Modeling Carbon and Nitrogen Dynamics for Soil Management

T0179208

Modeling Carbon and Nitrogen Dynamics for Soil Management

Edited by

M.J. Shaffer

USDA-ARS-NPA
Great Plains Systems Research Unit
Fort Collins, Colorado

Liwang Ma

USDA-ARS-NPA
Great Plains Systems Research Unit
Fort Collins, Colorado

S. Hansen

The Royal Veterinary and Agricultural University
Taastrup, Denmark

CRC Press
Taylor & Francis Group
Boca Raton London New York

CRC Press is an imprint of the
Taylor & Francis Group, an **informa** business

CRC Press
Taylor & Francis Group
6000 Broken Sound Parkway NW, Suite 300
Boca Raton, FL 33487-2742

First issued in paperback 2019

© 2001 by Taylor & Francis Group, LLC
CRC Press is an imprint of Taylor & Francis Group, an Informa business

No claim to original U.S. Government works

ISBN-13: 978-1-56670-529-5 (hbk)
ISBN-13: 978-0-367-39735-7 (pbk)

Visit the Taylor & Francis Web site at
http://www.taylorandfrancis.com

and the CRC Press Web site at
http://www.crcpress.com

Preface

Cycling of carbon and nitrogen is of paramount importance to agricultural productivity and global climate stability, and is strongly affected by human activities — especially intensive farming and deforestation. The balance between organic and inorganic forms is an indication of the well-being of ecosystems on Earth. On the other hand, any imbalance will cause disasters to the living sphere. In the last century, our understanding of carbon and nitrogen cycling has advanced tremendously at both the micro- and macro-scales. We have also experienced a transition in research from understanding the carbon/nitrogen processes to managing carbon/nitrogen resources. Such a transition requires not only new materials and methods in research, but also collaborations among scientists from different disciplines and between scientists and extension agents.

Due to the complexity of the carbon and nitrogen processes and their dynamic responses to environmental conditions, it is beyond our minds' ability to quantitatively synthesize all the knowledge. Mathematical modeling is an extension of our brains and helps us comprehend aspects in many scientific disciplines and is the key element in the systems approach. Mathematical models were not popular until the 1960s in agricultural sciences, and their applications for soil management have boomed due to advances in computer technology. There are numerous models simulating carbon and nitrogen processes, which differ in their objectives, levels of complexity, and relation to other agricultural components. Model applications — ranging from data interpretation, to technology transfer, to decision support — have not only advanced our understanding of carbon and nitrogen processes, but have also provided a theoretical basis for soil carbon and nitrogen management.

This book consists of 19 chapters authored by well-known model developers and users to update readers on the current status and future direction of carbon and nitrogen modeling. Its coverage ranges from a theoretical comparison of models to application of models for soil management problems, from laboratory applications to field and watershed scale applications, from short-term simulation to long-term prediction, and from DOS-based computer programs to object-oriented and graphical interface designs. The book will be useful to agricultural scientists, university professors, graduate-level students, environmental scientists, action agencies, consultants, and extension agents.

Chapter 1 is an introduction to the book and links the chapters by their contributions to understanding and managing carbon and nitrogen processes. Chapters 2 and 3 describe the carbon and nitrogen processes in upland and wetland soils, respectively. Chapter 3 also has a comprehensive description of models developed for wetland soils. Chapters 4 and 5 review and compare carbon and nitrogen models in the United States and Europe, respectively. Chapter 6 presents a Canadian ecosystem model called *ecosys*. Chapter 7 summarizes the application of the Root Zone Water Quality Model (RZWQM) for soil nitrogen management. Chapter 8 simulates the effect of crop rotation on long-term soil carbon, nitrogen gas emission, nitrate leaching, and crop yield using the DAYCENT model. Chapter 9 presents the use of NLEAP to evaluate soil water-quality problems as affected by manure and fertilizer applications in different cropping systems. Chapter 10 estimates soil nitrogen balance and crop nitrogen use efficiency under several cropping and management systems with NLEAP. Chapter 11 demonstrates the use of the Stella graphical interface for nitrogen modeling with NLEAP as an example. Chapter 12 presents the theoretical background of the NLEAP model and its Internet Web interface. Chapter 13 simulates soil nitrogen dynamics under different manure and fertilizer management in several soil and cropping systems using the SOILN model. Chapter 14 offers an optimization method to calibrate a soil nitrogen transport model (Expert-N). Chapters 15 and 16 discuss applications of the DAISY model for short- and long-term simulations at different scales of soil carbon and nitrogen dynamics. Chapter 17 presents a study of soil nitrogen dynamics under various agricultural management practices and at different scales using the HERMES/MINERVA (GIS) model. Chapter 18 evaluates the environmental impacts of fertilizer and crop management at the farm scale with a linkage to a GIS using a WAVE model. And finally, Chapter 19 illustrates the complexity of carbon and nitrogen processes and issues related to model parameterization and testing.

We sincerely thank all the authors for their fine contributions to the book and the CRC Press/Lewis Publishers staff for making this endeavor possible.

M.J. Shaffer
Liwang Ma
Søren Hansen

About the Editors

Dr. Marvin J. Shaffer is Soil Scientist, USDA-ARS, Great Plains Systems Research Unit, Fort Collins, CO, where he has been for the last 15 years. Previously, he was Soil Scientist for 6 years with the USDA-ARS, Soil and Water Management Unit in St. Paul, MN, and Hydrologist for 8 years with the U.S. Bureau of Reclamation in Denver, CO. He received the USDA Award for Distinguished Service and was 1992 Scientist of the Year for the USDA-ARS, Northern Plains Area. He currently serves as associate editor for the *Journal of Environmental Quality* (*JEQ*).

Dr. Shaffer was the primary developer of the NLEAP nitrate leaching model and served as lead scientist for development of the NTRM nitrogen, tillage, and residue management model. Most of his career has involved development of simulation models of soil-plant-aquifer systems, with emphasis on nitrogen and carbon cycling and soil chemistry. His scientific publications have dealt with model development, testing, and application; C/N cycling; nitrogen and crop management; nitrate leaching indices; groundwater quality; whole-farm simulation; effects of climate and erosion on crop productivity; crop–weed interactions; low-input agriculture; and soil salinity.

Dr. Liwang Ma is a Soil Scientist with USDA-ARS, Great Plains Systems Research Unit, Fort Collins, CO. Dr. Ma received his B.S. and M.S. in biophysics from Beijing Agricultural University (now China Agricultural University) in 1984 and 1987, respectively, and his Ph.D. in soil physics from Louisiana State University in 1993. He is a co-editor of two books and author or co-author of 50 publications. His research on agricultural system modeling extends from water and solute transport, to carbon and nitrogen dynamics, to plant growth. Dr. Ma is the recipient of the 1994 Prentiss E. Schilling Outstanding Dissertation Award in the College of Agriculture, Louisiana State University.

Søren Hansen is an associate professor at the Laboratory for Agrohydrology and Bioclimatology, Department of Agricultural Sciences, The Royal Veterinary and Agricultural University of Denmark (KVL). He has been with KVL for 23 years, first as a research assistant, later as an assistant professor, and since 1993 as an associate professor. His research comprises transport processes in the soil-plant-atmosphere continuum and agricultural system modeling, including modeling of carbon and nitrogen cycling and plant growth. He is the principal developer of the agro-ecosystem model DAISY. His scientific publications related to modeling have dealt with topics about model development, testing, and application, including up-scaling from field to catchment level and the assessment of uncertainty model results related to uncertainty on input data. Hansen is a member of the editorial board of *Applied Soil Ecology*.

Contributors

S. Achatz
GSF — National Research Center for
  Environment and Health
Institute of Soil Ecology
Oberschleissheim, Germany

L.R. Ahuja
USDA-ARS-NPA
Great Plains Systems Research Unit
Fort Collins, CO

A.J. Beblik
Centre of Agricultural Landscape
  Research
Institute for Landscape System Analysis
Muencheberg, Germany

Shabtai Bittman
Pacific Agri-Food Research Centre
Agriculture and Agri-Food Canada
Agaassiz, British Columbia

J. Brenner
Natural Resource Ecology Laboratory
Colorado State University
Fort Collins, CO

Sander Bruun
Department of Agricultural Sciences
The Royal Veterinary and Agricultural
  University
Taastrup, Denmark

W.F. DeBusk
Wetland Biogeochemistry Laboratory
Soil and Water Science Department
University of Florida
Gainesville, FL

Jorge A. Delgado
USDA-ARS-Soil Plant Nutrient
  Research Unit
Fort Collins, CO

S.J. Del Grosso
Natural Resource Ecology Laboratory,
Colorado State University
Fort Collins, CO

R. Flynn
Department of Soil and Crop Sciences
Colorado State University
Fort Collins, CO

Juan David Piñeros Garcet
Department of Environmental Sciences
  and Land Use Management
Université Catholique de Louvain
Louvain-la-Neuve, Belgium

Robert F. Grant
Department of Renewable Resources
University of Alberta
Edmonton, Alberta

Søren Hansen
Department of Agricultural Sciences
The Royal Veterinary and Agricultural
  University
Taastrup, Denmark

M.D. Hartman
Natural Resource Ecology Laboratory
Colorado State University
Fort Collins, CO

Derek E. Hunt
Pacific Agri-Food Research Centre
Agriculture and Agri-Food Canada
Agaassiz, British Columbia

Mathieu Javaux
Department of Environmental Sciences
  and Land Use Management
Université Catholique de Louvain
Louvain-la-Neuve, Belgium

Lars S. Jensen
Department of Agricultural Sciences
The Royal Veterinary and Agricultural
  University
Taastrup, Denmark

Milena Kercheva
N. Poushkarov Institute of Soil Science
  and Agroecology
Sofia, Bulgaria

K.C. Kersebaum
Centre of Agricultural Landscape
  Research
Institute for Landscape System Analysis
Muencheberg, Germany

K. Lasnik
Department of Soil and Crop Sciences
Colorado State University
Fort Collins, CO

D.R. Lewis
Environment Division
Scottish Agricultural College (SAC)
Edinburgh, Scotland

Liwang Ma
USDA-ARS-NPA
Great Plains Systems Research Unit
Fort Collins, CO

Malcolm B. McGechan
Environment Division
Scottish Agricultural College (SAC)
Edinburgh, Scotland

I.P. McTaggart
Environment Division
Scottish Agricultural College (SAC)
Edinburgh, Scotland

A.R. Mosier
USDA-ARS-NPA
Soil-Plant-Nutrient Research Unit
Fort Collins, CO

Torsten Mueller
Department of Agricultural Sciences
The Royal Veterinary and Agricultural
  University
Taastrup, Denmark

D.S. Ojima
Natural Resource Ecology Laboratory
Colorado State University
Fort Collins, CO

X. Ou
Department of Soil and Crop Sciences
Colorado State University
Fort Collins, CO

W.J. Parton
Natural Resource Ecology Laboratory
Colorado State University
Fort Collins, CO

E. Priesack
GSF – National Research Center for
  Environment and Health
Institute of Soil Ecology
Oberschleissheim, Germany

K.R. Reddy
Wetland Biogeochemistry Laboratory
Soil and Water Science Department
University of Florida
Gainesville, FL

Jens C. Refsgaard
Department of Agricultural Sciences
The Royal Veterinary and Agricultural
  University
Taastrup, Denmark

D.S. Schimel
Max Planck Institut für Biogeochemie
Jena, Germany

M. J. Shaffer
USDA-ARS-NPA
Great Plains Systems Research Unit
Fort Collins, CO

R. Stenger
Lincoln Environmental
Hamilton, New Zealand

Dimitar Stoichev
N. Poushkarov Institute of Soil Science
  and Agroecology
Sofia, Bulgaria

Dimitranka Stoicheva
N. Poushkarov Institute of Soil Science
  and Agroecology
Sofia, Bulgaria

Christian Thirup
WaterTech a/s
Roskilde, Denmark

Amaury Tilmant
Department of Environmental Sciences
  and Land Use Management
Université Catholique de Louvain
Louvain-la-Neuve, Belgium

Marnik Vanclooster
Department of Environmental Sciences
  and Land Use Management
Université Catholique de Louvain
Faculte des Sciences Agronomiques
Departement des Sciences du Milieu et
  de l'Amenagement du Territoire
Louvain-la-Neuve, Belgium

J.R. White
Wetland Biogeochemistry Laboratory
Soil and Water Science Department
University of Florida
Gainesville, FL

L. Wu
Environment Division
Scottish Agricultural College (SAC)
Edinburgh, Scotland

Contents

CHAPTER 1

Introduction to Simulation of Carbon and Nitrogen Dynamics in Soils

M.J. Shaffer, Liwang Ma, and Søren Hansen

CONTENTS

HISTORICAL PERSPECTIVES

Soil carbon and nitrogen dynamics directly affects soil quality and productivity and has been a focus of study since the very beginnings of soil science. With improvements in experimental methodology, scientists have accumulated valuable information on soil carbon and nitrogen dynamics and enhanced our ability to manage soil organic C and N (Shaffer and Ma, 2001; DeBusk et al., 2001). At the same time, conceptual models of carbon and nitrogen (C/N) processes in soils have been under development (Molina and Smith, 1998). Early researchers focused on individual processes, but lacked the tools and knowledge needed to address the complex soil environment as an integrated whole. To some extent this may still be true today, but recent major advances in computer technology, tracer techniques, remote sensing, field sampling procedures, and laboratory analytical methods have allowed significant improvements in our knowledge of the C/N processes themselves and their interactions (Shaffer and Ma, 2001; DeBusk et al., 2001).

A classic example of an early model dealing with an aspect of C/N cycling is the potential mineralization work of Stanford and Smith (1972). Other modeling examples include work by Jenny (1941), Henin and Dupuis (1945), and Olson (1963). Among the biggest revolutions in the study of soil C/N cycling in the last 30 years has been the development of comprehensive simulation models of integrated C/N and related processes within the ecosystem. This was made possible by the introduction of computers with enough speed and memory to allow simulation of complete integrated soil systems and subsystems.

The origin of computer simulation modeling of carbon and nitrogen dynamics in soils dates back to the early 1970s with the first integrated soil-system models containing C/N cycling reported by Dutt et al. (1972) in the United States and Beek and Frissel (1973) in Europe. These models were the first to combine C/N and related sub-processes of a soil-crop-nutrient system into an integrated model. In the late 1970s, other soil nitrogen and carbon models began to appear, such as the United States Bureau of Reclamation (USBR) irrigation return flow model (Shaffer et al., 1977); Tanji's nitrogen model (Tanji and Gupta, 1978); and C/N modeling reported by Hunt (1978), Watts and Hanks (1978), and Anderson (1979). The integrated approaches seen in this group of models were built upon later and included the multiple soil organic matter pools in the PHOENIX model (McGill et al., 1981), the PAPRAN model (Seligman and van Keulen, 1981), and Frissel and van Veen's (1981) C/N cycling model.

In the early to mid-1980s, interest in agricultural modeling surged due to the environmental movement, and a range of soil carbon and nitrogen submodels were developed in conjunction with crop- and soil-focused models such as CENTURY (Parton et al., 1983), NCSOIL (Molina et al., 1983), EPIC (Williams and Renard, 1985), CERES (Ritchie et al., 1986), NTRM (Shaffer and Larson, 1987), and LEACHM (Wagenet and Hutson, 1987) in the United States; and PAPRAN by Seligman and van Keulen (1981), SOILN (Johnsson et al., 1987), and ANIMO (Berghuijs van Dijk et al., 1985) in Europe. Attempts were already being made to simplify some existing models, while other authors concentrated on increased levels of detail (Grant et al., 2001a,b).

In the late 1980s to mid-1990s, several additional models were developed along with refinements in existing code. New models appearing included a nitrogen dynamics model reported by Bergstrom and Johnsson (1988), HERMES by Kersebaum (1989), DAISY by Hansen et al. (1990), NLEAP by Shaffer et al. (1991), SUNDIAL by Bradbury et al. (1993), GLEAMS by Knisel (1993), CANDY by Franko (1996), ECOSYS by Grant (1997), and ICBM by Andren and Katterer (1997); along with refined versions of NTRM (Radke et al., 1991); LEACHM (Hutson and Wagenet, 1992); CENTURY (Parton and Rasmussen, 1994); SOILN (Eckersten et al., 1996); and ANIMO (Groenendijk and Kroes, 1997). Detailed reviews of these models are included in this book (Ma and Shaffer, 2001; McGechan and Wu, 2001; Grant, 2001a).

By the late 1990s, the information age had arrived with the advent of the Internet and emphasis was beginning to shift to object-based or object-oriented modeling with the introduction of C++ and Java computer languages (Abrahamsen and Hansen, 2000; Shaffer et al., 2000a; 2001); the development of Windows-based user interfaces (Rojas et al., 2000); an increased focus on process interactions with the environment (Shaffer et al., 2000b; Grant, 2001a); application of the Stella modeling system (Bittman et al., 2001); and the introduction of models interactive on the Internet (Shaffer et al., 2001).

LEVELS OF COMPLEXITY IN C/N MODELS

With C/N cycling models, there has been a strong tendency to develop process-based models that emphasize the underlying biological, chemical, and physical mechanisms (Shaffer et al., 2000b; Grant, 2001a). These range from purely research-oriented models (e.g., *ecosys*) through applications-oriented models (e.g., NTRM) and screening tools (e.g., NLEAP). Scale of application focus ranges from short-term seasonal estimation of soil N and C status (e.g., NLEAP) to long-term levels of soil organic matter and associated carbon sequestration (e.g., CENTURY). Indeed, two classes of C/N cycling models emerged with different objectives, but similar technology — the so-called short-term models with the objective of predicting C/N cycling over a single year or several years of a crop rotation (e.g., CERES) and the longer-term models designed to estimate the carbon status of soils over decades or longer (e.g., CENTURY). The development of faster computers and the demands of new technology in agricultural and environmental sciences have caused both types of model to move toward a common objective (e.g., the DAYCENT model derived from the CENTURY model [Del Grosso et al., 2001]).

From the spatial viewpoint, the more complex research models tend to be point based, while broad-scale regional modeling is left to simpler screening tools. The recent increase in computing power of accessible computers and linkage of process-based models to GIS (Geographic Information Systems) has changed the definition of research vs. application tools, but the availability of data to support research models across broad regions continues to be a major issue (Garnier et al., 1998; Beinroth et al., 1998; Shaffer et al., 1996). Development of generic parameter data-bases for models across a wide range of soil and land management scenarios has been suggested as a means of extending their usability (Shaffer et al., 1991), but these databases do not exist for most models.

The advent of multiple soil organic matter, microbial, and plant residue pools during the late 1970s and early 1980s represented a major change in research direction that resulted in a significant increase in model complexity. Model complexity develops in terms of processes simulated, simulated rate of each process, role of soil microbes, and interaction among the processes (Ma and Shaffer, 2001; McGechan and Wu, 2001; Grant, 2001a). Model development is still being done using FORTRAN, although there have been some attempts at object-based or object-oriented simulation (Shaffer et al., 2000a). The use of programming languages such as Stella and C++ will probably increase in the 21st century, along with interactive applications that use Java and run over the Internet (Shaffer et al., 2001; Bittman et al., 2001). The significant increase in computing power has stimulated the development of ever more complex models along with the challenge of making these and existing models more useful to end users (Grant, 2001a).

MAJOR ISSUES IN C/N MODEL PARAMETERIZATION

C/N models have a history of being difficult to parameterize, especially after the introduction of multiple pools. Pool sizes and associated rate parameters are difficult to measure and usually require some model pre-runs for self-parameterization and initialization (Ma et al., 1998). Numerous attempts have been made to develop laboratory and field methods for initializing model pools for soil organic matter, crop residues, manure, municipal wastes, and microbial biomass, but to date these have not been particularly successful (Parton et al., 1994; Knisel, 1993; Williams, 1995).

Another difficulty with C/N models is determining the rate coefficients used for various process simulations, which may be first-order, zero-order, Michaelis-Menten kinetics, or Monod kinetics (Hansen et al., 1995). The rate coefficients may be constants, but usually are modified for soil moisture, temperature, pH, and O_2 effects. Most models are still using a 0–1 index for each environmental factor, but there is a trend of using integrated response curves for the combined environmental effects (Shaffer et al., 2000b; Ma et al., 2001; Grant, 2001a). The associated rate constants can be derived from literature values or calibrated from available data sets (Shaffer et al., 2000b). An additional problem with model parameterization is how to optimize parameters for a given experimental site. Most times, for a system model, a trial-and-error method is used and the obtained parameters may not be optimized in a strict mathematical sense. Occasionally, a mathematical algorithm is used for some simpler models (Priesack et al., 2001).

The relationship to other components in a system model also determines the ease of model parameterization, such as water movement, nitrate transport, plant growth, and soil heat transfer. Interactions among the sub-processes, a long-neglected subject area, also play an important role in carbon and nitrogen modeling. An inaccurate parameterization of these components will directly affect the validity and transferability of parameters controlling soil C/N dynamics. Another issue is the scale at which parameters are calibrated. Even if a model has the capability of predicting C/N dynamics at one scale (e.g., field plot) under all possible soil, climate, and management conditions, it may still fail at other scales (smaller or larger). Scaling of either derived model parameters or modeling results is needed for soil management, as illustrated in several chapters of this book (i.e., 16, 17, and 18) (Hansen et al., 2001; Kersebaum and Beblik, 2001; Garcet et al., 2001). The lack of sufficiently accurate (spatial and temporal) field measurements for model development, calibration, and validation may be a significant problem.

Indeed, the entire topic of model validation and testing continues to be challenging. Although soil carbon/nitrogen models cannot ever be absolutely validated because not all cases can ever be tested, there exists a considerable body of test data for most of the major models, as shown in Chapters 5, 7, 9, and 10 (Ma et al., 2001; McGechan et al., 2001; Stoichev et al., 2001; Delgado, 2001). A close look at the degree of accuracy being achieved seems to indicate that the general trends are being simulated along with some intermediate details, but the finer details are not being adequately reproduced. This statement applies to the more detailed research models as well as the application-level tools. The use of more detailed process models appears to improve the accuracy for a particular condition, but their predictability is still below expectation and could be improved.

CURRENT STATUS OF C/N MODEL APPLICATION

Given the state of knowledge and confidence for most C/N models, some degree of regional or local calibration is almost always required and prudent before making extrapolations in time or space with these tools. Once this has been accomplished, sufficient field data with enough detail to meet project objectives for model accuracy is needed. Uncertainty in field data measurements continues to be an issue with recognition that some type of Monte Carlo approach — in conjunction with statistical distributions for input data rather than fixed values — is needed (Shaffer, 1988; Ma et al., 2000; Hansen et al., 2001).

Although C/N models are still premature in terms of making regulatory management decisions and they are mostly at the research stage, they have been used to analyze real-world problems, such as manure and fertilizer management (McGechan et al., 2001; Ma et al., 2001), soil water quality (Stoichev et al., 2001; Shaffer et al., 2001; Hansen et al., 2001), air quality (Del Grosso et al., 2001; Shaffer et al., 2001), soil quality (Delgado, 2001; Del Grosso et al., 2001), and cropping system evaluation (Hansen et al., 2001; Del Grosso et al., 2001; Jensen et al., 2001; Delgado, 2001; Stoichev et al., 2001; Garcet et al., 2001) at various scales.

Applications of C/N sub-models also depend on other processes in the whole system model. Most models simulate a one-dimensional water flow (e.g., RZWQM, CERES), but a few models have the capability to simulate three-dimensional flow (e.g., *ecosys*). Plant growth modeling ranges from a logistic curve (e.g., NLEAP), a series of functional equations (e.g., CENTURY, EPIC), to a fully environmental controlled growth module (e.g., CERES, RZWQM). There has been a trend to modularize simulated components so that model users can generate their own model for their specific purposes.

Development and implementation of these models have had a major impact on nutrient management and on projections of management effects on the environment. During the 1970s, the early models showed that complex microbiological processes in soils could be simulated, and this generated considerable interest in the subject. Later refinements, extensions, and testing provided tools for nitrogen management to reduce nitrate leaching to groundwater and nitrogen losses in surface runoff, and provided tools for long-term estimation of N and C trends in soils (carbon sequestration). Subsequent models have provided tools for use in estimation of nitrogen fertilizer requirements, nitrogen use efficiency, excess acidification, and greenhouse gas emissions from soils. Farmers can be readily shown the impacts of their management on the environment and given answers to what-if questions regarding crop response to nitrogen, leaching of nitrates, and greenhouse gases. However, because farmers may not be able to use models directly (no matter how simple they are), a model-based information-decision support database is needed for extension and technology transfer.

FUTURE DIRECTIONS OF C/N MODELS

Given the current status of C/N models, the question needs to be asked — where are C/N models headed? From the research side, a more mechanistic, process-focused approach to C/N modeling is emerging with help from object-oriented languages and tools such as C++ and Stella. There is a trend away from the index approach to simulating environmental stresses to directly including these impacts in rate process equations. There is also more emphasis being placed on microbial biomass growth and dynamics (Ma et al., 2001; Grant, 2001a), and future progress will depend on correctly addressing interactions between processes and developing methods to more effectively parameterize all levels of models. Finally, recognition must be given to the fact that applications models are not necessarily research-level models with sophisticated interfaces and extensive databases; rather, they are models specifically designed for use in a user environment to accomplish a specific set of objectives.

On the applications side, we have entered the information age with the rise of the Internet and instant access possibilities with respect to data and models. Users are demanding models that are easy to use and that have accessible databases. This will soon mean models that can be accessed and run on the Internet with data found on the Internet. The linkage of process-level models to geographic information systems (GIS) will be the next generation of model application for spatially distributed fields or watersheds (Shaffer et al., 1996; Kersebaum and Beblik, 2001). The use of C/N models with real-time remote sensing data will enhance our ability to predict regional and/or global C/N dynamics and plant biomass productivity.

With the introduction of crop yield monitors and global positioning systems (GPS), interest has been generated in the potential use of models for site-specific or precision management. However, models suitable to optimize the amount, method, and timing of carbon and nitrogen management are almost completely lacking. The ability of C/N cycling models to make projections of future outcomes of management scenarios with respect to key environmental areas such as nitrate leaching, carbon sequestration, and emissions of greenhouse gases has generated intense interest in using them as regulatory tools. Model calibration, validation, and testing prior to their use in this arena are essential. Criteria need to be developed for model certification for use in regulation (Shaffer et al., 1991).

REFERENCES

Abrahamsen, P., and S. Hansen. 2000. DAISY: an open soil-crop-atmosphere system model. *Environmental Modelling and Software,* 15:113-330.

Anderson, D.W. 1979. Processes of humus formation and transformation in soils of the Canadian Great Plains. *Journal of Soil Science,* 30:77-84.

Andren, O., and T. Katterer. 1997. ICBM: the introductory carbon balance model for exploration of soil carbon balances. *Ecological Applications,* 7:1226-1236.

Beek, J., and M.J. Frissel. 1973. Simulation of Nitrogen Behavior in Soils, Centre for Agricultural Publishing and Documentation. Wageningen, The Netherlands, 67 pp.

Beinroth, F.H., J.W. Jones, E.B. Knapp, P. Papajorgji, and J. Luyten. 1998. Evaluation of land resources using crop models and a GIS. In G.Y. Tsuji, G. Hoogenboom, P.K. Thornton. (Eds.). *Understanding Options for Agricultural Production.* Kluwer Academic Publishers, London, 293-310.

Berghuijs van Dijk, J.T., P.E. Rijtema, and C.W.J. Roest. 1985. ANIMO Agricultural Nitrogen Model. NOTA 1671, Institute for Land and Water Management Research, Wageningen, The Netherlands.

Bergstrom, L., and H. Johnsson. 1988. Simulated nitrogen dynamics and nitrate leaching in a perennial grass ley. *Plant Soil,* 105:273-281.

Bittman, S., D.E. Hunt, and M.J. Shaffer. 2001. NLOS (NLEAP on STELLA) — a nitrogen cycling model with a graphical interface: implications for model developers and users. In M.J. Shaffer, L. Ma, and S. Hansen (Eds.) *Modeling Carbon and Nitrogen Dynamics for Soil Management.* Lewis Publishers, Boca Raton, FL, 383-402.

Bradbury, N.J., A.P. Whitmore, P.B.S. Hart, and D.S. Jenkinson. 1993. Modelling the fate of nitrogen in crop and soil in the years following application of ^{15}N-labeled fertilizer to winter wheat. *Journal of Agricultural Science, Cambridge,* 121:363.

DeBusk, W.F., J.R. White, and K.R. Reddy. 2001. Carbon and nitrogen dynamics in wetland soils. In M.J. Shaffer, L. Ma, and S. Hansen (Eds.). *Modeling Carbon and Nitrogen Dynamics for Soil Management.* Lewis Publishers, Boca Raton, FL, 27-53.

Delgado, J. 2001. Use of simulations for evaluation of best management practices on irrigated cropping systems. In M.J. Shaffer, L. Ma, and S. Hansen (Eds.). *Modeling Carbon and Nitrogen Dynamics for Soil Management.* Lewis Publishers, Boca Raton, FL, 355-381.

Del Grosso, S.J., W.J. Parton, A.R. Moser, M.D. Hartman, J. Brenner, D.S. Ojima, and D.S. Schimel. 2001. Simulated interaction of soil carbon dynamics and nitrogen trace gas fluxes using the DAYCENT model. In M.J. Shaffer, L. Ma, and S. Hansen (Eds.). *Modeling Carbon and Nitrogen Dynamics for Soil Management.* Lewis Publishers, Boca Raton, FL, 303-332.

Dutt, G.R., M.J. Shaffer, and W.J. Moore. 1972. Computer Simulation Model of Dynamic Bio-physicochemical Processes in Soils. Tech. Bulletin 196. Arizona Agricultural Experiment Station. University of Arizona, Tucson, AZ. 128 pp.

Eckersten, H., P-E. Jansson, and H. Johnsson. 1996. The SOILN model user's manual, Department of Soil Sciences, Swedish University of Agricultural Sciences, Uppsala.

Franko, U. 1996. Modelling approaches of soil organic matter turnover within the CANDY system. In D.S. Powlson, P. Smith, and J.U. Smith (Eds.). *Evaluation of Soil Organic Matter Models.* NATO ASI series, Vol. 38, Springer, Berlin.

Frissel, M.J., and J.A. van Veen. 1981. Simulation model for nitrogen immobilization and mineralization. In I.K. Iskandar (Ed.). *Modeling Wastewater Renovation.* John Wiley & Sons, New York, 359-381.

Garcet, J.D.P., A. Thilmant, M. Javaux, and M. Vanclooster. 2001. Modeling N behavior in the soil and vadose environment supporting fertilizer management at the farmer scale. In M.J. Shaffer, L. Ma, and S. Hansen (Eds.). *Modeling Carbon and Nitrogen Dynamics for Soil Management.* Lewis Publishers, Boca Raton, FL, 573-598.

Garnier, M., A. Lo Porto, R. Marini, and A. Leone. 1998. Integrated use of GLEAMS and GIS to prevent groundwater pollution caused by agricultural disposal of animal waste. *Environmental Management,* 22:747-756.

Grant, R.F. 1997. Changes in soil organic matter under different tillage and rotation: mathematical modelling in *ecosys. Soil Sci. Soc. Am. J.* 61:1159-1174.

Grant, R.F. 2001a. A review of the Canadian ecosystem model — *ecosys.* In M.J. Shaffer, L. Ma, and S. Hansen (Eds.). *Modeling Carbon and Nitrogen Dynamics for Soil Management.* Lewis Publishers, Boca Raton, FL, 173-264.

Grant, R.F. 2001b. Modeling transformations of soil organic carbon and nitrogen at differing scales of complexity. In M.J. Shaffer, L. Ma, and S. Hansen (Eds.). *Modeling Carbon and Nitrogen Dynamics for Soil Management*. Lewis Publishers, Boca Raton, FL, 599-632.

Groenendijk, P., and J.G. Kroes. 1997. Modelling the Nitrogen and Phosphorus Leaching to Groundwater and Surface Water; ANIMO 3.5. Report 144, DLO Winand Staring Centre, Wageningen, The Netherlands.

Hansen, S., H.E. Jensen, N.E. Nielsen, and H. Svendsen. 1990. *DAISY: Soil Plant Atmosphere System Model*. NPO Report No. A 10. The National Agency for Environmental Protection, Copenhagen, 272 pp.

Hansen, S., H.E. Jensen, and M.J. Shaffer. 1995. Developments in modeling nitrogen transformations in soil. In P.E. Bacon (Ed.). *Nitrogen Fertilization in the Environment*. Marcel Dekker, New York, 83-107.

Hansen, S., C. Thirup, J.C. Refsgaard, and L.S. Jensen. 2001. Modeling nitrate leaching at different scales — application of the DAISY model. In M.J. Shaffer, L. Ma, and S. Hansen (Eds.). *Modeling Carbon and Nitrogen Dynamics for Soil Management*. Lewis Publishers, Boca Raton, FL, 513-549.

Henin, S., and M. Dupuis. 1945. Essai de bilan de la matiere organique du soi. *Ann. Agron.* 15:17-29.

Hunt, H.W. 1978. A simulation model for decomposition in grasslands. *Ecology,* 58:469-484.

Hutson, J.L., and R.J. Wagenet. 1992. LEACHM: Leaching Estimation and Chemistry Model: a process-based model of water and solute movement, transformations, plant uptake and chemical reactions in the unsaturated zone. Version 3.0. Department of Soil, Crop and Atmospheric Sciences, Research Series No. 93-3, Cornell University, Ithaca, NY.

Jenny, H. 1941. *Factors of Soil Formation. A System of Quantitative Pedology.* McGraw-Hill, New York.

Jensen, L.S., T. Mueller, S. Bruun, and S. Hansen. 2001. Application of the DAISY model for short and long-term simulation of soil carbon and nitrogen dynamics. In M.J. Shaffer, L. Ma, and S. Hansen (Eds.). *Modeling Carbon and Nitrogen Dynamics for Soil Management*. Lewis Publishers, Boca Raton, FL, 485-511.

Johnsson H., L. Bergström, P-E. Jansson, and K. Paustian. 1987. Simulated nitrogen dynamics and losses in a layered agricultural soil, *Agriculture, Ecosystems and Environment,* 18, 333-356.

Kersebaum, K.C. 1989. Die Simulation der Stickstoffdynamik von Ackerböden, Ph.D. thesis, University Hannover. 141p.

Kersebaum, K.C., and A.J. Beblik. 2001. Performance of a nitrogen dynamics model applied to evaluate agricultural management practices. In M.J. Shaffer, L. Ma, and S. Hansen (Eds.). *Modeling Carbon and Nitrogen Dynamics for Soil Management*. Lewis Publishers, Boca Raton, FL, 551-571.

Knisel, W.G. (Ed.). 1993. GLEAMS, Groundwater Loading Effects of Agricultural Management Systems. Version 2.10. UGA-CPES-BAED, Publication No. 5. 259pp.

Ma, L., M.J. Shaffer, J.K. Boyd, R. Waskom, L.R. Ahuja, K.W. Rojas, and C. Xu. 1998. Manure management in an irrigated silage corn field: experiment and modeling. *Soil Sci. Soc. Am. J.,* 62:1006-1017.

Ma, L., J.C. Ascough II, L.R. Ahuja, M.J. Shaffer, J.D. Hanson, and K.W. Rojas. 2000. Root zone water quality model sensitivity analysis using Monte Carlo simulation. *Trans. ASAE,* 43:883-895.

Ma, L., and M.J. Shaffer. 2001. A review of carbon and nitrogen processes in nine U.S. soil nitrogen dynamics models. In M.J. Shaffer, L. Ma, and S. Hansen (Eds.). *Modeling Carbon and Nitrogen Dynamics for Soil Management*. Lewis Publishers, Boca Raton, FL, 55-102.

Ma, L., M.J. Shaffer, and L.R. Ahuja. 2001. Application of RZWQM for nitrogen management. In M.J. Shaffer, L. Ma, and S. Hansen (Eds.). *Modeling Carbon and Nitrogen
Dynamics for Soil Management.* Lewis Publishers, Boca Raton, FL, 265-301.

McGechan, M.B. and L. Wu. 2001. A review of carbon and nitrogen processes in European
soil nitrogen dynamics models. In M.J. Shaffer, L. Ma, and S. Hansen (Eds.). *Modeling
Carbon and Nitrogen Dynamics for Soil Management.* Lewis Publishers, Boca Raton,
FL, 103-171.

McGechan, M.B., D.R. Lewis, L. Wu, and I.P. McTaggart. 2001. Modeling the effects of
manure and fertilizer management options on soil carbon and nitrogen processes. In M.J.
Shaffer, L. Ma, and S. Hansen (Eds.). *Modeling Carbon and Nitrogen Dynamics for Soil
Management.* Lewis Publishers, Boca Raton, FL, 427-460.

McGill, W.B., H.W. Hunt, R.G. Woodmansee, and J.O. Reuss. 1981. PHOENIX: A model of
carbon and nitrogen dynamics in grassland soils. In F.E. Clark and T. Rosswall (Eds.).
*Terrestrial Nitrogen Cycles: Processes, Ecosystem Strategies, and Management Impacts.
Ecological Bulletin,* Vol. 33, Stockholm, 49-115.

Molina, J.A.E., C.E. Clapp, M.J. Shaffer, F.W. Chichester, and W.E. Larson. 1983. NCSOIL,
a model of nitrogen and carbon transformations in soil: description, calibration, and
behavior. *Soil Sci. Soc. Am. J.,* 47:85-91.

Molina, J.A.E., and P. Smith. 1998. Modeling carbon and nitrogen processes in soils. *Adv.
Agron.,* 62:253-298.

Olson, J.S. 1963. Energy storage and the balance of producers and decomposers in ecological
systems. *Ecology,* 44:322-331.

Parton, W. J., and P. E. Rasmussen. 1994. Long-term effects of crop management in wheat/fal-
low: II. CENTURY model simulations, *Soil Sci. Soc. Am. J.,* 58, 530-536.

Parton, W.J., J. Persson, and D.W. Anderson. 1983. Simulation of organic matter changes in
Swedish soils. In W.K. Lauenroth, G.V. Skogerboe, and M. Flug (Eds.). *Analysis of Ecological Systems: State-of-the-Art in Ecological Modelling.* Elsevier, Amsterdam, 511-516.

Parton, W.J., D.S. Ojima, C.V. Cole, and D.S. Schimel, 1994. A general model for soil organic
matter dynamics: sensitivity to litter chemistry, texture and management. In *Quantitative
Modeling of Soil Forming Processes,* SSSA, Spec. Pub. 39, Soil Science Society of
America, Madison, WI, 147-167.

Priesack, E., S. Achatz, and R. Stenger. 2001. Parameterisation of soil nitrogen transport models
by use of laboratory and field data. In M.J. Shaffer, L. Ma, and S. Hansen (Eds.). *Modeling
Carbon and Nitrogen Dynamics for Soil Management.* Lewis Publishers, Boca Raton, FL,
461-484.

Radke, J.K., M.J. Shaffer, and J. Saponara. 1991. Application of the Nitrogen-Tillage-Residue-
Management (NTRM) model to low-input and conventional agricultural systems. *Ecol.
Modelling,* 55:241-255.

Ritchie, J.T., D.C. Godwin, and S. Otter-Nacke. 1986. CERES-Wheat: A Simulation Model
of Wheat Growth and Development, CERES Model Description. Department of Crop
and Soil Science, Michigan State University, East Lansing.

Rojas, K.W., L. Ma, J.D. Hanson, and L.R. Ahuja. 2000. RZWQM98 User Guide. In L.R.
Ahuja, K.W. Rojas, J.D. Hanson, M.J. Shaffer, and L. Ma (Eds.). *Root Zone Water
Quality Model.* Water Resources Publications LLC, Highlands Ranch, CO, 327-364.

Seligman, N.G., and H. van Keulen. 1981. PAPRAN: a simulation model of annual pasture
production limited by rainfall and nitrogen. In M.J. Frissel and J.A. van Veen (Eds.).
Simulation of Nitrogen Behavior of Soil-Plant Systems, Proc. Workshop, Wageningen,
The Netherlands, 192-221.

Shaffer, M.J. 1988. Estimating confidence bands for soil-crop simulation models. *Soil Sci.
Soc. Am. J.,* 52:1782-1789.

Shaffer, M.J., R.W. Ribbens, and C.W. Huntley. 1977. Prediction of mineral quality of irrigation return flow. Volume V. Detailed return flow salinity and nutrient simulation model. EPA 600/2-77-179e. U.S. Environmental Protection Agency, Corvallis, OR. 229 pp.

Shaffer, M.J., and W.E. Larson (Eds.). 1987. NTRM, A Soil-Crop Simulation Model for Nitrogen, Tillage, and Crop-Residue Management. U.S. Department of Agriculture, Conservation Research Report No. 34-1. 103pp.

Shaffer, M.J., A.D. Halvorson, and F.J. Pierce. 1991. Nitrate leaching and economic analysis package (NLEAP): model description and application. In R.F. Follett, et al. (Eds.). *Managing Nitrogen for Groundwater Quality and Farm Profitability.* Soil Science Society of America, Madison, WI, chap. 13, 285-322.

Shaffer, M.J., M.D. Hall, B.K. Wylie, and D.G. Wagner. 1996. NLEAP/GIS approach for identifying and mitigating regional NO3-N leaching. In D.L. Corwin and K. Loague (Eds.). *Application of GIS to the Modeling of Non-point Source Pollutants in the Vadose Zone.* SSSA Special Publication, American Society of Agronomy, Madison, WI, 283-294.

Shaffer, M.J., P.N.S. Bartling, and J.C. Ascough, II. 2000a. Object-oriented simulation of integrated whole farms: GPFARM framework. *Computers and Electronics in Agriculture,* 28:29-49.

Shaffer, M.J., K. Rojas, D.G. DeCoursey, and C.S. Hebson. 2000b. Nutrient Chemistry Processes — OMNI. In *Root Zone Water Quality Model.* Water Resources Publc., Inc., Englewood, CO, chap. 5, 119-144.

Shaffer, M.J., and L. Ma. 2001. Carbon and nitrogen dynamics in upland soils. In M.J. Shaffer, L. Ma, and S. Hansen (Eds.). *Modeling Carbon and Nitrogen Dynamics for Soil Management.* Lewis Publishers, Boca Raton, FL, 11-26.

Shaffer, M.J., K. Lasnik, X. Ou, and R. Flynn. 2001. NLEAP Internet tools for estimating NO$_3$-N leaching and N$_2$O emissions. In M.J. Shaffer, L. Ma, and S. Hansen (Eds.). *Modeling Carbon and Nitrogen Dynamics for Soil Management.* Lewis Publishers, Boca Raton, FL, 403-426.

Stanford, G., and S.J. Smith. 1972. Nitrogen mineralization potentials of soils. *Soil Sci. Soc. Am. Proc.,* 36:465-472.

Stoichev, D., M. Kercheva, and D. Stoicheva. 2001. NLEAP water quality applications in Bulgaria. In M.J. Shaffer, L. Ma, and S. Hansen (Eds.). *Modeling Carbon and Nitrogen Dynamics for Soil Management.* Lewis Publishers, Boca Raton, FL, 333-354.

Tanji, K.K., and S.K. Gupta. 1978. Computer simulation modeling for nitrogen in irrigated croplands. In D.R. Nielsen and J.G. MacDonald (Eds.). *Nitrogen in the Environment.* Academic Press, New York, Vol. 1, 79-120.

Wagenet, R.J., and J.L. Hutson. 1987. LEACHM: Leaching Estimation and Chemistry Model. A process based model of solute movement, transformations, plant uptake and chemical reactions in the unsaturated zone. Continuum Vol. 2, Water Resources Institute, Cornell University, Ithaca, NY.

Watts, D.G., and J.R. Hanks. 1978. A soil-water-nitrogen model for irrigated corn on sandy soils. *Soil Sci. Soc. Am. J.,* 42:492-499.

Williams, J.R. 1995. The EPIC model. In V.P. Singh (Ed.). *Computer Models of Watershed Hydrology.* Water Resources Publications. Highlands Ranch, CO, 909-1000.

Williams, J.R., and K.G. Renard. 1985. Assessment of soil erosion and crop productivity with process models (EPIC). In R.F. Follett and B.A. Stewart (Eds.). *Soil Erosion and Crop Productivity.* American Society of Agronomy, Madison, WI, chap. 5, 68-102.

CHAPTER 2

Carbon and Nitrogen Dynamics in Upland Soils

M.J. Shaffer and Liwang Ma

CONTENTS

INTRODUCTION

Upland soils constitute the majority of soils in farm and ranch areas throughout the world. They also dominate forested areas. Because these soils tend to be better aerated than wetland soils, oxygen presence plays a key role in the process dynamics and management responses of these systems, and generally promotes higher decomposition rates of organic materials compared to wetland soils. Periodic wet and dry cycles on a seasonal or more frequent basis cause temporary shifts in oxygen availability that in turn affect process rates and directions. The landscape position of upland soils promotes drainage and movement of nutrients and soil particles to lower lying areas, and aspect exposure can have significant effects on temperature and water retention regimes. Location of the soils on the Earth relative to latitude and elevation also plays an important role in the accumulation of soil organic matter (SOM) and associated C/N dynamics (Povirk et al., 2000).

Particularly in farm and ranchland areas, upland soils have experienced dramatic changes since the introduction of agriculture (Lowdermilk, 1948; Dregne, 1982). Losses of SOM in the surface horizons from erosion and accelerated decomposition rates have been pronounced worldwide, and continue to occur especially in areas converted from the native condition within the past 50 to 100 years (Miller et al., 1985). Examples can be readily found in North America and developing countries throughout the world (Sears, 1980; Dregne, 1982). In addition to reducing SOM, erosion also removes soil minerals containing plant nutrients, exposes subsoil material with low fertility or high acidity or salinity, and degrades soil hydrologic conditions along with decreases in plant-available water holding capacity (Frye et al., 1985; Larson et al., 1985). In general, disturbance of the soil-vegetation complex by the introduction of cultivated agriculture and animal grazing has caused erosive soil movement to lower soil positions on the landscape and to wetlands, rivers, estuaries, and the oceans. Upland soils often experience pronounced losses of SOM in higher landscape positions with accumulations in lower areas where buried organic horizons are not uncommon. Also, increased mobilization of the soil carbon pools to atmospheric CO_2 has occurred due to accelerated decomposition rates. Thus, there has been a general lowering of the C:N ratios of the SOM in upland soils.

CARBON CYCLING IN UPLANDS

Organic carbon in upland soil occurs as SOM, dead and living plant material (tops and roots), microflora biomass (dead and living), and soil micro-, meso-, and macrofauna (dead and living). Fractionation of carbon into various designated pools for purposes of analysis by empirical methods and simulation modeling has been attempted for many years. Definitions for various physically and chemically fractionated pools, and a microbial-biomass carbon pool have been reviewed by Follett (2000). Carbon pools used in the simulation models are mostly conceptual (Ma and Shaffer, 2001; McGechan et al., 2001; Grant, 2001) and are only approximately correlated with the physical/chemical/biomass pools described above. Organic carbon in soils is a continuous (time and space) product of microbial processes, and fractionation into pools probably should be entirely microbial based. In general, the physical carbon pools are described based on size and include litter or easily recognizable residues (>2 mm in size), a light fraction that can be floated off using dense liquids (0.25 to 2 mm in size), and particulate organic matter that is separated using dispersion and physical separation techniques (0.05 to 2 mm in size). Soil microbial biomass carbon is determined using the chloroform fumigation-incubation (CFI) method (Jenkinson and Powlson, 1976); and more resistant forms of SOM, including humic acid, fulvic acid, and humin, are fractionated using chemical techniques such as the HCl and NaOH methods (Ping et al., 1999). As an approximation, most model-based pools can be compared to the physical/chemical/microbial pools as follows. The fast SOM model pools are similar to the light and/or particulate pools, the microbial biomass pools are similar to each other, and the slow SOM model pool is similar to the resistant SOM pool. In some cases, the slow residue model pool may be similar to the light SOM pool. The model pools are highly dynamic and initialization depends on the design of the model, the rate coefficients, and the relationships of model C/N components to other parts of the model.

The relative contributions of above- and below-ground higher plant parts to SOM carbon vary depending on the ecosystem involved. Agricultural and forested systems receive the majority of their SOM carbon from above-ground plant parts, while grassland and tundra systems receive SOM primarily from root decay (Reeder et al., 2000; Povirk et al., 2000). A comparison of standing organic carbon residues for various ecosystems such as forests, grasslands, tundra/desert, and croplands is presented by Sobecki et al. (2000). Forests have the most above- and below-ground biomass carbon, with 177 MT/ha, followed by grasslands with 19.8 MT/ha, croplands with 16.4 MT/ha, and tundra/desert with 4.2 MT/ha. These same authors also present a map for the distribution of estimated soil organic carbon in the coterminous United States. The highest accumulations of soil carbon occur in the upper Midwest, the Northeast, and coastal areas under wetland soil conditions as described by DeBusk et al. (2001). Because upland agricultural systems were usually derived either from forested or grassland systems, their origin and location (primarily latitude, altitude, and landscape position) are important for carbon pool comparisons, especially total organic carbon, although soil erosion and other losses (or gains) of SOM carbon may have altered these relationships. Living microorganisms mediate the vast majority of the C/N cycling processes in soils (Tate, 1987), although these organisms constitute only about 0.3 to 5% of the SOM carbon (Reeder et al., 2000). Microbial activity in soils is highly dependent on the presence of carbon substrates, which is why most activity and associated microbial biomass occurs near the soil surface and immediately after introduction of residues.

Soil inorganic carbon in the form of CO_2, bicarbonates, and carbonic acid, and soil and rock carbonates constitute most of the remaining carbon (Monger and Martinez-Rios, 2000). Given a long enough time period, carbon cycles among all of these pools. Cycling times vary from a few hours for CO_2 and plant materials, to a few weeks for plant material and fresh SOM, to several decades for labile SOM, to thousands of years for stable SOM, and to millions of years for carbonate rocks. Of most concern to agricultural managers and modelers is SOM with short turnover times up to several decades. This material supplies nutrients to the crops and can be managed and replaced, while the older material needs to be protected but is less easily managed. Cycling of carbon is closely tied to cycles involving other soil nutrients, such as nitrogen, phosphorus, sulfur, potassium, and trace constituents (McGill and Cole, 1981).

Upland agricultural soils have been studied extensively over the years under a range of management and climate regimes. For example, long-term plots such as the Sanborn Field site at Columbia, Missouri (Buyanovsky et al., 1997), the High Plains Agricultural Laboratory site at Sidney, Nebraska (Lyon et al., 1997), and the Morrow Plots at Urbana, Illinois (Darmondy and Peck, 1997), provide evidence of the effects of long-term agricultural management on soil carbon, fertility, and crop productivity. The decrease in soil carbon in these plots relative to the native state is quite evident and ranges from 40 to 80%. Disturbances resulting from cultivation (biological oxidation), removal of crop residues, and soil erosion are factors in these losses (Lyon et al., 1997; Buyanovsky and Wagner, 1997; Darmondy and Peck,

1997). In the case of biological oxidation, disturbance of the soil from tillage has been shown to release significant amounts of CO_2 gas (Reicosky et al., 1995). Buyanovsky and Wagner (1997) have demonstrated that decomposition of crop residues such as corn, wheat, and soybean returned to the soil makes only a minimal contribution to the more stable (long-term) soil carbon pool. Most of the carbon cycling from crop residues takes place in the faster, less stable pools. However, the return of crop residues to the land, rather than removal, has produced significant increases in organic carbon levels (Buyanovsky et al., 1997).

The introduction of reduced tillage and no-till has had significant impacts on C/N cycling. This is due primarily to decreased mixing of residues in the plow layer and a decrease in aeration plus an increase in soil water content. Soil organic carbon levels are significantly higher under no-till, especially near the surface (Kern and Johnson, 1993; Donigian et al., 1994; Peterson and Westfall, 1997). Nitrifier populations have decreased by 16 to 56% in reduced and no-till plots relative to conventional tillage, while denitrifier populations have increased (Lyon et al., 1997). Microbial biomass levels of no-till soils averaged 54% higher in the surface layer compared to plowed soils (Doran, 1987).

NITROGEN CYCLING IN UPLANDS

Considerable work has been published over the past several decades on C/N dynamics in upland soils (Minderman, 1968; Stanford and Smith, 1972; Jenkinson and Rayner, 1977; Zeikus, 1981; Campbell et al., 1995; Nakane et al., 1997; Janzen et al., 1998). In general, the nitrogen and carbon cycles are closely linked and cannot be effectively studied or modeled separately (McGill and Cole, 1981). This is especially true with upland soils where good aeration coupled with periodic wet-dry and temperature cycles direct the microbial and plant biological processes and significantly influence the physical processes. In fact, most models simulate nitrogen cycling from carbon cycling. Nitrogen occurs along with carbon in all of the crop residue and SOM pools and also occurs as inorganic NO_3^- and NH_4^+, urea, and as N_2, N_2O, NO_x, and NH_3 gases. The relative amounts of nitrogen found in the soil generally are SOM-N > residue-N > NO_3-N > NH_4-N > gaseous-N forms. From the standpoints of mobility and negative effects on the environment, NO_3-N and the gaseous-N forms (N_2O, NO_x, and NH_3) are of most concern.

The nitrogen cycle in upland soils is shown in Figure 2.1. Nitrogen enters the soil from precipitation and plant residues (tops and roots), from the atmosphere where symbiotic plants (e.g., legumes) fix nitrogen, from animal manures, and from the addition of commercial fertilizers and other organic materials such as sewage sludges and compost. Soil losses of nitrogen include removal in crops, nitrate leaching, denitrification and gaseous losses as NH_3, and losses in surface runoff and erosion (Romkens et al., 1973). The relative nitrogen loss amounts are highly variable and any of these major loss components can dominate given the right conditions. These may occur rapidly and major soil nitrogen changes can occur within a few hours or days (Shaffer et al., 1994).

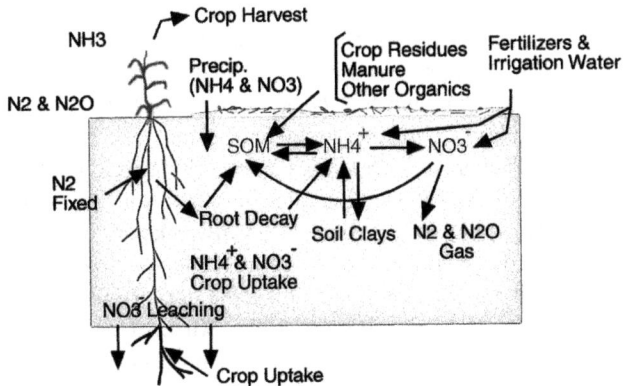

Figure 2.1 The nitrogen cycle in upland soils.

Figure 2.1 also depicts nitrogen cycling or turnover within the soil where, under aerobic conditions, the microbial processes of mineralization (or ammonification) decompose residues and SOM to NH_4^+ and more stable forms of SOM. CO_2 is the primary by-product, although temporary or localized anaerobic conditions in upland soils can also result in ammonification with the by-product CH_4 (methane) being produced instead of CO_2.

The heterotrophic microorganisms (primarily soil fungi along with lesser bio-mass quantities for bacteria and actinomyces) (Paul et al., 1979) utilize carbon and nitrogen from the decomposing residues and nitrogen from the soil NO_3^- and NH_4^+ pools to produce microbial biomass. This biomass generally ranges from about 250 to 900 kg C/ha for a range of agricultural soils (Doran, 1987). Rangeland biomass levels in the top 30 cm range from 65 g/m^2 to 635 g/m^2, depending on the type of prairie ecosystem (Reeder et al., 2000). With carbon-rich residues such as wheat and barley straw (C:N = 80+), insufficient nitrogen is available in the residues to supply the demand of the biomass growth (C:N = 6), resulting in temporary net uptake or immobilization of mineral soil NO_3^- and NH_4^+. This nitrogen is then released from the microbial biomass when the microbes die and are recycled. However, crops will not have access to this nitrogen source while it is immobilized. In general, decay of residues with C:N ratios greater than about 25 or 30 will cause temporary net immobilization, while lower C:N ratios result in net mineralization. Systems decomposing residues with C:N ratios greater than 30 may become deficient in mineral nitrogen and become self-limiting in their ability to decompose these residues. Temperature and water content are important state variables in the miner-alization process and control the microbial rates of growth and death (Marion and Black, 1987; Melillo et al., 1989).

Nitrogen has the capacity to cycle fairly rapidly between the inorganic and organic forms within the soil and a few weeks can see the complete turnover of some inorganic/organic nitrogen forms. Conversely, some forms of organic nitrogen, such as soil humus, may be very stable and have an age of several thousand years. Studies to identify the relative ages for SOM pools have been performed using carbon isotope ratios (Paul et al., 1997; Follett et al., 1997).

Conversion of NH_4^+ to NO_3^- is called nitrification and is mediated by specific genera of aerobic autotrophic microorganisms, namely *Nitrosomonas* and *Nitrobacter* (Alexander, 1977). These organisms use CO_2 as a carbon source, with *Nitrosomonas* oxidizing NH_4^+ to nitrite (NO_2) and *Nitrobacter* converting NO_2^- to NO_3^-. Nitrification state variables in upland soils include temperature and water content with a strong dependency on pH (McLaren, 1969; Sabey et al., 1969; Malhi and McGill, 1982; McInnes and Fillery, 1989). Nitrification rates in upland soils are generally medium relative to other C/N cycling processes, and 2 to 4 weeks are usually needed in the warmer months to nitrify an application of commercial ammonia-based fertilizer. The nitrification process is not 100% efficient and some conversion of NO_2^- to N_2O and NO_x greenhouse gases can occur (Grant, 1995; Xu et al., 1998). Nitrification is sensitive to soil pH, with significant reduction in process rates below pH 6.0 and above pH 8.0. Nitrification inhibitors such as nitrapyrin have been developed that slow the process by enzyme inhibition of *Nitrosomonas* (Peoples et al., 1995).

Denitrification involves the conversion of NO_3^- and NO_2 to N_2, N_2O, and NO_x. The process is mediated primarily by facultative anaerobic bacteria that oxidize organic carbon and reduce nitrogen oxides in the absence of oxygen. The key to understanding and simulating denitrification in upland soils is the transient and localized nature of the anaerobic conditions. Complete reduction of nitrogen oxides to N_2 is an anaerobic process, while partial or transient anaerobosis results in the production of N_2O and NO_x. Denitrification tends to be a very rapid pulse process with major losses of soil nitrogen possible in a single day given the right conditions (Xu et al., 1998). Ideal pulse conditions involve the occurrence of significant amounts of soil NO_3^-, a fresh carbon source such as manure or other residue SOM, warm temperatures, a precipitation or irrigation event, or temporary flooding (Nommick, 1956; Bremner and Shaw, 1958; Cady and Bartholomew, 1961; Smid and Beauchamp, 1976; Firestone et al., 1979; Firestone, 1982; Tiedje, 1988; Meisinger and Randall, 1991). These conditions are common in the early spring or summer in upland soils and apply to production of the greenhouse gases and N_2.

Gaseous loss of NH_3 from upland soils is primarily a soil surface process. In addition, it is one of the few soil nitrogen processes that is abiotic and not directly mediated by microorganisms, although they play a key role in NH_4^+ production. NH_3 gas exists in chemical equilibrium with the NH_4^+ ion in soil water, and the reaction is highly dependent on soil pH. Applications of ammonia fertilizers and manure to the soil surface set the stage for major losses of nitrogen as NH_3 gas. The process is enhanced by high pH levels (>8.5) and by wind and high temperatures (Smith et al., 1996; Shaffer et al., 2000). Incorporation of fertilizers and manure into the soil can significantly reduce or eliminate NH_3 volatilization.

Nitrate leaching below the crop root zone of upland soils is a primary mechanism for loss of nitrogen from these soils and contributes to pollution of underlying shallow aquifers. Nitrate leaching amounts from upland soils vary greatly, but values from less than 25 to greater than 300 kg/ha/year are common depending on management, soil, and climate. Leaching of NO_3^- is a pulse process keyed to precipitation and irrigation events, but tied closely to previous management of nitrogen and water, crop nitrogen uptake, and soil texture (Shaffer et al., 1994). Nitrates are highly soluble in water and move quickly along with the water because they are not usually

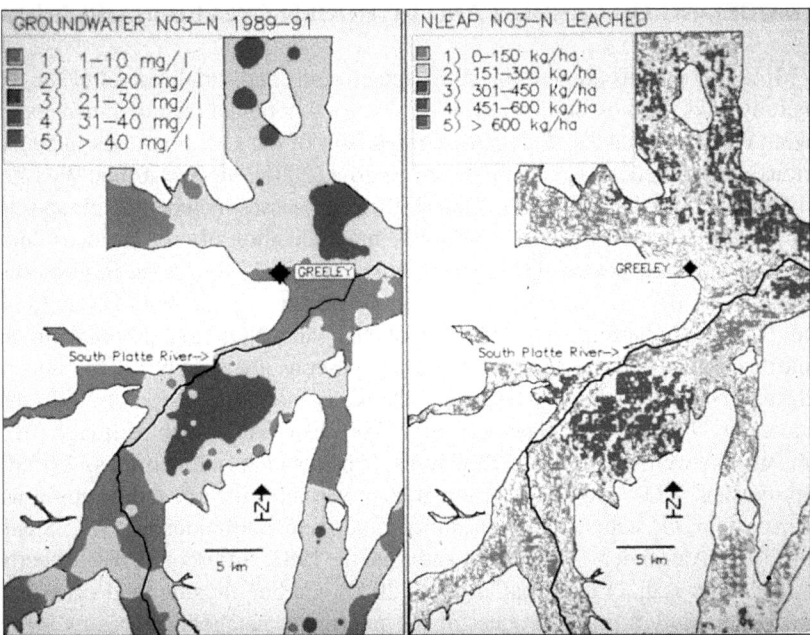

Figure 2.2   Regional spatial patterns for groundwater NO_3-N and the NLEAP NL index.

adsorbed by soil particles. Once leached below the root zone, recovery or reduction of NO_3^- is difficult and further downward movement over time can result in contamination of the groundwater (Shaffer et al., 1995). Groundwater pollution from agricultural NO_3^- generally does not occur uniformly, but rather in localized "hotspot" regions within shallow underlying aquifers (Figure 2.2). Combinations of coarse soil texture and poor management are the primary reasons for this effect (Shaffer et al., 1995). Application of a NO_3^- leaching (NL) index for annual mass of NO_3^- leached below the crop root zone can be used to help identify potential "hotspot" regions in aquifers (Wylie et al., 1995; Shaffer et al., 1995). Leaching of NO_3^- can be controlled by applying fertilizers and manure in smaller increments as needed by the crop to reduce the potential leaching risk from precipitation and irrigation, and by managing water applications where appropriate to reduce leaching (Follett, 1995; Shaffer et al., 1994; Hall et al., in press). Other mitigation techniques such as planting deep-rooted cover crops can be beneficial (Delgado, 1999; 2000).

Plant uptake of NO_3^- and NH_4^+ represents an important sink process in the nitrogen cycle and proper accounting is essential for simulation of C/N cycling and NO_3^- leaching in soils (Follett, 1995). For agricultural crops on upland soils, typical crop uptake of nitrogen is from about 50 to over 400 kg N/ha/year. Proper plant nutrition requires that nutrients such as nitrogen be available when the crop needs them. Producers are tempted to pre-apply large doses of commercial fertilizers or manure to help ensure that the crop will not become nutrient deficient. However, this practice runs the risk of losing significant amounts of nitrogen to the major sinks such as NO_3^- leaching, denitrification, and NH_3 volatilization (Hall et al., in press).

MODELING OF CARBON AND NITROGEN CYCLES IN UPLANDS

Most C/N models are developed for upland soils, and simulation of C/N cycling generally links the most significant processes as a function of their dominant external driving or state variables. A detailed comparison of the C/N processes included in various current models is given in Ma and Shaffer (2001), McGechan and Wu (2001), and Grant (2001). On an overview level, these processes include growth and death of microbial biomass; associated decay or mineralization of crop residues, animal manures, and other organic materials such as municipal sewage and food processor wastes; decay or accumulation of soil organic matter; nitrification of NH_4^+ to NO_3^- (and N_2O production), immobilization of NH_4^+ and NO_3^- to microbial biomass, denitrification of NO_3^- to N_2, N_2O, and NO_x; gaseous losses of NH_3; crop uptake of NO_3^- and NH_4^+; leaching of NO_3^- and soluble organics; adsorption of NH_4^+ (and in some cases NO_3^-) by soil clays; effects of soil disturbance such as tillage, freeze-thaw, shrink-swell, and soil fauna and roots; and erosional and runoff losses of SOM, crop residues, NO_3^- and NH_4^+. External state variables include temperature, wind, water content, O_2 status, pH, and salinity (Cady and Bartholomew, 1961; Stanford et al., 1973; Alexander, 1977; Malhi and McGill, 1982; Addiscott, 1983; Robertson, 1994; Hansen et al., 1995; Shaffer et al., 2000). Carbon dioxide (and occasionally CH_4) is generally a major product of the microbial metabolic processes involved and is often included as a sink in the models.

The form and detail of the rate equations and pools describing these processes are where the various models differ the most (Ma and Shaffer, 2001; McGechan and Wu, 2001; Grant, 2001). For example, most models treat decay of organic residues and soil humus as being first order, usually with respect to carbon (Hansen et al., 1995; Ma et al., 1998). However, the number and definition of the pools describing the various forms of residue, soil organic matter, and other constituents differ some-what (Ma and Shaffer, 2001; McGechan and Wu, 2001; Grant, 2001; Andrén and Paustian, 1987). The rapid process of urea hydrolysis to form NH_4^+ is simulated directly in some models and ignored in others (Reynolds et al., 1985; Shaffer et al., 2000; Ma and Shaffer, 2001). Nitrification of NH_4^+ to form NO_3^- is usually taken as being zero order with respect to NH_4^+, although other forms such as Michaelis-Menten and first order in NH_4^+ are also used (Hansen et al., 1995; Shaffer et al., 2000). Denitrification is generally simulated as being first order with respect to NO_3^-, but Michaelis-Menten and other approaches are also used (Firestone, 1982; Hansen et al., 1995; Shaffer et al., 1991). Ammonia volatilization is also generally treated as a first-order process relative to the NH_4^+ concentration on or near the soil surface, together with consideration of application management effects (Shaffer et al., 2000). Some of the models are more mechanistic relative to the chemical and physical processes involved (Ma and Shaffer, 2001; McGechan and Wu, 2001; Grant, 2001). Simulation techniques for the growth and death of microbial biomass (if included) are also quite variable, along with the manner in which the population (or biomass) influences the rates of the other processes (Shaffer et al., 2000; Grant, 2001). Inorganic nitrogen pools usually include NH_4^+ and NO_3^- and some models also add nitrite (NO_2^-), and N_2O and NO_x gases (Shaffer et al., 2000; Grant, 2001; Del Grosso et al., 2001). The models also differ considerably in the way external state variables are used to simulate effects on process rates. The predominant technique has been to develop some type of stress function usually with a 0–1 range. A good example of this technique is the soil water function based on water-filled pore space (Linn and Doran, 1984). These functions are often allowed to interact in some manner or the greatest stress value is used. More recent models have attempted to directly tie the state variables to the process rate equations (Grant, 1997; Grant, 2001; Shaffer et al., 2000).

SUMMARY

Upland soils dominate agricultural and ranchland regions throughout the world and are extremely important from the standpoints of food and fiber production globally. These soils have undergone significant reductions in SOM and fertility caused by soil erosion and losses due to accelerated decomposition rates as a result of human disturbance. Replacement of nutrients by the introduction of commercial fertilizers, the use of machinery, improved crop varieties, and pesticides have compensated for most potential yield losses. However, off-farm degradation from sedimentation and water pollution in groundwater, wetlands, rivers, estuaries, and the oceans continues to be a problem.

C/N dynamics in upland soils is dominated by aerobic processes mediated by soil microorganisms. These microbes (primarily soil fungi) produce NH_4^+ and live biomass by decomposing plant residues and other organic materials such as manure, sewage sludge, and food processor wastes. These organisms also perform nitrogen transformations on commercial fertilizer applications. Soil autotrophic nitrifying bacteria oxidize NH_4^+ to NO_2^- and NO_3^- with some leakage as N_2O. Facultative heterotrophic microbes operating in anaerobic or partially anaerobic conditions are capable of utilizing CO_2 and reducing NO_3^- to N_2, N_2O, and NO_x gases. Plants utilize CO_2 from the atmosphere (some also symbiotically fix atmospheric nitrogen), along with uptake of inorganic nitrogen (NO_3^- and NH_4^+) and other nutrients from the soil, to produce the vast majority of all biomass. This biomass from higher plants then becomes the raw material for recycling by the soil microorganisms, or is consumed by fauna (including humans). Fauna are eventually recycled by the soil microorganisms as well.

Cycling of carbon and nitrogen between the various organic and inorganic pools occurs naturally and correct accounting for these processes and the time scales involved is the key to simulation models for C/N dynamics and management. Key external driving variables for these models include temperature, wind, water content, O_2 status, pH, and salinity. The models simulate decay and assimilation of commercial fertilizers, crop residues and other organic soil amendments, and production/uptake of various soil gases such as CO_2, NH_3, CH_4, N_2, and N_2O. Plant uptake of NH_4^+ and NO_3^- is simulated to account for this important sink for inorganic nitrogen. The models simulate infiltration and redistribution of water and handle associated solute movement and leaching of soluble soil constituents such as NO_3^-, urea, and soluble organics.

Application of C/N cycling models to help make assessments and solve management problems has demonstrated the utility of these tools in various subject areas.

For example, crop rotations have been evaluated for their nitrogen status and yields (Radke et al., 1991; Wagner-Riddle et al., 1997; Diebel et al., 1995; Foltz et al., 1995); nitrogen cycling has been investigated in low-input agriculture (Radke et al., 1991); crop-weed interactions with nitrogen effects have been simulated (Ball and Shaffer, 1993); optimal crop transplantation dates have been determined (Saseendran et al., 1998); a simulation has determined the effect of planting date on nitrogen dynamics (Delgado, 1998); one study has compared irrigation scheduling methods (Steele et al., 1994); C/N modeling has been used to determine nitrogen fertilizer application rates and drain water quality (Paz et al., 1999; Watkins et al., 1998; Azevedo et al., 1997); erosion-productivity relationships have been defined with a C/N model (Shaffer, 1985); C/N modeling has been used for waste water management (Clanton et al., 1986); manure management effects have been simulated using C/N models (Wylie et al., 1995; Ma et al., 1998); tillage effects on NO_3-N transport to drains have been simulated (Singh and Kanwar, 1995); risk analysis for irrigation has been studied (Epperson et al., 1992; Martin et al., 1996); C/N modeling has been applied to decision support of nitrogen and crop management (Shaffer et al., 1994; Ford et al., 1993; Evers et al., 1998; Shaffer and Brodahl, 1998); screening tests have been run on C/N models (Khakural and Robert, 1993); simulation has been used to determine climate effects on crop production and soil sustainability (Swan et al., 1990; Phillips et al., 1996; Lee et al., 1996; Shaffer et al., 1994); simulation has been used in policy evaluation (Wu et al.,1996b); manure and fertilizer management have been investigated with C/N modeling (McGechan et al., 2001; Ma et al., 2001; Hall et al., in press); C/N modeling has been used to estimate nitrate leaching (Follett, 1995; Wu et al., 1996a; Delgado et al., 2000; Stoichev et al., 2001; Shaffer et al., 2001; Hansen et al., 2001); emissions of soil greenhouse gases have been modeled (Xu et al., 1998; Del Grosso et al., 2001; Shaffer et al., 2001); soil quality has been investigated using C/N models (Delgado, 2001; Del Grosso et al., 2001); C/N simulation has been used in cropping system evaluation (Hansen et al., 2001; Del Grosso et al., 2001; Jensen et al., 2001; Delgado, 2001; Stoichev et al., 2001; Garcet et al., 2001); and C/N modeling has been applied in conjunction with geographical information system (GIS) technology (Shaffer et al., 1995; Wu et al., 1996b; Kersebaum and Beblik, 2001; Garcet et al., 2001).

REFERENCES

Addiscott, T.M. 1983. Kinetics and temperature relationships of mineralization and nitrification in Rothamsted soils with differing histories. *J. Soil Sci.,* 34:343-353.

Alexander, M. 1977. *Introduction to Soil Microbiology.* John Wiley, New York. 251-271.

Andrén, O., and K. Paustian. 1987. Barley straw decomposition in the field: a comparison of models. *Ecology,* 68:1190-1200.

Azevedo, A.S., P. Singh, R.S. Kanwar, and L.R. Ahuja. 1997. Simulating nitrogen management effects on subsurface drainage water quality. *Agric. Systems,* 55:481-501.

Ball, D.A., and M.J. Shaffer. 1993. Simulating resource competition in multispecies agricultural plant communities. *Weed Research,* 33:299-310.

Bremner, J.M., and K. Shaw. 1958. Denitrification in soils. II. Factors affecting denitrification, *J. Agric. Sci.,* 51:40-52.

Broder, M.W., J.W. Doran, G.A. Peterson, and C.R. Fenster. 1984. Fallow tillage influence on spring populations of soil nitrifiers, denitrifiers, and available nitrogen. *Soil Sci. Soc. Am. J.,* 48:1060-1067.

Buyanovsky, G.A., and G.H. Wagner. 1997. Crop residue input to soil organic matter on Sanborn Field. In E.A. Paul et al. (Eds.). *Soil Organic Matter in Temperate Agroecosystems.* CRC Press, Boca Raton, FL, 73-83.

Buyanovsky, G.A., J.R. Brown, and G.H. Wagner. 1997. Sanborn Field: effect of 100 years of cropping on soil parameters. In E.A. Paul et al. (Eds.). *Soil Organic Matter in Temperate Agroecosystems.* CRC Press, Boca Raton, FL, 205-225.

Cady, F.B., and W.V. Bartholomew. 1961. Influence of low pO_2 on denitrification processes and products. *Soil Sci. Soc. Am. Proc.,* 25:362-365.

Campbell, C.A., B.G. McConkey, R.P. Zentner, F.B. Dyck, F. Selles, and D. Curtin. 1995. Carbon sequestration in a Brown Chernozem as affected by tillage and rotation. *Can. J. Soil Sci.,* 75:449-458.

Clanton, C.J., D.C. Slack, and M.J. Shaffer. 1986. Evaluation of the NTRM model for land application of wastewater. *Trans. ASAE,* 29(5):1307-1313.

Darmondy, R.G., and T.R. Peck, 1997. Soil organic carbon changes through time at the University of Illinois Morrow Plots. In E.A. Paul et al. (Eds.). *Soil Organic Matter in Temperate Agroecosystems.* CRC Press, Boca Raton, FL, 161-169.

DeBusk, W.F., J.R. White, and K.R. Reddy. 2001. Carbon and nitrogen dynamics in wetland soils. In M.J. Shaffer, L. Ma, and S. Hansen (Eds.). *Modeling Carbon and Nitrogen Dynamics for Soil Management.* Lewis Publishers, Boca Raton, FL, 27-53.

Delgado, J.A. 1998. Sequential NLEAP simulations to examine effect of early and late planted winter cover crops on nitrogen dynamics. *J. Soil & Water Conservation,* 53:241-244.

Delgado, J. 2001. Use of simulations for evaluation of best management practices on irrigated cropping systems. In M.J. Shaffer, L. Ma, and S. Hansen (Eds.). *Modeling Carbon and Nitrogen Dynamics for Soil Management.* Lewis Publishers, Boca Raton, FL, 355-381.

Delgado, J.A., R.T. Sparks, R.F. Follett, J.L. Sharkoff, and R.R. Riggenbach. 1999. Use of winter cover crops to conserve soil and water quality in the San Luis Valley of South Central Colorado. In R. Lal (Ed.). *Soil Quality and Soil Erosion.* CRC Press, Boca Raton, FL, 125-142.

Delgado, J.A., R.F. Follett, and M.J. Shaffer. 2000. Simulation of NO_3^--N dynamics for cropping systems with different rooting depths. *J. Soil Sci Soc Am.,* 64:1050-1054.

Delgado, J.A., M. Shaffer, and M.K. Brodahl. 1998. New NLEAP for shallow and deep rooted rotations. *J. Soil and Water Cons.,* 53:338-340.

Del Grosso, S.J., W.J. Parton, A.R. Moser, M.D. Hartman, J. Brenner, D.S. Ojima, and D.S. Schimel. 2001. Simulated interaction of soil carbon dynamics and nitrogen trace gas fluxes using the DAYCENT model. In M.J. Shaffer, L. Ma, and S. Hansen (Eds.). *Modeling Carbon and Nitrogen Dynamics for Soil Management.* Lewis Publishers, Boca Raton, FL, 303-332.

Diebel, P.L., J.R. Williams, and R.V. Llewelyn. 1995. An economic comparison of conventional and alternative cropping systems for a representative northeast Kansas farm. *Rev. of Agric. Econ.,* 17:323-335.

Donigian, A.S., Jr., T.O. Barnwell, R.B. Jackson, A.S. Patwardhan, K.B. Weinreich, A.L. Rowell, R.V. Chinnaswamy, and C.V. Cole. 1994. Assessment of Alternative Management Practices and Policies Affecting Soil Carbon in Agroecosystems of the Central United States. Pub. Number EPS/600/R-94/067, U.S. EPA, Athens, GA.

Doran, J.W. 1987. Microbial biomass and mineralizable nitrogen distributions in no-tillage and plowed soils. *Biol. Fertil. Soils,* 5:68-75.

Dregne, H.E. 1982. Historical perspective of accelerated erosion and effect on world civilization. In B.L. Schmidt et al. (Eds.). *Determinants of Soil Loss Tolerance.* Spec. Pub. 45. American Society of Agronomy, Madison, WI, 1-14.

Epperson, J.E., H.E. Hook, and Y.R. Mustafa. 1992. Stochastic dominance analysis for more profitable and less risky irrigation of corn. *J. Prod. Agric.,* 5:243-247.

Evers, A.J.M., R.L. Elliott, E.W. Stevens. 1998. Integrated decision making for reservoir, irrigation, and crop management. *Agricultural Systems,* 58:529-554.

Firestone, M.K. 1982. Biological denitrification. In Stevenson F.J. (Ed.). *Nitrogen in Agricultural Soils.* American Society of Agronomy, Inc.; Crop Science Society of America, Inc.; Soil Science Society of America, Inc., Madison, WI, 289-326.

Firestone, M.K., M.S. Smith, R.B. Firestone, and J.M. Tiedje. 1979. The influence of nitrate, nitrite and oxygen on the composition of the gaseous products of denitrification in soil. *Soil Sci. Soc. Am. J.,* 43:1140-1144.

Follett, R.F. 2000. Organic carbon pools in grazing land soils. In R.F. Follett et al. (Eds.). *The Potential of U.S. Grazing Lands to Sequester Carbon and Mitigate the Greenhouse Effect.* Lewis Publishers, Boca Raton, FL, chap. 3, 65-86.

Follett, R.F. 1995. NLEAP model simulation of climate and management effects on N leaching for corn grown on sandy soil. *J. Contaminant Hydrol.,* 20:241-252.

Follett, R.F., E.A. Paul, S.W. Leavitt, A.D. Halvorson, D. Lyon, and G.A. Peterson. 1997. Carbon isotope ratios of Great Plains soils in wheat-fallow systems. *Soil Sci. Soc. Am. J.,* 61:1068-1077.

Foltz, J.C., J.G. Lee, M.A. Martin, and P.V. Preckel. 1995. Multiattribute assessment of alternative cropping systems. *Am. J. Agric. Econ.,* 77:408-420.

Ford, D.A., A.P. Kruzic, and R.L. Doneker. 1993. Using GLEAMS to evaluate the agricultural waste application rule-based decision support (AWARDS) computer program. *Water Science and Technology,* 28:625-634.

Frye, W.W., O.L Bennett, and G.J. Buntley. 1985. Restoration of crop productivity on eroded or degraded soils. In R.F. Follett and B.A. Stewart (Eds.). *Soil Erosion and Crop Productivity.* American Society of Agronomy, Inc., Madison, WI. 339-356.

Garcet, J.D.P., A. Tilmant, M. Javaux, and M. Vanclooster. 2001. Modeling N behavior in the soil and vadose environment supporting fertilizer management at the farm scale. In M.J. Shaffer, L. Ma, and S. Hansen (Eds.). *Modeling Carbon and Nitrogen Dynamics for Soil Management.* Lewis Publishers, Boca Raton, FL, 573-598.

Grant, R.F. 1995. Mathematical modelling of nitrous oxide evolution during nitrification. *Soil Biol. Biochem.,* 27:1117-1125.

Grant, R.F. 1997. Changes in soil organic matter under different tillage and rotation: mathematical modelling in ecosys. *Soil Sci. Soc. Am. J.,* 61:1159-1174.

Grant, R.F. 2001. A review of the Canadian ecosystem model — *ecosys.* In M.J. Shaffer, L. Ma, and S. Hansen (Eds.). *Modeling Carbon and Nitrogen Dynamics for Soil Management.* Lewis Publishers, Boca Raton, FL, 173-264.

Hall, M.D., M.J. Shaffer, R.M. Waskom, and J.A. Delgado. In press. Regional Nitrate Leaching Variability: What Makes a Difference in Northeastern Colorado. *J. Am. Water Resources Assoc.*

Hansen, S., M.J. Shaffer, and H.E. Jensen. 1995. Developments in modelling nitrogen transformations in soil, In *Nitrogen Fertilization and the Environment.* Marcel Dekker, New York, 83-107.

Hansen, S., C. Thirup, J.C. Refsgaard, and L.S. Jensen. 2001. Modelling nitrate leaching at different scales — application of the DAISY model. In M.J. Shaffer, L. Ma, and S. Hansen (Eds.). *Modeling Carbon and Nitrogen Dynamics for Soil Management.* Lewis Publishers, Boca Raton, FL, 513-549.

Janzen, H.H., C.A. Campbell, E.G. Gregorich, and B.H. Ellert. 1998. Soil carbon dynamics in Canadian agroecosystems. In R. Lal, J.M. Kimble, R.F. Follet and B.A. Stewart (Eds.). *Soil Processes and the Carbon Cycle*. CRC Press, Boca Raton, FL, 57-80.

Jenkinson, D.S., and D.S. Powlson. 1976. The effects of biocidal treatments on metabolism in soil. V.A. method for measuring soil biomass. *Biochemistry*, 8:209-213.

Jenkinson, D.S., and J.H. Rayner. 1977. The turnover of soil organic matter in some of the Rothamsted classical experiments. *Soil Sci.*, 123:298-305.

Jensen, L.S., T. Mueller, S. Bruun, and S. Hansen. 2001. Application of the DAISY model for short and ling-term simulation of soil carbon and nitrogen dynamics. In M.J. Shaffer, L. Ma, and S. Hansen (Eds.). *Modeling Carbon and Nitrogen Dynamics for Soil Management*. Lewis Publishers, Boca Raton, FL, 485-511.

Kern, J.S., and M.G. Johnson. 1993. Conservation tillage impacts on national soil and atmospheric carbon levels. *Soil Sci. Soc. Am. J.*, 57:200-210.

Kersebaum, K.C., and A.J. Beblik. 2001. Performance of a nitrogen dynamics model applied to evaluate agricultural management practices. In M.J. Shaffer, L. Ma, and S. Hansen (Eds.). *Modeling Carbon and Nitrogen Dynamics for Soil Management*. Lewis Publishers, Boca Raton, FL, 551-571.

Khakural, B.R., and P. C. Robert. 1993. Soil nitrate leaching potential indices: using a simulation model as a screening system. *J. Environ. Qual.*, 22:839-845.

Larson, W.E., T.E. Fenton, E.L. Skidmore, and C.M. Benbrook. 1985. Effects of soil erosion on soil properties as related to crop productivity and classification. In R.F. Follett and B.A. Stewart (Eds.). *Soil Erosion and Crop Productivity*. American Society of Agronomy, Inc., Madison, WI, 189-211.

Lee, J.J., D.L. Phillips, and R.F. Dodson. 1996. Sensitivity of the U.S. corn belt to climate change and elevated CO2. II. Soil erosion and organic carbon. *Agricultural Systems*, 52:503-521.

Linn, D.M., and J.W. Doran. 1984a. Aerobic and anaerobic microbial populations in no-till and plowed soils. *Soil Sci. Soc. Am. J.*, 48:794-799.

Linn, D.M., and J.W. Doran. 1984b. Effect of water-filled pore space on carbon dioxide and nitrous oxide production in tilled and non-tilled soils. *Soil Sci. Soc. Am. J.*, 48:1267-1272.

Lowdermilk, W.C. 1948. Conquest of the Land Through 7000 years. SCS-MP-32. USDA-SCS. U.S. Govt. Printing Office, Washington, D.C.

Lyon, D.L., C.A. Monz, R.E. Brown, and A.K. Metherell. 1997. Soil organic matter changes over two decades of winter wheat-fallow cropping in western Nebraska. In E.A. Paul et al. (Eds.). *Soil Organic Matter in Temperate Agroecosystems*. CRC Press, Boca Raton, FL, 343-351.

Ma, L., M.J. Shaffer, and L. Ahuja. 2001. Application of RZWQM for nitrogen management. In M.J. Shaffer, L. Ma, and S. Hansen (Eds.). *Modeling Carbon and Nitrogen Dynamics for Soil Management*. Lewis Publishers, Boca Raton, FL, 265-301.

Ma, L., and M.J. Shaffer. 2001. A review of carbon and nitrogen processes in nine U.S. soil nitrogen dynamics models. In M.J. Shaffer, L. Ma, and S. Hansen (Eds.). *Modeling Carbon and Nitrogen Dynamics for Soil Management*. Lewis Publishers, Boca Raton, FL, 55-102.

Ma, L., M.J. Shaffer, L.R. Ahuja, K.W. Rojas, and C. Xu. 1998. Manure management in an irrigated silage corn field: experiment and modeling. *Soil Sci. Soc. Am. J.*, 62:1006-1017.

Malhi, S.S., and W.B. McGill. 1982. Nitrification in three Alberta soils: effect of temperature, moisture and substrate concentration. *Soil Biol. Biochem.*, 14:393-399.

Marion, G.W., and C.H. Black. 1987. The effect of time and temperature on nitrogen mineralization in arctic tundra soils. *Soil Sci. Soc. Am. J.*, 51:1501-1508.

Martin, E.C., J.T. Ritchie, and B.D. Baer. 1996. Assessing investment risk of irrigation in humid climate. *J. Prod. Agric.*, 9:228-233.

McGechan, M.B., and L. Wu. 2001. A review of carbon and nitrogen processes in European soil nitrogen dynamics models. In M.J. Shaffer, L. Ma, and S. Hansen (Eds.). *Modeling Carbon and Nitrogen Dynamics for Soil Management.* Lewis Publishers, Boca Raton, FL, 103-171.

McGechan, M.B., D.R. Lewis, L. Wu, and I.P. McTaggart. 2001. Modelling the effects of manure and fertilizer management options on soil carbon and nitrogen processes. In M.J. Shaffer, L. Ma, and S. Hansen (Eds.). *Modeling Carbon and Nitrogen Dynamics for Soil Management.* Lewis Publishers, Boca Raton, FL, 427-460.

McGill, W.B., and C.V. Cole. 1981. Comparative aspects of cycling of organic C, N, S, and P through soil organic matter. *Geoderma,* 26:267-286.

McInnes, K.J., and I.R.P. Fillery. 1989. Modeling and field measurements of the effect of nitrogen source on nitrification. *Soil Sci. Soc. Am. J.,* 53:1264-1269.

McLaren, A.D. 1969. Steady state studies of nitrification in soil: theoretical considerations. *Proc. Soil Sci. Soc. Am.,* 33:273-275.

Meisinger, J.J., and G.W. Randall. 1991. Estimating nitrogen budgets for soil-crop systems. In R.F. Follett, D.R. Keeney, and R.M. Cruse (Eds.). *Managing Nitrogen for Groundwater Quality and Farm Profitability,* Soil Science Society of America, Inc., Madison, WI. 85-124.

Melillo, J.M., J.D. Aber, A.E. Linkins, A. Ricca, B. Fry, and K.J. Nadelhoffer. 1989. Carbon and nitrogen dynamics along the decay continuum: plant litter to soil organic matter. *Plant and Soil,* 115:189-198.

Miller, F.P., W.D. Rasmussen, and L.D. Meyer. 1985. Historical perspective of soil erosion in the United States. In R.F. Follett and B.A. Stewart (Eds.). *Soil Erosion and Crop Productivity.* ASA/CSSA/SSSA, Madison, WI, 23-48.

Minderman, G. 1968. Addition, decomposition and accumulation of organic matter in forests. *J. Ecol.,* 56:355-362.

Monger, H.C., and J.J. Martinez-Rios. 2000. Inorganic carbon sequestration in grazing lands. In R.F. Follett et al. (Eds.). *The Potential of U.S. Grazing Lands to Sequester Carbon and Mitigate the Greenhouse Effect.* Lewis Publishers, Boca Raton, FL, 87-118.

Nakane, K., T. Kohno, T. Horikoshi, and T. Nakatsubo. 1997. Soil carbon cycling at a black spruce (*Picea mariana*) forest stand in Saskatchewan, Canada. *J. Geophys. Res.,* 102:785-793.

Nommick, N. 1956. Investigations of denitrification in soil. *Acta Agric. Scand.,* 6:195-228.

Paul, E.A., R.F. Follett, S.W. Leavitt, A.D. Halvorson, G.A. Peterson, and D. Lyon. 1997. Carbon isotope ratios of Great Plains soils in wheat-fallow systems. *Soil Sci. Soc. Am. J.,* 61:1058-1067.

Paul, E.A., F.E. Clark, and V.O. Biederbeck. 1979. Microorganisms. In R.T. Coupland (Ed.). *Grassland Ecosystems of the World: Analysis of Grasslands and Their Uses.* International Biological Programme 18, Cambridge University Press, Cambridge, 87-97.

Paz, J.O., W.D. Batchelor, B.A. Babcock, T.S. Colvin, S.D. Logsdon, T.C. Kaspar, D.L. Karlen. 1999. Model based technique to determine variable rate nitrogen for corn. *Agricultural Systems,* 61:69-75.

Peoples, M.B., J.R. Freney, and A.R. Mosier. 1995. Minimizing gaseous losses of nitrogen. In P.E. Bacon (Ed.). *Nitrogen Fertilization in the Environment.* Marcel Dekker, New York, 565-602.

Peterson, G.A., and D.G. Westfall. 1997. Management of dryland agroecosystems in the central Great Plains of Colorado. In E.A. Paul et al. (Eds.). *Soil Organic Matter in Temperate Agroecosystems.* CRC Press, Boca Raton, FL, 371-380.

Phillips, D.L., J.J. Lee, and R.F. Dodson. 1996. Sensitivity of the U.S. corn belt to climate change and elevated CO2. I. Corn and the soybean yields. *Agricultural Systems,* 52:481-502.

Ping, C.L., G.L. Michaelson, X.Y. Dai, and R.J. Candler. 1999. Characterization of soil organic matter. In J.M. Kimble et al. (Eds.). *Methods of Assessment of Soil Carbon*. CRC Press, Boca Raton, FL.

Povirk, K.L., J.M. Welker, and G.F. Vance. 2000. Carbon sequestration in arctic and alpine tundra and mountain meadow ecosystems. In R.F. Follett et al. (Eds.). *The Potential of U.S. Grazing Lands to Sequester Carbon and Mitigate the Greenhouse Effect*. Lewis Publishers, Boca Raton, FL, 189-228.

Radke, J.K., M.J. Shaffer, and J. Saponara. 1991. Application of the Nitrogen-Tillage-Residue-Management (NTRM) model to low-input and conventional agricultural systems. *Ecol. Modelling*, 55:241-255.

Reeder, J.D., C.D. Franks, and D.G. Milchunas. 2000. Root biomass and microbial processes. In R.F. Follett et al. (Eds.). *The Potential of U.S. Grazing Lands to Sequester Carbon and Mitigate the Greenhouse Effect*. Lewis Publishers, Boca Raton, FL, 139-166.

Reicosky, D.C., W.D. Kemper, G.W. Langdale, C.L. Douglas Jr., and P.E. Rasmussen. 1995. Soil organic matter changes resulting from tillage and biomass production. *J. Soil Water Cons.*, 253-261.

Reynolds, C.M., D.C. Wolf, and J.A. Armbruster. 1985. Factors related to urea hydrolysis in soils. *Soil Sci. Soc. Am. J.*, 49:104-108.

Robertson, K. 1994. Nitrous oxide emission in relation to soil factors at low to intermediate moisture levels. *J. Environ. Qual.*, 23: 805-809.

Romkens, M.J.M, D.W. Nelson, and J.V. Mannering. 1973. Nitrogen and phosphorus composition of surface runoff as affected by tillage method. *J. Environ. Quality*, 2:292-295.

Sabey, B.R., L.R. Frederick, and W.V. Bartholomew. 1969. The formation of nitrate from ammonium nitrogen in soils. IV. Use of the delay and maximum rate phases for making quantitative predictions. *Proc. Soil Sci. Soc. Am.*, 33:276-278.

Saseendran, S.A., K.G. Hubbard, K.K. Singh, N. Mendiratta, L.S. Rathore, and S.V. Singh. 1998. Optimum transplanting dates for rice in Kerala, India, determined using both CERES v3.0 and *Clim. Prob. Agron. J.*, 90:185-190.

Sears, P.B. 1980. *Deserts on the March*, 4th ed. Oklahoma University Press, Norman.

Shaffer, M.J. 1985. Simulation model for soil erosion-productivity relationships. *J. Environ. Qual.*, 14:144-150.

Shaffer, M.J. and M.K. Brodahl. 1998. Rule-based management for simulation in agricultural decision support systems. *Computers and Electronics in Agriculture*, 21:135-152.

Shaffer, M.J., and S.C. Gupta. 1981. Hydrosalinity models and field validation. In I.K. Iskandar (Ed.). *Modeling Wastewater Renovation*. Wiley, New York, 136-181.

Shaffer, M.J., A.D. Halvorson, and F.J. Pierce, 1991. Nitrate Leaching and Economic Analysis Package (NLEAP): model description and application. In R.F. Follett, D.R. Keeney, and R.M. Cruse (Eds.). *Managing Nitrogen for Groundwater Quality and Farm Profitability*, Soil Science Society of America, Inc., Madison, WI, 285-322.

Shaffer, M.J., K. Lasnik, X. Ou, and R. Flynn. 2001. NLEAP internet tools for estimating NO_3-N leaching and N_2O emissions. In M.J. Shaffer, L. Ma, and S. Hansen (Eds.). *Modeling Carbon and Nitrogen Dynamics for Soil Management*. Lewis Publishers, Boca Raton, FL, 403-426.

Shaffer, M.J., K.W. Rojas, D.G. DeCoursey, and C.S. Hebson. 2000. Nutrient chemistry processes — OMNI. In L.R. Ahuja et al. (Eds.). *Root Zone Water Quality Model*. Water Resources Publications, LLC. Highlands Ranch, CO. 119-144.

Shaffer, M.J., B.K. Wylie, R.F. Follett, and P.N.S. Bartling. 1994. Using climate/weather data with the NLEAP model to manage soil nitrogen. *Agricultural and Forest Meteorology*, 69:111-123.

Shaffer, M.J., B.K. Wylie, and M. Hall. 1995. Identification and mitigation of nitrate leaching hot spots using NLEAP/GIS technology. *J. Contaminant Hydrol.*, 20:253-263.

Singh, P., and R.S. Kanwar. 1995. Simulating NO3-N transport to subsurface drain flows as affected by tillage under continuous corn using modified RZWQM. *Trans. ASAE,* 38:499-506.

Smid, A.E., and E.G. Beauchamp. 1976. Effects of temperature and organic matter on denitrification in soil. *Can. J. Soil Sci.,* 56:385-391.

Smith, C.J., J.R. Freney, and W.J. Bond. 1996. Ammonia volatilization from soil irrigated with urban sewage effluent. *Aust. J. Soil Res.,* 34:789-802.

Smith, O.L. 1982. *Soil Microbiology: A Model of Decomposition and Nutrient Cycling.* CRC Press, Boca Raton, FL. 273pp.

Sobecki, T.M., D.L. Moffitt, J. Stone, C.D. Franks, and A.G. Mendenhall. 2001. A broadscale perspective on the extent, distribution, and characteristics of U.S. grazing lands. In R.F. Follett, et al. (Eds.). *The Potential of U.S. Grazing Lands to Sequester Carbon and Mitigate the Greenhouse Effect.* Lewis Publishers, Boca Raton, FL, 21-63.

Stanford, G., M.H. Frere, and D.H. Schwaninger. 1973. Temperature coefficient of soil nitrogen mineralization. *Soil Sci.,* 115:321-323.

Stanford, G., and S.J. Smith. 1972. Nitrogen mineralization potentials of soils. *Soil Sci. Soc. Am. Proc.,* 36:465-472.

Steele, D.D., E.C. Stegman, and B.L. Gregor. 1994. Field comparison of irrigation scheduling methods for corn. *Trans. ASAE,* 37:1197-1203.

Stoichev, D., M. Kercheva, and D. Stoicheva. 2001. NLEAP water quality applications in Bulgaria. In M.J. Shaffer, L. Ma, and S. Hansen (Eds.). *Modeling Carbon and Nitrogen Dynamics for Soil Management.* Lewis Publishers, Boca Raton, FL, 333-354.

Swan, J.B., M.J. Shaffer, W.H. Paulson, and A.E. Peterson. 1987. Simulating the effects of soil depth and climatic factors on corn yield. *Soil Sci. Soc. Am. J.,* 51:1025-1032.

Swan, J.B., J.A. Staricka, M.J. Shaffer, W.H. Paulson, and A.E. Peterson. 1990. Corn yield response to water stress, heat units, and management: model development and calibration. *Soil Sci. Soc. Am. J.* 54:209-216.

Tate, R.L. III. 1987. *Soil Organic Matter — Biological and Ecological Effectors.* John Wiley & Sons, New York.

Tiedje, J.M. 1988. Ecology of denitrification and dissimilatory nitrate reduction to ammonium. In A.J.B. Zehnder (Ed.). *Biology of Anaerobic Microorganisms.* John Wiley & Sons, New York, 179-245.

Wagner-Riddle, C., T.J. Gillespie, L.A. Hunt, and C.J. Swanton. 1997. Modeling a ray cover crop and subsequent soybean yield. *Agron. J.,* 89:208-218.

Watkins, K.B., Y.C. Lu, and W.Y. Huang. 1998. Economic and environmental feasibility of variable rate nitrogen fertilizer application with carry over effects. *J. Agricultural and Resource Economics,* 23:401-426.

Wu, Q.J., A.D. Ward, and S.R. Workman. 1996a. Using GIS in simulation of nitrate leaching from heterogeneous unsaturated soils. *J. Environ. Qual.,* 25:526-534.

Wu, J.J., H.P. Mapp, and D.J. Bernardo. 1996b. Integrating economic and physical models for analyzing water quality impacts of agricultural policies in the High Plains. *Review of Agricultural Economics,* 18:353-372.

Wylie, B.K., M.J. Shaffer, and M.D. Hall. 1995. Regional assessment of NLEAP NO_3-N leaching indices. *Water Resour. Bull.,* 31:399-408.

Xu, C., M.J. Shaffer, and M. Al-Kaisi. 1998. Simulating the impact of management practices on nitrous oxide emissions. *Soil Sci. Soc. Amer. J.,* 62:736-742.

Zeikus, J.G. 1981. Lignin metabolism and the carbon cycle. In M. Alexander (Ed.). *Advances in Microbial Ecology,* Plenum Press, New York, Vol. 5, 211-243.

CHAPTER 3

Carbon and Nitrogen Dynamics in Wetland Soils

W. F. DeBusk, J. R. White, and K. R. Reddy

CONTENTS

INTRODUCTION

Wetlands are an ecologically important part of the landscape. Many species of animals depend on wetlands for successful completion of their life cycle, and most species require, or benefit from, proximity to aquatic habitats. By virtue of their low-lying position in the landscape, wetlands receive hydrologic and nutrient inputs from adjacent upland areas. The retention of floodwater in wetlands, originating from surface and subsurface inflows, results in decreased rate of decomposition and, consequently, increased accumulation of organic matter. Because of their relatively high primary production and low decomposition rate, relative to upland ecosystems, wetlands are considered net sinks for organic carbon (C) and nitrogen (N).

In addition to C and N storage, wetlands serve a number of ecologically important and unique biogeochemical functions. One of the most notable characteristics of wetlands is the presence of aerobic/anaerobic interfaces in the water column (e.g., surfaces of plant litter), at the soil/water interface, and in the root zone (root/soil interface) of aquatic macrophytes. The juxtaposition of aerobic and anaerobic zones in wetlands supports a wide range of microbial populations with specialized metabolic functions, with oxygen reduction occurring in the aerobic interface of the substrate, and reduction of alternate electron acceptors in the anaerobic zone (Reddy and D'Angelo, 1994). Under continuously saturated soil conditions, vertical layering of different metabolic activities can be present, with oxygen reduction occurring at and just below the soil/floodwater interface. Much of the aerobic decomposition of plant detritus occurs in the water column; however, the supply of oxygen might be insufficient to meet demands and might drive certain microbial groups to utilize alternate electron acceptors (e.g., nitrate, oxidized forms of iron and manganese, sulfate, and bicarbonate).

Oxygen replenishment in a wetland soil is limited by the rate of diffusion through the water column and saturated soil matrix, while other inorganic electron acceptors may be added through hydraulic loading to the system. Draining of a wetland soil accelerates organic matter decomposition due to the introduction of oxygen deeper into the profile. In many wetlands, the influence of NO_3^-, Mn^{4+}, and Fe^{3+} on organic matter decomposition is minimal, as the concentration of these electron acceptors is usually low. The demand for electron acceptors of greater reduction potential (NO_3^-, Mn^{4+}, and Fe^{3+}) is high; therefore, they are rapidly depleted from the system. Long-term sustainable microbial activity is then supported by electron acceptors of lower reduction potential (SO_4^{2-} and HCO_3^-). Methanogenesis is often viewed as the terminal step in anaerobic decomposition in freshwater wetlands, whereas sulfate reduction is viewed as the dominant process in coastal wetlands. However, both processes can function simultaneously in the same ecosystem and compete for available substrates.

CARBON STORAGE IN WETLANDS

Organic C is stored in wetlands in living (vegetation, fauna, and microbial biomass) and non-living (dead plant tissue, plant litter, peat, or SOM) forms (Figure 3.1). Non-living storage of organic C is proportionally large in wetlands in relation to other ecosystems; this storage provides a substantial energy reserve to the ecosystem, which is gradually released through the detrital food web. Long-term storage of organic matter in the form of peat deposits is characteristic of a number of wetland types, and is largely a function of hydrology and climate.

Organic C accumulation in wetlands represents the mass balance between net primary production (C fixation) and heterotrophic metabolism (C mineralization). Organic C in plant litter, peat, or soil organic matter (SOM) serves as the source of energy to drive the detrital food chain in wetlands. Most of the organic matter produced in wetlands is deposited directly in the detrital pool (Moran et al., 1989;

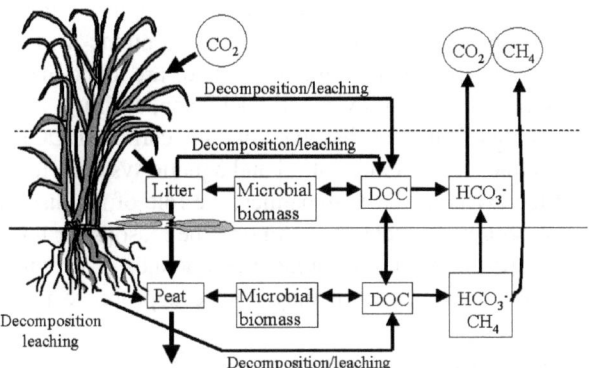

Figure 3.1 Schematic of the carbon cycle in wetlands.

Wetzel, 1992); thus, microbial decomposers play the major role in C cycling and energy flow in wetlands.

Burial of organic matter as peat provides a means for long-term storage of elements associated with organic C, such as nutrients and heavy metals (Clymo, 1983). Allochthonous compounds can be incorporated into peat and soil organic matter through plant uptake and senescence; immobilization within the soil organic matrix by physical/chemical processes such as adsorption, occlusion, and precipitation; or through uptake by microbial decomposers, with storage either within living cells or metabolic by-products. On a much broader scale, storage of organic C in wetland soil is an important component of the global carbon cycle and thus may impact large-scale processes such as global warming and ozone depletion (Happell and Chanton, 1993; Whiting, 1994).

Peat accumulation is the result of biological, chemical, and physical changes imposed on plant remains over an extended time period. Extent of decomposition, or humification, is qualitatively assessed by the extent to which plant structure is preserved (Given and Dickinson, 1975; Clymo, 1983). Numerous organic constituents have been isolated from peat, and many have been used in assessing the degree of decomposition, although these are generally classes and sub-classes of organic compounds rather than discrete compounds. For example, peat material soluble in nonpolar solvents is often termed "wax," but includes numerous compounds other than waxes (esters of fatty acids with alcohols other than glycerol) (Clymo, 1983). Although it is a small portion of total peat mass, this fraction is of interest when determining origins of the peat because fatty acids of lipids can be traced directly or indirectly to the original plant type (Given and Dickinson, 1975; Borga et al., 1994). Various types of acid or alkali hydrolysis procedures have been used to isolate fractions roughly equivalent to cellulose, hemicellulose, and lignin. The proportions of cellulose and hemicellulose in plant litter and peat tend to decrease with age (i.e., decomposition or humification), while lignin content increases with age; these three structural groups have all been used to characterize the degree of peat decomposition (Clymo, 1983; Brown et al., 1988; Bohlin et al., 1989). Analysis of peat has also revealed a large variety of phenolic compounds, many of which can be extracted in the "lignin" fraction. Concentration of cellulose and lignin is much higher in the fibrous fraction (remnant plant parts) of peat, while humic acids are much more prevalent in the humus fraction, although fulvic acids are found in somewhat greater amounts in the fibrous fraction (Given and Dickinson, 1975).

The ecological significance of dissolved organic carbon (DOC) in wetlands has not been clearly defined. Even in terrestrial and aquatic systems, for which a greater depth of knowledge exists for DOC dynamics, the role of the dissolved fraction of organic C has not been well established. Among the reasons for this is the fact that DOC represents a broad spectrum of organic compounds of varying environmental recalcitrance (Wetzel, 1984); thus, it may not be appropriate to treat DOC as a homogeneous category.

The DOC pool in aquatic ecosystems is a relatively stable component, both in terms of the size and quality of the pool (Wetzel, 1984). Decomposition of organic substrates in aquatic systems involves both labile and complex (recalcitrant) organic matter, the latter comprising the bulk of the DOC. Turnover of highly labile, energy-rich substrates may approach a rate of 5 to 10 times per day; thus, actual concentration (storage) of labile organic C is generally extremely low. Despite the fact that recalcitrant DOC and particulate organic C (POC) are slow to mineralize, these pools represent a major portion of the organic C processed by the heterotrophic community due to the relatively massive size of these pools.

During transport, selective removal of organic compounds occurs due to microbial utilization and chemical adsorption or precipitation; thus, recalcitrance of DOC increases downgradient. Decomposition of dissolved organic compounds (resulting from partial decomposition of particulate organic matter) occurs primarily at surfaces in the litter layer, soil/sediment, and among epiphytic microflora.

Most of the organic C fixed in wetlands and aquatic systems (by both phytoplankton and macrophytes) is processed and recycled entirely by bacteria, without entering the food web (i.e., higher animals) (Wetzel, 1984). Decomposition of organic matter is the primary ecological role of the heterotrophic microflora in soils, as it provides for mineralization of growth-limiting nutrients and formation of recalcitrant organic compounds (e.g., humus), which contribute to the chemical stability of the system (Swift, 1982). Microbial biomass comprises only a small fraction of the non-living organic matter, yet most of the net ecosystem production passes through the microbial component at least once and typically several times (Elliott et al., 1984; Heal and Ineson, 1984; Van Veen et al., 1984). Microbial decomposers derive their energy and C for growth from detrital organic C and facilitate recycling of energy and C within and external to the wetland ecosystem (Wetzel, 1984; 1992). Soil microbes can exert a significant influence on ecosystem energy flow in the form of feedback, because mineralization of organically bound nutrients is a regulator of nutrient availability for both primary production and decomposition (Elliott et al., 1984). Soil microbial biomass represents a significant fraction of the total C storage, dependent to a large extent on C substrate quality and nutrient availability (discussed later in this chapter). For example, in a northern Everglades marsh, microbial biomass C in peat (0 to 10 cm depth) ranged from 3.5 to 18.7 mg C/kg soil, representing 0.8 to 4.3% of the total soil organic C pool (DeBusk and Reddy, 1998).

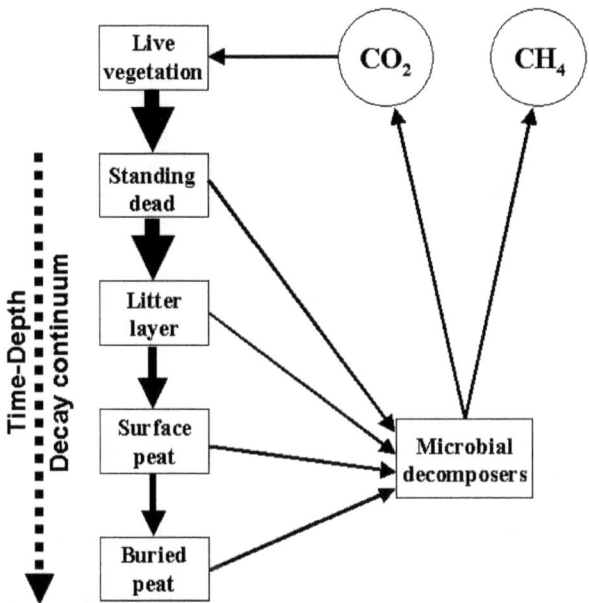

Figure 3.2 Schematic of the time-depth continuum of organic matter decomposition and accretion in wetlands.

CARBON CYCLING IN WETLANDS

Organic C undergoes complex cycling in wetlands, and the fate of C depends on the specific type of compound, as well as environmental conditions. In general, easily degradable (labile) fractions are decomposed to inorganic constituents, while recalcitrant pools are accreted as new peat layers (Figure 3.2). Organic matter associated with above- and below-ground plant, algal, and microbial biomass in wetlands consists of complex mixtures of non-humic substances, including particulate pools (cellulose, hemicellulose, tannins and lignins, proteins), water-soluble components (amino acids, sugars, and nucleotide bases), and ether-extractable components (lipids, waxes, oils). As plants and microorganisms senesce and die, for example during frost or natural seasonal die-off, non-humic substances are deposited in the water column and soil surface. Labile fractions undergo multistep conversion to inorganic constituents including (1) abiotic leaching of water-soluble components and fragmentation of tissues into small pieces (<1 mm), (2) extracellular enzyme hydrolysis of biopolymers (e.g., nucleic acids, proteins, and cellulose) into monomers, and (3) aerobic and anaerobic catabolism of monomers by heterotrophic microorganisms (Figure 3.3). Microbial biomass has been found to make up a significant amount of C in wetland substrates. Nutrients that are released are also available for plant uptake and growth.

Abiotic leaching of algal and plant biomass is largely complete within days to months after deposition into water, but depends on the amount of particulate and

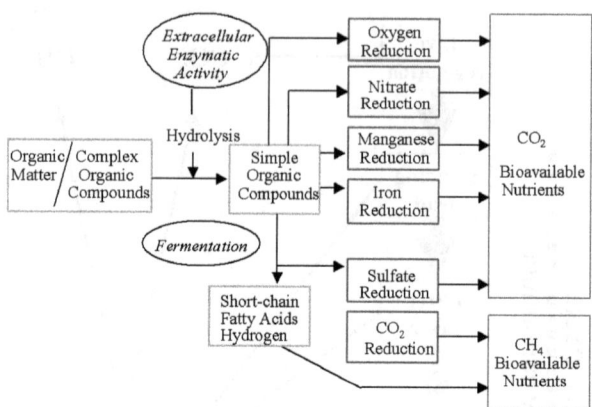

Figure 3.3   Schematic showing the biochemical pathways for breakdown of soil organic matter in wetland ecosystems.

structural material. For example, Benner et al. (1985) measured only 10 to 20% losses of lignocellulosic components of plants in one month. Particulate materials remaining after initial leaching are typically fragmented by meiofauna before undergoing further decomposition. This step is critical because surface area is increased, thus allowing microorganisms and enzymes to penetrate tissues.

Extracellular enzymes such as cellulases and proteases are excreted by fungi and bacteria, and hydrolyze high-molecular-weight biopolymers into oligomers and monomers that can then be taken up by microorganisms. Enzyme hydrolysis is generally considered the rate-limiting step in organic matter decomposition (Sinsabaugh et al., 1993). Oxidoreductase enzymes, such as peroxidases and phenoloxidases, are involved in lignin oxidation. Because these enzymes require O_2, degradation of lignin is most prevalent in aerobic zones such as soil/water and root/soil interfaces. Production and activity of extracellular enzymes is affected by a number of environmental factors, including nutrients, pH, O_2 supply, humic substances, and inorganic cations. For example, Sinsabaugh and Moorhead (1994) and Sinsabaugh et al. (1993) found that production of proteases and phosphatases was enhanced in N- and P-limited wetland systems, and lignocellulase activity and organic C mineralization were enhanced under nutrient non-limiting conditions. Kim and Wetzel (1993) demonstrated that humic substances complexed and curtailed the activity of extracellular enzymes, which was relieved by the presence of divalent cations such as Ca^{2+} and Mg^{2+}. They speculated that this phenomenon might explain the reduced decomposition in soft water compared to hard water aquatic systems. Activities of glucosidase, protease, and phosphatases were found to decrease with redox potential, exhibiting high activity under aerobic conditions and low activity under anaerobic conditions (McLatchy and Reddy, 1998).

After enzyme hydrolysis, low-molecular-weight compounds are taken up and utilized as C and energy sources by heterotrophic microorganisms. A multitude of different microorganisms may be involved in this terminal step of decomposition, which depends largely on the availability of electron acceptors. Diverse types of microorganisms couple oxidation of organic C substrates (electron donors) and reduction of electron acceptors to energy (ATP) production required for growth. The most common electron acceptors include O_2, NO_3^-, Mn^{4+}, Fe^{3+}, SO_4^{2-}, and CO_2, which are gained through both internal and external inputs. For example, internal inputs include oxidation of chemical species (NH_4^+, H_2S, Fe^{2+}, and Mn^{2+}) that diffuse from anaerobic to aerobic soil zones. External inputs of electron acceptors include atmospheric O_2, seawater (SO_4^{2-}), surface water runoff (NO_3^-), and precipitation. Most wetlands contain several electron acceptors; thus, there exists competition for electron donors between microbial groups. Organisms that derive the most energy, and have the fastest degradation kinetics, outcompete other groups. For these reasons, aerobic bacteria outcompete anaerobes. For example, DeBusk and Reddy (1998) determined that aerobic C mineralization was about 3 times faster than mineralization under anaerobic conditions.

Detrital material derived from roots and above-ground plant and algal biomass that are buried undergo anaerobic decomposition. Under anaerobic conditions, microorganisms must utilize alternate electron acceptors to oxidize organic matter. Due to energy differences, electron acceptors are utilized in the order $O_2 > NO_3^- > Mn^{4+} > Fe^{3+} > SO_4^{2-} > HCO_3^-$. This phenomenon explains the stratification in microbial activity and chemicals often observed in wetland soil profiles (Reddy and D'Angelo, 1994). Slow decomposition rates under anaerobic conditions partially explains why wetlands accumulate organic matter more than upland systems (DeBusk and Reddy, 1998).

Carbon compounds that are recalcitrant to aerobic and anaerobic decomposition tend to accumulate in wetlands, either as undecomposed plant tissues (peat) or humic substances. Formation of humic substances is thought to involve condensation reactions between reactive phenolic groups of tannins and lignins with water-soluble non-humic substances, catalyzed by phenoloxidase enzymes in soils (Stevenson, 1994). This mechanism accounts for the high molecular weight, and heterogeneous humic substances contain significant amounts of N, P, and S in their structures. In the absence of O_2, humic substances are resistant to decomposition, and thus represent a significant carbon and nutrient storage in wetlands. Under drained conditions, humic substances are rapidly degraded, which releases nutrients to the bioavailable pool, thereby affecting downstream water quality.

REGULATORS OF ORGANIC CARBON DECOMPOSITION

The decomposition/mineralization process in wetlands differs from that in upland ecosystems in a number of ways (Reddy and D'Angelo, 1994). The predominance of aerobic conditions in upland soils generally results in rapid decomposition of organic matter such as plant and animal debris. Net retention of organic matter is minimal in this case, and consists of accumulation of highly resistant compounds that are relatively stable even under favorable conditions for decomposition (Jenkinson and Rayner, 1977; Paul, 1984). The decomposition process occurs at a significantly lower rate in wetland soils, due to frequent-to-occasional anaerobic conditions throughout the soil profile resulting from flooding. Because of this, significant accumulation of moderately decomposable organic matter occurs, in addition to lignin and other recalcitrant fractions (Clymo, 1983).

There is evidence that growth strategies of the heterotrophic community reflect those of the plant community through their direct response to variation in resource quality, which is a function of plant growth strategies, and their similar response to common environmental conditions (Heal and Ineson, 1984). Carbon and nutrient utilization by the microbial decomposer community responds to three main groups of factors: (1) substrate quality, (2) physicochemical environment, and (3) other organisms. Substrate quality is a general term that refers to the combination of physical and chemical characteristics which determine its potential for microbial growth. Substrate quality is not determined by any particular factor; however, nitrogen and lignin content have been suggested as indicators of biodegradability (Heal et al., 1981; Minderman, 1968; Andrén and Paustian, 1987; Melillo et al., 1989). Physicochemical factors include temperature, pH, exogenous nutrient supply, moisture content (for non-flooded conditions), and oxygen or alternate electron acceptor availability (Swift et al., 1979; Heal et al., 1981; Reddy and D'Angelo, 1994).

Although loading of anthropogenic nutrients stimulates the growth of aquatic vegetation in wetlands, a significant portion of the nutrient requirements may be met through remineralization during decomposition of organic matter. The rate of organic matter turnover and nutrient regeneration is influenced by hydroperiod (Happell and Chanton, 1993), characteristics of the organic substrates, (Webster and Benfield, 1986; DeBusk and Reddy, 1998), supply of electron acceptors (D'Angelo and Reddy, 1994a; b), and addition of growth-limiting nutrients (Button, 1985; McKinley and Vestal, 1992; Amador and Jones, 1995).

Nutrient availability affects the decomposition rate by limiting microbial growth. Growth-limiting nutrients can be obtained by the microbial decomposers from the organic substrate or from dissolved compounds in the water and porewater (Godshalk and Wetzel, 1978). Microbial decomposers colonizing nutrient-depleted substrates (e.g., high C:N or lignin:N ratio) tend to scavenge significant amounts of nutrients from the surrounding media, resulting in net immobilization of growth-limiting nutrients in the system (Melillo et al., 1984). Although nutrient loading is typically greater in wetlands than in uplands due to location within the landscape, nutrient availability may be low relative to the pool of available organic C in wetlands (Reddy and D'Angelo, 1994). Both nitrogen (N) and phosphorus (P) have been identified as microbial growth-limiting nutrients in wetlands (Westermann, 1993). Nitrogen, unlike phosphorous, may be lost from wetlands through microbial metabolism via denitrification, as well as through ammonia volatilization (Reddy and D'Angelo, 1994).

Substrate availability to microbial decomposers is determined by molecular size distribution, availability of nutrients in the substrate as well as in the environmental matrix, and the amount of exposed surface area (Heal et al., 1981). Preferential utilization of low-molecular-weight compounds by microbial decomposers alters the chemical character of the organic resource such that the proportion of complex, slowly degradable compounds increases over time. Thus, the substrate quality of resources over a wide range of initial compositions tends to converge to a more uniform composition, and hence, degradability (Melillo et al., 1989).

The basic cell wall construction in vascular plants includes a framework component of α-cellulose, a matric component of linear polysaccharides (hemicellulose), and an encrusting component composed of lignin (Zeikus, 1981). Lignin occurs in cells of conductive and supportive tissue, and thus is not found in algae and mosses. The presence of lignin is the ultimate limiting factor in the decomposition of vascular plant tissue (Zeikus, 1981).

The heterogeneous group of organisms known as white-rot fungi, found primarily in terrestrial habitats, are the most active and complete decomposers of lignin (Eriksson and Johnsrud, 1982). Partial chemical modification of lignin has been documented for other eukaryotic microbes, most notably the brown-rot and soft-rot fungi, as well as for certain species of prokaryotes (Zeikus, 1981). Certain species of the genera *Streptomyces, Norcardia, Bacillus, Azotobacter,* and *Pseudomonas* are included in the latter category. Numerous species of fungi capable of cellulose and lignin degradation, including Hyphomycetes, have been identified in oxidized zones of wetland soils (Westermann, 1993).

Lignin degradation by white-rot fungi is highly oxidative and may involve singlet oxygen and hydroxyl radicals in chemical oxidation of aromatic ring structures or intermonomeric linkages (Benner et al., 1984a). However, recent research has cast doubt on the unconditional requirement of molecular O_2 for lignin degradation. Wetland plant decomposition studies have demonstrated decomposition of lignin in anoxic salt marsh, freshwater marsh, and mangrove sediments using ^{14}C labeling techniques (Benner et al., 1984a). Bacterial degradation of lignin from *Spartina alterniflora* predominated over fungal degradation in studies of decomposition in salt marsh sediment (Benner et al., 1984b).

The capacity for cellulose depolymerization is shared by several species of bacteria, actinomycetes, and microfungi (Sagar, 1988). Cellulose degradation readily occurs under anaerobic conditions, although at a reduced rate, mediated primarily by bacteria of the genus *Clostridium* (Swift et al., 1979). The ratio of cellulose to lignin degradation rate has been shown to be similar under both aerobic and anaerobic conditions (Benner et al., 1984a).

Aside from its effects on lignin degradation and other extracellular depolymerization, O_2 depletion forces a major shift in microbial metabolism of monomeric C compounds (e.g., glucose and acetate), from aerobic to anaerobic pathways (Westermann, 1993). Catabolic energy yields for bacteria utilizing alternate electron acceptors (NO_3^-, Mn^{4+}, Fe^{3+}, SO_4^{2-}, CO_2) are lower than for O_2; thus, microbial growth rates are generally lower in anaerobic environments (Westermann, 1993; Reddy and D'Angelo, 1994). In addition, sulfate reducing and methanogenic bacteria must depend on fermenting bacteria (e.g., *Clostridium* spp.) to produce substrate in the form of short-chain C compounds, such as volatile fatty acids, from the breakdown of mono- and polysaccharides (Howarth, 1993). Thus, although C metabolism occurs in the absence of O_2, and even in the complete absence of electron acceptors, the decomposition process for plant litter and soil organic matter is often significantly curtailed.

NITROGEN STORAGE IN WETLANDS

Nitrogen exists in a variety of inorganic and organic forms in wetland soils, with the majority present in organic forms. Components of N storage in living matter consist of submerged and emergent vegetation, fauna, and the microbial biomass. Nitrogen storage in non-living matter includes standing dead and detrital plant matter, peat, and soil organic matter. The stores of organic N in detrital tissue and soil organic matter comprise the vast majority of N within wetland systems, and primarily consist of complex proteins and humic compounds with trace amounts of simple amino acids and amines.

Organic N stored within wetland soils is the most stable pool of N and, as such, is not readily available for internal cycling. As with C, the more refractory organic compounds become buried within the soil and slowly accrete over time. There are a variety of factors that control the stability of the accreted organic N in the soil, including water depth, hydrologic fluctuations, temperature, supply of electron acceptors, and microbial activity.

There exist stores of inorganic N forms, including ammonium (NH_4^+), nitrate (NO_3^-), and nitrite (NO_2^-). In flooded soil, NO_3^- and NO_2^- are commonly found only in trace amounts, with the vast majority of inorganic N present as NH_4^+. Nitrate concentrations are higher only within the aerobic portion of the wetland. Gaseous N products include NH_3, N_2O, and N_2 and are generally transient stores as they are released to the atmosphere. The inorganic stores of N are not very stable over time and generally comprise less than 1% of the total N within the wetland (Howard-Williams and Downes, 1994).

The microbial biomass N compartment is relatively small when compared with the total pool of organic N, yet the microbial activity is the single greatest regulator of the stability of organic N. Microbial decomposers utilize simple organic N compounds for cell growth, while a variety of functional microbial groups utilize inorganic N compounds as key components in the electron transport phosphorylation system. The percent of microbial biomass N with respect to total N is typically within the range of 0.5 to 3% of total N (White and Reddy, 2000).

NITROGEN CYCLING IN WETLANDS

Nitrogen reactions in wetlands can effectively process inorganic N through nitrification and denitrification, ammonia volatilization, and plant uptake (Figure 3.4). These processes aid in maintaining low levels of inorganic N in the water column. A significant portion of dissolved organic N is returned to the water column during breakdown of detrital plant tissue or soil organic matter, and the majority of this dissolved organic N is resistant to decomposition. Under these conditions, surface water exiting the wetland may contain elevated levels of N in organic forms. The relative rates of these reactions will depend on the optimal environmental conditions present in the soil and water column.

Ammonification is the primary step in organic N mineralization and is defined as the biological process by which organic forms of N are transformed to NH_4^+. Similar to organic C, complex polymeric organic N compounds are hydrolyzed to simple monomers through the activity of extracellular enzymes (Gardner et al., 1989; McLatchey and Reddy, 1998). Eventually, this process leads to the final breakdown of amino acids and results in the liberation of NH_4^+.

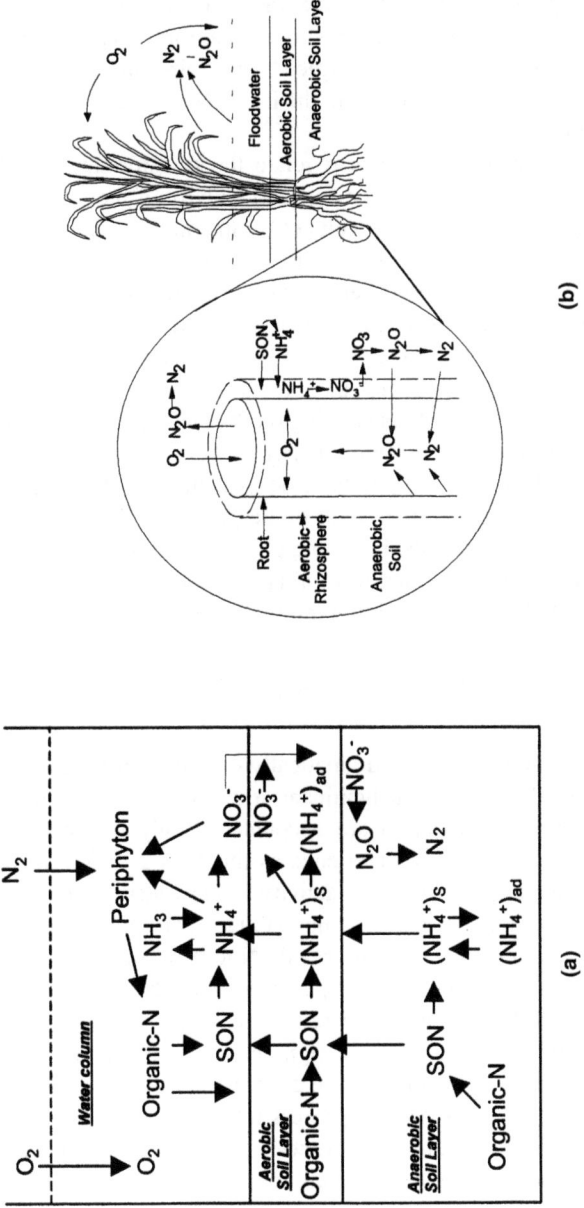

Figure 3.4 The wetland nitrogen cycle, depicting the N transformations within aerobic-anaerobic regions of the wetland soils (a) at the soil/floodwater interface and (b) the root/soil interface.

Ammonification is carried out by a wide variety of general-purpose heterotrophic microorganisms and occurs under both aerobic and anaerobic conditions. Ammonium concentrations increase under anaerobic conditions as a result of the low N requirements of anaerobic microorganisms. This increase is also due to the absence of oxygen, which prevents nitrification, the transformation of NH_4^+ to NO_3^-. Immobilization is the conversion of inorganic N to organic forms. The difference between immobilization and ammonification is termed "net N mineralization" and is generally a positive value in wetlands, again due to the lower N requirements of the anaerobic microbial community.

The soil microbial biomass helps regulate the transformation and storage of nutrients in soils (Martens, 1995). The size and activity of the microbial pool have been shown to significantly correlate with net N mineralization rates in wetland soils (Williams and Sparling, 1988; White and Reddy, 2000). Therefore, the activity of the microbial pool can regulate nutrient release, thereby affecting wetland surface water quality. Ammonification rates have been reported to range from 0.004 to 0.357 g N/m^2/d (mean = 0.111 + 0.124; n = 13) in wetland soils (Martin and Reddy, 1997). Ammonium N released through the mineralization process is readily available to plants and the microbial pool. Depending on the environmental conditions, NH_4^+ can be either lost through volatilization and/or oxidized to nitrate N.

Ammonia volatilization is an abiotic process controlled by the pH of the soil-water system. However, there are several biotic processes that can alter the pH of the soil-water system in wetlands, including photosynthesis and denitrification, and thus can indirectly control the volatilization rate. The following equation depicts the transformation of NH_4^+ to NH_3, highlighting the role of pH:

$$NH_4^+ + OH^- \rightarrow NH_3(aq) + H_2O$$

The rate of volatilization increases dramatically over the pH range from 8.5 to 10. Losses due to volatilization are insignificant at a pH < 7.5. In wetlands, the high pHs needed to drive volatilization are encountered only in the overlying water column. The water column may experience as much as 2 to 3 units of pH swing over a diel cycle due to photosynthetic activity in the water column.

Sources of NH_4^+ to the water column include: (1) addition of NH_4^+ through external sources, (2) mineralization of organic N to NH_4^+ in the water column, and (3) diffusion of NH_4^+ from the anaerobic soil layer to the water column. In most wetlands, losses of NH_4^+ through volatilization are controlled by low concentrations in the water column and may not be a significant process. However, in wetlands receiving external loads of N, this may be a significant process.

Nitrification is an obligate aerobic process involving oxidation of NH_4^+ to NO_3^-. Nitrification is mediated by autotrophic bacteria that couple the oxidation of NH_4^+ to electron transport phosphorylation while utilizing inorganic carbon (CO_2) to synthesize required cellular components. The process of nitrification is a two-step process that first involves the oxidation of NH_4^+ to NO_2^- by bacteria of the genus *Nitrosomonas*. The final step, the oxidation of NO_2^- to NO_3^-, is mediated by bacteria of the genus *Nitrobacter*. The oxidation state of N increases from –3 for NH_4^+ to +5 for NO_3^-.

$$NH_4^+ + 1\frac{1}{2} O_2 \rightarrow NO_2^- + 2H^+ + H_2O + \text{Energy}$$

$$NO_2^- + \frac{1}{2} O_2 \rightarrow NO_3^- + \text{Energy}$$

Nitrification occurs in three zones of a wetland soil-water profile: water column, aerobic soil layer, and the aerobic region of the rhizosphere. In all of these zones, the supply of NH_4^+ and availability of O_2 are the major limiting factors for nitrification. In most wetlands, NH_4^+ supply to aerobic portions of the wetland occurs through transport from anaerobic soil layers, as a result of concentration gradients across these two layers. Rates of nitrification have been reported to range from 0.01 to 0.161 g N/m²/d (mean = 0.048 ± 0.044; n = 9) in constructed wetlands (Martin and Reddy, 1997).

Nitrate, added to wetlands or derived through nitrification, rapidly diffuses into anaerobic soil layers, where it is utilized as an electron acceptor and reduced to gaseous end products of N_2O and N_2 or NH_4-N. The former pathway, known as denitrification, occurs at higher Eh levels (200 to 300 mV); while the latter process, known as dissimilatory NO_3 reduction to NH_4^+, occurs at low Eh levels (<0 mV).

Denitrification is the microbial-mediated reduction of nitrogenous oxides to N_2 gas. Denitrifiers are known to exist in almost all soils and come from a wide range of genera, including *Pseudomonas*, *Alcalignes*, *Flavobacterium*, *Paracoccus*, and *Bacillus* (Firestone, 1982; Tiedje, 1988). The facultative anaerobic bacteria are capable of using NO_3^- or NO_2^- as terminal electron acceptors, coupled with the oxidation of organic C in the absence of O_2. The oxidation state of N decreases from +5 for NO_3^- to 0 for N_2 gas.

Forms of N $\quad NO_3^- \rightarrow NO_2^- \rightarrow N_2O \rightarrow N_2$

Oxidation state $\quad$ +5 $\quad\quad$ +3 $\quad\quad$ +1 $\quad\quad$ 0

In general, the relatively high organic C content of wetlands soils provides ample substrate for heterotrophic microbial activity. Due to the high C content and paucity of O_2 in wetland soils, NO_3^- supply becomes the limiting factor in denitrification (Cooper, 1990; Gale et al., 1993; White and Reddy, 1999). Schipper et al. (1993) found that up to 77% of the variability in *in situ* denitrification rates in a riparian wetland could be explained by NO_3^- concentrations and the soil denitrifying enzyme activity. In addition, the authors noted that organic soils comprised only 12% of the total soils in the catchment, but were responsible for 56 to 100% of total NO_3^- consumption. Denitrification rates in wetlands on an areal basis have been reported to range from 0.003 to 1.02 g N/m²/d (mean = 0.097 + 0.139; n = 23; Martin and Reddy, 1997). Denitrification is the primary mechanism for inorganic N removal from the wetland systems and, as such, wetlands are utilized for their significant water quality enhancement capabilities.

Dissimilatory nitrate reduction to ammonium (DNRA) occurs in anaerobic soils and involves the reduction of NO_3^- to NH_4^+ rather than N_2 gas, as in denitrification. Conditions that favor DNRA include a highly reduced soil; therefore, DNRA could potentially compete with denitrification for available NO_3^-. Organisms responsible for DNRA are thought to be obligate anaerobes, as opposed to facultative anaerobes for denitrification (Tiedje, 1988). Isolated bacterial genera found to mediate DNRA are *Clostridium*, *Achromobacter*, and *Streptococcus* (Cole, 1990). The process of DNRA involves the consumption of eight electrons, as compared to five electrons for denitrification. Therefore, the more reduced the soil becomes, the more favorable the conditions are for DNRA. Assimilatory nitrate reduction (ANR) is favored to occur in environments where NH_4^+ availability is low and the N can then be incorporated directly into cellular components. Both these NO_3^- reduction processes are small when compared with denitrification losses in wetlands.

Biological N_2 fixation is the only mechanism available for conversion of inert, gaseous N_2, present in the atmosphere, to biologically available organic and inorganic forms. The process of N_2 fixation is driven by the nitrogenase enzyme produced by only a limited number of prokaryotic organisms, including several genera of bacteria and cyanobacteria (Vymazal and Richardson, 1995). These N_2-fixing organisms may occur in wetlands in the soil and water column, as plankton, filamentous mats, periphyton, or in symbiotic association with vegetation. For example, the aquatic fern Azolla has the ability to fix N_2 and meet its total N requirement, due to a symbiosis with the cyanobacteria *Anabaena azollae*, which grows in cavities on the fronds of the Azolla plant.

A wide range of biological N_2 fixation rates are reported for wetlands. For example, rates have been reported to range from 0.7 to 12 g N/m²/yr in the root zone, and from 0.4 to 46 g N/m²/yr within the photic zone of coastal wetlands (Buresh et al., 1980). The presence or absence of vegetation can affect the distribution and N_2 fixation rates. Well-developed cyanobacterial mats have exhibited rates ranging from 1.2 to 76 g N/m²/yr, while soils devoid of macrophyte coverage or cyanobacterial mats have demonstrated rates in the range from 0.002 to 1.6 g N/m²/yr (Howarth et al., 1988a). Nitrogen fixation rates were found to range from 1.8 to 18 mg N/m²/yr in a northern Everglades marsh (Inglett, 2000).

The overall cycling of organic N to inorganic N and through to removal from the wetland follows a path beginning with the mineralization of organic matter to NH_4^+. Ammonium produced in the primarily anaerobic soil will be conserved until diffusion into an aerobic water column or soil volume. Nitrification, constrained to these aerobic zones, is responsible for the transformation of NH_4^+ to NO_3^-. Nitrate is then conserved until diffusion into the anaerobic zones of the soil, where it becomes converted to either N_2O or N_2 gas, leading to removal from the wetland system. The time for cycling of N through the wetland is short in comparison with the residence time of N_2 gas in the atmosphere. However, the transformations of N from N_2 gas through the various bioavailable (inorganic) forms are critical in maintaining wetland ecosystem productivity.

REGULATORS OF NITROGEN CYCLING IN WETLANDS

There are a few systemwide regulators that can affect the overall cycling of N in wetlands, including hydrology linked to electron acceptor availability, temperature, the presence of macrophytes, and substrate quality. The vast majority of N, stored in organic forms, is generally unavailable for biological uptake. The presence of water-filled pore spaces in the soil creates anoxia due to high soil oxygen demand and the relatively slow diffusion of O_2 in water compared with air (~10,000 times slower). The existence of aerobic and anaerobic zones in flooded soils and the different oxidation states of N forms play a critical role in N transformations (D'Angelo and Reddy, 1994). For example, the reduction potential of the wetland soil can have marked control over the rate of ammonium production. The redox potential (Eh) of the soil decreases, from aerobic (+600mV) to nitrate reducing (+300mV), sulfate reducing (–100mV), and then finally to methanogenic conditions (–250mV). The mineralization rate of alanine in an organic wetland soil was found to significantly sequentially decrease under each of these four decreasing redox states (McLatchey and Reddy, 1998). The same result was seen for decomposition of the native soil organic matter measured by CO_2 evolution (Reddy and Graetz, 1988). Organic N mineralization can be controlled as a function of the C:N ratio, extracellular enzyme (such as protease), microbial biomass, and soil redox conditions. For example, N mineralization was highly correlated with microbial biomass N of wetland soil maintained under different redox conditions (McLatchey and Reddy, 1998).

There are a variety of regulating factors of nitrification, including NH_3 concentration, pH, alkalinity, wind velocity, density of emergent macrophytes, photosynthetic activity in the water column, and the soil cation exchange capacity. Many of these regulators are interrelated. The physical controls on volatilization losses include wind velocity and density of emergent macrophytes. The movement of aqueous NH_3 into the atmosphere across the surface water boundary is regulated by the difference in concentration of NH_3 in air and in water under a concentration gradient. Therefore, these factors can be combined to increase volatilization losses, with high surface wind speeds and an absence of emergent macrophytes. The factors can work in reverse to decrease losses where surface wind speeds are low and the emergent macrophyte density is high, thereby retarding air movement.

Nitrification is primarily confined to the water column and a very thin (<1 cm) surface layer of soil due to the obligate aerobic requirement. Very little nitrification occurs within flooded soil and then only within the rhizosphere of some wetland plants that create an aerobic microclimate (Reddy et al., 1989).

The soil aeration status can have an effect on denitrification. In general, denitrifiers are facultative aerobes capable of utilizing O_2 as well as nitrogenous oxides as respiratory electron acceptors. When O_2 is available, the overall return of energy to the microbial pool is greater for O_2 and its use is therefore preferred. In addition, the presence of O_2 has also been observed to have an inhibitory effect on the enzymes, in essence, turning them off in aerobic environments (Martin et al., 1988).

Temperature can also mediate rates of N transformation processes in wetland systems. Higher temperatures have been demonstrated to have a stimulatory effect on net N mineralization rates. In general, there is a doubling of the mineralization rate with a 10°C increase in temperature (Reddy and Patrick, 1984); others have shown a doubling of the ammonification rates when the temperature was raised from 15 to 40°C (Cho and Ponnamperuma, 1971).

Increasing the temperature from 10 to 25°C was shown to increase N use efficiency by *Typha* spp. from 5 to 38% (Reddy and Portier, 1987). In temperate climates, much of the N assimilation occurs during the growing season. During winter months, above-ground biomass is killed and accumulates as detrital tissue. At the same time, a significant portion of N is translocated into below-ground biomass. Nitrogen release during decomposition of detrital tissue *Typha* spp. was more rapid during summer months than during the winter.

The presence of vegetation can play a significant role in N cycling by: (1) assimilating inorganic N into plant tissue, and (2) providing environment in the root zone for nitrification-denitrification to occur. The efficiency of N utilization (defined as the increase in plant N per unit mass of available N) by aquatic vegetation is highly variable, depending on the type of wetlands (forested vs. herbaceous). The N use efficiency of aquatic vegetation and the C:N ratio of the plant litter decreased with nutrient loading (Shaver and Melillo, 1984; Reddy and Portier, 1987; Koch and Reddy, 1992). Plants derive most of their N from soil porewater, with only a small amount of the floodwater N being directly utilized. Floodwater N is rapidly utilized by algae and microbial communities growing on the plant litter substrates or lost through ammonia volatilization and nitrification-denitrification reactions (Reddy and Patrick, 1984; DeBusk and Reddy, 1987). The extent of these processes also increases with N loading. Nitrogen assimilation by herbaceous vegetation is usually short-term and usually cycled rapidly within the system.

The substrate composition of wetland detritus coupled with the absence of oxygen, can dramatically affect the amount of inorganic N released during decomposition. For example, under aerobic conditions that persist in drained soils, plant detritus containing a C:N ratio of approximately 25 will lead to net immobilization of N. However, under anaerobic conditions, plant detritus exhibiting a C:N ratio greater than 80 will lead to immobilization, while plant tissue below this threshold will lead to net N release (Damman, 1988; Williams and Sparling, 1988; Humphrey and Pluth, 1996). Therefore, under flooded conditions, most decaying organic matter acts as a source of inorganic N.

There exist several proximal regulators on N_2 fixation in wetlands. Positive influences include low N:P ratios of inputs (wastewater, plant residues), high redox potentials, molybdenum (Howarth and Cole, 1985), iron (Paerl et al., 1994), and dissolved organic matter (Howarth et al., 1988b). The controls that provide a negative or inhibitory effect include oxygen, dissolved inorganic N, and sulfate concentrations (Howarth et al., 1988b).

BIOGEOCHEMICAL ROLES OF CARBON
AND NITROGEN IN WETLANDS

Wetlands contain vast numbers of organic-substrate-utilizing microorganisms adapted to the aerobic surface waters and anaerobic soils. Organic C in wetlands is broken down into CO_2 and methane (CH_4), both of which are lost to the atmosphere. Similarly, organic N is converted to NH_4^+ and NO_3^- and subsequently to N_2O and N_2. Wetlands also store and recycle copious amounts of organic C and N contained in plants and animals, dead plant material (litter), microorganisms, and peat. Therefore, wetlands tend to be natural exporters of organic C and N as a result of the decomposition of organic matter into fine particulate matter and dissolved compounds.

In wetlands, as in many terrestrial ecosystems, plant litter accumulates at the soil surface. Some of the nutrients, metals, or other elements previously removed from the water by plant uptake are lost from the plant detritus by leaching and decomposition, and recycled back into the water and soil. Leaching of water-soluble contaminants may occur rapidly upon the death of the plant or plant tissue, while a more gradual loss of contaminants occurs during decomposition of detritus by bacteria and other organisms. Recycled contaminants can be flushed from the wetland in the surface water, or removed again from the water by biological uptake or other means.

Nutrients may be held tightly within the microbial biomass component of a low-nutrient ecosystem, reflecting efficient recycling of remineralized organic compounds (Melillo et al., 1984). Microbes and plants often compete for nutrients in ecosystems with limited nutrient input and tight nutrient cycles (Lodge et al., 1994). Under favorable conditions for organic matter decomposition, stored nutrients or contaminants can be released through mineralization and then recycled in the ecosystem or exported from the system (Reddy and D'Angelo, 1994). The rate of net organic matter accumulation is a critical determinant of how a wetland functions as an ecological unit within the landscape. The storage function is equally important for natural wetlands, especially those that represent an ecotone between terrestrial and aquatic ecosystems, and created wetlands, which may be used for treatment of wastewater or runoff (Howard-Williams, 1985).

Although microorganisms can provide a measurable amount of contaminant uptake and storage, it is their metabolic processes that play the most significant role in removal of organic compounds. Microbial decomposers, primarily soil bacteria, utilize the carbon (C) in organic matter as a source of energy, converting it to CO_2 or CH_4 gases. This provides an important biological mechanism for removal of a wide variety of organic compounds, including those found in municipal wastewater, food processing wastewater, pesticides, and petroleum products. The efficiency and rate of organic C degradation by microorganisms are highly variable for different types of organic compounds. Because of the limited supply of electron acceptors, organic matter decomposition is driven primarily by methanogenesis. Wetlands are a major source of carbonaceous gases, primarily CH_4, to the atmosphere, while the limited supply of inorganic electron acceptors in wetlands creates a sink for nitrogenous gases such as N_2O.

A wetland can function in the watershed as either a nutrient sink or source, providing net retention or release of nutrients to downstream waters. Whether a wetland serves as sink or source depends on the biogeochemical characteristics, including soil and vegetation, within the wetland, as well as the rate of nutrient input from surface or groundwater sources. If the export of nutrients from the wetland is lower than the incoming nutrient load, the wetland is considered a net sink for nutrients. On the other hand, if the export of nutrients is greater than the nutrient inflow, the wetland will be a net source of nutrients. Net export of nutrients can occur as a result of high nutrient loading rate to a wetland, followed by reduced loading rate. In many cases, some of the nutrients that accumulated under high loading conditions continue to be exported from the wetland after nutrient loading is decreased. Thus, a wetland that previously had functioned as a nutrient sink may become a nutrient source as a result of chronic nutrient loading.

Wetlands can also function as net transformers of nutrients from inorganic to organic forms. Nutrients frequently enter wetlands as dissolved inorganic compounds (e.g., nitrate, ammonium, phosphate). In contrast, nutrients exported from wetlands are predominantly in organic form, a consequence of the copious production of dissolved and particulate organic matter within a wetland. This net transformation of nutrients by wetlands is ecologically significant because organic forms of nutrients must undergo decomposition to inorganic forms prior to biological utilization. Because a large portion of the wetland nutrient export is not immediately bioavailable, nutrient-related impacts on downstream surface waters, such as excessive growth of algae, may be greatly reduced or eliminated. The nutrient transformation function of wetlands, coupled with their ability to buffer pulses of nutrients in the watershed by storing and slowly releasing nutrients to downstream waters, provides a significant measure of ecological stability to contiguous aquatic systems.

MODELING CARBON AND NITROGEN DYNAMICS IN WETLANDS

A wide array of conceptual and simulation models have been developed for wetland ecosystems (Costanza and Sklar, 1985; Mitsch and Reeder, 1991; Mitsch et al., 1988). Models describing C and N dynamics in wetlands are commonly used for evaluating or predicting nutrient removal efficiency for wastewater treatment and general water quality improvement functions. Other common themes for wetland C and N models include prediction of flux of gaseous forms of these elements to the atmosphere and, more generically, providing assessments of wetland function or ecological condition.

Wetland nutrient dynamics modelers have frequently utilized a "black-box" approach to describing soil biogeochemical processes. For example, nutrient removal in a natural or constructed treatment wetland can be depicted simply as a single first-order decay term incorporated into a nutrient transport model. Nevertheless, for many applications, such as treatment wetland performance evaluation, this approach is the most efficient and provides a suitable, albeit empirical description of nutrient removal. In this brief overview of wetland C and N modeling, the discussion is limited to models that incorporate, to various extents, biogeochemical processes controlling C and N cycling.

Models of wetland soil biogeochemistry are often adapted from terrestrial-based models of soil organic matter dynamics. For the most part, the microbial and biochemical processes governing C and N cycling are nearly identical in both terrestrial and wetland systems. The fundamental difference in modeling C and N dynamics in wetlands is the need to account for the availability of oxygen, which is highly variable in the wetland soil-water profile (both spatially and temporally), as well as the supply of alternate electron acceptors, all of which serve as controls of organic matter decomposition and C and N metabolism.

Conceptual models describing decomposition of heterogeneous substrates, such as plant detritus and soil organic matter, have generally employed similar approaches for both terrestrial and wetland ecosystems. Decomposition models generally fall into one of four major groups: single homogeneous compartment, two-compartment, multi-compartment, and non-compartmental heterogeneous models (Jenkinson, 1990). Compartmental models separate organic matter (or specifically organic C and N) into discrete compartments with significantly different turnover times. One approach attempts to quantitatively separate certain chemical fractions of organic matter presumed to control decomposition kinetics of the heterogeneous substrate, or that can be described as functions of some readily measured chemical parameter or environmental variable (Minderman, 1968). Many researchers have used a more operationally defined approach to group different fractions of organic matter. A common conceptual scheme for organic matter fractions involves grouping according to empirically derived measures of biodegradability. Resulting categories may simply be termed "labile," "resistant," "stable," etc. This approach has been most commonly used for terrestrial ecosystem C and N models (Van Veen and Paul, 1981; Parton et al., 1987; Jenkinson, 1990; Grant et al., 1993a; b). Another approach to dealing with heterogeneous substrates treats the substrate as a single component of variable quality, or overall biodegradability (Godshalk and Wetzel, 1978; Moran et al., 1989). This is more often seen as a theoretical approach to the description of decomposition, and has not been widely implemented in soil C and N models.

A single-compartment exponential decay model using an exponentially decreasing rate coefficient (k) was used by Godshalk and Wetzel (1978) to describe decomposition of five species of aquatic macrophytes. The form of the model was

$$dW/dt = -k \times W,$$

where $k = a \times exp(-b \times t)$ and W is percent of initial weight remaining, t is time in days, k is the first-order rate coefficient, and a and b are rate parameters. The exponentially decreasing rate coefficient reflected the increasing overall recalcitrance of the substrate as the composition shifted to a greater proportion of highly resistant compounds. Decay rates were negatively correlated to the total fiber content of the substrate, but were not well correlated to individual components such as cellulose, hemicellulose, and lignin.

A study by Moran et al. (1989) provided a critical evaluation of single- and multi-compartment exponential decay models for plant litter, based on chemical components of the litter. The authors measured *in situ* and laboratory decomposition rates of whole litter and the lignocellulose components of litter for the emergent macrophytes *Spartina alterniflora* and *Carex walteriana*. Following an initial rapid phase of mass loss from leaching and decomposition of non-lignocellulosic components, losses of C due to decomposition were attributable mainly to the lignocellulose fraction. By tracking the mass loss over time for individual chemical components of substrate, the authors were able to mechanistically construct a composite exponential decay model of whole-litter decomposition, with separate representation of soluble material, hemicellulose, cellulose, and lignin in the substrate.

However, evaluation of alternative decomposition models revealed that the whole-litter decomposition process (at least for a half-year period) was more accurately described by the decaying-coefficient model (Godshalk and Wetzel, 1978). The main shortcoming of the composite exponential model was the relatively poor fit (underestimation) during later stages of the experimental decomposition period, that is, after about 4 months. This was attributed, in large part, to an apparent decrease in the specific decomposition rate over time for individual components of the substrate. The authors suggested that selective degradation of less resistant fractions within each component (e.g., various phenolic sub-units of the lignin component), along with the increasing importance over time of physical protection and humification, were probable causes for the observed decrease in component decay rates. It was concluded that the enormous number of biochemically distinct components, each with a characteristic biodegradability (and distinct rate constant), presents an inherent limitation on the use of the composite exponential model for accurate mechanistic descriptions of decomposition.

Ecosystem-scale models of organic C and N cycling and organic matter accumulation in wetlands have also borrowed from terrestrial ecosystem models, most notably those describing grasslands and agricultural systems. Many of these models employ the concept of discrete compartments for organic C or N pools with different turnover times. The CENTURY model of soil organic matter (C and N) in Great Plains grasslands (Parton et al., 1987) has been used to simulate effects of climatic gradients and grazing on soil organic matter levels and plant productivity over a wide geographic region. This model assigns soil organic C and N to three separate compartments based on turnover time: active (including microbial biomass), slow, and passive soil organic matter. In addition, plant residue is divided into structural (slow) and metabolic (rapid) pools.

The multi-compartment representation of soil and sediment organic matter has also been employed in wetland models. The SEMIDEC model (Morris and Bowden, 1986) was developed to describe nutrient (N and P) mineralization and export in tidal marsh ecosystems. It was originally developed for simulation of a freshwater tidal marsh system of the North River in Massachusetts. The primary processes described by the model are sedimentation, that is, accretion of organic and inorganic sediments, organic matter decomposition, above- and below-ground plant biomass, and mineralization of N and P. Separate refractory and labile pools of sediment organic matter are described in the model. In contrast to terrestrial models, for which organic matter is incorporated into existing soil horizons, the SEMIDEC model treats sediment horizons as year-class cohorts, which are modified over time following burial. The SEMIDEC model is a steady-state model for which changes in cohorts are used to simulate depth profiles of total sediment C, N, and P.

Chen and Twilley (1999) developed a model (NUMAN), based on the SEMIDEC model, to simulate organic matter and nutrient distribution in mangrove wetland soils. As with the SEMIDEC model, the NUMAN model partitions leaf litter into labile and refractory soil organic matter pools, based on the lignin content of the litter. These pools are also analogous to the metabolic and structural pools for plant residue in the CENTURY model. The NUMAN model is designed to predict the effects of forest production, litter export, decomposition, and sedimentation on organic matter and N and P distribution in the soil. However, because NUMAN is a steady-state model, it does not account for catastrophic events such as hurricanes, which can cause significant redistribution of sediment profiles.

A process-based research model of N dynamics in wetland soil and water was developed by Martin and Reddy (1997). This wetland model provides a spatially distributed simulation of N transformations and transport in the soil-water profile, which incorporated three discrete layers, or compartments: floodwater, aerobic soil, and anaerobic soil. Transfers of N among layers occur through diffusion of solutes and settling of particulates. Biogeochemical processes described by this model include enzymatic hydrolysis, mineralization, nitrification, denitrification, ammonium adsorption/desorption, ammonia volatilization, vegetative uptake, and decomposition.

A simulation model of organic C mineralization was developed for management of cultivated Histosols in the Everglades (Browder and Volk, 1978). Organic C was partitioned into a non-living pool, according to degradability or chemical structure, and an active living pool containing microbial biomass. Water table (which controlled soil moisture and O_2 availability), temperature, and soil organic C content were used as effects to predict rates of soil subsidence and release of N compounds and organic acids to surface and groundwater.

Models of C and N cycling in wetlands have also been developed for the purpose of predicting the role and magnitude of gaseous emissions to the atmosphere. For example, the WMEM (Wetland Methane Emission Model) addresses methane flux from wetlands on a regional basis (Cao et al., 1996). Simulation of spatial and temporal variability of methane emissions is based on rates of net primary production, organic C decomposition, methane production in the soil, and methanotrophic oxidation in the soil and floodwater.

A recently developed ecosystem model of nutrient cycling in wetlands (van der Peijl and Verhoeven, 1999) addresses the evaluation of functional characteristics of European wetland ecosystems. The primary emphasis of this model is on the role of biogeochemical processes within soil, vegetation, and associated fauna, as governed by temperature, hydrology, and anthropogenic influences. The model explicitly addresses C and N in soil organic matter, ammonium and nitrate content of the soil, decomposition, denitrification, N mineralization and immobilization, and ammonia volatilization. The authors have also demonstrated the utility of this wetland model in simulating the effects of restoration, in the form of reestablishment of natural hydrological characteristics (i.e., removal of ditches), on a floodplain wetland system (van der Peijl and Verhoeven, 2000).

Numerous other process-oriented models of wetland C and N dynamics have been developed for research and management purposes. These models vary widely in their scope and degree of complexity. Some models address only individual processes, while others encompass entire ecosystems. A biogeochemical process such as decomposition can be modeled as a number of discrete component processes, both physical and biochemical, or as a single aggregated process.

Costanza and Sklar (1985) determined from a review of 87 wetland models that there is a significant trade-off between accuracy (goodness-of-fit) and articulation (size and complexity). Maximum model effectiveness, or explanatory power, is generally achieved at a moderate level of articulation, or complexity. Therefore, to achieve a high degree of accuracy in a process model, the scope of the model is necessarily limited; in other words, only a limited number of processes can be accommodated in the model. A model of broad scope, such as an ecosystem model, can generally be effective only if the degree of resolution, with respect to describing biogeochemical processes, is minimized. Thus, it is often the case that effective models for resource management are only moderately complex, or articulated, in their description of biogeochemical processes (e.g., the CENTURY model). Regardless of the level of aggregation of component processes, effective wetland biogeochemical models invariably account for, either implicitly or explicitly, the key microbial and biochemical processes governing nutrient and organic matter dynamics in the model system.

CONCLUSION

Wetlands are "transitional" ecosystems, sharing characteristics of both upland and aquatic ecosystems. Accordingly, wetlands possess a number of unique biogeochemical attributes, especially with regard to C and N cycling. A combination of external factors, such as basin morphology, position in the landscape, and widely fluctuating hydroperiod, strongly influences the biogeochemical functioning of wetland soils. Gradients of nutrients and electron acceptors are a significant feature in wetland soils, especially near the wetland/upland interface. The wide range of soil microenvironments created by nutrient and electron acceptor gradients results in a high diversity of microbial populations with specialized metabolic functions.

Of particular significance to N cycling are the O_2 gradients occurring at the aerobic/anaerobic interfaces. The position of these interfaces is strongly affected by hydrologic fluctuations, especially drying and reflooding cycles. As a result, the metabolic pathways and process rates for organic C and N mineralization in wetlands may vary considerably over time. Given the additional environmental variability imparted by lateral gradients of nutrients and electron acceptors, C and N cycling in wetland soils can be highly variable, even within a relatively small wetland. It follows that a fundamental understanding of the physical, biological, and chemical regulators of C and N cycling is essential for predicting their interactive effects on organic matter and nutrient dynamics and for developing effective management models for these highly complex systems.

REFERENCES

Amador, J.A., and R.D. Jones. 1995. Carbon mineralization in the pristine and phosphorus-enriched peat soils of the Florida Everglades. *Soil Sci.,* 159:129-141.
Andrén, O., and K. Paustian. 1987. Barley straw decomposition in the field: a comparison of models. *Ecology,* 68:1190-1200.

Benner, R., A.E. Maccubbin, and R.E. Hodson. 1984a. Anaerobic biodegradation of the lignin and polysaccharide components of lignocellulose and synthetic lignin by sediment microflora. *Appl. Environ. Microbiol.,* 47:998-1004.

Benner, R., M.A. Moran, and R.E. Hodson. 1985. Effect of pH and plant source on lignocellulose biodegradation rates in two wetland ecosystems, the Okefenokee Swamp and Georgia Salt Marsh. *Limnol. Oceonogr.,* 30:489-499.

Benner, R., S.Y. Newell, A.E. Maccubbin, and R.E. Hodson. 1984b. Relative contributions of bacteria and fungi to rates of degradation of lignocellulosic detritus in salt-marsh sediments. *Appl. Environ. Microbiol.,* 48:36-40.

Bohlin, E., M. Hämäläinen, and T. Sundén. 1989. Botanical and chemical characterization of peat using multivariate methods. *Soil Sci.,* 147:252-263.

Borga, P., M. Nilsson, and A. Tunlid. 1994. Bacterial communities in peat in relation to botanical composition as revealed by phospholipid fatty acid analysis. *Soil Biol. Biochem.,* 26:841-848.

Browder, J.A., and B.G. Volk. 1978. Systems model of carbon transformations in soil subsidence. *Ecol. Modelling,* 5:269-292.

Brown, A., S.P. Mathur, T. Kauri, and D.J. Kushner. 1988. Measurement and significance of cellulose in peat soils. *Can. J. Soil Sci.,* 68:681-685.

Buresh, R.J., M.E. Casselman, and W.H. Patrick. 1980. Nitrogen fixation in flooded soil systems, A review. *Adv. Agronomy,* 33:149-192.

Button, D.K. 1985. Kinetics of nutrient-limited transport and microbial growth. *Microbiol. Rev.,* 49:270-297.

Cao, M., S. Marshall, and K. Gregson. 1996. Global carbon exchange and methane emissions from natural wetlands: application of a process-based model. *J. Geophys. Res.,* 101:399-414.

Chen, R., and R.R. Twilley. 1999. A simulation model of organic matter and nutrient accumulation in mangrove wetland soils. *Biogeochemistry,* 44:93-118.

Cho, D.Y., and F.N. Ponnamperuma, 1971. The influence of soil temperature on the chemical kinetics of flooded soils and growth of rice. *Soil Sci.,* 112:184-189.

Clymo, R.S. 1983. Peat. In A.J.P. Gore (Ed.). *Mires: Swamp, Bog, Fen and Moor.* Elsevier, Amsterdam, 159-224.

Cole, J.A. 1990. Physiology, biochemistry and genetics of nitrate dissimilation to ammonium. In N.P. Reusbech and J. Sorensen (Eds.). *Denitrification in Soils and Sediment.* Plenum Press, New York. 213-228.

Cooper, A.B. 1990. Nitrate depletion in the riparian zone and stream channel of a small headwater catchment. *Hydrobiologia,* 202:13-26.

Costanza, R., and F.H. Sklar. 1985. Articulation, accuracy and effectiveness of mathematical models: a review of freshwater wetland applications. *Ecol. Modelling,* 27:45-68.

D'Angelo, E.M. and K.R. Reddy. 1994a. Diagenesis of organic matter in a wetland receiving hypereutrophic lake water. I. Distribution of dissolved nutrients in the soil and water column. *J. Environ. Qual.,* 23:937-943.

D'Angelo, E.M. and K.R. Reddy. 1994b. Diagenesis of organic matter in a wetland receiving hypereutrophic lake water. II. Role of inorganic electron acceptors in nutrient release. *J. Environ. Qual.,* 23:928-936.

Damman, A.W.H. 1988. Regulation of nitrogen removal and retention in Sphagnum bogs and other peatlands. *OIKOS,* 51:291-305.

DeBusk, W.F., and K.R. Reddy. 1987. Removal of floodwater nitrogen in a cypress-mixed hardwood swamp receiving primary sewage effluent. *Hydrobiologia,* 153:79-86.

DeBusk, W.F. and K.R. Reddy. 1998. Turnover of detrital organic carbon in a nutrient-impacted Everglades marsh. *Soil Sci. Soc. Am. J.,* 62:1460-1468.

Elliott, E.T., D.C. Coleman, R.E. Ingham, and J.A. Trofymow. 1984. Carbon and energy flow through microflora and microfauna in the soil subsystem of terrestrial ecosystems. In M.J. Klug and C.A. Reddy (Eds.). *Current Perspectives in Microbial Ecology.* American Society for Microbiology, Washington, D.C., 424-433.

Eriksson, K.E., and S.C. Johnsrud. 1982. Mineralisation of carbon. In R.G. Burns and J.H. Slater (Eds.). *Experimental Microbial Ecology.* Blackwell Scientific, Oxford, 134-153.

Firestone, M.K. 1982. Biological denitrification In F.J. Stevenson (Ed.). *Nitrogen in Agricultural Soils.* American Society of Agronomy, Inc.; Crop Science Society of America, Inc.; Soil Science Society of America, Inc. Madison, WI, 289-326.

Gale, P.M., I. Devai, K.R. Reddy, and D.A. Graetz. 1993. Denitrification potential of soils from constructed and natural wetlands. *Ecol. Engin.,* 2:119-130.

Gardner, W.S., J.F. Chandler, and G.A. Laird. 1989. Organic nitrogen mineralization and substrate limitation of bacteria in Lake Michigan. *Limnol. Oceanogr.,* 34:478-485.

Given, P.H., and C.H. Dickinson. 1975. Biochemistry and microbiology of peats. In E.A. Paul and A.D. McLaren (Eds.). *Soil Biochemistry,* Marcel Dekker, New York, Vol. 3, 123-212.

Godshalk, G.L., and R.G. Wetzel. 1978. Decomposition of aquatic angiosperms. II. Particulate components. *Aquat. Bot.,* 5:301-327.

Grant, R.F., N.G. Juma, and W.B. McGill. 1993a. Simulation of carbon and nitrogen transformations in soil: mineralization. *Soil Biol. Biochem.,* 25:1317-1329.

Grant, R.F., N.G. Juma, and W.B. McGill. 1993b. Simulation of carbon and nitrogen transformations in soil: microbial biomass and metabolic products. *Soil Biol. Biochem.,* 25:1331-1338.

Happell, J.D., and J.P. Chanton. 1993. Carbon remineralization in a north Florida swamp forest: effects of water level on the pathways and rates of soil organic matter decomposition. *Global Biogeochemical Cycles,* 7:475-490.

Heal, O.W., P.W. Flanagan, D.D. French, and S.F. MacLean, Jr. 1981. Decomposition and accumulation of organic matter. In L.C. Bliss, O.W. Heal, and J.J. Moore (Eds.). *Tundra Ecosystems: A Comparative Analysis.* Cambridge University Press, Cambridge, 587-633.

Heal, O.W., and P. Ineson. 1984. Carbon and energy flow in terrestrial ecosystems: relevance to microflora. In M.J. Klug and C.A. Reddy (Eds.). *Current Perspectives in Microbial Ecology.* American Society for Microbiology, Washington, D.C., 394-404.

Howard-Williams, C. 1985. Cycling and retention of nitrogen and phosphorus in wetlands: a theoretical and applied perspective. *Freshwater Biology,* 15:391-431.

Howard-Williams, C., and M.T. Downes. 1994. Nitrogen cycling in wetlands. In T.P. Burt, A.L. Heatherwaite, and S.T. Trudgill (Eds.). *Nitrate: Processes, Patterns and Management.* John Wiley & Sons, New York, 141-167.

Howarth, R.W. 1993. Microbial processes in salt-marsh sediments. In T.E. Ford (Ed.). *Aquatic Microbiology.* Blackwell Scientific, Oxford, 239-259.

Howarth, R.A. and J.J. Cole. 1985. Molybdenum availability, nitrogen limitation, and phytoplankton growth in natural waters. *Science,* 229:653-655.

Howarth, R.W., R. Marino, J. Lane, and J.J. Cole. 1988a. Nitrogen fixation in freshwater, estuarine, and marine ecosystems 1. Rates and importance. *Limnol. Oceanogr.,* 33:669-687.

Howarth, R.W., R. Marino, J. Lane, and J.J. Cole. 1988b. Nitrogen fixation in freshwater, estuarine, and marine ecosystems 2. Biogeochemical controls. *Limnol. Oceanogr.,* 33:688-701.

Humphrey, W.D. and D.J. Pluth. 1996. Net nitrogen mineralization in natural and drained fen peatlands in Alberta, Canada. *Soil Sci. Soc. Am. J.,* 60:932-940.

Inglett, P.W. 2000. Spatial and Temporal Patterns of Periphyton N2 Fixation in a Nutrient Impacted Everglades Ecosystem. M.S. thesis, University of Florida, Gainesville, FL. 67 pp.

Jenkinson, D.S. 1990. The turnover of organic carbon and nitrogen in soil. *Phil. Trans. R. Soc. Lond. B,* 329:361-368.

Jenkinson, D.S., and J.H. Rayner. 1977. The turnover of soil organic matter in some of the Rothamsted classical experiments. *Soil Sci.,* 123:298-305.

Kim, B., and R.G. Wetzel. 1993. The effect of dissolved humic substances on the alkaline phosphatase and the growth of microalgae. *Int. Ver. Theor. Angew. Limnol. Verh.,* 25:129-132.

Koch, M.S. and K.R. Reddy. 1992. Distribution of soil and plant nutrients along a trophic gradient in the Florida Everglades. *Soil. Sci. Soc. Am. J.,* 56:1492-1499.

Lodge, D.J., W.H. McDowell, and C.P. McSwiney. 1994. The importance of nutrient pulses in tropical forests. *Tree,* 9:384-387.

Martens, R. 1995. Current methods for measuring microbial biomass C in soils: potentials and limitations. *Biol. Fertil. Soils,* 19:87-99.

Martin, J.F., and K.R. Reddy. 1997. Interaction and spatial distribution of wetland nitrogen processes. *Ecological Modelling,* 105:1-21.

Martin, K., L.L. Parsons, R.E. Murray, and M.S. Smith. 1988. Dynamics of soil denitrifier populations: relationships between enzyme activity, most-probable-number counts, and actual N loss. *Appl. Environ. Microbiol.,* 54:2711-2716.

McKinley, V. L., and J. R. Vestal. 1992. Mineralization of glucose and lignocellulose by four arctic freshwater sediments in response to nutrient enrichment. *Appl. Environ. Microbiol.,* 58:1554-1563.

McLatchey, G.P. and K.R. Reddy. 1998. Regulation of organic matter decomposition and nutrient release in a wetland soil. *J. Environ. Qual.,* 27:1268-1274.

Melillo, J.M., J.D. Aber, A.E. Linkins, A. Ricca, B. Fry, and K.J. Nadelhoffer. 1989. Carbon and nitrogen dynamics along the decay continuum: plant litter to soil organic matter. *Plant and Soil,* 115:189-198.

Melillo, J.M., R.J. Naiman, J.D. Aber, and A.E. Linkins. 1984. Factors controlling mass loss and nitrogen dynamics of plant litter decaying in northern streams. *Bull. Mar. Sci.,* 35:341-356.

Minderman, G. 1968. Addition, decomposition and accumulation of organic matter in forests. *J. Ecol.,* 56:355-362.

Mitsch, W.J., and C. Reeder. 1991. Modelling nutrient retention of a freshwater coastal wetland: estimating the roles of primary productivity, sedimentation, resuspension and hydrology. *Ecol. Modelling,* 54:151-187.

Mitsch, W.J., M. Sraskraba, and S.E. Jørgensen (Eds.). 1988. *Wetland Modeling: Developments in Environmental Modeling,* 12. Elsevier, Amsterdam, 227.

Moran, M.A., R. Benner, and R.E. Hodson. 1989. Kinetics of microbial degradation of vascular plant material in two wetland ecosystems. *Oecologia,* 79:158-167.

Morris, J.T., and W.B. Bowden. 1986. A mechanistic, numerical model of sedimentation, mineralization, and decomposition for marsh sediments. *Soil Sci. Soc. Am. J.,* 50:96-105.

Paerl, H.W., L.E. Prufert-Bebout, and C. Gui. 1994. Iron-stimulated N2 fixation and growth in natural and cultured populations of the planktonic marine cyanobacteria *Trichodesium* spp. *Appl. Environ. Microbiol.,* 60:1044-1047.

Parton, W.J., D.S. Schimel, C.V. Cole, and D.S. Ojima. 1987. Analysis of factors controlling soil organic matter levels in Great Plains grasslands. *Soil Sci. Soc. Am. J.,* 51:1173-1179.

Paul, E.A. 1984. Dynamics of organic matter in soils. *Plant and Soil,* 76:275-285.

Ponnamperuma, F.N. 1972. The chemistry of submerged soils. *Advances in Agronomy,* 24:29-96.

Reddy, K.R., and E.M. D'Angelo. 1994. Soil processes regulating water quality in wetlands. In W.J. Mitsch (Ed.). *Global Wetlands: Old World and New.* Elsevier Science, Amsterdam, 309-324.

Reddy, K.R., and D.A. Graetz. 1988. Carbon and nitrogen dynamics in wetland soil. In D.D. Hook et al. (Eds.). *The Ecology and Management of Wetlands.* Timber Press, Portland, OR, 307-318.

Reddy, K.R., and W.H. Patrick. 1984. Nitrogen transformations and loss in flooded soils and sediments. *CRC Critical Reviews in Environ. Control,* 13:273-309.

Reddy, K.R., W.H. Patrick, Jr., and C.W. Lindau. 1989. Nutrification-denitrification at the plant root-sediment interface in wetlands. *Limnol. Oceanogr.,* 34:1004-1013.

Reddy, K.R., and K.M Portier. 1987. Nitrogen utilization by Typha latifolia L. as affected by temperature and rate of nitrogen application. *Aquatic Bot.,* 27:127-138.

Sagar, B.F. 1988. Microbial cellulases and their action on cotton fibres. In A.F. Harrison, P.M. Latter, and D.W.H. Walton (Eds.). *Cotton Strip Assay: An Index of Decomposition in Soils. ITE Symposium no. 24.* Institute of Terrestrial Ecology. Grange-Over-Sands, U.K.

Schipper, L.A., A.B. Cooper, C.G. Harfoot, and W.J. Dyck. 1993. Regulators of denitrification in an organic riparian soil. *Soil Biol. Biochem.,* 25:925-933.

Shaver G.R., and J.M. Melillo. 1984. Nutrient budgets of marsh plants: Efficiency concepts and relation to availability. *Ecology,* 65:1491-1510.

Sinsabaugh, R.L., R.K. Antibus, A.E. Linkins, C.A. McClaugherty, L. Rayburn, D. Repert, and T. Weiland. 1993. Wood decomposition: nitrogen and phosphorus dynamics in relation to extracellular enzyme activity. *Ecology,* 74:1586-1593.

Sinsabaugh, R.L., and D.L. Moorhead. 1994. Resource allocation to extracellular enzyme production: a model for nitrogen and phosphorus control of litter decomposition. *Soil Biol. Biochem.,* 26:1305-1311.

Stevenson, F.J. 1994. *Humus Chemistry.* John Wiley & Sons, Inc. New York. pp 496.

Swift, M.J. 1982. Microbial succession during the decomposition of organic matter. In R.G. Burns and J.H. Slater (Eds.). *Experimental Microbial Ecology.* Blackwell Scientific, Oxford, 164-177.

Swift, M.J., O.W. Heal, and J.M. Anderson. 1979. *Decomposition in Terrestrial Ecosystems.* Univ. of California Press, Berkeley.

Tiedje, J.M. 1988. Ecology of denitrification and dissimilatory nitrate reduction to ammonium. In A.J.B. Zehnder (Ed.). *Biology of Anaerobic Microorganisms.* John Wiley & Sons. New York, 179-245.

van der Peijl, M.J., and J.T.A. Verhoeven. 1999. A model of carbon, nitrogen and phosphorus dynamics and their interactions in river marginal wetlands. *Ecol. Modelling,* 118:95-130.

van der Peijl, M.J., and J.T.A. Verhoeven. 2000. Carbon, nitrogen and phosphorus cycling in river marginal wetlands; a model examination of landscape geochemical flows. *Bio-geochemistry,* 50:45-71.

Van Veen, J.A., J.N. Ladd, and M.J. Frissel. 1984. Modelling C and N turnover through the microbial biomass in soil. *Plant and Soil,* 76:257-274.

Van Veen, J.A., and E.A. Paul. 1981. Organic carbon dynamics in grassland soils. I. Background information and computer simulation. *Can. J. Soil Sci.,* 61:185-201.

Vymazal, J., and C.J. Richardson. 1995. Species composition, biomass, and nutrient content of periphyton in the Florida Everglades. *Algol Studies,* 73:75-97.

Webster, J.R., and E.F. Benfield. 1986. Vascular plant breakdown in freshwater ecosystems. *Annu. Rev. Ecol. Syst.,* 17:567-594.

Westermann, P. 1993. Wetland and swamp microbiology. In T.E. Ford (Ed.). *Aquatic Microbiology.* Blackwell Scientific Publications, Oxford, 215-238.

Wetzel, R.G. 1984. Detrital dissolved and particulate organic carbon functions in aquatic ecosystems. *Bull. Mar. Sci.,* 35:503-509.

Wetzel, R.G. 1992. Gradient-dominated ecosystems: sources and regulatory functions of dissolved organic matter in freshwater ecosystems. *Hydrobiol.,* 229:181-198.

White, J.R., and K.R. Reddy. 1999. Influence of nitrate and phosphorus loading on denitrifying enzyme activity in Everglades wetland soils. *Soil Sci. Soc. Am. J.,* 63:1945-1954.

White, J.R., and K.R. Reddy. 2000. The effects of phosphorus loading on organic nitrogen mineralization of soils and detritus along a nutrient gradient in the northern Everglades, Florida. *Soil Sci. Soc. Am. J.,* 64:1525-1534.

Whiting, G.J. 1994. CO_2 exchange in the Hudson Bay lowlands: community characteristics and multispectral reflectance properties. *J. Geophys. Res.,* 99:1519-1528.

Williams, B.L., and G.P. Sparling. 1988. Microbial biomass carbon and readily mineralized nitrogen in peat and forest humus. *Soil Biol. Biochem.,* 20:579-581.

Zeikus, J.G. 1981. Lignin metabolism and the carbon cycle. In M. Alexander (Ed.). *Advances in Microbial Ecology,* Plenum Press, New York, Vol. 5, 211-243.

A Review of Carbon and Nitrogen Processes in Nine U.S. Soil Nitrogen Dynamics Models

Liwang Ma and M.J. Shaffer

CONTENTS

INTRODUCTION

Carbon and nitrogen processes are essential components of agricultural systems, and correct simulation of these processes is essential for ecosystem models. Although modeling of soil carbon and nitrogen dynamics has been in the literature since the 1960s, it is still one of the least understood areas. The concept of carbon/nitrogen (C/N) dynamics in soils originated from chemistry, physical and enzyme chemistry, and biology and is applied to agricultural sciences without considering the limitations of many of the theories. For example, many of process rate equations apply only to ideal solutions or populations, which seldom exist in the soil environment. In addition, the soil matrix is highly heterogeneous in nature, compared to pure materials used in chemistry and biology laboratories. Although recent advances in chemistry and biology have helped to better understand heterogeneous materials and non-ideal solution reactions involving living organisms, their strict application to soil science is limited and may be difficult due to less controlled experimental conditions, ill-defined soil composition, and lack of measured parameters.

First-order kinetics remains the most commonly used approach to quantify reaction rates for many of the carbon and nitrogen processes in soils, such as decomposition of soil organic matter, nitrification, denitrification, and urea hydrolysis. The corresponding rate coefficients are then modified for effects of temperature, water, microorganisms, pH, and oxygen, depending on individual authors of the various models. Occasionally, the enzyme-based Michaelis-Menten equation or zero-order kinetics is used for selected processes (e.g., plant nitrogen uptake, nitrification, etc.). The basic assumption of first-order kinetics is that the reaction rate is controlled by the amount of one primary substrate and is proportional to that substrate. However, in soil systems, many of the biological processes involve several substrates, and reaction rates are limited by different substrates under different conditions.

The effects of environmental factors on soil carbon and nitrogen processes are difficult to simulate. Most U.S. models use a 0 to 1 modifier to adjust rate coefficients for each environmental factor (e.g, water and temperature), while others use the Arrhenius equation for temperature effects and proportional (or inverse) relationships for pH and oxygen (Shaffer et al., 2000a). Again, these functions or relationships were derived from basic chemistry or biology and should only be used with caution. Many inconsistencies in applying these functions between, for example, pure enzyme chemistry and soil chemistry are settled by calibrating one or more parameters in the equations. Therefore, it is not surprising that most of the derived model parameters are site or time dependent. There is an urgent need to (1) look more closely at implications of applying principles or theories from other scientific disciplines to soils, and (2) develop fundamental theoretical equations specifically for soils.

The role of microorganisms is very important in carbon and nitrogen dynamics. Their population responds so dynamically to environmental changes that most models cannot provide adequate simulation. Microbes are not only part of the carbon and nitrogen pools, but also catalyze most of the processes. Therefore, it is vitally important to correctly simulate the role of microbes in the soil environment. The ability to simulate the dynamics of soil microbes is primarily limited by an inability to monitor microbes experimentally under various field conditions and to assess the roles of each type of microbes. It is often assumed that there are maximal equilibrium microbial populations for a given biological reaction. In cases where time is needed for the microbial population to build up, a lag time is sometimes assumed rather than simulating the actual growth of the microbial population. In addition, biological processes usually have some sort of feedback mechanism to approach a quasi-equilibrium status, directly or indirectly. Most models, however, fail to adequately account for such a feedback mechanism. For example, if a model does not have representation of effects of microorganisms on each reaction rate, the model does not have a self-controlling mechanism for these reactions because many key processes are mediated through microbial activities.

Organic carbon and nitrogen in the soil are heterogeneous and cannot be treated with one lumped reaction. Most models differentiate between the various carbon and nitrogen pools, based on their physical, chemical, and/or biological characteristics. Physically, carbon and nitrogen can be mobile or immobile. Chemically, they can be organic (with different C:N ratios and compositions) or inorganic. Biologically, they can be transformed to other forms by various types of microbes and have different turnover rates. Thus far, very few models have dealt with mobile organic carbon and nitrogen (Grant, 2001; Parton et al., 1994). This is a particular problem under acid soil conditions where soluble forms of organic matter may be present in significant quantities. Organic carbon and nitrogen are also assumed to be coupled together by their C:N ratio, which may be different for different organic sources. Inorganic soil carbon is mostly in the gaseous form, such as CO_2 and CH_4, and as carbonates and carbonic acid. Inorganic nitrogen forms are NH_4, NO_3, urea-N, and various gases (e.g., NH_3, N_2O, NO_x).

A review of European carbon and nitrogen models is available in McGechan and Wu (2001) and a Canadian model (ecosys) in Grant (2001). Several other reviews of selected models are reported in the literature with different emphases and levels of detail (Molina and Smith, 1998; McGill, 1996; Myrold, 1998; Hansen et al., 1995). The purpose of this chapter is to compare nine major carbon and nitrogen models from the United States. These nine models have each developed a carbon/nitrogen component in the context of a system approach and have applications in agricultural research. Also, most of them were developed for upland ecosystems, with a few having modified versions for wetland soils (Godwin and Singh, 1998). The book by Smith (1982) remains a valid and a good reference for those who want to acquire a comprehensive knowledge of mathematical modeling of carbon and nitrogen dynamics in soils. Readers can also consult the book by Koch et al. (1998) for more information on mechanistic modeling of microbial ecology. To preserve the identity of each model, this chapter uses symbols and notation as near as possible to those in the original documentation. Also, the models reviewed in this chapter are the typical ones released by their developers, not including those later modified by individual users for specific purposes.

GENERAL DESCRIPTION OF THE REVIEWED MODELS

The nine models simulating carbon and nitrogen dynamics in the United States are:

1. NTRM (Nitrogen, Tillage, and crop-Residue Management) (Shaffer and Larson, 1987)
2. NLEAP (Nitrate Leaching and Economic Analysis Package) (Shaffer et al., 1991)
3. RZWQM (Root Zone Water Quality Model) (Ahuja et al., 2000a)

4. CENTURY (Parton et al., 1994)
5. CERES (Crop Estimation through Resource and Environment Synthesis) (Hanks and Ritchie, 1991)
6. GLEAMS (Groundwater Loading Effects of Agricultural Management Systems) (Leonard et al., 1987; Knisel, 1993)
7. LEACHM (Leaching Estimation and Chemistry Model) (Hutson 2000)
8. NCSOIL (Molina et al., 1983)
9. EPIC (Erosion/Productivity Impact Calculator) (Williams, 1995)

All of them simulate multiple components in an agricultural system with various objectives and degrees of complexity. Model maintenance and support vary greatly among models, along with their applicability and documentation. Some of the modeling concepts are shared by all of them, while others are unique to a particular model. Differences also lie in relationship to other components of the agricultural systems, such as plant growth and water movement. Although all the models have achieved various degrees of success in applications, they all have their weakness and fail under certain circumstances. Authors of the models seldom clearly define limitations of their models and ranges of applications.

The CERES family of crop growth models shares the same carbon and nitrogen simulation subroutine (Godwin and Singh, 1998; Godwin and Jones, 1991) and water movement (Ritchie, 1998). The crop growth models, such as CERES-maize, CERES-wheat, CERES-rice, and CROPGRO (for legumes) are integrated into the DSSAT (Decision Support System for Agrotechnology Transfer) framework (Jones et al., 1998). A group of scientists from the University of Florida, University of Hawaii, University of Georgia, International Fertilizer Development Center in Alabama, Michigan State University, USDA-ARS, and others are actively supporting these models. Copies of the model package are available from the University of Hawaii. This group has an e-mail listserver and a Web site at http://www.icasanet.org/. Some information is also available from Michigan State University at Web address http://nowlin.css.msu.edu/.

The GLEAMS model is an extension of the CREAMS (Chemicals, Runoff, and Erosion from Agricultural Management Systems) model with enhanced hydrology, pesticide, and plant nutrient components (Knisel, 1993; Leonard et al., 1987). GLEAMS was initially developed at the University of Georgia and has not been modified since 1993 with the release of version 2.10 (Knisel, 1993). Minimum support is provided by USDA-ARS, Grassland Soil and Water Research Laboratory in Temple, Texas (http://arsserv0.tamu.edu/). The model can be downloaded from the Web site http://sacs.cpes.peachnet.edu/sewrl/. GLEAMS was developed to evaluate the impact of management practices on potential pesticide and nutrient leaching within, through, and below the root zone. It can also be used to assess the effect of farm-level management decisions on water quality.

The EPIC model is one of the earliest agricultural simulation tools, and it was developed to assess the effect of soil erosion on productivity. It is maintained by Texas A & M University (http://www.brc.tamus.edu/epic/). The model was described in detail by Williams (1995). Two new models were developed based on EPIC. One is ALMANAC (Agricultural Land Management with Alternative Numerical Assessment Criteria), which contains improvements regarding plant growth, especially plant competition, and is maintained by USDA-ARS Grassland Soil and Water Research Laboratory in Temple, Texas (http://arsserv0.tamu.edu/). The other is APEX (Agricultural Policy Environmental Extender), which adds channel flow and can be used for multiple fields in a watershed; this model is also maintained by Texas A & M University. EPIC and GLEAMS are very similar in simulating soil nutrient dynamics (N and P).

The CENTURY model was developed to simulate long-term effects on soil and carbon dynamics (minimum 10 years), with a monthly simulation time step. The model was developed by the Natural Resource Ecology Laboratory at Colorado State University in Fort Collins, Colorado (Parton et al., 1994). A user guide for version 4.0 of the model was documented by Metherell et al. (1993), and a few modifications have been made since then. Recently, the model has been extended to a version that runs with a daily time step (DAYCENT) and also includes nitrogen gas flux from Li's model (Li et al., 1992; Parton et al., 1996; 1998). CENTURY is maintained and available from Colorado State University at http://www.nrel.colostate.edu/PROGRAMS/MODELING/CENTURY/CENTURY.html. It has been used to simulate carbon and nutrient dynamics for different types of ecosystems, including grasslands, agricultural lands, forests, and savannas. The model has the capability of simulating several nutrients (C, N, P, and S). Because the model uses a monthly time step and is designed for long-term carbon sequestration, the nitrogen sub-model is very simply represented. The model does not really differentiate between ammonium and nitrate. Therefore, nitrification, denitrification, NH_3 volatilization, and urea hydrolysis are not simulated.

The NCSOIL model was developed at the University of Minnesota and is maintained on the Web at http://soils.umn.edu/research/ncswap-ncsoil/. It started as a sub-model of NTRM (Molina et al., 1983) and was later incorporated into NCSWAP (N and C flows in the Soil-Water-Air-Plant System) (Molina, 1996). Different from other models, NCSOIL has the capability of simulating C and N tracers simultaneously with non-traced soil organic C and N to study the mechanisms of N and C cycling. The NCSOIL model can be used as a stand-alone model as well. A more detailed description of NCSOIL is available in Molina et al. (1987).

The NLEAP model was developed as part of a national effort to quantify mechanisms for nitrate leaching from agriculture (Shaffer et al., 1991) and has been widely used for water quality assessment. A Stella version of NLEAP called NLOS (NLEAP on Stella) was developed in Canada (Bittman et al., 2001). A Java-based Internet version is available on the Web at http://www.nleap.usda.gov that has GIS capabilities and linkages to NRCS (Natural Resources Conservation Services) databases (Shaffer et al., 2001). The model is well documented (Shaffer et al., 1991; Shaffer et al., 2001) and supported by USDA-ARS, Great Plains Systems Research Unit, in Fort Collins, Colorado. The DOS versions can be downloaded from two Web sites: http://www.gpsr.colostate.edu/gpsr/products/nleap/nleap.htm and ftp://ftp.nrcs.usda.gov/centers/itc/applications/wqmodels/. The NLEAP model has been applied in GIS studies identifying nitrate leaching at a regional scale (Shaffer et al., 1995; 1996). The N and C processes are also incorporated into GPFARM (Great Plains Framework for Agricultural Resource Management), which is a new model maintained by the USDA-ARS, Great Plains Systems Research Unit at http://www.gpsr.colostate.edu/gpsr/products/gpfarm.htm.

The RZWQM is a fully functional system model and has process-level simulations of soil water, soil temperature, plant growth, pesticide fate, and C and N soil dynamics (Ahuja et al., 2000a). The major feature of this model is its ability to simulate various agricultural management practices and their effects on water quality and crop production with great detail. Other distinct features include macropore flow and water table fluctuation. A Microsoft Windows-based interface has been developed and distributed along with the technique documentation by the Water Resources Publications in Colorado (Ahuja et al., 2000a). The USDA-ARS, Great Plains Systems Research Unit, is fully supporting the model. An effort to develop linkage to GIS and use for decision purposes is being undertaken. A review of RZWQM applications is available in Ma et al. (2000). Information on the model can be obtained at http://www.gpsr.colostate.edu/gpsr/products/rzwqm.htm.

The C and N soil processes simulated in the NTRM model present a precedent to NLEAP and RZWQM. The complexity of C and N processes increases from NTRM, to NLEAP, and to RZWQM. Detailed C and N process simulations were documented by Shaffer et al. (1969) and Dutt et al. (1972). Other components simulated in NTRM are available in Shaffer and Larson (1987). The model has been used to simulate the effects of soil erosion on soil productivity (Shaffer, 1985; Shaffer et al., 1995), crop production (Swan et al., 1987; Radke et al., 1991), water quality (Clanton et al., 1986), and plant competition (Ball and Shaffer, 1993). NTRM is fully supported by USDA-ARS, Great Plains Systems Research Unit, in Fort Collins, Colorado. Web information is available at http://www.wiz.uni-kassel.de/model_db/mdb/ntrm.html or http://www.wcc.nrcs.usda.gov/water/quality/common/h2oqual.html.

Another water and solute transport model, LEACHM, was developed at Cornell University. It is now fully supported in Australia and information on the model is available at http://www.es.flinders.edu.au/LeachmDownload/LEACHWEB.HTM. The LEACHM suite consists of four simulation models and several utilities. The simulation models utilize similar numerical solution schemes to simulate vertical water and chemical movement. They differ in their description of chemical equilibrium, transformation, and degradation pathways. LEACHW describes the water regime only. The other simulations describe pesticides (LEACHP), nitrogen and phosphorus (LEACHN), microbial population dynamics (LEACHB), and salinity in calcareous soils (LEACHC) (Hutson, 2000). The C/N process simulation was modified from the SOILN model (Johnsson et al., 1987).

SIMULATED CARBON AND NITROGEN PROCESSES
IN THE MODELS

The Partitioning of Surface Residue and Soil Organic Matter Pools

Because surface residues and soil organic matter are heterogeneous in their chemical and physical compositions, it is necessary to differentiate between the various groups (pools) of organic C and N in these modeling approaches. The immediate difficult task of dealing with soil organic matter and surface residue is how to partition organic materials among different groups, not to mention how to

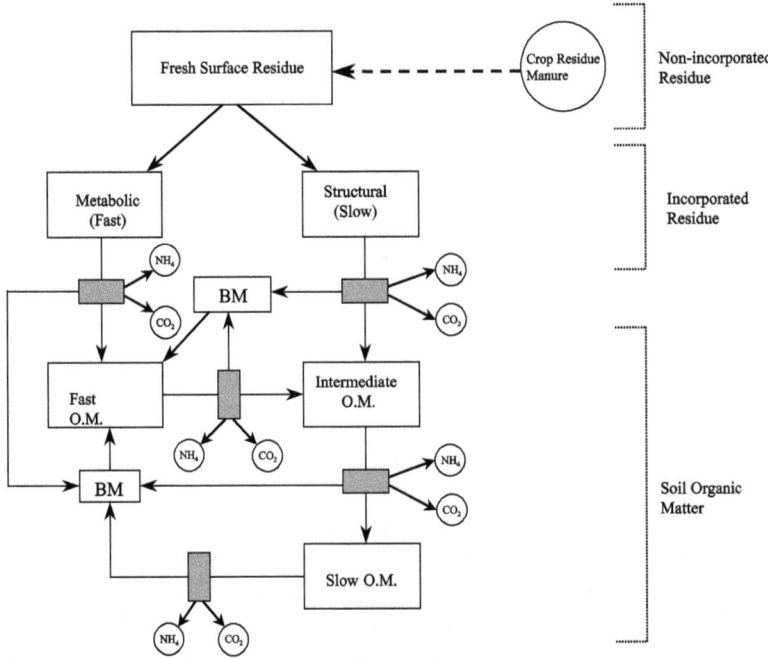

Figure 4.1   Conceptual structure of organic C/N pools used in the RZWQM model. C and N flow corresponding to immobilization process is not shown.

assign rate coefficients to each pool. One way to partition the pools is to make a best judgment on the size of the pools and then stabilize the pools with known management practices, as suggested in RZWQM (Ma et al., 1998). Another way is to calibrate the pools as used in the DAISY (Jensen et al., 2001) and NCSOIL models (Nicolardot et al., 1994). Although guidance or default values are provided in some of the models (e.g., CENTURY based on residue lignin content and EPIC based on years of cultivation), it is generally not an easy task to correctly partition the pools, especially with dynamic simulation of the microbial population.

RZWQM has a total of five organic C and N pools: the two surface residue pools and three soil organic matter pools (Figure 4.1). Differences among the pools include C:N ratios and decomposition rate constants (Shaffer et al., 2000a; Ma et al., 2001). A guideline is provided in the user interface on how to partition these pools. However, the authors of the model suggest a 10- to 12-year stabilization run prior to the targeted simulation period (Ma et al., 1998). Dead soil microbial biomass is recycled through the fast soil organic matter (O.M.) pools (or humus pools). First-order decay rate kinetics is used for all the pools, but with different rate coefficients. Inter-pool transfer coefficients can be adjusted for different organic C and N sources (Ma et al., 1998).

The NLEAP model is a relatively simpler model compared to RZWQM (Figure 4.2). It has two soil organic matter pools and one surface residue pool. Surface residue may be crop residue, applied manure, or other organic materials (Hansen et al., 1995). Each pool has its own C:N ratio and is subject to first-order

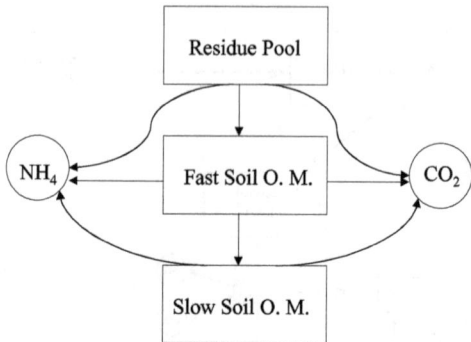

Figure 4.2 Conceptual structure of organic C/N pools used in the NLEAP model. C and N flow corresponding to immobilization process is not shown.

decomposition. The model assumes net immobilization at C:N ratios above 30 and net mineralization at C:N ratios below 30.

In the NTRM model, only one organic C pool is assumed, and the model keeps track of its C:N ratio throughout the mineralization process by assuming that 30 carbon atoms are released for every N atom released (Dutt et al., 1972). The pool's C:N ratio is also adjusted when residue is added into the soil. Mineralization and immobilization rates are calculated using linear regression equations. A zero mineralization-immobilization rate is assumed to occur at a C:N ratio of 23 (Dutt et al., 1972).

The NCSOIL model has four pools (Figure 4.3): a residue pool, a microbial pool (Pool I), an intermediate Pool II with half-life of 115 days and C:N ratio of 6, and a more stable Pool III (stable humus) with a C:N ratio of 12 and half-life of over 150 years (Molina, 1996). Pool I was shown to correspond to the soil microbial

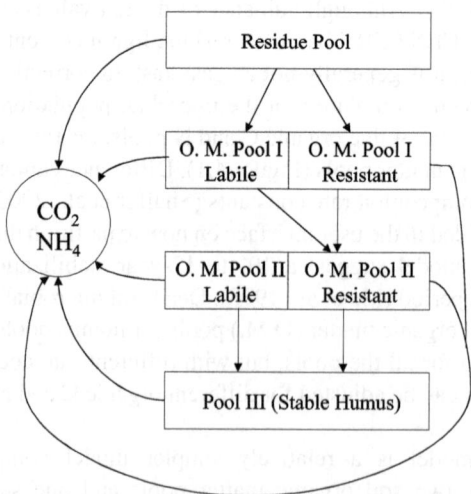

Figure 4.3 Conceptual structure of organic C/N pools used in the NCSOIL model. C and N flow corresponding to immobilization process is not shown.

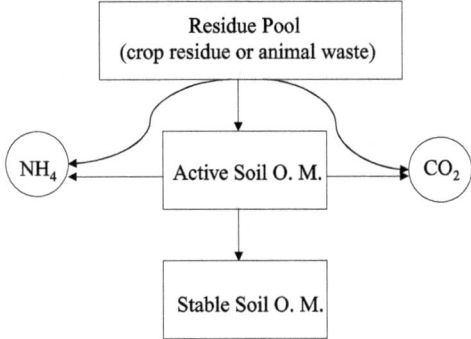

Figure 4.4 Conceptual structure of organic C/N pools used in the EPIC model. C and N flow corresponding to immobilization process is not shown.

biomass as measured by the fumigation method (Nicolardot et al., 1994). Each pool may be further divided into labile and resistant fractions. The labile and resistant Pool I fractions have half-lives of 2 and 17 days, respectively (Molina et al., 1997). Agricultural management practices such as tillage can change the resistant Pool II fraction to labile (Molina et al., 1983). Decomposition of each pool may be first-order or Monod kinetics. The stable humus pool has a much slower decay rate and is only important for long-term simulation.

The carbon and nitrogen processes in EPIC are modifications of the PAPRAN (Production of Arid Pastures limited by Rainfall And Nitrogen) model (Seligman and van Keulen, 1981). It has three organic C/N pools (Figure 4.4): the fresh organic pool of crop residue and microbial biomass, an active soil humus pool, and a stable soil organic humus pool (Williams, 1995). Only the fresh residue and the active humus pool are subjected to mineralization. The active humus pool has a half-life of a few years and a C:N ratio in the range of 12 to 25. The stable humus pool is not mineralized at all, and has a C:N ratio of less than 12. Williams (1995) presented a method to partition the active and stable soil humus pools. The active pool fraction (*RTN*) is calculated by:

$$RTN = 0.4\exp(-0.0277\ YC) + 0.1 \qquad (4.1)$$

where *YC* is years in cultivation. The EPIC model also estimates a nitrogen flow from the active to stable pools (*RON*) as:

$$RON = 1 \times 10^{-5}\left[\frac{ONA}{RTN} - ONS\right] \qquad (4.2)$$

where *ONA* and *ONS* are the organic N in active and stable pools, respectively. The same pool partitioning is used in the GLEAMS model (Knisel, 1993). GLEAMS uses a pool structure similar to that in the EPIC model (Knisel, 1993), except for *RTN* being calculated directly from the fraction of active pool, rather than estimated from *YC*.

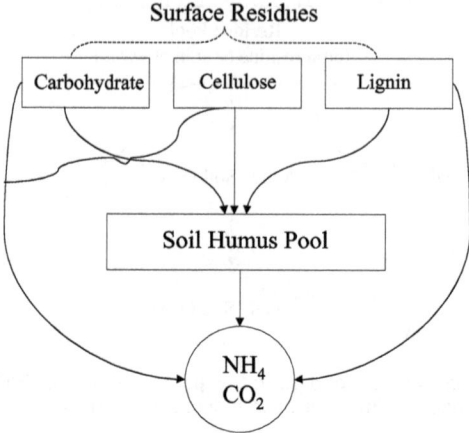

Figure 4.5 Conceptual structure of organic C/N pools used in the CERES model. C and N flow corresponding to immobilization process is not shown.

Another C and N process model that originated from PAPRAN is the CERES model (Godwin and Jones, 1991; Godwin and Singh, 1998). CERES divides the fresh organic pool (e.g., crop residue and manure) into three sub-pools of carbohydrate, cellulose, and lignin (Figure 4.5). Each has its own degradation rate and C:N ratio. There is only one soil humus pool, which has a much lower turnover rate than the fresh organic pool. The humus pool has a C:N ratio of 10. The model requires partitioning of the fresh organic pools with initial values of 20% carbohydrate, 70% cellulose, and 10% lignin. A 20% portion of the gross N released from the fresh organic pool is transferred to the soil humus pool.

The CENTURY model also has a multiple pool structure (Parton et al., 1987; 1994; Parton, 1996; Parton and Rasmussen, 1994). As in RZWQM, surface residue and root litter are divided into structural (slow) and metabolic (fast) pools, and soil organic matter is partitioned among active, slow, and passive pools (Figure 4.6). The

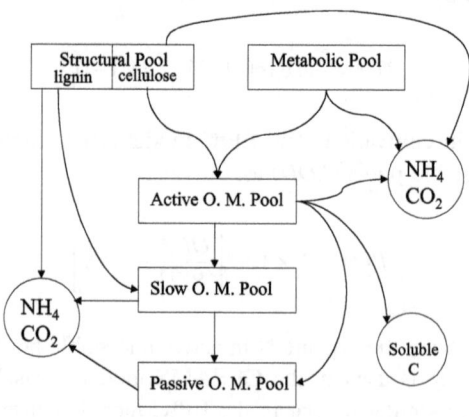

Figure 4.6 Conceptual structure of organic C/N pools used in the CENTURY model. C and N flow corresponding to immobilization process is not shown.

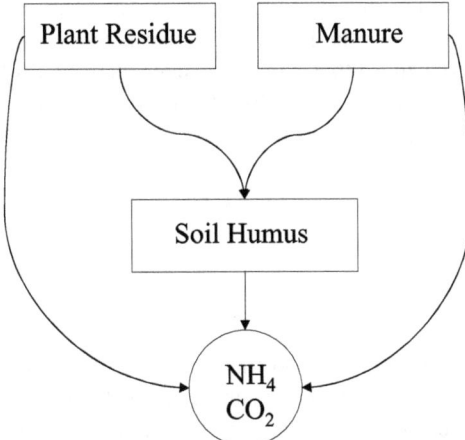

Figure 4.7 Conceptual structure of organic C/N pools used in the LEACHM model. C and N flow corresponding to immobilization process is not shown.

active O.M. pool consists of live microbes and microbial products along with soil O.M. with a short turnover time of 1 to 5 years and C:N in the range 3 to 14. The slow O.M. pool is physically or chemically protected, with an intermediate turnover time of 20 to 40 years and C:N ratio of 12 to 20. The passive O.M. pool is chemically recalcitrant or strongly physically protected, with the longest turnover time of 200 to 1500 years and a C:N ratio of between 11 and 12. The structural litter (residue) pool has a turnover time of 1 to 5 years and a C:N ratio of 150. The metabolic pool has a 0.1 to 1 year turnover time and a C:N ratio of 10 to 30. Partitioning of plant residue into structural and metabolic pools is based on its lignin-to-N ratio (LN). The fraction of surface residue or root litter partitioned into the metabolic pool (F_m) is:

$$F_m = 0.85 - 0.012 LN \tag{4.3}$$

In the LEACHM model, three organic pools are differentiated: plant residue, manure, and soil humus (Figure 4.7) (Hutson, 2000). Transformation of the organic pools and inorganic N forms is modified from Johnsson et al. (1987). The reason for having separate pools for plant residue and manure is due to their differences in C:N ratios and decomposition rates.

Mineralization and Immobilization Processes

Based on the above-mentioned pool partitioning, each model has explicitly defined a mineralization rate for each organic C pool. The role of microbes in the decomposition processes may or may not be explicitly expressed. With more complex models, Monod kinetics can be used (Hansen et al., 1995):

$$r_i = \frac{\mu_m C_i}{K_s + C_i} B \tag{4.4}$$

where r_i is decomposition rate, C_i is the carbon content in pool i, B is the size of the microbial biomass, μ_m is the maximum mineralization rate, and K_s is the half-saturation constant. For those that do not consider microbial roles, μ_m and B can be combined to give the form of Michaelis-Menten kinetics. The above equation can be further simplified as first-order kinetics:

$$r_i = k_1 C_i \tag{4.5}$$

or zero-order kinetics:

$$r_i = k_0 \tag{4.6}$$

where k_1 and k_0 are the first- and zero-order rate coefficients, respectively, and which can be modified for temperature, pH, O_2, and microbial effects. The inherent assumption for first- and zero-order kinetics in this model is that there are unlimited microorganisms for a biological process. Zero-order kinetics assume rates that are independent of carbon pool size. Decayed organic C from one pool can be transferred to another pool, assimilated into microbial biomass, or released as CO_2. The amount of nitrogen released is usually determined by C:N ratios of the individual pools (Molina and Smith, 1998).

The NLEAP model has a relatively simple approach to mineralization and immobilization processes. Net nitrogen mineralization of the residue pool depends on its C:N ratio and decomposition rate (Hansen et al., 1995):

$$M_r = r_r \left(\frac{1}{(C/N)_r} - \frac{1}{30} \right) \tag{4.7}$$

where M_r is the net nitrogen mineralization rate of the residue pool, $(C/N)_r$ is the C:N ratio of the residue pool, and r_r is the residue decomposition rate. Once $(C/N)_r$ reaches 6.5 to 12.0, depending on the type of residues (Shaffer et al., 2001), the remaining residue carbon and nitrogen are transferred to the fast soil organic matter pool (Hansen et al., 1995). Its decomposition rate is calculated with first-order kinetics (Shaffer et al., 1991):

$$r_r = k_r T_f W_f C_r \tag{4.8}$$

where k_r is a rate constant and can be adjusted based on C:N ratio (Shaffer et al., 2001), C_r is C content of the residue, and T_f and W_f are the temperature and water modifiers, respectively. The W_f value is the same as the one used in RZWQM (Ma et al., 2001), and T_f is in the form of an Arrhenius equation similar to that in RZWQM except for using constant values for the Arrhenius coefficients (Shaffer et al., 1991; Ma et al., 2001):

$$\begin{aligned} W_f &= 0.0075 \, WFP & WFP \le 20 \\ W_f &= -0.253 + 0.0203 \, WFP & 20 < WFP < 59 \\ W_f &= 41.1 e^{-0.0625 \, WFP} & WFP \ge 59 \end{aligned} \tag{4.9}$$

and

$$T_f = 1.68 \times 10^9 \; e^{\frac{-13.0}{1.99 \times 10^{-3}(T_s + 273)}} \tag{4.10}$$

where *WFP* is water-filled space and T_s is soil temperature. A similar decay equation is used for mineralization of soil organic matter pools except for using smaller rate constants (Shaffer et al., 1991; 2001). No soil microbial growth is simulated in the NLEAP model.

NTRM uses a regression approach for mineralization-immobilization processes, based on the C:N ratio of the organic matter pool (Dutt et al., 1972; Shaffer et al., 1969). Ammonium-N is the mineralized product. To monitor the C:N ratio of the single C pool, the model assumes that 30 C atoms are released for every N atom consumed by microbes. In addition, the model separates immobilization from NH_4-N and NO_3-N using different regression equations. Table 4.1 lists all the N processes and their regression equations. No microbial effects on mineralization-immobilization and microbial growth are simulated in NTRM.

In RZWQM, mineralization and immobilization are determined by the decay of organic pools and the growth of microbes. There is no predefined C:N ratio to control net mineralization and immobilization as in NLEAP and NTRM. Organic matter decay in each pool is simulated by a first-order equation, and the first-order rate coefficients are then modified for effects of soil temperature, oxygen, hydrogen, population of aerobic heterotrophic microbes, aerobic condition, and ionic strength

Table 4.1   Regression Variables and Constants for Various N Processes in NTRM

Urea Hydrolysis Equation

Urea hydrolysis rate (ppm/day) = Con + b_1 $\log_{10}(T_s)$ + b_2 $\log_{10}$ (Urea-N)
Con = 4.13 10^2 b_1 = −1.56 10^2 b_2 = −1.53 10^2

Mineralization-Immobilization Equation

Mineralization-immobilization rate (ppm/day) = Con + b_1 T_s + b_2 (Organic-N) + b_3 $\log_{10}$ (NH_4-N)
Con = 8.92 10^{-1} b_1 = 2.16 10^{-3}
b_2 = 2.70 10^{-2} b_3 = 3.92 10^{-1}

Nitrification Equation

Nitrification rate (ppm/day) = Con + b_1 T_s (NH_4-N) + b_2 $\log_{10}(NH_4$-N) + b_3 $\log_{10}(NO_3$-N)
Con = 4.64 b_1 = 1.62 10^{-3}
b_2 = 2.38 10^{-1} b_3 = −2.51

Nitrate-N Immobilization Equation

Nitrate-N immobilization rate (ppm/day) = Con + b_1 T_s/(organic-N)^2 + b^2 exp(T_s) +
b_3(T_s (organic-N) − (NO_3-N))/(organic-N)
Con = 0.0 b_1 = 1.52
b_2 = 3.23 10^{-15} b_3 = −4.90 10^{-3}

From Dutt, G.R., M.J. Shaffer, and W.J. Moore. 1972. Computer Simulation Model of Dynamic Bio-physicochemical Processes in Soils. Tech. Bulletin 196. Arizona Agricultural Experiment Station. 128 pp.

(Ma et al., 2001; Shaffer et al., 2000a). Microbes can grow via various soil carbon/nitrogen processes, and their growth is calculated differently for each process. The aerobic heterotrophs obtain their energy and growth by decay of the organic pools. RZWQM assumes that a fraction of decayed organic C not transferred into other organic pools is assimilated into microbial biomass. Net assimilation of N is calculated from the C:N ratios of the microbial biomass and the respective pools. Total growth of the aerobic heterotrophs is the summation of growth via the five organic pools. The autotrophs simulated in RZWQM obtain their energy from nitrification, and the model assumes a fraction of the total nitrified NH_4-N being assimilated into microbial biomass. The facultative heterotrophs grow via the denitrification process, and RZWQM uses an efficiency factor to partition a fraction of the denitrified NO_3-N to assimilated microbial biomass. Microbial death is simulated as first-order kinetics with respect to the individual microbial population.

The CERES model uses an approach very similar to that of the PAPRAN model of Seligman and van Keulen (1981). The mineralization rates of the three fresh residue pools simulated in the CERES model are (Godwin and Singh, 1998):

$$G1(j) = TF \times MF \times CNRF \times RDECR(j) \qquad (4.11)$$

where $G1(j)$ is the proportion of the pool j ($j = 1$ for carbohydrate, $j = 2$ for cellulose, and $j = 3$ for lignin) that decays in 1 day; $RDECR(j)$ is the decay rate constant of pool j; TF, MF, and $CNRF$ are temperature, moisture, and C:N ratio factors:

$$TF = \frac{T_s - 5.0}{30} \qquad (4.12)$$

$$MF = \frac{SW - AD}{DUL - AD} \quad if \ SW \le DUL$$

$$MF = 1.0 - 0.5 \frac{SW - DUL}{SAT - DUL} \quad if \ SW > DUL \qquad (4.13)$$

$$CNRF = e^{\left[-0.693 \frac{(CNR-25)}{25} \right]} \qquad (4.14)$$

where T_s is soil temperature; SW is current volumetric water content; SAT, DUL, and AD are saturated water content, drained upper limit, and air dry moisture content, respectively; and CNR is the C:N ratio based on nitrogen contents both in the organic and inorganic fractions. The gross mineralization of N (GRNOM) is then calculated from the pool size and its C:N ratio. The net N mineralized from the humus pool ($RHMIN$) is estimated from (Godwin and Singh, 1998; Godwin and Jones, 1991):

$$RHMIN = NHUM \times DMINR \times TF \times MF \times DMOD \qquad (4.15)$$

where $NHUM$ is the N content in the humus pool including the 20% of N released from the three fresh organic pools; $DMINR$ is the decay rate constant; and $DMOD$

is a factor to account for difference in soil cultivation history. *DMOD* equals 1.0 for most soils, 2.0 for newly cultivated soil, and 0.2 for Andisols (volcanic soils). Immobilization of N (*RNAC*) into microbial biomass is estimated by assuming [0.02 – fraction of N in the fresh organic material] grams of N assimilated for each gram of fresh organic material decomposed (Seligman and van Keulan, 1981). The net mineralization rate (*NNOM*) is the summation of:

$$NNOM = 0.8\,GRNOM + RHMIN - RNAC \qquad (4.16)$$

The CERES model does not explicitly simulate microbial biomass and its effect on soil C and N dynamics.

Nitrogen mineralization in EPIC is also a modification of the PAPRAN model (Williams, 1995). For each soil layer, a first-order mineralization rate (*RMN*) of the fresh organic N pool is written as:

$$RMN = DCR \times FON \qquad (4.17)$$

where *FON* is the amount of fresh organic N, and *DCR* is the decay rate coefficient:

$$DCR = 0.05\,CNP\,\sqrt{\frac{SW}{FC}}\,TFN \qquad (4.18)$$

where *SW* is soil water content, *FC* is water content at field capacity, *TFN* is a temperature factor, and *CNP* is C:N and C:P factor. *CNP* and *TFN* are calculated from:

$$CNP = minimum\left(1.0, e^{-0.693\frac{CNR-25}{25}}, e^{-0.693\frac{CPR-200}{200}}\right) \qquad (4.19)$$

and

$$TFN = maximum\left(0.1, \frac{T_s}{T_s + \exp(9.93 - 0.312\,T_s)}\right) \qquad (4.20)$$

where *CNR* and *CPR* are the C:N and C:P ratios of each soil layer. Some 20% of *RMN* goes to the active humus pool and 80% goes to the mineral N pool. Only the active humus pool in the EPIC model is subject to mineralization, and the mineralization rate (*HMN*) is:

$$HMN = CMN \times ONA \times \sqrt{\frac{SW}{FC}}\,TFN\left(\frac{BDS}{BDP}\right)^2 \qquad (4.21)$$

where *CMN* is a rate constant, *BDS* is the settled soil bulk density, and *BDP* is the current bulk density as affected by tillage. Immobilization is calculated the same way as in the PAPRAN model except for assuming that [0.016 – fraction of N in the fresh organic material] grams of N are assimilated for each gram of fresh organic material decomposed (Williams, 1995). When immobilization occurs, N is taken from the soil NO_3 pool. If N or P is limiting for microbial growth, the decomposition rate is adjusted accordingly.

The first-order mineralization rate of the active humus pool in the GLEAMS model is very similar to that in the EPIC model except for omitting the bulk density effects (Knisel, 1993):

$$HMN = CMN \times ONA \times \sqrt{\frac{SW - WP}{FC - WP}}\, TFN \qquad (4.22)$$

where WP is the water content at wilting point. The first-order decomposition rate coefficient (DCR) in GLEAMS is also similar to that in EPIC except for the use of a residue composition factor (RC) rather than a constant of 0.05 as in the EPIC model:

$$DCR = RC \times CNP \sqrt{\frac{SW - WP}{FC - WP}}\, TFN \qquad (4.23)$$

where RC is 0.8 for the first 20% of residue decomposed, 0.05 for the next 70% of residue decomposed, and 0.0095 for the last 10% decomposed (Knisel, 1993). The immobilization process is simulated in the same way as in EPIC except for taking N from both the NH_4 and NO_3 pools (Williams, 1995).

In the CENTURY model, surface residue is partitioned into a metabolic pool and a structural pool. The latter is further subdivided into lignin and cellulose components. The metabolic pool and cellulose C pool are transferred into the fast organic C pool (i.e., microbial biomass), whereas the lignin C is converted into the slow C pool directly (Parton et al., 1994). The fast soil microbial pool (active C pool) has four fates during decomposition: (1) to the slow C pool; (2) to the passive C pool; (3) to the soluble C pool (which can leach); and (4) to CO_2. Decomposition of all the pools is assumed to be first order (Parton et al., 1994). The rate coefficients are different for each of the C pools. The decay coefficient for the structural C pool (k) is written as (Parton et al., 1994):

$$k = k_c\, e^{-3L_s}\, A_t\, A_w \qquad (4.24)$$

where L_s is the lignin content in the structural C pool, k_c is a rate constant, and A_t and A_w are the temperature and water content factors, respectively:

$$A_t = t1^{0.2}\, t2 \qquad (4.25)$$

and

$$\begin{aligned} A_w &= \left(\frac{1}{1 + 30\, e^{-8.5 RAT}}\right) & \text{if } RAT \leq 1.5 \\ A_w &= 1.0 - 0.7\, \frac{RAT - 1.5}{1.5} & \text{if } RAT > 1.5 \end{aligned} \qquad (4.26)$$

where *RAT* is the water stress factor defined by [soil stored water + rainfall]/potential ET; and $t1$ and $t2$ are functions of soil temperature (T_s):

$$t1 = \frac{45 - T_s}{45 - 35} \quad \text{and} \quad t2 = e^{0.076\left(1 - t1^{2.63}\right)} \tag{4.27}$$

For the active SOM pool, the decay rate coefficient is affected by soil texture (T_m):

$$k = k_c A_w A_t T_m \tag{4.28}$$

where

$$T_m = 1 - 0.75(FSILT + FCLAY) \tag{4.29}$$

where *FSILT* and *FCLAY* are the silt and clay fractions of the soil, respectively. For all the other organic C pools (i.e., the metabolic C, slow C, and passive C pools), the decay rate coefficients are only functions of A_w and A_t (Parton et al., 1994). Transfer of C between pools is calculated from a series of linear functions of soil texture (clay, silt, and sand contents) (Parton et al., 1994). Nitrogen flow is linked to C flow by C:N ratios of the respective pools (Parton et al., 1993). It is worthwhile mentioning that the fraction of soluble O.M. leached from the active C pool (*CAL*) is estimated from monthly water leached below the top 30-cm soil depth (*MWL30*) and sand fraction (*FSAND*):

$$CAL = \frac{MWL30}{80}(0.01 + 0.04\,FSAND) \tag{4.30}$$

In CENTURY, a separate surface microbial pool (active C pool) is simulated, and its turnover rate is independent of soil texture (Parton et al., 1993; 1994).

In NCSOIL, each pool (residue, Pool I, and Pool II) is treated as two components: one labile and the other resistant. Both components decompose exponentially and independently from each other (Molina et al., 1983). The decay rate of each pool is:

$$\frac{dC_i}{dt} = minimum(TRED, WRED)\mu_N\left[S_L k_L C_i + (1 - S_L)h_R C_i\right] \tag{4.31}$$

where C_i is the carbon concentration of the substrate; S_L and $(1 - S_L)$ are the time invariant proportions of the labile and resistant components of the substrate, respectively; k_L and h_R are the corresponding rate constants; *TRED* and *WRED* are the temperature and water effects on mineralization, respectively; and μ_N is the reduction factor due to C:N ratio (Molina et al., 1983; 1987). The temperature reduction factor is:

$$\begin{aligned} TRED &= 0.0333\,T_s & T_s \le 30 \\ TRED &= 1.0 & 30 < T_s < 40 \\ TRED &= 2.6 - 0.04\,T_s & T_s \ge 40 \end{aligned} \tag{4.32}$$

where T_s is the soil temperature in °C. The water reduction factor is written as:

$$WRED = 0.002 \ PERSAT \qquad\qquad PERSAT \geq 10$$

$$WRED = 0.0195 \ PERSAT - 0.176 \qquad\qquad 10 < PERSAT \leq WHCPSAT$$

$$WRED = 0.52 + \frac{(1.0 - 0.52)(PERSAT - 80)}{WHCPSAT - 80} \qquad WHCPSAT < PERSAT \leq 80$$

$$WRED = 0.37 + \frac{(0.52 - 0.37)(PERSAT - 100)}{80 - 100} \qquad 80 < PERSAT < 100$$

$$WRED = 0.37 \qquad\qquad PERSAT = 100 \qquad (4.33)$$

where *PERSAT* is current soil water content (expressed as the percentage of water saturation) and *WHCPSAT* is soil water holding capacity (expressed as the percentage of saturation). The stable humus pool (Pool III) is assumed to decay with a much slower rate in the NCSWAP model.

A fraction of the decomposed Pool I is recycled into Pool I and partitioned between the labile and resistant components. However, the fraction of decomposed Pool I feeds directly into the resistant component of Pool II (Molina et al., 1983). When tillage practice is triggered in the NCSOIL model, a fraction of the resistant Pool II is transferred into labile Pool II to increase the mineralization rate. Based on C:N ratios of residue and microbial biomass, NCSOIL also determines whether the decomposition processes result in mineralization or immobilization. If immobilization occurs, the model will take ammonium and nitrate equally in proportion to their respective soil concentrations (Molina et al., 1983). NCSOIL also has the option of using Monod kinetics with Pool I as microbial biomass (Molina, 1996).

Decomposition of the organic C pools (residue, manure, and humus) in the LEACHM model is first order, and decayed C is partitioned into humus C, CO_2, and microbial biomass C (Hutson, 2000). The biomass pool is immediately recycled into its parent pool. If mineralized N is less than N assimilated into the biomass, net immobilization results. The first-order rate constants can be adjusted to reflect water content and temperature effects. The temperature factor (T_{cf}) is based on a Q_{10} type response:

$$T_{cf} = Q_{10}^{0.1(T_s - T_{base})} \qquad (4.34)$$

where Q_{10} is the change in rate constant for every 10°C temperature interval, and T_{base} is the base temperature at which the rate constant is specified. The water content factor (W_{cf}) is:

$$W_{cf} = W_{cfsat} + \frac{(1 - W_{cfsat})(\theta_s - \theta)}{\theta_s - \theta_{max}} \qquad \theta \geq \theta_{max}$$

$$W_{cf} = 1.0 \qquad\qquad \theta_{min} < \theta < \theta_{max} \qquad (4.35)$$

$$W_{cf} = \frac{\max(\theta, \theta_{wp}) - \theta_{wp}}{\theta_{min} - \theta_{wp}} \qquad \theta \leq \theta_{min}$$

where W_{cfsat} is the relative mineralization rate at saturation; θ is the current water content; θ_s and θ_{wp} are the saturated and wilting point water contents, respectively; and θ_{min} and θ_{max} are the minimum and maximum optimal water contents for mineralization, respectively.

Nitrification Process

Similar to the simulation of mineralization and immobilization, three types of kinetics (Monod, first-order, and zero-order) models are used to describe the nitrification processes (Hansen et al., 1995). In addition, some of the simple models assume that the mineralization process proceeds at a much slower rate than the nitrification process; therefore, the first inorganic N form is nitrate (NO_3), rather than ammonium (NH_4). This assumption also assumes that applied fertilizer ammonium is instantaneously converted into nitrate, rather than over several days or weeks as observed in the field (Hansen et al., 1995). One example of this assumption is the CENTURY model. Because the model uses a monthly time step, nitrification is not important. In fact, the model does not differentiate between ammonium and nitrate in the soil. The DAYCENT model (a daily version of CENTURY) assumes that nitrification is proportional to soil N turnover rate (Parton et al., 1996). N_2O production from nitrification is calculated from both N turnover rate and excess NH_4 in the soil (>3 µg N g^{-1}) in the DAYCENT model (Parton et al., 1996):

$$R_{N_2O} = N_{H_2O} N_{pH} N_t \left(K_{mx} + N_{mx} N_{NH_4} \right) \tag{4.36}$$

where R_{N_2O} is N_2O gas flux; N_{H_2O}, N_{pH}, N_t are the water, pH, and temperature factors, respectively; K_{mx} is the N turnover coefficient; N_{mx} is maximum N_2O gas flux due to excess soil NH_4; and N_{NH_4} is the effect of soil NH_4 on nitrification. The coefficients are written as:

$$N_{H_2O} = \left(\frac{WFP - b}{a - b} \right)^{d\left(\frac{b-a}{a-c}\right)} \left(\frac{WFP - c}{a - c} \right)^d \tag{4.37}$$

$$N_{pH} = 0.56 + \frac{\arctan\left(\pi \times 0.45 \times (-5 + pH)\right)}{\pi} \tag{4.38}$$

$$N_t = -0.06 + 0.13 e^{0.07 T_s} \tag{4.39}$$

$$N_{NH_4} = 1 - e^{-0.0105\, NH_4} \tag{4.40}$$

where WFP is water-filled pore space; a, b, c, and d are constants with values of 0.55, 1.70, –0.007, and 3.22 for sandy texture soil, respectively, and 0.60, 1.27, 0.0012, and 2.84 for medium texture soils; and T_s is the soil temperature.

   The NLEAP model uses the zero-order kinetic approach to simulate nitrification. The rate coefficient is estimated by a rate constant multiplied by the temperature (T_f) and water content (W_f) factors, as for the mineralization processes (Shaffer et al., 1991; 2001). The rate constant may be lower when a nitrification inhibitor is used. A fraction of nitrified N is released as N_2O. RZWQM initially used first-order kinetics for NH_4 concentrations greater than 3.0 millimoles N per liter of pore water (Hansen et al., 1995); but a recent version uses a simple zero-order function because the first-order assumption has little effect on C and N soil dynamics (Ma et al., 2001). The zero-order rate coefficient is a function of soil temperature, oxygen concentration, soil pH, and population of the autotrophs (Ma et al., 2001). A fraction of nitrified NH_4-N is assimilated into the autotroph's biomass. When anhydrous ammonia is applied, the model postpones the nitrification process for a period and then gradually restores the activity of the autotrophs. If a nitrification inhibitor is used, nitrification is further delayed.

   NTRM, on the other hand, uses a regression equation to calculate the nitrification rate without directly considering the mechanisms (Table 4.1). An optional transition state equation for nitrification is included, which has the form (Shaffer, 1985):

$$\frac{dNH_4}{dt} = k_{1/2} \frac{(NH_4)^{1/2} O_2}{H^+} \exp\left(\frac{E_a}{k_b T_s}\right) \tag{4.41}$$

where $k_{1/2}$ is a half-order rate constant, E_a is the apparent activation energy, and k_b is the Boltzmann constant.

   The nitrification process in the CERES model is simulated by the Michaelis-Menten equation (Godwin and Singh, 1998; Godwin and Jones, 1991). The rate coefficients are modified for water (WFD), temperature (TF), and pH effects (PHN). In a different manner from other models, the CERES model uses the Michaelis-Menten equation to calculate a percentage of total ammonium nitrified in 1 day:

$$\frac{dNH_4}{dt} = \frac{A\ 40\ NH_4}{NH_4 + 90} SNH_4 \tag{4.42}$$

where SNH_4 is the total ammonium-N in the soil layer and (NH_4) is the concentration of ammonium. The coefficient A is the minimum of the temperature factor (TF), water factor (WFD), pH factor (PHN), and nitrification potential index ($RP2$). The TF factor is the same as the one used for mineralization. PHN and WFD factors are:

$$PHN = \frac{pH - 4.5}{1.5} \quad \text{if } pH < 6.0$$

$$PHN = 1.0 \quad \text{if } 6.0 \leq pH \leq 8.0 \tag{4.43}$$

$$PHN = 9.0 - pH \quad \text{if } pH > 8.0$$

$$WFD = \frac{SW - AD}{DUL - AD} \qquad if \ SW \le DUL$$

$$WFD = 1.0 - \frac{SW - DUL}{SAT - DUL} \qquad if \ SW > DUL \qquad (4.44)$$

The $RP2$ index is derived from:

$$RP2 = CNI \ \exp(2.302 \ ELNC) \qquad (4.45)$$

where CNI is the nitrification potential of the previous day, and $ELNC$ is the environmental limit on nitrification capacity. $RP2$ is between 0.01 and 1.0.

$$ELNC = minimum(TF, WFD, SANC) \qquad (4.46)$$

and

$$SANC = 1.0 - \exp(-0.01363 \ SNH_4) \qquad (4.47)$$

The nitrification process is simulated as first-order kinetics in the EPIC model (Williams, 1995). Different from other models, EPIC uses the equivalent weight of ammonia ($WNH3$), rather than ammonium. The first-order rate ($RNIT$) is written as:

$$RNIT = WNH3[1.0 - \exp(TFNIT \times SWFNIT \times PHFNIT)] \qquad (4.48)$$

where $TFNIT$, $SWFNIT$, and $PHFNIT$ are the temperature, water, and pH factors for nitrification, respectively, and are estimated by:

$$TFNIT = 0.41(T_s - 5.0)/10.0 \qquad T_s > 5.0 \qquad (4.49)$$

$$SWFNIT = \frac{SW - WP}{SW25 - WP} \qquad SW < SW25$$

$$SWFNIT = 1.0 \qquad SW25 < SW < FC \qquad (4.50)$$

$$SWFNIT = 1.0 - \frac{SW - FC}{PO - FC} \qquad SW > FC$$

$$PHFNIT = 0.307 \ pH - 1.269 \qquad pH < 7.0$$

$$PHFNIT = 1.0 \qquad 7.0 < pH < 7.4 \qquad (4.51)$$

$$PHFNIT = 5.367 - 0.599 \ pH \qquad pH > 7.4$$

where PO is soil porosity (or saturated soil water content), WP is soil water content at the wilting point, and $SW25$ is the water content given by $WP + 0.25(FC - WP)$.

Zero-order kinetics is used in the GLEAMS model for nitrification, and the rate coefficient ($NIT0$) is adjusted by temperature and water factors (Knisel, 1993):

$$NIT0 \propto \frac{TFN0 \times SWFN0}{SOILMS} \tag{4.52}$$

where $SOILMS$ is the soil mass, and $TFN0$ and $SWFN0$ are temperature and water factors:

$$TFN0 = 0 \qquad\qquad\qquad T_s < 0°C$$

$$TFN0 = 0.496\,T_s \qquad\qquad 0 < T_s \leq 10°C \tag{4.53}$$

$$TFN0 = \exp\left(22.64 - \frac{5956.4}{T_s + 273}\right) \quad T_s > 10°C$$

and

$$SWFN0 = 0 \qquad\qquad\qquad SW \leq WP$$

$$SWFN0 = \frac{SW - WP}{FC - WP} \qquad\qquad WP < SW \leq FC \tag{4.54}$$

$$SWFN0 = 1.0 - \frac{SW - FC}{SAT - FC} \quad FC < SW \leq SAT$$

where SAT is the soil water content at saturation.

In the NCSOIL model, zero-order kinetics is assumed for nitrification. A user-defined maximum rate constant is modified by a factor, which is a constant in NCSOIL. However, in the NCSWAP model, the factor is the lower of the temperature ($TRED$) and water content ($XARENI$) reduction factors (Molina et al., 1983):

$$XAERNI = WRED \qquad\qquad\qquad\qquad PERSAT < WHCPSAT$$

$$XAERNI = AERNI + \frac{(1.0 - AERNI)(PERSAT - WHINFL)}{WHCPSAT - WHCCST - WHINFL} \qquad WHCPSAT \leq PERSAT < WHINFL$$

$$XAERNI = AERNI + \frac{(-AERNI)(PERSAT - WHINFL)}{100 - WHINFL} \qquad WHINFL \leq PERSAT < 100$$

$$XAERNI = 0.0 \qquad\qquad\qquad\qquad PERSAT = 100 \tag{4.55}$$

where $AERNI$ is a reduction factor at a water content (expressed as a percentage of saturation) of $WHINFL$, and $WHCCST$ is a constant to account for soil texture.

WHCCST is set to 10 for clay, silty clay, silty clay loam, clay loam, and sandy loam, and zero for other soil texture classes.

Nitrification in the LEACHM model is adapted from the SOILN model (Johnsson et al., 1987; McGechan and Wu, 2001) and is first-order with consideration for the nitrate-to-ammonium ratio (Hutson, 2000):

$$\frac{dNH_4}{dt} = -\mu_{nit}\left(NH_4 - \frac{NO_3}{r_{max}}\right) \tag{4.56}$$

where r_{max} is the maximum nitrate-to-ammonium ratio and μ_{nit} is the nitrification rate coefficient that can be modified by T_{cf} and W_{cf} as used in a mineralization process.

Denitrification Process

Simulation of the denitrification process is extremely empirical, not only because of the unknown nature of the process itself, but also because of the spatial and temporal variability of anaerobic conditions in the soil. Soil anaerobic conditions can be determined by soil oxygen or soil water content. Again, the mathematical simulation of denitrification can be zero-order, first-order, or Michaelis-Menten kinetics (Hansen et al., 1995). NTRM does not simulate denitrification. A fraction of N is lost as gas in the CENTURY model. Denitrification in the DAYCENT model (a daily time step version of CENTURY) is simulated to calculate N_2 and N_2O emission (Parton et al., 1996). Total N ($N_2 + N_2O$) gas flux is estimated from:

$$D_t = min\left(F_d(NO_3), F_d(CO_2)\right) F_d(WFP) \tag{4.57}$$

where D_t is total gas flux; $F_d(NO_3)$ and $F_d(CO_2)$ are the maximum total N gas flux for a given NO_3 and soil respiration rate, respectively; and $F_d(WFP)$ is the effect of water on denitrification. These coefficients are given by (Parton et al., 1996):

$$F_d(WFP) = \frac{a}{b^{\left(\frac{c}{b^{(d \times WFP)}}\right)}} \tag{4.58}$$

$$F_d(NO_3) = 11000 + \frac{40000 \times atan\left(\pi \times 0.002 \times (NO_3 - 180)\right)}{\pi} \tag{4.59}$$

$$F_d(CO_2) = \frac{24000}{1 + \frac{200}{e^{0.35 \times CO_2}}} - 100 \tag{4.60}$$

where *a*, *b*, *c*, and *d* are constants with values of 1.56, 12.0, 16.0, and 2.01, respectively, for sandy texture soil; 4.82, 14.0, 16.0, and 1.39 for medium texture soil; and 60.0, 18.0, 22.0, and 1.06 for fine texture soil. Simulated total N gas flux from denitrification is then partitioned between N_2 and N_2O based on an $N_2:N_2O$ ratio (R_{N_2/N_2O}) (Parton et al., 1996):

$$R_{N_2/N_2O} = \min\left(F_r(NO_3), F_r(CO_2)\right) F_r(WFP) \qquad (4.61)$$

where $F_r(NO_3)$, $F_r(CO_2)$, and $F_r(WFPS)$ are the effects of the soil NO_3, soil respiration rate, and water, respectively, on the $N_2:N_2O$ ratio (Parton et al., 1996):

$$F_r(WFP) = \frac{1.4}{13^{\left(\frac{17}{13^{(2.2 \times WFP)}}\right)}} \qquad (4.62)$$

$$F_r(NO_3) = 1 - \left(0.5 + \frac{1 \times atan\left(\pi \times 0.01 \times (NO_3 - 190)\right)}{\pi}\right) \times 25 \qquad (4.63)$$

$$F_r(CO_2) = 13 + \frac{30.78 \times atan\left(\pi \times 0.07 \times (CO_2 - 13)\right)}{\pi} \qquad (4.64)$$

Both NLEAP and RZWQM use first-order kinetics to simulate denitrification (Hansen et al., 1995). However, the rate coefficients are determined differently. The rate coefficient used in RZWQM is affected by soil anaerobic conditions, soil temperature, soil pH, soil carbon substrate, and population of the denitrifiers (Ma et al., 2001). A fraction of the denitrified NO_3-N is assimilated into the biomass of facultative heterotrophs in RZWQM. The denitrification rate coefficient in NLEAP is calculated from:

$$k_{dnit} = k_{dc} T_f \left[N_{wet} + W_{fd}\left(t - N_{wet}\right)\right] \qquad (4.65)$$

where k_{dnit} is the first-order rate coefficient, k_{dc} is the rate constant, t is the time step in days, N_{wet} is the number of days with precipitation or irrigation, and T_f is the same temperature effect for mineralization. W_{fd} is the water effect on denitrification calculated from water-filled pore space (WFP) (Shaffer et al., 1991; 2001):

$$W_{fd} = 0.000304 e^{0.0815 WFP} \qquad (4.66)$$

Denitrified N is then partitioned between N_2 and N_2O (Shaffer et al., 2001).

Denitrification occurs when soil water content (SW) exceeds the drained upper limit (DUL) in the CERES model (Godwin and Singh, 1998; Godwin and Jones, 1991). The 0–1 water index (FW) is calculated from:

$$FW = 1.0 - \frac{SAT - SW}{SAT - DUL} \qquad (4.67)$$

A temperature effect (FT) is also used to modify the denitrification rate:

$$FT = 0.1 \exp(0.046 T_s) \tag{4.68}$$

A third factor affecting denitrification in soil is the water-extractable carbon in soil organic matter (CW):

$$CW = 24.5 + 0.0031 SOILC \tag{4.69}$$

where $SOILC$ is the soil C content, assumed to be 58% of the stable humus pool (Godwin and Jones, 1991). A slightly different CW formula is used in the DSSAT framework (Godwin and Singh, 1998). Thus, the first-order denitrification rate is estimated from:

$$DNRATE = 6.0 \times 10^{-5} \times CW \times FW \times FT \times DLAYR \times NO_3 \tag{4.70}$$

where $DLAYR$ is the thickness of the soil layer.

Denitrification is assumed to occur when water content reaches 95% of field capacity in the EPIC model (Williams, 1995). The first-order denitrification rate (DN) is:

$$DN = WNO3[1 - \exp(-1.4 \times TFN \times CP)] \tag{4.71}$$

where $WNO3$ is weight of nitrate, CP is percentage of soil organic carbon content, and TFN is the same temperature factor as used in the mineralization process above.

A similar formula is used in the GLEAMS model (Knisel, 1993):

$$DNI = SNO3[1 - \exp(-DK \times TFN \times SWFD)] \tag{4.72}$$

where the daily decay rate (DK) is a linear function of active soil carbon, and $SWFD$ is a water content factor defined by:

$$SWFD = \frac{SW - [FC + 0.1(SAT - FC)]}{SAT - [FC + 0.1(SAT - FC)]} \tag{4.73}$$

The denitrification rate is calculated from the rates of decaying residue and soil organic matter pools in the NCSOIL model, which is based on the amount of energy available (Molina et al., 1983). It is generally assumed that 6 µg C have to be decayed from residue or soil O.M. to denitrify 1 µg NO_3-N. However, the NCSOIL model does not automatically switch from nitrification to denitrification mode. The user must either make a change to the source code to include the effect of soil water content or use the NCSWAP model where the water content effect on nitrification/denitrification is simulated. In NCSWAP, such a stoichiometric relationship is modified by temperature ($TRED$) and water ($XAERDN$) factors. The water content reduction factor in NCSWAP is written as:

$XAERDN = 0.0$ $PERSAT < WHCPSAT$

$$XAERDN = AERDN + \frac{(-AERDN)(PERSAT - WHINFL)}{WHCPSAT - WHCCST - WHINFL} \qquad WHCPSAT \leq PERSAT < WHINFL$$

$$XAERDN = AERDN + \frac{(1.0 - AERDN)(PERSAT - WHINFL)}{100 - WHINFL} \qquad WHINFL \leq PERSAT < 100$$

$XAERDN = 1.0$ $PERSAT = 100$ (4.74)

where $AERDN$ is a water reduction factor for denitrification at a water content of $WHINFL$ (expressed as percentage of saturation).

In the LEACHM model, the denitrification rate is simulated with Michaelis-Menten kinetics (Hutson, 2000):

$$\frac{dNO_3}{dt} = -\frac{\mu_{denit} NO_3}{NO_3 + C_{sat}} \qquad (4.75)$$

where C_{sat} is the half-saturation constant and μ_{denit} is a potential rate that can be adjusted by T_{cf} and W_{cfdn}:

$$W_{cfdn} = \left[\frac{\max(0, \theta - \theta_b)}{\theta_s - \theta_b} \right]^2 \qquad (4.76)$$

where θ_b is the water content at which W_{cfdn} decreases from 1 at saturation (θ_s) to zero. T_{cf} is the same as the one used for the mineralization process.

Urea Hydrolysis

Not all the models simulate urea hydrolysis. The NLEAP, GLEAM, CENTURY, NCSOIL, and EPIC models assume that urea hydrolysis is so rapid that applied urea can be treated as NH_4. First-order kinetics is used in RZWQM for urea hydrolysis (Ma et al., 2001). The rate coefficient is a function of soil temperature and soil aerobic conditions. Neither microbial population nor soil water content is simulated because they do not affect urease activity in soils (Shaffer et al., 2000a). Similarly, first-order kinetics is used in the LEACHM model, and the rate coefficient is modified by temperature (T_{cf}) and water content (W_{cf}) effects in a similar manner to that in the other processes (Hutson, 2000). Urea hydrolysis is simulated in NTRM using a regression equation (Table 4.1).

First-order kinetics is used for urea hydrolysis in the CERES model (Godwin and Jones, 1991). The hydrolysis rate ($UHYDR$) is calculated by:

$$UHYDR = AK \times minimum(SWF, STF) \times UREA \qquad (4.77)$$

where *UREA* is the soil urea concentration; *SWF* and *STF* are the water content and temperature effects on urea hydrolysis, respectively; and *AK* is given by a regression equation based on organic carbon (*OC*) and soil pH with minimum value of 0.25:

$$AK = -1.12 + 1.31\,OC + 0.203\,pH - 0.155\,OC \times pH \qquad (4.78)$$

The water content (*SWF*) and temperature (*STF*) effects are:

$$SWF = MF + 0.2$$

$$STF = \frac{T_s}{40} + 0.2 \qquad (4.79)$$

The *MF* factor is the same as that used for the mineralization process.

Ammonia Volatilization

Considerable quantities of ammonia may be lost when there is a high concentration of ammonium close to the soil surface at high soil pH. Many models use a first-order kinetic approach. In the NLEAP model, the rate coefficient is calculated from a rate constant modified by a temperature factor (T_f), and table values of the rate constant are provided according to fertilizer application method, occurrence of precipitation, CEC, and residue cover (Hansen et al., 1995; Shaffer et al., 1991; 2001). In RZWQM, the rate coefficient is a function of the pressure gradient of NH_3 between the soil and air, temperature, and wind speed (Ma et al., 2001). However, ammonia volatilization is not simulated in NTRM or NCSOIL. Volatilization is represented as a first-order loss of NH_4 from the soil surface in LEACHM, and the rate coefficient is constant in the current released version (Hutson, 2000).

Ammonia volatilization is not considered for upland soils in the CERES model (Godwin and Jones, 1991). However, for rice paddy soils, ammonia volatilization is simulated at high pH levels (Godwin and Singh, 1998). The CERES version for flooded soil simulates volatilization in two steps. First an ammonia-N concentration in the aqueous phase (*FLDH3C*) is calculated from:

$$FLDH3C = \frac{FLDH4C}{1.0 + 10.0^{\,0.009018 + \frac{2729.92}{TK} - FPH}} \qquad (4.80)$$

where *FLDH4C* is the ammonium concentration in floodwater; *TK* is the floodwater temperature (K); *FPH* is the floodwater pH. The second step is to calculate ammonia loss (*AMLOS*) based on the hourly water evaporation rate (*HEF*) and partial pressure of ammonia-N (*FNH3P*) derived from *FLDH3C*:

$$AMLOS = 0.036 \times FNH3P + 0.05863 \times HEF + 0.000257$$

$$\times FNH3P \times 2.0 \times HEF \times FLOOD \qquad (4.81)$$

where *FLOOD* is the depth of floodwater.

Ammonia volatilization in the EPIC model is simulated as first-order kinetics (Williams, 1995). Differently from other models, EPIC assumes ammonia can be volatilized from the entire soil profile, not only the first soil layer. The first-order volatilization rate (*RVOL*) is written as:

$$RVOL = WNH3\left[1.0 - \exp(AKV)\right] \tag{4.82}$$

where *AKV* is a coefficient determined by multiplying the temperature and wind speed factors for the uppermost soil layer or by multiplying the temperature, CEC (cation exchange capacity), and soil depth factors for deeper layers (Williams, 1995). The temperature effect is the same as that for nitrification (*TFNIT*). The wind speed effect (*WNF*) is estimated from:

$$WNF = 0.335 + 0.161\mathrm{n}(V) \tag{4.83}$$

where *V* is wind speed. The CEC factor (*CECF*) is:

$$CECF = 1.0 - 0.038\,CEC \tag{4.84}$$

and the soil depth factor (*DPF*) is a function of soil depth (*DEPTH*):

$$DPF = 1.0 - \frac{DEPTH}{DEPTH + \exp(4.706 - 0.0305\,DEPTH)} \tag{4.85}$$

Ammonia volatilization is simulated only for surface-applied solid, slurry, and liquid animal waste in GLEAMS (Knisel, 1993). A first-order volatilization rate (*VOLN*) is a function of air temperature (*ATP*) only:

$$VOLN = AWNH\,\exp(-K_v t) \tag{4.86}$$

where *AWNH* is ammonia in the animal waste, *t* is time, and K_v is a rate constant defined by:

$$K_v = 0.409(1.08)^{ATP-20} \tag{4.87}$$

Volatilization loss of N is assumed to be 5% of total N mineralized in the CENTURY model (Parton et al., 1987). The model also assumes a fraction of the N loss via volatilization from feces and urine. However, the current version of the model makes no attempt to conduct a mechanistic simulation of ammonia volatilization.

Plant Nitrogen Uptake

Plant N uptake is one of the most unknown yet most important areas in agricultural system models because it relates to plant growth, root distribution, C and N dynamics, water movement, and N transport. Any error in simulating the above processes will affect the accuracy of plant N uptake. Thus far, plant N uptake in most of the models is driven by plant N demand, which can be estimated from a logistic curve or by interaction with a plant growth model (Hansen et al., 1995). Supply of soil N to roots is determined by plant available total N (NH_4 and NO_3), rate of water transpiration, and possibly by diffusion of soil N to root surfaces. Most models assume N concentration at the root surface to be the same as in the bulk soil solution (Hansen et al., 1995).

RZWQM has a plant growth sub-model to estimate N demand at various growth stages. Plant available N in a soil layer is the sum of all inorganic forms of N. Nitrogen uptake from each soil layer is first estimated from the transpiration water stream (called passive uptake). If total N uptake from all the soil layers (based on root density) is less than the N demand, an active uptake mechanism in the form of Michaelis-Menten kinetics is invoked to meet the demand until all the plant available N is exhausted (Hanson, 2000). If an N-fixing legume is planted, RZWQM assumes that all the N demand can be met, regardless of the amount of N in the soil. In addition, the model assumes uptake of NH_4 and NO_3, with no preference for which. The amount of N uptake from each soil mineral pool is calculated from its relative percentage in the soil layers. NLEAP, on the other hand, uses a logistic curve to estimate N demand based on yield goal and relative growth stage (Shaffer et al., 1991). Actual uptake is determined by available soil mineral N. In the case of legumes, all N demand can be met through N fixation. NTRM uses a plant growth simulation model with separate but interrelated modules for the roots and tops. The plant growth model is driven by air and soil temperatures and solar radiation, and environmental constraints include water stress and nitrogen stress (Shaffer and Larson, 1987).

Plant N uptake in the CERES model is determined by plant N demand and soil N supply (Godwin and Jones, 1991; Godwin and Singh, 1998). Plant N demand is composed of the N deficiency of existing biomass plus N required for new growth. Nitrogen supply is calculated from root length density, soil water content, and available soil NH_4 and NO_3 concentrations. Actual N uptake is the lower of demand and supply. If potential N supply is greater than crop N demand, N uptake from each soil layer is reduced proportionally to the level of demand. The model also allows luxury N uptake and organic N exudation from the plant. In the case of N exudation, the exuded N is added into the fresh organic N pool (Godwin and Jones, 1991; Godwin and Singh, 1998). Originally, different availability functions were used to calculate soil NH_4 and NO_3 supply (Godwin and Jones, 1991). However, an average function is now used for both NH_4 and NO_3 in the DSSAT version of CERES, except that plants cannot take up N when the soil NH_4 concentration is below 0.5 µg N/g soil (Godwin and Singh, 1998).

Plant N uptake is also based on a supply and demand approach in the EPIC model (Williams, 1995). Nitrogen demand is estimated from the difference between current plant N content and optimal N content. Nitrogen supply is based on water uptake with adjustments made to ensure that actual N uptake does not exceed demand at high soil nitrate concentration and to increase N uptake at low soil nitrate concentration. In the EPIC model, only NO_3 is taken up by plants. It has a functional approach to N fixation. Daily N fixation is proportional to daily plant N demand. The proportional coefficient is affected by the soil water content, soil nitrate concentration, and plant growth stage (Williams, 1995).

Plant N uptake in the GLEAMS model is a modification of the EPIC model (Knisel, 1993). The major difference is that GLEAMS assumes the uptake of both NH_4 and NO_3. In a similar manner to that in RZWQM, the quantity of N taken up from each species is proportional to their respective masses in the soil layer. Legumes will not fix nitrogen from the atmosphere if the total ammonia and nitrate concentration exceeds 5 mg/L. Otherwise, N fixation is equal to daily N demand. Fixed N is not added into the soil directly, except through crop residues or roots returned to the soil (Knisel, 1993).

In the CENTURY model, plant N uptake is also based on the demand and supply concept (Parton et al., 1987). Nitrogen demand is based on the C:N ratio of monthly biomass increments, whereas N supply is based on soil N concentration and root biomass. Actual N uptake is the lower of the two. Initially, potential plant biomass is calculated from annual rainfall (Parton et al., 1987). Later, monthly biomass growth is estimated from fixed monthly potential growth, multiplied by temperature, moisture content, and plant shading factors (Parton et al., 1993). CENTURY assumes that solar radiation is not a limiting factor; therefore, it is not a control variable in the plant biomass simulation. (Parton et al., 1993). Nitrogen fixation is a function of annual precipitation (Parton et al., 1987).

NCSOIL itself simulates only N and C processes for each soil layer. Plant N uptake is part of the NCSWAP model. A logistic reference growth curve is used to simulate daily plant growth and modified according to water, N, and temperature stresses as required (Molina et al., 1985; 1997). An N demand is calculated from the percentage of N in plant biomass. The plant is assumed to take both NH_4-N and NO_3-N based on their respective concentrations.

Plant N uptake in the LEACHM model is based on the approach of Watts and Hanks (1978) where a potential N uptake is calculated from the growth stage (in the range 0 to 1). Nitrate and ammonium are assumed to be taken up through the transpiration stream. However, if the transpiration stream cannot meet N demand, a diffusive equation is used to calculate additional N uptake by plants (Hutson, 2000).

Soil Horizon Differentiation

The various models define the soil profile differently, based on model complexity. NLEAP has only two soil layers (called the upper and lower horizons) because two horizons are sufficient for the objective of the model (although the newest version of NLEAP has three soil horizons). RZWQM can define up to 10 soil horizons and can be expanded if needed. Each horizon can be subdivided into computational layers. NTRM is coded to accept four soil horizons, which are then divided into seven segments for chemical transport or 25 nodes for water movement (Shaffer, 1987). The CERES model divides the soil profile into up to 20 layers. GLEAMS allows for a maximum of five soil horizons, with up to 12 possible computational layers for plant water uptake (Knisel, 1993). A soil profile in EPIC can be divided into up to 10 layers. The CENTURY model has set a maximum of 10 soil layers, with suggested thicknesses of 15 cm for the first two layers. Soil organic matter dynamics are simulated only in the first 20 cm of the soil profile in CENTURY. NCSWAP uses the same soil horizon structure as NTRM, and users can specify up to 10 (minimum of three) soil horizons. Each horizon can be subdivided into segments. The LEACHM model uses segments of equal thickness, with 13 to 25 segments recommended (Hutson, 2000).

Other Related Components Simulated in the Models

Carbon and nitrogen processes are generally coupled with other processes in agricultural systems, such as water movement, chemical transport, plant growth, and management practices. In all the models reviewed in this chapter, we have seen considerable differences among the models in terms of C and N processes. In addition to these differences, depending on the objectives, each individual model is linked with a wide range of approaches to simulating water movement, chemical transport, plant growth, and agricultural management. Therefore, it is very difficult (if not impossible) to evaluate the C and N process modeling in the various models because the C and N processes are strongly affected by the way other processes are simulated.

RZWQM was developed to provide a system tool for understanding agricultural production and its impact on soil and water quality, by synthesizing the state-of-the-science knowledge from literature sources. Therefore, in addition to comprehensive modeling of C and N processes, it has a detailed simulation approach to water movement, plant growth, and chemical transport. Water movement is simulated by the Green-Ampt equation during infiltration and the Richards equation during redistribution (Ahuja et al., 2000b). Potential evapotranspiration (ET) is estimated from the extended Shuttleworth-Wallace model (Ahuja et al., 2000a). Plant water uptake is evaluated using the approach of Nimah and Hanks (1973). In addition, the model simulates heat transport, tile drainage, water table fluctuation, macropore flow, and surface runoff (Ahuja et al., 2000b). A generic plant growth is used in RZWQM and has been tested under various experimental conditions (Hanson, 2000). RZWQM also has a chemical equilibrium module to estimate soil pH and other soil cations if required (Shaffer et al., 2000b). A pesticide module is included to estimate pesticide fates in agricultural soils (Ahuja et al., 2000a). Various agricultural management practices and their effects on the system are simulated in details also (Rojas and Ahuja, 2000).

Because NLEAP was designed primarily for estimating N leaching potential, it uses a very simple water balance routine. For each of the two soil layers, water available for leaching is determined by water input (e.g., precipitation, irrigation, or leaching from the layer above), ET losses, water holding capacity, and water content at the previous time step. ET is calculated from pan evaporation data modified by a crop coefficient (Shaffer et al., 1991). Nitrate-N leached during a simulation time step is computed using an exponential relationship based on water available for leaching and soil porosity. No plant growth is simulated; instead, the model uses a logistic growth curve adjusted for target yield (Shaffer et al., 1991). Fairly complete soil, climate, and crop databases have been developed for different regions in the United States.

NTRM is a system model with all the major components, such as water and chemical movement, plant growth, root development, management effects (e.g., crop residue, tillage, irrigation, fertilization), soil C/N dynamics, soil temperature simulation, runoff, and soil equilibrium chemistry (Shaffer and Larson, 1987). The Richards equation is used for water movement (Shaffer, 1985; Shaffer and Gupta, 1987). Surface runoff is calculated using a regression equation as a function of soil slope and surface residue mass (Shaffer, 1985). NTRM also uses the partial differential equation of heat flow to simulate soil temperature, and the diffusion equation for solute transport (Shaffer, 1985). A functional approach is used to calculate photo-synthesis, biomass accumulation, leaf area, and root distribution. Those functions are extended to include 20 crops using appropriate coefficients (Shaffer and Larson, 1987).

CERES is one of the most widely used models thus far. Its C and N simulation sub-model has been linked into a family of 16 crop growth models under the DSSAT framework (barley, maize, millet, rice, sorghum, wheat, dry bean, soybean, peanut, chickpea, sugarcane, tomato, sunflower, pasture, cassava, and potato) (Tsuji et al., 1998). The soil water balance is estimated using a relatively simple approach com-pared to some of the recently developed models (e.g., RZWQM). Surface runoff is estimated using the USDA-Soil Conservation Service (SCS) curve number (Ritchie, 1998). Water redistribution in the soil is dependent on water holding capacity, which is the water contained between saturation (SAT) and drained upper limit (DUL). Plant water uptake is calculated from root length density and water content above a predefined lower limit water content (Ritchie, 1998). Root growth is affected by soil bulk density, soil temperature, soil water content, as well as Al toxicity and Ca deficiency (Jones et al., 1991). A similar root growth model is used by RZWQM (Hanson, 2000).

The EPIC model is a fully functional agricultural system model and all its components are documented by Williams (1995). Surface runoff is calculated from a USDA-SCS curve number. Water percolation occurs only if a soil layer exceeds field capacity. The model also calculates upward water flow if a lower soil layer exceeds field capacity. It can also estimate lateral flow in the case of a hillside landscape. EPIC estimates soil erosion caused by water and wind. Another feature of EPIC is the simulation of P dynamics in addition to N dynamics. Daily soil temperature is calculated from a series of empirical equations, rather than solving the heat equation. The plant growth simulation is not mechanistic; rather, it uses a series of conversion or scaling factors to estimate biomass, leaf area index (LAI), plant height, root growth, and yield. Biomass is synthesized at a rate related to the solar radiation level and is used to calculate yield using a harvest index. LAI and plant height are estimated directly from daily heat units (degree days) up to given maximum values. Root growth is related to both heat units and biomass by a simple equation (Williams, 1995). These growth rates may be affected by water, tempera-ture, and nutrient stresses. In the case of root growth, it is affected by soil bulk density, soil texture, and Al toxicity. EPIC also simulates tillage effects on soil bulk density, soil nutrient mixing, crop residues, and surface roughness (Williams, 1995).

GLEAMS has four major components: hydrology, erosion/sediment yield, pes-ticide transport, and nutrients (N and P) (Knisel, 1993). The nutrient component is a modified version of EPIC. Soil erosion is simulated in a manner similar to that in EPIC. Its main objective was to estimate non-point source pollution from agricultural land. Potential ET can be estimated by the Penman-Monteith or Priestly-Taylor equation. Soil temperature is simulated the same way as in EPIC. Water movement through the soil profile is based on a simple piston flow and field capacity concept (Knisel, 1993). Runoff is calculated from the USDA-SCS curve number, and erosion is based on the USLE (Universal Soil Loss Equation) model. Plant LAI and biomass are based on logistic curves with given maxima.

The CENTURY model has a monthly time step and most of the simulated processes are simplified for that reason. The newer version, DAYCENT, has more detailed processes (Del Grosso et al., 2001; Parton et al., 1996). CENTURY calculates a monthly water budget with various empirical functions (Parton et al., 1993). Water movement from layer to layer is determined by draining excess water above the field capacity. Field capacity is calculated from soil bulk density, soil texture, and organic matter content as developed by Gupta and Larson (1979). Nitrate leaching is a function of monthly saturated water flow and percentage of sand (Parton et al., 1993). A potential monthly biomass growth is estimated from temperature, moisture, and plant shading factors, which may be reduced if there is insufficient N, P, and S. In addition, the model also simulates P and S based on C:P and C:S ratios (Parton et al., 1988).

NCSOIL can be a stand-alone model or serve as a sub-model of NCSWAP, where water flow and plant growth are simulated. NCSWAP uses a simple piston flow for water movement without allowing for upward flow. Plant growth is based on a reference logistic curve with modifications for water, temperature, and N stresses (Molina et al., 1985; 1997).

The LEACHM model is a water and solute transport model (Hutson, 2000). Plant growth is simply represented by empirical equations so that ET and nutrient uptake can be estimated. There is no feedback between soil condition and plant behavior. The model provides both simple (tipping-bucket approach) and complicated approaches (convective-dispersive equation and Richards equation) to water and solute transport for different users. Pesticide fates and soil P dynamics are simulated. Soil temperature is modeled with the heat flow equation (Hutson, 2000). In addition, the model has a runoff component based on the USDA-SCS curve number and a detailed pesticide process simulation.

PARAMETERIZATION AND APPLICATION OF THE MODELS

Data requirements vary considerably among the nine models. Most models need daily weather information, while others need only monthly averages (e.g., CENTURY). To drive C/N cycling in a system model, methods to estimate ET, water and nitrogen uptake, soil water content, and soil temperature are needed, and associated input parameters vary from method to method. For example, when a logistical plant growth curve is used to estimate biomass production, only the maximum yield and growth period are needed, with possible environmental stress factors (e.g., NLEAP). However, if a fully interactive plant growth model is implemented, detailed plant growth parameters are required (e.g., RZWQM). Also, if a piston flow is used to estimate soil water balance, only a few critical soil water contents (e.g., drained upper limit, saturated soil water content) are needed (e.g., in CERES). However, if a fully explicit Richards equation is used for water balance, much more detailed soil hydrological properties are required (e.g., in RZWQM).

How to obtain model parameters is one of the major obstacles in model applications. There are three types of model parameters governing C and N processes. One type includes the experimentally measurable/recordable parameters (e.g., weather data, soil data, management practices). Another type includes internal parameters that cannot be measured (e.g., nitrification rate constant, denitrification rate constant, mineralization rate constant). The third type involves the parameters that depend on other processes in an ecosystem (e.g., soil water content). Uncertainty exists for all the parameters, including spatial and temporal variability of measured data, non-uniqueness of fitted model parameters, and interaction among various processes. In addition, it is difficult to assess the significance of this uncertainty on model performance. Most often, a model user is dealing with more parameters than experimental data points, and has to select only a few parameters to optimize simulation results. Furthermore, very few users are applying a mathematically sound optimization algorithm to obtain model parameters (Priesack et al., 2001). Parameters derived by one user (either from field or laboratory data) are seldom applicable to other users or data without additional calibration (Nicolardot et al., 1994). Thus, databases for model parameters may be needed and improved by extensive application of a model (Shaffer et al., 1991).

Due to difficulty in model parameterization, applications of the mathematical models are limited to the model developers and their collaborators unless the model is promoted by a governmental agency, such as the EPA and USDA. The extent of model use also depends on the objectives of the model, ease-of-use, and model support availability. Very few applications include more than one model, unless they are for comparison purposes (Hansen et al., 1995; Smith et al., 1997). Model users tend to first try the models that are readily available, and then look for others if these models are not adequate for their purposes. A typical model application is to describe an experimental phenomenon and then make conclusions or recommendations using the calibrated model. Another application of a model is to use it as a building block in another large-scale model, such as in a linkage to GIS (geographic information systems) (e.g., the GLEAMS-GIS linkage [Garnier et al., 1998; Wu et al., 1996a]; CERES-GIS [Beinroth et al., 1998]; and NLEAP-GIS [Shaffer and Wylie, 1995; Shaffer et al., 1996]) and economics (e.g., EPIC economic component [Williams, 1995]).

Most model applications are at the stage of research and model evaluation. However, the ultimate goal of developing models is to use them for technology transfer and decision-making. Although none of the above-mentioned models is perfect, each has been used for soil management purposes, such as crop rotation (Wagner-Riddle et al., 1997; Diebel et al., 1995; Foltz et al., 1995); crop transplanting (Saseendran et al., 1998); planting date (Delgado, 1998); irrigation scheduling (Steele et al., 1994); nitrogen application (Paz et al., 1999; Watkins et al., 1998; Azevedo et al., 1997); manure management (Ma et al., 1998); tillage effects (Singh and Kanwar, 1995); risk analysis (Epperson et al., 1992; Martin et al., 1996); decision support (Ford et al., 1993; Evers et al., 1998); screening tests (Khakural and

Robert, 1993); climate effects on crop production and soil sustainability (Phillips et al., 1996; Lee et al., 1996); policy evaluation (Wu et al., 1996b; Chowdhury and Lacewell, 1996); pest management (Teng et al., 1998); and carbon sequestration (Lee and Dodson, 1996). Conclusions drawn from these applications are qualitative and indecisive, at best.

SUMMARY AND CONCLUSION

The vast differences among the nine models reflect our understanding of the soil C and N processes. There are no right or wrong answers to the approaches used in the models. All of them are based on a general understanding of the heterogeneity of soil organic matter and residues, supported by limited experimental data. Some models may choose to have more surface residue pools (e.g., CERES), whereas others attempt differentiation of soil humus pools (e.g., RZWQM and CENTURY). Some models use more sophisticated temperature, water, and pH factors to modify the rate coefficients (e.g., RZWQM), while others use a series of 0–1 factors (e.g., CERES and EPIC) or constant values for rate coefficients (e.g., NTRM). However, the effects of these environmental factors on C and N processes are simulated in ways that differ so much from model to model that very few model users can understand them (or bother to ask such details). For example, why does CERES assume no pH effects at pH 6.0 to 8.0, whereas EPIC assumes no pH effects at pH 7.0 to 7.4 for nitrification? In addition, how do these environmental factors interact to affect the rate coefficients? Should one calculate the effect of each factor independently (e.g., CERES, EPIC) or dependently (e.g., RZWQM)? In the case of independent calculation, should one take the lower of the factors (e.g., CERES, NCSOIL) or multiply them (e.g., EPIC, GLEAMS)? Both approaches have some experimental background. The most important element for a good model is to provide users with step-by-step information on how to obtain the conceptual model parameters. Because of misunderstanding between model developers and users, there is often misuse of simulation models, either by applying them beyond their capability or having a wrong set of model parameters. Therefore, it is strongly suggested that model users consult model documentation or developers whenever possible so that models are applied correctly and as intended.

The partitioning of soil organic C and N differs significantly among the nine models, ranging from a single pool model (e.g., NTRM) up to five pools (e.g., RZWQM and CENTURY). The purpose of multiple pools is to capture the heterogeneity of soil organic materials, so that derived rate coefficients from one soil can be applied to others. Such a wish has not yet been fulfilled in practice in model application because (1) there are no experimental methods to adequately differentiate the pools, and the pools are more conceptual than real; (2) the physical and chemical properties of the pools have yet to be sufficiently examined experimentally, and the parameters associated with the pools remain empirical or unknown; and (3) the distribution of soil organic C and N pools in the soil is continuous, and any discretion of the pools is arbitrary. As a result, multiple pool models may not have significant advantages in terms of solving real-world problems, except for research purposes.

On the other hand, a simple pool model such as NTRM, once calibrated for a given soil condition, may do just as good a job as the multiple pool models.

Mineralization and immobilization processes are simulated very differently among the nine U.S. models, ranging from a purely empirical approach in NTRM to a highly complex approach in RZWQM. Although most models include a pool containing soil microbial biomass (e.g., NCSOIL and CENTURY), only RZWQM explicitly simulates the growth and death of microbial populations. Simple models (e.g., NTRM and NLEAP) use a C:N ratio to determine net mineralization or immobilization, and more sophisticated models estimate an immobilization rate based on decomposition rate (e.g., CERES and EPIC). The most complicated U.S. models (e.g., RZWQM) have a highly dynamic soil microbial growth model to calculate the immobilization rate. Decomposition of organic C can be simulated using a regression equation (e.g., NTRM), first-order kinetics (e.g., RZWQM and NLEAP), or Monod kinetics (e.g., NCSOIL). Rate coefficients are affected by soil temperature and soil water content in most models. However, they can also be affected by soil pH (e.g., RZWQM), C:N ratio (e.g., CERES), C:P ratio (e.g., EPIC), bulk density (e.g., EPIC), lignin content (e.g., CENTURY), soil texture (e.g., CEN-TURY), soil ionic strength (e.g., RZWQM), and soil microbial population (e.g., RZWQM). Among the nine models, only the CENTURY model assumes a fraction of the decomposed organic C to be mobile and leaching out of the soil profile. CERES is the only one that explicitly simulates C/N dynamics in floodwater with algal growth. When immobilization occurs, inorganic N can be taken from both NH_4 and NO_3 pools (e.g., NCSOIL and NLEAP), or only the NO_3 pool (e.g., EPIC). The NTRM model uses different regression equations for immobilization from NH_4 and NO_3. Transfer of mass among organic C pools during decomposition is usually assumed to be a constant fraction of those decomposed (e.g., RZWQM, CERES, EPIC, GLEAMS, LEACHM, and NCSOIL) or modified for soil texture (e.g., CEN-TURY) or determined by the C:N ratio of the decomposing pool (e.g., NLEAP).

The nitrification process is simulated in all the models except CENTURY, which uses a monthly time step so that nitrification is not important. In fact, CENTURY does not differentiate among inorganic N forms. Nitrification kinetics may be zero order (e.g., GLEAMS, NCSOIL, and RZWQM), half order (e.g., NTRM), first order (e.g., EPIC, LEACHM), or Michaelis-Menten (e.g., CERES). The RZWQM model simulates the nitrifier population, which applies a feedback mechanism to nitrification. LEACHM, on the other hand, assumes a maximum nitrate-to-ammonium ratio to control the nitrification rate. Rate coefficients are generally modified by soil water, pH, and temperature.

By the same token, the denitrification process is also simulated differently among the nine models. There is no mechanistic simulation of the denitrification processes as in the NTRM and CENTURY models. Thus far, only the CERES model has a special version for wetland soils when it is used in the CERES-rice model. Denitrification can be triggered by days with precipitation (e.g., NLEAP) or by soil water content (e.g., CERES, RZWQM, and EPIC). The denitrification rate can be zero order (e.g., NCSOIL), first order (e.g., RZWQM and NLEAP), or Michaelis-Menten (e.g., LEACHM). In addition to soil water content, the denitrification rate is also affected by soil temperature and carbon substrate availability as an energy source for denitrifiers (e.g., CERES, NCSOIL, RZWQM, EPIC, NLEAP, and DAYCENT).

Urea hydrolysis is not simulated in five out of the nine models (e.g., NLEAP, GLEAMS, CENTURY, NCSOIL, and EPIC). Applied urea is directly added as its equivalent into the soil NH_4 pool or inorganic N pool. NTRM uses a regression equation to predict urea hydrolysis, and the rest assume first-order kinetics (e.g., RZWQM, LEACHM, and CERES). The first-order rate coefficient is modified by soil temperature and water content in the same way as for other C and N processes (e.g., mineralization, nitrification, and denitrification) in the LEACHM model. However, the CERES and RZWQM models use different water and temperature effects for urea hydrolysis as compared to other processes. In addition, CERES includes a rate constant that is a linear regression function of organic carbon and pH effects.

Ammonia volatilization is assumed to take place only in the top soil layer in most models, but it can also be simulated for all the soil layers (e.g., EPIC). Two models, NTRM and NCSOIL, do not simulate ammonia volatilization. The CERES model for upland crops does not simulate volatilization either. However, the rice version of CERES has the capability of estimating ammonia losses through volatilization. Ammonia volatilization is usually treated as a first-order process and the rate coefficient is affected by temperature (e.g., RZWQM, GLEAMS); wind speed (e.g., RZWQM and EPIC); the cation exchange capacity (e.g., EPIC); soil depth (e.g., EPIC and RZWQM); and water evaporation rate (e.g., CERES-rice). CENTURY does not explicitly simulate ammonia volatilization; rather, it takes a fraction of mineralized N as volatilization loss. The CERES-rice model calculates an ammonia-N concentration in the aqueous phase and then uses a regression equation to predict volatilization loss from hourly water evaporation, partial pressure of ammonia-N, and floodwater depth.

Plant N uptake is very much dependent on crop growth. All the models use a plant demand and soil supply concept. Most models do not differentiate between NH_4 and NO_3 for plant N uptake, except for EPIC where only NO_3 can be taken up by the plant. Plant N demand is calculated from growth stage (e.g., LEACHM) or plant biomass. The latter can be simulated in extensive detail (e.g., RZWQM and CERES) or using a logistic curve (e.g., NLEAP and NCSOIL). Most models assume N uptake first through the transpiration stream. However, if such a passive uptake cannot meet plant N demand and there is enough inorganic soil N, the plant may take additional soil N by means of diffusion (e.g., LEACHM) or Michaelis-Menten kinetics (e.g., RZWQM). In the case of N fixation, the quantity of N fixed is estimated from annual precipitation (e.g., CENTURY) or plant N demand (e.g., RZWQM and EPIC). Partitioning of N uptake among soil layers is determined by the plant root distribution, and this can be calculated from empirical functions (e.g., NTRM, NCSOIL, and LEACHM) or detailed root growth with carbon allocation (e.g., RZWQM).

In addition to the differences in simulating C and N processes among the nine models, the models also vary in simulating other components, such as water movement, chemical transport, pesticide fates, and plant growth. Therefore, each model is unique and co-exists with others. For example, RZWQM has its own strengths (e.g., macropore flow), but it does not simulate erosion and P dynamics. LEACHM and RZWQM have similar strengths with respect to water movement and pesticide transport, but the plant growth component in LEACHM is very empirical and it is weak in simulating management effects. CERES family models are strong in their plant growth component, but their water balance component is simple, and it has no macropore flow or tile drainage flow. EPIC and GLEAMS both simulate N and P in soils, soil erosion, and surface runoff, but use simple hydrology and plant growth approaches. The CENTURY model has N, P, and S components, but all the components are extremely simple to accommodate the monthly time step. NLEAP uses simple approaches to all the components, but it does have a nationwide database, and it is very easy to use.

Although the different models were initially developed to fill some knowledge gaps in previous models, they were expanded into a system model by incorporating more and more components. For example, GLEAMS was initially a chemical transport, soil erosion, and runoff model. Now, it has subsurface drainage flow, macropore flow, and nutrient dynamics (Morari and Knisel, 1997; Reyes et al., 1994). Very often, model developers borrow modules from each other. For example, GLEAMS's C and N module has its origin in EPIC, and EPIC's pesticide module is from GLEAMS (Knisel, 1993; Sabbagh et al., 1991). In addition, model developers make modifications to modules even for those that are borrowed. Therefore, it is very difficult to know the details of the model being used without consulting the model owners or the computer code. Thus, the same model name in the literature may contain different approaches to some processes. A future direction of model development would be a modular modeling framework, where users can assemble their own system models from well-defined modules for their specific applications.

REFERENCES

Ahuja, L.R., K.W. Rojas, J.D. Hanson, M.J. Shaffer, and L. Ma. (Eds.). 2000a. *Root Zone Water Quality Model: Modeling Management Effects on Water Quality and Crop Production.* Water Resources Publications, LLC, Highlands Ranch, CO. 372pp.

Ahuja, L.R., K.E. Johnsen, and K.W. Rojas. 2000b. Water and chemical transport in soil matrix and macropores. In L.R. Ahuja, K.W. Rojas, J.D. Hanson, M.J. Shaffer, and L. Ma (Eds.). *Root Zone Water Quality Model: Modeling Management Effects on Water Quality and Crop Production.* Water Resources Publications, LLC, Highlands Ranch, CO. 13-50.

Azevedo, A.S., P. Singh, R.S. Kanwar, and L.R. Ahuja. 1997. Simulating nitrogen management effects on subsurface drainage water quality. *Agricultural Systems,* 55:481-501.

Ball, D.A., and M.J. Shaffer. 1993. Simulating resource competition in multispecies agricultural plant communities. *Weed Research,* 33:299-310.

Beinroth, F.H., J.W. Jones, E.B. Knapp, P. Papajorgji, and J. Luyten. 1998. Evaluation of land resources using crop models and a GIS. In G.Y. Tsuji, G. Hoogenboom, and P.K. Thornton (Eds.). *Understanding Options for Agricultural Production.* Kluwer Academic. London, 293-310.

Bittman, S., D.E. Hunt, and M.J. Shaffer. 2001. NLOS (NLEAP on STELLA) — A nitrogen cycling model with a graphical interface: implications for model developers and users. In M.J. Shaffer, L. Ma, and S. Hansen (Eds.). *Modeling Carbon and Nitrogen Dynamics for Soil Management.* Lewis Publishers, Boca Raton, FL, 383-402.

Chowdhury, M.E., and R.D. Lacewell. 1996. Implications of alternative policies on nitrate contamination of groundwater. *J. Agric. Resour. Econ.,* 21:82-95.

Clanton, C.J., D.C. Slack, and M.J. Shaffer. 1986. Evaluation of the NTRM model for land application of wastewater. *Trans. ASAE,* 29:1307-1313.

Delgado, J.A. 1998. Sequential NLEAP simulations to examine effect of early and late planted winter cover crops on nitrogen dynamics. *J. Soil & Water Conservation,* 53:241-244.

Del Grosso, S.J., W.J. Parton, A.R. Mosier, M.D. Hartman, J. Brenner, D.S. Ojima, and D.S. Schimel. 2001. Simulated interaction of soil carbon dynamics and nitrogen trace gas fluxes using the DAYCENT model. In M.J. Shaffer, L. Ma, and S. Hansen (Eds.). *Modeling Carbon and Nitrogen Dynamics for Soil Management.* Lewis Publishers, Boca Raton, FL, 303-332.

Diebel, P.L., J.R. Williams, and R.V. Llewelyn. 1995. An economic comparison of conventional and alternative cropping systems for a representative northeast Kansas farm. *Rev. Agric. Econ.,* 17:323-335.

Dutt, G.R., M.J. Shaffer, and W.J. Moore. 1972. Computer Simulation Model of Dynamic Bio-physicochemical Processes in Soils. Tech. Bulletin 196. Arizona Agricultural Experiment Station. 128 pp.

Epperson, J.E., H.E. Hook, and Y.R. Mustafa. 1992. Stochastic dominance analysis for more profitable and less risky irrigation of corn. *J. Prod. Agric.,* 5:243-247.

Evers, A.J.M., R.L. Elliott, and E.W. Stevens. 1998. Integrated decision making for reservoir, irrigation, and crop management. *Agricultural Systems,* 58:529-554.

Foltz, J.C., J.G. Lee, M.A. Martin, and P.V. Preckel. 1995. Multiattribute assessment of alternative cropping systems. *Am. J. Agric. Econ.,* 77:408-420.

Ford, D.A., A.P. Kruzic, and R.L. Doneker. 1993. Using GLEAMS to evaluate the agricultural waste application rule-based decision support (AWARDS) computer program. *Water Science and Technology,* 28:625-634.

Garnier, M., A. Lo Porto, R. Marini, and A. Leone. 1998. Integrated use of GLEAMS and GIS to prevent groundwater pollution caused by agricultural disposal of animal waste. *Environmental Management,* 22:747-756.

Godwin, D.C., and C.A. Jones. 1991. Nitrogen dynamics in soil-plant systems. In J. Hanks and J.T. Ritchie (Eds.). *Modeling Plant and Soil Systems.* American Society of Agronomy, Inc. Madison, WI, 287-321.

Godwin, D.C., and U. Singh. 1998. Nitrogen balance and crop response to nitrogen in upland and lowland cropping systems. In G.Y. Tsuji, G. Hoogenboom, and P.K. Thornton (Eds.). *Understanding Options for Agricultural Production.* Kluwer Academic, Boston, 55-77.

Grant, R.F. 2001. A review of Canadian ecosystem model — *ecosys.* In M.J. Shaffer, L. Ma, and S. Hansen (Eds.). *Modeling Carbon and Nitrogen Dynamics for Soil Management.* Lewis Publishers, Boca Raton, FL, 173-264.

Gupta, S.C., and W.E. Larson. 1979. Estimating soil water retention characteristics from particle size distribution, organic matter content, and bulk density. *Water Resour. Res.,* 15:1633-1635.

Hanks, J., and J.T. Ritchie. 1991. *Modeling Plant and Soil Systems.* American Society of Agronomy, Inc., Madison, WI, 545pp.

Hansen, S., H.E. Jensen, and M.J. Shaffer. 1995. Developments in modeling nitrogen transformations in soil. In P.E. Bacon (Ed.). *Nitrogen Fertilization in the Environment.* Marcel Dekker, New York, 83-107.

Hanson, J.D. 2000. Generic Crop Production. In L.R. Ahuja, K.W. Rojas, J.D. Hanson, M.J. Shaffer, and L. Ma. (Eds.). *Root Zone Water Quality Model: Modeling Management Effects on Water Quality and Crop Production.* Water Resources Publications, LLC, Highlands Ranch, CO, 81-118.

Hutson, J.L. 2000. LEACHM: *Model Description and User's Guide.* School of Chemistry, Physics, and Earth Sciences, The Flinders University of South Australia, Adelaide, South Australia.

Jensen, L.S., T. Mueller, S. Bruun, and S. Hansen. 2001. Application of the DAISY model for short and long-term simulation of soil carbon and nitrogen dynamics. In M.J. Shaffer, L. Ma, and S. Hansen (Eds.). *Modeling Carbon and Nitrogen Dynamics for Soil Management.* Lewis Publishers, Boca Raton, FL, 485-511.

Johnsson, H., L. Bergström, P.-E. Janson, and K. Paustian. 1987. Simulated nitrogen dynamics and losses in a layered agricultural soil. *Agriculture, Ecosystems and Environment,* 18:333-356.

Jones, C.A., W.L. Bland, J.T. Ritchie, J.R. Williams. 1991. Simulation of root growth. In J. Hanks and J.T. Ritchie (Eds.). *Modeling Plant and Soil Systems.* American Society of Agronomy, Inc., Madison, WI, 91-123.

Jones, J.W., G.Y. Tsuji, G. Hoogenboom, L.A. Hunt, P.K. Thornton, P.W. Wilkens, D.T. Imamura, W.T. Bowen, and U. Singh. 1998. Decision support system for agrotechnology transfer; DSSAT v3. In G.Y. Tsuji, G. Hoogenboom, and P.K. Thornton (Eds.). *Understanding Options for Agricultural Production.* Kluwer Academic, London, 157-177.

Khakural, B.R., and P.C. Robert. 1993. Soil nitrate leaching potential indices: using a simulation model as a screening system. *J. Environ. Qual.,* 22:839-845.

Knisel, W.G. (Ed.). 1993. GLEAMS, Groundwater Loading Effects of Agricultural Management Systems. Version 2.10. UGA-CPES-BAED, publication No. 5. 259pp.

Koch, A.L., J.A. Robinson, and G.A. Milliken (Eds.). 1998. *Mathematical Modeling in Microbial Ecology.* Chapman & Hall. Microbiology Series, New York. 273pp.

Lee, J.J., and R.F. Dodson. 1996. Potential carbon sequestration by afforestation of pasture in the south central United States. *Agron. J.,* 88:381-384.

Lee, J.J., D.L. Phillips, R.F. Dodson. 1996. Sensitivity of the US corn belt to climate change and elevated CO2. II. Soil erosion and organic carbon. *Agricultural Systems,* 52:503-521.

Leonard, R.A., W.G. Knisel, and D.S. Still. 1987. GLEAMS: Groundwater Loading Effects of Agricultural Management Systems. *Trans. ASAE,* 30:1403-1418.

Li, C., S. Frolking, and T.A. Frolking. 1992. A model of nitrous oxide evolution from soil driven by rainfall events. 1. Model structure and sensitivity. *J. Geophys. Res.,* 97:9777-9796.

Ma, L., M.J. Shaffer, J.K. Boyd, R. Waskom, L.R. Ahuja, K.W. Rojas, and C. Xu. 1998. Manure management in an irrigated silage corn field: experiment and modeling. *Soil Sci. Soc. Am. J.,* 62:1006-1017.

Ma, L., L.R. Ahuja, J.C. Ascough, II, M.J. Shaffer, K.W. Rojas, R.W. Malone, and M.R. Cameira. 2000. Integrating system modeling with field research in agriculture: applications of the Root Zone Water Quality Model (RZWQM). *Adv. Agron.,* 71:233-292.

Ma, L., M.J. Shaffer, and L.R. Ahuja. 2001. Application of RZWQM for soil nitrogen management. In M.J. Shaffer, L. Ma, and S. Hansen (Eds.). *Modeling Carbon and Nitrogen Dynamics for Soil Management.* Lewis Publishers, Boca Raton, FL, 265-301.

Martin, E.C., J.T. Ritchie, and B.D. Baer. 1996. Assessing investment risk of irrigation in humid climate. *J. Prod. Agric.,* 9:228-233.

McGechan, M.B., and L. Wu. 2001. A review of carbon and nitrogen processes in European soil nitrogen dynamics models. In M.J. Shaffer, L. Ma, and S. Hansen (Eds.). *Modeling Carbon and Nitrogen Dynamics for Soil Management.* Lewis Publishers, Boca Raton, FL, 103-171.

McGill, W.B. 1996. Review and classification of ten soil organic matter (SOM) models. In D.S. Powlson, P. Smith, and J.U. Smith (Eds.). *Evaluation of Soil Organic Matter Models Using Existing Long-term Datasets.* NATO ASI Series I, Vol. 38. Springer-Verlag, Heidelberg, 111-132.

Metherell, A.K., L.A. Harding, C.V. Cole, and W.J. Parton. 1993. CENTURY: Soil Organic Matter Model Environment. USDA-ARS, Great Plains Systems Research Unit, Technical Report No. 4. Fort Collins, CO.

Molina, J.A.E. 1996. Description of the model NCSOIL. In D.S. Powlson, P. Smith, and J.U. Smith (Eds.). *Evaluation of Soil Organic Matter Models Using Existing Long-term Datasets.* NATO ASI Series I, Vol. 38. Springer-Verlag, Heidelberg, 269-274.

Molina, J.A.E. and P. Smith. 1998. Modeling carbon and nitrogen processes in soils. *Adv. Agron.,* 62:253-298.

Molina, J.A.E., C.E. Clapp, M.J. Shaffer, F.W. Chichester, and W.E. Larson. 1983. NCSOIL, A model of nitrogen and carbon transformations in soil: description, calibration, and behavior. *Soil Sci. Soc. Am. J.,* 47:85-91.

Molina, J.A.E., M.J. Shaffer, R.H. Dowdy, and J.F. Power. 1985. Simulation of tillage, residue, and nitrogen management. In R.F. Follett and B.A. Stewart (Eds.). *Soil Erosion and Crop Productivity.* Soil Science Society of America, Madison, WI, 413-430.

Molina, J.A.E., C.E. Clapp, M.J. Shaffer, F.W. Chichester, and W.E. Larson. 1987. Carbon and nitrogen transformations in soil, Submodel NCSOIL. In M.J. Shaffer and W.E. Larson (Eds.). *NTRM, A Soil-Crop Simulation Model for Nitrogen, Tillage, and Crop-Residue Management.* U.S. Department of Agriculture, Conservation Research Report No. 34-1. 38-50.

Molina, J.A.E., G.J. Crocker, P.R. Grace, J. Klir, M. Korschems, P.R. Poulton, and D.D. Richter. 1997. Simulating trends in soil carbon in long-term experiments using the NCSOIL and NCSWAP models. *Geoderma,* 81:91-107.

Morari, F., and W.G. Knisel. 1997. Modification of the GLEAMS model for crack flow. *Trans. ASAE,* 40:1337-1348.

Myrold, D.D. 1998. Modeling nitrogen transformations in soil. In A.L. Koch, J.A. Robinson, and G.A. Milliken (Eds.). *Mathematical Modeling in Microbial Ecology.* Chapman & Hall. Microbiology Series, New York, 142-161.

Nicolardot, B., J.A.E. Molina, and M.R. Allard. 1994. C and N fluxes between pools of soil organic matter. Model calibration with long-term incubation data. *Soil Biology & Biochemistry,* 26:235-243.

Nimah, M., and R.J. Hanks. 1973. Model for estimating soil-water-plant-autospheric interrrelation. I. Description and sensitivity. *Soil Sci. Soc. Am. Proc.,* 37:522-527.

Parton, W.J. 1996. The CENTURY model. In D.S. Powlson, P. Smith, and J.U. Smith (Eds.). *Evaluation of Soil Organic Matter Models Using Existing Long-term Datasets.* NATO ASI Series I, Vol. 38. Springer-Verlag, Heidelberg, 284-291.

Parton, W.J., D.S. Schimel, C.V. Cole, and D.S. Ojima. 1987. Analysis of factors controlling soil organic matter levels in the Great Plains grasslands. *Soil Sci. Soc. Am. J.,* 51:1173-1179.

Parton, W.J., J.W.B. Stewart, and C.V. Cole. 1988. Dynamics of C, N, P, and S in grassland soils: a model. *Biogeochemistry,* 5:109-131.

Parton, W.J., J.M.O. Scurlock, D.S. Ojima, T.G. Gilmanov, R.J. Scholes, D.S. Schimel, T. Kirchner, J.-C. Menaut, T. Seastedt, E. Garcia Moya, Apinan Kamnalrut, and J.I. Kinyamario. 1993. Observations and modeling of biomass and soil organic matter dynamics for the grassland biome worldwide. *Global Biogeochemical Cycles,* 7:785-809.

Parton, W.J., D.S. Ojima, C.V. Cole, and D.S. Schimel, 1994. A general model for soil organic matter dynamics: sensitivity to litter chemistry, texture and management. In *Quantitative Modeling of Soil Forming Processes,* SSSA Spec. Pub. 39, Madison, WI, 147-167.

Parton, W.J., and P.E. Rasmussen. 1994. Long-term effects of crop management in wheat-fallow: II CENTURY model simulations. *Soil Sci. Soc. Am. J.,* 58:530-536.

Parton, W.J., A.R. Mosier, D.S. Ojima, D.W. Valentine, D.S. Schimel, K. Weier, and A.E. Kulmala. 1996. Generalized model for N_2 and N_2O production from nitrification and denitrification. *Global Biogeochemical Cycles,* 10:401-412.

Parton, W.J., M. Hartman, D.S. Ojima, and D.S. Schimel, 1998. DAYCENT: its land surface submodel: description and testing, *Global Planetary Change,* 19:35-48.

Paz, J.O., W.D. Batchelor, B.A. Babcock, T.S. Colvin, S.D. Logsdon, T.C. Kaspar, and D.L. Karlen. 1999. Model based technique to determine variable rate nitrogen for corn. *Agricultural Systems,* 61:69-75.

Phillips, D.L., J.J. Lee, R.F. Dodson. 1996. Sensitivity of the US corn belt to climate change and elevated CO2. I. Corn and the soybean yields. *Agricultural Systems,* 52:481-502.

Priesack, E., S. Achatz, and R. Stenger. 2001. Parameterisation of soil nitrogen transport models by use of laboratory and field data. In M.J. Shaffer, L. Ma, and S. Hansen (Eds.). *Modeling Carbon and Nitrogen Dynamics for Soil Management.* Lewis Publishers, Boca Raton, FL, 461-484.

Radke, J.K., M.J. Shaffer, and J. Saponara. 1991. Application of the Nitrogen-Tillage-Residue-Management (NTRM) model to low-input and conventional agricultural systems. *Ecol. Modelling,* 55:241-255.

Reyes, M.R., R.L. Bengston, and J.L. Fouss. 1994. GLEAMS-WT hydrology submodel modified to include subsurface drainage. *Trans. ASAE,* 37:1115-1120.

Ritchie, J.T. 1998. Soil water balance and plant water stress. In G.Y. Tsuji, G. Hoogenboom, and P.K. Thornton (Eds.). *Understanding Options for Agricultural Production.* Kluwer Academic, Boston, 41-54.

Rojas, K.W., and L.R. Ahuja. 2000. Management practices. In L.R. Ahuja, K.W. Rojas, J.D. Hanson, M.J. Shaffer, and L. Ma (Eds.). *Root Zone Water Quality Model, Modeling Management Effects on Water Quality and Crop Production.* Water Resources Publications, Highlands Ranch, CO. 245-280.

Sabbagh, G.J., S. Geleta, R.L. Elliott, J.R. Williams, and R.H. Griggs. 1991. Modification of EPIC to simulate pesticide activities: EPIC-PST. *Trans. ASAE,* 34:1683-1692.

Saseendran, S.A., K.G. Hubbard, K.K. Singh, N. Mendiratta, L.S. Rathore, S.V. Singh. 1998. Optimum transplanting dates for rice in Kerala, India, determined using both CERES v3.0 and ClimProb. *Agron. J.,* 90:185-190.

Seligman, N.G., and H. van Keulen. 1981. PAPRAN: a simulation model of annual pasture production limited by rainfall and nitrogen. In M.J. Frissel and J.A. van Veen (Eds.). *Simulation of Nitrogen Behavior of Soil-Plant Systems, Proc. Workshop.* Wageningen, The Netherlands, 192-221.

Shaffer, M.J., G.R. Dutt, and W.J. Moore. 1969. Predicting Changes in Nitrogenous Compounds in Soil-Water Systems. Water Pollution Control Research Series 13030 ELY, 15-28.

Shaffer, M.J. 1985. Simulation model for soil erosion-productivity relationships. *J. Environ. Qual.,* 14:144-150.

Shaffer, M.J. 1987. Model structure and submodel interactions. In Shaffer, M.J., and W.E. Larson (Eds.). NTRM, A Soil-Crop Simulation Model for Nitrogen, Tillage, and Crop-Residue Management. U.S. Department of Agriculture, Conservation Research Report No. 34-1, 6-14.

Shaffer, M.J. and S.C. Gupta. 1987. Unsaturated flow submodel. In Shaffer, M.J., and W.E. Larson (Eds.). NTRM, A Soil-Crop Simulation Model for Nitrogen, Tillage, and Crop-Residue Management. U.S. Department of Agriculture, Conservation Research Report No. 34-1, 51-56.

Shaffer, M.J., and W.E. Larson (Eds.). NTRM, A Soil-Crop Simulation Model for Nitrogen, Tillage, and Crop-Residue Management. U.S. Department of Agriculture, Conservation Research Report No. 34-1, 103pp.

Shaffer, M.J., and B.K. Wylie. 1995. Identification and mitigation of nitrate leaching hot spots using NLEAP/GIS technology. *Journal of Contaminant Hydrology,* 20:253-263.

Shaffer, M.J., A.D. Halvorson, and F.J. Pierce. 1991. Nitrate leaching and economic analysis package (NLEAP): model description and application. In R.F. Follett et al. (Eds.). *Managing Nitrogen for Groundwater Quality and Farm Profitability*. Soil Science Society of America, Inc., Madison, WI, 285-322.

Shaffer, M.J., T.E. Schumacher, and C.L. Ego. 1995. Simulating the effects of erosion on corn productivity. *Soil Sci. Soc. Am. J.*, 59:672-676.

Shaffer, M.J., M.D. Hall, B.K. Wylie, and D.G. Wagner. 1996. NLEAP/GIS approach for identifying and mitigating regional NO$_3$-N leaching. In D.L. Corwin and K. Loague (Eds.). *Application of GIS to the Modeling of Non-point Source Pollutants in the Vadose Zone*. SSSA Special Publication, American Society of Agronomy, Madison, WI, 283-294.

Shaffer, M.J., K.W. Rojas, D.G. DeCoursey, and C.S. Hebson. 2000a. Nutrient chemistry processes — OMNI. In L.R. Ahuja, K.W. Rojas, J.D. Hanson, M.J. Shaffer, and L. Ma (Eds.). *Root Zone Water Quality Model, Modeling Management Effects on Water Quality and Crop Production*. Water Resources Publications, Highlands Ranch, CO, 119-144.

Shaffer, M.J., K.W. Rojas, D.G. DeCoursey, and C.S. Hebson. 2000b. The equilibrium soil chemistry process — SOLCHEM. In L.R. Ahuja, K.W. Rojas, J.D. Hanson, M.J. Shaffer, and L. Ma (Eds.). *Root Zone Water Quality Model, Modeling Management Effects on Water Quality and Crop Production*. Water Resources Publications, Highlands Ranch, CO, 145-161.

Shaffer, M.J., K. Lasnik, X. Ou, and R. Flynn. 2001. NLEAP Internet tools for estimating NO$_3$-N leaching and N$_2$O emission. In M.J. Shaffer, L. Ma, and S. Hansen (Eds.). *Modeling Carbon and Nitrogen Dynamics for Soil Management*. Lewis Publishers, Boca Raton, FL, 403-426.

Singh, P., and R.S. Kanwar. 1995. Simulating NO3-N transport to subsurface drain flows as affected by tillage under continuous corn using modified RZWQM. *Trans. ASAE*, 38:499-506.

Smith, O.L. 1982. *Soil Microbiology: A Model of Decomposition and Nutrient Cycling*. CRC Press, Boca Raton, FL, 273pp.

Smith, P., J.U. Smith, D.S. Powlson, W.B. McGill, J.R.M. Arah, O.G. Chertov, K. Coleman, U. Franko, S. Frolking, D.S. Jenkinson, L.S. Jensen, R.H. Kelly, H. Klein-Gunnewiek, A.S. Komarov, C. Li, J.A.E. Molina, T. Mueller, W.J. Parton, J.H.M. Thornley, and A.P. Whitmore. 1997. A comparison of the performance of nine soil organic matter models using datasets from seven long-term experiments. *Geoderma*, 81:153-225.

Steele, D.D., E.C. Stegman, and B.L. Gregor. 1994. Field comparison of irrigation scheduling methods for corn. *Trans. ASAE*, 37:1197-1203.

Swan, J.B., M.J. Shaffer, W.H. Paulson, and A.E. Peterson. 1987. Simulating the effects of soil depth and climatic factors on corn yield. *Soil Sci. Soc. Am. J.*, 51:1025-1032.

Teng, P. S., W.D. Batchelor, H.O. Pinnschmidt, and G.G. Wilkerson. 1998. Simulation of pest effects on crops using coupled pest-crop models: the potential for decision support. In G.Y. Tsuji, G. Hoogenboom, P.K. Thornton (Eds.). *Understanding Options for Agricultural Production*. Kluwer Academic, London, 221-265.

Tsuji, G.Y., G. Hoogenboom, and P.K. Thornton (Eds.). 1998. *Understanding Options for Agricultural Production*. Kluwer Academic, Boston, 399pp.

Wagner-Riddle, C., T.J. Gillespie, L.A. Hunt, and C.J. Swanton. 1997. Modeling a ray cover crop and subsequent soybean yield. *Agron. J.*, 89:208-218.

Watkins, K.B., Y.C. Lu, and W.Y. Huang. 1998. Economic and environmental feasibility of variable rate nitrogen fertilizer application with carry over effects. *J. Agricultural and Resource Economics*, 23:401-426.

Watts, D.G., and R.J. Hanks. 1978. A soil-water-nitrogen model for irrigated corn on sandy soils. *Soil Sci. Soc. Am. J.*, 42:492-499.

Williams, J.R. 1995. The EPIC model. In V.P. Singh (Ed.). *Computer Models of Watershed Hydrology.* Water Resources Publications, Highlands Ranch, CO, 909-1000.

Wu, Q.J., A.D. Ward, and S.R. Workman. 1996a. Using GIS in simulation of nitrate leaching from heterogeneous unsaturated soils. *J. Environ. Qual.,* 25:526-534.

Wu, J.J., H.P. Mapp, and D.J. Bernardo. 1996b. Integrating economic and physical models for analyzing water quality impacts of agricultural policies in the High Plains. *Review of Agricultural Economics,* 18:353-372.

GLOSSARY OF SYMBOLS

A:	Environmental factor for nitrification (0–1) (CERES)
A_t:	Temperature effect on mineralization rate (0–1) (CENTURY)
A_w:	Water effect on mineralization rate (0–1) (CENTURY)
AD:	Air dry soil moisture content (cm^3 cm^3) (CERES)
AERDN:	Water reduction constant for denitrification (dimensionless) (NCSOIL)
AERNI:	Water reduction constant for nitrification (dimensionless) (NCSOIL)
AK:	A regression constant for urea hydrolysis (dimensionless) (CERES)
AKV:	An environmental factor for ammonia volatilization (dimensionless) (EPIC)
AMLOS:	Ammonia volatilization loss rate (kg N/ha/h) (CERES)
ATP:	Air temperature (°C) (GLEAMS)
AWNH:	Ammonia in animal waste (kg/ha) (GLEAMS)
b_1, b_2, b_3:	Regression coefficient (dimensionless) (NTRM)
B:	Microbial biomass used in the Monod kinetic equation (general)
BDP:	Current bulk density (t/m^3) (EPIC)
BDS:	Settled soil bulk density (t/m^3) (EPIC)
C:	Carbon (general)
C_i:	Carbon concentration in pool i (general)
C_r:	Carbon content of surface residue (lb/acre) (NLEAP)
C_{sat}:	Half-saturation constant for denitrification (mg N/L) (LEACHM)
CAL:	Fraction of soluble O.M. leached below the top 30-cm soil depth (dimensionless) (CENTURY)
CEC:	Cation exchange capacity (cmol/kg) (EPIC)
CECF:	Effect of CEC on ammonia volatilization (0–1) (EPIC)
CMN:	Mineralization rate constant for the active humus pool (d^{-1}) (EPIC, GLEAMS)
CNI:	Nitrification potential index of previous day (0.01–1.0) (CERES)
CNP:	A C:N or C:P factor for mineralization (0–1) (EPIC)
CNR:	C:N ratio of organic pool (dimensionless) (CERES, EPIC)
CNRF:	C:N ratio effect on residue pool decomposition (0–1) (CERES)
CO_2:	Carbon dioxide concentration (kg C/ha) (DAYCENT)
Con:	Regression constant (dimensionless) (NTRM)
CP:	Percentage of soil organic carbon content (%) (EPIC)
CPR:	C:P ratio effect on mineralization (dimensionless) (EPIC)
CW:	Water extractable carbon (µg C/g soil) (CERES)
$(C/N)_r$:	C:N ratio of the surface residue pool (dimensionless) (NLEAP)
DCR:	Decay rate coefficient for fresh organic N pool (d^{-1}) (EPIC)
DEPTH:	Depth of soil layer (mm) (EPIC)
DK:	Daily denitrification rate coefficient (d^{-1}) (GLEAMS)
DLAYR:	Thickness of a soil layer (cm) (CERES)
DMINR:	Decay rate constant of the humus pool (d^{-1}) (CERES)

DMOD:	Soil factor on humus mineralization (dimensionless) (CERES)
DN:	First-order denitrification rate (kg N/ha/d) (EPIC)
DNRATE:	First-order denitrification rate (kg N/ha/d) (CERES)
DNI:	First-order denitrification rate (kg N/ha/d) (GLEAMS)
DPF:	Soil depth effect on ammonia volatilization (0–1) (EPIC)
DUL:	Upper drain limit (cm^3/cm^3) (CERES)
E_a:	Activation energy for nitrification (kcal/mole) (NTRM)
ELNC:	Environmental limit on nitrification capacity (0–1) (CERES)
ET:	Evapotranspiration (general)
$F_d(NO_3)$:	Maximum denitrification rate for a given soil NO_3 (g N/ha/d) (DAYCENT)
$F_d(CO_2)$:	Maximum denitrification rate for a given soil respiration rate (g N/ha/d) (DAYCENT)
$F_d(WFP)$:	Effect of soil water on denitrification rate (dimensionless) (DAYCENT)
$F_r(NO_3)$:	Effect of soil NO_3 on $N_2:N_2O$ ratio (dimensionless) (DAYCENT)
$F_r(CO_2)$:	Effect of soil CO_2 on $N_2:N_2O$ ratio (dimensionless) (DAYCENT)
$F_r(WFP)$:	Effect of soil water on $N_2:N_2O$ ratio (dimensionless) (DAYCENT)
F_m:	Fraction of residue partitioned into the metabolic pool (0–1) (CENTURY)
FC:	Water content at field capacity (mm) (EPIC, GLEAMS)
FCLAY:	Clay fraction of the soil (0–1) (CENTURY)
FLDH3C:	Ammonia-N concentration in the aqueous phase ($\mu g\ N/m^3$) (CERES)
FLDH4C:	Ammonium-N concentration in the floodwater ($\mu g\ N/m^3$) (CERES)
FLDH3P:	Ammonia-N partial pressure in the aqueous phase (Pascals) (CERES)
FLOOD:	Depth of floodwater (m) (CERES)
FON:	N in fresh organic pool (kg N/ha) (EPIC)
FPH:	Floodwater pH (CERES)
FSAND:	Sand fraction of the soil (0–1) (CENTURY)
FSILT:	Silt fraction of the soil (0–1) (CENTURY)
FT:	Temperature effect on denitrification (0–1) (CERES)
FW:	Water effect on denitrification (0–1) (CERES)
G1:	Proportion of a residue pool that decays in one day (0–1) (CERES)
GRNOM:	Gross mineralization of N (kg N/ha/d) (CERES)
h_R:	First-order decay rate constant for the resistant component (d^{-1}) (NCSOIL)
H:	Hydrogen concentration in soil (NTRM)
HEF:	Hourly water evaporation rate (mm/h) (CERES)
HMN:	Mineralization rate from the active humus pool (kg N/ha/d) (EPIC, GLEAMS)
k:	First-order decay rate coefficient of the structural C pool (kg m^2/yr) (CENTURY)
k_0:	Zero-order mineralization rate coefficient (general)
$k_{1/2}$:	Half-order nitrification rate constant (NTRM)
k_1:	First-order mineralization rate coefficient (general)
k_b:	Boltzmann constant (J $°K^{-1}$) (NTRM)
k_c:	Decay rate constant of the structural C pool (yr^{-1}) (CENTURY)
k_{dc}:	Denitrification rate constant (d^{-1}) (NLEAP)
k_{dnit}:	First-order denitrification rate coefficient (d^{-1}) (NLEAP)
k_L:	First-order decay rate constant for the labile component (d^{-1}) (NCSOIL)
k_r:	Surface residue decomposition rate constant (d^{-1}) (NLEAP)
K_{mx}:	N turnover coefficient used to calculate nitrification N_2O flux (g N/ha/d) (DAYCENT)
K_s:	Half-saturation constant used in the Monod equation (general)
K_v:	Rate coefficient for ammonia volatilization (d^{-1}) (GLEAMS)
L_s:	Fraction of lignin content in the structure C pool (0–1) (CENTURY)

LAI:	Leaf area index (general)
LN:	Lignin-to-N ratio (dimensionless) (CENTURY)
M_r:	Mineralization-immobilization rate (lb/acre/d) (NLEAP)
MF:	Moisture effect on mineralization rate (0–1) (CERES)
MWL30:	Water leached below the top 30-cm soil depth (cm/mo) (CENTURY)
N:	Nitrogen (general)
N_2:	Nitrogen gas (general)
N_{H_2O}:	Effect of water on nitrification (0–1) (DAYCENT)
N_{pH}:	Effect of soil pH on nitrification (0–1) (DAYCENT)
N_t:	Effect of soil temperature on nitrification (0–1) (DAYCENT)
N_{NH_4}:	Effect of soil NH_4 on nitrification (0–1) (DAYCENT)
N_{mx}:	Maximum nitrification N_2O gas flux (g N/ha/d) (DAYCENT)
N_2O:	Nitrous oxide (general)
N_{wet}:	Number of days with precipitation or irrigation (d) (NLEAP)
NH_4:	Ammonium concentration or name (general)
NHUM:	N content in the humus pool (kg N/ha) (CERES)
NIT0:	Zero-order nitrification rate (kg N/ha/d) (GLEAMS)
NO_3:	Nitrate concentration or name (general)
NNOM:	Net mineralization (kg N/ha/d) (CERES)
O_2:	Oxygen concentration in soil (NTRM)
OC:	Organic carbon content (%) (CERES)
ONA:	Organic N in the active pool (kg N/ha) (EPIC, GLEAMS)
ONS:	Organic N in the stable pool (kg N/ha) (EPIC, GLEAMS)
PERSAT:	Water content expressed as percentage of saturation (%) (NCSOIL)
PHFNIT:	pH effect on nitrification (0–1) (EPIC)
PHN:	pH effect on nitrification (0–1) (CERES)
PO:	Soil porosity (mm) (EPIC)
Q_{10}:	Change in rate constant for every 10°C temperature interval (dimensionless) (LEACHM)
r_i:	Mineralization rate (general)
r_{max}:	Maximum nitrate to ammonium ratio (dimensionless) (LEACHM)
r_r:	Residue decomposition rate (lb/acre/d) (NLEAP)
R_{N_2O}:	N_2O gas flux from nitrification (g N/ha/d) (DAYCENT)
RAT:	Water stress factor (dimensionless) (CENTURY)
RC:	Residue composition factor (dimensionless) (GLEAMS)
RDECR:	Residue decay rate constant (d^{-1}) (CERES)
RHMIN:	Net N mineralization from the humus pool (kg N/ha/d) (CERES)
RMN:	First-order mineralization rate of the fresh organic pool (kg N/ha/d) (EPIC)
RNAC:	Immobilization of N into microbial biomass (kg N/ha/d) (CERES)
RNIT:	First-order nitrification rate (kg N/ha/d) (EPIC)
RON:	Nitrogen flow from active to stable pool (kg N/ha/d) (EPIC, GLEAMS)
RP2:	Nitrification potential index (0.01–1.0) (CERES)
RTN:	Fraction of active soil organic N pool (0–1) (EPIC, GLEAMS)
RVOL:	First-order volatilization rate (kg N/ha/d) (EPIC)
S_L:	Labile fraction of the C pool (0–1) (NCSOIL)
SANC:	Soil ammonium concentration factor (0–1) (CERES)
SAT:	Saturated soil water content (cm^3/cm^3) (CERES, GLEAMS)
SNH_4:	Total ammonium-N in a soil layer (kg N/ha) (CERES, DAYCENT)
SOILC:	Soil C in the stable humus pool (μg C/g soil) (CERES)

SOILMS:	Soil mass (mg/ha) (GLEAMS)
SW:	Volumetric soil water content (mm or cm^3/cm^3) (CERES, GLEAMS, EPIC)
SW25:	Water content = WP + 0.25(FC − WP), (mm) (EPIC)
SWF:	Water effect on urea hydrolysis (0–1) (CERES)
SWFD:	Soil water effect on denitrification (0–1) (GLEAMS)
SWFN0:	Water effect on nitrification (0–1) (GLEAMS)
SWFNIT:	Soil water effect on nitrification (0–1) (EPIC)
STF:	Soil temperature effect on urea hydrolysis (0–1) (CERES)
t:	Time (d) (NLEAP, GLEAMS)
t1:	A derived function from soil temperature (dimensionless) (CENTURY)
t2:	A derived function from soil temperature (dimensionless) (CENTURY)
T_{base}:	Soil temperature when rate constant is specified (ºC) (LEACHM)
T_{cf}:	Temperature effect on mineralization rate (dimensionless) (LEACHM)
T_f:	Temperature effect on mineralization rate (dimensionless) (NLEAP)
T_m:	Soil texture effect on mineralization of the active SOM pool (0–1) (CENTURY)
T_s:	Soil temperature (°C) (CERES, NCSOIL,GLEAMS, DAYCENT)
TF:	Temperature effect on mineralization rate (0–1) (CERES)
TFN:	Temperature effect on mineralization rate (dimensionless) (EPIC)
TFN0:	Temperature effect on nitrification (0–1) (GLEAMS)
TFNIT:	Temperature effect on nitrification (dimensionless) (EPIC)
TK:	Floodwater temperature (°K) (CERES)
TRED:	Temperature effect on mineralization rate (0–1) (NCSOIL)
UHYDR:	First-order urea hydrolysis rate (kg/ha/d) (CERES)
UREA:	Urea concentration in soil (kg/ha) (CERES)
V:	Wind speed (m/s) (EPIC)
VOLN:	First-order ammonia volatilization rate (kg N/ha/d) (GLEAMS)
W_{cf}:	Water effect on mineralization rate (dimensionless) (LEACHM)
W_{cfdn}:	Water effect on denitrification rate (dimensionless) (LEACHM)
W_{cfsat}:	Relative mineralization rate constant at saturated water content (dimensionless) (LEACHM)
W_f:	Water effect on mineralization rate (dimensionless) (NLEAP)
W_{fd}:	Water effect on denitrification (dimensionless) (NLEAP)
WFP:	Water-filled pore space (0–1) (NLEAP, DAYCENT)
WHCCST:	Soil texture effect on nitrification (dimensionless) (NCSOIL)
WHCPSAT:	Soil water holding capacity as percentage of saturation (%) (NCSOIL)
WHINFL:	Percentage water content at which AERNI and AERDN are measured (%) (NCSOIL)
WNF:	Wind effect on ammonia volatilization (dimensionless) (EPIC)
WNH3:	Ammonia equivalent ammonium content (kg/ha) (EPIC)
WNO3:	Nitrate weight in a soil layer (kg/ha) (EPIC)
WRED:	Water effect on mineralization rate (dimensionless) (NCSOIL)
WP:	Soil water content at wilting point (mm) (GLEAMS, EPIC)
XAERDN:	Water reduction factor for denitrification (0–1) (NCSOIL)
XAERNI:	Water reduction factor for nitrification (0–1) (NCSOIL)
YC:	Years in cultivation (yr) (EPIC)
μ_m:	Maximum mineralization rate used in the Monod equation (general)
μ_N:	C:N ratio effect on mineralization rate (dimensionless) (NCSOIL)
μ_{denit}:	Denitrification rate coefficient (g $N/m^2/d$) (LEACHM)
μ_{nit}:	Nitrification rate constant (d^{-1}) (LEACHM)

θ: Soil water content (cm^3/cm^3) (LEACHM)
θ_b: Zero effect water content for denitrification (cm^3/cm^3) (LEACHM)
θ_{min}:   Minimum optimal soil water content for mineralization (cm^3/cm^3) (LEACHM)
θ_{max}:   Maximum optimal soil water content for mineralization (cm^3/cm^3) (LEACHM)
θ_s: Saturated soil water content (cm^3/cm^3) (LEACHM)
θ_{wp}:   Wilting point soil water content (cm^3/cm^3) (LEACHM)
π: A constant 3.14159... (general)

CHAPTER 5

A Review of Carbon and Nitrogen Processes in European Soil Nitrogen Dynamics Models

Malcolm B. McGechan and L. Wu

CONTENTS

INTRODUCTION AND SCOPE

Requirement for Models to Study Environmental Pollution by Nitrogen Compounds

For several years there has been concern about the contribution of agricultural activities to environmental pollution by nitrogen compounds. Particular concerns are nitrate leached to watercourses and deep groundwater, leading to contaminated water supplies, and atmospheric pollution by nitrous oxide (a greenhouse gas) and ammonia (which contributes to acid rain). Simulation models of the dynamic processes that nitrogen compounds undergo in the soil have been developed in a number of countries. Up to now, they have been used mainly to study processes in arable land receiving mineral fertilizer under various cropping and fertilizer regimes.

All-year housing of pigs and poultry and over-winter housing of cattle result in large quantities of animal manure and slurry that is traditionally disposed of by land spreading. Municipal and industrial wastes are also increasingly disposed of on agricultural land. Spreading these wastes can cause serious water and air pollution problems, but they are also valuable as plant nutrients and, if managed efficiently, can substantially reduce chemical fertilizer requirements. However, poor utilization in current farming practice, resulting from inappropriate land spreading technologies and unsuitable timing of spreading operations, represents both a threat to the environment and an economic loss of a valuable resource. The use of soil nitrogen dynamics models is now being extended to the study of environmental pollution from land spreading of manures and wastes, and the benefits of measures to mitigate such pollution. As manure from ruminant livestock is a product of pastoral agriculture, the application of the models is also being extended to cover processes concerned with pasture and forage as well as arable crops.

While environmental pollution is primarily concerned with nitrogen, many soil nitrogen dynamic processes depend on a supply of carbon and thus are closely linked with the dynamics of carbon in soil organic matter. This chapter discusses and compares a number of European models of soil process dynamics; all these models include representation of soil carbon dynamics in parallel with soil nitrogen dynamics. The constituent processes are analyzed with particular reference to the equations used. Sources of information for model parameters are also reviewed.

Previous Comparisons of Models

A workshop titled "Nitrogen Turnover in the Soil-Crop Ecosystem: Modelling of Biological Transformations, Transport of Nitrogen and Nitrogen Use Efficiency" was held in The Netherlands in June 1990. Fourteen simulation models of nitrogen turnover in the soil-crop system (predominantly from European countries) were presented, and de Willigen (1991) compared and appraised these models in terms of the physical, biological, and chemical processes they describe. The uptake of nitrogen and water was also considered. The results of simulations carried out with the same data set (from three arable cropped sites in The Netherlands) were compared. More recently, at a workshop entitled "Validation of Agroecosystem Models," a similar range of models was tested against data sets from arable cropped land in northern Germany. The results were summarized by Diekkrüger et al. (1995), who concluded that, for each model, there was a variation in the accuracy of its representation between different processes. However, both workshops represented collections of independent studies using different models on the same data sets, and lacked a detailed comparison of the equations representing the constituent processes modeled. Of the models considered in these two workshops, three models plus one more recently developed model were selected for review by Wu and McGechan (1998a). They were considered to be the most suitable candidate models for use in studies of environmental pollution from waste spreading. These four models were all physically based and thus had different characteristics compared with some other more empirical models presented at the workshops.

Models Considered for Review

The four models reviewed by Wu and McGechan include:

1. ANIMO, set up by the Winand Staring Centre for Integrated Land Soil and Water Research, Wageningen, The Netherlands (Berghuijs-van Dijk et al., 1985; Rijtema and Kroes, 1991; Groenendijk and Kroes, 1997)
2. SOILN, developed by the Swedish University of Agricultural Sciences, Sweden (Johnsson et al., 1987; Bergström et al., 1991; Eckersten et al., 1998)
3. DAISY, developed by The Royal Veterinary and Agricultural University, Denmark (Hansen et al., 1991; 1993a; b)
4. SUNDIAL, the Rothamstead Nitrogen Turnover Model (Bradbury et al., 1993).

This chapter reviews these four models in detail, plus one additional similar model, CANDY, developed at the Leipzig-Halle Environment Research Centre (UFZ), Germany (Franko et al., 1995; Franko, 1996; Franko and Oelschlägel, 1996). One other model, developed by Verberne et al. (1990) at the DLO Haren Institute of Soil Fertility Research, The Netherlands, covers only carbon processes associated with organic matter decomposition plus the associated nitrogen processes of mineralization or immobilization, and is discussed only in the chapter section on these processes.

Three further models are adaptations of SOILN and are mentioned only where they differ significantly from the standard SOILN. These are SOILNNO from the Norwegian Agricultural University, Ås (Vold, 1997); SWATNIT from Belgium (Vereecken et al., 1990; 1991; Vanclooster et al., 1995); and a Dutch version of SOILN (van Grinsven and Makaske, 1993). Mention should also be made of one further model developed in the United Kingdom, the Hurley Pasture Model (Thornley, 1998). This is primarily a plant growth model, but it does include carbon and nitrogen sub-routines with pools and flows broadly similar to the soil nitrogen dynamics models.

Run-time versions of all the models discussed here are distributed by their authors, together with documentation. A fee is charged for ANIMO. All were originally designed to operate on a standard PC under DOS, and SOIL/SOILN is now available under Windows within COUP. Some models can also be downloaded from Web sites, with source code as well as run-time version and documentation in each case, as follows: http://www.dina.kvl.dk/~abraham/daisy/for DAISY, http://www.mv.slu.se/bgf/for the DOS version of SOILN, and http://amov.ce.kth.se/coup.htm for the Windows version of SOIL/SOILN in COUP.

Main Soil Nitrogen Dynamic Processes

All the models consider the main soil nitrogen dynamic processes, namely, surface application (as fertilizer, manure, or slurry, atmospheric deposition, and deposition or incorporation of dead plant material), mineralization/immobilization (between organic and inorganic forms), nitrification (from ammonium to nitrate), nitrate leaching, denitrification (to N_2O and N_2), and uptake by plants. The level of detail in the treatment of some of the processes varies between models. The process of anaerobiosis, the root cause of denitrification, is only modeled in ANIMO. Ammonia volatilization is an important nitrogen process, but this is modeled only in ANIMO and DAISY, and then only in a limited way.

Other Related Processes

Soil Water

Most soil nitrogen dynamic processes are dependent on soil water content. Denitrification in particular is very dependent on the lack of oxygen in the soil atmosphere and thus rises rapidly with increasing soil wetness. Nitrate transport through the soil profile and out into field drains or deep groundwater is controlled by water movement. Any model of soil nitrogen dynamics is therefore very dependent on an accurate description of soil water content and soil water movement. The original version of SOILN was designed to carry out simulations using the output from previously run simulations with the soil water and heat model SOIL (Jansson, 1995). However, there is now a new version called COUP in which SOIL and SOILN are merged to simultaneously simulate soil water, heat, carbon, and nitrogen processes. ANIMO is designed to operate in conjunction with alternative Dutch soil water simulation models, the multilayer field-scale model SWATRE (Feddes et al., 1978), the two-layer field-scale model WATBAL (Berghuijs-van Dijk, 1990), or the regional-scale model SIMGRO (Querner, 1988). SOIL (and the soil water routines in COUP), SWATRE, and WATBAL all include representation of water flow to field drains as well as downward through the soil layers. Van Grinsven and Makaske (1993) ran their Dutch version of SOILN with their own simple empirical soil water model known as SOLWAT. The most recent version of DAISY has a soil water sub-model based on Richards (1931) equation (as SOIL and SWATRE), with various lower boundary conditions, including the option of field drains. SUNDIAL models water flow as a so-called "piston process," with water filling each layer up to its available water holding capacity before draining to the layer below, evaporation from the uppermost layer containing water, and bypass flow if rainfall in a particular week exceeds a certain value.

Soil Heat

All transformations concerned with soil nitrogen dynamics are very temperature dependent; thus, it is important to have an accurate description of soil heat processes. In addition, evaporation links the soil water and soil heat processes. The SOIL model, on which the SOILN carbon and nitrogen processes depend, is a linked soil water and heat model with a sophisticated treatment of the heat processes, including freezing and thawing of soil water (which change the soil hydraulic properties) and effects of snow cover. All these processes are incorporated into the new COUP model. DAISY models soil temperature by solving a heat flow equation, and also includes freezing and thawing of soil water. ANIMO includes a simple soil heat sub-model independent of its associated soil water model. In SUNDIAL, the temperature dependence of soil nitrogen processes is based on air temperature rather than soil temperature.

Crop Growth

SUNDIAL and ANIMO assume that the rate of extraction of nitrogen from the soil by the growing crop is dependent on date by a simple equation or curve. Early versions of both SOILN and DAISY assumed extraction of nitrogen by the crop dependent on date, and this is still available as an option in each case. Recent versions of SOILN (including that in the COUP model) include daily interaction with a weather-driven cereal crop growth sub-model as described by Eckersten and Jansson (1991), and optionally with a forest growth sub-model as described by Eckersten and Slapokas (1990). A third alternative crop growth sub-model for the SOILN model was developed by Wu and McGechan (1998b; 1999) for grass and grass/clover crops, based on photosynthesis equations in the grass/clover growth model described by Topp and Doyle (1996). The latest version of DAISY includes weather-driven simulations alternatively for cereals, rape, beet, potatoes, peas (with N fixation), or cereal undersown with grass. An option in the DAISY crop growth sub-model is to simulate photosynthesis, assimilate partitioning, respiration, and net production (in a manner similar to the Topp and Doyle grass/clover model) on an hourly basis, with a daily interaction between soil nitrogen and crop growth. For mixed crops (cereals and grass in DAISY, or grass and clover in SOILN), species compete for light, water, and nitrogen.

Subdivision of Compartments and Processes in Models

The main flows and states for the soil nitrogen and soil carbon dynamic processes represented in the models are shown diagramatically in Figures 5.1 through 5.5, with pools and flows as shown in the source paper for each model but rearranged as far as possible to a common layout. There is some variation between models in the definition and subdivision of pools, and in the names given to the flows. Soil organic matter in particular is divided into several sub-pools, generally according to the rate at which material flows out of each sub-pool. Flows from fast cycling pools to the main slow cycling pool of organic matter are variously described as decomposition, humification, and decay.

Layer Subdivision in Soil Profile

All the models divide the soil profile into layers to simulate vertical movement of nitrogen, carbon, and water between layers, as well as transformation within each layer and uptake by plants from each layer at various rates. The SOILN, ANIMO, and DAISY models can divide the profile into 20 or more layers, with thicknesses specified by the user. SUNDIAL divides the soil profile into four fixed layers, of thicknesses 250 mm for the upper two, and 500 mm for the lower two. The upper two layers are then further subdivided into five slices, each 50 mm thick. In CANDY, the profile is divided into 20 layers of 100-mm thickness. SUNDIAL and CANDY only consider one-dimensional (vertical) transport processes, whereas SOILN, ANIMO, and DAISY can also consider two-dimensional transport with horizontal flows along layers to the field drainage system or ditches.

Application of Animal Manure and Slurry

SOILN, DAISY, ANIMO, and CANDY make provision for application of animal manure or slurry. SOILN allows for straw-bedded farmyard manure by providing for an input of N in straw in manure (which is added to the plant litter pool), as well as N in feces and N in urine which need to be specified for both solid manure and for slurry. These nitrogen inputs are initially added to the appropriate pools in the upper layer of the profile; but after plowing (at a date also specified in the input),

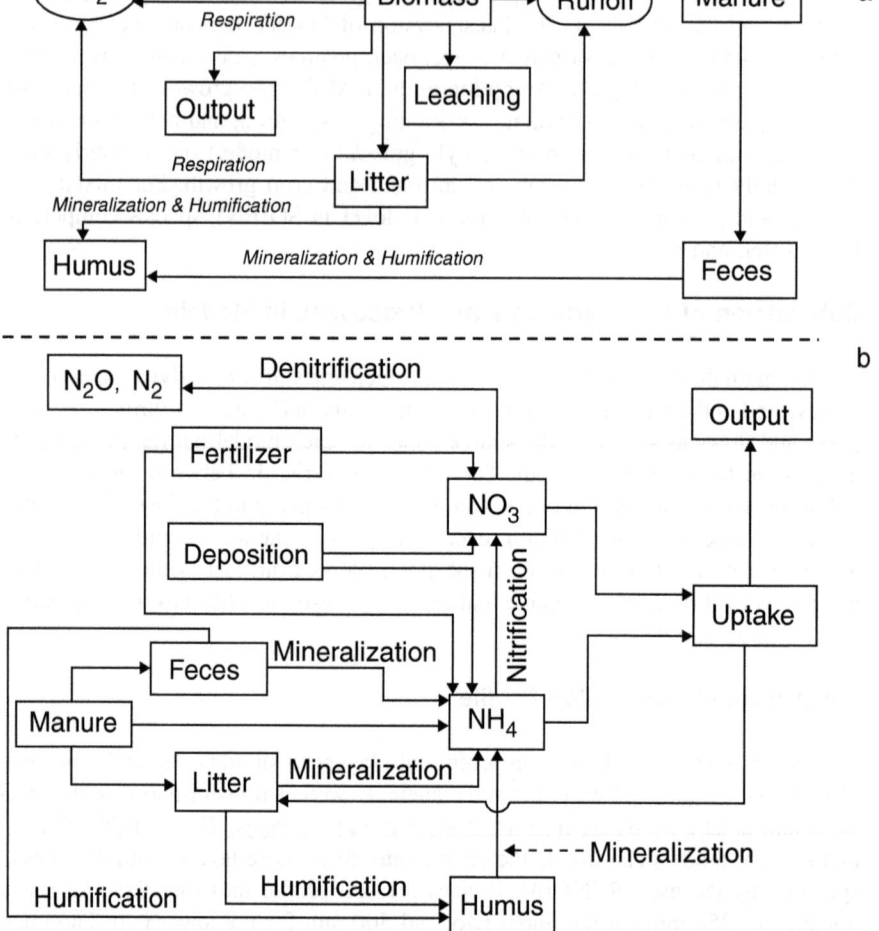

Figure 5.1   Block diagram of (a) carbon cycling and (b) nitrogen cycling in the SOILN model. (From Wu, L., and M.B. McGechan. 1998a. A review of carbon and nitrogen processes in four soil nitrogen dynamic models. *Journal of Agricultural Engineering Research,* 69:279-305. With permission.)

the material is distributed throughout the profile down to the plowing depth. DAISY allows for applications of farmyard manure and slurry characterized by dry matter content, carbon and nitrogen concentrations in dry matter, and the fraction of nitrogen present as ammonia. ANIMO allows for applications of cattle slurry, specified as mineral N (assumed to be 100% ammonium N), plus three categories of organic material each with its own N content. CANDY allows for applications of up to six categories of "fresh" organic matter, including manure or slurry (or more than one category of manure if their differing characteristics are known) as well as plant litter such as plant root material.

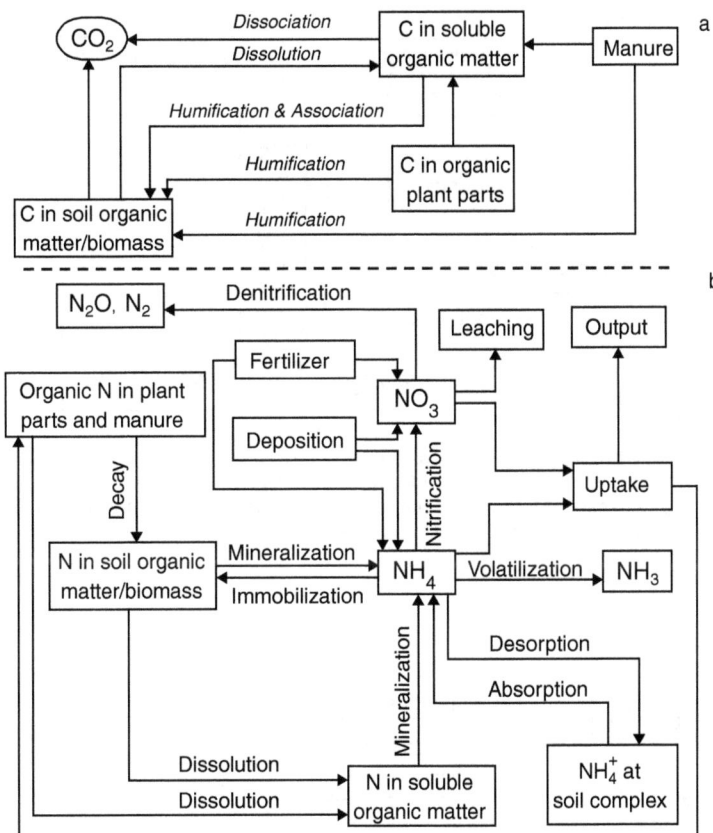

Figure 5.2   (a) Carbon and (b) nitrogen transport and transformation processes in the soil for the ANIMO model. (Adapted from Rijtema, P.E., and J.G. Kroes. 1991. Some results of nitrogen simulations with the model ANIMO. *Fertilizer Research,* 27:189-198; Wu, L., and M.B. McGechan. 1998a. A review of carbon and nitrogen processes in four soil nitrogen dynamic models. *Journal of Agricultural Engineering Research,* 69:279-305. With permission.)

Dynamic Simulation

All the models incorporate weather-driven dynamic simulations. ANIMO and the original SOILN do not use weather data directly, as they use output from previously run, weather-driven simulations with their associated soil water model; they operate at the same time step as the soil water simulations (usually daily). DAISY and the new COUP version of SOIL/SOILN use weather data with a daily or hourly interval to drive their soil water and temperature routines, while SUNDIAL operates on a weekly time step. There is some loss of realism and accuracy of model representation when operating with a weekly rather than a daily time step, particularly regarding processes influenced by rainfall such as high soil wetness, denitrification, drainflows and nitrate leaching, because rainfall occurs in distinct events rather than having an average intensity over a weekly period.

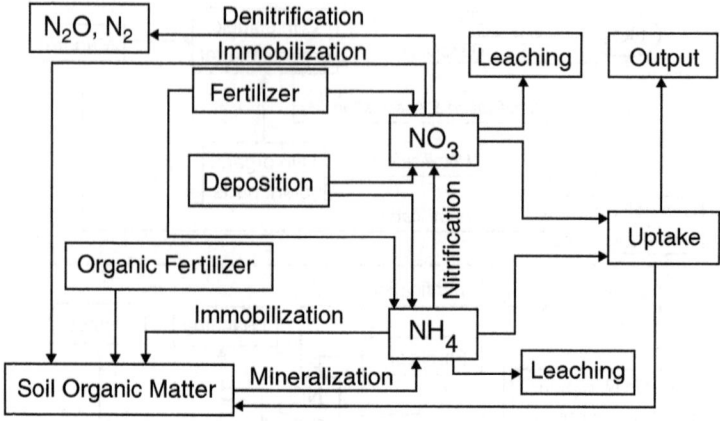

Figure 5.3 Nitrogen transport and transformation processes in the soil for the DAISY model. (From Wu, L., and M.B. McGechan. 1998a. A review of carbon and nitrogen processes in four soil nitrogen dynamic models. *Journal of Agricultural Engineering Research,* 69:279-305. With permission.)

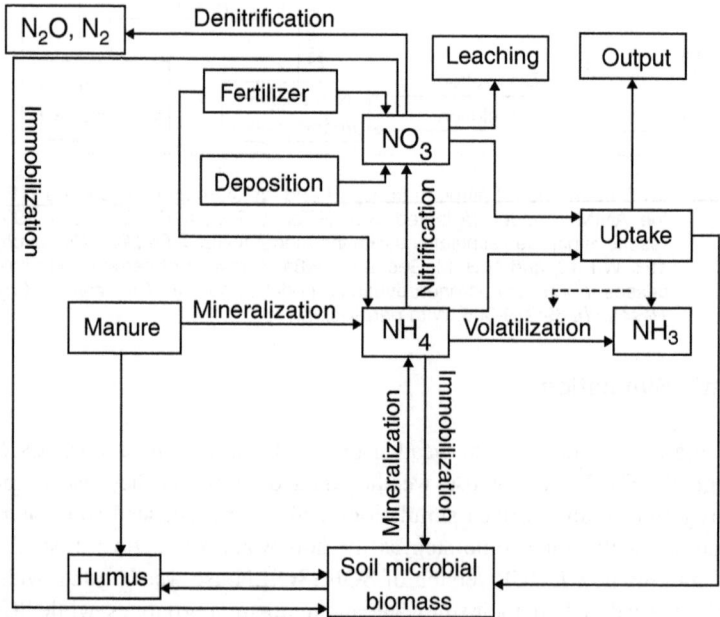

Figure 5.4 Flow diagram for nitrogen in the SUNDIAL model. (From Wu, L., and M.B. McGechan. 1998a. A review of carbon and nitrogen processes in four soil nitrogen dynamic models. *Journal of Agricultural Engineering Research,* 69:279-305. With permission.)

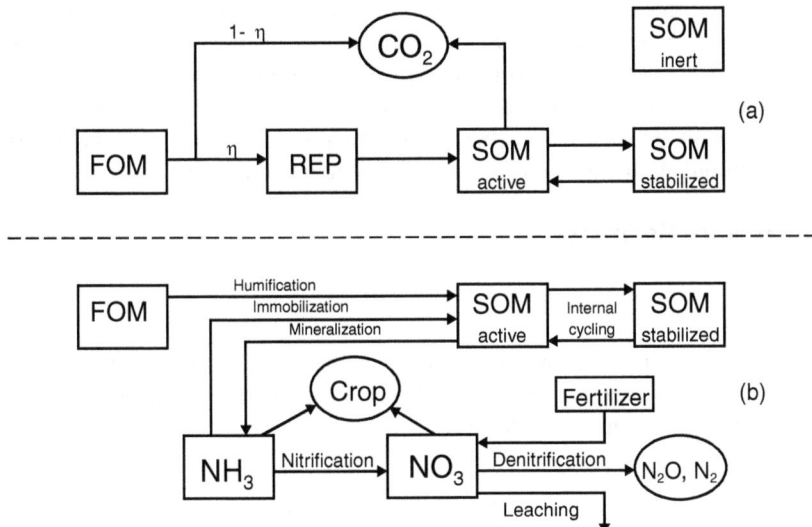

Figure 5.5   Block diagram of (a) carbon cycling and (b) nitrogen cycling in the CANDY model.

THEORETICAL DESCRIPTION OF MATHEMATICS OF PROCESSES

First-Order Rate Processes

In all the models considered, a transformation or flow between pools is represented in most instances as a first-order rate process, which means that the flow out of the first pool to the second is proportional to the quantity q of material remaining in the first pool. This is commonly represented in differential equation form as:

$$\frac{dq}{dt} = -Kq \qquad (5.1)$$

Alternatively, this can be expressed as the solution to the differential equation. The ANIMO model in particular has equations described in this form, which also requires specification of the initial value q_0 at the beginning of the time period or time step:

$$q = q_0 \exp(-K\delta t) \qquad (5.2)$$

A further alternative is to express the change δq over a time step δt. The form of equation used in simulations over a short time step in all the models is:

$$\delta q = -q_0 \left\{ 1 - \exp(-K\delta t) \right\} \qquad (5.3)$$

To account for the sensitivity of such transformations to environmental factors such as temperature and soil wetness, the transformation rate coefficient K in Equations (5.1) to (5.3) for a particular process is multiplied by a response function for each of these factors. Because these response functions are generally the same for all processes in each model (but differ between models), they will be described before introducing the actual process transformation equations.

Temperature Response

SOILN

The soil temperature response function e_t in SOILN is based on a Q_{10} expression, a commonly used function for chemical reactions. Alternative equations are used, depending on the temperature range.

$$e_t = \begin{cases} Q_{10}^{\frac{T-T_b}{10}} & (T \geq T_{lin}) \\ \dfrac{T}{T_{lin}} Q_{10}^{\frac{T-T_b}{10}} & (T < T_{lin}) \end{cases} \tag{5.4}$$

where e_t is the temperature response function, Q_{10} is the response to a 10°C soil temperature change, T is soil temperature (°C), T_b is the base temperature (°C) at which $e_t = 1$, and T_{lin} is a threshold temperature (°C) below which the temperature response is a linear function of temperature.

ANIMO

On the basis of literature sources, the authors of ANIMO concluded that temperature has the greatest effect on organic matter decomposition processes. Other processes follow a similar pattern, but the influence of temperature is less important.

It is assumed that a reaction rate, $r(T)$, increases with soil temperature (T) in the range 0 to 26°C in a manner described by the Arrhenius equation:

$$r(T) = Z \exp\left(-\frac{a}{T+273}\right) \tag{5.5}$$

where Z and a are constants. A relative temperature approach is adopted to incorporate this equation form in the temperature response function e_t, comparing the actual soil temperature in a layer (T) with the mean temperature in that layer ($\bar{T}$). This is expressed as:

$$e_t = \exp\left\{-9000\left(\frac{1}{T+273} - \frac{1}{\bar{T}+273}\right)\right\} \tag{5.6}$$

When T is below zero, e_t is set to zero.

DAISY

The temperature effect in DAISY is expressed as a linear or exponential equation dependent on temperature range, mainly because soil temperatures in Denmark are predominantly in the range 0 to 20°C, which is well below the optimum temperature for nitrogen mineralization. The following is the expression for the effect of temperature on decomposition rates (based on experimental data for the temperature range 0 to 40°C).

$$
e_t = \begin{cases} 0 & (T < 0°C) \\ 0.10T & (0°C < T \le 20°C) \\ \exp(0.47 - 0.027T + 0.00193T^2) & (T > 20°C) \end{cases} \tag{5.7}
$$

There is a similar expression for nitrification with different range boundaries.

$$
e_t = \begin{cases} 0 & (T \le 2°C) \\ 0.15(T - 2) & (2°C < T \le 6°C) \\ 0.10T & (6°C < T \le 20°C) \\ \exp(0.47 - 0.027T + 0.00193T^2) & (20°C < T < 40°C) \end{cases} \tag{5.7a}
$$

SUNDIAL

Three environmental factors affect the decomposition rates of the compartments in SUNDIAL. The temperature function is expressed as a first-order process:

$$
e_t = \frac{47.9}{1 + \exp\left(\dfrac{106}{T_w + 18.3}\right)} \tag{5.8}
$$

where T_w is the weekly mean air temperature (°C).

CANDY

Three environmental factors affect the decomposition rates of the compartments in CANDY. The temperature function (which is also used for the denitrification equation) is expressed as a Q_{10} expression:

$$
e_t = \begin{cases} Q_{10}^{\frac{T-35}{10}} & (T \le 35°C) \\ 1 & (T > 35°C) \end{cases} \tag{5.9}
$$

Soil Water Response

SOILN

The effect of soil water content differs between processes in SOILN. The general response function for all processes except denitrification is based on the assumption that the activity decreases on either side of an optimal range of soil water contents.

$$
e_m = \begin{cases}
\left(\dfrac{\theta - \theta_w}{\theta_{lo} - \theta_w} \right)^m & \left(\theta_w \le \theta < \theta_{lo} \right) \\
1 & \left(\theta_{lo} \le \theta \le \theta_{hi} \right) \\
e_s + (1 - e_s) \left(\dfrac{\theta_s - \theta}{\theta_s - \theta_{hi}} \right)^{m_1} & \left(\theta_{hi} \le \theta \le \theta_s \right)
\end{cases}
\tag{5.10}
$$

where θ_s (%) is the saturated water content; θ_{hi} and θ_{lo} (%) are the high and low water contents, respectively, for which the soil water factor is optimal (equal to unity); and θ_w (%) is the minimum water content for process activity. A coefficient (e_s) defines the soil water response function at saturation, and m is an empirical constant. The two thresholds, defining the optimal range, are calculated as:

$$
\theta_{lo} = \theta_w + \Delta\theta_1
\tag{5.10a}
$$

$$
\theta_{hi} = \theta_s - \Delta\theta_2
\tag{5.10b}
$$

where $\Delta\theta_1$ is the volumetric range of water content where the response increases and $\Delta\theta_2$ is the corresponding range where the response decreases.

For denitrification, the function is:

$$
e_m = \begin{cases}
0 & \left(\theta \le \theta_d \right) \\
\left(\dfrac{\theta - \theta_d}{\theta_s - \theta_d} \right)^{d_1} & \left(\theta > \theta_d \right)
\end{cases}
\tag{5.11}
$$

where θ_d (%) is the threshold water content for denitrification and d_1 is an empirical constant.

ANIMO

In ANIMO, rate constants are reduced by a factor dependent on soil water tension (matric potential) expressed as pF ($\log_{10}$ of the soil water tension in centimeters of water).

DAISY

For decomposition and nitrification, the soil water content function is expressed in terms of soil water tension (matric potential):

$$
e_m = \begin{cases}
A_1 & \left(\psi \geq -10^{-1}\right) \\
A_1 + \dfrac{A_2 \log(10\psi)}{1.5} & \left(-10^{-1} > \psi > -10^{0.5}\right) \\
1.0 & \left(-10^{0.5} \geq \psi \geq -10^{1.5}\right) \\
A_3 - \dfrac{\log(10\psi)}{A_4} & \left(-10^{1.5} > \psi > -B_1\right) \\
0.0 & \left(\psi \leq -B_1\right)
\end{cases}
\tag{5.12}
$$

where ψ is the soil water tension (matric potential, mH_2O), and A_1 to A_4 as well as B_1 (the threshold soil water tension) are constants.

For denitrification, the soil water content function is expressed in terms of the degree of water saturation x_w (the ratio of soil water content to soil water content at saturation):

$$
e_m = \begin{cases}
0 & \left(x_w \leq x_1\right) \\
x \dfrac{x_w - x_1}{x_2 - x_1} & \left(x_1 < x_w < x_2\right) \\
x + (1-x)\dfrac{x_w - x_2}{1 - x_2} & \left(x_2 \leq x_w \leq 1\right)
\end{cases}
\tag{5.13}
$$

where x, x_1, and x_2 are constants.

SUNDIAL

The soil water content effect function is given as:

$$
e_m = 1 - (1 - e_{m0}) \frac{\gamma_c - \gamma_i}{\gamma_f - \gamma_i}
\tag{5.14}
$$

where γ_c is the calculated water deficit in a certain layer (mm), γ_i is the water held between field capacity and a tension of 100 kPa (mm) in that layer, γ_f is the available water holding capacity of the layer (mm), and e_{m0} is the function value at wilting point (1500 kPa tension), which is set to 0.6.

CANDY

The effect of soil wetness differs between processes in CANDY. For all processes except denitrification, there are separate functions for soil water content e_m and for soil aeration e_a:

$$e_m = \begin{cases} 4 \cdot \dfrac{\theta}{\theta_s} \cdot \left(1 - \dfrac{\theta}{\theta_s}\right) & \left(\theta/\theta_s \leq 0.5\right) \\ 1 & \left(\theta/\theta_s \geq 0.5\right) \end{cases} \tag{5.15}$$

$$e_a = \exp\left(-10000 \cdot z \cdot \sqrt{\dfrac{S_t \cdot e_t \cdot e_m}{\varepsilon_a \cdot \left(\varepsilon_a - \varepsilon_p\right)}}\right) \tag{5.16}$$

where θ_s (%) is the saturated water content, z is the depth of the soil layer (m), ε_a is the air-filled pore space (%), and ε_p is a constant. S_t is a texture-dependent parameter equal to the proportion by weight (%) of particles less than 6.3 μm in diameter.

For denitrification the function is:

$$e_m = \begin{cases} 0 & \left(\theta \leq \theta_d\right) \\ \dfrac{\theta - \theta_d}{\theta_{fc} - \theta_d} & \left(\theta > \theta_d\right) \end{cases} \tag{5.17}$$

where θ is the actual soil water content, θ_{fc} is the soil water content at field capacity, and ε_{pv} is the relative pore volume. θ_d (%) is the threshold water content for denitrification given by:

$$\theta_d = 0.627 \cdot \theta_{fc} - 0.0267 \cdot \theta_{fc} \cdot \varepsilon_{pv} \tag{5.18}$$

Decomposition, Mineralization, and Immobilization

While there are some detailed differences between the models in their treatment of decomposition, mineralization, and immobilization processes, they all operate according to the same general principles. Mineralization and immobilization are nitrogen transformations that are controlled by carbon processes in organic matter. Organic matter is divided into several components or pools, roughly categorized as fast cycling pools (e.g., plant litter or manure) and slow cycling pools, mainly of soil humus. These are shown in a simplistic way in Figure 5.6. Transformation processes (flows) important for nitrogen dynamics models are from the fast cycling pools to the slow cycling pool and mineralization (or the reverse process of immobilization) from each organic pool to ammonium. Carbon transformation processes that control nitrogen transformations are commonly described as decomposition,

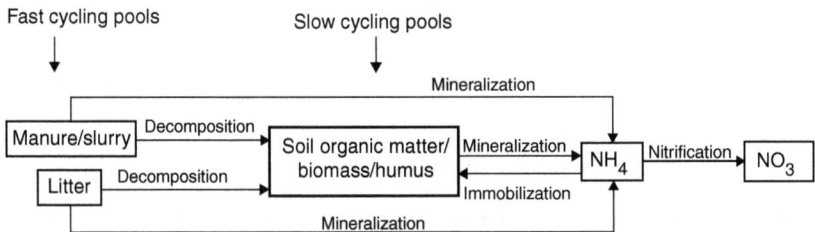

Figure 5.6   Simplified common representation of decomposition and mineralization. (From Wu, L., and M.B. McGechan. 1998a. A review of carbon and nitrogen processes in four soil nitrogen dynamic models. *Journal of Agricultural Engineering Research,* 69:279-305. With permission.)

although other terms, such as humification, are used for specific flows. Decomposition in the fast cycling pools transfers some material to the slow cycling humus pool (humification), the proportion going there being described by the assimilation factor, and the remaining carbon is lost as CO_2. The slow cycling humus pool can be thought of as the final destination for organic carbon; thus, decomposition of carbon in this pool generally produces CO_2 alone, except in the ANIMO, DAISY, and the Verberne models, which do allow some carbon to return from humus to soluble or microbial pools. However, some of the models also allow for a decomposition process in one or more organic pool which recycles some material back to the same pool with some carbon loss as CO_2. It might be expected that where the carbon component of an organic pool is lost by decomposition to CO_2, the corresponding nitrogen component will be mineralized to ammonium. However, while this happens in some instances, in the case of transfer from fast to slow cycling pools, the fast cycling material generally has a higher C:N ratio than slow cycling material. Thus, unless most of the carbon in the fast cycling pool is lost as CO_2 due to a low assimilation factor, soluble inorganic nitrogen must be converted back to organic form or "immobilized" to satisfy the higher nitrogen requirement of the slow cycling pool. Individual models vary in their definitions of the organic pools, with pools additional to those shown in Figure 5.6 for microbial biomass (considered to be fast cycling material, slow cycling, or both), and in the case of ANIMO a soluble organic matter pool. In all the models, decomposition from each carbon pool is a first-order rate process [Equations (5.1)–(5.3)], with an adjustment to the rate constant K for temperature and soil water content, plus an adjustment for soil clay content in one model only.

SOILN with No Microbial Biomass Pool

In the earlier versions of SOILN (an option in the current version), microbial biomass is not considered as a separate pool but rather as a component of each fast cycling pool. Organic material pools correspond to those in Figure 5.6, with two fast cycling pools representing manure-derived feces and litter (undecomposed plant material), and a slow cycling humus pool consisting of stabilized decomposed material. Decomposition from the two fast cycling organic carbon pools (at a rate

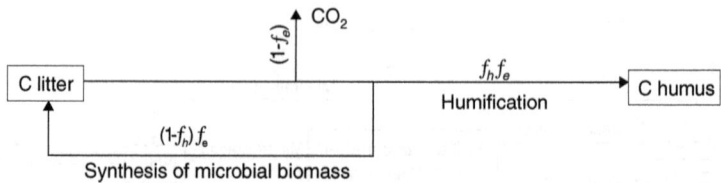

Figure 5.7   Details of organic matter pools and flows in SOILN. (From Wu, L., and M.B. McGechan. 1998a. A review of carbon and nitrogen processes in four soil nitrogen dynamic models. *Journal of Agricultural Engineering Research*, 69:279-305. With permission.)

C_{di}, g C/m²/d) follows the standard first-order rate process with the decomposition rate coefficient K adjusted by the soil temperature and water content functions given by Equations (5.4) and (5.10). The destination of the carbon is determined by two assimilation factors, the synthesis efficiency constant (f_e) and humification fraction (f_h) (Figure 5.7). The first fraction $(1 - f_e)$ is lost to the atmosphere as CO_2. Carbon in the second fraction $f_e (1 - f_h)$ becomes part of the microbial biomass, which is also considered to be a fast cycling material held in the original litter or feces pool; this fraction is therefore recycled within the same pool. The third fraction $f_e f_h$ is transferred to the slow cycling, stabilized humus pool.

The rate of transfer of nitrogen from the fast cycling pool into the humus pool N_{hi} (g N m²/d) is then:

$$N_{hi} = \frac{f_e\, f_h\, C_{di}}{r_0} \tag{5.19}$$

and the net mineralization rate of nitrogen from the fast cycling pool N_{ml} (g N m²/d) is:

$$N_{ml} = \left(\frac{N_i}{C_i} - \frac{f_e}{r_0} \right) C_{di} \tag{5.20}$$

where r_0 is an assumed C:N ratio of decomposed biomass and humification products, C_i is the carbon content in the pool (g C/g dry weight), N_i is the nitrogen mass in the pool (g N/m²), and C_{di} is the carbon decomposition rate in the pool (g C/m²/d).

When $N_i/C_i < f_e/r_0$, net immobilization will occur, but the immobilization rate is limited to a maximal fraction of mineral-N in the soil. Both ammonium and nitrate can be immobilized but preference is made for available ammonium.

Decomposition of organic matter (carbon) and mineralization of nitrogen in humus follow the standard first-order rate equation, both with the same rate coefficient K_h adjusted by the soil temperature and water content functions given by Equations (5.4) and (5.10).

SOILN with a Separate Microbial Biomass Pool

The latest version of SOILN provides an option to account for microbial biomass dynamics in the decomposition process by including a separate microbial biomass

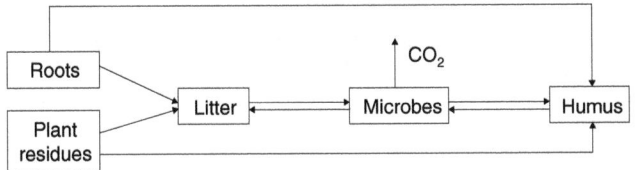

Figure 5.8   SOILN with separate microbial biomass pool. (From Wu, L., and M.B. McGechan. 1998a. A review of carbon and nitrogen processes in four soil nitrogen dynamic models. *Journal of Agricultural Engineering Research*, 69:279-305. With permission.)

pool (Figure 5.8). This allows the C:N ratio of microbial biomass to be specified (rather than assuming it has the same value as humus as in the simpler version of SOILN); and this is particularly important when modeling forest land where the C:N ratio of humus is higher than in agricultural land (Jansson et al., 1999). With this option, the material in the litter pool or feces pool is transferred into microbial biomass before it is decomposed and the allocation of the decomposed carbon is changed. From decomposed carbon, the same fraction $(1 - f_e)$ is lost to the atmosphere as CO_2 as without the microbial biomass option, and f_h is stabilized as humus.

The rate of transfer of nitrogen from microbial biomass into humus N_{hm} (g N m²/d) is:

$$N_{hm} = f_h \, N_m \tag{5.21}$$

and the mineralization rate of nitrogen from microbial biomass N_{mm} (g N m²/d) is:

$$N_{mm} = C_{im} \frac{N_i}{C_i} - f_h N_m - \frac{f_e C_{im} - f_h C_m}{r_0} \tag{5.22}$$

where C_{im} is microbial gain from a pool (g C/m²/d), and C_m is carbon content of microbial biomass (g C/m²).

SOILNNO

Organic matter pools in SOILNNO (Vold et al., 1997) are similar to those in SOILN with the separate microbial biomass pool, but in addition the fast cycling litter pool is subdivided into a readily decomposable fraction and a more slowly decomposable fraction. Each litter sub-pool has a different decomposition rate constant, based on the work of Verberne et al. (1990) as discussed in the section "Verberne Model." There is a flow representing recycling of microbial biomass back into its own pool, which is absent from the Swedish version of SOILN shown in Figure 5.8 (Jansson et al., 1999). Representation of humus mineralization also differs from the standard SOILN in which the same equation with the same decomposition rates was applied to both carbon and nitrogen. In SOILNNO, mineralization of humus carbon and nitrogen is simulated in a manner similar to that for the fast cycling pools. Decomposition of carbon in humus is first calculated according to the first-order rate equation with the rate constant adjusted according to the temperature

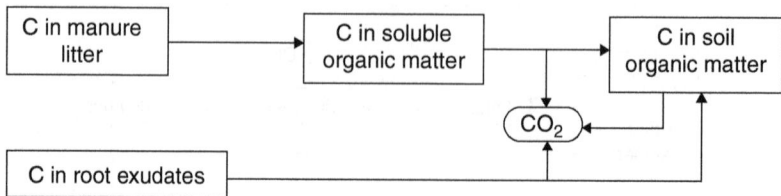

Figure 5.9   Organic pools and flow in ANIMO. (From Wu, L., and M.B. McGechan. 1998a. A review of carbon and nitrogen processes in four soil nitrogen dynamic models. *Journal of Agricultural Engineering Research,* 69:279-305. With permission.)

and water content functions, with a proportion f_e transferring to the humus pool and the remainder $(1 - f_e)$ being lost to the atmosphere as CO_2. Next, the net loss of carbon from the humus pool is estimated as carbon mineralization minus carbon entering the pool by humification from the fast cycling pools. Finally, nitrogen mineralization is calculated taking into account the C:N ratio of humus.

ANIMO

In the ANIMO model, organic matter is divided into four categories: organic plant parts and manure (sometimes also called fresh organic material), root exudates, soil organic matter/biomass, and soluble organic material (see Figure 5.2). The single fast cycling fresh organic material pool (corresponding to a combination of the litter and manure/slurry pools in Figure 5.6) consists of root and other crop residues after harvesting plus the organic parts of manure. Root exudates, an additional fast cycling pool not considered in other models and not shown in Figure 5.6, are organic products excreted by living roots plus dead root cells discarded by the plant. The slow cycling soil organic matter/biomass pool contains material formed from part of the available fresh organic material and root exudates, and consists of both dead organic soil material and living biomass. The fourth pool, soluble organic matter, again not considered in other models, is part of soil organic matter from other pools that has been rendered soluble. Unlike other models, ANIMO also considers manure or slurry to contain a soluble organic component, and this is added to the soluble organic matter pool when applied. Because the authors of ANIMO were uncertain about which transformation products would become soluble, they chose the representation of the transformations between pools as shown in Figure 5.9.

DAISY

In the DAISY model, organic matter in soil is described in the overall flow diagram (see Figure 5.3) as a single organic matter pool, but this is later subdivided into six sub-pools. Organic matter is first divided into three main categories, described as added organic matter, microbial biomass, and dead native soil organic matter, but each main category is further subdivided into a slower cycling (subscript 1) and a faster cycling component (subscript 2) with flows between sub-pools as shown in Figure 5.10. The added organic matter category, which includes farmyard manure,

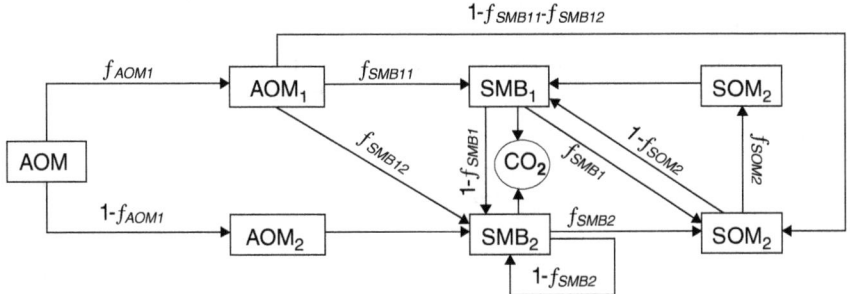

Figure 5.10 DAISY organic matter pools and flows (AOM: added organic matter, SMB: microbial biomass, SOM: soil organic matter; subscript 1 for slower cycling and 2 for
faster cycling sub-pools in each case). (From Wu, L., and M.B. McGechan. 1998a.
A review of carbon and nitrogen processes in four soil nitrogen dynamic models.
Journal of Agricultural Engineering Research, 69:279-305. With permission.)

slurry, green crop manure, or crop residues, is treated in a manner similar to that in
ANIMO, with two sub-pools of material considered to have different decomposition
rates. The two sub-pools of microbial biomass are considered to be the more stable
part and the more dynamic part of this material. Dead native soil organic matter (the
slow cycling pool in other models) is subdivided into two sub-pools, consisting of
chemically stabilized organic matter that is decomposed at a slow rate, and physically
stabilized organic matter with a higher decomposition rate.

All the flows shown in Figure 5.10 are first-order rate processes with a specified
rate coefficient for material leaving each sub-pool. There is also a specified partition
coefficient where material from one sub-pool goes to more than one destination sub-
pool. For each microbial biomass sub-pool, the rate coefficient K is the sum of a
maintenance coefficient m^* and a death rate coefficient d^*. All the rate coefficients
are multiplied by the temperature and soil water content functions given by
Equations (5.7) and (5.12). The rate coefficients for the two dead native soil organic
matter sub-pools and for the slower cycling microbial biomass sub-pool are also
multiplied by a function of clay content e_c, representing chemical as well as a physical
protection of organic matter against decomposition. The expression is given as:

$$e_c = \begin{cases} 1 - \sigma X_c & \left(0 < X_c < X_c^*\right) \\ 1 - \sigma X_c^* & \left(X_c \geq X_c^*\right) \end{cases} \tag{5.23}$$

where X_c is the clay content (fraction), X_c^* is the limit for effect of clay content,
and σ is a coefficient.

Net mineralization of nitrogen is related to the reserves of nitrogenous organic
matter in soil, the nature of the organic matter, and its C:N ratio, and is expressed
in terms of the sum of the flows between organic matter sub-pools:

$$N_{NH_4^+ - N} = -\Sigma \frac{\Delta C_i}{(C:N)_i} \tag{5.24}$$

where ΔC_i is the increment of carbon in the ith pool, and $(C{:}N)_i$ is the C:N ratio of the ith pool.

If net mineralization is negative so that immobilization occurs, then the upper limit for the immobilization rate is determined by the availability of inorganic nitrogen:

$$N^{im}_{NH_4^+} = K^{im}_{NH_4^+} \, N_{NH_4^+} \qquad\qquad (5.25)$$

$$N^{im}_{NO_3^-} = K^{im}_{NO_3^-} \, N_{NO_3^-} \qquad\qquad (5.26)$$

where $N_{NH_4^+}$ is the concentration of ammonium-N in the soil (kg N/m³), $N_{NO_3^-}$ is the concentration of nitrate-N in the soil (kg N/m³), $K^{im}_{NH_4^+}$ is the immobilization rate coefficient for ammonium-N (s⁻¹), and $K^{im}_{NO_3^-}$, the immobilization rate coefficient for nitrate-N (s⁻¹).

SUNDIAL

Decomposition in SUNDIAL is represented by flows between three organic carbon pools, for stubble chaff and straw (RO), microbial biomass (BIO), and humus (HUM), as shown in Figure 5.11. Each flow is a first-order process with a specified rate constant for flow from each pool. Each flow divides, with fraction α converted to BIO, fraction β converted to HUM, and the remainder going to CO_2. For the BIO and HUM pools, one such fraction represents recycling back to the same pool. Rate coefficients are multiplied by the temperature and soil water content functions given by Equations (5.8) and (5.14). The effect of soil texture on decomposition is obtained by adjusting the fractions α and β according to the clay content of the soil, according to the following equation:

$$\frac{1-\alpha-\beta}{\alpha+\beta} = 0.714\left\{1.85 + 1.60\exp\left(-0.997\,X_c\right)\right\} \qquad\qquad (5.27)$$

where X_c is the clay content of the 0–50 cm layer (%) and α and β are constants.

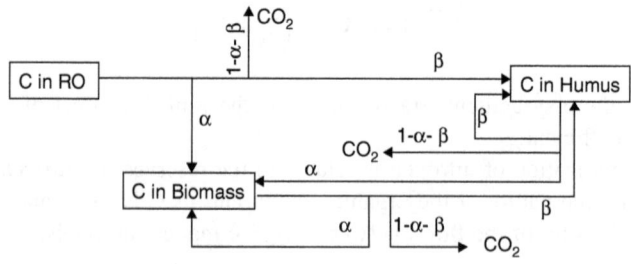

Figure 5.11   Organic matter pools and flows in SUNDIAL. (From Wu, L., and M.B. McGechan. 1998a. A review of carbon and nitrogen processes in four soil nitrogen dynamic models. *Journal of Agricultural Engineering Research*, 69:279-305. With permission.)

The nitrogen content of each compartment is calculated from the appropriate C:N ratio, and the processes of mineralization and immobilization are expressed in a similar manner to that in DAISY.

CANDY

In CANDY, organic matter (carbon) is held in three sets of pools — FOM, REP, and SOM — as shown in Figure 5.5. Carbon moves out of each pool according to the first-order rate equation, with rate constants K_F, K_R, and K_S, adjusted by the environmental factors given by Equations (5.9), (5.15), and (5.16). Fast cycling fresh organic matter (FOM) is divided into several sub-pools, each representing different organic material (plant residues, roots, manures, etc.) added to the soil. The synthesis coefficient η indicates the proportion of carbon from FOM transferring to the reproductive organic matter (REP, microbial biomass) pool, the remainder being released as CO_2. Slow cycling soil organic matter (SOM) is divided into sub-pools described as biologically activated or stabilized, with internal turnover movements between each. There is also a third SOM sub-pool described as inert, which takes no part in the dynamic processes. Nitrogen mineralization or immobilization takes place to fit in with transfer of carbon between pools, taking into account the differing C:N ratios of each pool.

Verberne Model

Verberne et al. (1990) have described a model of soil organic matter dynamics that has a particular emphasis on the concept of "protection" of some components (as already mentioned for DAISY in relation to the clay content of soils [Equation (5.23)]). Protected organic matter cycles at a slow rate when the soil is undisturbed, but becomes unprotected with a much faster cycling rate when soil is disturbed by plowing or other cultivations. The structure of the Verberne model has also been programmed and tested within a modeling framework called MOTOR (Whitmore et al., 1997). Hassink and Whitmore (1997) used the MOTOR framework to compare the protected pools structure with an alternative representation of protection based on the kinetics of sorption and desorption. The soil organic matter sub-model of the Hurley Pasture model (Thornley, 1998) is also based on the Verberne model with protected and unprotected pools.

In the Verberne model, pools of organic matter are described as shown in Figure 5.12. Input organic matter (plant residues, etc.) is subdivided into three compartments: decomposable material (DPM) consists of carbohydrates and proteins, and stabilized material (SPM) consists of cellulose and hemicellulose, both becoming components of the substrate carbon on which microbial biomass feeds; resistant material (RPM) consists of lignified plant components, and carbon leaving this pool enters one of the active organic matter sub-pools. Both the microbial biomass pool (MIC) and the active organic matter pool (OM) are divided into protected and unprotected sub-pools (NMIC, PMIC, NOM, and POM, respectively). The final destination of carbon is the slow cycling stabilized organic matter (SOM) pool (humus). Transfer of carbon out of pools generally follows the first-order rate

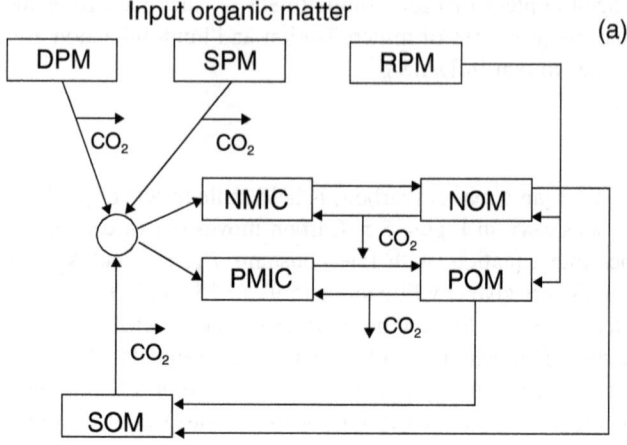

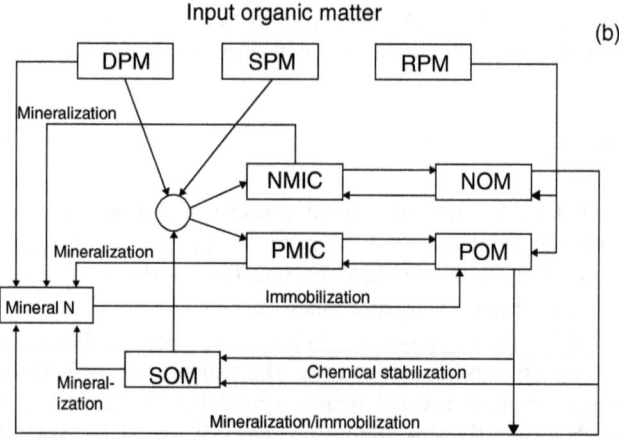

Figure 5.12   (a) Carbon and (b) nitrogen flows around soil organic matter pools in the Verberne (1990) model.

equation, with an associated efficiency factor indicating how much carbon reaches the destination pool, the remainder being released as CO_2. Adjustment of the rate coefficients for environmental factors is based on an earlier model by van Veen and Paul (1981). On the basis of work which shows that lignin influences decomposition of (hemi)cellulose (Parton et al., 1987; Swift et al., 1979), the decomposition rate constants for the SPM and RPM pools are adjusted by the factor f_l according to the proportion of resistant material, as follows:

$$f_l = \exp\left(-3.0 \times \frac{RPM}{SPM + RPM}\right)$$

(5.28)

Allocation of microbial biomass between the protected and unprotected sub-pools is determined as follows:

$$\left.\begin{array}{l} C_{m,p} = C_{m,\max} \\ C_{m,n} = C_m - C_{m,\max} \end{array}\right\} \quad \left(C_m \geq C_{m,\max}\right)$$

$$\left.\begin{array}{l} C_{m,p} = C_m \\ C_{m,n} = 0 \end{array}\right\} \qquad \left(C_m \leq C_{m,\max}\right) \tag{5.29}$$

where C_m is the total microbial biomass, $C_{m,\max}$ is the maximum value of total microbial biomass, $C_{m,n}$ is the non-protected microbial biomass, and $C_{m,p}$ is protected microbial biomass.

Nitrogen mineralization or immobilization takes place to fit in with transfer of carbon between pools, taking into account the differing C:N ratios of each pool or sub-pool.

Nitrification

SOILN

Nitrification of ammonium to nitrate is considered in SOILN to be a first-order rate process, driven by the excess of ammonium above an assumed equilibrium ammonium:nitrate ratio. In addition to the standard functions for soil temperature and soil water content given by Equations (5.4) and (5.10), the rate constant can (optionally) also be multiplied by a function e_{pH} for pH:

$$e_{pH} = \frac{pH - pH_{min}}{pH_{max} - pH_{min}} \tag{5.30}$$

where pH is the real pH value of the soil, pH_{max} is the pH value at which nitrification is not affected by acidity, and pH_{min} is the value at which nitrification is zero.

ANIMO

In ANIMO, the following equation is used to calculate the net production rate of nitrate N_{Nitri} in a specific soil layer in the model, taking into account the effect of anaerobic conditions on the decomposition rate of NH_4.

$$N_{Nitri} = 0.01 A_e K_N \theta_{av} N_{NH_4^+} - N_{NO_3 \rightarrow N_2O + N_2} \tag{5.31}$$

where A_e is the aerated fraction of the soil layer, K_n is a specific nitrification rate under aerobic conditions (d^{-1}), θ_{av} is the water content in the layer (%), and $N_{NH_4^+}$ is the concentration of ammonium-N (kg/m^3).

The aerated fraction of the layer is given by:

$$A_e = 1 - \left(1 - \pi R^2\right)^{\frac{\theta_r}{1047r}}$$ (5.32)

where r is the average pore radius in a soil layer (m) and θ_r is the difference in water content between that at the air entry tension corresponding to the largest pore radius in the layer and the average tension in the layer (%). The distance R from the soil pore center to the position where the concentration of oxygen in water is equal to zero (m) is related to the equation for gas horizontal transport from a soil pore system derived by Hoeks (1972).

DAISY and CANDY

Michaelis-Menten kinetics has been adopted in DAISY to calculate the nitrification rate:

$$N_{Nitri} = \frac{K_n^* N_{NH_4^+}}{c_{sh} + N_{NH_4^+}}$$ (5.33)

where K_n^* is the maximum nitrification rate at 10°C and optimum soil water conditions (d^{-1}), c_{sh} is the half saturation constant (kg NH_4/m³), and $N_{NH_4^+}$ is an ammonium concentration in soil (kg NH_4/m³). CANDY calculates nitrification according to a Michaelis-Menten type equation almost identical to Equation (5.33), based on Ritchie et al. (1986).

SUNDIAL

Nitrification is expressed by a first-order rate process in SUNDIAL, with the rate constant multiplied by the standard temperature and soil water content functions given by Equations (5.8) and (5.14).

Denitrification

SOILN

In SOILN, denitrification is described by a "macroscale" response function that integrates the effects of processes taking place at the microscale level. It is calculated as a zero-order rate process from a potential denitrification rate, accounting for the effects both of different soils and different cropping systems, and also response functions to incorporate the effects of soil aeration status, soil nitrate concentration, and soil temperature.

The denitrification rate (g N/m²/d) is expressed as:

$$N_{NO_3 \rightarrow N_2O + N_2} = K_d^* e_t e_m e_{NO_3}$$ (5.34)

where K_d^* is the potential denitrification rate (g N m²/d) and e_{NO_3} is a dimensionless response function of nitrate content. Easily metabolizable organic matter is assumed freely available.

The effect of nitrate is controlled by the half-saturation constant c_s, and soil nitrate concentration, $N_{NO_3^-}$.

$$e_{NO_3^-} = \frac{N_{NO_3^-}}{N_{NO_3^-} + c_s} \tag{5.35}$$

The soil water content function e_m [Equation (5.11)] is used as an indirect expression of the soil oxygen status influence on the denitrification rate, increasing from a threshold point θ_d, to a maximum at saturation (θ_s).

SOILNNO

In SOILNNO, an additional multiplicative response function e_r is included in Equation (5.34) to allow for the effect of carbon substrate availability. This accounts for the close correlation between the denitrification rate and the concentration of readily decomposable soil carbon, as shown by Li et al. (1994). This function is given by:

$$e_r = \frac{C_{res}}{C_{res} + e_t e_m c_r} \tag{5.36}$$

where C_{res} is the respiration rate and c_r is the apparent half-saturation constant with respect to respiration rate (g CO_2/kg soil/d).

ANIMO

The treatment of anoxia and denitrification in ANIMO is based on a description of the pore size distribution in the soil, and this in turn is derived from the water retention curve. The rate of denitrification is determined by both the amount of decomposable organic matter (D_{om}, kg N/m²) and the presence of oxygen. The anaerobic oxidation reaction of glucose is assumed to calculate denitrification in a layer.

$$N_{NO_3 \rightarrow N_2O + N_2} = \frac{0.541 D_{om} (1 - A_e)}{L t_d} \tag{5.37}$$

where L is the soil layer thickness (m) and t_d is the day number from the start of growth.

DAISY

Denitrification is simulated by a simple index model in DAISY. The rate is determined either by the transport of NO_3-N to anaerobic microsites or the actual microbial activity at these sites.

$$N_{NO_3 \to N_2O+N_2} = \min\left(K_a, K_t\right) \text{ and}$$

$$K_t = \beta_d \, N_{NO_3^-}, \quad K_a = e_m K_d^* = e_m \alpha_d^* \xi_{CO_2} \tag{5.38}$$

where K_t is the denitrification rate determined by the transport of nitrate-N to anaerobic microsites (kg N/m³/s), K_a is the rate determined by actual microbial activity at anaerobic microsites (kg N/m³/s), $N_{NO_3^-}$ is the concentration of nitrate-N in the soil (kg N/m³), ξ_{CO_2} is the CO_2 evolution rate (kg C/m³/s) derived from decomposition of organic matter, K_d^* is the potential denitrification rate, α_d^* is an empirical constant (kg N/kg C), and β_d another empirical constant (s⁻¹).

SUNDIAL

In SUNDIAL, the quantity of N denitrified in a particular layer in a particular week is assumed to be proportional to the quantity of CO_2 produced by that layer during that week, and also to its NO_3-N content, provided that denitrification occurs only in the 0 to 0.25 m layer and that it does not reduce the NO_3-N content of a particular layer below the residual quantity of NO_3-N that cannot be removed from the layer. The rate of loss of nitrogen by denitrification from a particular 0.05 m layer is expressed as:

$$N_{NO_3 \to N_2O+N_2} = f_{df} \, N_{NO_3^-} \, \frac{W\left(\gamma_f - \gamma_c\right)}{5\gamma_f} \tag{5.39}$$

where f_{df} is the denitrification factor, W is the combined evolution of carbon in CO_2 during that week by the RO, BIO, and HUM compartments in the 0 to 25 cm layer (kg/ha), $N_{NO_3^-}$ is the nitrate-N in a specified soil layer at beginning of week, and γ_f and γ_c are defined as for Equation (5.14).

CANDY

Denitrification in CANDY is represented as follows:

$$N_{NO_3 \to N_2O+N_2} = e_t e_m k_{den} \, N_{NO_3^-} \, C_{SOM} \tag{5.40}$$

where k_{den} is the denitrification reaction rate coefficient, $N_{NO_3^-}$ is the nitrate-N in the soil layer, C_{SOM} is the quantity of carbon in the SOM fraction (kg/ha), and the temperature and water content functions e_t and e_m are given by Equations (5.9) and (5.17), respectively.

Further Models of the Denitrification Process

Two models of the denitrification process described by Arah and Vinten (1995) currently operate in stand-alone mode, but perform a function similar to the denitrification routines of the soil nitrogen dynamics models reviewed in this chapter, and each could be adapted to become such a sub-model. Vinten et al. (1996) tested these two models, linked in each case to the simple solute leaching model SLIM (Addiscott and Whitmore, 1991).

The first is described as a "simple-structure (randomly distributed pore) model." It is based on principles similar to the denitrification sub-model of ANIMO as described by Rijtema and Kroes (1991), although the details differ.

The second is called an "aggregate assembly model," and is described in detail in an earlier paper by Arah and Smith (1989). It calculates steady-state values of the anoxic fraction and denitrification rate in spherical aggregates for any combination of aggregate radii, reaction potential, nitrate concentration in interaggregate pore water, and oxygen concentration in interaggregate pore water.

Ammonia Volatilization

Ammonia can be volatilized from the soil ammonium pool and from manure, slurry, fertilizer, or plant parts deposited on the soil surface. ANIMO is the only model considered here that includes representation of ammonia volatilization from slurry, and then only in a crude way; DAISY is the only model to consider volatilization from fertilizer or plant parts; and no models consider volatilization from the soil ammonium pool.

Ammonia Volatilization from Slurry

Volatilization of ammonia from slurry or manure lying on the soil surface is a process that strongly depends on short-term weather conditions. During dry and warm or windy weather, a major part of this material may be lost; whereas if the material is incorporated in the soil or precipitation falls directly after application, the major part will be saved.

None of the models considered here treat ammonia volatilization as a weather-dependent soil transformation process. ANIMO makes some adjustments without weather dependence as follows. For additions on top of the soil, a certain percentage, defined beforehand, of the ammonia applied is immediately lost and the remaining material is added to the uppermost soil layer. For additions incorporated in the first layer, no volatilization occurs. With other models (e.g., SOILN), manure applications must be specified in terms of the N contents of urine, feces, and bedding, and the C contents of feces and bedding, that enter the uppermost soil layer. Ammonia volatilization must be allowed for when calculating the N content of urine from the analysis of slurry leaving the store, either using tabulated adjustment values (according to conditions) such as those listed by Dyson (1992), or from a simulation with an ammonia volatilization model. One such model, described by van der Molen et al. (1990), uses a transfer function approach to simulate ammonia volatilization, either from the soil as used in the Dutch version of SOILN described by van Grinsven and Makaske (1993), or from slurry spread on the soil surface. The rate of transfer depends on a combination of the difference in ammonia concentration between the soil (or slurry) air spaces and the boundary layer of air immediately above the soil surface, and three resistance terms which depend on weather parameters and characteristics of vegetation on the soil surface. McGechan and Wu (1998), for an application of the SOILN model, used another weather-driven ammonia volatilization model from Hutchings et al. (1996) to estimate the quantity of ammonia N entering the soil surface. In a manner similar to that of DAISY, Jensen et al. (1994) used an ammonia volatilization model from Hansen et al. (1990).

Volatilization from Fertilizer and Plant Parts in SUNDIAL

Volatilization following the application of fertilizer or the senescencing components of plants is estimated in SUNDIAL. Volatilization from soil following fertilizer application V_s (kg N/ha/d) is given by the following equation:

$$V_s = \phi_s F_A \qquad (5.41)$$

where ϕ_s is the fraction of the fertilizer nitrogen volatilized and F_A is the quantity of fertilizer (kg/ha). It only occurs in the week following application if the rainfall is less than 5 mm in that week and the fertilizer is applied as ammonium sulfate or urea. This is the only weather-dependent volatilization process in any of the models considered.

Losses of nitrogen by volatilization during senescence of the above-ground part of the crop V_c (kg N/ha/d) is assumed to occur during the last 5 weeks before harvest. The rate of loss by volatilization is given by:

$$V_c = \frac{\phi_c U_T}{35} \qquad (5.42)$$

where ϕ_c is the fraction of the above-ground crop nitrogen lost by volatilization and U_T is the nitrogen content of the above-ground part of the crop at harvest (kg N/ha).

Interaction with Plants

SOILN

The actual growth of plants in SOILN is considered as the potential growth reduced by non-optimal temperature, leaf nitrogen concentration, and transpiration. The nitrogen demand is proportional to the daily growth. Actual uptake is the lowest value of demand and the quantity available in soil, which is a fraction of the mineral nitrogen.

In early versions of SOILN, growth, and hence uptake of nitrogen by the growing crop, was represented by a logistic curve with parameters specified by the user (Johnsson et al., 1987). This procedure is retained in the Dutch version of SOILN described by van Grinsven and Makaske (1993). This approach has been developed into a slightly more complex crop growth sub-model in the Norwegian SOILNNO (Vatn et al., 1996; Vold and Søreng, 1997). In more recent versions and applications of the Swedish SOILN, an alternative option is provided with a plant growth sub-model operating interactively with the soil nitrogen routines in the main model (Eckersten and Jansson, 1991; Eckersten et al., 1998). The quantity of available nitrogen in the soil simulated by SOILN (as well as water availability given by the SOIL simulation) is an input variable to the plant growth sub-model, and in turn the nitrogen uptake estimated by the sub-model is considered as one component in the nitrogen balance in the soil by the main routines of SOILN. The accuracy of the plant growth estimation affects the accuracy of the entire model and the flows of nitrogen and carbon to different pools. The standard SOILN model (also incorporated in the COUP model) includes alternative solar radiation-driven crop growth sub-models representing cereal crops (Eckersten and Jansson, 1991) and forest trees (Eckersten and Slopokas, 1990; Eckersten et al., 1995). Wu and McGechan (1998b; 1999) have developed alternative interactive crop growth sub-models for SOILN representing grass and grass/clover crops, based on a weather-driven grass and grass/clover model developed by Topp and Doyle (1996), which in turn is based on photosynthesis equations described by Johnson and Thornley (1984). Plant biomass and nitrogen are divided into pools for roots, stem, and leaf, with a further pool for the cereal crop for grain. Nitrogen uptake is dependent on total plant growth and nitrogen availability, although always limited to a certain fraction of the soil mineral nitrogen. If nitrogen deficiency occurs in one layer, compensatory uptake takes place from other layers with excess nitrogen. The nitrogen taken up is divided between root, leaf, and stem, following the idea that roots receive nitrogen first until they reach their maximum concentration, then similarly for the stem and finally for the leaf.

Plant uptake of nitrogen is calculated empirically from the crop root distribution in the soil profile and a total potential uptake rate for each component of the crop at the current stage of growth. To account for limited root exploitation of the soil, a rate of nitrogen uptake less than the potential rate occurs when the demand exceeds the quantity available. Eckersten and Jansson (1991) have described this process in detail.

At harvest, the total crop nitrogen is divided into three fractions: harvest output, and above- and below-ground residues. The below-ground residue is incorporated into the litter pool in the respective soil layers, together with a corresponding amount of carbon calculated from an assumed C:N ratio of the roots. Similarly, at plowing, above-ground residue is incorporated into the litter pool in the plow layer, assuming a specific C:N ratio for above-ground residues.

ANIMO

In ANIMO, only mineral nitrogen is taken up in the form of nitrate or ammonium. It is assumed that the mineral nitrogen requirement of the crop is related to the growth rate, which in turn is related to the quantity of water transpired. A maximal crop production will occur in a warm year with optimal N supply. An optimal mineral N concentration in the evapotranspiration flux C_{opt} (kg N/m²) is defined as

$$C_{opt} = \frac{N_{max}}{E_{pot}}$$
(5.43)

where N_{max} is the total maximal nitrogen uptake (kg N/m^2) and E_{pot} is the total evapotranspiration during the growing season in a warm year (m). Because the nitrogen requirement varies with the development of the crop, two periods are defined to distinguish requirements: the first period from the beginning of crop growth to the time when the mineral nitrogen requirement begins to decrease; the second period thereafter until the end of the growing season. The maximal nitrogen uptake rate N_{mtake} (kg N m^2/d) can be determined by:

$$N_{mtake} = \sum_{i=1}^{n_r} fe_i \, C_{opt} \qquad (5.44)$$

where n_r is number of layers in the root zone, and fe_i is the actual evapotranspiration flux from layer i (m).

If the evapotranspiration during the growing season is reduced, the total nitrogen requirement of the plant will decrease. The actual nitrogen uptake rate for the layers of the root zone, which must be corrected for the nitrogen excreted via root exudates, can be calculated as:

$$N_{rtake} = \sum_{i=1}^{n_r} \left\{ S_i \left[fe_i \, N_{NO_3^-} + N_{NH_4^+} \right] - f_{ifre} \, ke_i \right\} \qquad (5.45)$$

where S_i is a selectivity constant (fraction), $N_{NO_3^-}$ and $N_{NH_4^+}$ are average concentrations of nitrate-N and ammonium-N in layer i (kg N/m^3), respectively, f_{ifre} is the N-fraction of exudates, and ke_i is the root exudate production rate in layer i (kg m^2/d).

If the actual nitrogen uptake rate is greater than the maximal nitrogen uptake, the selectivity constant for the next time step can be approximated by $S_i = N_{mtake}/N_{rtake}$; otherwise, $S_i = 1$. If no crop is present in the field, $S_i = 0$.

DAISY

Nitrogen uptake by plants is determined by the potential crop nitrogen content (sink) and nitrogen status in the soil (source) in DAISY. Potential nitrogen uptake is estimated from the growth stage given by the crop growth sub-model, in which photosynthesis is first estimated from LAI, temperature, and radiation, and then reduced if there is water stress or nitrogen stress. The nitrogen stress and the potential nitrogen content of the crop depend on dry matter production as well as target nitrogen concentrations for the particular crop at the given stage of growth (with the stage of growth based on a temperature sum in early versions of the model). The crop nitrogen demand, depending on the actual compared to the potential crop nitrogen content and interacting with the soil nitrogen content, is temporarily distributed over each hour of the day. The uptake is spatially distributed over the entire root zone, taking account of distributions of root density, moisture, and nitrogen throughout the zone. For nitrogen movement from the bulk soil to the root surfaces, the assumption is made that only radial movement takes place and only from the cylindrical soil volume surrounding the root. The mass conservation equation in cylindrical coordinates is used to calculate the uptake rate of substrates per unit length of root:

$$
N_{rtake} = \begin{cases} 4\pi D(\overline{C} - C_r)\left[\dfrac{\beta_c^2 \ln \beta_c^2}{\beta_c^2 - 1} - 1\right]^{-1} & (\mu = 0) \\[3ex] q_w \dfrac{(\beta_c^2 - 1)\overline{C} - C_r \ln \beta_c^2}{(\beta_c^2 - 1) - \ln \beta_c^2} & (\mu = 2) \\[3ex] q_w \dfrac{(\beta_c^2 - 1)\left(1 - \dfrac{\mu}{2}\right)\overline{C} - (\beta_c^{2-\mu} - 1)C_r}{(\beta_c^2 - 1)\left(1 - \dfrac{\mu}{2}\right) - (\beta_c^{2-\mu} - 1)} & (\mu \neq 0 \text{ and } \mu \neq 2) \end{cases}
\qquad (5.46)
$$

$$
\mu = \frac{q_w}{2\pi D}, \qquad \beta_c = \left(r_r^2 \pi \rho_r\right)^{-\frac{1}{2}}
$$

where D is the diffusion coefficient of nitrogen (m²/s), $\overline{C}$ is the average concentration in solution of nitrogen within a cylindrical soil volume (kg/m³), C_r is the nitrogen concentration at the root surface (kg/m³), assumed to be constant along all root surfaces; q_w is the water flux toward the root (m³/m/s); ρ_r is the root density (m/m³); and r_r is the radius of root (m).

The crop growth sub-model in the most recent version of DAISY accounts for shedding of senescent leaves and root death, leading to transfer of carbon and nitrogen to the soil layers. Also, a "bio-incorporation" sub-model mimics the role of earth-living fauna in incorporation of organic material lying on the soil surface.

SUNDIAL

In SUNDIAL, the cumulative nitrogen uptake over a particular time period is given by:

$$
N_{rtake} = \left\{U_m^{-1/P_1} + \exp(-P_2 d)\right\}
\qquad (5.47)
$$

where P_1 is a shape factor relating the rate of uptake to the point of inflection of the uptake curve, d is the number of cumulative week-degrees since sowing, P_2 is a rate constant, and U_m is the final nitrogen target of the crop and empirically calculated as:

$$
U_m = (1 + \phi_c)\left\{\exp(0.075G) - 1\right\} + 60\left\{1 - \exp(-0.5G)\right\}
\qquad (5.48)
$$

where G is the grain yield (kg/ha) at 85% dry matter content and ϕ_c is the fraction of the above-ground crop nitrogen lost by volatilization.

Solute Transport

Removal of nitrogen from the soil/water system by leaching of solutes (mainly nitrate) is one of the most important components of the overall balance represented by the models, and the study of its environmental consequences is the main reason for the existence of the models. Representation of solute transport and leaching is dependent on the associated soil water model or sub-routine (see "Other Related Processes, Soil Water"). SOILN, ANIMO, and DAISY can represent leaching to both field drains and deep groundwater with the balance between the two determined by the soil water model, whereas SUNDIAL represents leaching to deep groundwater only. Solute transport of both nitrate and ammonium, determined by numerical solution of convection or convection/dispersion equations, is represented with a complexity that varies between the models.

SUMMARY OF MODEL TESTING AND VALIDATION

General Approach

Models should be developed and parameterized (calibrated) using one set of data, then tested or "validated" using an independent set of data. With complex interactive models as reviewed in this chapter, an appropriate approach is to select model parameters for individual processes on the basis of small-scale laboratory experiments, and use field-measured data only for validation, perhaps applying "fine-tuning" to a few parameter values only at this stage. Field measurements of leached nitrate and gas emissions are ideal data for validation of these models. This approach works well for some processes such as decomposition, for which adequate laboratory measured data are available. However, for denitrification, parameterization has in most instances been carried out on the basis of measurements from field experiments, although there are reports of attempts to reproduce the process in the laboratory using flooded soil columns (e.g., Clough et al., 1999). There is also a conflict here in relation to the complexity of representation of processes, because a simpler model, particularly if based on sound theoretical principles, lends itself better to parameterization using laboratory measurements so field measurements can be used for validation only. The complex pool structure in DAISY and numerous slurry components in ANIMO make it difficult to select parameter values without field measurements. However, the complex treatment of nitrogen uptake in SOILN can be dealt with by further studies of plant growth leading to more reliable estimates of the selected plant growth sub-model.

Testing Individual Models

All the models discussed in this chapter have been tested and validated using data sets from the countries in which they were developed, with some further testing in other countries. Testing of SOILN in Sweden on measured soil nitrogen pool data has been reported by Johnsson et al. (1987); on soil nitrogen and crop biomass pool data by Eckersten and Jansson (1991), Eckersten et al. (1995), and Käterrer and Andrén (1996); and on denitrification data by Johnsson et al. (1991). The crop growth sub-model in SOILN has been further tested by Käterrer et al. (1997) against data on biomass and nitrate content of crop components from Käterrer et al. (1993). The performance of the procedure for specifying microbial biomass as a separate pool in SOILN has been tested by Blombäck et al. (2000), with parameter values similar to those for the same procedure in SOILNNO (Jansson et al., 1999). The ability of SOILN to simulate changes in soil N pools and organic matter turnover has been tested by comparing simulated results with data from long-term experiments (Persson and Kirchmann, 1994) by Käterrer et al. (1997), Blombäck and Eckersten (1997), Blombäck (1998), and Blombäck et al. (2001). ANIMO has been tested by Rijtema and Kroes (1991) against mineral nitrogen pool and crop nitrogen uptake data reported by Groot and Verberne (1991). Measured nitrate concentrations in water flowing out through tile drains at Scottish sites (Vinten et al., 1991; 1992; 1994) has been used to test simulations with SOILN by Wu et al. (1998), and with ANIMO by Vinten (1999). The soil water, carbon, and nitrogen, as well as crop processes, in DAISY have been extensively tested against field data. The hydrological processes have been tested by Jensen et al. (1994a) and Djurhuus et al. (1999) using measured soil water contents and nitrate concentrations in soil water. Representation of the C and N transformation processes in the soil has been tested by Mueller et al. (1997; 1998) against data representing organic pools (including soil microbial biomass, SMB), soil mineral N concentrations, and CO_2 emissions. The performance of the crop routines (for various cereals, rape, potatoes, and beet) have been tested against crop biomass and N offtake data by Jensen et al. (1994a; b; 1996) and Petersen et al. (1995). SUNDIAL was tested against data representing soil mineral nitrogen pools and crop uptake using labeled N by Bradbury et al. (1993). CANDY was tested against German data describing long-term changes to soil organic matter pools by Franko et al. (1997).

Comparing Models Using Common Data Sets

There have been several attempts to compare the performance of models on common data sets. These data have included, in each case, soil characteristic parameters for calibration of the models, weather data for the sites, and measured time series data representing soil nitrogen pool sizes, nitrate leaching, or crop uptake corresponding to variables simulated by the models (as required for validation).

Soil and crop data from three sites in The Netherlands during 1983 and 1984 (Groot and Verberne, 1991) were used to test SOILN (Bergström et al., 1991; Eckersten and Jansson, 1991), SWATNIT containing the nitrogen transformation equations of SOILN (Vereecken et al., 1991), ANIMO (Rijtema and Kroes, 1991), DAISY (Hansen et al., 1991), and a precursor to the SUNDIAL model (Whitmore et al., 1991). Data from several sites in one catchment in Germany (McVoy et al., 1995) were used to test SOILN (Blombäck et al., 1995), SWATNIT (Vanclooster et al., 1995), DAISY (Svendsen et al., 1995), and SUNDIAL (Whitmore, 1995). A comparison of nine models, including DAISY (Jensen et al., 1997), CANDY (Franko et al., 1997), the Verberne model (Whitmore et al., 1997), and the Hurley pasture model (Arah et al., 1997), in terms of their performance at representing long-term changes in soil organic matter, has been reported by Smith et al. (1997). This comparison was carried out using 12 data sets from 7 long-term experiments in several countries, covering grassland, arable cropped land, and woodland. Further details of the data sets at European sites were presented by Körschens (1996) and Smith et al. (1996).

DATA REQUIREMENTS AND AVAILABILITY

Requirements for Calibration Values for Model Parameterization

Some calibration parameter values are included as coefficients in the equations presented in the chapter section "Theoretical Description of Mathematics of Processes," but many other parameters need to be determined by field or laboratory experiments. Information about such parameter values is presented in this section. These include standard or default values suggested by the authors of the models, plus some additional information from related experimental work. Some function values are also presented graphically, enabling comparisons between models to be made. A number of sets of time series data representing nitrate leaching, soil nitrogen pool sizes, and denitrification, as required for model validation, were described in the previous chapter section "Summary of Model Testing and Validation."

Temperature Response

SOILN

The temperature response function in SOILN is illustrated in Figure 5.13 for a range of values of the parameters Q_{10}, T_b, and T_{lin}. While the facility is provided to specify different parameter values for each different process, the authors of SOILN suggest the same values in every case as listed in Table 5.1, with the sources of data indicated.

ANIMO

Figure 5.14 illustrates the temperature function in ANIMO, assuming a standard mean temperature value. In practice, the equation is implemented in ANIMO with the growth period divided into several distinct phases with a different mean temperature for each. This is too complex to express as a simple set of curves.

Comparison of Functions Between Models

Because weekly mean air temperature (T_w) is used in SUNDIAL, it was not possible to make a comparison with the other three models that work with daily soil

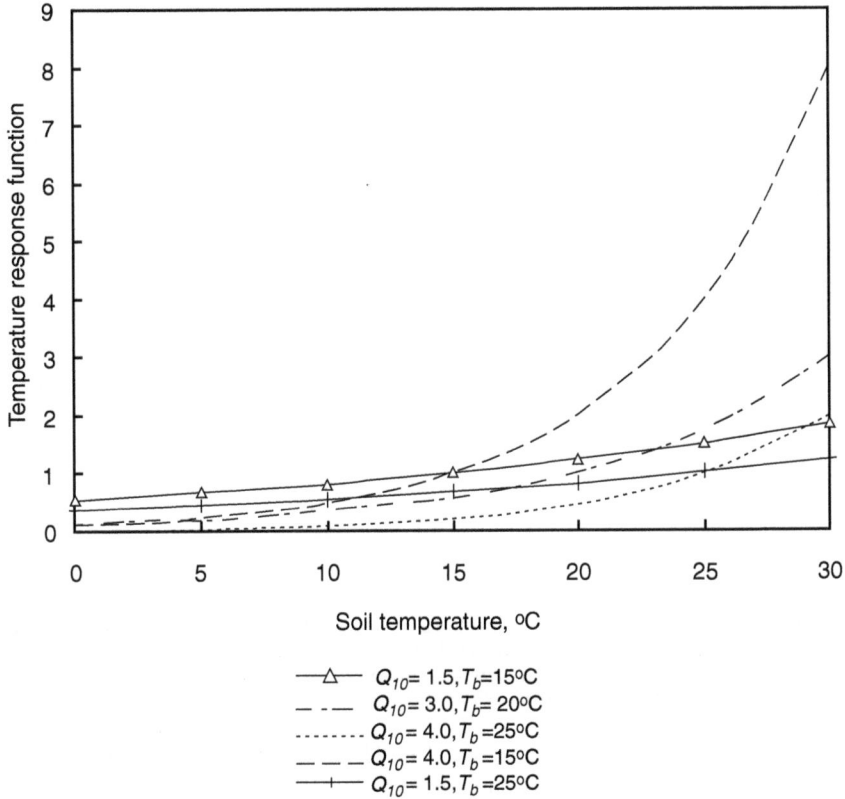

Figure 5.13 Response functions to soil temperature with different values of Q_{10} and T_b in SOILN. (From Wu, L., and M.B. McGechan. 1998a. A review of carbon and nitrogen processes in four soil nitrogen dynamic models. *Journal of Agricultural Engineering Research*, 69:279–305. With permission.)

Table 5.1   Parameters in Temperature Function for All Processes in SOILN

Parameter	Suggested Coefficient	Ref.
Q_{10} for the mineralization-immobilization process, fraction	3	Campbell et al. (1994)
Base temperature at which $e_t = 1$, T_b, (°C)	20	Eckersten et al. (1998)
Threshold temperature below which the temperature response is a linear function, T_{lin}, (°C)	5	Eckersten et al. (1998)

From Wu, L., and M.B. McGechan. 1998a. A review of carbon and nitrogen processes in four soil nitrogen dynamic models. *Journal of Agricultural Engineering Research*, 69:279–305. With permission.

temperature (T). There is some difference between SOILN and DAISY in the response function e_t at a particular temperature if the suggested parameter values are used in SOILN (20°C and 3 for T_b and Q_{10}, respectively). However, close

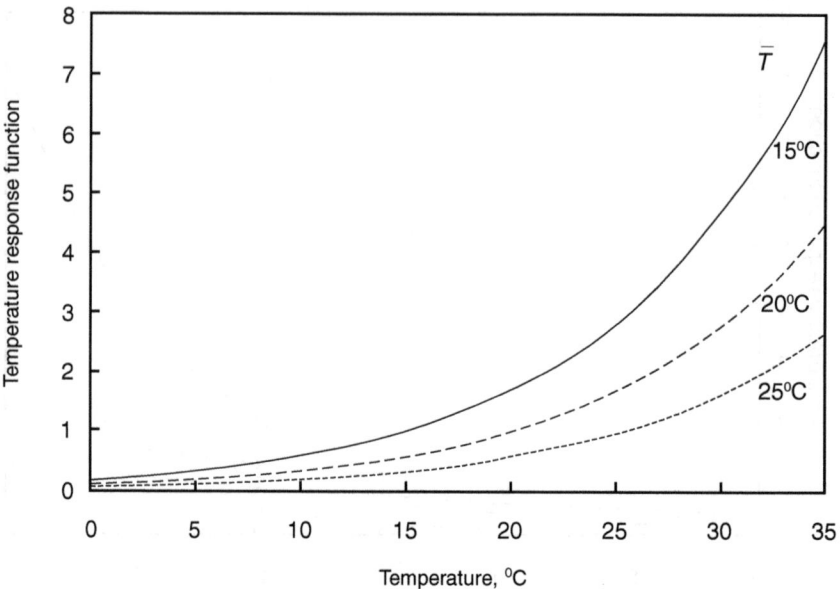

Figure 5.14   Effect of different mean soil temperature $\bar{T}$ values on processes in ANIMO. (From Wu, L., and M.B. McGechan. 1998a. A review of carbon and nitrogen processes in four soil nitrogen dynamic models. *Journal of Agricultural Engineering Research*, 69:279-305. With permission.)

agreement over the temperature range 7 to 31°C is given if T_b and Q_{10} are set to 10°C and 2, respectively, as shown in Figure 5.15. ANIMO considers the temperature effect as a relative value by comparing the current temperature (T) with its long-term average ($\bar{T}$), which makes it less straightforward to compare with the other models, although for $\bar{T}$ = 15°C (a typical average soil temperature in summer), the function is similar to that in SOILN and DAISY over the temperature range 5 to 25°C. Verberne et al. (1990) used a temperature response function from an earlier model of van Veen and Paul (1981). This was a plotted curve rather than mathematical equations, and also operates in a different manner from the other models in that it shows a reduction effect from a maximum value of 1. This is also shown in Figure 5.15, but adjusted to a maximum value of 2.72 at 25°C to correspond with the other models.

Soil Water Response

SOILN

The general soil water content response function for all processes in SOILN except denitrification given by Equation (5.10) is illustrated in Figure 5.16, and suggested parameter values (Johnsson et al., 1987) are listed in Table 5.2. For denitrification, the function with different values of d_1 (always ≥ 0) is shown in Figure 5.17, and suggested parameter values (Eckersten et al., 1998) are listed in Table 5.2.

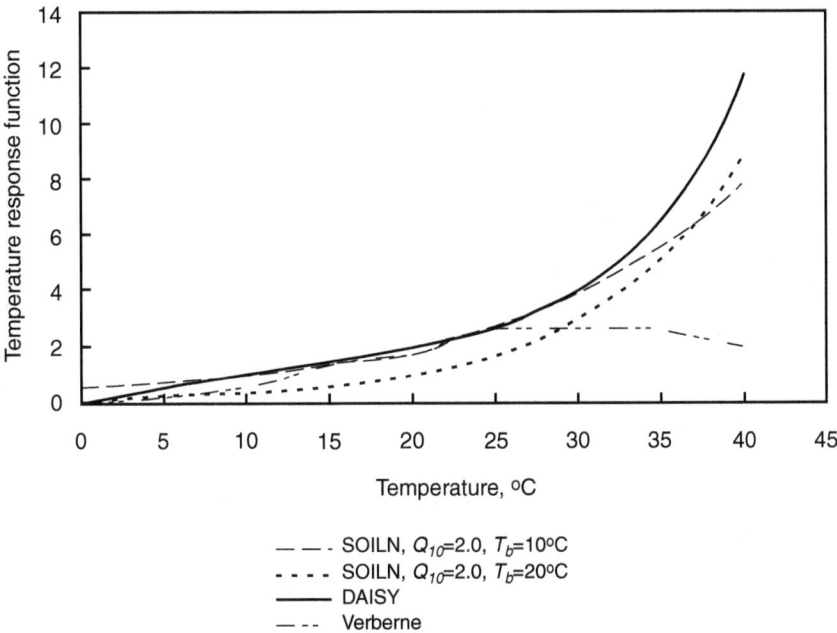

Figure 5.15   Comparison of temperature functions in SOILN and DAISY.

ANIMO

The soil water content response function in ANIMO takes the form of a reduction factor for the rate constant as shown in Table 5.3 for a number of soil water tension values. Reduction factors for intermediate tension values are estimated by interpolation.

DAISY

Different values for the soil water response function in DAISY for decomposition and nitrification are listed in Table 5.4. For denitrification, values of the constants x, x_1, and x_2 in Equation (5.12) are listed in Table 5.4.

Comparison of Functions Between Models

While all models operate according to similar principles, there is some variation in detail between them regarding the assumed soil water response function. ANIMO and SUNDIAL assume the same function for all processes, while for SOILN and DAISY, the function varies between processes. In some cases, functions are defined in terms of water content; in others according to tension (matric potential). The alternative functions are illustrated in Figure 5.18 for a clay loam soil, assuming a typical water release (soil water tension) curve to indicate the increase in tension (pF) value with decrease in water content. Decomposition and nitrification, and in

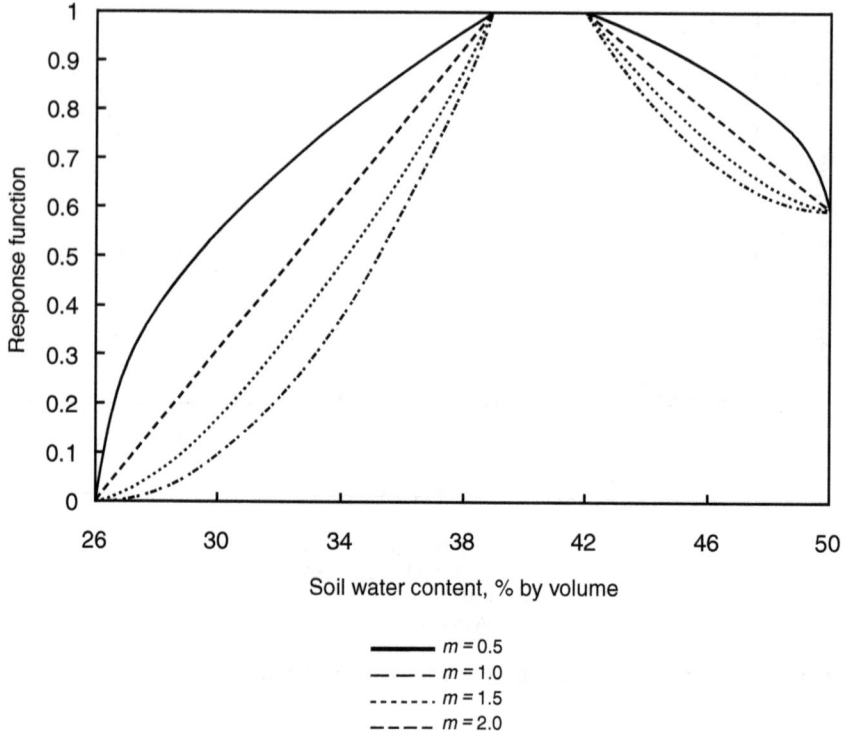

Figure 5.16 Soil water content response function for nitrification in SOILN with different values of m, assuming $\theta_W = 26$ (% by volume), $\theta_S = 50$, $\theta_{lo} - \theta_w = 13$, $\theta_s - \theta_{hi} = 8$, and $e_s = 0.6$. (From Wu, L., and M.B. McGechan. 1998a. A review of carbon and nitrogen processes in four soil nitrogen dynamic models. *Journal of Agricultural Engineering Research*, 69:279-305. With permission.)

Table 5.2 Parameters of Soil Water Response Function in SOILN

Parameter	Suggested Coefficient	Ref.
Water content intervals in general soil water response function, θ_{hi}, θ_{lo}	10, 8	Johnsson et al. (1987)
Coefficient in general soil water content response function, m	1	Johnsson et al. (1987)
Saturation activity in general soil water content response function, e_s	0.6	Johnsson et al. (1987)
Coefficient in function for soil water/aeration effect on denitrification, d_1	2	Eckersten et al. (1998)
Threshold water content for soil water/aeration effect on denitrification, θ_d	17	Eckersten et al. (1998)

From Wu, L., and M.B. McGechan. 1998a. A review of carbon and nitrogen processes in four soil nitrogen dynamic models. *Journal of Agricultural Engineering Research*, 69:279-305. With permission.

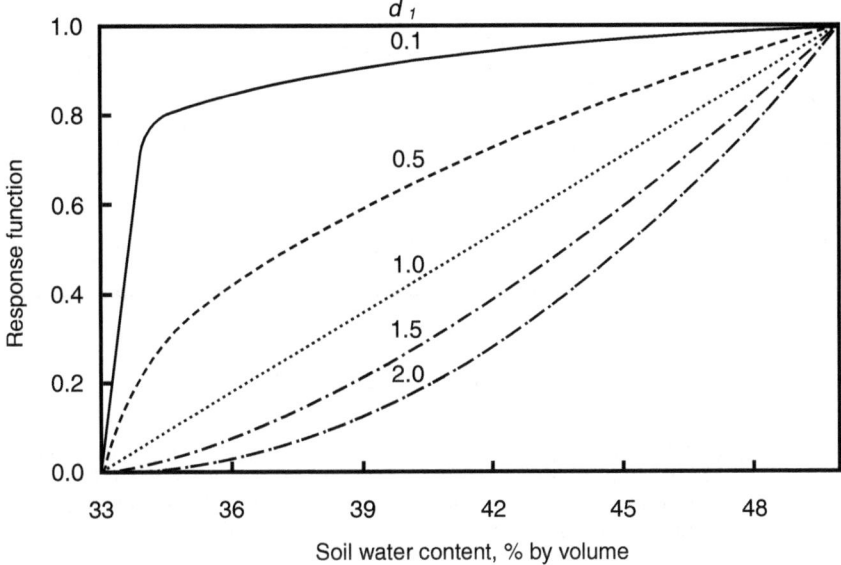

Figure 5.17 Soil water content response function for denitrification in SOILN with different values of d_1. (From Wu, L., and M.B. McGechan. 1998a. A review of carbon and nitrogen processes in four soil nitrogen dynamic models. *Journal of Agricultural Engineering Research,* 69:279-305. With permission.)

Table 5.3 Soil Water Tension (pF) Values with Corresponding Rate Reduction Factors in ANIMO

Soil water tension (pF)	≤2.4	2.7	2.9	3.2
Rate reduction factor	1.00	0.57	0.36	0.32

From Wu, L., and M.B. McGechan. 1998a. A review of carbon and nitrogen processes in four soil nitrogen dynamic models. *Journal of Agricultural Engineering Research,* 69:279-305. With permission.

Table 5.4 Parameter and Boundary Values for Soil Water Content Function in DAISY

Process	A_1	A_2	A_3	A_4	B_1
Decomposition	0.6	0.4	1.625	4.0	$10^{5.5}$
Nitrification	0.0	1.0	2.0	2.5	10^4

Process	x	x_1	x_2
Denitrification	0.2	0.8	0.9

From Hansen, S., H.E. Jensen, N.E. Nielsen, and H. Svendsen. 1993a. *Description of the Soil Plant System Model DAISY — Basic Principles and Modelling Approach.* Jordbrugsforlaget, The Royal Veterinary and Agricultural University, Copenhagen, Denmark. With permission.

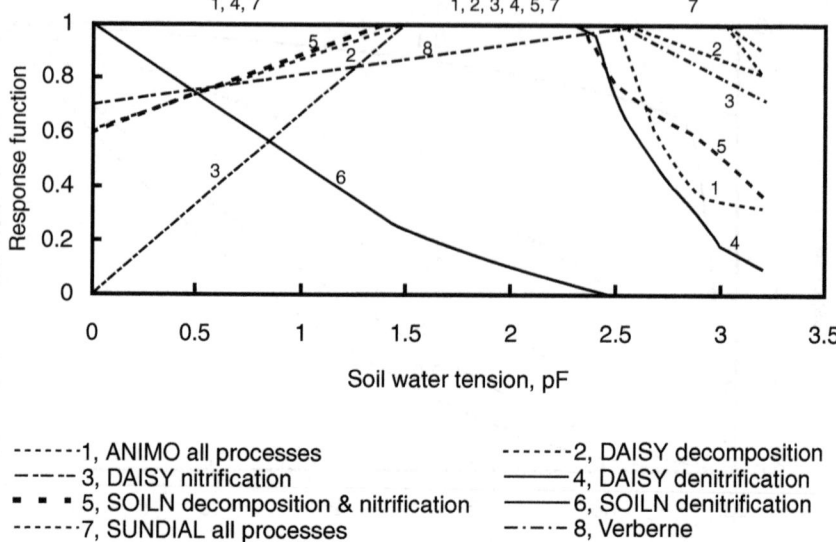

Figure 5.18 Comparison of soil water content response functions in SOILN, ANIMO, DAISY, SUNDIAL, and Verberne. For SOILN assuming θ_w = 26 (% by volume), θ_s = 50, $\theta_{lo} - \theta_w$ = 13, $\theta_s - \theta_{hi}$ = 8, and m = 1.5, and for SUNDIAL assuming a clay loam soil.

most cases denitrification also, are assumed to take place at the maximum rate (e_m = 1) in the middle of the water content range, declining at low water contents, and also declining at high water contents for DAISY and SOILN only. The plotted soil water content response curve used by Verberne et al. (1990), again from the earlier model of van Veen and Paul (1981), is also shown in Figure 5.18. The only function that differs by a large degree from the others is that for denitrification in SOILN, where it takes a value of unity only at saturation and declines continuously with decreasing water content; whereas in other cases the decline commences at a pF value of around 2.5 to 3.0.

Decomposition, Mineralization, and Immobilization

ANIMO

The option is provided in ANIMO to divide each category of fresh organic material into several fractions, each with its own rate coefficient for the first-order rate decomposition process. The solution of the equation representing decomposition of fresh organic material to soluble organic matter is then the sum of all the fractions with different rate coefficients. Decomposition dynamics of a number of fresh organic materials, each consisting of several fractions with different rate coefficients, are shown in Figure 5.19. The rate of root exudate production is assumed to be 0.41 times the rate of root biomass production. Rate coefficients must

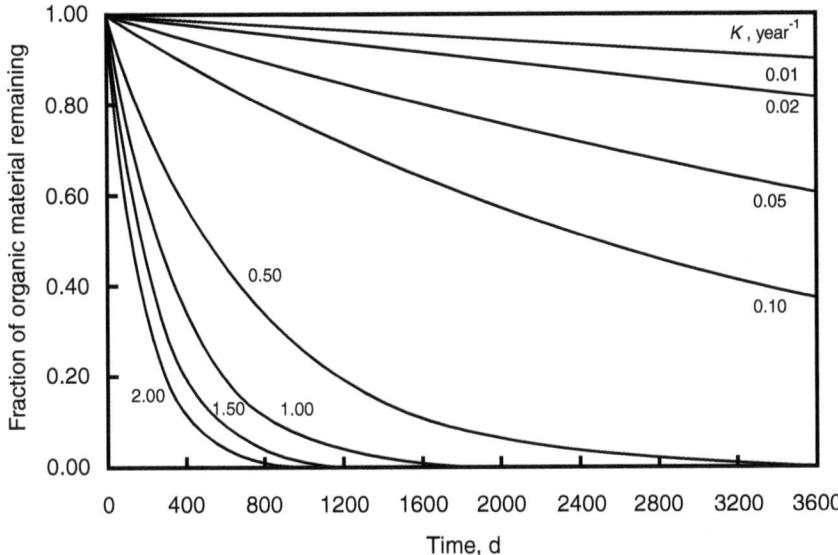

Figure 5.19 Decomposition of fresh organic material in ANIMO with different values of decomposition rate constant, representing the decomposition characteristics of a particular organic material. (From Wu, L., and M.B. McGechan. 1998a. A review of carbon and nitrogen processes in four soil nitrogen dynamic models. *Journal of Agricultural Engineering Research*, 69:279-305. With permission.)

be specified for the first-order rate processes representing flows of carbon out of the soluble organic matter pool and out of the root exudate pool, both with the same assimilation factor to indicate the fraction going to the slow cycling humus pool, the remainder going to CO_2. As in SOILN, corresponding mineralization or immobilization flows of nitrogen are calculated from the assimilation factor and the C:N ratios of the source and destination pools. Again as in SOILN, flows of carbon (to CO_2) and of nitrogen (mineralization) from the slow cycling pool are first-order rate processes, both with the same specified rate coefficient. The rate coefficients for all the first-order rate decomposition processes are multiplied by the same factors for temperature [Equations (5.5) and (5.6)], soil water content (Table 5.3), and pH of the soil. For the effect of pH, a single function has been introduced as shown in Figure 5.20.

DAISY

In DAISY, X_c^* and σ take values of 0.25 and 2.0, respectively in Equation (5.23) representing the protection effect as a function of clay content. Both $K_{NH_4^+}^{im}$ in Equation (5.25) and $K_{NO_3^-}^{im}$ in Equation (5.26) (representing the upper limit for the immobilization rate where net mineralization occurs) are set to 0.5 (from Hansen et al., 1993).

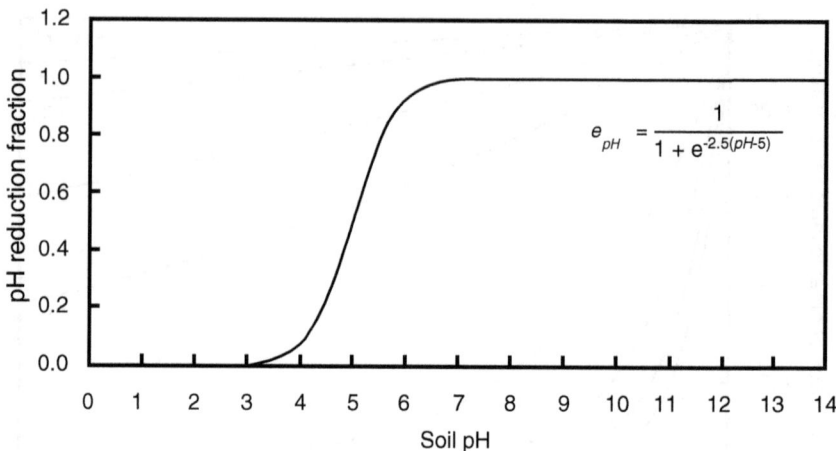

The equation shown in the figure:

$$e_{pH} = \frac{1}{1 + e^{-2.5(pH-5)}}$$

Figure 5.20   Response function for pH in ANIMO. (From Wu, L., and M.B. McGechan. 1998a. A review of carbon and nitrogen processes in four soil nitrogen dynamic models. *Journal of Agricultural Engineering Research,* 69:279-305. With permission.)

Comparison of Decomposition Parameters Between Models

Rate coefficients for the first-order rate processes of flows out of each organic matter pool assumed in the models are listed in Table 5.5, with the literature sources indicated. Separate maintenance m^* and death d^* components of the rate coefficient are shown for the microbial biomass sub-pools in DAISY (added together to give K). Assumed assimilation factors (partition constants) are listed in Table 5.6 and C:N ratios in Table 5.7, again with the literature sources indicated.

Decomposition rate constants for some pools were estimated from incubation experiments, including incubation of litter in bags used in SOILN (Andrén and Paustian, 1987), and in DAISY (Lind et al., 1990). Results from similar experiments at SAC, reported by Vinten et al. (1996), were used as source data for decomposition rates of litter and humus. Experimental data for material added to the soil is mainly for plant parts (litter), with little data for manure or slurry. The general assumption made is that the feces component of manure and slurry consists of plant parts similar to those in litter, and that the same decomposition rate values can be assumed. Only the authors of ANIMO have attempted to estimate values specific to slurry, showing generally slightly slower decomposition rates than for litter. For DAISY and SUN-DIAL, which have a more complex treatment of organic matter with many compartments, decomposition rate parameter selection has been carried out partly by adjusting parameters to improve fits of simulated results to field experiments. For both models, long-term experiments at Rothamstead Experimental Station [referred to in Jenkinson et al. (1987) by the authors of DAISY and in Powlson et al. (1986) and Hart et al. (1993) by the authors of SUNDIAL] have been used for this purpose.

A comparison between models of the suggested coefficients for the first-order rate process representing decomposition from similar pools is complicated somewhat

Table 5.5 Decomposition Rate Coefficients

Model	Material/Pool	Suggested Coefficient (d⁻¹)	Ref.
	Fast Cycling Pools		
SOILN	Litter and internally recycled microbial biomass	3.5×10^{-2}	Andrén and Paustian (1987)
	Feces and internally recycled microbial biomass (dependent on type of manure)	3.5×10^{-2}	Eckersten et al. (1998)
	Litter (where separate microbial biomass pool specified)	2.0×10^{-2}	Blombäck and Eckersten (1997); Blombäck et al. (2000)
	Microbial biomass death rate (overall)	2.7×10^{-3}	Blombäck and Eckersten (1997); Blombäck et al. (2000)
	Microbial biomass death rate (to litter pool)	2.05×10^{-3}	Blombäck and Eckersten (1997); Blombäck et al. (2000)
	Microbial biomass death rate (to humus pool)	6.85×10^{-4}	Blombäck and Eckersten (1997); Blombäck et al. (2000)
SOILNNO	Readily decomposable litter	1.3×10^{-1}	Verberne et al. (1990); Breland (1994)
	Slowly decomposable litter	5.0×10^{-2}	Verberne et al. (1990); Breland (1994)
ANIMO	Slurry Fraction 1 (soluble) and soluble part of Fraction 2	8.2×10^{-2}	[a]
	Slurry Fraction 2 (rapidly decomposing) "fresh" part	2.7×10^{-3}	[a]
	Slurry Fraction 3 (slowly decomposing)	3.3×10^{-4}	[a]
	Crop residues (mainly roots) Rapidly decomposing part (0.9 of total)	5.5×10^{-3}	[a]
	Slowly decomposing part (0.1 of total)	6.0×10^{-4}	[a]
	Root exudates	1.0	[a]
DAISY	Root material Faster cycling part	7.0×10^{-2}	Hansen et al. (1991)
	Slower cycling part	7.0×10^{-3}	Hansen et al. (1991)
	Plant residues from ryegrass, straw, and pig slurry Faster cycling part	5.0×10^{-2}	Lind et al. (1990); Jenkinson et al. (1987)
	Slower cycling part	5.0×10^{-3}	

Table 5.5 (continued)   Decomposition Rate Coefficients

Model	Material/Pool	Suggested coefficient (d^{-1})	Ref.
SUNDIAL	Stubble, chaff, and straw (RO)	2.3×10^{-2}	Powlson et al. (1986); Hart et al. (1993)
SAC experiments	Litter		Vinten et al. (1996)
	Clay loam soil (topsoil)	2.4×10^{-2}	
	Sandy loam soil (topsoil)	2.6×10^{-2}	
CANDY	Lucerne roots	0.2	Franko (1997)
Verberne	DPM	$0.2\ (0.1)^{b}$	Verberne et al. (1990)
	SPM	$0.1\ (0.01)^{b}$	
	RPM	$0.02\ (0.01)^{b}$	
	NOM→MIC	$0.01\ (0.015)^{b}$	
	NOM→SOM	$1.0 \times 10^{-6}\ (5.0 \times 10^{-5})^{b}$	
	POM→MIC	$0.0003\ (0.0009)^{b}$	
	POM→SOM	$1.0 \times 10^{-6}\ (4.0 \times 10^{-5})^{b}$	

Microbial Biomass

Model	Material/Pool	Suggested coefficient (d^{-1})	Ref.
SOILN	Death rate (overall)	2.7×10^{-3}	Blombäck and Eckersten (1997); Blombäck et al. (2000)
	(to litter pool)	2.05×10^{-3}	Blombäck and Eckersten (1997); Blombäck et al. (2000)
	(to humus pool)	6.85×10^{-4}	Blombäck and Eckersten (1997); Blombäck et al. (2000)
SOILNNO		$1.0 \times 10^{-3} - 3.0 \times 10^{-1}$	Vold et al. (1997)
DAISY	More dynamic part — Maintenance	1.0×10^{-2}	Hansen et al. (1991); Lind et al. (1990); Jenkinson et al. (1987)
	Death	1.0×10^{-2}	
	Less dynamic part — Maintenance	1.0×10^{-2}	
		1.0×10^{-3}	
	Death	1.8×10^{-3}	
SUNDIAL	NMIC→NOM	$0.5\ (0.045)^{b}$	Powlson et al. (1986); Hart et al. (1993)
Verberne	PMIC→POM	$0.005\ (0.008)^{b}$	Verberne et al. (1990)

Slow Cycling (Humus) Material

Model	Material/soil	Value	Reference
SOILN		7.0×10^{-5}	Johnsson et al. (1987)
SOILNNO	Humus carbon	7.2×10^{-5}	Breland (1994)
ANIMO		$4.1\text{–}5.5 \times 10^{-5}$	Kortleven, J. (1963)
DAISY	Physically stabilized	1.4×10^{-4}	Jenkinson et al. (1987)
	Chemically stabilized	2.7×10^{-6}	Jenkinson et al. (1987)
SUNDIAL		5.7×10^{-5}	Powlson et al. (1986); Hart et al. (1993)
SAC experiments	Clay loam soil (topsoil)	1.0×10^{-5}	Vinten et al. (1996)
	Sandy loam soil (topsoil)	3.3×10^{-5}	
Verberne	SOM	8.0×10^{-7} (2.0×10^{-5})[b]	Verberne et al. (1990)

[a] See chapter section "Subdivision of Components of Slurry in ANIMO."
[b] Values in parentheses are alternative values for the Verberne model quoted by Whitmore et al. (1997).

Table 5.6 Assimilation Factors (Partition Coefficients)

Model	Factor		Suggested Value		Ref.
SOILN	Synthesis efficiency constant f_e		0.5		Kjøller and Struwe (1982)
	Humification fraction f_h		0.2		Johnsson et al. (1987)
SOILNNO	Microbial growth yield efficiency f_e		0.2–0.5		Vold et al. (1997)
	Humification fraction (litter and microbial biomass) f_h		0.1–0.5		Vold et al. (1997)
ANIMO	Assimilation factor		0.25		[a]

| | | Plant Residues | Farmyard manure | | |
			(Broadbalk)	(Hoosfield)	
DAISY	f_{AOM1}	0.80	0.90	0.95	Hansen et al. (1991); Jenkinson et al. (1987)
	f_{SMB11}	0.40	0.375	0.33	
	f_{SMB12}	0.40	0.375	0.33	
			Microbial Biomass		
	f_{SMB1}		0.40		Hansen et al. (1993a); Jenkinson et al. (1987)
			Soil Organic Matter		
	f_{SOM2}		0.10		Hansen et al. (1993a); Jenkinson et al. (1987)
	f_{SMB2}		0.40		Hansen et al. (1993a); Jenkinson et al. (1987)

		Saxmunden (40.0% clay)	Rothamstead (23.5% clay)	Woburn (10.0% clay)	
SUNDIAL	$\alpha+\beta$	0.42	0.40	0.35	Powlson et al. (1986); Hart et al. (1993)
	α/β	1.1	1.1	1.1	
	α	0.20	0.19	0.167	
	β	0.22	0.21	0.183	
SAC experiments	Assimilation factor		0.25		Vinten et al. (1996)
CANDY	Synthesis coefficient	Farmyard manure	0.60		Franko (1997)
		Cattle slurry	0.60		
		Pig slurry	0.60		
		Fodder beet leaves	0.65		
		Cereal straw	0.45		
		Green manure	0.35		
Verberne	Efficiency factor	DPM	0.4		Verberne et al. (1990)
		SPM	0.3		
		RPM	1.0		
		NOM→MIC	0.25		
		NOM→SOM	1.0		
		POM→MIC	0.2 (0.25)[b]		
		POM→SOM	1.0		
		NMIC→NOM	1.0		
		PMIC→POM	1.0		
		SOM	0.2 (0.25)[b]		

[a] See chapter section "Subdivision of Components of Slurry in ANIMO."
[b] Values in parentheses are alternatives for the Verberne model quoted by Whitmore et al. (1997).

Table 5.7 C:N Ratios

Model	Material/Pool	Suggested C:N ratio	Ref.
	Fast Cycling Material		
SOILN	Litter (harvest residues)	50	Johnsson et al. (1987)
	Roots	25	Johnsson et al. (1987)
	Litter (plant residues and roots) directly humified, where separate microbial biomass pool specified as in Figure 5.8.	60	Blombäck et al. (2000)
	Feces	50	Eckersten et al. (1998)
SOILNNO	Newly synthesized microorganisms and humified products when nitrogen is not limiting	3–15	Vold et al. (1997)
ANIMO	Slurry Fraction 1 (soluble) and soluble part of Fraction 2	8.3	[a]
	Slurry Fraction 2 (rapidly decomposing) "fresh" part	11.6	[a]
	Slurry Fraction 3 (slowly decomposing)	58	[a]
	Arable crop residues (mainly roots) Rapidly decomposing part (0.9 of tot)	58	[a]
	Slowly decomposing part (0.1 of total)	39	[a]
	Grassland crop residues (mainly roots) Rapidly decomposing part (0.9 of tot)	58	[a]
	Slowly decomposing part (0.1 of total)	58	[a]
	Root exudates	23	[a]
DAISY	Added organic material slower cycling part	100	Hansen et al. (1991)
SUNDIAL	Roots	5–70	Powlson et al. (1986); Hart et al. (1993)
CANDY	Farmyard manure	15	Franko (1997)
	Cattle slurry	8	
	Pig slurry	4.5	
	Fodder beet leaves	10.5	
	Cereal straw	60	
	Green manure	12	
Verberne	DPM	6	Verberne et al. (1990)
	SPM	150	
	RPM	100	
	NOM	15	
	POM	10	

Microbial Biomass

Model	Description	Value	Reference
SOILN	In litter/feces pool	10	McGill et al. (1981)
	In separate pool	7	Blombäck et al. (2000); Sjödahl-Svensson et al. (1994)
DAISY	More dynamic part	10	Hansen et al. (1991)
	Less dynamic part	6	Hansen et al. (1991)
SUNDIAL		8.5	Powlson et al. (1986); Hart et al. (1993)
Verberne	MIC (NMIC)[b]	8	Verberne et al. (1990)
	(PMIC)[b]	(4)[b]	Whitmore et al. (1997)

Slow Cycling (Humus) Material

Model	Description	Value	Reference
SOILN		10	McGill et al. (1981)
ANIMO	Soil organic matter (humus) alone	11	Stanford and Smith (1972)
			[a]
DAISY	Humus plus microbial biomass	14	
	Physically stabilized	12	Jenkinson et al. (1987)
	Chemically stabilized	10	Jenkinson et al. (1987)
SUNDIAL		8.5	Powlson et al. (1986), Hart et al. (1993)
Verberne	SOM Clay soil	10	Verberne et al. (1990)
	SOM Sandy soil	20 (10)[b]	

[a] See chapter section "Subdivision of Components of Slurry in ANIMO."
[b] Values in parentheses are alternative values for the Verberne model quoted by Whitmore et al. (1997).

by differing definitions of pools, differing flows, and differing destination pools. Meaningful comparisons are best made in relation to the simplified common description of the pools as illustrated in Figure 5.6. A useful comparison in terms of flows out of pools representing plant litter and feces in manure or slurry can be made because this is an important constraint in the system, whereas decomposition from an intermediate pool such as the soluble organic material pool in ANIMO is very fast and imposes little constraint. Where a material is subdivided into fractions with different decomposition rates, a mean decomposition rate weighted according to the proportion of the total in that fraction can be used for comparison. Suggested decomposition rates for litter and manure components are of similar magnitude for all the models. The value of the decomposition rate parameter for litter and manure suggested for SOILN (after adjusting by a factor of 0.6 to allow for material recycled back to the same pool in this case) is similar to the mean of all fractions in the more complex treatments in some of the other models, and also similar to the values measured for litter by Vinten et al. (1996). Decomposition of litter, with the suggested rate for optimal conditions for each of the four models, is illustrated in Figure 5.21. Suggested values for the decomposition rate of humus are of similar magnitude for all models. However, the work of Vinten et al. (1996) suggests that different values should be selected for different soil types, although an adjustment for clay content on the rate coefficient or recycling partition factor in DAISY and SUNDIAL attempts to account for this.

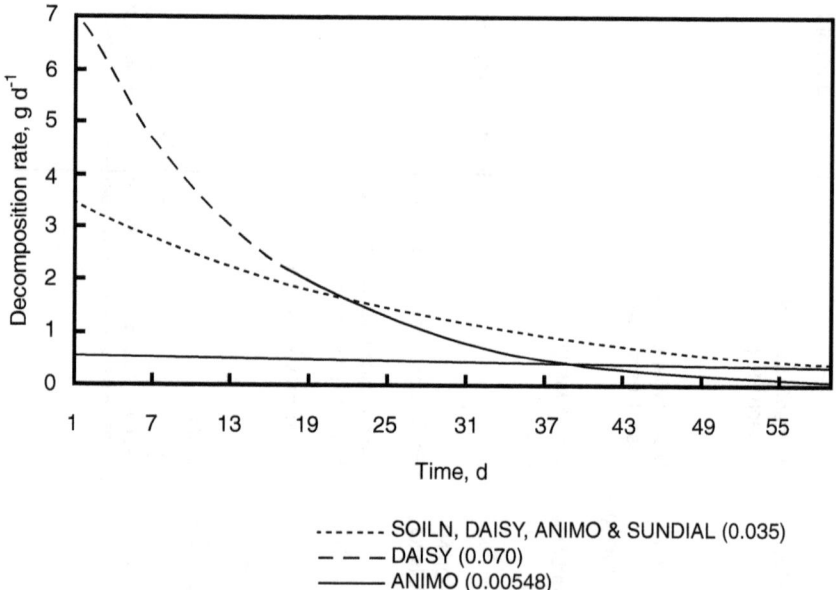

Figure 5.21   Dynamic decomposition with different values of the decomposition rate coefficient (d⁻¹, in parenthesis) for four models. Assuming 100 g dry mass in litter pool in each soil layer. (From Wu, L., and M.B. McGechan. 1998a. A review of carbon and nitrogen processes in four soil nitrogen dynamic models. *Journal of Agricultural Engineering Research*, 69:279-305. With permission.)

Net Mineralization or Immobilization

Whether net mineralization or immobilization occurs from or to the fast cycling pools can be illustrated with reference to the SOILN model. With the suggested values of 10 for the C:N ratio of the humus pool and 0.5 for f_e, net immobilization according to Equation (5.20) occurs if the C:N ratio of the fast cycling pool (litter or feces plus recycled microbial biomass) exceeds 20. Net immobilization is most likely after harvest or after plowing when there is a large input of litter with high C:N ratio to the pool. At other times, the pool will tend to be dominated by recycled microbial biomass (C:N = 10) to give a combined C:N ratio below 20, so net mineralization will occur. A similar reversal in direction for these processes occurs with the other models in accordance with dominant C:N ratios; but due to the more complex definition of pools, this is less easy to visualize.

Subdivision of Components of Slurry in ANIMO

Compared with the other models, ANIMO has a much more detailed treatment of animal slurry as a material being added to the soil. The other models can only consider slurry to be a combination of insoluble feces and soluble urine (which rapidly converts to ammonium in the soil and thus is added straight into the ammonium pool in the models). This subdivision corresponds to a standard analysis of slurry in which nitrogen is specified as "total nitrogen" and "available nitrogen." In contrast, ANIMO considers slurry to consist of a number of main components, with the organic part subdivided into a number of smaller components that can be further subdivided into insoluble (described as "fresh") and insoluble sub-components, each with different decomposition rates. ANIMO is also the only model that includes a separate soluble organic material pool, receiving inputs both as components of added slurry and by decomposition of solid components of added slurry, plant parts (mainly roots), and root exudates. Because the material in this pool is soluble, it moves with water in the soil and is subject to losses by leaching. Details of the proportions of organic material (dry matter) allocated to each of the components for several types of animal slurry are listed in Table 5.8 from Berghuijs-van Dijk et al. (1985). From these, the nitrogen content of each component as a percentage of total slurry dry matter can be calculated, as listed in Table 5.9. Decomposition rates for each component (which are the same for all animal types) are listed in Table 5.5.

Berghuijs-van Dijk et al. (1985) also describe the procedure adopted to estimate the size of each fraction and their decomposition rates. The organic matter content of each slurry type was based on the work of Lammers et al. (1983), and the proportions of soluble material in each fraction were based on an analysis of cattle slurry before and after centrifuging, as reported by Oosterom and Steenvoorden (1980). The decomposition rate of humus was selected as 5.5×10^{-5}/d from Kortleven (1963) and that for root exudates as 1.0/d, assuming this material would decompose at a very fast rate. An iterative procedure was then adopted to estimate the quantity of material in each fraction, the decomposition rate of each fraction, the nitrogen content of each fraction, and the proportion of material that is already in solution at the time of field application. The procedure comprised repeated simulations with

Table 5.8 Subdivision of Slurry Organic Material into Components, as Described for ANIMO Model

Slurry Component	N Content (%)	Proportion of Total in Each Component				
		Cattle	Calf	Pig	Poultry	Dry Poultry Manure
Component 1, 100% dissolved, rapidly decomposing	7	0.10	0.10	0.10	0.10	0.10
Component 2, part dissolved, rapidly decomposing, total	5	0.70	0.80	0.80	0.80	0.40
Subdivided into: "Fresh" sub-component	5	0.65	0.75	0.75	0.75	0.35
Dissolved sub-component	5	0.05	0.05	0.05	0.05	0.05
Component 3, 100% "fresh," slowly decomposing	1	0.20	0.10	0.10	0.10	0.50

From Berghuijs-van Dijk, J.T., P.E. Rijtema, and C.W.J. Roest. 1985. *ANIMO Agricultural Nitrogen Model.* NOTA 1671, Institute for Land and Water Management Research, Wageningen. With permission.

Table 5.9 Nitrogen Contents of Slurry Components, as Described for ANIMO Model

Slurry Component	N Content of Component (% of dry matter)	N Content of Each Component (% of slurry dry matter)				
		Cattle	Calf	Pig	Poultry	Dry Poultry Manure
Mineral N (100% NH_4-N)		0.22	0.21	0.275	0.63	0.95
Total organic matter		6.0	1.5	6.3	9.5	37.0
Organic component 1, 100% dissolved, rapidly decomposing	7	0.042	0.0105	0.0441	0.0665	0.259
Organic component 2, part dissolved, rapidly decomposing, total	5	0.210	0.060	0.252	0.380	0.740
Subdivided into: "Fresh" sub-component	5	0.195	0.0563	0.236	0.356	0.648
Dissolved sub-component	5	0.015	0.00375	0.0158	0.0238	0.0925
Organic component 3, 100% "fresh," slowly decomposing	1	0.012	0.0015	0.0063	0.0095	0.185

From Berghuijs-van Dijk, J.T., P.E. Rijtema, and C.W.J. Roest. 1985. *ANIMO Agricultural Nitrogen Model.* NOTA 1671, Institute for Land and Water Management Research, Wageningen. With permission.

Table 5.10   Parameters in Nitrification Functions

Model	Parameter	Value	Ref.
SOILN	Specific nitrification rate (d⁻¹)	0.2	Johnsson et al. (1987)
	Nitrate-ammonium ratio in nitrification function	8	Johnsson et al. (1987)
SOILNNO	Specific nitrification rate (d⁻¹)	0.05–0.5	Vold et al. (1997)
	Minimum value of $N_{NO3}:N_{NH4}$ ratio for nitrification	50–1000	Vold et al. (1997)
ANIMO	Specific nitrification rate (d⁻¹)	1.0	Huet, H. van (1983)
DAISY	Specific nitrification rate (d⁻¹)	5×10^{-3}	Svendsen et al. (1995)
	Half-saturation constant (kg N/m³)	5.0×10^{-2}	Svendsen et al. (1995)

HISTOR, a simplified subset of the ANIMO model, adjusting parameters to give results that agreed with information in a number of literature sources. These included decomposition curves for various organic materials from Kolenbrander (1969) as illustrated in Figure 5.19. Parameter selection took into account the movement of soluble material by leaching to reproduce profiles of total organic material in solution in a lysimeter experiment reported by Steenvoorden (1983) in which large quantities of animal slurry were added to the soil. Other results to be reproduced included the buildup of organic material in layers (as reported by McGrath, 1981) and the reduction of organic material from slurry to 50% of its initial value by mineralization over the first year after spreading (as reported by van Dijk and Sturm, 1983).

Nitrification Rate Coefficient Data and Comparison Between Models

The nitrification of ammonium to nitrate is determined by a specific nitrification rate, ammonium content, and environmental functions in all models. Theoretically, the rate should be similar under optimal conditions because in every case the process is described as a first-order rate controlled by the same factors. However, the suggested rate coefficients (Table 5.10) vary from model to model (Johnsson et al., 1987; van Huet, 1983; Svendsen et al., 1995). A range of 0.08 to 0.2 varying with fertilizer treatment is suggested in a paper on an application of SOILN by Borg et al. (1990). An even greater spread would have arisen if environmental factors had been ignored from the listed values.

Denitrification

Data Sources

Vinten et al. (1996) carried out field measurements of gas emissions and nitrate leaching using a ¹⁵N tracer to test the two models discussed in the chapter section "Further Models of the Denitrification Process" (i.e., the "randomly distributed pore model" and the "aggregate assembly model"). Results varied with different soils and management practices, with better agreement between simulations and measurements for leaching than for gas emissions. In particular, significant denitrification was measured in a sandy loam soil with few aggregates where both models predicted

Table 5.11   Parameters in Denitrification Functions

Model	Process	Suggested Coefficient	Ref.
SOILN	Half-saturation constant for N concentration effect on denitrification (mg N/L), c_s	10	Klemedtsson et al. (1991)
	Potential rate of denitrification (g N m²/d), K_d^*	0.027	Johnsson et al. (1991)
		0.135	Johnsson et al. (1991)
		0.2	Janssen and Andersson (1988)
SOILNNO	Half-saturation constant for N concentration effect on denitrification (mg N/L), c_s	10	Johnsson et al. (1991)
	Half-saturation constant for respiration in denitrification function (g C/kg soil/d), c_s	$5.0 \times 10^{-4} - 1.0 \times 10^{-6}$	Vold et al. (1997)
ANIMO	Average denitrification rate (d⁻¹)	1.6×10^{-4}	Kroes (1994)
DAISY	Empirical constant, α_d^*	0.1	Hansen et al. (1991)

almost zero denitrification. This discrepancy was attributed to denitrification in deep soil layers with low permeability and high water contents.

Comparison of Denitrification Rates Between Models

Representation of denitrification fundamentally differs among the denitrification sub-models of the four nitrogen dynamics models, and between them and the two stand-alone denitrification models considered in the chapter section "Further Models of the Denitrification Process." Denitrification is linked to the quantity of decomposable organic matter in ANIMO, to the quantity of nitrate in the soil in SOILN, and depends on both nitrate quantity and CO_2 evolution rate from decomposition of organic matter in DAISY. From the expressions used, no coefficient can be directly compared between the models. Suggested values for denitrification parameters (Hansen et al., 1991; Klemedtsson et al., 1991; Johnsson et al., 1991; Jansson and Andersson, 1988; Kroes, 1994) are listed in Table 5.11.

Unlike the other processes considered in this chapter for which rate coefficients measured in laboratory studies could be compared, an adequate comparison of predicted denitrification rates would require simulations with each of the complete soil nitrogen dynamics models that could be compared with field measurements. This has not been attempted here. A detailed comparison of models of denitrification is beyond the scope of this review and could be the subject of a further paper or book chapter on its own.

Soil Hydraulic Parameters

The hydrological routines in DAISY, as well as the soil water models SOIL and SWATRE (or, alternatively, the simpler model WATBAL), in conjunction with which the soil nitrogen models SOILN and ANIMO operate (see chapter section "Other Related Process, Soil Water"), require specification of hydraulic functions, the water release curve (water content against matric tension), and hydraulic conductivity (also at a range of matric tension values). These relationships are expressed as mathematical equations, as specified either by the Brooks and Corey (1964) equation, or the van Genuchten (1980) equation; thus, parameters of these functions need to be specified. Parameter values for some Swedish soils have been reported by Johnsson and Jansson (1991) and Stähli et al. (1996); for Norwegian soils by Vold et al. (1997); for Scottish soils by McGechan et al. (1997); and for Dutch soils by Kabat and Hack-ten Broeke (1988). Hydrological functions for Danish soils as used in DAISY were presented graphically by Jensen et al. (1994) and Djurhuus et al. (1999). SUN-DIAL, CANDY, and the Verberne model incorporate within them their own much simpler sub-models indicating soil water content variation, which do not require these hydraulic parameters.

SUMMARY AND CONCLUSION

A number of European soil nitrogen dynamics models have been reviewed and compared. These models are SOILN from Sweden, SWATNIT from Belgium, SOILNNO from Norway, ANIMO and the Verberne model from The Netherlands, DAISY from Denmark, SUNDIAL from the United Kingdom, and CANDY from Germany. The constituent processes have been analyzed with particular reference to the equations used in each of the models. Processes considered are surface application (as fertilizer, manure or slurry, atmospheric deposition, and deposition or incorporation of dead plant material); mineralization/immobilization (between organic and inorganic forms); nitrification (from ammonium to nitrate); nitrate leaching; denitrification (to N_2O and N_2); and uptake by plants. Sources of information for model parameters have also been reviewed and compared with values assumed in the models.

All the models considered here have strengths and weaknesses, and they vary in the complexity of the treatment of individual processes. However, mention will be made of some important features of particular models that distinguish them from the others. SOILN and DAISY have the most detailed treatment of plant uptake, each with the option of a weather-driven plant growth sub-model operating interactively with the soil nitrogen dynamic processes. DAISY has options for a wide range of crops, while for SOILN a grass growth sub-model has recently been provided in addition to the original cereal and forest tree options. Although all models consider manure in some form, ANIMO has the most complex treatment of animal slurry with the inclusion of a soluble organic material component and several sub-components of solid feces material. ANIMO also has the most mechanistic representation of denitrification, probably the most difficult process to adequately represent. The Verberne model includes the concept of "protected organic matter," which becomes important when soil is disturbed by plowing. Protected pools also feature in a more limited way in DAISY. The ideal model would incorporate all of these features.

Regarding individual processes taking place within the soil profile, the following comments can be made.

1. The temperature response functions are broadly similar in SOILN, ANIMO, DAISY, and CANDY if appropriate parameter values are chosen (not always the same as those suggested by their authors); however, the function in SUNDIAL is of a different form, which makes it difficult to make comparisons with the others.
2. For the soil water response, there is a common form of function that applies to most processes in all the models. This takes a value of unity (no reduction in the rate coefficient) in the middle of the water content range, with a decline at high and low water contents. However, the water content at which the decline starts and the rate of decline varies between models. In addition, the function is defined in terms of water content in some cases and tension (matric potential) in others, and some models assume the same function for all processes, while in others parameter values vary between processes. The function for denitrification in SOILN is of a different form altogether.
3. There are additional response functions for clay content influencing the decomposition rate only in DAISY and SUNDIAL, and for pH influencing the nitrification rate only in ANIMO, which do not appear in the other functions or models.
4. A comparison of decomposition, mineralization, and immobilization is complicated by differing definitions of organic matter pools, although all the models have the same broad pattern of a mixture of fast cycling (plant litter and slurry/manure) and slow cycling (humus) pools. There is general agreement that the processes are controlled by the decomposition rate of soil organic matter and the C:N ratios of materials in the pools. These processes have sufficiently similar broad representation in all the models to draw conclusions about parameter values (rate coefficients), by comparing values suggested for the different models with those from experimental measurements. The suggested decomposition rates for fast cycling material (plant litter and manure) in SOILN (even allowing for some material being recycled back into the source pool) are similar to those suggested by the other model authors and in laboratory measurements at SAC (Vinten et al., 1996), although they do not account for "protection" as discussed above. There is also reasonable agreement between models and laboratory measurements about decomposition rates for slow cycling soil humus material.
5. The need for and benefits of splitting the fast cycling litter pool into very fast and slightly slower cycling components has been identified by the authors of DAISY. This is discussed further by the Swedish authors of SOILN (Jansson et al. [1999], but thus far only implemented in the Norwegian version, SOILNNO (Vold et al., 1997).
6. One or more soil microbial biomass categories are identified as C and N pools in DAISY, SOILNNO, and a recent version of SOILN, with transformation flows linking them with other C and N pools. However, their presence or magnitude has no influence or role in other transformations, other than flows in or out of their own pools.
7. Nitrification is represented by the same form of equation in all the models, but the rate constants suggested vary considerably between models.
8. Denitrification is a complex process that has not been examined in depth in this chapter. Different forms of equations are suggested for each of the models considered, although they all assume that denitrification is stimulated by high soil water contents restricting soil aeration. Of all the processes considered in this chapter,

denitrification in particular will require further modeling and experimental work to give adequate representation and estimates of parameter values.

9. Volatilization of plant litter and fertilizer have been considered only in SUNDIAL. Ammonia volatilization following slurry spreading has been considered only in ANIMO, and only in a very simple way. This is an important process requiring further study or links with a separate model.

The models have been applied and tested for a wide range of applications and tasks, but three main categories can be identified: (1) sizes of soil mineral nitrogen pools, and hence susceptibility to leaching of nitrate to field drains or deep ground-water (aquifers); (2) long-term trends in sizes of soil organic matter pools; and (3) crop performance, with biomass and N offtakes and balances. The emphasis in this chapter (similar to that in the paper by Wu and McGechan, 1998a) has been on the processes and the ability of each model to represent them. The comparisons discussed in the chapter section "Theoretical Description of Mathematics of Processes" (summarized by de Willigen, 1991; Diekkrüger et al., 1995) also emphasize the models' ability to represent processes, rather than their overall performance at representing the variables in the common data sets used. In fact, there is very little objective evidence with which to compare the overall performance at carrying out tasks that might lead to favoring one particular model. The comparison reported by Smith et al. (1997) was confined to assessing performance in representing long-term trends in soil organic matter (9 models, 12 data sets, but not all combinations, and only 4 models tested with all 12 data sets), showing no one model performing better than others over all data sets. There was one group of models including CANDY and DAISY with roughly the same overall level of performance, and another group including the Hurley Pasture model and the Verberne model with significantly greater errors. All the models considered in this chapter operate according to the same fundamental principles, with roughly the same mix of components and comparable equation forms, as well as values of parameters such as rate coefficients for processes other than denitrification. Their performance at the above range of tasks is likely to be similar, in each case very dependent on identifying appropriate calibration parameters for the particular site or conditions. However, from the groups of tasks, DAISY and SOILN with their more complex interactive crop growth sub-model may be expected to perform better at simulating C and N flows to crops.

REFERENCES

Addiscott, T.M., and A.P. Whitmore. 1991. Simulation of solute leaching in different soil of differing permeabilities. *Soil Use Management*, 7:94-102.

Andrén, O., and K. Paustian. 1987. Barley straw decomposition in the filed — A comparison of models. *Ecology*, 68:1190-1200.

Arah, J.R.M., and K.A. Smith. 1989. Steady-state denitrification in aggregated soils: a mathematical model. *Journal of Soil Science*, 40:139-149.

Arah, J.R.M., and A.J.A. Vinten. 1995. Simplified models of anoxia and denitrification in aggregated and simple-structured soils. *European Journal of Soil Science*, 46:507-517.

Arah, J.R.M., J.H.M. Thornley, P.R. Poulton, and D.D. Richter. 1997. Simulating trends in soil organic carbon in long-term experiments using the ITE (Edinburgh) Forest and Hurley Pasture Ecosystems models. *Geoderma*, 81:61-74.

Berghuijs-van Dijk, J.T., P.E. Rijtema, and C.W.J. Roest. 1985. *ANIMO Agricultural Nitrogen Model.* NOTA 1671, Institute for Land and Water Management Research, Wageningen, The Netherlands.

Berghuijs-van Dijk, J.T. 1990. *WATBAL — Water Balance Model for the Unsaturated and Saturated Zone,* Winand Staring Centre, Wageningen, The Netherlands.

Bergström, L., H. Johnsson, and G. Torstensson. 1991. Simulation of soil nitrogen dynamics using the SOILN model. *Fertilizer Research,* 27:181-188.

Blombäck, K., M. Stähli, and H. Eckersten. 1995. Simulation of nitrogen flows and plant growth for a winter wheat stand in central Germany. *Ecological Modelling,* 81:157-167.

Blombäck, K., and H. Eckersten. 1997. Simulated growth and nitrogen dynamics of a perennial rye grass. *Agricultural and Forest Meteorology,* 8:37-45.

Blombäck, K. 1998. *Carbon and Nitrogen in Catch Crop Systems: Modelling of Seasonal and Long-term Dynamics in Plant and Soil.* Agraria 134 (Doctoral thesis), Swedish University of Agricultural Sciences, Uppsalla.

Blombäck, K., H. Eckersten, L. Lewan, and H. Aronsson. 2001. Simulations of carbon and nitrogen dynamics during seven years in a catch crop experiment, *Agricultural Systems,* (in press).

Borg, G.C., P.-E. Jansson, and B. Lindén. 1990. Simulated and measured nitrogen conditions in a manured and fertilized soil. *Plant and Soil,* 121:251-267.

Bradbury, N. J., A.P. Whitmore, P.B.S. Hart, and D.S. Jenkinson. 1993. Modelling the fate of nitrogen in crop and soil in the years following application of ^{15}N-labelled fertilizer to winter wheat. *Journal of Agricultural Science, Cambridge,* 121:363-379.

Breland, T.A. 1994. Enhanced mineralization and denitrification as a result of the heterogeneous distribution of clover residues in the soil. *Plant and Soil,* 166:1-12.

Brooks, R.H., and A.T. Corey. 1964. Hydraulic Properties of Porous Media. Hydrology paper No. 3, Colorado State University, Fort Collins, CO, 27 pp.

Campbell, C.A., Y.W. Jame, and G.E. Winkleman. 1994. Mineralization rate constants and their use for estimating nitrogen mineralization in some Canadian prairie soils. *Canadian Journal of Soil Science,* 64:333-343.

Clough, T.J., S.C. Jarvis, E.R. Dixon, R.J. Stevens, R.J. Laughlin, and D.J. Hatch. 1999. Carbon induced subsoil denitrification of ^{15}N-labelled nitrate in 1 mm deep soil columns. *Soil Biology and Biochemistry,* 31:31-41.

Diekkrüger, B., D. Söndgerath, K.C. Kersebaum, and C.W. McVoy, 1995. Validity of agroecosystem models. A comparison of results of different models applied to the same data set. *Ecological Modelling,* 83:3-29.

Dijk, T.A. van, and H. Sturm. 1983. Fertiliser Values of Animal Manures on the Continent. Proceedings No. 220, The Fertiliser Society, London.

Djurhuus, J., S. Hansen, K. Schelde, and O.H. Jacobsen. 1999. Modelling the mean nitrate leaching from spatially variable fields using effective parameters. *Geoderma,* 87:261-279.

Dyson, P.W. 1992. Fertiliser Allowances for Manures and Slurries. Fertiliser Series No. 14, Technical Note, Scottish Agricultural College, Edinburgh, U.K.

Eckersten, H., and P.-E. Jansson. 1991. Modelling water flow, nitrogen uptake and production for wheat. *Fertilizer Research,* 27:313-329.

Eckersten, H., and T. Slapokas, 1990. Biomass production and nitrogen turnover in an irrigated short rotation forest. *Agricultural and Forest Meteorology,* 50:99-123.

Eckersten, H., A. Gärdenäs, and P.-E. Jansson. 1995. Modelling seasonal nitrogen, carbon, water and heat dynamics of the Solling spruce stand. *Ecological Modelling*, 83:119-129.

Eckersten, H., P.-E. Hansson, and H. Johnsson. 1998. SOILN Model User's Manual, Version 9.2. Division of Hydrotechnics, Communications 98:6, Swedish University of Agricultural Sciences, Uppsala, 113pp.

Feddes, R.A., P.J. Kowalik, and H. Zaradny. 1978. *Simulation of Field Water Use and Crop Yield*. Pudok, Wageningen, The Netherlands.

Franko, U., B. Oelschlägel, and S. Schenk. 1995. Simulation of temperature, water, and nitrogen dynamics using the model CANDY. *Ecological Modelling*, 81:213-222.

Franko, U. 1996. Modelling approaches of soil organic matter turnover within the CANDY system, In D.S. Powlson, P. Smith, and J.U. Smith (Eds.). *Evaluation of Soil Organic Matter Models*, Nato ASI Series, Vol. 38, Springer, Berlin.

Franko, U., and B. Oelschlägel. 1996. *Das Bodenprozess modell CANDY (The soil process model CANDY)*. In H. Muhle and S. Claus, (Eds.). *Reaktionsverhalten von agrarischen Ökosystemen homogener Areale*. Teubner, Leipzig.

Franko U., G.J. Crocker, P R. Grace, J. Klír, M. Körschens, P.R. Poulton, and D.D. Richter. 1997. Simulating trends in soil organic carbon in long-term experiments using the CANDY model. *Geoderma*, 81:109-120.

Franko, U. 1997. Modellierung des Umsatzes der organischen Bodensubstanz, (Modelling of Soil Organic Matter Turnover). *Archiv. Acker-Planzenbau und Boden*, 41:527-547.

Genuchten, M. Th. van. 1980. A closed-form equation for predicting the hydraulic conductivity of unsaturated soils. *Soil Science Society of America Journal*, 44:892-898.

Grinsven, J.J.M. van, and G.B. Makaske, 1993. A one dimensional-model for transport and accumulation of water and nitrogen, based on the Swedish model "SOILN." Report No. 714908001.

Groenendijk, P., and J.G. Kroes, 1997. Modelling the Nitrogen and Phosphorus Leaching to Groundwater and Surface Water; ANIMO 3.5. Report 144, DLO Winand Staring Centre, Wageningen, The Netherlands.

Groot, J.J.R., and E.J.L. Verberne. 1991. Response of wheat to nitrogen fertilization, a data set to validate simulation models for nitrogen dynamics in crop and soil. *Fertilizer Research*, 27:349-383.

Hansen, J.F., J.E. Olesen, I. Munk, U. Henius, J. Høy, S. Rude, M. Steffensen, T. Huld, F. Guul-Simonsen, A. Danfaer, S. Boisen, and S. Mikkelsen. 1990. Kvaelstof I husdyrgødning. Statusredegørelse og systemanalyse vedrørende kvaelstofudnyttelsen. Tidsskrift for Planteavl.

Hansen, S., Jensen, H.E., Nielsen, N.E. and Svendsen, H. 1991. Simulation of nitrogen dynamics and biomass production in winter wheat using the Danish simulation model DAISY, *Fertilizer Research*, 27, 245-259.

Hansen, S., H.E. Jensen, N.E. Nielsen, and H. Svendsen. 1993a. *Description of the Soil Plant System Model DAISY — Basic Principles and Modelling Approach*. Jordbrugsforlaget, The Royal Veterinary and Agricultural University, Copenhagen, Denmark.

Hansen, S., H.E. Jensen, N.E. Nielsen, and H. Svendsen. 1993b. *Users Guide to the DAISY Simulation Model*. The Royal Veterinary and Agricultural University, Copenhagen, Denmark.

Hart, P.B.S., D.S. Powlson, P.R. Poulton, A.E. Johnston, and D.S. Jenkinson. 1993. The availability of the nitrogen in the crop residues of winter wheat to subsequent crops. *Journal of Agricultural Science, Cambridge*, 121:355-362.

Hassink, J., and A.P. Whitmore. 1997. A model of the physical protection of organic matter in soils. *Soil Science Society of America Journal*, 61:131-139.

Hoeks, J. 1972. *Effect of Leaking Natural Gas on Soil and Vegetation in Urban Area*. PUDOC, Wageningen, The Netherlands.

Huet, H. van. 1983. Kwantificering en modellering van de stikstofhuishouding in bodem en grondwater na bemesting. ICW-nota 1426, Wageningen, The Netherlands.

Hutchings, N.J., S.G. Sommer, and S.C. Jarvis. 1996. Model of ammonia volatilization from a grazing livestock farm. *Atmospheric Environment*, 30(4):589-599.

Jansson, P.-E. 1995. Simulation Model for Soil Water and Heat Conditions. Report 165 (Revised), Swedish University of Agricultural Sciences, Uppsala.

Jansson, P.-E., and R. Andersson. 1988. Simulation of runoff and nitrate leaching from an agricultural district in Sweden. *Journal of Hydrology*, 99:33-47.

Jansson, P.-E., T. Persson, and T. Kätterer. 1999. *Nitrogen Processes in Arable and Forest Soils in the Nordic Countries — Field Scale Modelling and Experiments*. TemaNord 1999:560. Nordic Council of Ministers, Copenhagen, 203pp.

Jenkinson, D.S., P.B.S. Hart, J.H. Rayner, and L.C. Parry. 1987. Modelling the turnover of organic matter in long-term experiments at Rothamstead. *Intecol Bulletin*, 15:1-18.

Jensen, C., B. Stougaard, and P. Olsen. 1994a. Simulation of nitrogen dynamics at three Danish locations by use of the DAISY model. *Acta Agriculturae Scandinavica, Sect. B, Soil and Plant Science*, 44:75-83.

Jensen, C., B. Stougaard, and H.S. Østergaard. 1994b. Simulation of the nitrogen dynamics in farm land areas in Denmark 1989–1993. *Soil Use and Management*, 10:111-118.

Jensen, C., B. Stougaard, and H.S. Østergaard. 1996. The performance of the Danish simulation model DAISY in prediction of N_{min} at spring. *Fertilizer Research*, 44:79-85.

Jensen, L.S., T. Mueller, N.E. Nielsen, S. Hansen, G.J. Crocker, P.R. Grace, J. Klír, M. Körschens, and P.R. Poulton. 1997. Simulating trends in soil organic carbon in long-term experiments using the CANDY model. *Geoderma*, 81:109-120.

Johnson, I.R., and J.H.M. Thornley. 1984. A model of instantaneous and daily canopy photosynthesis. *Journal of Theoretical Biology*, 107:531-545.

Johnsson H., L. Bergström, P.-E. Jansson, and K. Paustian. 1987. Simulated nitrogen dynamics and losses in a layered agricultural soil. *Agriculture, Ecosystems and Environment*, 18:333-356.

Johnsson H., and P.-E. Jansson. 1991. Water balance and soil moisture dynamics of field plots with barley and grass ley. *Journal of Hydrology*, 129:149-173.

Johnsson, H., L. Klemedtsson, Å. Nilsson, and B. Svensson. 1991. Simulation of field scale denitrification losses from soils under grass ley and barley. *Plant and Soil*, 138:287-302.

Kabat, P., and M.J.D. Hack-ten Broeke. 1988. Input data for agrohydrological models: some parameter estimation techniques. In H.A.J. Lanen and A.K. Bregt (Eds.). *Proceedings of the EC-Workshop on Application of Computerized Soil Maps and Climate Data*. Wageningen, The Netherlands.

Kätterer, T., A-C. Hansson, and O. Andrén. 1993. Wheat root biomass and nitrogen dynamics — effects of daily irrigation and fertilization. *Plant and Soil*,151:21-30.

Kätterer, T., and O. Andrén. 1996. Measured and simulated nitrogen dynamics in winter wheat and a clay soil subject to daily irrigation and fertilization. *Fertilizer Research*, 44:51-63.

Kätterer, T., H. Eckersten, O. Andrén, and R. Pettersson. 1997. Winter wheat biomass and nitrogen dynamics under different fertilization and water regimes: application of a crop growth model. *Ecological Modelling*, 102:302-314.

Kjøller, A., and S. Struwe. 1982. Microfungi in ecosystems: fungal occurrence and activity in litter and soil. *Oikos*, 39:389-422.

Klemedtsson, L., S. Simkins, B.H. Vensson, H. Johnsson, and T. Rosswall. 1991. Soil denitrification in three cropping systems characterized by differences in nitrogen and carbon supply, II. The water and NO_3 effects on the denitrification process. *Plant and Soil*, 138:273-286.

Kolenbrander, G.J. 1969. De bepaling van de waarde van verschillende soorten organische stof ten aanzien van hun effect op het humusgehalte bij bouwland. Report C6988, Instituut voor Bodemvruchbaarheid, Haren.

Körschens, M. 1996. Long-term datasets from Germany and Eastern Europe. In D.S. Powlson, P. Smith, and J.U. Smith (Eds.). *Evaluation of Soil Organic Matter Models*. NATO ASI Series, Vol. 38, Springer, Berlin.

Kortleven, J. 1963. Kwantitatieve aspecten van humusopbouw en humusafbraak, Dissertatie Landbouwhogenschool Wageningen, Verslagen van Landbouwkundige onderzoekingen, 69, Pudoc, Wageningen, The Netherlands.

Kroes, J.G. 1994. *ANIMO Version 3.4, User's Guide*. Interne mededeling 102, Wageningen, The Netherlands.

Lammers, H.W., T.A. van Dijk., C.H. Henkens, G.J. Kolenbrander, P.E. Rijtema, and K.W. Smilde. 1983. *Gevolgen van het gebruik van organische mest op bouwland*. Consulentschap voor Bodemaangelegenheden in de Landbouw.

Li, C., S. Frolking, R.C. Harris, and R.E. Terry. 1994. Modelling nitrous oxide emissions from agriculture: a Florida case study. *Chemosphere*, 28:1401-1415.

Lind, A-M., K. Debosz, J. Djurhuus, and M. Maag. 1990. *Kvaelstofomsaetning og — transport i to dyrkede jorde*, NOO-forskning fra Miljostyrelsen, Report A9.

McGechan, M.B., R. Graham, and A.J.A. Vinten, J.T. Douglas, and P.S. Hooda. 1997. Parameter selection and testing the soil water model SOIL. *Journal of Hydrology*, 195:312-334.

McGechan, M.B., and L. Wu. 1998. Environmental and economic implications of some slurry management options. *Journal of Agricultural Engineering Research*, 71:273-283.

McGill, W.B., H.W. Hunt, R.G. Woodmansee, and J.O. Reuss. 1981. Phoenix, a model of the dynamics of carbon and nitrogen in grassland soils. In *Terrestrial Nitrogen Cycles*. *Ecological Bulletin*, 33:49-115.

McGrath, D. 1981. Accumulation of organic carbon and nitrogen in soil as a consequence of pig slurry application to grassland. In J.C. Brogan (Ed.). *Nitrogen Losses and Surface Run-off from Land Spreading of Manures*. Nijhoff/Junk, The Hague, The Netherlands.

McVoy, C.W., K.C. Kersebaum, M. Arnonng, P. Kleeberg, H. Othmer, and U. Schröder. 1995. A deterministic evaluation analysis applied to an integrated soil-crop model. *Ecological Modelling*, 81:265-300.

Molen, J. van der, A.C.M. Beljaars, W.J. Chardon, and H.G. van Faasen. 1990. Ammonia volatilization from arable land after application of cattle slurry. 2. Derivation of a transfer model. *Netherlands Journal of Agricultural Science*, 38:239-254.

Mueller, T., L.S. Jensen, J. Magid, and N.E. Nielsen. 1997. Temporal variation of C and N turnover in soil after oilseed rape straw incorporation in the field: simulations with the soil-plant-atmosphere model DAISY. *Ecological Modelling*, 99:247-262.

Mueller, T., J. Magid, L.S. Jensen, H.S. Svendsen, and N.E. Nielsen. 1998. Soil C and N turnover after incorporation of chopped maize, barley straw and bluegrass in the field; Evaluation of the DAISY soil-organic-matter submodel. *Ecological Modelling*, 111:1-15.

Oosterom, H.P., and J.H.A.M. Steenvoorden. 1980. Chemische samenstelling van oppervlakkig afstromend water (Proefveld onderzoek te Achterveld). ICW-nota 1237, Wageningen, Instituut voor Cultuurtechniek en Waterhuishouding.

Parton, W.J., D.S. Schimmel, C.V. Cole, and D.S. Ojima. 1987. Analysis of factors controlling soil organic matter levels in Great Plains grasslands. *Soil Science Society of America Journal*, 51:1173-1179.

Persson, J., and H. Kirchmann. 1994. Carbon and nitrogen in arable soils as affected by supply of N fertilisers and organic manures. *Agriculture, Ecosystems and Environment*, 51:249-255.

Petersen, C.T., U. Jørgensen, H. Svendsen, H.E. Hansen, H.E. Jensen, and N.E. Nielsen. 1995. Parameter assessment for simulation of biomass production and nitrogen uptake in winter rape. *European Agronomy Journal*, 4:77-79.

Powlson, D.S., G. Pruden, A.E. Johnston, and D.S. Jenkinson. 1986. The nitrogen cycle in the Broadbalk Wheat Experiment: recovery and losses of ¹⁵N-labelled fertilizer applied in spring and inputs of nitrogen from the atmosphere. *Journal of Agricultural Science, Cambridge*, 106:591-609.

Querner, E.P. 1998. Description of a regional groundwater flow model. *Agricultural Water Management*, 14:209-218.

Richard, L.A. 1931. Capillary conductivity of liquids in porous mediums. *Physics*, 1:318-333.

Rijtema, P.E., and J.G. Kroes. 1991. Some results of nitrogen simulations with the model ANIMO. *Fertilizer Research*, 27:189-198.

Ritchie, J.T., D.C. Godwin, and S. Otter-Nacke. 1986. CERES-Wheat: A Simulation Model of Wheat Growth and Development. CERES Model description, Department of Crop and Soil Science, Michigan State University, East Lansing.

Sjödahl-Svensson, K., L. Lewan, and M. Clarholm. 1994. Effects of a ryegrass catch crop on microbial biomass and mineral nitrogen in an arable soil during winter. *Swedish Journal of Agricultural Research*, 24:31-38.

Smith, P., D.S. Powlson, and M.J. Glendinning. 1996. Establishing a European GCTE soil organic matter network (SOMNET), In D.S. Powlson, P. Smith, and J.U. Smith (Eds.). *Evaluation of Soil Organic Matter Models*, NATO ASI Series, Vol. 38, Springer, Berlin.

Smith, P., J.U. Smith, D.S. Powlson, W.B. McGill, J.H.M. Arah, O.G. Chertov, K. Coleman, U. Franko, S. Frolking, D.S. Jenkinson, L.S. Jensen, R.H. Kelly, H. Klein-Gunnewiek, A.S. Komarov, C. Li, J.A.E. Molina, T. Mueller, W.J. Parton, J.H.M. Thornley, and A.P. Whitmore. 1997. A comparison of performance of nine soil organic matter models using datasets from seven long-term experiments. *Geoderma*, 81:153-225.

Stanford, G., and S. J. Smith. 1972. Nitrogen mineralization potentials of soils. *Soil Science Society of America Proceedings*, 36:465-472.

Steenvoorden, J.H.A.M. 1983. Nitraatbelasting van het grondwater in zandgebieten; denitrificatie in de ondergrond, ICW-nota 1435 Instituut voor Cultuurtecniek en Waterhuishouding, Wageningen, The Netherlands.

Svendsen, H., S. Hansen, and H.E. Jensen. 1995. Simulation of crop production, water and nitrogen balances in two German agro-ecosystems using the DAISY model. *Ecological Modelling*, 81:197-212.

Swift, M.J., O.W. Heal, and J.M. Anderson. 1979. Decomposition in terrestrial ecosystems. *Studies in Ecology*, Vol. 5, Blackwell, Oxford, 220-226.

Thornley, J.H.M. 1998. *Grassland Dynamics, An Ecosystem Simulation Model*. CAB International, Wallingford, U.K.

Topp, C.F.E., and C.J. Doyle. 1996. Simulating the impact of global warming on milk and forage production in Scotland. 1. The effects on dry-matter yield of grass and grass-white clover swards. *Agricultural Systems*, 52:213-242.

Vanclooster, M., P. Viaene, J. Diels, and J. Feyen. 1995. A deterministic evaluation analysis applied to an integrated soil-crop model. *Ecological Modelling*, 81:183-195.

Vatn, A., L.R. Bakken, M.A. Bleken, P. Botterweg, H. Lundeby, E. Romstad, P.K. Rørstad, and A. Vold. 1996. Policies for reduced nutrient losses and erosion from Norwegian agriculture. *Norwegian Journal of Agricultural Sciences, Supplement No. 23*.

Veen, J.A. van, and E.A. Paul. 1981. Organic carbon dynamics in grassland soils. 1. Background information and computer simulation. *Canadian Journal of Soil Science*, 61:185-201.

Verberne, E.L.J., J. Hassink, P. de Willigen, J.J.R. Groot, and J.A. van Veen. 1990. Modelling organic matter dynamics in different soils. *Netherlands Journal of Agricultural Science*, 38:221-238.

Vereecken, H., M. Vanclooster, and M. Swerts. 1990. A model for the estimation of nitrogen leaching with regional applicability. *Fertilizer Research*, 27:250-263.

Vereecken, H., M. Vanclooster, M. Swerts, and J. Diels. 1991. Simulating water and nitrogen behavior in soils cropped with winter wheat. *Fertilizer Research,* 27:233-243.

Vinten, A.J.A., R.S. Howard, and M.H. Redman. 1991. Measurement of nitrate leaching losses from arable plots under different nitrogen input regimes. *Soil Use and Management,* 7:3-14.

Vinten, A.J.A., B.J. Vivian, and R.S. Howard. 1992. The effect of nitrogen fertiliser on the nitrogen cycle of two upland arable soils of contrasting textures. *Proceedings of International Conference, Cambridge. The Fertiliser Society.* Peterborough, U.K.

Vinten, A.J.A., B.J. Vivian, F. Wright, and R.S. Howard. 1994. A comparative study of nitrate leaching from soils of differing textures under similar climatic and cropping conditions. *Journal of Hydrology,* 159:197-213.

Vinten, A.J.A., K. Castle, and J.R.M. Arah. 1996. Field evaluation of models of denitrification linked to nitrate leaching for aggregated soil. *European Journal of Soil Science,* 47:305-317.

Vinten, A.J.A. 1999. Predicting nitrate leaching from drained arable soils derived from glacial till. *Journal of Environmental Quality,* 28:988-996.

Vold, A. 1997. Development and Evaluation of a Mathematical Model for Plant N-uptake and N-leaching from Agricultural Soils, Ph.D. thesis, Agricultural University of Norway.

Vold, A., and J.S. Søreng. 1997. Optimization of dynamic plant nitrogen uptake, using *a priori* information of plant nitrogen content. *Biometrics Journal,* 39:707-718.

Vold, A., T.A. Breland, and J.S. Søreng. 1997. Multiresponse estimation of parameter values in a model of soil carbon and nitrogen dynamics. In A. Vold (Ed.). *Development and Evaluation of a Mathematical Model for Plant N-uptake and N-leaching from Agricultural Soils.* Agricultural University of Norway, Ph.D. thesis.

Whitmore, A.P., N.J. Coleman, N.J. Bradbury, and T.M. Addiscott. 1991. Simulation of nitrogen in soil and winter wheat crops: modelling nitrogen turnover through organic matter. *Fertilizer Research,* 27:283-291.

Whitmore, A.P. 1995. Modelling the mineralization and leaching of nitrogen from soil residues during three successive growing seasons. *Ecological Modelling,* 81:233-241.

Whitmore, A.P. 1996. Modelling the release and loss of nitrogen after vegetable crops. *Netherlands Journal of Agricultural Science,* 44:73-86.

Whitmore, A.P., J. Klein-Gunnewiek, G.J. Crocker, J. Klír, M. Körschens, and P.R. Poulton. 1997. Simulating trends in soil organic carbon in long-term experiments using the Verberne/MOTOR model. *Geoderma,* 81:137-151.

Willigen, P. de. 1991. Nitrogen turnover in the soil-crop system; comparison of fourteen simulation models. *Fertilizer Research,* 27:141-149.

Wu, L., M.B. McGechan, D.R. Lewis, P.S. Hooda, and A.J.A. Vinten. 1998. Parameter selection and testing the soil nitrogen dynamics model SOILN. *Soil Use and Management,* 14:170-181.

Wu, L., and M.B. McGechan. 1998a. A review of carbon and nitrogen processes in four soil nitrogen dynamic models. *Journal of Agricultural Engineering Research,* 69:279-305.

Wu, L., and M.B. McGechan. 1998b. Simulation of biomass, carbon and nitrogen accumulation in grass to link with a soil nitrogen dynamics model. *Grass and Forage Science,* 53:233-249.

Wu, L., and M.B. McGechan. 1999. Simulation of nitrogen uptake, fixation and leaching in a grass/white clover mixture. *Grass and Forage Science,* 54:30-41.

GLOSSARY

A_1, A_2, A_3, A_4	Coefficients in Eq. (5.12)
A_e	Aerated fraction of soil layer
a	Constant in Eq. (5.5), °C
B_1	Soil water tension threshold, mH_2O, Eq. (5.12)
$\overline{C}$	Average concentration in solution of nitrogen within a cylindrical soil volume, kg/m^3, Eq. (5.46)
C_{di}	Carbon decomposition rate, $g\ C/m^2/d$
C_i	Carbon content of a pool, $g\ C/g$ dry weight
C_{im}	Microbial biomass gain from fast cycling pools, $g\ C/m^2/d$, Eq. (5.22)
C_m	Carbon content of microbial biomass, $g\ C/m^2$
$C_{m,max}$	Maximum value of total microbial biomass, $g\ C/m^2$, Eq. (5.29)
$C_{m,n}$	Carbon content of non-protected microbial biomass, $g\ C/m^2$, Eq. (5.29)
$C_{m,p}$	Carbon content of protected microbial biomass, $g\ C/m^2$, Eq. (5.29)
C_{opt}	Optimal mineral N concentration in evapotranspiration flux, $kg\ N/m^3$
C_{SOM}	Carbon content of stabilized organic matter (SOM) pool, $kg\ C/m^2$, Eq. (5.40)
C_r	Nitrogen concentration at root surface, kg/m^3, Eq. (5.46)
C_{res}	Respiration rate, $g\ CO_2/kg$ soil/d, Eq. (5.36)
ΔC_i	Carbon increment in pool i, $kg\ C/m^3/d$, Eq. (5.24)
$(C{:}N)_i$	C:N ratio in pool i, Eq. (5.24)
c_r	Half-saturation constant with respect to respiration rate, $g\ CO_2/kg$ soil/d, Eq. (5.36)
c_s	Half-saturation constant for nitrate-N, $mg\ N/L$
c_{sh}	Half saturation constant, $kg\ NH_4/m^3$, Eq. (5.33)
D	Diffusion coefficient of nitrogen, m^2/s, Eq. (5.46)
D_{om}	Decomposable organic matter, $kg\ N/m^2$, Eq. (5.37)
d	Cumulative week-degrees since sowing, Eq. (5.47)
d_1	Constant in Eq. (5.11)
d^*	Death rate coefficient for microbial biomass, s^{-1}
E_{pot}	Total evapotranspiration during growing season in a warm year, m
e_a	Soil aeration response function, Eq. (5.16)
e_c	Clay content response function
e_m	Soil water content response function
e_{m0}	Soil water response at wilting point (1500 kPa tension) in Eq. (5.14)
e_r	Carbon substrate availability response function, Eq. (5.36)
e_s	Soil water response function value at saturation, Eq. (5.10)
e_{NO_3}	Response function for nitrate content
e_{pH}	pH response function
e_t	Temperature response function
F_A	Fertilizer applied, $kg\ N/ha$, Eq. (5.41)
f_{AOM1} f_{SMB11} f_{SMB12} f_{SMB1} f_{SOM2} f_{SMB2}	Assimilation factors (partition coefficients) in DAISY, as shown in Figure 5.10
f_{df}	Denitrification factor, Eq. (5.39)

f_e — Synthesis efficiency constant

fe_i — Actual evapotranspiration flux from layer over growing season, m

f_h — Humification fraction

f_{ifre} — N-fraction of exudates, Eq. (5.45)

G — Grain yield at 85% dry matter, kg dry matter/ha, Eq. (5.48)

K — Rate coefficient for first-order rate process

K_a — Denitrification rate determined by actual microbial activity at anaerobic microsites, kg N/m³/s, Eq. (5.38)

K_d^* — Potential denitrification rate, g N m²/d

K_h — Decomposition rate coefficient of humus, d⁻¹

K_n — Specific nitrification rate under aerobic conditions, d⁻¹, Eq. (5.31)

K_n^* — Maximum nitrification rate coefficient at 10°C and optimum soil water conditions, d⁻¹, Eq. (5.33)

$K_{NH_4^+}^{im}$ — Ammonium immobilization rate coefficient, s⁻¹, Eq. (5.25)

$K_{NO_3^-}^{im}$ — Nitrate immobilization rate coefficient, s⁻¹, Eq. (5.26)

K_t — Denitrification rate determined by the transport of nitrate-N to anaerobic microsites, kg N/m³/s, Eq. (5.38)

k_{den} — Denitrification rate coefficient, kg/m²/d, Eq. (5.40)

ke_i — Root exudate production rate in a soil layer, kg/m²/d, Eq. (5.45)

L — Soil layer thickness, m

m — Constant (power) in Eq. (5.10) (with a value greater than zero to indicate shape of function curve)

m^* — Maintenance coefficient for microbial biomass, s⁻¹

N_{hi} — Humus nitrogen transformation rate, g N m²/d, Eq. (5.19)

N_{hm} — Rate of transfer of nitrogen from microbial biomass into humus, g N m²/d, Eq. (5.21)

N_i — Nitrogen mass in a pool, g N/m², Eq. (5.20)

N_m — Nitrogen content of microbial biomass, g N/m², Eqs. (5.21) and (5.22)

N_{max} — Total maximal nitrogen uptake, kg N/m², Eq. (5.43)

N_{ml} — Net mineralization rate from fast cycling pool, g N m²/d, Eq. (5.20)

N_{mtake} — Maximal nitrogen uptake rate, kg N m²/d, Eq. (5.44)

N_{mm} — Mineralization rate from microbial biomass, g N m²/d, Eq. (5.22)

$N_{NH_4^+}$ — Concentration of ammonium-N, kg N/m³

N_{Nitri} — Nitrification rate

$N_{NH_4^+}^{im}$ — Ammonium immobilization rate, kg N/m³/s, Eq. (5.25)

N_{NH_4-N} — Total net mineralization rate, g N m²/d

$N_{NO_3^-}$ — Concentration of nitrate-N

$N_{NO_3^-}^{im}$ — Nitrate immobilization rate, kg N/m³/s, Eq. (5.26)

$N_{NO_3 \rightarrow N_2O + N_2}$ — Denitrification rate, d⁻¹

N_{rtake} — Nitrogen uptake rate from root zone, kg N m²/d

n_r — Number of layers in the root zone, Eq. (5.44)

P_1 — Shape factor in Eq. (5.47) relating the rate of uptake to the point of inflection of the uptake curve

P_2 — Rate constant in Eq. (5.47)

pH — pH of the soil in Eq. (5.30)

pH_{max} — pH value at which nitrification is not affected by acidity, Eq. (5.30)

pH_{min} — pH value at which nitrification is zero, Eq. (5.30)

Q_{10} — Response to a 10°C soil temperature change, Eq. (5.4)

q — Quantity of material in a pool, kg

q_0 — Initial value of quantity of material in pool

q_w	Water flux toward the root, m³/m/s, Eq. (5.46)
R	Distance from the soil pore center to the position where the oxygen concentration in water is equal to zero, m, Eq. (5.32)
RPM	Size of resistant material (RPM) pool, kg/ha, Eq. (5.28)
r	Average pore radius, m, Eq. (5.32)
r_0	C:N ratio of decomposed microbial biomass and humification products, Eq. (5.20)
r_r	Radius of root, m, Eq. (5.46)
$r(T)$	Reaction rate at temperature T, Eq. (5.5)
S_i	Selectivity constant in Eq. (5.45)
S_t	Texture dependent parameter, Eq. (5.16)
SPM	Size of stabilized material (SPM) pool, kg/ha, Eq. (5.28)
T	Soil temperature, °C
$\bar{T}$	Mean temperature, °C, Eq. (5.6)
T_b	Base temperature at which $e_t = 1$, °C, Eq. (5.4)
T_{lin}	Threshold temperature below which the temperature response is a linear function of temperature, °C, Eq. (5.4)
T_w	Mean weekly air temperature, °C, Eq. (5.8)
t	Time, d
t_d	Day number from the start of growth, day, Eq. (5.37)
U_m	Empirically calculated target N uptake of crop, kg N/ha, Eqs. (5.47) and (5.48)
U_T	Nitrogen content of the above-ground part of the crop at harvest, kg N/ha, Eq. (5.42)
V_c	Volatilization of senescent above ground crop material, kg N/ha/d, Eq. (5.42)
V_s	Volatilization of fertilizer in soil, kg N/ha, Eq. (5.41)
W	Combined loss of carbon in CO_2 in 0–0.25 m layer from the RO, biomass, and Humus compartments, kg C/ha/d, Eq. (5.39)
X_c	Clay content, fraction
X_c^*	Threshold for effect of clay content, fraction, Eq. (5.23)
x, x_1, x_2	Constants in Eq. (5.13)
x_w	Degree of water saturation, Eq. (5.13)
Z	Constant in Arrhenius' equation, Eq. (5.5)
z	Depth in soil
α	Fraction of decomposing organic carbon transferred to microbial biomass, Eq. (5.27)
α_d^*	Empirical constant in Eq. (5.38), kg N/kg C
β	Fraction of decomposing organic carbon transferred to humus, Eq. (5.27)
β_c	Parameter in Eq. (5.46)
β_d	Empirical constant in Eq. (5.38), s⁻¹
ε_a	Air-filled pore space, Eq. (5.16)
ε_p	Constant in Eq. (5.16)
γ_c	Water deficit in soil layer, mm, Eqs. (5.14), (5.39)
γ_f	Available water (difference between water held at field capacity and at wilting point, 1500 kPa tension) in soil layer, mm, Eqs. (5.14), (5.39)
γ_i	Water held in layer between field capacity (5 kPa tension) and 100 kPa tension, mm, Eq. (5.14)

ϕ_c	Fraction of the above-ground crop nitrogen lost by volatilization, day^{-1}, Eqs. (5.42), (5.48)
ϕ_s	Fraction of fertilizer nitrogen volatilized during week following application, Eq. (5.41)
μ	Parameter in Eq. (5.46)
θ	Soil water content, % by volume
θ_{av}	Water content per unit thickness, %, Eq. (5.31)
θ_d	Threshold water content for denitrification in Eqs. (5.11), (5.17), and (5.18), % by volume
θ_{fc}	Soil water content at field capacity, % by volume
θ_{lo}	Lowest water content at which the function e_m has a value of unity, %, Eq. (5.10)
θ_{hi}	Highest water content at which the soil water response function has a value of unity, %, Eq. (5.10)
θ_r	Water content difference between that at the air entry tension corresponding to the largest pore radius in the layer and the average tension in the layer, %, Eq. (5.32)
θ_s	Porosity, % by volume
θ_w	Wilting point, % by volume, Eq. (5.10)
$\Delta\theta_1$	Range of volumetric soil water contents where the soil water response increases, Eq. (5.10a)
$\Delta\theta_2$	Range of volumetric soil water contents where the soil water response decreases, Eq. (5.10b)
ρ_r	Root density, m/m^3, Eq. (5.46)
σ	Coefficient in Eq. (5.23)
ξ_{CO_2}	CO_2 evolution rate from decomposition or organic matter, kg C/m^3/s, Eq. (5.38)
ψ	Soil water tension (matric potential), kPa, Eq. (5.12)

CHAPTER **6**

A Review of the Canadian Ecosystem Model — *ecosys*

Robert F. Grant

CONTENTS

INTRODUCTION

The *ecosys* modeling program is dedicated to the construction and testing of a comprehensive mathematical model (*ecosys*) of natural and managed terrestrial ecosystems. The long-term objectives of this program are to provide a means to anticipate ecosystem behavior under different environmental conditions (soils, climates, and land use practices). The design and scope of *ecosys* are described in general terms below, and further documentation may be found in the cited literature. The governing equations in the model are listed in the attached appendices.

The development of *ecosys* has been based on the following guidelines.

1. *Ecosys* parameters have a defined physical or biological meaning, and are capable of evaluation independently of the model.
2. These *ecosys* parameters function at spatial and temporal scales smaller than those at which the model is tested.
3. Each ecosystem process is represented in the model at a level of detail that allows well-constrained tests of the process.
4. *Ecosys* integrates temporal scales from seconds to centuries, allowing validation vs. data from experiments that range from short-term laboratory incubations to long-term field studies.
5. *Ecosys* integrates spatial scales ranging from mm to km in 1, 2, or 3 dimensions, as required, allowing the scaling up of microscale phenomena to the landscape level.
6. *Ecosys* integrates biological scales (both plant and microbial) from the organ to the community, allowing the representation of complex biomes.
7. *Ecosys* simulates the transport and transformation of heat, water, carbon, oxygen, nitrogen, phosphorus, and ionic solutes through soil-plant-atmosphere systems with the atmosphere as the upper boundary and soil parent material as the lower boundary.
8. *Ecosys* is constructed entirely in FORTRAN 77, allowing portability among different computers. Although *ecosys* has been run on high-performance computing facilities, recent developments in computing technology now allow the model to be run on desktop computers.

Options are provided in *ecosys* to introduce a full range of management practices into model simulations, including tillage (defined by dates, depths, degrees of soil mixing); fertilization (dates, application methods and depths, types and amounts of N, P lime, gypsum, residue, manure); irrigation (dates, times, amounts, chemical composition); planting (species, dates, densities); and harvesting (dates, parts of plants removed, heights of cutting, fraction of plants harvested, and efficiency of removal). Options are also provided to introduce changes in atmospheric boundary conditions (incremental or step changes in C_a, radiation, temperature, precipitation, humidity, wind speed, and changes in chemical composition of precipitation). These options allow the effects of a wide range of disturbances to be simulated when studying disturbance effects on ecosystem function.

The *ecosys* source code is constructed such that all flux equations are solved in three dimensions for each cell of a matrix defined by row, column, and layer position. Model users can thus construct simulated ecosystems in one, two, or three dimensions with any length, width, depth, azimuth, and slope, and any number of competing plant populations. The capability of *ecosys* to function in three dimensions addresses the need to scale high-resolution ecosystem processes up to the landscape level. Provision is made for soil properties to vary in three dimensions through the matrix. All loops are constructed to ensure independence of each cell during code execution. This code structure can be implemented in a massively parallel computing environment, such that computations for each cell can be carried out concurrently by different processors. Such implementation will eventually enable routine use of the model at the landscape level and will make an important contribution to the scaling up of *ecosys* performance to represent soil productivity and atmospheric gas exchange at the regional scale.

INPUT DATA

Weather

Hourly

Irradiance (W m^{-2} or kJ m^{-2} h^{-1})
Air temperature (°C)
Wind speed (m s^{-1} or km h^{-1})
Humidity (RH %, vapor pressure kPa, or dewpoint °C)
Precipitation (mm as snow or rain)

or Daily

Radiation (MJ m^{-2} d^{-1})
Max. temperature (°C)
Min. temperature (°C)
Wind travel (km d^{-1})
Precipitation (mm as snow or rain)
Humidity (RH %, vapor pressure kPa, or dewpoint °C)

Site

General

Latitude (°)
Altitude (m)
Average annual air temperature (°C)
Solar noon (h)
Ambient N_2, O_2, CO_2, CH_4, N_2O, and NH_3 concentration (μmol mol^{-1})
Type of experiment (phytotron, field)
Previous crop (legume or non-legume)
Depth to water table (m)
pH and mineral ion concentration in rain (NH_4^+, NO_3^-, PO_4^{2-}, Al^{3+}, Fe^{3+}, Ca^{2+}, Mg^{2+}, K^+, Na^+, SO_4^{2-}, Cl^-, in g Mg^{-1})

For Each Topographic Position in Site

Soil profile name
NW, SE location
Slope inclination (°)
Slope aspect (°)
Initial surface roughness (m)
Initial depth of snowpack (m)

Soil

For Each Soil Profile

Water potential defined as field capacity (MPa)
Water potential defined as wilting point (MPa)
Wet soil albedo

For Each Layer in the Soil Profile:

Depth (m)
Bulk density (Mg m^{-3})
Water content at field capacity (m^3 m^{-3})
Water content at wilting point (m^3 m^{-3})
Vertical saturated hydraulic conductivity (mm h^{-1})
Horizontal saturated hydraulic conductivity (mm h^{-1})
Sand (g kg^{-1})
Silt (g kg^{-1})
Macropores (% vol.)
Coarse fragments (% vol.)
Organic C (g kg^{-1})
Particulate organic C (g kg^{-1})
Organic N (g Mg^{-1})
Organic P (g Mg^{-1})
pH
Cation exchange capacity (cmol kg^{-1})
Anion exchange capacity (cmol kg^{-1})
NH_4^+ (g Mg^{-1})
NO_3^- (g Mg^{-1})
PO_4^{2-} (g Mg^{-1})
Al^{3+} (g Mg^{-1})
Fe^{3+} (g Mg^{-1})
Ca^{2+} (g Mg^{-1})
Mg^{2+} (g Mg^{-1})
Na^+ (g Mg^{-1})
K^+ (g Mg^{-1})
Cl^- (g Mg^{-1})
SO_4^{2-} (g Mg^{-1})
Initial water content (m^3 m^{-3})
Initial ice content (m^3 m^{-3})
Initial plant residue content (g C m^{-2})
Initial animal manure content (g C m^{-2})

Plant

Plant Descriptors

Photosynthetic type: C_3 or C_4
Monocot or dicot
Annual or perennial
Growth habit: determinate or indeterminate
Legume or non-legume
Spring or winter (vernalization requirement)
Photoperiod sensitivity: short day or long day
Plant architecture: crown or trunk
Storage organs: above-ground or below-ground
Mycorrhizal or non-mycorrhizal

CO_2 Fixation Kinetics

Specific carboxylation activity of rubisco (μmol g^{-1} s^{-1})
Specific oxygenation activity of rubisco (μmol g^{-1} s^{-1})
Michaelis-Menten K_m for carboxylation (μmol)
Michaelis-Menten K_m for oxygenation (μmol)
Fraction of leaf protein in rubisco
Specific activity of chlorophyll (μmol g^{-1} s^{-1})
Fraction of leaf protein in chlorophyll
Albedo of leaf surfaces for shortwave radiation
Albedo of leaf surfaces for photosynthetic radiation
Transmissivity of leaf surfaces for shortwave radiation
Transmissivity of leaf surfaces for photosynthetic radiation

Phenology

Rate of primordial advance (h^{-1})
Rate of leaf appearance (h^{-1})
Chilling temperature (°C)
Vernalization requirement (h)
Number of primordial nodes in seed
Critical photoperiod (h)
Photoperiod sensitivity of floral initiation (node h^{-1})
Critical temperature of floral initiation (°C)
Temperature sensitivity of floral initiation (node °C^{-1})

Morphology

Fraction of leaf area inclined 0°–22.5°, 22.5°–45°, 45°–67.5°, 67.5°–90° from horizontal
Clumping factor for non-uniform horizontal distribution of leaf area (unitless)
Angle of tillers (monocot) or branches (dicot) from horizontal (°)
Angle of sheaths (monocot) or petioles (dicot) from horizontal (°)

Grain Characteristics

Maximum number of fruiting sites per reproductive node
Maximum number of grain kernels per fruiting site
Maximum mass of grain kernel (g C)
Mass of seed at planting (g C)
Maximum rate of kernel filling (g C kernel h^{-1})

Root Characteristics

Radius of primary roots (m)
Radius of secondary roots (m)
Porosity of roots (m^3 m^{-3})
Root radial resistivity (MPa h m^{-2})
Root axial resistivity (MPa h m^{-4})
Branching frequency of secondary roots on primary axes (m^{-1})
Branching frequency of tertiary roots on secondary axes (m^{-1})

Ammonium Uptake Parameters:

Maximum rate of ammonium-N uptake (g N m^{-2} root h^{-1})
Michaelis-Menten K_m for ammonium-N uptake (g N m^{-3})
Minimum concentration for ammonium-N uptake (g N m^{-3})

Nitrate Uptake Parameters:

Maximum rate of nitrate-N uptake (g N m^{-2} root h^{-1})
Michaelis-Menten K_m for nitrate-N uptake (g N m^{-3})
Minimum concentration for nitrate-N uptake (g N m^{-3})

Phosphate Uptake Parameters:

Maximum rate of phosphate-P uptake (g P m^{-2} root h^{-1})
Michaelis-Menten K_m for phosphate P uptake (g P m^{-3})
Minimum concentration for phosphate-P uptake (g P m^{-3})

Plant Water Relations

Osmotic potential when water potential = 0 (MPa)
Shape parameter for stomatal sensitivity to plant turgor
Leaf cuticular resistance to vapor diffusion (s m^{-1})

Organ Characteristics

Growth yield of leaf, sheath or petiole, stalk, reserve, husk or pod, ear or rachis, grain,
  root and nodule (if legume) (g C g C^{-1})
Nitrogen:carbon ratio of leaf, sheath or petiole, stalk, reserve, husk or pod, ear or
  rachis, grain, root and nodule (if legume) (g N g C^{-1})
Phosphorus:carbon ratio of leaf, sheath or petiole, stalk, reserve, husk or pod, ear or
  rachis, grain, root and nodule (if legume) (g P g C^{-1}).

Temperature Sensitivity

Arrhenius function for temperature sensitivity of carboxylation: intercept, energy of activation ($J mol^{-1}$), energy of enthalpy ($J mol^{-1}$), energy of low-temperature inactivation ($J mol^{-1}$), energy of high-temperature inactivation ($J mol^{-1}$)

Arrhenius function for temperature sensitivity of oxygenation: intercept, energy of activation ($J mol^{-1}$), energy of enthalpy ($J mol^{-1}$), energy of low-temperature inactivation ($J mol^{-1}$), energy of high-temperature inactivation ($J mol^{-1}$)

Arrhenius function for temperature sensitivity of growth respiration: intercept, energy of activation ($J mol^{-1}$), energy of enthalpy ($J mol^{-1}$), energy of low-temperature inactivation ($J mol^{-1}$), energy of high-temperature inactivation ($J mol^{-1}$)

Management

Tillage

Date of each tillage event
Implement (SCS standard)
Depth (m)

Fertilizer

Date of each fertilizer event
NH_4^+ broadcast ($g N m^{-2}$)
NH_3 broadcast ($g N m^{-2}$)
Urea broadcast ($g N m^{-2}$)
NO_3^- broadcast ($g N m^{-2}$)
NH_4^+ banded ($g N m^{-2}$)
NH_3 banded ($g N m^{-2}$)
Urea banded ($g N m^{-2}$)
NO_3^- banded ($g N m^{-2}$)
PO_4^{2-} broadcast ($g P m^{-2}$)
PO_4^{2-} banded ($g P m^{-2}$)
Rock PO_4^{2-} broadcast ($g P m^{-2}$)
Animal manure ($g C m^{-2}$)
Plant residue ($g C m^{-2}$)
Distance between band rows (m)

Irrigation

Date of each irrigation event
Amount (mm)
Time started (h)
Time finished (h)
pH and mineral ion concentration (NH_4^+, NO_3^-, PO_4^{2-}, Al^{3+}, Fe^{3+}, Ca^{2+}, Mg^{2+}, K^+, Na^+, SO_4^{2-}, Cl^- in $g Mg^{-1}$) in irrigation water

Plant Management

Number of plant species or cohorts

For Each Plant Species:

Plant density (m^{-2})
Maturity type (early, medium, late)
Date of planting
Date(s) of harvest
Type of harvest (remove whole plant, grain, or nothing)
Height of harvest cut (m)
Fraction of population harvested
Efficiency of harvest removal

OUTPUT DATA

Each Plant Species or Cohort

Energy

Net radiation $(W\ m^{-2})$
Latent heat $(W\ m^{-2})$
Sensible heat $(W\ m^{-2})$
Canopy temperature $(^\circ C)$

Water

Transpiration $(mm\ h^{-1})$
Canopy water potential (MPa)
Canopy turgor (MPa)
Boundary layer resistance $(s\ m^{-1})$
Canopy stomatal resistance $(s\ m^{-1})$

Growth

CO_2 fixation of leaf and canopy $(\mu mol\ m^{-2}\ s^{-1}$ or $g\ m^{-2}\ h^{-1})$
CO_2 respiration $(\mu mol\ m^{-2}\ s^{-1}$ or $g\ m^{-2}\ h^{-1})$
Soluble CH_2O concentration $(g\ g^{-1})$
C $(g\ m^{-2})$ and length (m) of each leaf, sheath or petiole, and internode on each tiller or branch
C $(g\ m^{-2})$ and length (m) of each primary and secondary root axis
C $(g\ m^{-2})$ of each spike or pod, and seed on each tiller or branch and of root nodules in each soil layer
Canopy height (m)

Nutrients

NH_4^+ uptake (g m^{-2} h^{-1}) from each soil layer
NO_3^- uptake (g m^{-2} h^{-1}) from each soil layer
Soluble N concentration (g g^{-1})
N (g m^{-2}) of each leaf, sheath or petiole, and internode on each tiller or branch
N (g m^{-2}) of each primary and secondary root axis
N (g m^{-2}) of each spike or pod, and seed on each tiller or branch and of root nodules
    in each soil layer
PO_4^{2-} uptake (g m^{-2} h^{-1}) from each soil layer
Soluble P concentration (g g^{-1})
P (g m^{-2}) of each leaf, sheath or petiole, and internode on each tiller or branch
P (g m^{-2}) of each primary and secondary root axis
P (g m^{-2}) of each spike or pod, and seed on each tiller or branch and of root nodules
    in each soil layer
O_2 uptake (g m^{-2} h^{-1}) from each soil layer
CO_2 evolution (g m^{-2} h^{-1}) into each soil layer
CO_2 concentration in root axis and rhizosphere (g m^{-3})
O_2 concentration in root axis and rhizosphere (g m^{-3})
NH_4^+ concentration in rhizosphere (g m^{-3})
NO_3^- concentration in rhizosphere (g m^{-3})
PO_4^{2-} concentration in rhizosphere (g m^{-3})

Snow, Residue, and Soil Surfaces

Energy

Net radiation (W m^{-2})
Latent heat (W m^{-2})
Sensible heat (W m^{-2})
Temperature (°C)

Water

Evaporation (mm h^{-1})
Boundary layer resistance (s m^{-1})
Snowpack depth (m)
Snowpack composition (snow, water, ice) (m^3 m^{-3})

Gas Fluxes

CO_2 flux (g m^{-2} h^{-1})
O_2 flux (g m^{-2} h^{-1})
NH_3 flux (g m^{-2} h^{-1})
N_2O flux (g m^{-2} h^{-1})
N_2 flux (g m^{-2} h^{-1})
CH_4 flux (g m^{-2} h^{-1})

Each Soil Layer

Physical

Temperature (°C)
Water content ($m^3\ m^{-3}$)
Ice content ($m^3\ m^{-3}$)
Water potential (matric and osmotic) (MPa)

Biochemical

CO_2 concentration in gaseous and aqueous phases ($g\ m^{-3}$)
O_2 concentration in gaseous and aqueous phases ($g\ m^{-3}$)
NH_3 concentration in gaseous and aqueous phases ($g\ m^{-3}$)
N_2O concentration in gaseous and aqueous phases ($g\ m^{-3}$)
N_2 concentration in gaseous and aqueous phases ($g\ m^{-3}$)
CH_4 concentration in gaseous and aqueous phases ($g\ m^{-3}$)
NH_4^+ concentration in aqueous and solid phases ($g\ m^{-3}$)
NO_3^- concentration in aqueous phase ($g\ m^{-3}$)
PO_4^{2-} concentration in aqueous and solid phases ($g\ m^{-3}$)
Al^{3+} concentration in aqueous and solid phases ($mol\ m^{-3}$)
Fe^{3+} concentration in aqueous and solid phases ($mol\ m^{-3}$)
Ca^{2+} concentration in aqueous and solid phases ($mol\ m^{-3}$)
Mg^{2+} concentration in aqueous and solid phases ($mol\ m^{-3}$)
Na^+ concentration in aqueous and solid phases ($mol\ m^{-3}$)
K^+ concentration in aqueous and solid phases ($mol\ m^{-3}$)
Cl^- concentration in aqueous and solid phases ($mol\ m^{-3}$)
SO_4^{2-} concentration in aqueous and solid phases ($mol\ m^{-3}$)

Biological

Microbial biomass ($g\ m^{-3}$)
Microbial C substrates ($g\ m^{-3}$)
Soluble C concentration ($g\ m^{-3}$)
Root C and density ($g\ m^{-3}$ and $m\ m^{-3}$)
O_2 reduction ($g\ m^{-2}\ h^{-1}$)
NO_3^- reduction ($g\ m^{-2}\ h^{-1}$)
NO_2^- reduction ($g\ m^{-2}\ h^{-1}$)
N_2O reduction ($g\ m^{-2}\ h^{-1}$)
NH_4^+ immobilization-mineralization ($g\ m^{-2}\ h^{-1}$)
NO_3^- immobilization ($g\ m^{-2}\ h^{-1}$)
NH_4^+ oxidation ($g\ m^{-2}\ h^{-1}$)
NO_2^- oxidation ($g\ m^{-2}\ h^{-1}$)
CH_4 oxidation ($g\ m^{-2}\ h^{-1}$)
CH_4 reduction ($g\ m^{-2}\ h^{-1}$)

MODEL DEVELOPMENT

Ecosystem-Atmosphere Energy Exchange

Model Description

Separate first-order closure schemes are used to simulate hourly energy exchange between the atmosphere and each of several terrestrial surfaces in *ecosys* including the canopy of each species in the plant community ([Equations (A.1) to (A.15)]; also Grant et al., 1999b; Grant and Baldocchi, 1992; Grant et al., 1993e; 1995a,c; 1999a), and snow, residue, and soil surfaces ([Equations (A.16) to (A.23)]; also Grant et al., 1999b; Grant et al., 1995b). Surface energy exchanges are coupled to subsurface conductive, convective, and latent heat transfers using a forward differencing scheme with heat capacities and thermal conductivities calculated from de Vries (1963) [Equations (A.24) to (A.27)]. Total energy exchange between the atmosphere and terrestrial surfaces is calculated as the sum of the exchanges with the canopy of each plant species, and each snow, detritus, and soil surface. A summary of ecosystem-atmosphere exchanges simulated in *ecosys* is shown in Figure 6.1.

Canopy energy exchange in *ecosys* is calculated from a two-stage convergence solution for the transfer of water and heat through a multi-specific, multi-layered soil-root-canopy system. The first stage of this solution requires convergence to a canopy temperature at which the first-order closure of the energy balance is achieved for the canopy of each plant species [Equations (A.1) to (A.15)]. The energy balance requires first that the absorption, reflection, and transmission of both shortwave and photosynthetically active radiation be calculated for each leaf and stem surface in a multi-layered canopy. Radiation includes direct and diffuse sources, defined by solar and sky angles, as well as forward and backscattering within the canopies. Each leaf and stem surface is defined by species, height, azimuth, inclination, exposure (sunlit vs. shaded), and optical properties (reflection and transmission). Non-uniformity in the horizontal distribution of leaf surfaces within each canopy layer (clumping) is represented by a species-specific interception fraction between zero and one applied to each leaf and stem surface [Equation (A.7)]. This fraction describes its fractional exposure to direct and diffuse irradiance (vs. self-shading) (e.g., Chen et al., 1997), and is assumed constant through the canopy. The fraction of photosynthetic photon flux density absorbed by each canopy is used to partition the exchange of longwave radiation emitted by sky, ground, and canopy surfaces on the assumption that leaves are similarly opaque to both radiation types [Equation (A.8)]. The net values of longwave exchange are added to those of shortwave radiation absorbed by all leaf and stem surfaces to calculate canopy net radiation [Equation (A.2)].

The energy balance then requires solutions for latent and sensible heat fluxes at the temperature of each canopy. If intercepted precipitation (calculated from leaf and stem areas according to Waring and Running, 1998) is present on leaf or stem surfaces, latent heat flux is calculated from evaporation determined by the canopy-atmosphere vapor density gradient and aerodynamic conductance [Equation (A.4)]. If intercepted precipitation is not present, latent heat flux is calculated from transpiration, which is also determined by stomatal conductance [Equation (A.3)]. Sensible heat flux is

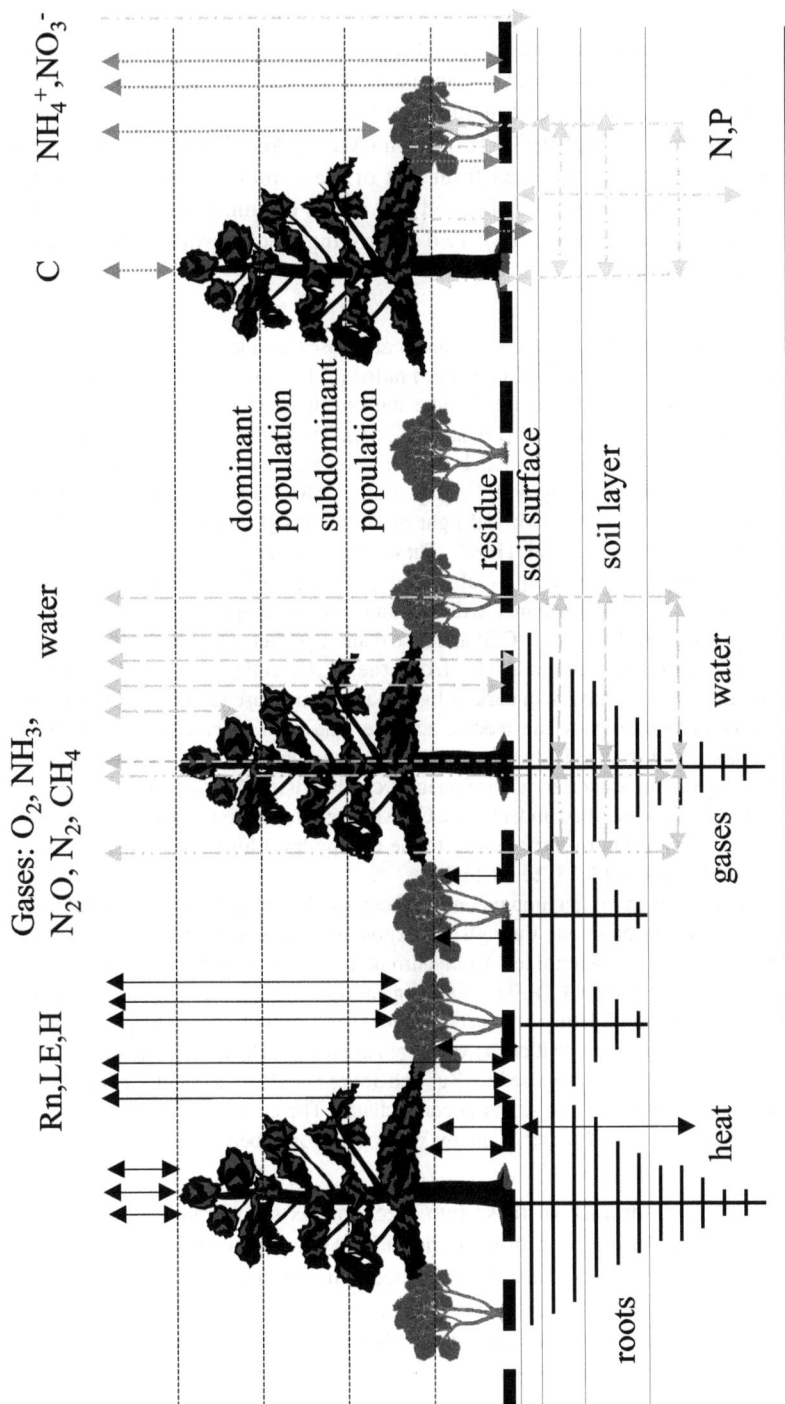

Figure 6.1 Ecosystem-atmosphere exchanges and subsurface transfers of heat, gases, water, C, N, and P simulated in *ecosys*.

calculated from the canopy-atmosphere temperature gradient and aerodynamic conductance [Equation (A.5)]. Canopy heat storage is calculated from changes in canopy temperature and from the heat capacities of leaves, twigs, and stems [Equation (A.6)].

Aerodynamic conductance in energy balance solutions is calculated from zero plane displacement and surface roughness of the dominant canopy derived from its height and leaf area (Perrier, 1982) [Equation (A.12)]. Aerodynamic conductance of non-dominant canopies is reduced from that of the dominant canopy according to the differences between their heights and that of the dominant canopy as proposed by Choudhury and Monteith (1988). Two controlling mechanisms are postulated for stomatal conductance:

1. At the leaf scale, maximum conductances are those that allow an assumed constant intercellular to atmospheric CO_2 concentration ($C_i:C_a$) ratio of 0.7 to be maintained at carboxylation rates calculated under ambient irradiance, temperature, C_a, and non-limiting water potential [Equation (C.14)]. The assumption of constant $C_i:C_a$ is implicit in other stomatal models in which conductance varies directly with CO_2 assimilation and inversely with C_a (e.g., Ball, 1988). Carboxylation rates are calculated as the lesser of dark and light reaction rates [Equations (C.1) to (C.10)] according to Farquhar et al. (1980). These rates are driven by irradiance (rectangular hyperbolic function of absorbed photosynthetically active radiation) [Equation (C.7)], temperature (Arrhenius function of canopy temperature from the energy balance) [Equations (C.3) and (C.8)] and C_i (Michaelis-Menten function of the aqueous equivalents of C_i from the $C_i:C_a$ ratio and of that of O_2) [Equation (C.1)]. Maximum dark or light reaction rates used in these functions are assumed proportional to the specific activities and areal concentrations of rubisco or chlorophyll, respectively. These activities and concentrations may be reduced from maximum values by environmental conditions (radiation, temperature, C_a, water, N, P) as described under "Plant Growth" below. Maximum leaf conductances are then aggregated by surface area to the canopy level for energy balance calculations [Equations (A.13) to (A.15)].

2. At the canopy scale, maximum conductances are then reduced from those at non-limiting water potential through an exponential function of turgor potential [Equation (A.13)] determined from current total and osmotic water potentials [Equations (B.1) and (B.2)]. This function causes conductance to be largely insensitive to turgor at high turgor, to become more sensitive to turgor as turgor declines, and to reach minimum (cuticular) values at zero turgor. The function of turgor used here is based on that proposed by Zur and Jones (1981) to account for the effects of osmotic adjustment on stomatal conductance. The calculation of canopy water potential is described under "Canopy Water Relations" below.

The relative importance of these two mechanisms for stomatal conductance in the model depends on site conditions. In humid ecosystems such as the black spruce-moss forest (Grant et al., 2001a), the first mechanism is the dominant controller of conductance and the second is of limited influence. In water-limited ecosystems such as underirrigated wheat (Grant et al., 1999b), the second mechanism is more important, especially during periods of water stress.

Model Testing

The sensitivity of the energy exchange algorithm to C_a and soil water content has been tested with results from a Free Air CO_2 Enrichment experiment (FACE) on wheat. The algorithm simulated reductions in midday latent heat fluxes of 50 to 100 W m^{-2} and increases in canopy temperatures of up to 1°C when C_a was raised from 355 to 550 μmol mol^{-1} over fully irrigated and fertilized wheat. These modeled changes were within the standard errors of measured changes (Grant et al., 1995a, 1999b) (Figure 6.2). Changes in energy exchange when C_a was raised from 355 to 550 μmol mol^{-1} were greater under limiting N fertilization (Grant et al., 2001c) (Figure 6.3). The changes in energy exchange were modeled from differing sensitivities to C_a of leaf carboxylation rates and $C_a - C_i$ differences (from the C_i:C_a ratio described above) that caused leaf stomatal conductance to decrease under higher C_a and lower N. The energy exchange algorithm simulated reductions in leaf stomatal conductances of up to 0.2 mol m^{-2} s^{-1} and increases in canopy temperature of up to 5°C in the FACE experiment when irrigation was reduced from non-limiting rates to one half of evapotranspirational demand (Figure 6.4). These changes, which were within the standard errors of those measured (Grant et al., 1999b), were modeled from the effects of lower canopy water potentials and turgors on stomatal conductances.

The sensitivity of the energy exchange algorithm to diurnal changes in weather has also been tested against eddy correlation data measured over deciduous and coniferous forests in the southern study area of BOREAS. The algorithm explained 80% of diurnal variation in ecosystem latent heat fluxes measured by Black et al. (1996) over a 70-year-old mixed aspen-hazelnut stand (Grant et al., 1999a) (Figure 6.5). The algorithm also simulated the much lower latent heat fluxes measured by Jarvis et al. (1997) over a nearby 115-year-old black spruce-moss stand (Grant et al., 2001a) (Figure 6.6). Variation in energy exchange at the two sites was modeled from changes in leaf carboxylation rates and hence in leaf stomatal conductances driven by diurnal changes in radiation and temperature and assuming constant C_i:C_a.

The soil heat and water transfer scheme caused modeled soil thawing and warming in spring to be delayed by 1 week under reduced vs. conventional tillage at a site in central Alberta. It also caused soil under clear fallow to become as much as 10°C warmer than that under grass during summer at the same site. These changes in soil temperature were modeled from changes in surface energy exchange caused by differing plant and detritus cover, and were supported by field measurements (Grant et al., 1995b).

Canopy Water Relations

Model Description

Following first-order closure of the energy balance, a convergence solution is sought for the canopy water potential of each plant population at which the difference between its transpiration [Equation (A.3)] and total root water uptake [Equation (B.9)] equals the difference between its water contents at the previous and

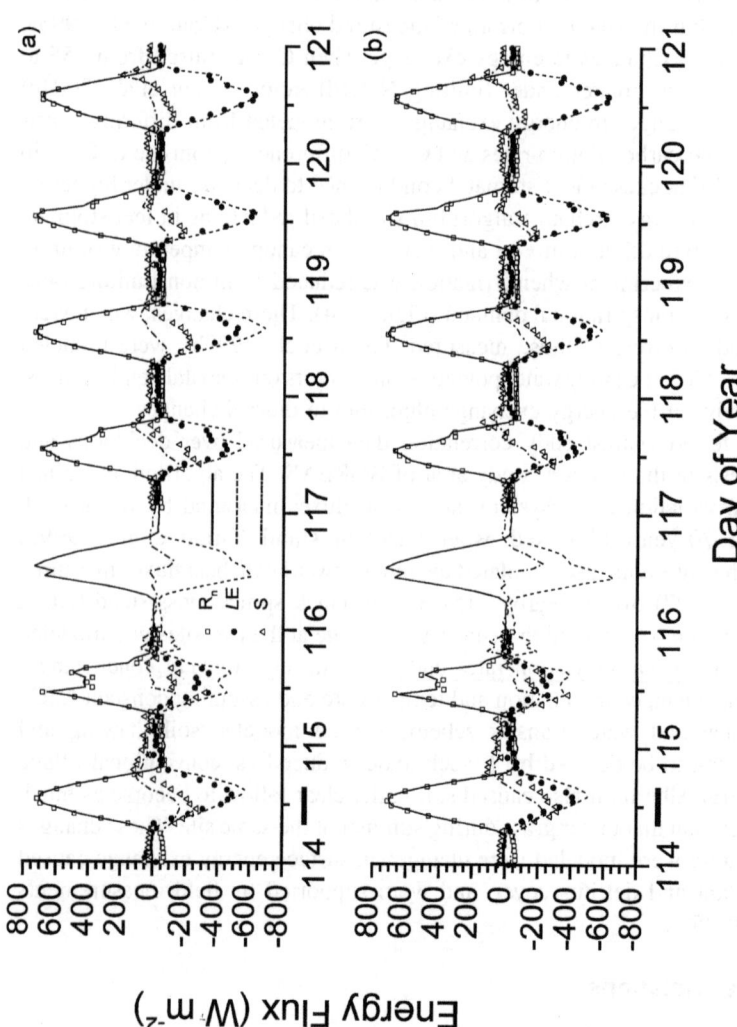

Figure 6.2 Net radiation (R_n), latent heat flux (LE), and sensible heat flux (H) simulated (lines) and measured (symbols) in the Free Air CO_2 Enrichment (FACE) experiment at Phoenix, AZ, from 25 April to 1 May 1994 (days 115 to 121) inclusive under (a) 355 and (b) 550 μmol mol^{-1} atmospheric CO_2 concentration and high ($1 \times$ PET) irrigation. Bold line on x-axis indicates irrigation event. (From Grant, R.F., G.W. Wall, B.A. Kimball, K.F.A. Frumau, P.J. Pinter Jr., D.J. Hunsaker, and R.L. Lamorte. 1999b. Crop water relations under different CO_2 and irrigation: testing of ecosys with the Free Air CO_2 Enrichment (FACE) experiment. *Agric. For. Meteorol.*, 95:27-51. With permission.)

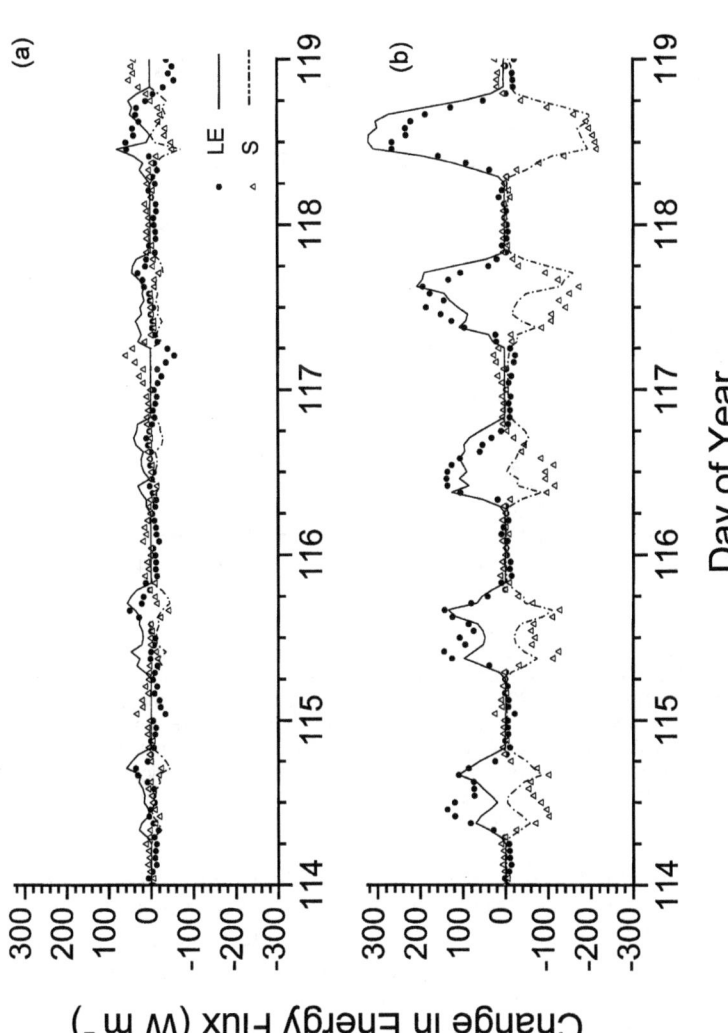

Figure 6.3 Differences in latent and sensible heat fluxes under 548 vs. 363 μmol mol⁻¹ CO_2 with (a) 35 g N m⁻² and (b) 7 g N m⁻² fertilization simulated (lines) and measured (symbols) in the Free Air CO_2 Enrichment (FACE) experiment at Phoenix, AZ, from 19 to 28 April 1996 (DOY 110 to 119). Positive values represent reductions in upward fluxes. (From Grant, R.F., B.A. Kimball, T.J. Brooks, G.W. Wall, P.J. Pinter Jr., D.J. Hunsaker, F.J. Adamsen, R.L. Lamorte, S.J. Leavitt, T.L. Thompson, and A.D. Matthias. 2001c. Interactions among CO_2, N and climate on energy exchange of wheat: model theory and testing with a free air CO_2 enrichment (FACE) experiment. *Agron. J.*, in press. With permission.)

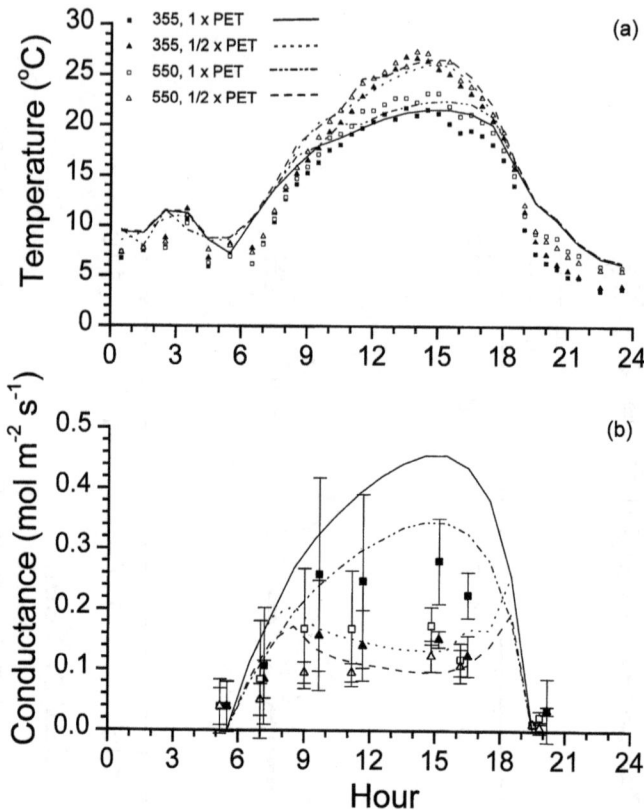

Figure 6.4 (a) Canopy temperature and (b) leaf stomatal conductance simulated (lines) and measured (symbols) in the Free Air CO_2 Enrichment (FACE) experiment at Phoenix, AZ, during 5 April 1994 under 355 vs. 550 µmol mol[-1] atmospheric CO_2 concentration and high (1 × PET) vs. low (½ × PET) irrigation. (From Grant, R.F., G.W. Wall, B.A. Kimball, K.F.A. Frumau, P.J. Pinter Jr., D.J. Hunsaker, and R.L. Lamorte. 1999b. Crop water relations under different CO_2 and irrigation: testing of *ecosys* with the Free Air CO_2 Enrichment (FACE) experiment. *Agric. For. Meteorol.*, 95:27-51. With permission.)

current water potentials (Figure 6.7). Canopy water potential controls transpiration by determining canopy turgor (calculated as the difference between canopy total and osmotic potentials from Equations [(B.1) to (B.2)] which affects stomatal conductance (as described under "Ecosystem-Atmosphere Energy Exchange" above) and thereby canopy temperature, vapor pressure, and conductance to vapor transfer. Canopy water potential also controls soil-root water uptake by determining canopy-root-soil water potential gradients to which uptake in the model is directly proportional [Equations (B.9) to (B.10)]. Soil water potential is the sum of matric, gravimetric, and osmotic fractions, the last of which is determined from solute concentrations (from [Equations (D.1) to (D.55)]; also Grant, 1995b). Soil-root water uptake is also directly proportional to soil-root and root-canopy hydraulic conductances in each rooted soil layer [Equations (B.11) to (B.15)]. Soil-root conductance is calculated from root length given by a root growth sub-model driven by shoot-root C

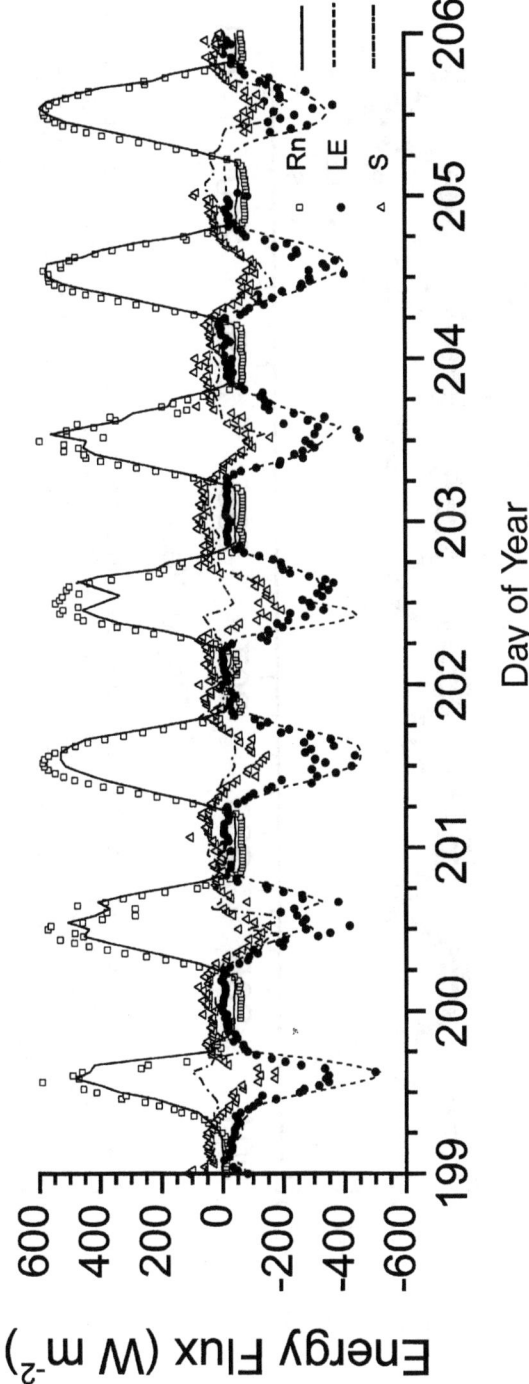

Figure 6.5 Net radiation (Rn), latent (LE) and sensible (S) heat fluxes simulated (lines) and measured (symbols) over the aspen overstory of a mixed aspen-hazelnut stand in the southern study area of BOREAS during late July 1994. (From Grant, R.F., T.A. Black, G. den Hartog, J.A. Berry, S.T. Gower, H.H. Neumann, P.D. Blanken, P.C. Yang, and C. Russell. 1999a. Diurnal and annual exchanges of mass and energy between an aspen-hazelnut forest and the atmosphere: testing the mathematical model *ecosys* with data from the BOREAS experiment. *J. Geophys. Res.*, 104:27,699-27,717. With permission.)

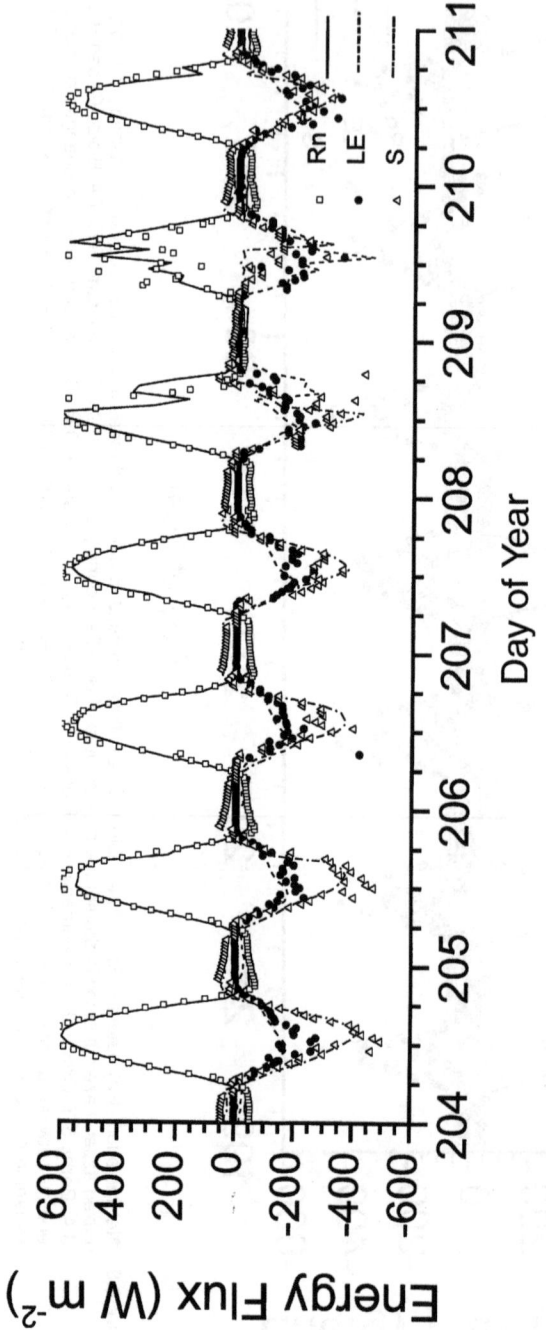

Figure 6.6 Net radiation (Rn), latent (LE) and sensible (S) heat fluxes simulated (lines) and measured (symbols) over a black spruce-moss stand in the southern study area of BOREAS during late July 1994. (From Grant, R.F., P.G. Jarvis, J.M. Massheder, S.E. Hale, J.B. Moncrieff, M. Rayment, S.L. Scott, and J.A. Berry. 2001a. Controls on carbon and energy exchange by a black spruce — moss ecosystem: testing the mathematical model *ecosys* with data from the BOREAS experiment. *Global Biogeochem. Cycles*, in press. With permission.)

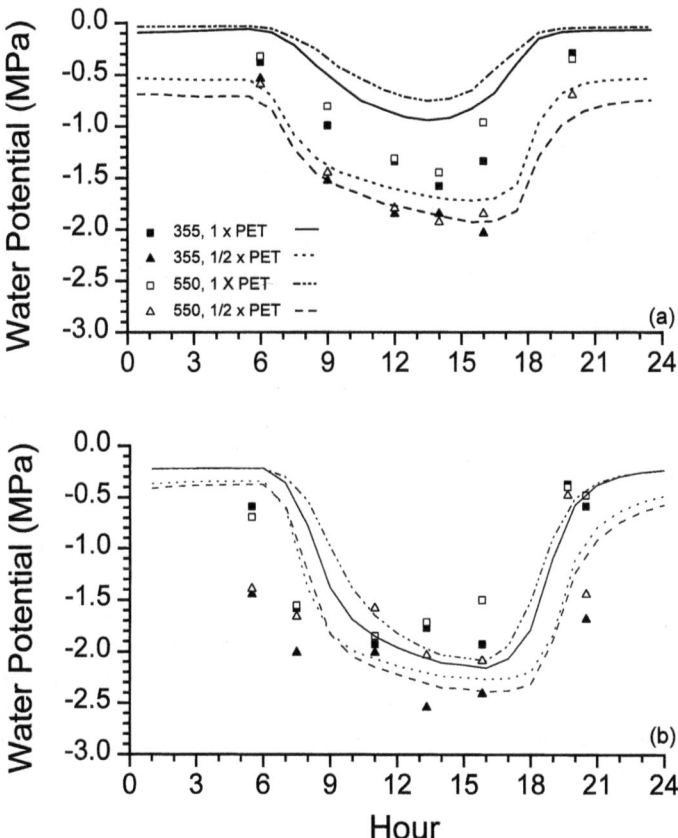

Figure 6.7 Canopy water potential simulated (lines) and measured (symbols) in the Free Air CO₂ Enrichment (FACE) experiment at Phoenix, AZ, during (a) 5 April 1994 and (b) 30 April 1994 under 355 vs. 550 µmol mol⁻¹ atmospheric CO_2 concentration and high (1 × PET) vs. low (½ × PET) irrigation. (From Grant, R.F., G.W. Wall, B.A. Kimball, K.F.A. Frumau, P.J. Pinter Jr., D.J. Hunsaker, and R.L. Lamorte. 1999b. Crop water relations under different CO_2 and irrigation: testing of *ecosys* with the Free Air CO₂ Enrichment (FACE) experiment. *Agric. For. Meteorol.,* 95:27-51. With permission.)

transfers ([Equations (F.1) to (F.47)], also Grant, 1993a,b; Grant 1998b; Grant and Robertson, 1997), and from soil-root hydraulic conductivity calculated according to Cowan (1965) for an assumed horizontally uniform root distribution in each soil layer [Equation (B.11)]. Root-canopy conductance is calculated from radial and axial root conductances (Reid and Huck, 1990) and from lengths of primary and secondary roots [Equations (B.12) to (B.14)] as described in Grant (1998b). If the convergence criterion for difference between canopy transpiration and uptake vs. change in canopy water content is not met, the canopy energy balance is solved again using an adjusted value for canopy water potential. The convergence is then tested again using new values for transpiration, uptake, and change in canopy water content calculated from the adjusted water potential.

The stomatal model used in the convergence solution does not include an explicit relationship between conductance and atmospheric humidity as included in other such models (e.g., Ball, 1988). However, lower humidity in *ecosys* causes more rapid transpiration, and therefore lower canopy water and turgor potentials and lower stomatal conductance. The sensitivity of the stomatal response to humidity in this model will be higher at lower soil water potentials because conductance is more sensitive to turgor when turgor is lower. This response of conductance to humidity is consistent with the hypothesis of Monteith (1995) that stomatal conductance responds to transpiration rate rather than to atmospheric humidity.

Model Testing

The sensitivity of the water relations algorithm to C_a and soil water content has been tested with results from a Free Air CO_2 Enrichment experiment (FACE) on wheat. The algorithm simulated rises in canopy water potential of 0.2 to 0.3 MPa in wheat when C_a was raised by 200 µmol mol^{-1} above ambient concentrations (Figure 6.7). These rises in water potential, which were within the standard error of measured rises (Grant et al., 1999b), were modeled from reduced stomatal conductances and hence reduced soil-canopy water fluxes simulated with constant C_i:C_a under elevated C_a. The algorithm simulated declines in wheat canopy water potential of 1 MPa in the FACE experiment when irrigation was reduced from non-limiting rates to one half of evapotranspirational demand (Figure 6.7). These declines, which were consistent with measured declines (Grant et al., 1999b), were modeled from the effects of lower soil water potentials and of lower soil and root hydraulic conductivities on soil-root-canopy water uptake when coupled to canopy-atmosphere transpiration.

Canopy C Fixation

Model Description

After successful convergence solutions for current canopy temperature and water potential, leaf carboxylation rates are adjusted from those calculated under non-limiting water potential to those under current water potential. This adjustment is required by the decrease in current stomatal conductance from its maximum value [calculated in Equation (A.14) of "Ecosystem-Atmosphere Energy Exchange" above] to that at current turgor [calculated in Equation (A.13) from Equation (B.1) of "Canopy Water Relations" above]. The adjustment is achieved through a convergence solution for current C_i and its aqueous equivalent at which the diffusion rate of gaseous CO_2 [Equations (C.11) to (C.16)] equals the carboxylation rate of aqueous CO_2 [Equations (C.1) to (C.10)]. The diffusion rate is calculated from the $C_a - C_i$ concentration difference multiplied by the current stomatal conductance (calculated under "Canopy Water Relations" above). The carboxylation rate is calculated in the same way as that under non-limiting water potential described above, but using C_i from the convergence solution [Equation (C.17)] rather than that from the C_i:C_a ratio. Current C_i will converge at values close to the C_i:C_a ratio when decreases in current conductance from maximum values are small (e.g., well-watered conditions), and at lower values when they are larger (e.g., water deficit conditions). The CO_2 fixation rate of each leaf surface at convergence is added to arrive at a value for gross canopy CO_2 fixation by each branch of each plant species in the model.

Model Testing

The sensitivity of the C fixation algorithm to wide ranges of C_a, O_2, temperature and irradiance has been tested at the leaf level against phytotron (Grant, 1989; 1992a) and field (Grant et al., 1989; 1999a,b) data. At the canopy scale the algorithm simulated an increase in net CO_2 exchange from 35 to 65 $\mu mol\ m^{-2}\ s^{-1}$ and a decrease in transpiration from 14 to 11 $mmol\ m^{-2}\ s^{-1}$ when C_a was raised from 330 to 800 $\mu mol\ mol^{-1}$ over irrigated soybean. These changes were consistent with those measured in outdoor controlled environment chambers (Grant, 1992b). The C fixation algorithm also simulated midday increases in CO_2 fixation by wheat of 5 and 10 $\mu mol\ m^{-2}\ s^{-1}$ under non-limiting and limiting irrigation rates respectively in the FACE experiment when C_a was raised by 200 $\mu mol\ mol^{-1}$. These increases agreed with increases measured in a flux chamber (Grant et al., 1995a) (Figure 6.8). In forests the same C fixation algorithm simulated 0.8 of the diurnal variation in net CO_2 exchange measured by Black et al. (1996) with eddy correlation over a mixed aspen-hazelnut stand (Grant et al., 1999a) (Figure 6.9) and by Jarvis et al. (1997) over a

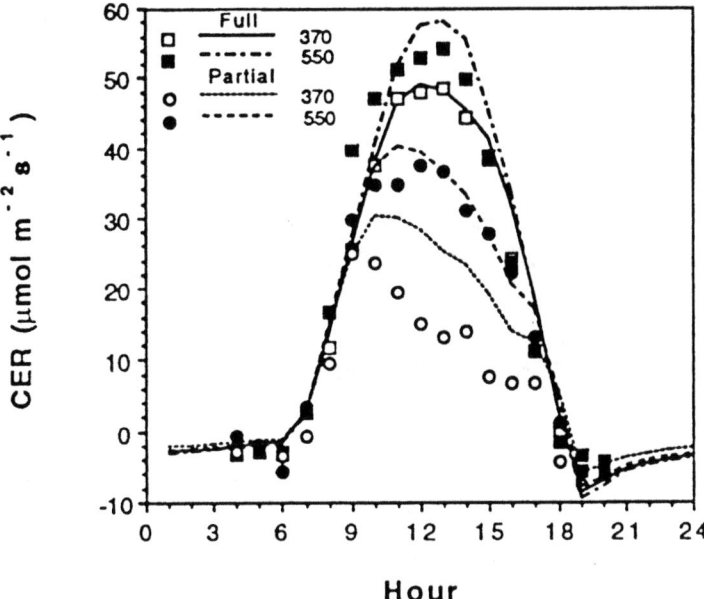

Figure 6.8 Canopy CO_2 fixation simulated (lines) and measured (symbols) in the Free Air CO_2 Enrichment (FACE) experiment at Phoenix, AZ, during 16 March 1993 under 355 vs. 550 $\mu mol\ mol^{-1}$ atmospheric CO_2 concentration and full (1 × PET) vs. partial (½ × PET) irrigation. (From Grant, R.F., R.L. Garcia, P.J. Pinter Jr., D. Hunsaker, G.W. Wall, B.A. Kimball, and R.L. LaMorte. 1995a. Interaction between atmospheric CO_2 concentration and water deficit on gas exchange and crop growth: Testing of *ecosys* with data from the Free Air CO_2 Enrichment (FACE) experiment. *Global Change Biol.*, 1:443-454. With permission.)

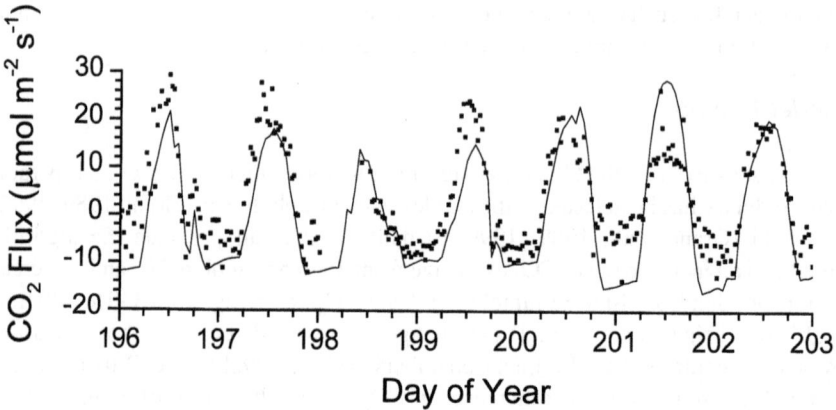

Figure 6.9 CO_2 fluxes simulated (lines) and measured (symbols) over the aspen overstory of a mixed aspen-hazelnut stand in the southern study area of BOREAS during late July 1994. (From Grant, R.F., T.A. Black, G. den Hartog, J.A. Berry, S.T. Gower, H.H. Neumann, P.D. Blanken, P.C. Yang, and C. Russell. 1999a. Diurnal and annual exchanges of mass and energy between an aspen-hazelnut forest and the atmosphere: testing the mathematical model *ecosys* with data from the BOREAS experiment. *J. Geophys. Res.*, 104:27,699-27,717. With permission.)

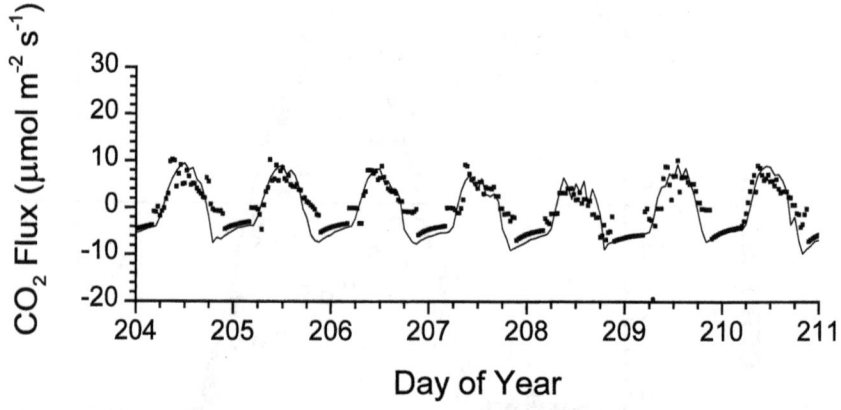

Figure 6.10 CO_2 fluxes simulated (lines) and measured (symbols) over a black spruce-moss stand in the southern study area of BOREAS during late July 1994. (From Grant, R.F., P.G. Jarvis, J.M. Massheder, S.E. Hale, J.B. Moncrieff, M. Rayment, S.L. Scott, and J.A. Berry. 2001a. Controls on carbon and energy exchange by a black spruce — moss ecosystem: testing the mathematical model *ecosys* with data from the BOREAS experiment. *Global Biogeochem. Cycles*, in press. With permission.)

nearby black spruce-moss stand (Grant et al., 2001a) (Figure 6.10). As for those in energy exchange described above, these changes in C fixation were modeled from differing sensitivities to C_i of leaf carboxylation rates and $C_a - C_i$ differences, radiation, and temperature that caused leaf stomatal conductance to decrease under higher C_a, lower radiation, and lower temperature.

Canopy Respiration and Senescence

Model Description

The product of CO_2 fixation is added to a mobile pool of C stored each branch of each plant species from which C is oxidized to meet maintenance respiration requirements using a first-order function of storage C [Equations (B.3) and (B.4)]. If the C storage pool is depleted, the C oxidation rate may be less than the maintenance respiration rate, in which case the difference is made up through respiration of remobilizable C [Equation (B.6)], considered to be 0.5 of protein C in leaves and supporting organs (twigs, petioles, or sheaths) (Kimmins, 1987). Protein N and P associated with remobilized protein C are withdrawn from leaves and supporting organs, and added to mobile pools of N and P stored in each branch of the plant from which N and P can be withdrawn to support new leaf and twig growth as described under "Plant Growth" below. Protein N and P can also be remobilized when ratios of storage N or P to storage C become lower than those required for synthesis of new phytomass. Remobilization starts at the lowest node at which leaves and twigs are present, and proceeds upward. Upon exhaustion of the remobilizable C, N, and P in each leaf or twig, the remaining C, N, and P are dropped from the branch and added to detritus at the soil surface from which they undergo decomposition as described under "Soil Microbial Activity" below. Environmental constraints such as nutrient, heat or water stress that reduce C fixation and hence C storage with respect to maintenance respiration will therefore accelerate the loss of leaf and twig C, N, and P from the branch. Storage C oxidized in excess of maintenance respiration requirements is used as growth respiration [Equation (B.5)] to drive the formation of new biomass [Equation (B.7)] as described under "Plant Growth" below. Net CO_2 fixation is calculated for each branch of each plant species as the difference between gross fixation and the sum of maintenance, growth, and senescence respiration [Equation (B.8)].

Model Testing

Both elevated C_a and reduced N fertilization caused lower ratios of storage N to storage C, and thereby more rapid remobilization of protein C, N, and P from leaves and sheaths in the model. This in turn caused more rapid late-season leaf senescence that was consistent with that measured under elevated C_a and reduced N fertilization in a FACE experiment (Grant et al., 2001c) (Figure 6.11).

Nutrient Uptake

Model Description

Nutrient (N and P) uptake is calculated for each plant species by solving for aqueous concentrations at its root and mycorrhizal surfaces in each soil layer at which radial transport by mass flow and diffusion from the soil solution to the surfaces equals active uptake by the surfaces [Equation (F.25)]. This solution dynamically

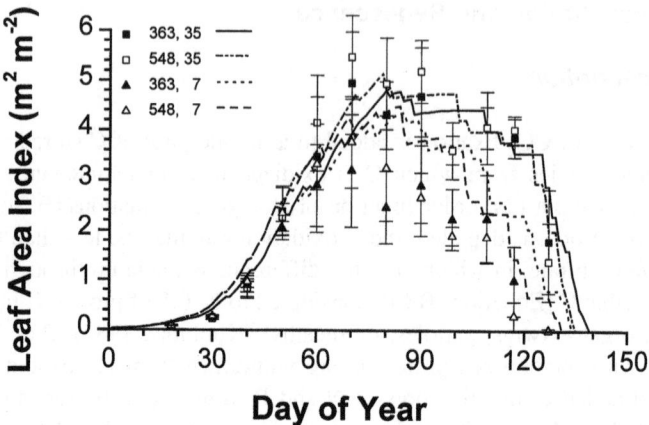

Figure 6.11   LAI under 548 vs. 363 µmol mol⁻¹ CO_2 with (a) 35 g N m⁻² and (b) 7 g N m⁻² fertilization simulated (lines) and measured (symbols) in the Free Air CO_2 Enrichment (FACE) experiment at Phoenix, AZ, during 1996. (From Grant, R.F., B.A. Kimball, T.J. Brooks, G.W. Wall, P.J. Pinter Jr., D.J. Hunsaker, F.J. Adamsen, R.L. Lamorte, S.J. Leavitt, T.L. Thompson, and A.D. Matthias. 2001c. Interactions among CO_2, N and climate on energy exchange of wheat: model theory and testing with a free air CO_2 enrichment (FACE) experiment. *Agron. J.,* in press. With permission.)

links rates of soil nutrient transformations with those of root and mycorrhizal nutrient uptake. These transformations control the aqueous concentrations of N and P in each soil layer through thermodynamically driven precipitation, adsorption, and ion pairing reactions ([Equations (D.1) to (D.55); also Grant and Heaney, 1997], convective-dispersive solute transport [shown in Equations (F.39) and (F.40)] for O_2, but also for all aqueous and gaseous solutes] and microbial mineralization-immobilization [Equation (E.23)]. Solution N and P concentrations determine mass flow during root water uptake [Equation (F.25)] described under "Canopy Water Relations" above. They also determine diffusion through concentration gradients between the soil solution and root and mycorrhizal surfaces that are generated by active root and mycorrhizal uptake [Equation (F.25)]. Active uptake is calculated from length densities and surface areas (Itoh and Barber, 1983) [Equation (F.26)] given by a root and mycorrhizal growth sub-model [Equations (F.1) to (F.47)], also Grant, 1993a,b; Eqs. (1) to (17) in Grant 1998a; Grant and Robertson, 1997) assuming uniform horizontal distributions within each soil layer. Active nutrient uptake is constrained by O_2 uptake, by solution N and P concentrations, and by root and mycorrhizal C, N, and P storage [Equations (F.25) to (F.29)]. The products of N and P uptake are added to mobile pools of N and P stored in each root and mycorrhizal layer. Growth respiration drives the combination of N and P from these pools with storage C to form new plant biomass as described under "Plant Growth" below.

The solution concentrations of N and P that occur in the model depend on site conditions. Under the nutrient-limited conditions found in most natural ecosystems, rates of nutrient transport to and uptake by root and mycorrhizal surfaces are constrained by rates of nutrient mineralization, desorption, and dissolution from organic and inorganic sources. These conditions can cause nutrient concentrations to be drawn down to very low values ($\ll 1$ g m^{-3} as in Grant and Robertson, 1997). Under the nutrient-rich conditions found in heavily fertilized agricultural crops, larger nutrient concentrations occur, hastening nutrient transport and uptake.

Model Testing

The root uptake algorithm simulated seasonal N uptake in wheat that increased from 13 to 20 and 25 g m^{-2} when fertilizer N was increased from 0 to 6 and 16 g m^{-2}, which did not differ significantly from measured uptake (Grant, 1991). The root uptake algorithm also simulated seasonal P uptake in barley that increased from 0.8 to 1.5 g m^{-2} when fertilizer P was increased from 0 to 2 g m^{-2}, which was in both cases about 0.2 g m^{-2} higher than measured uptake (Grant and Robertson, 1997) (Figure 6.12). Increased uptake with fertilizer was simulated from higher soil nutrient concentrations that drove more rapid mass flow and diffusion to, and active uptake by, root and mycorrhizal surfaces as described under "Nutrient Uptake" above.

Plant Growth

Model Description

Growth respiration from "Canopy Respiration and Senescence" above [Equation (B.5)] drives expansive growth of vegetative and reproductive organs through mobilization of storage C, N, and P in each branch of each plant species according to phenology-dependent partitioning coefficients and biochemically based growth yields [Equation (B.7)]. This growth is used to simulate the lengths, areas, and volumes of individual internodes, twigs, petioles, sheaths, and leaves (Grant, 1994b; Grant and Hesketh, 1992), from which heights and areas of leaf and stem surfaces are calculated for irradiance interception and aerodynamic conductance algorithms as described in "Ecosystem-Atmosphere Energy Exchange" above. Growth respiration also drives extension of primary and secondary root axes and of mycorrhizal axes of each plant species in each soil layer [Equation (F.15)] through mobilization of storage C, N, and P in each root zone of each plant species [Equations (F.10) and (F.11)] as described in Grant (1993a; 1998b). This growth is used to calculate lengths and areas of root and mycorrhizal axes [Equation (F.16)] from which root uptake of water [Equation (B.10)]; Grant et al., 1999b) and nutrients [Equation (F.25)] (Grant, 1991; Grant and Robertson, 1997) is calculated as described above.

The growth of different branch organs and root axes in the model depends on transfers of storage C, N, and P among branches, roots, and mycorrhizae. These transfers are driven from concentration gradients within the plant that develop from different rates of C, N, or P acquisition and consumption by its branches, roots, or mycorrhizae [Equations (F.18) to (F.24)]. Storage C:N:P ratios are thus determined by the comparative availability of ecological resources for C fixation (C$_a$, radiation, heat, water) and N or P uptake (soil mineral N or P concentrations). Very low ratios

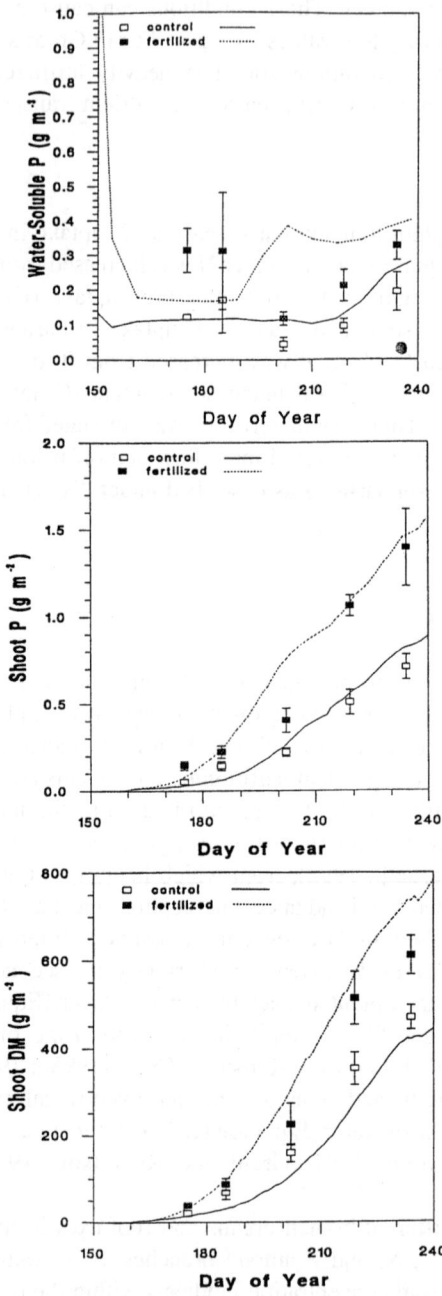

Figure 6.12   (a) Soluble phosphate-P concentrations, (b) P uptake, and (c) plant growth simulated (lines) and measured (symbols) in soil fertilized with 0 vs. 2 g P m⁻² as superphosphate at Ellerslie, Alberta. (From Grant, R.F., and J.A. Robertson. 1997. Phosphorus uptake by root systems: mathematical modelling in *ecosys*. *Plant Soil,* 188:279-297. With permission.)

of storage N or P to storage C in branches indicate that N or P uptake, rather than C fixation, is constraining branch growth. Such ratios in the model have two effects:

1. They reduce the specific activities of rubisco and chlorophyll from set maximum values through a product inhibition function (Thornley, 1995) according to the extent that these ratios are below the ratios of N or P to C required for growth. This function simulates the deactivation of rubisco under low N or high C_a caused by leaf carbohydrate accumulation (reviewed by Stitt, 1991).
2. They reduce N:C and P:C ratios at which leaf growth occurs from set maximum values, thereby simulating the reduction in foliar nutrient concentrations observed under nutrient-limited conditions. This reduction occurs in the model because storage C, N, and P are the direct substrates for leaf growth in each branch, and hence their ratios affect those of the leaves during growth. Lower N:C and P:C ratios in leaves result in lower areal concentrations and hence in areal activities of rubisco and chlorophyll because rubisco and chlorophyll N are assumed to be constant fractions of total leaf N. Lower areal activities of rubisco and chlorophyll cause lower leaf C fixation rates [Equations (C.3) and (C.8)] and hence lower stomatal conductances in order to conserve the $C_i:C_a$ ratio.

Low ratios of storage N or P to storage C in branches cause less storage C, N, and P to be used for leaf growth, and thereby cause root-shoot concentration gradients of N and P to become smaller, and those of storage C to become larger. These changes result in smaller root-shoot transfers of N and P and in larger shoot-root transfers of C, thereby allowing more plant resources to be used in root growth. Such use has been observed experimentally under conditions of inadequate N or P uptake and excess C fixation (Makino et al., 1997; Rogers et al., 1993). The consequent increase in root:shoot ratios and thus in N and P uptake, coupled with the decrease in C fixation rate, redresses to some extent the storage C:N:P imbalance under limiting N or P. The model thus implements the algorithm for functional equilibrium between roots and shoots proposed by Thornley (1995).

For perennial deciduous plant species, soluble C, N, and P are withdrawn from storage pools in branches into a long-term storage pool in the crown during autumn, causing leaf senescence. Soluble C, N, and P are remobilized from this pool to drive leaf and twig growth the following spring. The timing of withdrawal and remobilization is determined by duration of exposure to low temperatures (between 3 and 8°C) under shortening and lengthening photoperiods, respectively.

Model Testing

The plant growth algorithm simulated increases in peak growth rates of well-fertilized wheat from 11.9 to 14.2 g C m^{-2} d^{-1} and from 8.3 to 10.1 g C m^{-2} d^{-1} when C_a in a FACE experiment was increased from 363 to 548 µmol mol^{-1} under non-limiting and limiting irrigation, respectively (Grant et al., 1999b) (Figure 6.13a). The plant growth algorithm also simulated increases in peak growth rates of well-irrigated wheat from 11.4 to 12.9 g C m^{-2} d^{-1} and from 9.1 to 9.3 g C m^{-2} d^{-1} when C_a in a FACE experiment was increased from 363 to 548 µmol mol^{-1} under non-limiting and limiting N fertilization, respectively (Grant et al., 2001c) (Figure 6.13b).

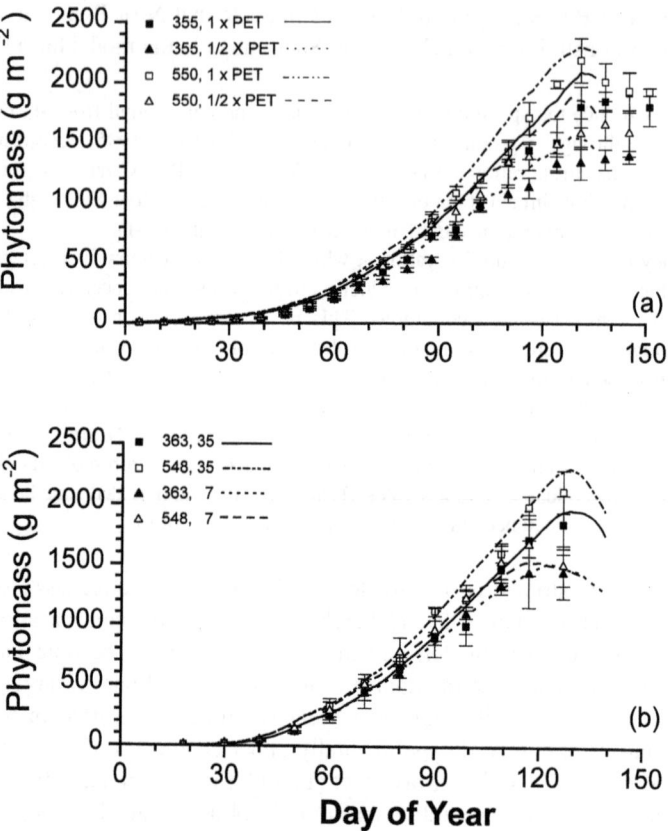

Figure 6.13 Phytomass under 548 vs. 363 μmol mol⁻¹ CO_2 with (a) high (1 × PET) vs. low (½ × PET) irrigation and (b) 35 vs. 7 g N m⁻² fertilization simulated (lines) and measured (symbols) in the Free Air CO_2 Enrichment (FACE) experiment at Phoenix, AZ, during (a) 1994. (From Grant, R.F., T.A. Black, G. den Hartog, J.A. Berry, S.T. Gower, H.H. Neumann, P.D. Blanken, P.C. Yang, and C. Russell. 1999a. Diurnal and annual exchanges of mass and energy between an aspen-hazelnut forest and the atmosphere: testing the mathematical model *ecosys* with data from the BOREAS experiment. *J. Geophys. Res.*, 104:699-717. With permission.) (b) 1996. (From Grant, R.F., B.A. Kimball, T.J. Brooks, G.W. Wall, P.J. Pinter Jr., D.J. Hunsaker, F.J. Adamsen, R.L. Lamorte, S.J. Leavitt, T.L. Thompson, and A.D. Matthias. 2001c. Interactions among CO_2, N and climate on energy exchange of wheat: model theory and testing with a free air CO_2 enrichment (FACE) experiment. *Agron. J.*, in press. With permission.)

These rates were within the standard errors of those measured. Over a longer time scale, the same algorithm simulated growth rates of 96 and 59 g C m⁻² y⁻¹ by nutrient-limited boreal aspen (Figure 6.14a) and black spruce (Figure 6.14b), respectively, in Saskatchewan which was within the range estimated from allometric studies (Grant et al., 1999a; 2001a). These growth rates were modeled from C fixation and respiration, and from the consequent growth of leaves, stems, reproductive organs, reserves, and roots. *Ecosys* was then used to project changes in wood and soil C accumulation in these forests under 150 years of IS92a climate change.

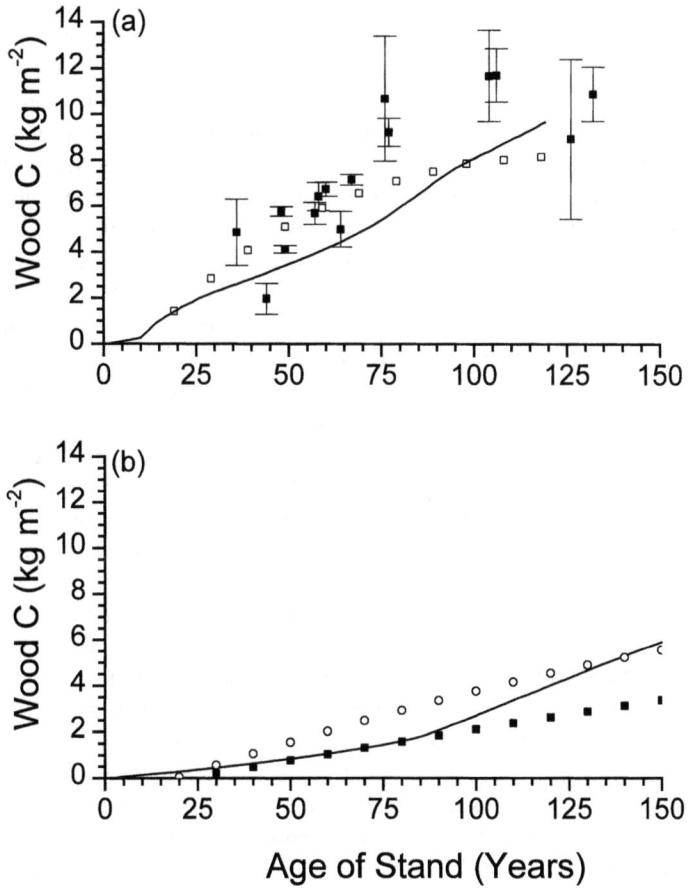

Figure 6.14 (a) Growth of aspen wood simulated in the southern study area of BOREAS (line) and calculated from allometric measurements of differently aged stands in Prince Albert National Park, Saskatchewan. (From Grant, R.F., and I.A. Nalder. 2000. Climate change effects on net carbon exchange of a boreal aspen-hazelnut forest: estimates from the ecosystem model *ecosys. Global Change Biol.,* 6:183-200. With permission.) (b) Growth of black spruce wood simulated in the southern study area of BOREAS (line) and calculated by the Alberta Forest Service (1985) at fair and medium sites (symbols) in Alberta. (From Grant, R.F., P.G. Jarvis, J.M. Massheder, S.E. Hale, J.B. Moncrieff, M. Rayment, S.L. Scott, and J.A. Berry. 2001a. Controls on carbon and energy exchange by a black spruce — moss ecosystem: testing the mathematical model *ecosys* with data from the BOREAS experiment. *Global Biogeochem. Cycles,* in press. With permission.)

Soil Microbial Activity

Model Development

The modeling of microbial activity is based on six organic states: solid, soluble, sorbed, acetate (for methanogenesis), microbial biomass, and microbial residues. C, N, and P may move among these states within each of four organic matter-microbe complexes: plant litterfall, animal manure, particulate organic matter, and humus. Each complex is resolved into hierarchical levels of biological organization: organic matter-microbe complex, functional type within each complex (microbial populations of obligately aerobic heterotrophs, facultatively anaerobic heterotrophs (denitrifiers), obligately aerobic heterotrophs (fermenters), heterotrophic methanogens, aerobic diazotrophs, anaerobic diazotrophs, autotrophic nitrifiers, autotrophic methanogens, and autotrophic methanotrophs), structural or kinetic components within each complex or functional type, and elemental fraction within each structural or kinetic component. Transformations among organic states in each organic matter-microbe complex simulated in *ecosys* are shown in Figure 6.15.

Litterfall from "Canopy Respiration and Senescence" above is partitioned into carbohydrate, protein, cellulose, and lignin components according to Trofymow et al. (1995), each of which is of differing vulnerability to hydrolysis by heterotrophic decomposers (Table 3 in Grant et al., 1999a). Soil organic matter is partitioned into active and humus components of differing vulnerability to hydrolysis. The rate at which each component is hydrolyzed is a first-order function of the active biomass of obligately aerobic, facultatively anaerobic, and obligately anaerobic heterotrophic decomposers associated with each component ([Equations (E.1) and (E.2)], Grant et al., 1993a,b) as affected by the temperatures and water contents of surface detritus, and those of a spatially resolved soil profile (Grant, 1997; Grant and Rochette, 1994; Grant et al., 1998). Microbial biomass in *ecosys* is thus an active agent of organic matter transformation rather than a passive organic state as in most other ecosystem models. The rate at which each component is hydrolyzed is also a function of substrate concentration that approaches a first-order function at low concentrations, and a zero-order function at high concentrations [Equations (E.3) and (E.4)]. These rates are controlled by soil temperature through an Arrhenius function [Equation (E.5)] and by soil water content through its effect on microbial concentration [Equations (E.3) and (E.4)]. Soil temperatures and water contents are calculated from the surface energy balance and the heat and water transfer scheme through a snow-residue-soil profile described under "Ecosystem-Atmosphere Energy Exchange" above (Grant et al., 1995b). Residue hydrolysis products undergo humification at a rate that depends on that of residue lignin hydrolysis and on soil clay concentration [Equations (E.27) and (E.28)].

The concentration of the soluble hydrolysis products drives C oxidation by each decomposer population [Equations (E.11) and (E.12)], the total of which drives CO_2 emission from the soil surface. Heterotrophic oxidation rates may be constrained by O_2 [Equation (E.13)] and nutrients [Equation (E.16)]. This oxidation is coupled to the reduction of O_2 by all aerobic populations [Equations (H.1) and (H.2)]; Grant et al., 1993a,b; Grant and Rochette, 1994), to the sequential reduction of NO_3^-, NO_2^- and N_2O by heterotrophic denitrifiers) [Equations (H.3) to (H.6)]; Grant et al., 1993c,d; Grant and Pattey, 1999) and to the reduction of organic C by fermenters and heterotrophic methanogens ([Equations (G.1) to (G.11)]; Grant, 1998a). The energetics of these oxidation-reduction reactions [Equations (E.14), (G.4), (G.10), and (H.12)] determine heterotrophic growth yields and hence biomass growth [Equations (E.25), (G.6), (G.11), and (H.13)] from which heterotrophic decomposer activity is calculated. All soluble and gaseous reactants and products undergo

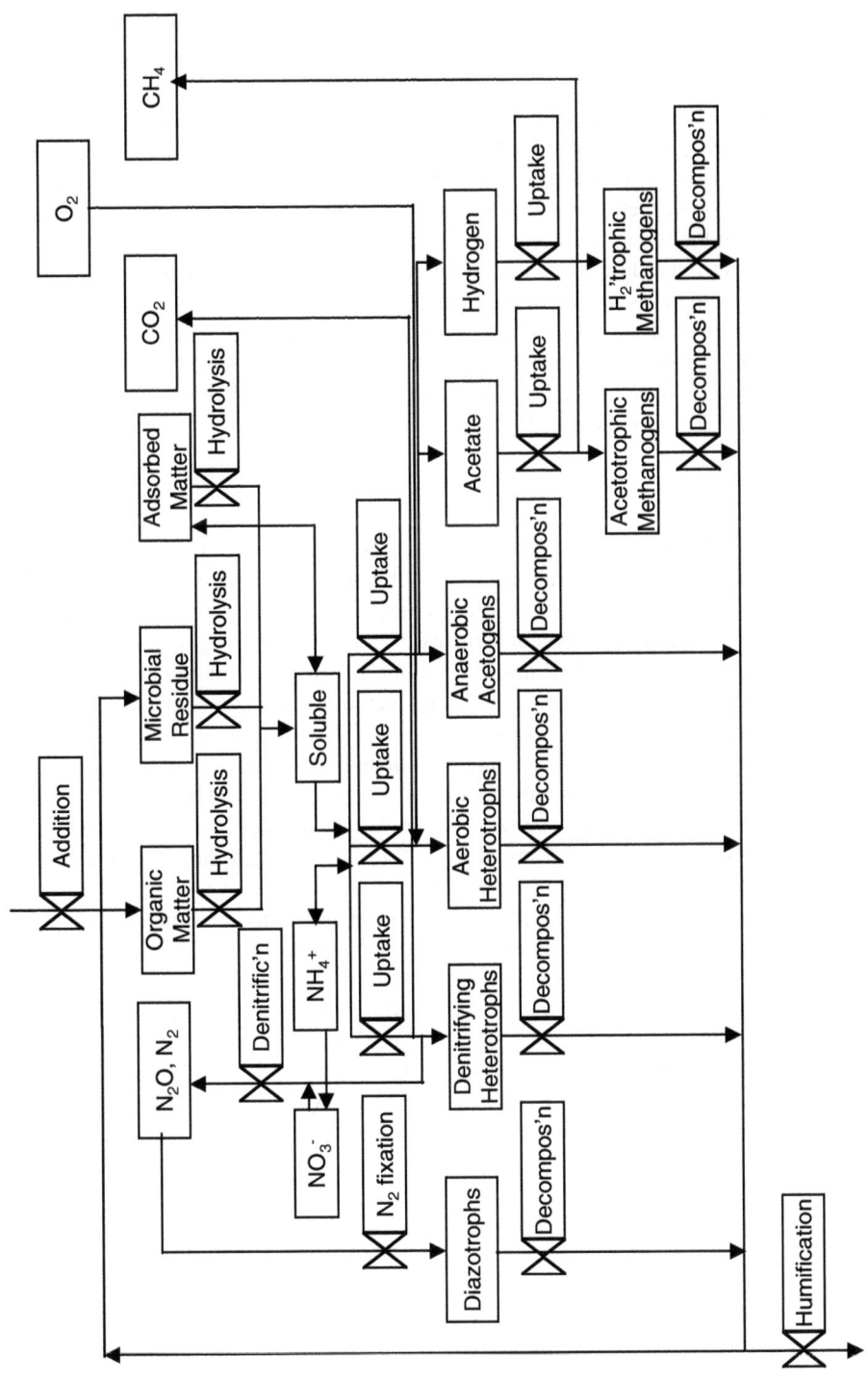

Figure 6.15 Organic transformations in each organic matter–microbe complex simulated in *ecosys*.

convective-dispersive transport through the soil profile (as shown for O_2 in ([Equations (F.37) to (F.47)]; Grant et al., 1993c,d; Grant and Heaney, 1997). In addition, autotrophic nitrifiers conduct NH_4^+ oxidation and NO_3^- production [Equations (I.1) to (I.16)]; Grant, 1994a), and N_2O evolution ([Equations (I.17) to (I.19)]; Grant, 1995a), and autotrophic methanogens and methanotrophs conduct CH_4 reduction ([Equations (G.12) to (G.17)]; Grant, 1998a) and oxidation ([Equations (G.18) to (G.27)]; Grant, 1999), the energetics of which determine autotrophic growth yields and hence autotrophic biomass and activity. Microbial populations in the model seek to maintain equilibrium ratios of biomass C:N:P by mineralizing or immobilizing mineral N and P [Equation (E.23)], thereby regulating solution concentrations used to drive N and P uptake by roots and mycorrhizae as described under "Nutrient Uptake" above. Microbial populations undergo first-order decomposition [Equations (E.21) and (E.22)], the products of which are partitioned between humus and microbial residue according to a function of soil clay content ([Equation (E.29)]; Grant et al., 1993a,b).

Model Testing

The energetics that drive heterotrophic microbial activity in the model allowed temporal changes in the mineralization, immobilization, and stabilization of C and N to be simulated within 10% of recorded changes over time scales of hours, days, and years following laboratory amendments of glucose, cellulose, lignin, and plant detritus (Grant et al., 1993a,b) (Figure 6.16). The sensitivity of microbial respiration in the model to incremental changes in soil temperature and water content was tested with laboratory data in Grant and Rochette (1994). This sensitivity allowed diurnal changes in soil CO_2 efflux to be accurately simulated with differing soil temperatures under fallow vs. field-grown barley (Grant and Rochette, 1994). The energetics that drive heterotrophic denitrifier activity in the model allowed diurnal changes in surface N_2O fluxes from 0 to 1 mg N m^{-2} h^{-1} to be simulated during spring thaw events (Grant and Pattey, 1999) (Figure 6.17).

These same energetics of heterotrophic microbial activity, and their sensitivity to soil temperature and water content, were used to simulate gains in soil C of 10 to 20 g m^{-2} y^{-1} under reduced vs. conventional tillage and continuous vs. fallow wheat cropping systems that were within the standard errors of measured gains in soil C over 15 years in semiarid (Grant, 1997) and boreal (Grant et al., 1998) climates. They were also used to simulate gains in soil C of 15 to 20 g m^{-2} y^{-1} under a 5-year cereal-forage rotation vs. a 2-year wheat-fallow rotation over 70 years in a boreal climate (Grant et al., 2001b) (Figure 6.18).

FUTURE DIRECTIONS

Involvement of ecosystem models in multidisciplinary ecosystem-level research projects is absolutely necessary for their development because the diversity and

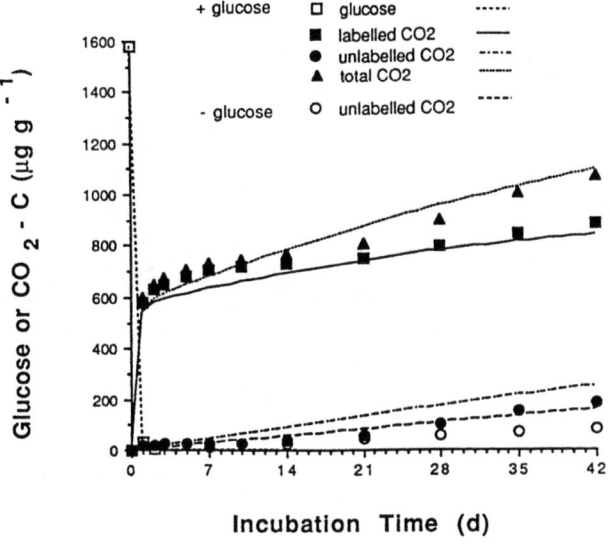

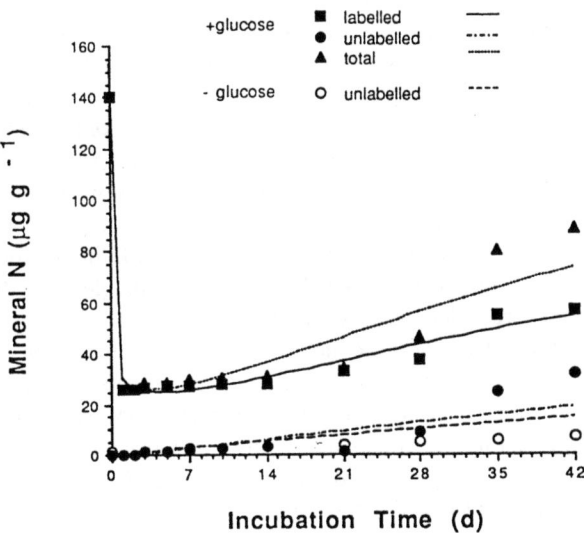

Figure 6.16 Simulated (lines) and measured (symbols) (a) accumulated CO_2-C evolved and glucose-C remaining and (b) mineral N remaining after additions of 0 and 1580 μg g^{-1} of glucose-^{14}C and 140 μg g^{-1} of $^{15}NO_3$-N to a Bradwell fine sandy loam. Measured data from Voroney and Paul (1984). (From Grant, R.F., N.G. Juma, and W.B. McGill. 1993a. Simulation of carbon and nitrogen transformations in soils. I. Mineralization. *Soil Biol. Biochem.*, 27:1317-1329. With permission.)

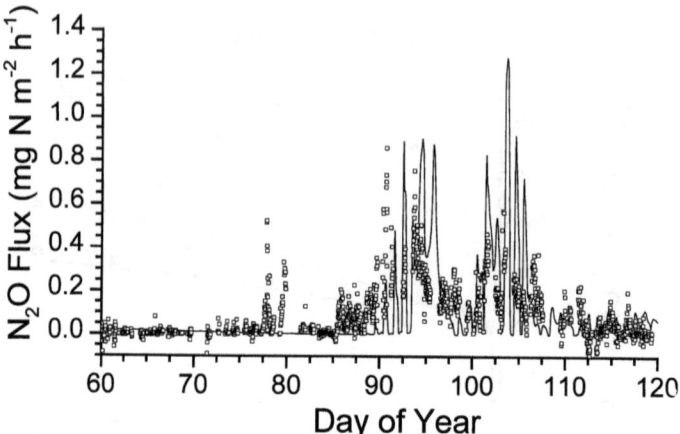

Figure 6.17 Surface N_2O fluxes simulated (lines) and measured (symbols) during March and April 1996 over an agricultural field near Ottawa, Ontario. (From Grant, R.F., and E. Pattey. 1999. Mathematical modelling of nitrous oxide emissions from an agricultural field during spring thaw. *Global Biogeochem. Cycles,* 13:679-694. With permission.)

temporal scale of process measurements in such projects enable well-constrained model tests of ecosystem function. The development of *ecosys* has greatly benefitted from its involvement in the FACE and BOREAS projects, some results from which are given here. Testing of C and energy exchange, and of net primary and ecosystem productivity, in *ecosys* is being extended to boreal wetlands as part of BOREAS, to arctic tundra and temperate coniferous ecosystems as part of the Ameriflux program, and to agricultural and forest ecosystems in tropical, temperate, semiarid, and arid zones as part of other climate change projects. This testing should not require any further model parameterization beyond that described above (in theory). Any changes to model performance in these tests should be confined to inputs for soil type, plant species, or land use practices for which values are available from the test site or in the literature that are independent of the model.

The next stage of this testing should involve the comparison of simulated vs. measured C and energy exchange in disturbed (cultivated, harvested, logged, or burned) vs. undisturbed ecosystems. Such testing is an essential prerequisite for model use in predicting the effects of land use practices on ecosystem C and water balances, and is currently planned as part of several proposed climate change projects. It is important that this testing be carried out over an extended period of time so that the impact of climate variation on C and energy exchange can be assessed. The prediction of disturbance effects at regional or national scales will require systematic linkages between *ecosys* and soil, climate, and land use databases in geographic information systems. These linkages are currently being studied.

Several important ecosystem processes are not yet represented in *ecosys*. Key among these are wind and water erosion, herbivory (except as preset partial harvests), and fire. Constraints on C cycling imposed by deficiencies of K and S are not yet simulated. Future consideration will have to be given to including these processes in the model.

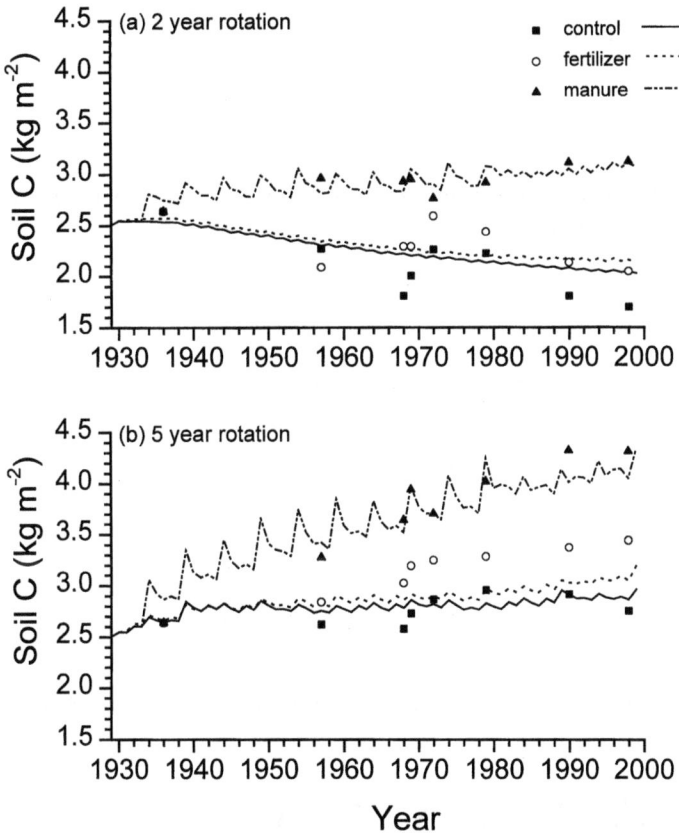

Figure 6.18 Soil C content measured (symbols) and simulated (lines) in the upper 0.15 m of the profile during 70 years of control, fertilizer and manure treatments in (a) a wheat-fallow (2-year) and (b) a wheat-oats-barley-forage-forage (5-year) rotation at Breton, Alberta. (From Grant, R.F., N.G. Juma, J.A. Robertson, R.C. Izaurralde, and W.B. McGill. 2001b. Long term changes in soil C under different fertilizer, manure and rotation: testing the mathematical model *ecosys* with data from the Breton plots. *Soil Sci. Soc. Am. J.*, in press. With permission.)

APPENDIX A: ENERGY EXCHANGE

Canopy

First-order closure of energy balance

$$R_{N_i} + LE_i + LV_i + H_i + S_i = 0 \qquad \text{(A.1)}$$

Net radiation from transfer of longwave and shortwave radiation

$$R_{N_i} = R_{S_i} + F_{P_i}\left(R_{L_a} + R_{L_g} - 2\,R_{L_i}\right) \qquad \text{(A.2)}$$

Latent heat (transpiration) from vapor pressure differences between mesophyll surfaces and the atmosphere

$$LE_i = F_{S_i} L(e_a - e_i) / (r_{A_i} + r_{C_i})$$
(A.3)

Latent heat (evaporation) from vapor pressure differences between wet canopy surfaces and the atmosphere

$$LV_i = F_{S_i} L(e_a - e_i') / r_{A_i}$$
(A.4)

Sensible heat from temperature differences between the plant canopy and the atmosphere

$$H_i = F_{S_i} c_a (T_a - T_i) / r_{A_i}$$
(A.5)

Change in heat stored in the canopy

$$S_i = c_i (T_{i(t-1)} - T_{i(t)}) / \Delta t$$
(A.6)

Shortwave radiation absorbed by canopy aggregated from that absorbed by each leaf surface

$$R_{S_i} = \sum_{j=1}^{J} \sum_{k=1}^{K} \sum_{l=1}^{L} \sum_{m=1}^{M} \sum_{n=1}^{N} \sum_{o=1}^{O} \left(R_{S_{i,l,m,n,o}} X_i A_{i,j,k,l,m,n,o} \right)$$
(A.7a)

PAR absorbed by canopy aggregated from that absorbed by each leaf surface

$$R_{P_i} = \sum_{j=1}^{J} \sum_{k=1}^{K} \sum_{l=1}^{L} \sum_{m=1}^{M} \sum_{n=1}^{N} \sum_{o=1}^{O} \left(R_{P_{i,l,m,n,o}} X_i A_{i,j,k,l,m,n,o} \right)$$
(A.7b)

Fraction of incident shortwave radiation absorbed by canopy

$$F_{S_i} = R_{S_i} / R_S$$
(A.8a)

Fraction of incident PAR absorbed by canopy

$$F_{P_i} = R_{P_i} / R_P$$
(A.8b)

Longwave radiation emitted from sky

$$R_{L_a} = \varepsilon_a \sigma T_a^4$$
(A.9)

Longwave radiation emitted from soil and residue surfaces

$$R_{L_g} = \Sigma_{g=s}^r F_g \left(\varepsilon_g \sigma T_g^4 \right) \tag{A.10}$$

Longwave radiation emitted from canopy

$$R_{L_i} = \varepsilon_i \sigma T_i^4 \tag{A.11}$$

Boundary layer resistance from wind speed and heights of zero plane displacement and surface roughness

$$r_{A_i} = \left\{ \left(\ln\left(\left(z_R - d_i + z_{C_i} \right) / z_{C_i} \right) \ln\left(\left(z_R - d_i + z_{V_i} \right) / z_{V_i} \right) / \left(K^2 u_a \right) \right\} \middle/ (1.0 - 10.0 \; Ri) \right. \tag{A.12}$$

Stomatal resistance to vapor diffusion at ambient water potential driven by canopy turgor

$$r_{C_i} = r_{C_i'} + \left(r_{C_{max}} - r_{C_i'} \right) e^{\left(-\beta \psi_{T_i} \right)} \tag{A.13}$$

Stomatal resistance to vapor diffusion at zero water potential driven by canopy CO_2 concentration difference and CO_2 fixation

$$r_{C_i'} = 0.64 \; F_{P_i} \left(C_{B_i} - C_{I_i'} \right) \middle/ V_{B_i'} \tag{A.14}$$

CO_2 fixation by canopy aggregated from that by all leaf surfaces

$$V_{B_i'} = \sum_{j=1}^J \sum_{k=1}^K \sum_{l=1}^L \sum_{m=1}^M \sum_{n=1}^N \sum_{o=1}^O \left(V_{B_{i,j,k,l,m,n,o}'} \; A_{i,j,k,l,m,n,o} \right) \tag{A.15}$$

Soil and Residue Surfaces

First-order closure of energy balance for both soil and residue surfaces

$$R_{N_g} + LE_g + H_g + S_g = 0 \tag{A.16}$$

Net radiation from transfer of longwave and shortwave radiation transmitted through canopy

$$R_{N_g} = F_g \left[R_{S_g} + \left(1 - F_{P_i} \right) R_{L_a} + F_{P_i} R_{L_i} \right] - R_{L_g} \tag{A.17}$$

Latent heat from vapor pressure differences between soil and residue surfaces and the atmosphere

$$LE_g = F_g \left[L\left(e_a - e_g\right) / r_{A_g} \right] \qquad (A.18)$$

Sensible heat from temperature differences between soil and residue surfaces and the atmosphere

$$H_g = F_g \left[c_a \left(T_a - T_g\right) / r_{A_g} \right] \qquad (A.19)$$

Change in heat stored in surface residue

$$S_r = c_r \left(T_{r(t-1)} - T_{r(t)}\right) / \Delta t + 2 \kappa_{S_{r,s}} \left(T_{s(t-1)} - T_{r(t)}\right) / \left(z_r + z_s\right) \qquad (A.20)$$

Change in heat stored in soil surface

$$S_s = c_s \left(T_{s(t-1)} - T_{s(t)}\right) / \Delta t + 2 \kappa_{S_{s,l}} \left(T_{l(t-1)} - T_{s(t)}\right) / \left(z_s + z_l\right) \qquad (A.21)$$

Boundary layer resistance over surface residue as affected by canopy

$$r_{A_r} = r_{A_i} + d_i / \left(D_{V_i} f_{D_i} c_{D_i}\right) \qquad (A.22)$$

Boundary layer resistance over soil surface as affected by canopy and surface residue

$$r_{A_s} = r_{A_r} + z_r / \left(D_{V_r} f_{D_r} c_{D_r}\right) \qquad (A.23)$$

Subsurface Heat and Water Transfer

Soil water + vapor flux between adjacent soil layers

$$Q_{l,l+1} = 2 \left(K_{l,l+1} \left(\psi_{S_l} - \psi_{S_{l+1}}\right) + D_{V_{l,l+1}} \left(e_l - e_{l+1}\right) \right) / \left(z_l + z_{l+1}\right) \qquad (A.24)$$

Soil vapor pressure within a soil layer

$$e_l = e_l' e^{\left(M \psi_{sl} / (R T_l)\right)} \qquad (A.25)$$

Soil heat flux between adjacent soil layers

$$G_{l,l+1} = 2 \kappa_{l,l+1} \left(T_l - T_{l+1}\right) / \left(z_l + z_{l+1}\right) + c_w T_l Q_{l,l+1} \qquad (A.26)$$

Soil latent heat (freeze-thaw) within a soil layer

$$G_{l-l,1} - G_{l,l+1} + LF_l + c_l \left(T_{l_{(t-1)}} - T_{l_{(t)}} \right) \Big/ \Delta t = 0 \qquad (A.27)$$

Glossary

Subscripts Used to Define Spatial Resolution of Variables

a Atmosphere
g Soil or residue surface (r = residue, s = soil)
i Plant species
j Tiller or branch of plant species
k Node of tiller or branch
l Canopy or soil layer
m Azimuth class of leaf in canopy layer
n Inclination class of leaf in azimuth class
o Irradiance class of leaf in inclination class (sunlit or shaded)
r Residue at ground surface
s Soil at ground surface
x Root axis order (1 = primary, 2 = secondary)
z root type (root or mycorrhizae)

Definition of Variables and Equations in Which They Are Used

$A_{i,j,k,l,m,n,o}$	Leaf surface area ($m^2\,m^{-2}$) [Eqs. (A.7, 15)]
β	Stomatal resistance parameter [Eq. (A.13)] †
C_{B_i}	[CO_2] in canopy air ($\mu mol\ mol^{-1}$) [Eq. (A.14)]
$C_{1'_i}$	[CO_2] in canopy leaves ($\mu mol\ mol^{-1}$) at ψ_{C_i} = 0 MPa [Eq. (A.14)]
c_a	Heat capacity of atmosphere ($J\ m^{-3}\ K^{-1}$) [Eqs. (A.5, 19)] †
c_{D_i}	Correction term for wind-driven dispersion of vapor and heat through canopy [Eq. (A.22)]
c_{D_r}	Correction term for wind-driven dispersion of vapor and heat through surface residue [Eq. (A.23)]
c_i	Heat capacity of canopy ($J\ m^{-2}\ K^{-1}$) [Eq. (A.6)]
c_l	Heat capacity of soil layer ($J\ m^{-2}\ K^{-1}$) [Eq. (A.27)]
c_r	Heat capacity of surface residue ($J\ m^{-2}\ K^{-1}$) [Eq. (A.20)]
c_s	Heat capacity of soil surface layer ($J\ m^{-2}\ K^{-1}$) [Eq. (A.21)]
c_w	Heat capacity of water ($J\ g^{-1}\ K^{-1}$) [Eq. (A.26)] †
D_{V_i}	Effective diffusion coefficient of vapor through canopy ($m^2\ s^{-1}$) [Eq. (A.22)]
D_{V_l}	Effective diffusion coefficient of vapor through soil ($m^2\ s^{-1}$) [Eq. (A.24)]
D_{V_r}	Effective diffusion coefficient of vapor through surface residue ($m^2\ s^{-1}$) [Eq. (A.23)]
d_i	Zero-plane displacement height (m) [Eqs. (A.12, 22)]

e_a Atmospheric vapor density (g m^{-3}) at T_a and ambient humidity [Eqs. (A.3, 4, 18)]

e_g Vapor density at ground surfaces (g m^{-3}) at T_g and ψ_{S_g} [Eq. (A.18)]

e_i Canopy vapor density (g m^{-3}) at T_i and ψ_{C_i} [Eq. (A.3)]

e'_i Saturation vapor density (g m^{-3}) at T_i [Eq. (A.4)]

e_l Vapor density in soil (m^3 m^{-3}) at ψ_{Sl} and T_l [Eqs. (A.24, 25)]

e'_l Saturation vapor density in soil (m^3 m^{-3}) at T_l [Eq. (A.25)]

ε_a Atmospheric emissivity [Eq. (A.9)]

ε_g Ground surface emissivity [Eq. (A.10)] [†]

ε_i Canopy emissivity [Eq. (A.11)] [†]

F_g Fraction of ground surface covered by residue r or soil s (m^2 m^{-2}) [Eqs. (A.10, 17–19)]

F_{P_i} Fraction of incident photosynthetic photon flux density absorbed by canopy (m^2 m^{-2}) [Eqs. (A.2, 8b, 14, 17)]

F_{S_i} Fraction of incident shortwave radiation absorbed by canopy (m^2 m^{-2}) [Eqs. (A.3–5, 8a)]

f_{D_i} Temperature sensitivity of D_{V_i} [Eq. (A.22)] [†]

f_{D_r} Temperature sensitivity of D_{V_r} [Eq. (A.23)] [†]

$G_{l,l+1}$ Heat flux between adjacent soil layers (W m^{-2}) [Eqs. (A.26, 27)]

H_g Sensible heat flux between ground surfaces and atmosphere (W m^{-2}) [Eqs. (A.16, 19)]

H_i Sensible heat flux between canopy and atmosphere (W m^{-2}) [Eqs. (A.1, 5)]

K von Karman's constant [Eq. (A.12)] [†]

$K_{l,l+1}$ Hydraulic conductivity between adjacent soil layers (m^2 s MPa^{-1}) [Eq. (A.24)]

$\kappa_{S_{r,s}}$ Thermal conductivity of residue-soil (W m^{-1} K^{-1}) [Eq. (A.20)]

$\kappa_{S_{s,l}}$ Thermal conductivity of surface soil (W m^{-1} K^{-1}) [Eq. (A.21)]

$\kappa_{l,l+1}$ Thermal conductivity between adjacent soil layers (W m^{-1} K^{-1}) [Eq. (A.26)]

L Latent heat of evaporation (J g^{-1}) [Eqs. (A.1, 3, 4, 18)] [†]

LE_g Latent heat flux between ground surfaces and atmosphere (W m^{-2}) [Eqs. (A.16, 18)]

LE_i Latent heat flux between canopy and atmosphere (W m^{-2}) [Eqs. (A.1, 3)]

LF_l Latent heat flux in soil layer (W m^{-2}) [Eq. (A.27)]

LV_i Latent heat flux between canopy surface and atmosphere (W m^{-2}) [Eqs. (A.1, 4)]

M Molecular mass of water (g mol^{-1}) [Eq. (A.25)] [†]

$Q_{l,l+1}$ Water flux between adjacent soil layers (g m^{-2} s^{-1}) [Eqs. (A.24, 26)]

R Gas constant (J mol^{-1} K^{-1}) [Eq. (A.25)] [†]

Ri Richardson number [Eq. (A.12)]

R_{L_a} Longwave radiation emitted from atmosphere (W m^{-2}) [Eqs. (A.2, 9, 17)]

R_{L_g} Longwave radiation emitted from ground surface (W m^{-2}) [Eqs. (A.2, 10, 17)]

R_{L_i} Longwave radiation emitted from canopy (W m^{-2}) [Eqs. (A.2, 11, 17)]

R_{N_g} Net radiation absorbed by ground surfaces (W m^{-2}) [Eqs. (A.16, 17)]

R_{N_i}	Net radiation absorbed by canopy (W m^{-2}) [Eqs. (A.1, 2)]
R_{P_i}	Photosynthetic photon flux density absorbed by canopy (μmol m^{-2} s^{-1}) [Eqs. (A.7b, 8)]
$R_{P_{i,l,m,n,o}}$	Photosynthetic photon flux density absorbed by leaf surface (μmol m^{-2} s^{-1}) [Eqs. (A.7b, 8)]
R_{S_g}	Shortwave radiation absorbed by ground surfaces (W m^{-2}) [Eq. (A.17)]
R_{S_i}	Shortwave radiation absorbed by canopy (W m^{-2}) [Eqs. (A.2, 7a)]
$R_{S_{i,l,m,n,o}}$	Shortwave radiation absorbed by leaf surface (W m^{-2}) [Eq. (A.7a)]
r_{A_g}	Aerodynamic resistance to vapor flux from ground surfaces (s m^{-1}) [Eqs. (A.18, 19)]
r_{A_i}	Aerodynamic resistance to vapor flux from canopy (s m^{-1}) [Eqs. (A.3–5, 12, 22)]
r_{A_r}	Aerodynamic resistance to vapor flux from surface residue (s m^{-1}) [Eqs. (A.22, 23)]
r_{A_s}	Aerodynamic resistance to vapor flux from soil surface (s m^{-1}) [Eq. (A.23)]
r_{C_i}	Canopy stomatal resistance to vapor flux (s m^{-1}) [Eqs. (A.3, 13)]
r_{C_i}'	Minimum r_{C_i} (s m^{-1}) at ψ_{C_i} = 0 MPa [Eqs. (A.13, 14)]
$r_{C\max}$	Canopy cuticular resistance to vapor flux (s m^{-1}) [Eq. (A.13)] [†]
S_g	Change in thermal energy stored in ground surfaces (W m^{-2}) [Eq. (A.16)]
S_i	Change in thermal energy stored in crop phytomass (W m^{-2}) [Eq. (A.1,6)]
S_r	Change in thermal energy stored in surface residue (W m^{-2}) [Eq. (A.20)]
S_s	Change in thermal energy stored in soil surface layer (W m^{-2}) [Eq. (A.21)]
σ	Stefan-Boltzmann constant (W m^{-2} K^{-4}) [Eqs. (A.9–11)] [†]
T_a	Air temperature (K) [Eqs. (A.5, 9, 19)]
T_g	Ground surface temperatures (K) [Eqs. (A.10, 19)]
T_i	Canopy temperature (K) [Eqs. (A.5, 6, 11)]
T_l	Temperature of soil layer below surface (K) [Eqs. (A.20, 21, 25–27)]
T_r	Surface residue temperature (K) [Eq. (A.20)]
T_s	Soil surface temperature (K) [Eqs. (A.20, 21)]
t	Current time step [Eq. (A.6)]
$t-1$	Previous time step [Eq. (A.6)]
u_a	Wind speed recorded at z_R (m s^{-1}) [Eq. (A.12)]
V_{B_i}'	Potential canopy carboxylation rate (μmol m^{-2} s^{-1}) at ψ_{C_i} = 0 MPa [Eqs. (A.14, 15)]
$V_{B_{i,j,k,l,m,n,o}}'$	Potential leaf carboxylation rate (μmol m^{-2} s^{-1}) at ψ_{C_i} = 0 MPa [Eq. (A.15)]
X_i	Clumping factor (0 = complete self-shading, 1 = no self-shading)
ψ_{Sl}	Soil water potential (MPa) [Eqs. (A.24, 25)]
ψ_{T_i}	Canopy turgor potential (MPa) [Eq. (A.13)]
z_{C_i}	Momentum roughness parameter (m) [Eq. (A.12)]
z_{V_i}	Vapor roughness parameter (m) [Eq. (A.12)]

z_R Height of wind speed measurement (m) [Eq. (A.12)] [†]
z_l Depth of soil layer below surface (m) [Eqs. (A.21, 24, 26)]
z_r Depth of surface residue (m) [Eqs. (A.20, 23)]
z_s Depth of soil surface layer (m) [Eqs. (A.20, 21)]

[†] Indicates values taken from the literature and provided to the ecosystem model.

APPENDIX B: WATER RELATIONS

Canopy

Turgor potential is the difference between total and osmotic potentials

$$\Psi_{T_i} = \Psi_{C_i} - \Psi_{\pi_i} \tag{B.1}$$

Osmotic potential is determined by canopy water content (passive adjustment) and stored C concentration (active adjustment)

$$\Psi_{\pi_i} = \Psi_{\pi'_i} F_{DM}/F_{DM'} - [X]_i \, F_{DM} \, RT_i/M \tag{B.2}$$

Total respiration from stored C in each branch or tiller

$$Q_{T_{i,j}} = Q'_T \, X_{i,j} \, f_{Q_i} \tag{B.3}$$

Maintenance respiration from N content of each branch or tiller

$$Q_{M_{i,j}} = Q'_M \, N_{i,j} \, f_{M_i} \tag{B.4}$$

Growth respiration is reduced by maintenance respiration and by turgor

$$Q_{G_{i,j}} = max\left\{0, \left(Q_{T_{i,j}} - Q_{M_{i,j}}\right) max\left\{0, \left(\Psi_{T_i} - \Psi_{T'_i}\right)\right\}\right\} \tag{B.5}$$

Maintenance respiration causes senescence if it exceeds total respiration from stored C

$$Q_{S_{i,j}} = max\left\{0, Q_{M_{i,j}} - Q_{T_{i,j}}\right\} \tag{B.6}$$

Phytomass growth is driven by growth respiration and efficiency of phytomass synthesis

$$\partial M_{i,j}/\partial t = Q_{G_{i,j}} \, F_{X_{i,j}} \Big/ \left(1.0 - F_{X_{i,j}}\right) - Q_{S_{i,j}} - Q_{D_{i,j}} \tag{B.7}$$

Net CO_2 exchange is the difference between CO_2 fixation and maintenance + growth respiration

$$V_{N_{i,j}} = V_{X_{i,j}} - min\left\{Q_{T_{i,j}}, Q_{M_{i,j}}\right\} - Q_{G_{i,j}} - Q_{S_{i,j}} \qquad (B.8)$$

Soil

Total water uptake from root system

$$U_i = \Sigma_{l=1}^{L} \Sigma_{r=1}^{2} U_{i,l,z} \qquad (B.9)$$

Water uptake from root system in each soil layer from soil-canopy water potential gradients

$$U_{i,l,z} = \left(\psi_{C_i} - \psi_{S_l}\right) \Big/ \left(\Omega_{S_{i,l,z}} + \Omega_{R_{i,l,z}} + \Sigma_{x=1}^{2} \Omega_{A_{i,l,z,x}}\right) \qquad (B.10)$$

Soil hydraulic resistance to root water uptake depends on root length and soil hydraulic conductivity

$$\Omega_{S_{i,l,z}} = \ln\left\{\left(d_{i,l,z}/r_{i,l,z}\right) \Big/ \left(2\pi\, L_{i,l,z}\, \kappa_{R_{i,l,z}}\right)\right\} \theta'_l / \theta_{i,l,z} \qquad (B.11)$$

Radial hydraulic resistance of primary and secondary root systems varies inversely with root length

$$\Omega_{R_{i,l,z}} = \Omega_{R_{i,z}}/L_{i,l,z} \qquad (B.12)$$

Axial hydraulic resistance of the primary root system varies with its depth and inversely with the number of axes and their radii

$$\Omega_{A_{i,l,z,1}} = \Omega_{A_{i,r}}\, z_l \Big/ \left\{n_{i,l,z,1}\left(r_{i,l,z1}/r'_{i,2,z}\right)^4\right\} \qquad (B.13)$$

Axial hydraulic resistance of the secondary root system varies with its average length and inversely with the number of axes and their radii

$$\Omega_{A_{i,l,z,2}} = \Omega_{A_{i,r}}\, 0.5\left(L_{i,l,z,2}/n_{i,l,z,2}\right) \Big/ \left\{n_{i,l,z,2}\left(r_{i,l,z,2}/r'_{i,2,z}\right)^4\right\} \qquad (B.14)$$

Root water potential is that at which water flux from the soil to the root water equals that from the root to the canopy

$$\psi_{R,l,z} = \frac{\left\{\psi_{S_l}\left(\Omega_{Ri,l,z} + \Omega_{Ai,l,2,z} + \Omega_{Ai,l,2,z}\right) + \psi_{C_i}\Omega_{Si,l,z}\right\}}{\left(\Omega_{Si,l,z} + \Omega_{Ri,l,z} + \Omega_{Ai,l,2,z} + \Omega_{Ai,l,2,z}\right)} \qquad (B.15)$$

Glossary

Subscripts Used to Define Spatial Resolution of Variables

a Atmosphere
g Soil or residue surface (r = residue, s = soil)
i Plant species
j Tiller or branch of plant species
k Node of tiller or branch
l Canopy or soil layer
m Azimuth class of leaf in canopy layer
n Inclination class of leaf in azimuth class
o Irradiance class of leaf in inclination class (sunlit or shaded)
r Residue at ground surface
s Soil at ground surface
x Root axis order (1 = primary, 2 = secondary)
z Root type (root or mycorrhizae)

Definition of Variables and Equations in Which They Are Used

$d_{i,l,z}$ Half distance between adjacent roots (m) [Eq. (B.11)]

F_{DM} Ratio of leaf+sheath dry mass to symplasmic water (g g^{-1}) [Eq. (B.2)]

$F_{DM'}$ F_{DM} when ψ_{C_i} = 0 (g g^{-1}) [Eq. (B.2)] †

$F_{X_{i,j}}$ Biosynthesis efficiency of $M_{i,j}$ (g g^{-1}) [Eq. (B.7)] †

f_{M_i} Temperature sensitivity of Q'_M [Eq. (B.4)] †

f_{Q_i} Temperature sensitivity of Q'_T [Eq. (B.3)] †

$\kappa_{R_{i,l,z}}$ Hydraulic conductivity between soil and root surface (m^2 MPa^{-1} h^{-1})
     [Eq. (B.11)]

$L_{i,l,z}$ Length of roots (m m^{-2}) [Eqs. (B.11, 12)]

$L_{i,l,z,2}$ Length of secondary roots (m m^{-2}) [Eq. (B.14)]

M Average molecular mass of X_i (g mol^{-1}) [Eq. (B.2)] †

$M_{i,j}$ Mass of tiller (branch) (g C m^{-2}) [Eq. (B.7)]

$N_{i,j}$ Nitrogen content of tiller (branch) (g N m^{-2}) [Eq. (B.4)]

$n_{i,l,z,1}$ Number of primary axes (m^{-2}) [Eq. (B.13)]

$n_{i,l,z,2}$ Number of secondary axes (m^{-2}) [Eq. (B.14)]

$Q_{D_{i,j}}$ Loss of plant material caused by senescence respiration of tiller (branch)
     (g C m^{-2} h^{-1}) [Eq. (B.7)]

$Q_{M_{i,j}}$ Maintenance respiration of tiller (branch) (g C m^{-2} h^{-1}) [Eqs. (B.4–6, 8)]

Q'_M Specific maintenance respiration (g C g N^{-1} h^{-1}) at T_i = 30 °C [Eq. (B.4)] †

$Q_{S_{i,j}}$ Senescence respiration of tiller (branch) (g C m^{-2} h^{-1}) [Eqs. (B.6–8)]

$Q_{T_{i,j}}$ Total respiration of $X_{i,j}$ (g C m^{-2} h^{-1}) [Eqs. (B.3, 5, 6, 8)]

Q'_T Specific respiration of $X_{i,j}$ (g C g X^{-1} h^{-1}) at T_i = 30°C [Eq. (B.3)] †

$\theta_{i,l,z}$ Soil water content at root surface (m^3 m^{-3}) [Eq. (B.11)]

θ'_l Soil porosity (m^3 m^{-3}) [Eq. (B.11)]

R Gas constant (J mol^{-1} K^{-1}) [Eq. (B.2)] †

$r_{i,l,z}$ Root radius (m) [Eq. (B.11)]

$r_{i,l,z,1}$ Primary root radius (m) [Eq. (B.13)]

$r_{i,l,z,2}$ Secondary root radius (m) [Eq. (B.14)]

$r'_{i,2,z}$ Secondary root radius (m) at $\psi_{T_{i_{l,z}}} = 1.0$ MPa [Eqs. (B.13, 14)] [†]

T_i Canopy temperature (K) [Eq. (B.2)]

U_i Water uptake by canopy root system (m^3 m^{-2} h^{-1}) [Eq. (B.9)]

$U_{i,l,z}$ Water uptake by canopy roots or mycorrhizae in layer l (m^3 m^{-2} h^{-1}) [Eqs. (B.9, 10)]

$V_{N_{i,j}}$ Tiller (branch) net CO_2 fixation (g C m^{-2} h^{-1}) [Eq. (B.8)]

$V_{X_{i,j}}$ Tiller (branch) gross CO_2 fixation (g C m^{-2} h^{-1}) [Eq. (B.8)]

$\Omega_{A_{i,l,z,x}}$ Axial resistance to water transport through root or mycorrhizal system (MPa h m^{-1}) [Eqs. (B.10, 13, 14)]

$V_{X_{i,j}}$ Tiller (branch) gross CO_2 fixation (g C m^{-2} h^{-1}) [Eq. (B.8)]

$\Omega_{A_{i,l,z,1}}$ Axial resistance to water transport along primary root axes (MPa h m^{-1}) [Eq. (B.13)]

$\Omega_{A_{i,l,z,2}}$ Axial resistance to water transport along secondary root axes (MPa h m^{-1}) [Eq. (B.14)]

$\Omega_{A_{i,r}}$ Axial resistivity to water transport along root axes (MPa h m^{-4}) [Eqs. (B.13, 14)] [†]

$\Omega_{R_{i,l,z}}$ Radial resistance to water transport from root surface to root axis (MPa h m^{-1}) [Eqs. (B.10, 12, 15)]

$\Omega_{R_{i,z}}$ Radial resistivity to water transport from surface to axis of root or mycorrhizae (MPa h m^{-2}) [Eq. (B.12)] [†]

$\Omega_{S_{i,l,z}}$ Radial resistance to water transport from soil to root surface (MPa h m^{-1}) [Eqs. (B.10, 11, 15)]

X Primary photosynthate storage pool (g C m^{-2}) [Eq. (B.3)]

$[X]_i$ Concentration of X in leaf+sheath (g g^{-1}) [Eq. (B.2)]

ψ_{C_i} Canopy water potential (MPa) [Eqs. (B.1, 10, 15)]

ψ_{π_i} Canopy osmotic potential (MPa) [Eqs. (B.1, 2)]

$\psi_{\pi'_i}$ ψ_{π_i} at $\psi_{C_i} = 0$ (MPa) [Eq. (B.2)] [†]

$\psi_{R,l,z}$ Root water potential (MPa) [Eq. (B.15)]

ψ_{S_l} Soil water potential (MPa) [Eqs. (B.10, 15)]

ψ_{T_i} Canopy turgor potential (MPa) [Eqs. (B.1, 5)]

$\psi_{T'_i}$ Canopy turgor potential (MPa) below which $\partial M_{i,j}/\partial t \leq 0$ [Eq. (B.5)]

z_l Depth of soil layer below surface (m) [Eq. (B.13)]

[†] Indicates values taken from the literature and provided to the ecosystem model.

APPENDIX C: CO_2 FIXATION

Dark Reactions

Leaf carboxylation rate — dependence on aqueous CO_2 concentration

$$V_{C_{i,j,k,l,m,n,o}} = V_{C\max_{i,j,k}}\left(C_{C_{i,j,k,l,m,n,o}} - \Gamma_{i,j,k}\right)\bigg/\left(C_{C_{i,j,k,l,m,n,o}} + K_{C_i}\right) \qquad (C.1)$$

Michaelis-Menten constant for carboxylation

$$K_{C_i} = K_{C_i'}\left(1 + O_C/K_{O_i'}\right) \tag{C.2}$$

Maximum leaf carboxylation rate from superficial density of rubisco as a fraction of leaf protein

$$V_{C\max_{i,j,k}} = V_i' f_{C_i} R_{C_i} 6.25\left(N_{i,j,k}/M_{i,j,k}\right)\left(M_{i,j,k}/A_{i,j,k}\right) \tag{C.3}$$

Leaf expansion rate driven by leaf growth rate and canopy turgor

$$\partial A_{i,j,k}/\partial t = \partial A_{i,j,k}/\partial M_{i,j,k}\, \partial M_{i,j,k}/\partial t\, max\left\{0,\left(\psi_{T_i} - \psi_{T_i'}\right)\right\} \tag{C.4}$$

CO_2 compensation point

$$\Gamma_{i,j,k} = 0.5\, V_{O\max_{i,j,k}} K_{C_i'} O_C \Big/ \left(V_{C\max_{i,j,k}} K_{O_i'}\right) \tag{C.5}$$

Light Reactions

Leaf carboxylation rate — dependence on electron transport

$$V_{J_{i,j,k,l,m,n,o}} = J_{i,j,k,l,m,n,o}\, Y_{i,j,k,l,m,n,o} \tag{C.6}$$

Electron transport rate — dependence on irradiance

$$J_{i,j,k,l,m,n,o} = \left(Q R_{P_{i,l,m,n,o}} + J_{\max_{i,j,k}} - \left(\left(Q R_{P_{i,l,m,n,o}} + J_{\max_{i,j,k}}\right)^2\right.\right.$$
$$\left.\left. - 4\,\alpha\, Q\, R_{P_{i,l,m,n,o}} J_{\max_{i,j,k}}\right)^{0.5}\right) \Big/ (2\,\alpha) \tag{C.7}$$

Maximum electron transport rate from superficial density of chlorophyll as a fraction of leaf protein

$$J_{\max_{i,j,k}} = J_i' f_{J_i} E_{C_i} 6.25\left(N_{i,j,k}/M_{i,j,k}\right)\left(M_{i,j,k}/A_{i,j,k}\right) \tag{C.8}$$

Leaf carboxylation efficiency

$$Y_{i,j,k,l,m,n,o} = \left(C_{C_{i,j,k,l,m,n,o}} - \Gamma_{i,j,k}\right) \bigg/ \left(4.5\, C_{C_{i,j,k,l,m,n,o}} + 10.5\, \Gamma_{i,j,k}\right) \tag{C.9}$$

Leaf carboxylation rate — combined dependence on CO_2 concentration and irradiance

$$V_{B_{i,j,k,l,m,n,o}} = min\left\{V_{C_{i,j,k,l,m,n,o}},\, V_{J_{i,j,k,l,m,n,o}}\right\} \tag{C.10}$$

Gaseous Flux

CO_2 diffusion through stomates driven by canopy air-leaf CO_2 concentration differences

$$V_{G_{i,j,k,l,m,n,o}} = \left(C_{B_i} - C_{I_{i,j,k,l,m,n,o}}\right) \bigg/ r_{L_{i,j,k,l,m,n,o}} \tag{C.11}$$

CO_2 concentration in canopy air depends on net CO_2 flux and boundary layer resistance

$$C_{B_i} = C_A - 1.33\, r_{A_i}\left(V_N + V_S\right) \times 10^6 \big/ \left(12 \times 3600 \times 44.6 \times 273.15 / T_i\right) \tag{C.12}$$

Leaf stomatal resistance at ambient water potential depends on leaf turgor

$$r_{L_{i,j,k,l,m,n,o}} = r_{L'_{i,j,k,l,m,n,o}} + \left(r_{L\max} - r_{L'_{i,j,k,l,m,n,o}}\right) e^{\left(-\beta \psi T_i\right)} \tag{C.13}$$

Leaf stomatal resistance at zero water potential depends on canopy air-leaf CO_2 concentration differences and leaf carboxylation rate

$$r_{L'_{i,j,k,l,m,n,o}} = \left(C_{B_i} - C_{I'_i}\right) \big/ V_{B'_{i,j,k,l,m,n,o}} \tag{C.14}$$

Leaf carboxylation rate at zero water potential is the minimum of dark and light reaction rates

$$V_{B'_{i,j,k,l,m,n,o}} = min\left\{V_{C'_{i,j,k}},\, V_{J'_{i,j,k,l,m,n,o}}\right\} \tag{C.15}$$

Leaf carboxylation rate at zero water potential — dependence on CO_2 concentration

$$V_{C'_{i,j,k}} = V_{C\max_{i,j,k}}\left(C_{C'_i} - \Gamma_{i,j,k}\right) \big/ \left(C_{C'_i} + K_{C_i}\right) \tag{C.16}$$

Combined Flux

Solution for leaf internal CO_2 concentration at which carboxylation equals diffusion

$$C_{I_{i,j,k,l,m,n,o}}\Big|V_{X_{i,j,k,l,m,n,o}} = V_{B_{i,j,k,l,m,n,o}} = V_{G_{i,j,k,l,m,n,o}} \qquad (C.17)$$

Carboxylation rate of canopy is aggregated from that of all leaf surfaces

$$V_{X_{i,j}} = \sum_{k=1}^{K}\sum_{l=1}^{L}\sum_{m=1}^{M}\sum_{n=1}^{N}\sum_{o=1}^{O}\left(V_{X_{i,j,k,l,m,n,o}} A_{i,j,k,l,m,n,o}\right) \qquad (C.18)$$

Glossary Of Terms

Subscripts Used to Define Spatial Resolution of Variables

a	Atmosphere
g	Ground surface (r = residue, s = soil)
i	Plant species
j	Tiller or branch of plant species
k	Node of tiller or branch
l	Canopy or soil layer
m	Azimuth class of leaf in canopy layer
n	Inclination class of leaf in azimuth class
o	Irradiance class of leaf in inclination class (sunlit or shaded)
r	Residue at ground surface
s	Soil at ground surface
x	Root axis order (1 = primary, 2 = secondary)
z	Root type (root or mycorrhizae)

Definition of Variables and Equations in Which They Are Used

$A_{i,j,k,l,m,n,o}$	Leaf surface area ($m^2\,m^{-2}$) [Eqs. (C.3, 4, 8, 18)]
α	Shape parameter for response of $J_{i,j,k,l,m,n,o}$ to $R_{P_{i,l,m,n,o}}$ [Eq. (C.7)] †
β	Stomatal resistance parameter [Eq. (C.13)] †
C_A	[CO_2] in atmosphere ($\mu mol\ mol^{-1}$) [Eq. (C.12)]
C_{B_i}	[CO_2] in canopy air ($\mu mol\ mol^{-1}$) [Eqs. (C.11, 12, 14)]
$C_{C_{i,j,k,l,m,n,o}}$	Chloroplast CO_2 concentration (μM) in soluble equilibrium with $C_{I_{i,j,k,l,m,n,o}}$ [Eqs. (C.1, 9)]
$C_{C_i'}$	Chloroplast CO_2 concentration (μM) in soluble equilibrium with $C_{I_i'}$ [Eq. (C.16)]
$C_{I_i'}$	[CO_2] in canopy leaves ($\mu mol\ mol^{-1}$) at $\psi_{C_i} = 0$ MPa [Eq. (C.14)]
$C_{I_{i,j,k,l,m,n,o}}$	[CO_2] in canopy leaves ($\mu mol\ mol^{-1}$) [Eqs. (C.11, 17)]

E_{C_i}	Specific leaf chlorophyll content (g g leaf protein^{-1}) [Eq. (C.8)] [†]
f_{C_i}	Temperature sensitivity of V_{C_i}' [Eq. (C.3)] [†]
f_{J_i}	Temperature sensitivity of J_i' [Eq. (C.8)] [†]
$\Gamma_{i,j,k}$	CO_2 compensation point (μM) at ambient O_C [Eqs. (C.1, 5, 9, 16)]
$J_{i,j,k,l,m,n,o}$	Electron transport rate (μmol e$^-$ m^{-2} s^{-1}) [Eqs. (C.6, 7)]
$J_{max\,i,j,k}$	Electron transport rate (μmol e$^-$ m^{-2} s^{-1}) at saturating R_{P_i} [Eqs. (C.7, 8)]
J_i'	Specific electron transport by chlorophyll (μmol e$^-$ g^{-1} s^{-1}) at $T_i = 30°C$ and saturating R_{P_i} [Eq. (C.8)] [†]
K_{C_i}	Michaelis-Menten constant (μM) for carboxylation under ambient O_2 [Eqs. (C.1, 2, 16)] [†]
K_{C_i}'	Michaelis-Menten constant (μM) for carboxylation in the absence of O_2 [Eqs. (C.2, 5)] [†]
K_{O_i}'	Michaelis-Menten constant (μM) for oxygenation in the absence of CO_2 [Eqs. (C.2, 5)] [†]
$M_{i,j,k}$	Mass of leaf (g C m^{-2}) [Eqs. (C.3, 4, 8)]
$N_{i,j,k}$	Nitrogen content of leaf (g N m^{-2}) [Eqs. (C.3, 8)]
O_C	Soluble O_2 concentration (μM) in chloroplasts [Eqs. (C.2, 5)]
Q	Quantum yield (μmol e$^-$ μmol quanta^{-1}) [Eq. (C.7)] [†]
R_{C_i}	Specific leaf rubisco content (g g leaf protein^{-1}) [Eq. (C.3)] [†]
$R_{P\,i,l,m,n,o}$	Photosynthetic photon flux density absorbed by leaf surface (μmol m^{-2} s^{-1}) [Eq. (C.7)]
r_{A_i}	Aerodynamic resistance to vapor flux from canopy (s m^{-1}) [Eq. (C.12)]
$r_{L_{i,j,k,l,m,n,o}}$	Leaf stomatal resistance to CO_2 flux (s m^{-1}) [Eqs. (C.11, 13)]
$r_{L_{i,j,k,l,m,n}}'$	Leaf stomatal resistance to CO_2 flux (s m^{-1}) at $\psi_{C_i} = 0$ MPa [Eqs. (C.13, 14)]
r_{Lmax}	Leaf cuticular resistance to CO_2 flux (s m^{-1}) [Eq. (C.13)]
T_i	Canopy temperature (K) [Eq. (C.12)]
V_i'	Specific carboxylation activity of rubisco (μmol g^{-1} s^{-1}) at $T_i = 30°C$ and saturating CO_2 in the absence of O_2 [Eq. (C.3)] [†]
$V_{B_{i,j,k,l,m,n,o}}$	Leaf carboxylation rate (μmol m^{-2} s^{-1}) [Eqs. (C.10, 17)]
$V_{B_{i,j,k,l,m,n,o}}'$	Potential leaf carboxylation rate (μmol m^{-2} s^{-1}) at $\psi_{C_i} = 0$ MPa [Eqs. (C.14, 15)]
$V_{C_{i,j,k,l,m,n,o}}$	CO_2-limited leaf carboxylation rate (μmol m^{-2} s^{-1}) at $C_{C_{i,j,k,l,m,n,o}}$ [Eqs. (C.1, 10)]
$V_{C_{i,j,k}}'$	Potential CO_2-limited leaf carboxylation rate (μmol m^{-2} s^{-1}) at C_{C_i}' and $\psi_{C_i} = 0$ MPa [Eqs. (C.15, 16)]
$V_{Cmax\,i,j,k}$	Leaf carboxylation rate (μmol m^{-2} s^{-1}) at saturating CO_2 in the absence of O_2 [Eqs. (C.1, 3, 5, 16)]
$V_{G_{i,j,k,l,m,n,o}}$	Leaf CO_2 diffusion rate (μmol m^{-2} s^{-1}) at $r_{L_{i,j,k,l,m,n,o}}$ [Eqs. (C.11, 17)]
$V_{J_{i,j,k,l,m,n,o}}$	Irradiance-limited leaf carboxylation rate (μmol m^{-2} s^{-1}) at $R_{P_{l,m,n,o}}$ [Eqs. (C.6, 10)]
$V_{J_{i,j,k,l,m,n,o}}'$	Potential irradiance-limited leaf carboxylation rate (μmol m^{-2} s^{-1}) at $R_{P_{l,m,n,o}}$ and $\psi_{C_i} = 0$ MPa [Eqs. (C.6, 10)]
$V_{Omax\,i,j,k}$	Leaf oxygenation rate (μmol m^{-2} s^{-1}) at saturating O_2 in the absence of CO_2 [Eq. (C.5)]

V_N	Community net CO_2 fixation (g C m^{-2} h^{-1}) [Eq. (C.12)]
V_S	Soil CO_2 flux (g C m^{-2} h^{-1}) [Eq. (C.12)]
$V_{X_{i,j,k,l,m,n,o}}$	Leaf gross CO_2 fixation rate (μmol m^{-2} s^{-1}) at ambient ψ_{C_i} [Eqs. (C.17, 18)]
$Y_{i,j,k,l,m,n,o}$	Carboxylation yield (μmol CO_2 μmol e^{--1}) [Eqs. (C.6, 9)]
ψ_{T_i}	Canopy turgor potential (MPa) [Eqs. (C.4, 13)]
ψ_{T_i}'	Canopy turgor potential (MPa) below which $\partial M_{i,j}/\partial t \leq 0$ [Eq. (C.4)]

† Indicates values taken from the literature and provided to the ecosystem model.

APPENDIX D: SOLUTE EQUILIBRIA

Precipitation — Dissolution

$$Al(OH)_{3(s)} \Leftrightarrow \left(Al^{3+}\right) + 3\left(OH^-\right) \qquad \left(\text{amorphous } Al(OH)_3\right) \qquad -33.0 \quad (\text{D.1})[1]$$

$$Fe(OH)_{3(s)} \Leftrightarrow \left(Fe^{3+}\right) + 3\left(OH^-\right) \qquad (\text{soil Fe}) \qquad -39.3 \quad (\text{D.2})$$

$$CaCO_{3(s)} \Leftrightarrow \left(Ca^{2+}\right) + \left(CO_3^{2-}\right) \qquad (\text{calcite}) \qquad -9.28 \quad (\text{D.3})$$

$$CaSO_{4(s)} \Leftrightarrow \left(Ca^{2+}\right) + \left(SO_4^{2-}\right) \qquad (\text{gypsum}) \qquad -4.64 \quad (\text{D.4})$$

$$AlPO_{4(s)} \Leftrightarrow \left(Al^{3+}\right) + \left(PO_4^{3-}\right) \qquad (\text{variscite}) \qquad -22.1 \quad (\text{D.5})[2]$$

$$FePO_{4(s)} \Leftrightarrow \left(Fe^{3+}\right) + \left(PO_4^{3-}\right) \qquad (\text{strengite}) \qquad -26.4 \quad (\text{D.6})$$

$$Ca(H_2PO_4)_{2(s)} \Leftrightarrow \left(Ca^{2+}\right) + 2\left(H_2PO_4^-\right) \text{ (monocalcium phosphate)} \quad -1.15 \quad (\text{D.7})[3]$$

$$CaHPO_{4(s)} \Leftrightarrow \left(Ca^{2+}\right) + \left(HPO_4^{2-}\right) \qquad (\text{monetite}) \qquad -6.92 \quad (\text{D.8})$$

$$Ca_5(PO_4)_3 OH_{(s)} \Leftrightarrow 5\left(Ca^{2+}\right) + 3\left(PO_4^{3-}\right) + \left(OH^-\right) \text{(hydroxyapatite)} \quad -58.2 \quad (\text{D.9})$$

[1] Round brackets denote solute activity. Numbers denote log K (precipitation-dissolution, ion pairs), Gapon coefficient (cation exchange), or log c (anion exchange).
[2] All equilibrium reactions involving N and P are calculated for both band and non-band volumes if a banded fertilizer application has been made. These volumes are calculated dynamically from diffusive transport of soluble N and P.
[3] May only be entered as fertilizer, not considered to be naturally present in soils.

Cation Exchange[1]

$$X\text{-}Ca + 2\left(NH_4^+\right) \Leftrightarrow 2\ X\text{-}NH_4 + \left(Ca^{2+}\right) \qquad 1.00 \qquad (D.10)$$

$$3\ X\text{-}Ca + 2\left(Al^{3+}\right) \Leftrightarrow 2\ X\text{-}Al + 3\left(Ca^{2+}\right) \qquad 1.00 \qquad (D.11)$$

$$X\text{-}Ca + \left(Mg^{2+}\right) \Leftrightarrow X\text{-}Mg + \left(Ca^{2+}\right) \qquad 0.60 \qquad (D.12)$$

$$X\text{-}Ca + 2\left(Na^+\right) \Leftrightarrow 2\ X\text{-}Na + \left(Ca^{2+}\right) \qquad 0.16 \qquad (D.13)$$

$$X\text{-}Ca + 2\left(K^+\right) \Leftrightarrow 2\ X\text{-}K + \left(Ca^{2+}\right) \qquad 3.00 \qquad (D.14)$$

$$X\text{-}Ca + 2\left(H^+\right) \Leftrightarrow 2\ X\text{-}H + \left(Ca^{2+}\right) \qquad 1.00 \qquad (D.15)$$

Anion Adsorption

$$X\text{-}OH_2^+ \Leftrightarrow X\text{-}OH + \left(H^+\right) \qquad -7.35 \qquad (D.16)$$

$$X\text{-}OH \Leftrightarrow X\text{-}O^- + \left(H^+\right) \qquad -8.95 \qquad (D.17)$$

$$X\text{-}H_2PO_4 + H_2O \Leftrightarrow X\text{-}OH_2^+ + \left(H_2PO_4^-\right) \qquad -2.80 \qquad (D.18)$$

$$X\text{-}H_2PO_4 + \left(OH^-\right) \Leftrightarrow X\text{-}OH + \left(H_2PO_4^-\right) \qquad 4.20 \qquad (D.19)$$

$$X\text{-}HPO_4^- + \left(OH^-\right) \Leftrightarrow X\text{-}OH + \left(HPO_4^{2-}\right) \qquad 2.60 \qquad (D.20)$$

Organic Acid

$$X\text{-}COOH \Leftrightarrow X\text{-}COO^- + \left(H^+\right) \qquad -5.00 \qquad (D.21)$$

Ion Pair

$$\left(NH_4^+\right) \Leftrightarrow \left(NH_3\right)_{(g)} + \left(H^+\right) \qquad -9.24 \qquad (D.22)$$

[1] X- denotes surface exchange site for cation or anion adsorption.

$$H_2O \Leftrightarrow (H^+) + (OH^-) \qquad\qquad -14.3 \qquad\qquad (D.23)$$

$$(CO_2)_{(g)} + H_2O \Leftrightarrow (H^+) + (HCO_3^-) \qquad\qquad -6.42 \qquad\qquad (D.24)$$

$$(HCO_3^-) \Leftrightarrow (H^+) + (CO_3^{2-}) \qquad\qquad -10.4 \qquad\qquad (D.25)$$

$$(AlOH^{2+}) \Leftrightarrow (Al^{3+}) + (OH^-) \qquad\qquad -9.06 \qquad\qquad (D.26)$$

$$(Al(OH)_2^+) \Leftrightarrow (AlOH^{2+}) + (OH^-) \qquad\qquad -10.7 \qquad\qquad (D.27)$$

$$(Al(OH)_3^0) \Leftrightarrow (Al(OH)_2^+) + (OH^-) \qquad\qquad -5.70 \qquad\qquad (D.28)$$

$$(Al(OH)_4^-) \Leftrightarrow (Al(OH)_3^0) + (OH^-) \qquad\qquad -5.10 \qquad\qquad (D.29)$$

$$(AlSO_4^+) \Leftrightarrow (Al^{3+}) + (SO_4^{2-}) \qquad\qquad -3.80 \qquad\qquad (D.30)$$

$$(FeOH^{2+}) \Leftrightarrow (Fe^{3+}) + (OH^-) \qquad\qquad -12.1 \qquad\qquad (D.31)$$

$$(Fe(OH)_2^+) \Leftrightarrow (FeOH^{2+}) + (OH^-) \qquad\qquad -10.8 \qquad\qquad (D.32)$$

$$(Fe(OH)_3^0) \Leftrightarrow (Fe(OH)_2^+) + (OH^-) \qquad\qquad -6.94 \qquad\qquad (D.33)$$

$$(Fe(OH)_4^-) \Leftrightarrow (Fe(OH)_3^0) + (OH^-) \qquad\qquad -5.84 \qquad\qquad (D.34)$$

$$(FeSO_4^+) \Leftrightarrow (Fe^{3+}) + (SO_4^{2-}) \qquad\qquad -4.15 \qquad\qquad (D.35)$$

$$(CaOH^+) \Leftrightarrow (Ca^{2+}) + (OH^-) \qquad\qquad -1.90 \qquad\qquad (D.36)$$

$$(CaCO_3^0) \Leftrightarrow (Ca^{2+}) + (CO_3^{2-}) \qquad\qquad -4.38 \qquad\qquad (D.37)$$

$$(CaHCO_3^+) \Leftrightarrow (Ca^{2+}) + (HCO_3^-) \qquad\qquad -1.87 \qquad\qquad (D.38)$$

$$\left(CaSO_4^0\right) \Leftrightarrow \left(Ca^{2+}\right) + \left(SO_4^{2-}\right) \qquad -2.92 \qquad \text{(D.39)}$$

$$\left(MgOH^+\right) \Leftrightarrow \left(Mg^{2+}\right) + \left(OH^-\right) \qquad -3.15 \qquad \text{(D.40)}$$

$$\left(MgCO_3^0\right) \Leftrightarrow \left(Mg^{2+}\right) + \left(CO_3^{2-}\right) \qquad -3.52 \qquad \text{(D.41)}$$

$$\left(MgHCO_3^+\right) \Leftrightarrow \left(Mg^{2+}\right) + \left(HCO_3^-\right) \qquad -1.17 \qquad \text{(D.42)}$$

$$\left(MgSO_4^0\right) \Leftrightarrow \left(Mg^{2+}\right) + \left(SO_4^{2-}\right) \qquad -2.68 \qquad \text{(D.43)}$$

$$\left(NaCO_3^-\right) \Leftrightarrow \left(Na^+\right) + \left(CO_3^{2-}\right) \qquad -3.35 \qquad \text{(D.44)}$$

$$\left(NaSO_4^-\right) \Leftrightarrow \left(Na^+\right) + \left(SO_4^{2-}\right) \qquad -0.48 \qquad \text{(D.45)}$$

$$\left(KSO_4^-\right) \Leftrightarrow \left(K^+\right) + \left(SO_4^{2-}\right) \qquad -1.30 \qquad \text{(D.46)}$$

$$\left(H_3PO_4\right) \Leftrightarrow \left(H^+\right) + \left(H_2PO_4^-\right) \qquad -2.15 \qquad \text{(D.47)}$$

$$\left(H_2PO_4^-\right) \Leftrightarrow \left(H^+\right) + \left(HPO_4^{2-}\right) \qquad -7.20 \qquad \text{(D.48)}$$

$$\left(HPO_4^{2-}\right) \Leftrightarrow \left(H^+\right) + \left(PO_4^{3-}\right) \qquad -12.4 \qquad \text{(D.49)}$$

$$\left(FeH_2PO_4^{2+}\right) \Leftrightarrow \left(Fe^{3+}\right) + \left(H_2PO_4^-\right) \qquad -5.43 \qquad \text{(D.50)}$$

$$\left(FeHPO_4^+\right) \Leftrightarrow \left(Fe^{3+}\right) + \left(HPO_4^{2-}\right) \qquad -10.9 \qquad \text{(D.51)}$$

$$\left(CaH_2PO_4^+\right) \Leftrightarrow \left(Ca^{2+}\right) + \left(H_2PO_4^-\right) \qquad -1.40 \qquad \text{(D.52)}$$

$$\left(CaHPO_4^0\right) \Leftrightarrow \left(Ca^{2+}\right) + \left(HPO_4^{2-}\right) \qquad -2.74 \qquad \text{(D.53)}$$

$$\left(CaPO_4^-\right) \Leftrightarrow \left(Ca^{2+}\right) + \left(PO_4^{3-}\right) \qquad -6.46 \qquad \text{(D.54)}$$

$$\left(MgHPO_4^0\right) \Leftrightarrow \left(Mg^{2+}\right) + \left(HPO_4^{2-}\right) \qquad -2.91 \qquad \text{(D.55)}$$

Calculation of Thermodynamic Equilibria

Calculation of precipitation-dissolution and ion pair reaction rates

$$\frac{\gamma_A[A-aQ]^a \gamma_B[B-bQ]^b \gamma_C[C-cQ]^c}{\gamma_D[D+dQ]^d \gamma_E[E+eQ]^e} = K_{ABCDE} \tag{D.56}$$

Calculation of cation exchange rates

$$\frac{\gamma_A[A-Q]^{1/a} b[X-B]}{a[X-A+Q]\gamma_B[B]^{1/b}} = g_{AB} \tag{D.57}$$

Calculation of anion exchange rates

$$\frac{\gamma_A[A-Q]\,\beta_B[X-B-Q]}{\beta_A[X-A+Q]\,\gamma_B[B+Q]} = c_{AB} \tag{D.58}$$

Glossary

A, B, C	Soluble products from precipitation-dissolution [Eqs. (D.1–55] (mol m^{-3}) [Eq. (D.56)]
a, b, c	Stoichiometric coefficients of A,B,C [Eqs. (D.56–58)]
β	Surface activity coefficients for anion exchange reactions [Eq. (D.58)]
D, E	Soluble reactants from precipitation-dissolution [Eqs. (D.1–55)] (mol m^{-3}) [Eqs. (D.56–58)]
d and e	Stoichiometric coefficients of D,E [Eqs. (D.56–58)]
g_{AB}	Gapon selectivity coefficient for cation exchange reaction [Eq. (D.57)]
c_{AB}	Equilibrium constant for anion exchange reaction [Eq. (D.58)]
γ	Solution activity coefficients from precipitation-dissolution [Eqs. (D.1-55); (D.56–58)]
K_{ABCDE}	Equilibrium constant for precipitation-dissolution or ion pairing reaction [Eq. (D.56)]
Q	Transformation rate of A, B, C, D, E (mol m^{-3} s^{-1})
X	Cation or anion exchange ligand [Eqs. (D.57, 58)]

APPENDIX E: HETEROTROPHIC MICROBIAL ACTIVITY

Hydrolysis of Organic Matter

Hydrolysis of soil organic C by active biomass in each substrate-microbe complex

$$D_{Si,j,c} = D'_{Si,j,c} M_{i,a,c} f_{tg} \tag{E.1}$$

Hydrolysis of microbial residue C by active biomass in each substrate-microbe complex

$$D_{Zi,j,c} = D'_{Zi,j,c} M_{i,a,c} f_{tg} \tag{E.2}$$

Concentrations of soil organic C and active microbial C determine specific microbial activity according to competitive enzyme kinetics

$$D'_{Si,j,c} = \left\{ D_{Si,j,c} [S_{i,c}] \right\} \Big/ \left\{ [S_{i,c}] + K_{mSi,c} \left(1.0 + [M_{i,a,c}]/K_{iSi,c}\right) \right\} \tag{E.3}$$

Concentrations of microbial residue C and active microbial C determine specific microbial activity according to competitive enzyme kinetics

$$D'_{Zi,j,c} = \left\{ D_{Zi,j,c} [Z_{i,c}] \right\} \Big/ \left\{ [Z_{i,c}] + K_{mZi,c} \left(1 + [M_{i,a,c}]/K_{iZi,c}\right) \right\} \tag{E.4}$$

Temperature affects biological process rates according to Arrhenius kinetics

$$f_{tg} = T_l \left\{ e^{[A-H_a/(RT_l)]} \right\} \Big/ \left\{ 1 + e^{[(H_{dl}-ST_l)/(RT_l)]} + e^{[(ST_l-H_{dh})/(RT_l)]} \right\} \tag{E.5}$$

Hydrolysis of soil organic N by active biomass is driven by that of soil C

$$D_{Si,j,n} = D_{Si,j,c} \left(S_{i,j,n}/S_{i,j,c}\right) \tag{E.6}$$

Hydrolysis of microbial residue N by active biomass is driven by that of soil C

$$D_{Zi,j,n} = D_{Zi,j,n} \left(Z_{i,j,n}/Z_{i,j,c}\right) \tag{E.7}$$

Sorption and Microbial Uptake of Hydrolysis Products

Sorption of soluble C hydrolysis products according to Freundlich isotherm

$$A_{i,c} = k_{ts} \left\{ a \, F_s [Q_{i,c}]^b - X_{i,c} \right\} \tag{E.8}$$

Adsorption of soluble N hydrolysis products is driven by that of soluble C

$$A_{i,n} = A_{i,c} \left(Q_{i,n}/Q_{i,c}\right) \tag{E.9}$$

Desorption of soluble N hydrolysis products is driven by that of soluble C

$$A_{i,n} = A_{i,c} \left(X_{i,n}/X_{i,c}\right) \tag{E.10}$$

Specific respiration by active microbial biomass is driven by the concentration of soluble hydrolysis products

$$R_{gi,c} = \left\{ R'_{gc}\left[Q_{i,c}\right]\right\} \Big/ \left\{K_{mPc} + \left[Q_{i,c}\right]\right\} f_{tg} \qquad (E.11)$$

Aerobic respiration by active microbial biomass under non-limiting O_2

$$R'_{gi,c} = R_{gi,c}\, M_{i,a,c} \qquad (E.12)$$

Aerobic respiration by active microbial biomass under ambient O_2 is constrained by microbial uptake of O_2 (see [Eqs. (H.1) and ([H.2)]

$$R_{gi,c} = R'_{gi,c}\, R_{O2i,c}/R'_{O2i,c} \qquad (E.13)$$

Total uptake of C hydrolysis products by active microbial biomass under non-limiting nutrient availability is driven by aerobic respiration and the energy yield from C oxidation

$$U'_{i,c} = R_{gi,c}\left(1 + \Delta G/E_m\right) \qquad (E.14)$$

Net uptake of C hydrolysis products by active microbial biomass under non-limiting nutrient availability

$$U'_{ni,c} = U'_{i,c} - R_{gi,c} \qquad (E.15)$$

Net uptake of C hydrolysis products by microbial biomass under limiting nutrient availability is constrained by microbial N content

$$U_{i,c} = U'_{i,c}\left[C_N/\left(C_N + K_{CN}\right)\right] \qquad (E.16)$$

Uptake of N hydrolysis products by active microbial biomass is driven by that of C

$$U_{i,n} = U_{i,c}\, Q_{i,n}/Q_{i,c} \qquad (E.17)$$

Respiration and Decomposition of Microbial Biomass

Maintenance respiration by microbial biomass depends on microbial N content

$$R_{mi,j,c} = R_m\, M_{i,j,n}\, f_{tm} \qquad (E.18)$$

Maintenance respiration increases exponentially with temperature

$$f_{tm} = x\, e^{\left(y\left(T_l - 273.16\right)\right)} \qquad (E.19)$$

Total CO_2 emission is the sum of growth and maintenance respiration

$$R_{i,c} = R_{gi,c} + \sum_{j=1}^{J} R_{mi,j,c} \qquad \text{(E.20)}$$

Microbial biomass decomposes according to first-order kinetics

$$D_{Mi,j,c} = \boldsymbol{D}_{Mi,j} \, M_{i,j,c} \, f_{tg} \qquad \text{(E.21)}$$

Release of microbial N is driven by that of C

$$D_{Mi,j,n} = \boldsymbol{D}_{Mi,j} \, M_{i,j,n} \, f_{tg} \qquad \text{(E.22)}$$

Mineralization-immobilization of N by microbial biomass is driven by microbial C:N ratios

$$I_{i,j,n} = \left(M_{i,j,c} \, C_{N_j} - M_{i,j,n} \right) \qquad \left(I_{i,j,n} < 0 \right) \qquad \text{(E.23a)}$$

$$I_{i,j,n} = \left(M_{i,j,c} \, C_{N_j} - M_{i,j,n} \right) [N]/([N] + K_N) \qquad \left(I_{i,j,n} > 0 \right) \qquad \text{(E.23b)}$$

Active biomass is calculated as labile biomass plus the amount of resistant biomass associated with it

$$M_{i,a,c} = M_{i,l,c} + M_{i,r,c} \, F_r/F_l \qquad \text{(E.24)}$$

Net growth of each heterotrophic population is the total uptake of hydrolysis products minus C oxidation for maintenance and growth respiration and for decomposition

$$\delta M_{i,j,c}/\delta t = F_j \, U_{i,c} - F_j \, R_{i,c} - D_{i,j,c} \qquad \left[R_{i,c} > R_{mi,j,c} \right] \qquad \text{(E.25a)}$$

$$\delta M_{i,j,c}/\delta t = F_j \, U_{i,c} - R_{mi,j,c} - D_{i,j,c} \qquad \left[R_{i,c} < R_{mi,j,c} \right] \qquad \text{(E.25b)}$$

$$\delta M_{i,c}/\delta t = \sum_{j=1}^{J} \delta M_{i,j,c}/\delta t \qquad \text{(E.25c)}$$

Net uptake or loss of N by each heterotrophic population is the total uptake of hydrolysis products plus immobilization minus decomposition

$$\delta M_{i,j,n}/\delta t = F_j \, U_{i,n} + I_{i,j,n} - D_{i,j,n} \qquad \text{(E.26)}$$

Humification of Products from Residue Hydrolysis and Microbial Decomposition

Humification of hydrolysis products from residue lignin depends upon hydrolysis rate and soil clay concentration

$$H_{Si=residue,\,j=lignin,\,c} = D_{Si=residue,\,j=lignin,\,c}\; F_H \qquad (E.27)$$

Humification of hydrolysis products from other residue components is driven by that of lignin

$$H_{Si=residue,\,j\neq lignin,\,c} = H_{Si=residue,\,j=lignin,\,c}\; L_{Hj} \qquad (E.28)$$

Humification of microbial decomposition products depends upon decomposition rate and soil clay concentration

$$H_{Mi,j,c} = D_{Mi,j,c}\; F_H \qquad (E.29)$$

Glossary

Subscripts Used to Define Spatial Resolution of Variables (decreasing spatial scale)

i Substrate-microbe complex
j Structural or kinetic components within each complex
k Elements such as C, N, and P within each component

Definition of Variables and Equations in Which They Are Used

A   Parameter such that $f_{tg} = 1.0$ at $T_l = 303.15K$ [Eq. (E.5)] [†]
$A_{i,c}$ Sorption of hydrolysis products (g C m^{-2} h^{-1}) [Eqs. (E.8–10)]
$A_{i,n}$ Sorption of hydrolysis products (g N m^{-2} h^{-1}) [Eqs. (E.9, 10)]
a Total substrate + residue C = $\displaystyle\sum_{i=1}^{I}\sum_{j=1}^{J}\left(\left[S_{i,j,c}\right]+\left[Z_{i,j,c}\right]\right)$ (g C Mg^{-1})
           [Eq. (E.8)]
b Freundlich exponent for sorption isotherm [Eq. (E.8)] [†]
C_N Ratio of $M_{i,a,n}$ to $M_{i,a,c}$ (g N g C^{-1}) [Eq. (E.16)]
C_{Nj} Maximum ratio of $M_{i,j,n}$ to $M_{i,j,c}$ maintained by $M_{i,j,c}$ (g N g C^{-1}) [Eq. (E.23)] [†]
$D_{Mi,j}$ Specific decomposition rate of $M_{i,j,c}$ at 30°C (g C g C^{-1} h^{-1}) [Eqs. (E.21, 22)] [†]
$D_{Mi,j,c}$ Decomposition rate of $M_{i,j,c}$ (g C m^{-2} h^{-1}) [Eqs. (E.21, 29)]
$D_{Mi,j,n}$ Decomposition rate of $M_{i,j,n}$ (g N m^{-2} h^{-1}) [Eq. (E.22)]
$D_{Si,j,c}$ Decomposition rate of $S_{i,j,c}$ by $M_{i,a,c}$ (g C m^{-2} h^{-1}) [Eqs. (E.1, 6, 27)]
$D_{Si,j,c}$ Specific decomposition rate of $S_{i,j,c}$ by $M_{i,a,c}$ at 30°C and saturating $[S_{i,c}]$ (g C g C^{-1} h^{-1}) [Eq. (E.3)] [†]

$D'_{Si,j,c}$	Specific decomposition rate of $S_{i,j,c}$ by $M_{i,a,c}$ at 30°C (g C g C^{-1} h^{-1}) [Eqs. (E.1, 3)]
$D_{Si,j,n}$	Decomposition rate of $S_{i,j,n}$ by $M_{i,a,c}$ (g N m^{-2} h^{-1}) [Eq. (E.6)]
$D_{Zi,j,c}$	Decomposition rate of $Z_{i,j,c}$ by $M_{i,a,c}$ (g C m^{-2} h^{-1}) [Eqs. (E.2, 7)]
$D'_{Zi,j,c}$	Specific decomposition rate of $Z_{i,j,c}$ by $M_{i,a,c}$ at 30°C and saturating [$Z_{i,c}$] (g C g C^{-1} h^{-1}) [Eq. (E.4)] [†]
$D_{Zi,j,n}$	Decomposition rate of $Z_{i,j,n}$ by $M_{i,a,c}$ (g N m^{-2} h^{-1}) [Eq. (E.7)]
ΔG	Energy yield from aerobic oxidation of $Q_{i,c}$ [Eq. (E.14)] [†]
E_m	Energy requirement for growth of $M_{i,a,c}$ [Eq. (E.14)] [†]
F_H	Fraction of products from lignin hydrolysis and microbial decomposition that are humified (function of clay content) [Eqs. (E.27, 29)] [†]
F_l	Fraction of microbial growth allocated to labile component $M_{i,l,c}$ [Eq. (E.24)] [†]
F_r	Fraction of microbial growth allocated to resistant component $M_{i,r,c}$ [Eq. (E.24)] [†]
F_s	Equilibrium ratio between $Q_{i,c}$ and $H_{i,c}$ [Eq. (E.8)]
f_{tg}	Temperature function for growth respiration (dimensionless) [Eqs. (E.1, 2, 11, 21, 22)] [†]
f_{tm}	Temperature function for maintenance respiration (dimensionless) [Eqs. (E.18, 19)] [†]
H_a	Energy of activation (J mol^{-1}) [Eq. (E.5)] [†]
H_{dl}	Energy of low-temperature deactivation (J mol^{-1}) [Eq. (E.5)] [†]
H_{dh}	Energy of high-temperature deactivation (J mol^{-1}) [Eq. (E.5)] [†]
$H_{Si,j,c}$	Humification rate of lignin hydrolysis products (g C m^{-2} h^{-1}) [Eqs. (E.27, 28)]
$H_{Mi,j,c}$	Humification rate of microbial decomposition products (g C m^{-2} h^{-1}) [Eq. (E.29)]
I	Number of substrate classes I [Eq. (E.1)]
$I_{i,j,n}$	Mineralization ($I_{i,j,n} < 0$) or immobilization ($I_{i,j,n} > 0$) of N by $M_{i,j,c}$ (g N m^{-2} h^{-1}) [Eq. (E.23)]
J	Number of components j in class I [Eq. (E.1)]
K_{CN}	Ratio of $M_{i,a,n}$ to $M_{i,a,c}$ at which $U_{i,c} = 1/2\ U'_{i,c}$ [Eq. (E.16)] [†]
$K_{iSi,c}$	Inhibition constant for $M_{i,a,c}$ on $S_{i,c}$ (g C Mg^{-1}) [Eq. (E.3)] [†]
$K_{iZi,c}$	Inhibition constant for $M_{i,a,c}$ on $Z_{i,c}$ (g C Mg^{-1}) [Eq. (E.4)] [†]
K_{mQc}	Michaelis-Menten constant for $R_{gi,c}$ on [$Q_{i,c}$] [Eq. (E.11)] [†]
$K_{mSi,c}$	Michaelis-Menten constant for $D_{Si,j,c}$ (g C Mg^{-1}) [Eq. (E.3)] [†]
$K_{mZi,c}$	Michaelis-Menten constant for $D_{Zi,j,c}$ (g C Mg^{-1}) [Eq. (E.4)] [†]
K_N	Michaelis-Menten constant for microbial uptake of solution N (g N m^{-3}) [Eq. (E.23)] [†]
k_{ts}	Equilibrium rate constant for sorption (h^{-1}) [†]
L_{Hj}	Ratio of non-lignin to lignin components in humified hydrolysis products [Eq. (E.28)] [†]
$M_{i,a,c}$	Active microbial C associated with $\sum_{j=1}^{J} \left(S_{i,j,c} + Z_{i,j,c} \right)$ (g C m^{-2}) [Eqs. (E.1, 2, 12, 24)]
[$M_{i,a,c}$]	Concentration of $M_{i,a,c}$ in soil water (g C m^{-3}) [Eqs. (E.3, 4)]
$M_{i,j,c}$	C content of microbial biomass $M_{i,j}$ (g C m^{-2}) [Eqs. (E.21, 23)]
$M_{i,j,n}$	N content of microbial biomass $M_{i,j}$ (g N m^{-2}) [Eqs. (E.18, 22, 23)]

$M_{i,l,c}$ — Labile microbial C $(j = l)$ associated with $\sum_{j=1}^{J}\left(S_{i,j,c}+Z_{i,j,c}\right)\left(g\ C\ m^{-2}\right)$ [Eq. (E.24)]

$M_{i,l,c}$ — Resistant microbial C $(j = r)$ associated with $\sum_{j=1}^{J}\left(S_{i,j,c}+Z_{i,j,c}\right)\left(g\ C\ m^{-2}\right)$ [Eq. (E.24)]

$[N]$ — $[NH_4^+]$ or $[NO_3^-]$ in soil solution (g N m^{-3}) [Eq. (E.23)]

$R_{O2i,c}$ — O_2 uptake by $M_{i,a,c}$ under ambient O_2 (g m^{-2} h^{-1}) [Eq. (E.13)]

$R'_{O2i,c}$ — O_2 uptake by $M_{i,a,c}$ under non-limiting O_2 (g m^{-2} h^{-1}) [Eq. (E.13)]

$Q_{i,c}$ — Hydrolysis products of $\sum_{j=1}^{J}\left(D_{Si,j,c}+D_{Zi,j,c}\right)\left(g\ C\ m^{-2}\right)$ [Eqs. (E.8, 17)]

$[Q_{i,c}]$ — Solution concentration of $Q_{i,c}$ (g C Mg^{-1}) [Eqs. (E.8, 11)]

$Q_{i,n}$ — Hydrolysis products of $\sum_{j=1}^{J}\left(D_{Si,j,n}+D_{Zi,j,n}\right)\left(g\ N\ m^{-2}\right)$ [Eqs. (E.9, 17)]

R — Gas constant (J mol^{-1} K^{-1}) [Eq. (E.5)] [†]

R'_{gc} — $R_{gi,c}$ at saturating $[Q_{i,c}]$ and 30°C (g C g C^{-1} h^{-1}) [Eq. (E.11)] [†]

$R_{gi,c}$ — Aerobic respiration of $M_{i,a,c}$ under ambient O_2 (g C m^{-2} h^{-1}) [Eqs. (E.13–15)]

$R_{gi,c}$ — Specific respiration of $M_{i,a,c}$ on $Q_{i,c}$ (g C g C^{-1} h^{-1}) [Eqs. (E.11, 12, 20)]

$R'_{gi,c}$ — Aerobic respiration of $M_{i,a,c}$ under non-limiting O_2 (g C m^{-2} h^{-1}) [Eq. (E.12)]

$R_{i,c}$ — Total C from growth and maintenance respiration (g C m^{-2} h^{-1}) [Eq. (E.20)]

R_m — Specific maintenance respiration at 30°C (g C g N^{-1} h^{-1}) [Eq. (E.18)] [†]

$R_{mi,j,c}$ — Maintenance respiration by $M_{i,j,c}$ (g C m^{-2} h^{-1}) [Eqs. (E.18, 20)]

S — Change in entropy (J mol^{-1} K^{-1}) [Eq. (E.5)] [†]

$S_{i,j,c}$ — Mass of solid or sorbed organic C in soil (g C m^{-2}) [Eq. (E.6)]

$S_{i,j,n}$ — Mass of solid or sorbed organic N in soil (g C m^{-2}) [Eq. (E.6)]

$[S_{i,c}]$ — Concentration of $\sum_{j=1}^{J}S_{i,j,c}$ in soil (g C Mg^{-1}) [Eq. (E.3)]

T_l — Soil temperature of layer l (K) [Eqs. (E.5, 19)]

$U_{i,c}$ — Uptake of $Q_{i,c}$ by $M_{i,a,c}$ under limiting nutrient availability (g C m^{-2} h^{-1}) [Eqs. (E.16, 17)]

$U'_{i,c}$ — Uptake of $Q_{i,c}$ by $M_{i,a,c}$ under non-limiting nutrient availability (g C m^{-2} h^{-1}) [Eqs. (E.14–16)]

$U_{i,n}$ — Uptake of $Q_{i,n}$ by $M_{i,a,c}$ under limiting nutrient availability (g N m^{-2} h^{-1}) [Eq. (E.17)]

$U'_{ni,c}$ — Net uptake of $Q_{i,c}$ by $M_{i,a,c}$ under non-limiting nutrient availability (g C m^{-2} h^{-1}) [Eq. (E.15)]

$X_{i,c}$ — Adsorbed C hydrolysis products (g C Mg^{-1}) [Eqs. (E.8, 10)]

$X_{i,n}$ — Adsorbed N hydrolysis products (g N Mg^{-1}) [Eq. (E.10)]

x — Selected to give a value for f_{tm} of 1.0 at 30°C [Eq. (E.19)] [†]

y — Selected to give a Q_{10} for f_{tm} of 2 [Eq. (E.19)] [†]

$Z_{i,j,c}$ — Mass of microbial residue C in soil (g C m^{-2}) [Eq. (E.7)]

$Z_{i,j,n}$ — Mass of microbial residue N in soil (g C m^{-2}) [Eq. (E.7)]

$[Z_{i,c}]$ — Concentration of $\sum_{j=1}^{J}Z_{i,j,c}$ in soil (g C Mg^{-1}) [Eq. (E.4)]

[†] Indicates values taken from the literature and provided to the ecosystem model.

APPENDIX F: ROOT SYSTEM, NUTRIENT AND OXYGEN UPTAKE

Root and Mycorrhizal Growth

Total root respiration at maximum turgor driven by storage C pool, temperature, O_2, and comparative access of growing points to storage C in each soil layer

$$R_{T_{s,a,l,x,z}} = R_R \, Z_{s,l,z,c} \, f_{T_{s,l,z}} \, f_{O_{s,l,z}} \, J_{s,a,l,x,z} / J_{s,l,z} \tag{F.1}$$

Conductance of storage C to primary root axes ($x = 1$, $z = r$) varies directly with their number and radii, and varies inversely with their lengths in the soil layer in which the root tip is located

$$J_{s,a,l,1,r} = n_{s,a,l,1,r} \, r_{s,l,1,r^4} \Big/ \big(d_{s,t} + d_{s,a,1,r} \big) \tag{F.2}$$

Conductance of storage C to secondary roots ($x = 2$, $z = r$) is calculated from the conductance of the primary and secondary roots in series for each soil layer through which each primary axis extends

$$J_{s,a,l,2,r} = J'_{s,a,l,1,r} \, J'_{s,a,l,2,r} \Big/ \big(J'_{s,a,l,1,r} + J'_{s,a,l,2,r} \big) \tag{F.3}$$

Conductance of storage C to secondary root axes varies directly with the number of secondary axes per unit area and their radii, and varies inversely with the lengths of the primary and secondary axes

$$J'_{s,a,l,1,r} = n_{s,a,l,1,r} \, r_{s,a,l,1,r^4} \Big/ \Big(d_{s,t} + z_{l-1} + 0.5 \, \max\{0, d_{s,a,1,1,r} - d\} \Big) \quad \big[d_{s,a,1,1,r} > d \big] \tag{F.4a}$$

$$J'_{s,a,l,1,r} = 0 \qquad\qquad\qquad \big[d_{s,a,1,1,r} \le d \big] \tag{F.4b}$$

and

$$J'_{s,a,l,2,r} = n_{s,a,l,2,r} \, r_{s,a,l,2,r^4} \Big/ d_{s,a,1,2,r} \qquad \big[d_{s,a,1,1,r} > d \big] \tag{F.5a}$$

$$J'_{s,a,l,2,r} = 0 \qquad\qquad\qquad \big[d_{s,a,1,1,r} \le d \big] \tag{F.5b}$$

Average length of the secondary root axes in each soil layer

$$d_{s,a,1,2,r} = L_{s,a,l,2,r} / n_{s,a,l,2,r} \tag{F.6}$$

Number of secondary root axes calculated from primary and secondary branching densities $\mathbf{n'}$ and $\mathbf{n''}$

$$n_{s,a,l,2,r} = \mathbf{n'}\, n_{s,a,l,1,r}\, \max\{0,\, d_{s,a,l,1,r} - \mathbf{d}\} + \mathbf{n''}\, L_{s,a,l,2,r} \tag{F.7}$$

Conductance of storage C to mycorrhizae ($x = 2$, $z = m$) varies directly with the number of mycorrhizal axes per unit area and their radii, and varies inversely with their lengths

$$J_{s,a,l,2,m} = n_{s,a,l,2,m}\, r_{s,l,2,m}^4 \big/ d_{s,a,l,2,m} \tag{F.8}$$

Radii of root and mycorrhizal axes depends upon their turgor calculated from root water potential through an explicit treatment of osmotic adjustment:

$$\psi_{\tau_{s,l,z}} = \psi_{s,l,z} - \psi_{\pi_{s,l,z}} \tag{F.9}$$

Total root or mycorrhizal respiration from Eq. (F.1) is partitioned into maintenance $R_{Ms,a,l,x,z}$ and growth $R_{Gs,a,l,x,z}$ components, with maintenance having the higher priority:

$$R_{M_{s,a,l,x,z}} = M_{s,a,l,x,z,n}\, \mathbf{R}_M\, f_{M_l} \tag{F.10}$$

The remainder of $R_{Ts,a,l,x,z}$ is available for growth, depending upon the more limiting of root water and nutrient status as represented by root turgor and nutrient storage pools, respectively

$$R_{G_{s,a,l,x,z}} = \tag{F.11}$$

$$\max\left\{0.0,\ \min\left\{\left(R_{T_{s,a,l,x,z}} - R_{M_{s,a,l,x,z}}\right)\min\left\{1.0,\ \max\left\{0.0, \psi_{\tau_{s,l,z}} - \psi_\tau - \omega_{s,a,l,x}\right\}\right\},\right.\right.$$

$$\left.\left. J_{s,a,l,x,z}\big/J_{s,l,z}\, \mathbf{R}_G\, \min\left\{Z_{s,l,z,n}\big/\left(C_{s,z,n}(1-\mathbf{R}_G)\right),\, Z_{s,l,z,p}\big/\left(C_{s,z,p}(1-\mathbf{R}_G)\right)\right\}\right\}\right\}$$

Root senescence occurs if total respiration from Eq. (F.1) is less than maintenance respiration from Eq. (F.10):

$$R_{D_{s,a,l,x,z,c}} = -\min\left\{0.0,\ R_{T_{s,a,l,x,z}} - R_{M_{s,a,l,x,z}}\right\} \tag{F.12}$$

with associated loss of N and P:

$$R_{D_{s,a,l,x,z,n}} = R_{D_{s,a,l,x,z,c}}\, C_{s,z,n}\, F_D \tag{F.13}$$

$$R_{D_{s,a,l,x,z,p}} = R_{D_{s,a,l,x,z,c}} C_{s,z,p} F_D \tag{F.14}$$

Growth respiration and senescence [Eqs. (F.11–12)] are used to calculate change in root or mycorrhizal mass

$$\delta M_{s,a,l,x,z,c}/\delta t = R_{G_{s,a,l,x,z}} (1 - R_G)/R_G - R_{D_{s,a,l,x,z,c}} \tag{F.15}$$

which is then used to calculate change in root or mycorrhizal length from their specific volumes, internal porosities, radii, and an assumed cylindrical geometry

$$\delta L_{s,a,l,x,z}/\delta t = \delta M_{s,a,l,x,z,c} / \left\{ \rho_r (1 - \theta_{Ps,z})(\pi r^2_{s,l,x,z}) \right\} \tag{F.16}$$

Respiration is also required for nutrient uptake by roots or mycorrhizae

$$R_{Q_{s,l,z}} = \Sigma_{(n,p)} R_{Q_{(n,p)}} Q_{Z_{s,l,z,(n,p)}} \tag{F.17}$$

Root-Shoot Nutrient Transfer

Potential transfer of storage C between each root and shoot axis is driven by concentration differences in storage C

$$Z_{s,t,c}/M_{s,t,c} - F'_{s,l,t,r,c} = Z_{s,l,r,c}/M_{s,l,r,c} + F'_{s,l,t,r,c} \tag{F.18}$$

Actual transfer is constrained by a rate constant from reaching equilibrium

$$F_{s,l,t,r,c} = K_{s,l,r} F'_{s,l,t,r,c} \tag{F.19}$$

The rate constant for equilibration is determined by the comparative conductance of storage C to all nodes of the root system in each soil layer (Eqs. [F.2–8])

$$K_{s,l,r} = L K_{s,r} J_{s,l,r}/J_{s,r} \tag{F.20}$$

Transfer of storage N and P between each root and shoot storage pool is driven by concentration differences of storage N and P with respect to storage C

$$Z_{s,t,(n,p)}/Z_{s,t,c} - F'_{s,l,t,r,(n,p)} = Z_{s,l,r,(n,p)}/Z_{s,l,r,c} + F'_{s,l,t,r,(n,p)} \tag{F.21}$$

Root-Mycorrhizal Nutrient Transfer

Storage pools in the root exchange C, N, and P with storage pools in the mycorrhizae according to concentration differences:

$$Z_{s,l,r,c}/M_{s,l,r,c} - F'_{s,l,r,m,c} = Z_{s,l,m,c}/M_{s,l,m,c} + F'_{s,l,r,m,c} \tag{F.22}$$

$$Z_{s,l,r,(n,p)} \big/ Z_{s,l,r,c} - F'_{s,l,r,m,(n,p)} = Z_{s,l,m,(n,p)} \big/ Z_{s,l,m,c} + F'_{s,l,r,m,(n,p)} \tag{F.23}$$

Actual transfer is constrained by a rate constant from reaching equilibrium

$$F_{s,l,r,m,k} = \mathbf{K}_{s,m} F'_{s,l,r,m,k} \tag{F.24}$$

Root and Mycorrhizal Nutrient Uptake

Uptake of N and P is calculated from solution concentrations at root and mycorrhizal surfaces at which radial transport by mass flow plus diffusion equals active uptake

$$Q_{Z_{s,l,z,(n,p)}} = \left\{ Q_{W_{s,l,z}} [S]_{l,(n,p)} + 2\pi L_{s,l,z} \ D_{E_{s,l,(n,p)}} \left([S]_{l,(n,p)} - [S]_{s,l,z,(n,p)} \right) \big/ \ln\!\left(d_{s,l,z} / r_{s,l,z} \right) \right\}$$

$$= Q_{Zmx_{s,l,z,(n,p)}} \left([S]_{s,l,z,(n,p)} - [S]_{s,z,(n,p)} \right) \big/ \left([S]_{s,l,z,(n,p)} - [S]_{s,z,(n,p)} + K_{Zs,z,(n,p)} \right) \tag{F.25}$$

Maximum active uptake rate of N or P depends on root or mycorrhizal area, temperature, O_2, and nutrient status

$$Q_{Zmx_{s,l,z,(n,p)}} = \mathbf{Q}_{Zmx_{s,z,(n,p)}} \ 2\pi r_{s,l,z} L_{s,l,z} \ f_{T_{s,l,z}} \ f_{O_{s,l,z}} \ f_{Q_{s,l,z,(n,p)}} \tag{F.26}$$

High root or mycorrhizal nutrient concentrations can constrain nutrient uptake

$$f_{Q_{s,l,z,(n,p)}} = Z_{s,l,z,c} \big/ \left(Z_{s,l,z,c} + \mathbf{x}_{(n,p)} Z_{s,l,z,(n,p)} \right) \tag{F.27}$$

Respiration and growth [Eqs. (F.1–17)], nutrient transfer [Eqs. (F.18–24)]), nutrient uptake [Eqs. (F.25–27)], and exudation ($E_{s,l,z,c}$) (Grant, 1993d) cause changes in storage pools $Z_{s,l,z,k}$

$$Z_{s,l,z,c} = Z'_{s,l,z,c} - \sum_{a=1}^{A} \sum_{x=1}^{X} \min\left\{ R_{T_{s,a,l,x,z}}, \ R_{M_{s,a,l,x,z}} \right\}$$

$$- \sum_{a=1}^{A} \sum_{x=1}^{X} \max\left\{ 0.0, \ \delta M_{s,a,l,x,z,c} \big/ \delta t + R_{G_{s,a,l,x,z}} \right\} \tag{F.28}$$

$$- R_{Q_{s,l,z}} + F_{s,l,t,r,c} + F_{s,l,r,m,c} - E_{s,l,z,c}$$

$$Z_{s,l,z,(n,p)} = Z'_{s,l,z,(n,p)} - \sum_{a=1}^{A} \sum_{x=1}^{X} \max\left\{ 0.0, \delta M_{s,a,l,x,z,c} \big/ \delta t \ C_{Rs,z,(n,p)} \right\} \tag{F.29}$$

$$+ F_{s,l,t,r,(n,p)} + F_{s,l,r,m,(n,p)} + Q_{Zs,l,z,(n,p)}$$

Root and Mycorrhizal Oxygen Uptake

Respiration of storage C in maintenance, growth and nutrient uptake processes under non-limiting O_2

$$R'_{s,l,z} = \sum_{a=1}^{A} \sum_{x=1}^{X} R_{T's,a,l,x,z} + R_{Q's,l,z} \qquad (F.30)$$

Demand for aqueous O_2 uptake under non-limiting O_2 depends on respiratory quotient

$$U_{O'_{s,l,z}} = R'_{s,l,z}/RQ \qquad (F.31)$$

Radial movement of O_2 from the aqueous phase in the soil to respiration sites in roots or mycorrhizae

$$U_{OS_{s,l,z}} = \left\{ -U_{W_{s,l,z}} \left[O_{AS_l} \right] + \right.$$
$$\left. 2\pi\, L_{s,l,z}\, D_{UO_{s,l,z}} \left(\left[O_{AS_l} \right] - \left[O_{AR_{s,l,z}} \right] \right) \middle/ \ln \left(\left(r_{s,l,z} + r_{W_l} \right) \middle/ r_{s,l,z} \right) \right. \qquad (F.32)$$

Dispersivity-diffusivity of aqueous O_2 during uptake

$$D_{UO_{s,l,z}} = \lambda \left| U_{W_{s,l,z}} \right| + D_{OA}\, f_{tDA_l}\, \tau_A\, \theta_{R_{s,l,z}}{}^{\upsilon}A \qquad (F.33)$$

Radial movement of O_2 from the aqueous phase in the root or mycorrhizae to respiration sites in roots or mycorrhizae

$$U_{OP_{s,l,z}} = \left\{ 2\pi\, L_{s,l,z}\, D_{UO_{s,l,z}} \left(\left[O_{AP_{s,l,z}} \right] - \left[O_{AR_{s,l,z}} \right] \right) \right\} \middle/ \ln \left(r_{s,l,z}/r_{P_{s,l,z}} \right) \qquad (F.34)$$

Active uptake of O_2 at respiration sites in the root or mycorrhizae

$$U_{OR_{s,l,z}} = U_{O'_{s,l,z}} \left[O_{AR_{s,l,z}} \right] \middle/ \left(\left[O_{AR_{s,l,z}} \right] + K_{O_s} \right) \qquad (F.35)$$

The O_2 uptake rate is that at which convergence for $[O_{AR_{s,l,z}}]$ in Eqs. (F.32, 34, and 35) is achieved. This rate is used with the demand for O_2 uptake under non-limiting O_2 from Eq. (F.31) to calculate O_2 constraints in Eqs. (F.1) and (F.26)

$$f_{O_{s,l,z}} = U_{O_{s,l,z}}/U_{O'_{s,l,z}} \qquad (F.36)$$

Transfer of Oxygen Through Soil, Roots, and Mycorrhizae

O_2 exchange between gaseous and aqueous phases in the soil depends on solubility and concentration gradients

$$E_{OS_l} = A_{AS_l} D_{E_o} \left(S_{O_l} \left[O_{GS_l} \right] - \left[O_{AS_l} \right] \right)$$ (F.37)

O_2 exchange between gaseous and aqueous phases in the root depends on solubility and concentration gradients

$$E_{OP_{s,l,z}} = A_{AP_{s,l,z}} D_{E_o} \left(S_{O_l} \left[O_{GP_{s,l,z}} \right] - \left[O_{AP_{s,l,z}} \right] \right)$$ (F.38)

Convective-diffusive vertical transport of O_2 in the aqueous phase of the soil

$$V_{OAS_l} = V_{W_l} \left[O_{AS_l} \right] + D_{VOA_l} \Delta \left[O_{AS_l} \right] / \Delta z_l$$ (F.39)

Dispersion-diffusion coefficient for vertical transport of O_2 in the aqueous phase of the soil

$$D_{VOA_l} = \lambda \left| V_{W_l} \right| + D_{OA} f_{tDA_l} \tau_A \theta_l {}^{\nu}A$$ (F.40)

Convective-diffusive vertical transport of O_2 in the gaseous phase of the soil

$$V_{OGS_l} = -\left(\sum_{s=1}^{S} \sum_{z=1}^{Z} U_{W_{s,l,z}} + V_{W_l} \right) \left[O_{GS_l} \right] + D_{VOGS_l} \Delta \left[O_{GS_l} \right] / \Delta z_l$$ (F.41)

Dispersion-diffusion coefficient for vertical transport of O_2 in the gaseous phase of the soil

$$D_{VOGS_l} = D_{OG} f_{tDG_l} \tau_G \theta_{GS_l} {}^{\nu}G$$ (F.42)

Diffusive vertical transport of O_2 in the gaseous phase of the root depends on root area/depth

$$V_{OGP_{s,l,z}} = D_{VOGP_{s,l,z}} \left(\left[O_{GP_{s,l,z}} \right] - \left[O_{GE} \right] \right) A_{GP_{s,l,z}} / z_{P_{s,l,z}}$$ (F.43)

Dispersion-diffusion coefficient for vertical transport of O_2 in the gaseous phase of the root

$$D_{VOGP_{s,l,z}} = D_{OG} f_{tDG_l} \tau_P \theta_{GP_s} {}^{\nu}P$$ (F.44)

The calculation of root area/depth in Eq. (F.43) is based on transport of O_2 through primary and secondary roots:

$$A_{GP_{s,l,z}} / z_{P_{s,l,z}} = A_{GP_{s,l,1,r}} A_{GP_{s,l,2,z}} / \left\{ A_{GP_{s,l,1,r}} L_{s,l,2,z} + A_{GP_{s,l,2,z}} z_{l-1} \right\} \quad \text{(F.45)}$$

Diffusive vertical transport of O_2 between the atmosphere and the aqueous phase of the soil surface layer depends on soil aqueous diffusivity and aerodynamic boundary layer conductance

$$V_{OAS_a} = \quad \text{(F.46)}$$

$$g_a \left\{ [O_{GE}] - \left\{ D_{VOA_l} [O_{AS_l}] \right\} / (0.5 \, z_l) + g_a [O_{GE}] \right\} / \left\{ D_{VOA_l} S_{O_l} f_{tDA_l} / (0.5 \, z_l) + g_a \right\}$$

Diffusive vertical transport of O_2 between the atmosphere and the gaseous phase of the soil surface layer depends on soil gaseous diffusivity and aerodynamic boundary layer conductance

$$V_{OGS_a} = \quad \text{(F.47)}$$

$$g_a \left\{ [O_{GE}] - \left\{ D_{VOGS_l} [O_{GS_l}] \right\} / (0.5 z_l) + g_a [O_{GE}] \right\} / \left\{ D_{VOGS_l} f_{tDG_l} / (0.5 \, z_l) + g_a \right\}$$

Glossary

Subscripts Used to Define Spatial Resolution of Variables

a Shoot node from which root axis originates
k Content of carbon c, nitrogen n, or phosphorus p
l Soil layer
s Plant species
t Shoot branch or tiller
x Root axis order (primary 1 or secondary 2)
z Root type (root r or mycorrhizae m)

Variables Used in Equations

$A_{AP_{s,l,z}}$ Air-water interfacial area in root (m² m⁻²) [Eq. (F.38)]
A_{AS_l} Air-water interfacial area in soil (m² m⁻²) [Eq. (F.37)]
$A_{GP_{s,l,z}}$ Cross-sectional area of root system (m² m⁻²) [Eqs. (F.43, 45)]
$C_{s,z,(n,p)}$ N or P concentration maintained by $M_{s,a,l,x,z,c}$ (g g⁻¹) [Eqs. (F.11, 13, 14, 29] [†]
D_{E_o} Diffusive transfer coefficient (m h⁻¹) [Eqs. (F.37, 38)]

$D_{E_{s,l,(n,p)}}$ — Effective dispersivity-diffusivity of n or p during root uptake ($m^2\ h^{-1}$) [Eq. (F.25)]

D_{OA} — Diffusivity of aqueous O_2 ($m^2\ h^{-1}$) [Eqs. (F.33, 40)]

D_{OG} — Diffusivity of gaseous O_2 ($m^2\ h^{-1}$) [Eqs. (F.42, 44)]

D_{VOA_l} — Dispersivity-diffusivity of O_{AS_l} between adjacent soil layers ($m^2\ h^{-1}$) [Eqs. (F.39, 40, 46)]

$D_{VOGP_{s,l,z}}$ — Dispersivity-diffusivity of $X_{GP_{s,l,z}}$ ($m^2\ h^{-1}$) [Eqs. (F.43, 44)]

D_{VOGS_l} — Dispersivity-diffusivity of X_{GS_l} between adjacent soil layers ($m^2\ h^{-1}$) [Eqs. (F.41, 42, 47)]

$D_{UO_{s,l,z}}$ — Dispersivity-diffusivity of O_{AS_l} during uptake ($m^2\ h^{-1}$) [Eqs. (F.32–34)]

d — Distance from primary root tip within which secondary growth is suppressed (m) [Eqs. (F.4, 5, 7)][†]

$d'_{s,l,z}$ — Half-distance between adjacent roots assumed equal to uptake path length (m) [Eq. (F.25)]

$d_{s,a,l,x,z}$ — Length of root axis in soil layer l (m) [Eqs. (F.4–8)]

$d_{s,a,x,z}$ — Length of root axis from ground surface to tip (m) [Eq. (F.2)]

$d_{s,t}$ — Height of upper leaves in plant canopy above ground surface (m) [Eqs. (F.2, 4)]

$E_{s,l,z,c}$ — Root exudation ($g\ m^{-2}\ h^{-1}$) [Eq. (F.28)]

$E_{OP_{s,l,z}}$ — O_2 exchange between gaseous and aqueous phases in the root ($g\ m^{-2}\ h^{-1}$) [Eqs. (F.38, 39)]

E_{OS_l} — O_2 exchange between gaseous and aqueous phases in the soil ($g\ m^{-2}\ h^{-1}$) [Eq. (F.37)]

F_D — Fraction of N and P not recovered from senesced root material [Eqs. (F.13, 14)]

$F'_{s,l,r,m,k}$ — Equilibrium transfer of C, N, or P between $Z_{s,l,m,k}$ and $Z_{s,l,r,k}$ ($g\ m^{-2}\ h^{-1}$) [Eqs. (F.22–24)]

$F'_{s,l,t,r,k}$ — Equilibrium transfer of C, N, or P between $Z_{s,t,k}$ and $Z_{s,l,r,k}$ ($g\ m^{-2}\ h^{-1}$) [Eqs. (F.18, 19, 21)]

$F_{s,l,r,m,k}$ — Actual transfer of C, N, or P between $Z_{s,l,m,k}$ and $Z_{s,l,r,k}$ ($g\ m^{-2}\ h^{-1}$) [Eqs. (F.24, 28, 29)]

$F_{s,l,t,r,k}$ — Actual transfer of C, N, or P between $Z_{s,t,k}$ and $Z_{s,l,r,k}$ ($g\ m^{-2}\ h^{-1}$) [Eqs. (F.19, 28, 29)]

f_{Ml} — Function for effect of temperature on root maintenance (dimensionless) [Eq. (F.10)]

$f_{O_{s,l,z}}$ — Function for effect of O_2 transfer on root activity (dimensionless) [Eqs. (F.1, 26, 36)]

$f_{Q_{s,l,z(n,p)}}$ — Function for feedback effect of root N or P concentration on root uptake [Eqs. (F.26, 27)]

$f_{T_{s,l,z}}$ — Function for effect of temperature on root activity (dimensionless) [Eqs. (F.1, 26)]

f_{tDA_l} — Temperature sensitivity of aqueous diffusivity [Eqs. (F.33, 40, 46)]

f_{tDG_l} — Temperature sensitivity of gaseous diffusivity [Eqs. (F.42, 44, 47)]

g_a — Aerodynamic conductance at the soil surface ($m\ h^{-1}$) [Eqs. (F.46, 47)]

$J_{s,a,l,x,z}$ — Comparative conductance of roots to $Z_{s,l,z,c}$ [Eqs. (F.1–3, 8, 11)]

$J'_{s,a,l,x,z}$ Comparative conductance of roots to $Z_{s,l,z,c}$ used to calculate $J_{s,a,l,2,r}$ [Eqs. (F.3–5)]

$J_{s,l,z}$ $= \sum_{a=1}^{A} \sum_{x=1}^{X} J_{s,a,l,x,z}$ [Eqs. (F.1, 11, 20)]

$J_{s,z}$ $= \sum_{l=1}^{L} J_{s,l,z}$ [Eq. (F.20)]

K_{O_s} Michaelis-Menten constant for uptake of O_2 [Eq. (F.35)]

$K_{s,l,r}$ Rate constant for $F_{s,l,t,r,k}$ in soil layer l (h^{-1}) [Eqs. (F.19, 20)]

$K_{s,m}$ Rate constant for $F_{s,l,r,m,k}$ (h^{-1}) [Eq. (F.24)] [†]

$K_{s,r}$ Rate constant for $F_{s,l,t,r,k}$ (h^{-1}) [Eq. (F.20)] [†]

$K_{Z_{s,z,(n,p)}}$ Michaelis-Menten constant for N or P uptake at root or mycorrhizal surface (g m^{-3}) [Eq. (F.25)] [†]

L Number of soil layer in which primary root tips are growing [Eq. (F.20)]

$L_{s,a,l,x,z}$ Root length (m m^{-2}) [Eqs. (F.6, 7, 16)]

$L_{s,l,x,z}$ $= \sum_{a=1}^{A} L_{s,a,l,x,z} \left(\text{m m}^{-2}\right)$ [Eq. (F.45)]

$L_{s,l,z}$ $= \sum_{x=1}^{X} L_{s,l,x,z} \left(\text{m m}^{-2}\right)$ [Eqs. (F.25, 26, 32, 34)]

λ Hydrodynamic dispersion coefficient (m)[†] [Eqs. (F.33, 40)]

$M_{s,a,l,x,z,k}$ Root mass of C, N, or P (g m^{-2}) [Eqs. (F.10, 15, 16, 28, 29)]

$M_{s,l,z,k}$ $= \sum_{a=1}^{A} \sum_{x=1}^{X} M_{s,a,l,x,z,k} \left(\text{g m}^{-2}\right)$ [Eqs. (F.18, 22, 23)]

$M_{s,t,k}$ Shoot mass of C, N, or P (leaves + sheaths or petioles) (g m^{-2}) [Eq. (F.18)]

n' Branching density of secondary roots on primary axis (m^{-1}) [Eq. (F.7)] [†]

n'' Branching density of secondary roots on secondary axis (m^{-1}) [Eq. (F.7)] [†]

$n_{s,a,l,x,z}$ Number of axes (m^{-2}) [Eqs. (F.2, 4–8)]

$O_{AP_{s,l,z}}$ Aqueous O_2 in the root (g m^{-2})

$[O_{AP_{s,l,z}}]$ Aqueous concentration of $O_{AP_{s,l,z}}$ (g m^{-3}) [Eqs. (F.34, 38)]

$[O_{AR_{s,l,z}}]$ Aqueous concentration of O_2 at root and mycorrhizal surfaces (g m^{-3}) [Eqs. (F.32, 34, 35)]

O_{AS_l} Aqueous O_2 in the soil (g m^{-2}) [Eq. (F.32)]

$[O_{AS_l}]$ Aqueous concentration of O_{AS_l} (g m^{-3}) [Eqs. (F.32, 37, 39, 46)]

$[O_{GE}]$ Concentration of O_2 in the atmosphere (g m^{-3}) [Eqs. (F.43, 46, 47)]

$[O_{GP_{s,l,z}}]$ Gaseous concentration of O_2 in the root (g m^{-3}) [Eqs. (F.38, 43)]

O_{GS_l} Gaseous O_2 in the soil (g m^{-2})

$[O_{GS_l}]$ Gaseous concentration of O_{GS_l} (g m^{-3}) [Eqs. (F.37, 41, 47)]

$Q_{W_{s,l,z}}$ Water uptake by $M_{s,l,z,c}$ (m^3 m^{-2} h^{-1}) [Eq. (F.25)]

$Q_{Zmx_{s,l,z,(n,p)}}$ Maximum N or P uptake by roots (g m^{-2} h^{-1}) [Eqs. (F.25, 26)]

$Q_{Zmx_{s,z,(n,p)}}$ Maximum N or P uptake by root surface (g m^{-2} h^{-1}) [Eq. (F.26)] [†]

$Q_{Z_{s,l,z,(n,p)}}$ N or P uptake by $M_{s,l,z,c}$ (g m^{-2} h^{-1}) [Eqs. (F.17, 25, 29)]

θ_{GP_s} Root air-filled porosity (m^3 m^{-3})[†] [Eq. (F.44)]

θ_{GS_l}	Soil air-filled porosity ($m^3\ m^{-3}$) [Eq. (F.42)]
$\theta_{P_{s,z}}$	Root or mycorrhizal porosity ($m^3\ m^{-3}$) [Eq. (F.16)] [†]
θ_l	Soil water content ($m^3\ m^{-3}$) [Eq. (F.40)]
$\theta_{R_{s,l,z}}$	Water content at the root or mycorrhizal surface ($m^3\ m^{-3}$) [Eq. (F.33)]
$R_{D_{s,a,l,x,z,k}}$	Senescence of $M_{s,a,l,x,z,k}$ ($g\ m^{-2}\ h^{-1}$) [Eqs. (F.12–15)]
$\mathbf{R_G}$	Specific growth respiration of $M_{s,a,l,x,z,c}$ ($g\ g^{-1}$) [Eqs. (F.11, 15)] [†]
$R_{G_{s,a,l,x,z}}$	Growth respiration of $M_{s,a,l,x,z,c}$ ($g\ m^{-2}\ h^{-1}$) [Eqs. (F.11, 15, 28)]
$\mathbf{R_M}$	Specific root maintenance respiration ($g\ g^{-1}N\ h^{-1}$) [Eq. (F.10)] [†]
$R_{M_{s,a,l,x,z}}$	Maintenance respiration of $M_{s,a,l,x,z,c}$ ($g\ m^{-2}\ h^{-1}$) [Eqs. (F.10–12, 28)]
RQ	Respiratory quotient for storage C oxidation ($g\ C\ g\ O_2^{-1}$)[†] [Eq. (F.31)]
$\mathbf{R_{Q(n,p)}}$	Specific root respiration during nutrient uptake ($g\ g^{-1}$) [Eq. (F.17)] [†]
$R_{Q_{s,l,z}}$	Root respiration for nutrient uptake at ambient $[O_2]$ ($g\ m^{-2}\ h^{-1}$) [Eqs. (F.17, 28)]
$R_{Q'_{s,l,z}}$	Root respiration for nutrient uptake at non-limiting $[O_2]$ ($g\ m^{-2}\ h^{-1}$) [Eq. (F.30)]
$\mathbf{R_R}$	Maximum specific respiration of $Z_{s,l,z,c}$ ($g\ g^{-1}\ h^{-1}$) [Eq. (F.1)] [†]
$R'_{s,l,z}$	Total respiration at maximum turgor and non-limiting $[O_2]$ [Eqs. (F.30, 31)]
$R_{T_{s,a,l,x,z}}$	Total respiration at maximum turgor and ambient $[O_2]$ ($g\ m^{-2}\ h^{-1}$) [Eqs. (F.1, 11, 12, 28)]
$R_{T'_{s,a,l,x,z}}$	Maintenance + growth respiration at maximum turgor and non-limiting $[O_2]$ ($g\ m^{-2}\ h^{-1}$) [Eq. (F.30)]
$r_{P_{s,l,z}}$	Radius of root aerenchyma (m) [Eq. (F.34)]
$r_{s,l,x,z}$	Root radius (m) [Eqs. (F.2, 4, 5, 8, 16)]
$r_{s,l,z}$	Average for all x of $r_{s,l,x,z}$ weighted for $L_{s,l,x,z}$ [Eqs. (F.25, 26, 32, 34)]
r_{W_l}	Thickness of water film (m) calculated from soil water potential [Eq. (F.32)]
ρ_r	Dry matter content of root biomass ($g\ g^{-1}$) [Eq. (F.16)] [†]
S_{O_l}	Temperature-corrected Ostwald solubility coefficient for gaseous O_2 [Eqs. (F.37, 38, 46)]
$[S]_{s,z,(n,p)}$	Solution concentration of N or P at root surface below which uptake = 0 ($g\ m^{-3}$) [Eq. (F.25)] [†]
$[S]_{l,(n,p)}$	Solution concentration of N or P ($g\ m^{-3}$) [Eq. (F.25)]
$[S]_{s,l,z,(n,p)}$	Solution concentration of N or P at root surface ($g\ m^{-3}$) [Eq. (F.25)]
τ_A	Tortuosity factor for aqueous diffusion[†] [Eqs. (F.33, 40)]
τ_G	Tortuosity factor for gaseous diffusion in soil[†] [Eq. (F.42)]
τ_P	Tortuosity factor for gaseous diffusion in roots[†] [Eq. (F.44)]
$U_{OP_{s,l,z}}$	Uptake of O_2 through root or mycorrhizae ($g\ m^{-2}\ h^{-1}$) [Eq. (F.34)]
$U_{OR_{s,l,z}}$	Active uptake of O_2 at respiration sites in the root or mycorrhizae ($g\ m^{-2}\ h^{-1}$) [Eq. (F.35)]
$U_{OS_{s,l,z}}$	Uptake rate of O_2 from soil ($g\ m^{-2}\ h^{-1}$) [Eq. (F.32)]
$U_{O_{s,l,z}}$	Uptake of O_2 under ambient O_2 ($g\ m^{-2}\ h^{-1}$) [Eq. (F.36)]
$U_{O'_{s,l,z}}$	Uptake of O_2 under non-limiting O_2 ($g\ m^{-2}\ h^{-1}$) [Eqs. (F.31, 35, 36)]
$U_{W_{s,l,z}}$	Water uptake (–ve) or exudation (+ve) ($m^3\ m^{-2}\ h^{-1}$) [Eqs. (F.32, 33, 41)]
V_{OAS_a}	Vertical O_2 transport between the aqueous phase of the soil and the atmosphere ($g\ m^{-2}\ h^{-1}$) [Eq. (F.46)]

V_{OAS_l}	Vertical O_2 transport in the aqueous phase of the soil (g m^{-2} h^{-1}) [Eq. (F.39)]
$V_{OGP_{s,l,z}}$	Vertical O_2 transport in the gaseous phase of the root (g m^{-2} h^{-1}) [Eq. (F.43)]
V_{OGS_a}	Vertical O_2 transport between the gaseous phase of the soil and the atmosphere (g m^{-2} h^{-1}) [Eq. (F.47)]
V_{OGS_l}	Vertical O_2 transport in the gaseous phase of the soil (g m^{-2} h^{-1}) [Eq. (F.41)]
V_{W_l}	Vertical water transport (+ve down, –ve up) (m^3 m^{-2} h^{-1}) [Eqs. (F.39–41)]
υ_A	Sensitivity of τ_A to $\theta_{W_{s,l,z}}$ [1] [Eqs. (F.33, 40)]
υ_G	Sensitivity of τ_G to θ_{GS_l} [1] [Eq. (F.42)]
υ_P	Sensitivity of τ_P to θ_{GP_s} [1] [Eq. (F.44)]
V_r	Specific volume of root biomass (m^3 g^{-1}) [Eq. (F.16)]
$\omega_{s,a,l,x}$	Soil resistance (MPa) [Eq. (F.11)]
$\mathbf{x}_{(n,p)}$	Ratio of $Z_{s,l,z,(n,p)}$:$Z_{s,l,z,c}$ at which root N or P uptake is 1/2 maximum (g g^{-1}) [Eq. (F.27)] [1]
$\psi_{\pi s,l,z}$	Root or mycorrhizal osmotic potential (MPa)
$\psi_{s,l,z}$	Root or mycorrhizal water potential (MPa) [Eq. (F.9)]
ψ_τ	Turgor below which shoot, root or mycorrhizal extension stops (MPa) [Eq. (F.11)] [1]
ψ_{τ_s}	Shoot turgor (MPa)
$\psi_{\pi s,l,z}$	Root or mycorrhizal osmotic potential (MPa) [Eq. (F.9)]
$\psi_{\tau s,l,z}$	Root or mycorrhizal turgor (MPa) [Eqs. (F.9, 11)]
$Z_{s,l,z,k}$	Storage C, N, or P pool in root or mycorrhizae (g m^{-2}) [Eqs. (F.1, 11, 18, 21–23, 27–29)]
$Z_{s,t,k}$	Storage C, N, or P pool in shoot (g m^{-2}) [Eqs. (F.18, 21)]
z_l	Depth to bottom of soil layer l (m) [Eqs. (F.4, 39, 41, 45–47)]
$z_{P_{s,l,z}}$	Depth from soil surface to roots in soil layer (m) [Eqs. (F.43, 45)]

† Indicates values taken from the literature and provided to the ecosystem model.

APPENDIX G: METHANE

Anaerobic Fermenters and H$_2$-Producing Acetogens

Respiration of soluble hydrolysis products by anaerobic fermenters is inhibited by O_2

$$R_{i,f} = \left\{ R'_f \, M_{i,f,a} \left[Q_{i,c} \right] \Big/ \left(K_f + \left[Q_{i,c} \right] \left(1 + \left[O_2 \right] / K_i \right) \right) \right\} f_t \tag{G.1}$$

Oxidation-reduction of hydrolysis products to acetate, CO_2, and H_2

$$Q_{i,c} \rightarrow 0.67 \, A_{i,c} + 0.33 \, CO_2\text{-}C + 0.11 \, H_2 \tag{G.2}$$

Uptake of hydrolysis products by anaerobic fermenters is driven by maintenance and growth respiration

$$U_{i,f,c} = R_{m^{i,f}} + \left(R_{i,f} - R_{m^{i,f}}\right)\left(1.0 + Y_f\right) \qquad \left[R_{i,f} > R_{m^{i,f}}\right] \qquad \text{(G.3a)}$$

$$U_{i,f,c} = R_{i,f} \qquad\qquad\qquad\qquad \left[R_{i,f} < R_{m^{i,f}}\right] \qquad \text{(G.3b)}$$

Growth yield of anaerobic fermentation depends on fermentation energy yield vs. biomass synthesis requirements

$$Y_f = -\Delta G_f / E_M \qquad \text{(G.4)}$$

Energy yield of anaerobic fermentation is affected by H_2 product concentration

$$\Delta G_f = \Delta G'_f + \left\{RT \ln\left(\left[H_2\right]/\left[H'_2\right]\right)^4\right\} \qquad \text{(G.5)}$$

Net biomass growth of anaerobic fermenters is the difference between uptake and respiration of hydrolysis products less decomposition

$$\delta M_{i,f,j,c}/\delta t = F_j U_{i,f,c} - F_j R_{i,f} - D_{i,f,j,c} \qquad \left[R_{i,f} > R_{m^{i,f}}\right] \qquad \text{(G.6a)}$$

$$\delta M_{i,f,j,c}/\delta t = F_j U_{i,f,c} - R_{mi,f,j} - D_{i,f,j,c} \qquad \left[R_{i,f} < R_{m^{i,f}}\right] \qquad \text{(G.6b)}$$

Acetotrophic Methanogens

Respiration of acetate by acetotrophic methanogens

$$R_{i,m} = \left\{R'_m M_{i,m,a}\left[A_{i,c}\right]/\left(K_m + \left[A_{i,c}\right]\right)\right\} f_t \qquad \text{(G.7)}$$

Oxidation-reduction of acetate to methane and CO_2

$$A_{i,c} \rightarrow 0.50\ CH_4\text{-}C + 0.50\ CO_2\text{-}C \qquad \text{(G.8)}$$

Uptake of acetate by acetotrophic methanogens is driven by maintenance and growth respiration

$$U_{i,m,c} = R_{m^{i,m}} + \left(R_{i,m} - R_{m^{i,m}}\right)\left(1.0 + Y_m\right) \qquad \left[R_{i,m} > R_{m^{i,m}}\right] \qquad \text{(G.9a)}$$

$$U_{i,m,c} = R_{i,m} \qquad\qquad\qquad\qquad \left[R_{i,m} < R_{m^{i,m}}\right] \qquad \text{(G.9b)}$$

Growth yield of acetotrophic methanogenesis depends on acetotrophic energy yield vs. biomass synthesis requirements

$$Y_m = -\Delta G'_m / E_M \tag{G.10}$$

Biomass growth of acetotrophic methanogens is the difference between uptake and respiration of acetate less decomposition

$$\delta M_{i,m,j,c} / \delta t = F_j U_{i,m,c} - F_j R_{i,m} - D_{i,m,j,c} \qquad \left[R_{i,m} > R_{m^{i,m}} \right] \tag{G.11a}$$

$$\delta M_{i,m,j,c} / \delta t = F_j U_{i,m,c} - R_{m^{i,m,j}} - D_{i,m,j,c} \qquad \left[R_{i,m} < R_{m^{i,m}} \right] \tag{G.11b}$$

Hydrogenotrophic Methanogens

Respiration from reduction of CO_2 by hydrogenotrophic methanogens

$$R_h = \left\{ R'_h M_{h,a} \left[H_2 \right] / \left(K_h + \left[H_2 \right] \right) \left[CO_2 \right] / \left(K_c + \left[CO_2 \right] \right) \right\} f_t \tag{G.12}$$

Reduction of CO_2 to methane

$$CO_2\text{-}C + 0.67\, H_2 \rightarrow CH_4\text{-}C + 3\, H_2O \tag{G.13}$$

Uptake of CO_2 by hydrogenotrophic methanogens is driven by maintenance and growth respiration

$$U_{h,c} = R_{m^h} + \left(R_h - R_{m^h} \right) \left(1.0 + Y_h \right) \qquad \left[R_h > R_{m^h} \right] \tag{G.14a}$$

$$U_{h,c} = R_h \qquad \left[R_h < R_{m^h} \right] \tag{G.14b}$$

Growth yield of hydrogenotrophic methanogenesis depends on hydrogenotrophic energy yield vs. biomass synthesis requirements

$$Y_h = -\Delta G_h / E_C \tag{G.15}$$

Energy yield of hydrogenotrophic methanogenesis affected by reactant H_2 concentration

$$\Delta G_h = \Delta G'_h - \left\{ RT \ln \left(\left[H_2 \right] / \left[H'_2 \right] \right)^4 \right\} \tag{G.16}$$

Biomass growth of hydrogenotrophic methanogens is the difference between uptake and respiration of CO_2 less decomposition

$$\delta M_{h,j,c}/\delta t = F_j\, U_{h,c} - F_j\, R_h - D_{h,j,c} \qquad \left[R_h > R_{mh}\right] \qquad \text{(G.17a)}$$

$$\delta M_{h,j,c}/\delta t = F_j\, U_{h,c} - R_{mh,j} - D_{h,j,c} \qquad \left[R_h < R_{mh}\right] \qquad \text{(G.17b)}$$

Autotrophic Methanotrophs

Oxidation of methane for energy and respiration by autotrophic methanotrophs at saturating O_2

$$X'_t = \left\{X'_t M_{t,a}\left[CH_4\right]/\left(K_t + \left[CH_4\right]\right)\right\} f_t \qquad \text{(G.18)}$$

Oxidation of methane for respiration by autotrophic methanotrophs at saturating O_2

$$R'_t = X'_t\, Y_{tR} \qquad \text{(G.19)}$$

Ratio of methane oxidized for respiration vs. methane oxidized for energy depends on methanotrophic energy yield vs. energy required to transform CH_4 into organic C for respiration

$$Y_{tR} = -\Delta G'_t / E_G \qquad \text{(G.20)}$$

Oxidation of methane for energy and respiration by autotrophic methanotrophs at ambient O_2 is constrained by microbial O_2 uptake

$$X_t = X'_t\, f_{O_2 t} \qquad \text{(G.21a)}$$

Oxidation of methane for respiration by autotrophic methanotrophs at ambient O_2 is constrained by microbial O_2 uptake

$$R_t = R'_t\, f_{O_2 t} \qquad \text{(G.21b)}$$

Oxidation of methane to CO_2

$$CH_4\text{-}C + 4.0\ O_2 \rightarrow CO_2\text{-}C + 1.5\ H_2O + 0.167\ H^+ \qquad \text{(G.22)}$$

Oxidation of methane to microbial storage C

$$CH_4\text{-}C + 1.33\ O_2 \rightarrow CO_2O\text{-}C + 0.167\ H^+ \qquad \text{(G.23)}$$

Oxidation of microbial storage C to CO_2

$$CH_2O\text{-}C + 2.67\ O_2 \rightarrow CO_2\text{-}C + 1.5\ H_2O \qquad (G.24)$$

Uptake of methane by autotrophic methanotrophs is driven by maintenance and growth respiration

$$U_{t,c} = R_{m^t} + \left(R_t - R_{m^t}\right)\left(1.0 + Y_{tG}\right) \qquad \left[R_t > R_{m^t}\right] \qquad (G.25a)$$

$$U_{t,c} = R_t \qquad \left[R_t < R_{m^t}\right] \qquad (G.25b)$$

Growth yield of autotrophic methanotrophy depends on C oxidation energy yield vs. biomass synthesis requirements

$$Y_{tG} = -\Delta G'_c / E_M \qquad (G.26)$$

Biomass growth of autotrophic methanotrophs is the difference between uptake and respiration of methane less decomposition

$$\delta M_{t,j,c}/\delta t = F_j U_{t,c} - F_j R_t - D_{t,j,c} \qquad \left[R_t > R_{m^t}\right] \qquad (G.27a)$$

$$\delta M_{t,j,c}/\delta t = F_j U_{t,c} - R_{m_{t,j}} - D_{t,j,c} \qquad \left[R_t < R_{m^t}\right] \qquad (G.27b)$$

Glossary

Subscripts Used to Define Spatial Resolution of Variables

a Descriptor for j = active component of M_i
f Descriptor for fermenters and acetogens in each M_i
h Descriptor for hydrogenotrophic methanogens in each M_i
i Descriptor for organic matter-microbe complex (i = plant residue, manure, particulate OM, or humus)
j Descriptor for structural or kinetic components for each functional type within each M_i (e.g., a = active)
k Descriptor for elemental fraction within each j (j = c, n or p)
m Descriptor for acetotrophic methanogens in each M_i
t Descriptor for autotrophic methanotrophs

Variables Used in Equations

A   Acetate (g C m^{-2}) [Eq. (G.2)]
[A]   Aqueous concentration of acetate (g C m^{-3}) [Eq. (G.7)]
[CH$_4$] Aqueous concentration of CH_4 (g C m^{-3}) [Eq. (G.18)]

$[CO_2]$ Aqueous concentration of CO_2 (g C m^{-3}) [Eq. (G.12)]

$D_{h,j,c}$ Decomposition of hydrogenotrophic methanogens (g C m^{-2} h^{-1}) [Eq. (G.17)]

$D_{i,f,j,c}$ Decomposition of fermenters and acetogens (g C m^{-2} h^{-1}) [Eq. (G.6)]

$D_{i,m,j,c}$ Decomposition of acetotrophic methanogens (g C m^{-2} h^{-1}) [Eq. (G.11)]

$D_{t,j,c}$ Decomposition of autotrophic methanotrophs (g C m^{-2} h^{-1}) [Eq. (G.27)]

E_C Energy required to construct new M from CO_2 (kJ g C^{-1}) [Eq. (G.15)] [†]

E_G Energy required to transform CH_4 into organic C (kJ g C^{-1}) [Eq. (G.20)] [†]

E_M Energy required to construct new M from organic C (kJ g C^{-1}) [Eqs. (G.4, 10, 26)] [†]

F_j Partitioning coefficient for j in $M_{i,n,j}$ [Eqs. (G.6, 11, 17, 27)] [†]

f_{o_2t} Ratio of O_2 uptake to O_2 requirement for CH_4 oxidation [Eqs. (G.21a, 21b)]

f_t Temperature function for growth-related processes (dimensionless) [Eqs. (G.1, 7, 12)] [†]

$\Delta G'_c$ Free energy change of C oxidation-O_2 reduction (kJ g C^{-1}) [Eq. (G.26)] [†]

ΔG_f Free energy change of fermentation plus acetogenesis (kJ g $Q_{i,c}^{-1}$) [Eqs. (G.4, 5)] [†]

$\Delta G'_f$ ΔG_f when $[H_2] = [H_2']$ (kJ g $Q_{i,c}^{-1}$) [Eq. (G.5)]

ΔG_h Free energy change of hydrogenotrophic methanogenesis (kJ g CO_2-C^{-1}) [Eqs. (G.15,16)] [†]

$\Delta G'_h$ Free energy change of hydrogenotrophic methanogenesis when $[H_2] = [H_2']$ (kJ g CO_2-C^{-1}) [Eq. (G.16)]

$\Delta G'_m$ Free energy change of acetotrophic methanogenesis (kJ g $A_{i,c}^{-1}$) [Eq. (G.10)] [†]

$\Delta G'_t$ Free energy change of CH_4 oxidation by methanotrophs (kJ g CH_4-C^{-1}) [Eq. (G.20)] [†]

$[H_2]$ Aqueous concentration of H_2 (g m^{-3}) [Eqs. (G.5, 12, 16)]

$[H_2']$ Aqueous concentration of H_2 when $\Delta G_h = \Delta G'_h$ and $\Delta G_f = \Delta G'_f$ (g H m^{-3}) [Eqs. (G.5, 16)]

K_c Michaelis-Menten constant for uptake of CO_2 by hydrogenotrophic methanogens (g C m^{-3}) [Eq. (G.12)] [†]

K_f Michaelis-Menten constant for uptake of $Q_{i,c}$ by fermenters and acetogens (g C m^{-3}) [Eq. (G.1)] [†]

K_h Michaelis-Menten constant for uptake of H_2 by hydrogenotrophic methanogens (g H m^{-3}) [Eq. (G.12)] [†]

K_m M-M constant for uptake of $A_{i,c}$ by acetotrophic methanogens (g C m^{-3}) [Eq. (G.7)] [†]

K_t Michaelis-Menten constant for uptake of CH_4 by methanotrophs (g C m^{-3}) [Eq. (G.18)] [†]

M   Microbial communities (g C m^{-2})

M_h Hydrogenotrophic methanogen community (g C m^{-2}) [Eqs. (G.12, 17)]

$M_{i,f}$ Fermenter and acetogenic community (g C m^{-2}) [Eqs. (G.1, 6)]

$M_{i,m}$ Acetotrophic methanogen community (g C m^{-2}) [Eqs. (G.7, 11)]

M_t Autotrophic methanotrophic community (g C m^{-2}) [Eqs. (G.18, 27)]

Q   Soluble organic matter (g C m^{-2})

$[Q]$ Aqueous concentration of soluble organic matter (g C m^{-3}) [Eq. (G.1)]

R   Gas constant (kJ mol^{-1} K^{-1}) [Eqs. (G.5, 16)] [†]

R'_f	Specific respiration by fermenters and acetogens at saturating $[Q_{i,c}]$, 30°C and high water potential (g C g $M_{i,f,a}^{-1}$ h^{-1}) [Eq. (G.1)] [†]
R_h	CO_2 reduction by hydrogenotrophic methanogens (g C m^{-2} h^{-1}) [Eqs. (G.12–14, 17, 18)]
R'_h	Specific CO_2 reduction by hydrogenotrophic methanogens at saturating $[H_2]$ and $[CO_2]$, and at 30°C and high water potential (g C g $M_{h,a}^{-1}$ h^{-1}) [Eq. (G.12)] [†]
$R_{i,f}$	Respiration of hydrolysis products by fermenters and acetogens (g C m^{-2} h^{-1}) [Eqs. (G.1–3, 6)]
$R_{i,m}$	Respiration of acetate by acetotrophic methanogens (g C m^{-2} h^{-1}) [Eqs. (G.7–9, 11)]
R'_m	Specific respiration by acetotrophic methanogens at saturating $[A_{i,c}]$, 30°C and high water potential (g C g $M_{i,m,a}^{-1}$ h^{-1}) [Eq. (G.7)] [†]
$R_{mh,j}$	Maintenance respiration by hydrogenotrophic methanogens (g C m^{-2} h^{-1}) [Eqs. (G.14, 17)]
$R_{mi,f,j}$	Maintenance respiration by fermenters and acetogens (g C m^{-2} h^{-1}) [Eqs. (G.3, 6)]
$R_{mi,m,j}$	Maintenance respiration by acetotrophic methanogens (g C m^{-2} h^{-1}) [Eqs. (G.9, 11)]
$R_{mt,j}$	Maintenance respiration by methanotrophs (g C m^{-2} h^{-1}) [Eqs. (G.25, 27)]
R_t	CH_4 oxidation by methanotrophs for respiration (g C m^{-2} h^{-1}) [Eqs. (G.21b, 23–25, 27a)]
R'_t	CH_4 oxidation by methanotrophs for respiration at saturating O_2 (g C m^{-2} h^{-1}) [Eqs. (G.19, 21b)]
T	Soil temperature (K) [Eqs. (G.5, 16)]
$U_{h,c}$	Rate of CO_2 uptake by M_h (g C m^{-2} h^{-1}) [Eqs. (G.14, 17, 18)]
$U_{i,f,k}$	Rate of $Q_{i,k}$ uptake by $M_{i,f}$ (g C m^{-2} h^{-1}) [Eqs. (G.3, 6)]
$U_{i,m,c}$	Rate of $A_{i,c}$ uptake by $M_{i,m}$ (g C m^{-2} h^{-1}) [Eqs. (G.9, 11)]
$U_{t,c}$	Rate of CH_4 uptake by M_t (g C m^{-2} h^{-1}) [Eqs. (G.25, 27)]
X_t	CH_4 oxidation by methanotrophs (g C m^{-2} h^{-1}) [Eqs. (G.21a, 22)]
X'_t	CH_4 oxidation by methanotrophs at saturating O_2 (g C m^{-2} h^{-1}) [Eqs. (G.1, 2, 4a)]
X'_t	Specific CH_4 oxidation by methanotrophs at saturating O_2, 30°C and high water potential (g C g^{-1} h^{-1}) [Eq. (G.18)] [†]
Y_f	Biomass yield from fermentation and acetogenic reactions (g $M_{i,f}$ g $Q_{i,c}^{-1}$) [Eqs. (G.3, 4)]
Y_h	Biomass yield from hydrogenotrophic methanogenic reaction (g M_h g CO_2-C^{-1}) [Eqs. (G.14, 15, 18)]
Y_m	Biomass yield from acetotrophic methanogenic reaction (g $M_{i,m}$ g $A_{i,c}^{-1}$) [Eqs. (G.9, 10)]
Y_{tG}	Biomass yield from methanotrophic growth respiration (g M_t-C g CH_4-C^{-1}) [Eqs. (G.25a, 26)]
Y_{tR}	Ratio of CH_4 respired vs. CH_4 oxidized by methanotrophs (g C g C^{-1}) [Eqs. (G.19, 20)]

[†] Indicates values taken from the literature and provided to the ecosystem model.

APPENDIX H: DENITRIFICATION

Heterotrophic Oxidation

Respiratory oxidation of soluble hydrolysis products by facultatively anaerobic heterotrophs creates a demand for electron acceptors, initially from O_2

$$R'_{i,h} = 32/12 \left\{ R'_h M_{i,h,a} \left[Q_{i,c} \right] / \left(K_{Rh} + \left[Q_{i,c} \right] \right) \right\} f_t \qquad \text{(H.1a)}$$

$$R'_{O_2i,h} = 32/12 \, R'_{i,h} \qquad \text{(H.1b)}$$

The demand for electron acceptors may be partially met by O_2 through radial diffusion to, and active uptake by, microbial microsites through a solution for the O_2 concentration at microbial microsites $[O_{2m}]$ at which diffusion = uptake

$$R_{O_2i,h} = 4\pi \, n \, M_{i,h,a} \, D_{S_{O_2}} \left(d_m d_w / (d_w - d_m) \right) \left(\left[O_{2s} \right] - \left[O_{2m} \right] \right) \qquad \text{(H.2a)}$$

$$R_{O_2i,h} = R'_{O_2i,h} \left[O_{2m} \right] / \left(\left[O_{2m} \right] + K_{O_2h} \right) \qquad \text{(H.2b)}$$

NO$_x$ Reduction

The demand for electron acceptors unmet by O_2 creates a demand from heterotrophic denitrifiers $(h = d)$ for other acceptors

$$R_e = 0.125 \left(R'_{O_2i,d} - R_{O_2i,d} \right) \qquad \text{(H.3)}$$

The unmet demand for electron acceptors may be transferred sequentially to NO_3^-, NO_2^-, and N_2O

$$R_{NO_3i,d} = 7 \, R_e \left[NO_3^- \right] / \left(\left[NO_3^- \right] + K_{NO_3d} \right) \qquad \text{(H.4)}$$

$$R_{NO_2i,d} = \left(7 \, R_e - R_{NO_3i,d} \right) \left[NO_2^- \right] / \left(\left[NO_2^- \right] + K_{NO_2d} \right) \qquad \text{(H.5)}$$

$$R_{N_2Oi,d} = 2 \left(7 \, R_e - R_{NO_3i,d} - R_{NO_2i,d} \right) \left[N_2O \right] / \left(\left[N_2O \right] + K_{N_2Od} \right) \qquad \text{(H.6)}$$

Microbial Growth

The oxidation of C [Eq. (H.1)] is coupled to the reduction of O_2 [Eq. (H.2)] by all heterotrophs

$$R_{i,h} = R'_{i,h} \, R_{O_2i,h} / R'_{O_2i,h} \qquad \text{(H.7)}$$

Additional oxidation of C by denitrifiers ($h = d$) is coupled to the sequential reduction of NO_3^-, NO_2^-, and N_2O

$$R_{i,d} = 0.429\, R_{NO_3i,d} + 0.429\, R_{NO_2i,d} + 0.214\, R_{N_2Oi,d} \tag{H.8}$$

Total oxidation of C by denitrifiers is therefore

$$R_i = R_{i,h} + R_{i,d} \tag{H.9}$$

The energy yields from aerobic and denitrifier growth respiration drive the uptake of additional hydrolysis products and their transformation into microbial biomass so that total uptake of hydrolysis products by obligate aerobes is

$$U_{i,h,c} = R_{Mi,h} + \left(R_{i,h} - R_{Mi,h}\right)\left(1.0 + Y_h\right) \qquad \left[R_{i,h} > R_{Mi,h}\right] \tag{H.10a}$$

$$U_{i,h,c} = R_{i,h} \qquad \left[R_{i,h} < R_{Mi,h}\right] \tag{H.10b}$$

and by facultative denitrifiers is

$$U_{i,h(h=d),c} = R_{Mi,h} + \left(R_{i,h} - R_{Mi,h}\right)\left(1.0 + Y_h\right) + $$
$$R_{i,d}\left(1.0 + Y_d\right) \qquad \left[R_{i,h} > R_{Mi,h}\right] \tag{H.11a}$$

$$U_{i,h(h=d),c} = R_{i,h} + R_{i,d}\left(1.0 + Y_d\right) \qquad \left[R_{i,h} < R_{Mi,h}\right] \tag{H.11b}$$

The energy yields are calculated by dividing the free energy change of the oxidation-reduction reactions by the energy required to construct new microbial biomass from hydrolysis products

$$Y_h = -\Delta G_h / G_M \tag{H.12a}$$

$$Y_d = -\Delta G_d / G_M \tag{H.12b}$$

Net growth of each heterotrophic population is the total uptake of hydrolysis products minus C oxidation for maintenance and growth respiration and for decomposition

$$\delta M_{i,h,j,c} / \delta t = F_j\, U_{i,h,c} - F_j\, R_{i,h} - D_{i,h,j,c} \qquad \left[R_{i,h} > R_{Mi,h}\right] \tag{H.13a}$$

$$\delta M_{i,h,j,c} / \delta t = F_j\, U_{i,h,c} - R_{Mi,h,j} - D_{i,h,j,c} \qquad \left[R_{i,h} < R_{Mi,h}\right] \tag{H.13b}$$

$$\delta M_{i,h,c} / \delta t = \sum_{j=1}^{J} \delta M_{i,h,j,c} / \delta t \tag{H.13c}$$

Glossary

$D_{i,h,j,c}$ Decomposition of heterotrophs (g C m^{-2} h^{-1}) [Eq. (H.13)]

$D_{S_{O_2}}$ Aqueous dispersivity-diffusivity of O$_2$ in soil (m^2 h^{-1}) [Eq. (H.2)]

d_m Radius of heterotrophic microsite (m) [Eq. (H.2)] †

d_w Radius of d$_m$ + water film at current water content (m) [Eq. (H.2)]

F_j Partitioning coefficient for j in M$_{i,h,j}$ [Eq. (H.13)] †

f_t Temperature function for microbial processes (dimensionless) [Eq. (H.1)] †

ΔG_d Free energy change of heterotrophic C oxidation–N reduction (kJ g C^{-1}) [Eq. (H.12)] †

ΔG_h Free energy change of heterotrophic C oxidation–O$_2$ reduction (kJ g C^{-1}) [Eq. (H.12)] †

G_M Energy required to construct new M from Q$_{i,k}$ (kJ g C^{-1}) [Eq. (H.12)] †

K_{NO_2d} Michaelis-Menten constant for reduction of NO$_2^-$ by heterotrophic denitrifiers (g N m^{-3}) [Eq. (H.5)] †

K_{NO_3d} Michaelis-Menten constant for reduction of NO$_3^-$ by heterotrophic denitrifiers (g N m^{-3}) [Eq. (H.4)] †

K_{N_2Od} Michaelis-Menten constant for reduction of N$_2$O by heterotrophic denitrifiers (g N m^{-3}) [Eq. (H.6)] †

K_{O_2h} Michaelis-Menten constant for reduction of O$_{2s}$ by heterotrophs (g O$_2$ m^{-3}) [Eq. (H.2)] †

K_{Rh} Michaelis-Menten constant for respiration of Q$_{i,c}$ by heterotrophs (g C m^{-3}) [Eq. (H.1)] †

$M_{i,h,a}$ Active biomass of heterotrophs (g C m^{-2}) [Eqs. (H.1, 2)]

$M_{i,h,j,c}$ Biomass of heterotrophs (g C m^{-2}) [Eq. (H.13)]

$[NO_2^-]$ Concentration of NO$_2^-$ in soil solution (g N m^{-3}) [Eq. (H.5)]

$[NO_3^-]$ Concentration of NO$_3^-$ in soil solution (g N m^{-3}) [Eq. (H.4)]

$[N_2O]$ Concentration of N$_2$O in soil solution (g N m^{-3}) [Eq. (H.6)]

n   Number of active heterotrophic microsites (g^{-1}) [Eq. (H.2)]

$[O_{2m}]$ O$_2$ concentration at heterotrophic microsites (g O$_2$ m^{-3}) [Eq. (H.2)]

$[O_{2s}]$ O$_2$ concentration in soil solution (g O$_2$ m^{-3}) [Eq. (H.2)]

$[Q_{i,c}]$ Concentration of soluble decomposition products of S$_{i,c}$ in soil solution (g C m^{-3}) [Eq. (H.1)]

R_e Electron transfer to N oxides by denitrifiers (mol e$^-$ m^{-2} h^{-1}) [Eqs. (H.3–6)]

R'_h Specific oxidation of Q$_{i,c}$ by heterotrophs at saturating [Q$_{i,c}$], 30°C and high water potential (g C g C^{-1} h^{-1}) [Eq. (H.1)] †

$R_{i,d}$ Oxidation of Q$_{i,c}$ coupled to reduction of N by denitrifiers (g C m^{-2} h^{-1}) [Eqs. (H.8–11)]

$R_{i,h}$ Oxidation of Q$_{i,c}$ coupled to reduction of O$_2$ by heterotrophs under ambient [O$_{2s}$] (g C m^{-2} h^{-1}) (H.7, 9–11, 13)]

$R'_{i,h}$ Oxidation of Q$_{i,c}$ coupled to reduction of O$_2$ by heterotrophs under saturating [O$_{2s}$] (g C m^{-2} h^{-1}) [Eqs. (H.1, 7)]

$R_{Mi,h}$ Maintenance respiration by heterotrophs (g C m^{-2} h^{-1}) [Eqs. (H.10, 11, 13)]

$R_{NO_2i,d}$ NO$_2^-$ reduction by heterotrophic denitrifiers (g N m^{-2} h^{-1}) [Eqs. (H.5, 6, 8)]

$R_{NO_3i,d}$ NO$_3^-$ reduction by heterotrophic denitrifiers (g N m^{-2} h^{-1}) [Eqs. (H.4–6, 8)]

$R_{N_2Oi,d}$ N_2O reduction by heterotrophic denitrifiers (g N m^{-2} h^{-1}) [Eqs. (H.6, 8)]

$R_{O_2i,h}$ O_2 reduction by heterotrophs under ambient $[O_{2s}]$ (g O_2 m^{-2} h^{-1}) [Eqs. (H.2, 3, 7)]

$R'_{O_2i,h}$ O_2 reduction by heterotrophs under saturating $[O_{2s}]$ (g O_2 m^{-2} h^{-1}) [Eqs. (H.1–3, 7)]

$U_{i,h,c}$ $Q_{i,c}$ uptake by $M_{i,h}$ (g C m^{-2} h^{-1}) [Eqs. (H.10, 11, 13)]

Y_d Biomass yield from heterotrophic reduction of N (g M g C^{-1}) [Eqs. (H.11, 12)]

Y_h Biomass yield from heterotrophic reduction of O_2 (g M g C^{-1}) [Eqs. (H.10, 12)]

† Indicates values taken from the literature and provided to the ecosystem model.

APPENDIX I: NITRIFICATION

Aerobic Oxidation — Reduction Reactions

Potential oxidation rate of $NH_{3(s)}$ to NO_2^- by ammonia oxidizers under non-limiting $O_{2(s)}$ depends on biomass and substrate concentrations

$$X'_{NH_3} = f_{tx} f_w \mu_{NH_3} M_{NH_3 a,c} \left\{ [CO_{2(s)}] / \left(K_{CO_2} + [CO_{2(s)}] \right) \right\}$$
$$\left\{ [NH_{3(s)}] / \left(K_{NH_3} + [NH_{3(s)}] \right) \right\}$$

(I.1)

Potential oxidation rate of NO_2^- to NO_3^- by nitrite oxidizers under non-limiting $O_{2(s)}$ depends on biomass and substrate concentrations

$$X'_{NO_2} = f_{tx} f_w \mu_{NO_2} M_{NO_2 a,c} \left\{ [CO_{2(s)}] / \left(K_{CO_2} + [CO_{2(s)}] \right) \right\}$$
$$\left\{ [NO_2] / \left(K_{NO_2} + [NO_2] \right) \right\}$$

(I.2)

Potential reduction rate of $CO_{2(s)}$ by ammonia oxidizers under non-limiting $O_{2(s)}$ is driven by potential oxidation rate of $NH_{3(s)}$

$$R'_{CO_2 - NH_3} = X'_{NH_3} r_{CO_2 - NH_3}$$

(I.3)

Potential reduction rate of $CO_{2(s)}$ by nitrite oxidizers under non-limiting $O_{2(s)}$ driven by potential oxidation rate of NO_2^-

$$R'_{CO_2 - NO_2} = X'_{NO_2} r_{CO_2 - NO_2}$$

(I.4)

Potential oxidation rate of reduced C for growth and maintenance processes by ammonia oxidizers under non-limiting $[O_{2(s)}]$ driven by their growth efficiency calculated from the energy yield of ammonia oxidation

$$X'_{CO_2-NH_3} = R'_{CO_2-NH_3}\left(1 - E_{O_2-NH_3}\right) \tag{I.5}$$

Potential oxidation rate of reduced C for growth and maintenance processes by nitrite oxidizers under non-limiting $[O_{2(s)}]$ driven by their growth efficiency calculated from the energy yield of nitrite oxidation

$$X'_{CO_2-NO_2} = R'_{CO_2-NO_2}\left(1 - E_{O_2-NO_2}\right) \tag{I.6}$$

Potential reduction rate of $O_{2(s)}$ by ammonia oxidizers under non-limiting $[O_{2(s)}]$ driven by potential oxidation rates of $NH_{3(s)}$ and reduced C

$$R'_{O_2-NH_3} = r_{O_2-NH_3} X'_{NH_3} + r_{O_2-CO_2} X'_{CO_2-NH_3} \tag{I.7}$$

Potential reduction rate of $O_{2(s)}$ by nitrite oxidizers under non-limiting $[O_{2(s)}]$ driven by potential oxidation rates of NO_2^- and reduced C

$$R'_{O_2-NO_2} = r_{O_2-NO_2} X'_{NO_2} + r_{O_2-CO_2} X'_{CO_2-NO_2} \tag{I.8}$$

Actual reduction rate of $O_{2(s)}$ by ammonia oxidizers under ambient $[O_{2(s)}]$ is calculated from $[O_{2(s)}]$ in microbial microsites at which spherical diffusion of $O_{2(s)}$ equals active uptake of $O_{2(s)}$

$$R_{O_2-NH_3} = 4\pi n M_{NH_3a,c} D_{O_2S}\left(d_m d_w/(d_w - d_m)\right)\left(\left[O_{2(s)}\right] - \left[O_{2(m)-NH_3}\right]\right) \tag{I.9a}$$

$$= R'_{O_2-NH_3}\left[O_{2(m)-NH_3}\right]\Big/\left(\left[O_{2(m)-NH_3}\right] + K_{O_2}\right) \tag{I.9b}$$

Actual reduction rate of $O_{2(s)}$ by nitrite oxidizers under ambient $[O_{2(s)}]$ is calculated from $[O_{2(s)}]$ in microbial microsites at which spherical diffusion of $O_{2(s)}$ equals active uptake of $O_{2(s)}$

$$R_{O_2-NO_2} = 4\pi n M_{NO_2a,c} D_{O_2S}\left(d_m d_w/(d_w - d_m)\right)\left(\left[O_{2(s)}\right] - \left[O_{2(m)-NO_2}\right]\right) \tag{I.10a}$$

$$= R'_{O_2-NO_2}\left[O_{2(m)-NO_2}\right]\Big/\left(\left[O_{2(m)-NO_2}\right] + K_{O_2}\right) \tag{I.10b}$$

The oxidation rate of $NH_{3(s)}$ and NO_2^- by ammonia and nitrite oxidizers under ambient $[O_{2(s)}]$ is calculated from the ratio of actual [Eqs. (I.9) and (I.10)] to potential [Eqs. (I.7) and (I.8)] reduction rates of $O_{2(s)}$

$$X_{NH_3} = X'_{NH_3} \, R_{O_2-NH_3} / R'_{O_2-NH_3} \tag{I.11}$$

$$X_{NO_2} = X'_{NO_2} \, R_{O_2-NO_2} / R'_{O_2-NO_2} \tag{I.12}$$

The reduction rate of $CO_{2(s)}$ by ammonia and nitrite oxidizers under ambient $[O_{2(s)}]$ is calculated from the oxidation rate of $NH_{3(s)}$ and NO_2^- under ambient $[O_{2(s)}]$

$$R_{CO_2-NH_3} = X_{NH_3} \, r_{CO_2-NH_3} \tag{I.13}$$

$$R_{CO_2-NO_2} = X_{NO_2} \, r_{CO_2-NO_2} \tag{I.14}$$

Oxidation rate of reduced C by ammonia oxidizers for growth and maintenance processes under ambient $[O_{2(s)}]$ is driven by the reduction rate of $CO_{2(s)}$ and their growth efficiency calculated from the energy yield of ammonia oxidation and the energy requirements for biomass synthesis

$$C_{CO_2-NH_3} = R_{CO_2-NH_3} \left(1 - E_{O_2-NH_3}\right) \tag{I.15}$$

Oxidation rate of reduced C by nitrite oxidizers for growth and maintenance processes under ambient $[O_{2(s)}]$ is driven by the reduction rate of $CO_{2(s)}$ and their growth efficiency calculated from the energy yield of nitrite oxidation and the energy requirements for biomass synthesis

$$X_{CO_2-NO_2} = R_{CO_2-NO_2} \left(1 - E_{O_2-NO_2}\right) \tag{I.16}$$

Anaerobic Oxidation — Reduction Reactions

Reduction of NO_2^- by ammonia oxidizers is driven by the demand for electron acceptors to oxidize reduced C (Eq. (I.5)] that was unmet by $O_{2(s)}$ (Eq. (I.15)]:

$$\mathbf{R}_{NO_2-NH_3} = \mathbf{r}_{NO_2-O_2} \, r_{O_2-CO_2} \left(X'_{CO_2-NH_3} - X_{CO_2-NH_3}\right)$$
$$\left\{\left[CO_{2(s)}\right] / \left(K_{CO_2} + \left[CO_{2(s)}\right]\right)\right\} \left\{\left[NO_2^-\right] / \left(K_{RNO_2} + \left[NO_2^-\right]\right)\right\} \tag{I.17}$$

The reduction of NO_2^- by ammonia oxidizers drives the reduction of $CO_{2(s)}$

$$\mathbf{R}_{CO_2-NH_3} = \mathbf{r}_{CO_2-NO_2} \, \mathbf{R}_{NO_2-NH_3} \tag{I.18}$$

Oxidation rate of C that is reduced through the reduction of NO_2^- by ammonia oxidizers depends on growth efficiency of ammonia oxidizers when reducing NO_2^-

$$X'_{CO_2-NH_3} = R_{CO_2-NH_3}\left(1-E_{NO_2}\right) \tag{I.19}$$

Growth of Nitrifiers

Growth of ammonia and nitrite oxidizers under ambient $[O_{2(s)}]$ is calculated from the difference between the reduction of $CO_{2(s)}$ [Eqs. (I.13), (I.14), and (I.18)] and the oxidation of C [Eqs. (I.15), (I.16), and (I.19)] minus losses from decomposition

$$\Delta M_{NH3a,c}/\Delta t = R_{CO_2-NH_3} + \mathbf{R}_{CO_2-NH_3} - X_{CO_2-NH_3} - X'_{CO_2-NH_3}$$

$$-\sum\nolimits_{j=1}^{J}\left\{f_{tm}\,\mathbf{X}_j\,M_{NH3a,c} + f_{tx}\,\mathbf{D}_j\,M_{NH3a,c}\right\} \tag{I.20}$$

$$\Delta M_{NO2a,c}/\Delta t = R_{CO2-NO2} - X_{CO2-NO2}$$

$$-\sum\nolimits_{j=1}^{J}\left\{f_{tm}\,\mathbf{X}_j\,M_{NO2a,c} + f_{tx}\,\mathbf{D}_j\,M_{NO2a,c}\right\} \tag{I.21}$$

Glossary

Subscripts Used to Define Spatial Resolution of Variables

a	Active component of M (Grant et al., 1993b)
c	Dry matter (DM) fraction of active component of M (Grant et al., 1993b)
j	Kinetic components of *a*, *c*
(s)	Soluble
(g)	Gaseous

Variables Used in Equations

$[CO_{2(s)}]$	CO_2 concentration in soil solution (μg mL^{-1}) [Eqs. (I.1, 2, 17)]
$\mathbf{D}_j$	Specific decomposition (g reduced C g $M_{j,c}$ $^{-1}$ h^{-1}) [Eqs. (I.20, 21)]
D_{O_2S}	Diffusivity of $O_{2(s)}$ at ambient temperature and water content (m^2 h^{-1}) [Eqs. (I.9, 10)]
d_m	Radius of microbial microsite for $M_{NH3a,c}$ and $M_{NO2a,c}$ (m) [Eqs. (I.9, 10)][†]
d_w	Radius of d_m + water film at current water content (m) [Eqs. (I.9, 10)]
$E_{O_2-NH_3}$	Growth efficiency of $M_{NH3a,c}$ using $O_{2(s)}$ (g C g C^{-1}) [Eqs. (I.5, 15))][†]
$E_{O_2-NO_2}$	Growth efficiency of $M_{NO2a,c}$ using $O_{2(s)}$ (g C g C^{-1}) [Eqs. (I.6, 16)][†]
E_{NO_2}	Growth efficiency of $M_{NH3a,c}$ using NO_2^- (g C g C^{-1}) [Eq. (I.19)][†]
f_{tm}	Temperature function for maintenance with a value of 1 at 30°C (dimensionless) [Eqs. (I.20,21)][†]

f_{tx} — Temperature function for oxidation with a value of 1 at 30°C (dimensionless) [Eqs. (I.1, 2, 20, 21)][†]

f_w — Water function for oxidation (dimensionless) [Eqs. (I.1, 2)][†]

K_{CO_2} — Michaelis-Menten constant for reduction of $CO_{2(s)}$ by $M_{NH3a,c}$ and $M_{NO2a,c}$ (µg ml⁻¹) [Eqs. (I.1, 2, 17)][†]

K_{RNO_2} — Michaelis-Menten constant for reduction of NO_2^- (µg ml⁻¹) [Eq. (I.17)][†]

K_{NH_3} — Michaelis-Menten constant for oxidation of $NH_{3(s)}$ (µg ml⁻¹) [Eq. (I.1)][†]

K_{NO_2} — Michaelis-Menten constant for oxidation of NO_2^- (µg ml⁻¹) [Eq. (I.2)][†]

K_{O_2} — Michaelis-Menten constant for reduction of $O_{2(s)}$ by $M_{NH3a,c}$ and $M_{NO2a,c}$ (µg ml⁻¹) [Eqs. (I.9, 10)][†]

$M_{NH3a,c}$ — Biomass of active $NH_{3(s)}$ oxidizers (µg C g⁻¹) [Eqs. (I.1, 9, 20)]

$M_{NO2a,c}$ — Biomass of active NO_2^- oxidizers (µg C g⁻¹) [Eqs. (I.2, 10, 21)]

μ_{NH3} — Specific oxidation rate of $NH_{3(s)}$ by $M_{NH3a,c}$ (g NH_3-N g⁻¹ $M_{NH3a,c}$ h⁻¹) at 30°C under non-limiting conditions [Eq. (I.1)][†]

μ_{NO2} — Specific oxidation rate of NO_2^- by $M_{NO2a,c}$ (g NO_2^--N g⁻¹ $M_{NO2a,c}$ h⁻¹) at 30°C under non-limiting conditions [Eq. (I.2)][†]

$[NH_{3(s)}]$ — NH_3 concentration in soil solution (µg ml⁻¹) [Eq. (I.1)]

$[NO_2]$ — NO_2^- concentration in soil solution (µg ml⁻¹) [Eqs. (I.2, 17)]

n — Number of active $M_{NH3a,c}$ or $M_{NO2a,c}$ (µg⁻¹) [Eqs. (I.9, 10)]

$[O_{2(m)-NH3}]$ — O_2 concentration at $M_{NH3a,c}$ microsite (µg ml⁻¹) [Eq. (I.9)]

$[O_{2(m)-NO2}]$ — O_2 concentration at $M_{NO2a,c}$ microsite (µg ml⁻¹) [Eq. (I.10)]

$[O_{2(s)}]$ — O_2 concentration in soil solution (µg ml⁻¹) [Eqs. (I.9, 10)]

$R'_{CO_2-NH_3}$ — Potential rate of $CO_{2(s)}$ reduction by $M_{NH3a,c}$ under non-limiting $[O_{2(s)}]$ (µg g⁻¹ h⁻¹) [Eqs. (I.3, 5)]

$R'_{CO_2-NO_2}$ — Potential rate of $CO_{2(s)}$ reduction by $M_{NO2a,c}$ under non-limiting $[O_{2(s)}]$ (µg g⁻¹ h⁻¹) [Eqs. (I.4, 6)]

$R_{CO_2-NH_3}$ — Actual rate of $CO_{2(s)}$ reduction by $M_{NH3a,c}$ from R_{O2-NH3} under ambient $[O_{2(s)}]$ (µg g⁻¹ h⁻¹) [Eqs. (I.13, 15, 20)]

$\mathbf{R}_{CO_2-NH_3}$ — Actual rate of $CO_{2(s)}$ reduction by $M_{NH3a,c}$ from $\mathbf{R}_{NO2-NH3}$ under ambient $[O_{2(s)}]$ (µg g⁻¹ h⁻¹) [Eqs. (I.18–20)]

$R_{CO_2-NO_2}$ — Actual rate of $CO_{2(s)}$ reduction by $M_{NO2a,c}$ from R_{O2-NO2} under ambient $[O_{2(s)}]$ (µg g⁻¹ h⁻¹) [Eqs. (I.14, 16, 21)]

$\mathbf{R}_{NO_2-NH_3}$ — Actual rate of NO_2^- reduction by $M_{NH3a,c}$ under ambient $[O_{2(s)}]$ (µg g⁻¹ h⁻¹) [Eqs. (I.17, 18)]

$R'_{O_2-NH_3}$ — Potential rate of $O_{2(s)}$ reduction by $M_{NH3a,c}$ under non-limiting $[O_{2(s)}]$ (µg g⁻¹ h⁻¹) [Eqs. (I.7, 9, 11)]

$R'_{O_2-NO_2}$ — Potential rate of $O_{2(s)}$ reduction by $M_{NO2a,c}$ under non-limiting $[O_{2(s)}]$ (µg g⁻¹ h⁻¹) [Eqs. (I.8, 10, 12)]

$R_{O_2-NH_3}$ — Actual rate of $O_{2(s)}$ reduction by $M_{NH3a,c}$ under ambient $[O_{2(s)}]$ (µg g⁻¹ h⁻¹) [Eqs. (I.9, 11)]

$R_{O_2-NO_2}$ — Actual rate of $O_{2(s)}$ reduction by $M_{NO2a,c}$ under ambient $[O_{2(s)}]$ (µg g⁻¹ h⁻¹) [Eqs. (I.10, 12)]

$r_{CO_2-NH_3}$ — Ratio of $CO_{2(s)}$ reduced to $NH_{3(s)}$ oxidized by $M_{NH3a,c}$ (g g⁻¹) [Eqs. (I.3, 13)][†]

$r_{CO_2-NO_2}$ — Ratio of $CO_{2(s)}$ reduced to NO_2^- oxidized by $M_{NO2a,c}$ (g g⁻¹) [Eqs. (I.4, 14)][†]

$\mathbf{r}_{CO_2-NO_2}$ — Ratio of $CO_{2(s)}$ reduced to NO_2^- reduced by $M_{NH3a,c}$ (g g⁻¹) [Eq. (I.18)][†]

$r_{NO_2-O_2}$	Ratio of NO_2^- reduced to O_2 reduced by $M_{NO_{2a,c}}$ (g g^{-1}) [Eq. (I.17)][†]
$r_{O_2-CO_2}$	Ratio of $O_{2(s)}$ reduced to C oxidized by $M_{NH_{3a,c}}$ and $M_{NO_{2a,c}}$ (g g^{-1}) [Eqs. (I.7, 8, 17)][†]
$r_{O_2-NH_3}$	Ratio of $O_{2(s)}$ reduced to $NH_{3(s)}$ oxidized by $M_{NH_{3a,c}}$ (g g^{-1}) [Eq. (I.7)][†]
$r_{O_2-NO_2}$	Ratio of $O_{2(s)}$ reduced to $NH_{3(s)}$ oxidized by $M_{NO_{2a,c}}$ (g g^{-1}) [Eq. (I.8)][†]
$X'_{CO_2-NH_3}$	Potential rate of C oxidation by $M_{NH_{3a,c}}$ under non-limiting [$O_{2(s)}$] (μg g^{-1} h^{-1}) [Eqs. (I.5, 7, 17)]
$X'_{CO_2-NO_2}$	Potential rate of C oxidation by $M_{NO_{2a,c}}$ under non-limiting [$O_{2(s)}$] (μg g^{-1} h^{-1}) [Eqs. (I.6, 8)]
$X_{CO_2-NH_3}$	Actual rate of C oxidation by $M_{NH_{3a,c}}$ from $O_{2(s)}$ reduction (μg g^{-1} h^{-1}) [Eqs. (I.15, 17, 20)]
$X'_{CO_2-NH_3}$	Actual rate of C oxidation by $M_{NH_{3a,c}}$ from NO_2^- reduction (μg g^{-1} h^{-1}) [Eqs. (I.19, 20)]
$X_{CO_2-NO_2}$	Actual rate of C oxidation by $M_{NO_{2a,c}}$ from $O_{2(s)}$ reduction (μg g^{-1} h^{-1}) [Eqs. (I.16, 21)]
X_j	Specific maintenance respiration (g reduced C g $M_{j,c}$ $^{-1}$ h^{-1}) [Eqs. (I.20, 21)]
X'_{NH_3}	Potential rate of NH_3 oxidation by $M_{NH_{3a,c}}$ under non-limiting [$O_{2(s)}$] (μg g^{-1} h^{-1}) [Eqs. (I.1, 3, 7, 11)]
X'_{NO_2}	Potential rate of NO_2^- oxidation by $M_{NO_{2a,c}}$ under non-limiting [$O_{2(s)}$] (μg g^{-1} h^{-1}) [Eqs. (I. 2, 4, 8, 12)]
X_{NH_3}	Actual rate of NH_3 oxidation by $M_{NH_{3a,c}}$ under ambient [$O_{2(s)}$] (μg g^{-1} h^{-1}) [Eqs. (I.11, 13)]
X_{NO_2}	Actual rate of NO_2^- oxidation by $M_{NO_{2a,c}}$ under ambient [$O_{2(s)}$] (μg g^{-1} h^{-1}) [Eqs. (I.12, 14)]

[†] Indicates values taken from the literature and provided to the ecosystem model.

REFERENCES

Alberta Forest Service. 1985. *Alberta Phase 3 Forest Inventory: Yield Tables for Unmanaged Stands*. Alberta Energy and Natural Resources. Edmonton, Alberta.

Ball, J.T. 1988. An Analysis of Stomatal Conductance. Ph.D. thesis, Stanford University.CA. 89 pp.

Black, T.A., G. Den Hartog, H.H. Neumann, P.D. Blanken, P.C. Yang, C. Russell, Z. Nesic, X. Lee, S.G. Chen, R. Staebler, and M.D. Novak. 1996. Annual cycles of water vapour and carbon dioxide fluxes in and above a boreal aspen forest. *Global Change Biol.*, 2:219-229.

Chen, J.M., P.M. Rich, S.T. Gower, J.M. Norman, and S. Plummer. 1997. Leaf area index of boreal forests: theory, techniques and measurements. *J. Geophys. Res.*, 102:429-443.

Choudhury, B.J., and J.L. Monteith. 1988. A four-layer model for the heat budget of homogenous land surfaces. *Quart. J. Royal Met. Soc.*, 114:373-398.

Cowan, I.R. 1965. Transport of water in the soil-plant-atmosphere system. *J. Appl. Ecol.*, 2:221-239.

de Vries, D.A. 1963. Thermal properties of soils. In R. van Wijk (Ed.). *Physics of Plant Environment*. North Holland Publishing, Amsterdam, The Netherlands, 210-235.

Farquhar, G.D., S. von Caemmerer, and J.A. Berry. 1980. A biochemical model of photosynthetic CO_2 assimilation in leaves of C_3 species. *Planta,* 149:78-90.

Grant, R.F. 1989. Test of a simple biochemical model for photosynthesis of maize and soybean leaves. *Agric. For. Meteorol.,* 48:59-74.

Grant, R.F. 1991. The distribution of water and nitrogen in the soil-crop system: a simulation study with validation from a winter wheat field trial. *Fert. Res.,* 27:199-214.

Grant, R.F. 1992a. Simulation of carbon dioxide and water deficit effects upon photosynthesis of soybean leaves with testing from growth chamber studies. *Crop Sci.,* 32:1313-1321.

Grant, R.F. 1992b. Simulation of carbon dioxide and water deficit effects upon photosynthesis and transpiration of soybean canopies with testing from growth chamber studies. *Crop Sci.,* 32:1322-1328.

Grant, R.F. 1993a. Simulation model of soil compaction and root growth. I. Model development. *Plant Soil,* 150:1-14.

Grant, R.F. 1993b. Simulation model of soil compaction and root growth. II. Model testing. *Plant Soil,*150:15-24.

Grant, R.F. 1994a. Simulation of ecological controls on nitrification. *Soil Biol. Biochem.,* 26:305-315.

Grant, R.F. 1994b. Simulation of competition between barley (*Hordeum vulgare* L.) and wild oat (*Avena fatua* L.) under different managements and climates. *Ecol. Modelling,* 71:269-287.

Grant, R.F. 1995a. Mathematical modelling of nitrous oxide evolution during nitrification. *Soil Biol. Biochem.,* 27:1117-1125.

Grant, R.F. 1995b. Salinity, water use and yield of maize: testing of the mathematical model ecosys. *Plant Soil,* 172:309-322.

Grant, R.F. 1996. ecosys. In *Global Change and Terrestrial Ecosystems Focus 3 Wheat Network. Model and Experimental Meta Data,* 2nd ed. GCTE Focus 3 Office, NERC Centre for Ecology and Hydrology, Wallingford, Oxon, U.K., 65-74.

Grant, R.F. 1997. Changes in soil organic matter under different tillage and rotation: mathematical modelling in ecosys. *Soil Sci. Soc. Am. J.,* 61:1159-1174.

Grant, R.F. 1998a. Simulation of methanogenesis in the mathematical model *ecosys. Soil Biol. Biochem.,* 30:883-896.

Grant, R.F. 1998b. Simulation in *ecosys* of root growth response to contrasting soil water and nitrogen. *Ecol. Modelling,* 107:237-264.

Grant, R.F. 1999. Simulation of methanotrophy in the mathematical model ecosys. *Soil Biol. Biochem.,* 31:287-297.

Grant, R.F., D.B. Peters, E.M. Larson, and M.G. Huck. 1989. Simulation of canopy photosynthesis in maize and soybean. *Agric. For. Meteorol.,* 48:75-92.

Grant, R.F., and D.D. Baldocchi. 1992. Energy transfer over crop canopies: simulation and experimental verification. *Agric. For. Meteorol.,* 61:129-149.

Grant, R.F., and J.D. Hesketh. 1992. Canopy structure of maize (*Zea mays* L.) at different populations: simulation and experimental verification. *Biotronics,* 21:11-24.

Grant, R.F., N.G. Juma, and W.B. McGill. 1993a. Simulation of carbon and nitrogen transformations in soils. I. Mineralization. *Soil Biol. Biochem.,* 27:1317-1329.

Grant, R.F., N.G. Juma, and W.B. McGill. 1993b. Simulation of carbon and nitrogen transformations in soils. II. Microbial biomass and metabolic products. *Soil Biol. Biochem.,* 27:1331-1338.

Grant, R.F., M. Nyborg, and J. Laidlaw 1993c. Evolution of nitrous oxide from soil. I. Model development. *Soil Sci.,* 156:259-265.

Grant, R.F., M. Nyborg, and J. Laidlaw 1993d. Evolution of nitrous oxide from soil. II. Experimental results and model testing. *Soil Sci.,* 156:266-277.

Grant, R.F., P. Rochette, and R.L. Desjardins. 1993e. Energy exchange and water use efficiency of crops in the field: validation of a simulation model. *Agron. J.*, 85:916-928.

Grant, R.F., and P. Rochette. 1994. Soil microbial respiration at different temperatures and water potentials: theory and mathematical modelling. *Soil Sci. Soc. Am. J.*, 58:1681-1690.

Grant, R.F., R.L. Garcia, P.J. Pinter Jr., D. Hunsaker, G.W. Wall, B.A. Kimball, and R.L. LaMorte. 1995a. Interaction between atmospheric CO_2 concentration and water deficit on gas exchange and crop growth: Testing of *ecosys* with data from the Free Air CO_2 Enrichment (FACE) experiment. *Global Change Biol.*, 1:443-454.

Grant, R.F., R.C. Izaurralde, and D.S. Chanasyk. 1995b. Soil temperature under different surface managements: testing a simulation model. *Agric. For. Meteorol.*, 73:89-113.

Grant, R.F., B.A. Kimball, P.J. Pinter Jr., G.W. Wall, R.L. Garcia, R.L. LaMorte, and D.J. Hunsaker. 1995c. CO_2 effects on crop energy balance: testing *ecosys* with a Free Air CO_2 Enrichment (FACE) experiment. *Agron. J.*, 87:446-457.

Grant, R.F., and D.J. Heaney. 1997. Inorganic phosphorus transformation and transport in soils: mathematical modelling in ecosys. *Soil Sci. Soc. Am. J.*, 61:752-764.

Grant, R.F., and J.A. Robertson. 1997. Phosphorus uptake by root systems: mathematical modelling in ecosys. *Plant Soil*, 188:279-297.

Grant, R.F., R.C. Izaurralde, M. Nyborg, S.S. Malhi, E.D. Solberg, and D. Jans-Hammermeister. 1998. Modelling tillage and surface residue effects on soil C storage under current vs. elevated CO_2 and temperature in *ecosys*. In R. Lal, J.M. Kimble, R.F. Follett, and B.A. Stewart (Eds.). *Soil Processes and the Carbon Cycle*. CRC Press, Boca Raton, FL, 527-547.

Grant, R.F., and E. Pattey. 1999. Mathematical modelling of nitrous oxide emissions from an agricultural field during spring thaw. *Global Biogeochem. Cycles*, 13:679-694.

Grant, R.F., T.A. Black, G. den Hartog, J.A. Berry, S.T. Gower, H.H. Neumann, P.D. Blanken, P.C. Yang, and C. Russell. 1999a. Diurnal and annual exchanges of mass and energy between an aspen-hazelnut forest and the atmosphere: testing the mathematical model *ecosys* with data from the BOREAS experiment. *J. Geophys. Res.*, 104:699-717.

Grant, R.F., G.W. Wall, B.A. Kimball, K.F.A. Frumau, P.J. Pinter Jr., D.J. Hunsaker, and R.L. Lamorte. 1999b. Crop water relations under different CO_2 and irrigation: testing of ecosys with the Free Air CO_2 Enrichment (FACE) experiment. *Agric. For. Meteorol.*, 95:27-51.

Grant, R.F., and I.A. Nalder. 2000. Climate change effects on net carbon exchange of a boreal aspen-hazelnut forest: estimates from the ecosystem model *ecosys*. *Global Change Biol.*, 6:183-200.

Grant, R.F., P.G. Jarvis, J.M. Massheder, S.E. Hale, J.B. Moncrieff, M. Rayment, S.L. Scott, and J.A. Berry. 2001a. Controls on carbon and energy exchange by a black spruce — moss ecosystem: testing the mathematical model *ecosys* with data from the BOREAS experiment. *Global Biogeochem. Cycles*, in press.

Grant, R.F., N.G. Juma, J.A. Robertson, R.C. Izaurralde, and W.B. McGill. 2001b. Long term changes in soil C under different fertilizer, manure and rotation: testing the mathematical model *ecosys* with data from the Breton plots. *Soil Sci. Soc. Am. J.*, in press.

Grant, R.F., B.A. Kimball, T.J. Brooks, G.W. Wall, P.J. Pinter Jr., D.J. Hunsaker, F.J. Adamsen, R.L. Lamorte, S.J. Leavitt, T.L. Thompson, and A.D. Matthias. 2001c. Interactions among CO_2, N and climate on energy exchange of wheat: model theory and testing with a free air CO_2 enrichment (FACE) experiment. *Agron. J.*, in press.

Itoh, S., and S.A. Barber. 1983. Phosphorus uptake by six plant species as related to root hairs. *Agron. J.*, 75:457-461.

Jarvis, P.G., J.M. Massheder, S.E. Hale, J.B. Moncrieff, M Rayment, and S.L. Scott. 1997. Seasonal variation in carbon dioxide, water vapor, and energy exchanges of a boreal black spruce forest. *J. Geophys. Res.*, 102:953-966.

Kimmins, J.P. 1987. *Forest Ecology*. Macmillan, New York, 531 pp.

Makino, A., M. Harada, T. Sato, H. Nakano, and T. Mae. 1997. Growth and N allocation in rice plants under CO_2 enrichment. *Plant Physiol.*, 115:199-203.

Monteith, J.L. 1995. A reinterpretation of stomatal response to humidity. *Plant, Cell Environ.*, 18:357-364.

Perrier, A. 1982. Land surface processes: vegetation. In *Atmospheric General Circulation Models*. P.S. Eagleson (Ed.). Cambridge Univ. Press. Cambridge, U.K., 395-448.

Reid, J.B., and M.G. Huck. 1990. Diurnal variation of crop hydraulic resistance: a new analysis. *Agron. J.*, 82:827-834.

Rogers, G.S., L. Payne, P. Milham, and J. Conroy. 1993. Nitrogen and phosphorus requirements of cotton and wheat under changing CO_2 concentrations. In *Plant Nutrition: from Genetic Engineering to Field Practice*. J. Barrow (Ed.). Kluwer, Dordrecht, The Netherlands, 257-260.

Stitt, M. 1991. Rising CO_2 levels and their potential significance for carbon flow in photosynthetic cells. *Plant Cell Environ.*, 14:741-762.

Thornley, J.H. 1995. Shoot:root allocation with respect to C, N and P: an investigation and comparison of resistance and teleonomic models. *Ann. Bot.*, 75:391-405.

Trofymow, J.A., C.M. Preston, and C.E. Prescott. 1995. Litter quality and its potential effect on decay rates of materials from Canadian forests. *Water Air Soil Poll.*, 82:215-226.

Voroney, R.P., and E.A. Paul. 1984. Determination of k_c and k_n *in situ* for calibration of the chloroform fumigation-incubation method. *Soil Biol. Biochem.*, 16:9-14.

Waring, R.H. and S.W. Running. 1998. *Forest Ecosystems,* 2nd ed. Academic Press, San Diego CA.

Zur, B., and Jones, J.W. 1981. A model for the water relations, photosynthesis and expansive growth of crops. *Water Resour. Res.*, 17:311-320.

CHAPTER 7

Application of RZWQM for Soil Nitrogen Management

Liwang Ma, M. J. Shaffer, and L. R. Ahuja

CONTENTS

INTRODUCTION

Carbon and nitrogen modeling is one of the most challenging areas in agricultural and environmental sciences. It involves processes related to chemical, physical, and biological aspects of the soil system. Correct simulation of carbon and nitrogen cycling also determines the success of other components of the agricultural systems (e.g., plant growth). Models of carbon and nitrogen in soil systems cover a wide spectrum ranging from upland agricultural systems to wetland ecosystems, from irrigated land to dry land, from agriforestry to farmed land, from laboratory to field studies, and from point source (e.g., landfill) to non-point source contamination studies. Thus far, carbon and nitrogen models have been used in many aspects of research, decision-making, and policy-making processes. Many management practices have been simulated for their effects on carbon and nitrogen cycling, including tillage, irrigation, fertilizer/manure applications, surface crop residue coverage, and fauna activity. Applications of the models range from short-term predictions of experimental results to long-term simulations of global warming.

For a model to be used as a tool for soil nitrogen management, it must simulate all the nitrogen processes in the context of agricultural systems and their responses to various environmental conditions (e.g., soil moisture, soil temperature, and agricultural management practices). In addition, and because most understanding of soil nitrogen processes is obtained from laboratory studies under isolated conditions, the assembled model needs to be reevaluated under field conditions. However, one difficulty with system model evaluation is that very few experiments have been designed to measure all the fates of nitrogen in an agricultural system so that a rigorous model calibration and evaluation can be carried out. As a result, agricultural system models are seldom tested as rigorously as single-process 'models. Furthermore, estimated model parameters are generally not unique and may not be transferable from location to location, or from time to time. A much more interactive collaboration between experimental scientists and model developers is needed in order to build better and practically usable models.

Despite problems with model applications to real-world scenarios, carbon/nitrogen models have achieved some success at various process levels of agricultural systems, ranging from functional models, such as simple regression equation on yield response to nitrogen application rate, to process-based whole system models, such as NTRM (Nitrogen, Tillage, and crop-Residue Management) (Shaffer and Larson, 1987); NLEAP (Nitrate Leaching and Economic Analysis Package) (Shaffer et al., 1991); RZWQM (Root Zone Water Quality Model) (Ahuja et al., 2000a); GPFARM (Great Plains Framework for Agricultural Resource Management) (Shaffer et al., 2000b); CENTURY (Parton et al., 1994); CERES (Hanks and Ritchie, 1991); GLEAMS (Groundwater Loading Effects of Agricultural Management Systems) (Leonard et al., 1987); EPIC (Erosion Productivity Impact Calculator) (Williams, 1995); ECOSYS (Grant 1995a; b); SUNDIAL (Bradbury et al., 1993); DAISY (Hansen et al., 1991); SOILN (Johnsson et al., 1987; Eckersten and Jansson, 1991); and ANIMO (Berghuijs van Dijk et al., 1985). Most of the models and their applications have been reviewed in other chapters (Ma and Shaffer, 2001; McGechan and Wu, 2001; Grant, 2001). The purpose of this chapter is to document RZWQM applications and provide further understanding of agricultural problems with the use of an agricultural system model.

The RZWQM is a whole-system model for simulating agricultural production and environmental quality. It includes all the major components of an agricultural system, such as plant growth, water movement and usage, heat transport, chemical transport/transformation, soil carbon/nitrogen dynamics, and agricultural management practices. A completed version of the model was released in 1992 to be used in the MSEA (Management Systems Evaluation Areas) projects. Tested results were published in 1999 (Wu et al., 1999; Ghidey et al., 1999; Jaynes and Miller, 1999; Martin and Watts, 1999; Landa et al., 1999). A more comprehensive summary of model applications is available in Ma et al. (2000a). This chapter focuses on nitrogen processes in RZWQM with example scenarios to demonstrate the applications of RZWQM in nitrogen management. Because detailed descriptions of model components are available in Ahuja et al. (2000a), only the carbon/nitrogen-related processes are presented here.

SIMULATED CARBON/NITROGEN PROCESSES IN RZWQM

Soil Carbon/Nitrogen Dynamics

Similar to other models, RZWQM has the capability of simulating both organic and inorganic carbon/nitrogen in the soil. Organic nitrogen is calculated from C:N ratios and organic carbon content. There are five organic carbon pools: two for surface residues and three for humus materials (known as soil organic matter). These five pools are dynamically linked together as shown in Figure 7.1. The fast residue pool has a C:N ratio of 80, and the slow one has a C:N ratio of 8 (modified to account for manure; Ma et al., 1998a). Partitioning of fresh residues between the fast residue and slow residue pools is based on their C:N ratio and N mass balance. The three

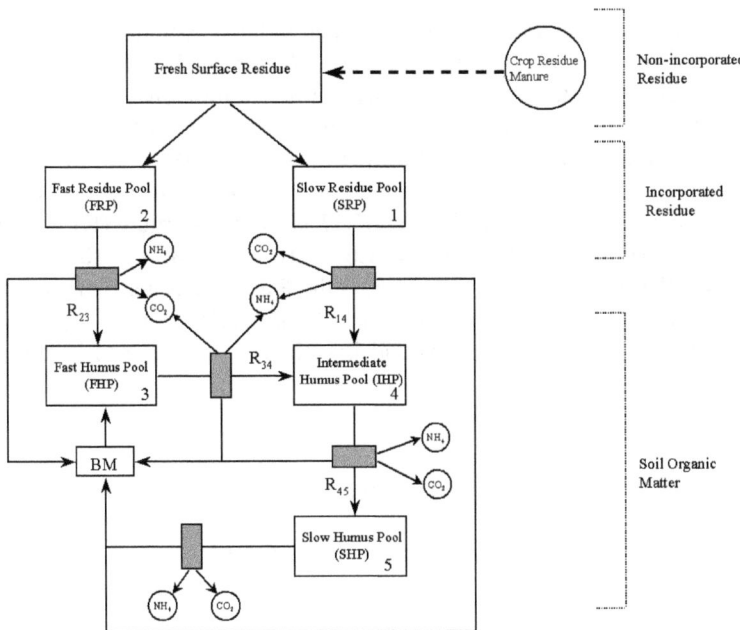

Figure 7.1   A schematic diagram of residue and soil organic matter pools. R_{14}, R_{23}, R_{34}, and R_{45} are inter-pool mass transfer coefficients. BM is microbial biomass. (Modified from Ma, L., M.J. Shaffer, J.K. Boyd, R. Waskom, L.R. Ahuja, K.W. Rojas, and C. Xu. 1998a. Manure management in an irrigated silage corn field: experiment and modeling. *Soil Sci. Soc. Am. J.*, 62:1006-1017.)

organic matter (OM) pools have C:N ratios of 8 (fast), 10 (intermediate), and 12 (slow), respectively. In addition, there are three types of soil microorganisms involved in carbon/nitrogen transformations: aerobic heterotrophs, autotrophs, and facultative heterotrophs. All three microbial pools have a C:N ratio of 8. For convenience, an index number was assigned to each carbon pool based on the notation of Shaffer et al. (2000a) [1 = slow residue pool (SRP), 2 = fast residue pool (FRP), 3 = fast humus pool (FHP), 4 = intermediate humus pool (IHP), 5 = slow humus pool (SHP), 6 = CO_2, 7 = aerobic heterotrophs (soil decomposers), 8 = autotrophs (nitrifiers), 9 = facultative heterotrophs (facultative anaerobes)].

Each residue or humus pool is subject to a first-order decay with respect to carbon content (Shaffer et al., 2000a):

$$r_i = -k_i C_i \qquad 1 < i < 5 \tag{7.1}$$

where r_i is the decay rate of pool i (μg – C/g/d), C_i is carbon concentration (μg – C/g soil), and k_i is a first-order decay rate (1/d) and is a function of soil environmental variables as follows:

$$k_i = f_{aer} \left(\frac{k_b T}{h_p} \right) A_i \ e^{-\frac{E_a}{R_g T}} \frac{[O_2]}{\left[H^{kh} \gamma_1^{kh} \right]} P_{het} \tag{7.2}$$

where $[O_2]$ is O_2 concentration in the soil water with the assumption that oxygen in soil air is not limited (moles O_2/liter pore water); H is the hydrogen ion concentration (moles H/liter pore water); γ_1 is the activity coefficient for monovalent ions ($1/\gamma_1^{kh}$ = 3.1573×10^3 if pH > 7.0, and $1/\gamma_1^{kh}$ = 1.0 if pH ≤ 7.0); kh is hydrogen ion exponent for decay of organic matter (= 0.167 for pH ≤ 7.0 and = –0.333 for pH > 7.0), P_{het} is the population of aerobic heterotrophic microbes [# organisms/g soil]; k_b is the Boltzmann constant (1.383×10^{-23} J/deg K); T is soil temperature (K); A_i is a pool-specific rate coefficient (s/d) with A_1 = 1.67×10^{-7}, A_2 = 8.14×10^{-6}, A_3 = 2.5×10^{-7}, A_4 = 5.0×10^{-8}, A_5 = 4.5×10^{-10}; h_p is the Planck constant (6.63×10^{-34} J.s); R_g is the universal gas constant (1.99×10^{-3} kcal/mole/deg K); and f_{aer} is a factor for the extent of aerobic conditions and is estimated from Figure 1 of Linn and Doran (1984):

$$f_{aer} = 0.0075 \ P_\theta \qquad\qquad P_\theta \leq 20$$

$$f_{aer} = -0.253 + 0.0203 \ P_\theta \qquad 20 < P_\theta < 59 \tag{7.3}$$

$$f_{aer} = 41.1 \ e^{-0.0625 \ P_\theta} \qquad\qquad P_\theta \geq 59$$

where P_θ is water-filled pore space, and E_a [= 15.1 + 12.3 U, U is ionic strength (mole)] is the apparent activation energy (kcal/mole). Obtained model parameters associated with soil carbon decay result in a turnover time of 5, 20, and 2000 yr for the fast, intermediate, and slow soil humus pools, respectively. Parameters related to surface residue pools are calibrated from wheat, corn (maize), and soybean residue data (Figure 7.2). A fraction of decayed organic materials is transferred between pools as denoted by R_{14}, R_{23}, R_{34}, and R_{45} in Figure 7.1. A set of calibrated R values by Ma et al. (1998a) were $R_{14} = 0.6$, $R_{23} = 0.1$, $R_{34} = 0.1$, and $R_{45} = 0.1$. These values were based on an experimental study of manure management in a corn field. Because the C:N ratios are different among pools, nitrogen conservation is observed during the transformations. The remainder of the carbon is released as CO_2, and the model uses CO_2 or CH_4 as a carbon sink or source.

Nitrogen is released as inorganic NH_4 during the decay processes and may be nitrified to NO_3 following a zero-order equation:

$$r_{nit} = k_n \tag{7.4}$$

with k_n being function of:

$$k_n = f_{aer}\left(\frac{k_b T}{h_p}\right) A_{nit}\, e^{-\frac{E_{an}}{R_g T}} \frac{[O_2]^{1/2}}{[H^{kh}\gamma_1^{kh}]} P_{aut} \tag{7.5}$$

where r_{nit} is the zero-order nitrification rate (moles N/liter/d); A_{nit} ($= 1.0 \times 10^{-9}$) is the nitrification rate coefficient (s/d/organism); P_{aut} is the autotrophic biomass population (nitrifiers) [# organisms/g soil]; and E_{an} [$= 12.64 + 12.3$ U] is the apparent activation energy (kcal/mole).

Nitrate from nitrification or applied commercial fertilizers is subject to denitrification under anaerobic conditions, and the denitrification rate is described using a first-order equation:

$$r_{den} = k_{den} C_{NO_3} \tag{7.6}$$

where C_{NO_3} is the NO_3 concentration in soil solution (mole NO_3/liter pore water); k_{den} is the first-order denitrification rate (mole N/liter/d) and can be expressed as:

$$k_{den} = f_{anaer}\left(\frac{k_b T}{h_p}\right) A_{den}\, e^{-\frac{E_{den}}{R_g T}} \frac{C_s}{[H^{kh}\gamma_1^{kh}]} P_{ana} \tag{7.7}$$

where C_s is the weighted carbon substrate concentration (μg C/g soil), A_{den} ($= 1.0 \times 10^{-13}$) is the denitrification rate coefficient (s/d/organism), E_{den} is the apparent activation energy for denitrification (45.0 kcal/mole), and P_{ana} is the anaerobic microbe biomass [# organisms/g soil]. f_{anaer} is anaerobic factor expressed as (Linn and Doran, 1984):

$$f_{anaer} = 0.000304\, e^{0.0815 P_\theta} \tag{7.8}$$

and C_s is weighted on the carbon contents of the five pools based on the relative contributions of each organic material pool:

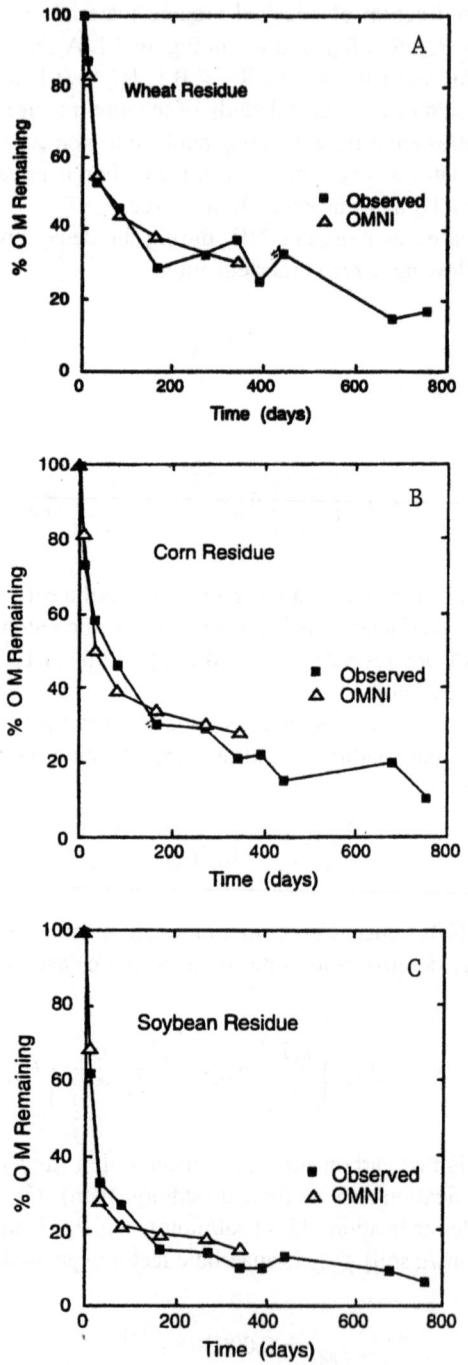

Figure 7.2   Calibrated vs. observed decay of surface crop residues. (From Shaffer, M.J., K.W. Rojas, D.G. DeCoursey, and C.S. Hebson. 2000a. Nutrient chemistry processes (OMNI). In L.R. Ahuja, K.W. Rojas, J.D. Hanson, M.J. Shaffer, and L. Ma (Eds.). *The Root Zone Water Quality Model.* Water Resources Publications LLC, Highlands Ranch, CO, 119-144.)

$$C_s = 0.234\,C_1 + C_2 + 0.0234\,C_3 + 0.00234\,C_4 + 0.000234\,C_5 \qquad (7.9)$$

As discussed, three microbial populations are involved in nitrogen transformations; that is, heterotrophic decomposers, nitrifiers, and facultative anaerobes. Minimum populations set in the RZWQM are 50,000, 500, and 5000 organisms/g soil, respectively. It is suggested to initialize the pools with historical management practices (Ma et al., 1998a). There is no further microbial death if populations are less than their respective minimum populations. The growth of heterotrophs (heterotrophic decomposers and facultative anaerobes) is not modeled directly with state equations; instead, their growth is calculated from organic matter decay by assigning a fraction of decayed organic matter to microbial biomass. Decayed OM-C has three possible fates: transfer to another OM pool, microbial biomass, and CO_2. The accumulated rates of the heterotrophs are:

$$r_g(7) = \sum_{i=1}^{5}\left[1 - f_t(i)\right]e_{max}\,f_r\,r(i)$$

$$r_g(9) = \sum_{i=1}^{5}\left[1 - f_t(i)\right]e_{max}\left[1 - f_r\right]r(i) \qquad (7.10)$$

where f_r is the fraction of soil decomposers over total heterotrophs, $f_t(i)$ is the fraction of decayed OM lost to other OM pools, r(i) is the decay rate of pool i given in Equation 7.1, and e_{max} is an efficiency factor (0.267). Autotroph (nitrifier) growth is dependent on nitrification rate:

$$r_g(8) = e_{nit}\,C_N\,r_{nit} \qquad (7.11)$$

where e_{nit} (= 0.0135) is an efficiency factor converting NH_4 to autotrophic growth rate, and C_N is the C:N ratio of autotrophic biomass. Denitrifiers (facultative anaerobes) can also grow under anaerobic conditions and decompose soil residue and soil OM. The total decay rate of the five pools under anaerobic condition can be expressed as:

$$r_{ad} = -a_{ad}\,r_{den} \qquad (7.12)$$

where r_{ad} is the total anaerobic decay rate of the five organic pools (μg C/g soil/day) and a_{ad} is a decay conversion factor from mole N/liter pore water to μg C/g soil/day ($1.4\theta/\rho$, where θ is volumetric water content and ρ is soil bulk density). The total decay rate is then partitioned into the five carbon pools assuming that the relative anaerobic decay rates of each pool are the same as the relative aerobic decay rates. Corresponding growth of facultative anaerobes can then be written as:

$$r_{ga}(9) = e_{ad}\,r_{ad} \qquad (7.13)$$

where $r_{ga}(9)$ is the growth rate of facultative anaerobes under anaerobic conditions and e_{ad} is the fraction of decayed soil carbon to biomass (= 0.1). Death rates for the three biomass populations are calculated as first-order equations:

$$r_{death}(i) = k_{death}(i)\, C_i \qquad i = 7,8,9 \tag{7.14}$$

where

$$k_{death}(7) = \frac{1}{f_{aer}}\left(\frac{k_b T}{h_p}\right) A_{death}(7)\, e^{-\frac{E_7}{R_g T}} \frac{H\gamma_1}{[O_2]C_s} \tag{7.15}$$

$$k_{death}(8) = \frac{1}{f_{aer}}\left(\frac{k_b T}{h_p}\right) A_{death}(8)\, e^{-\frac{E_8}{R_g T}} \frac{H\gamma_1}{[O_2]C_{NH_4}} \tag{7.16}$$

$$k_{death}(9) = \frac{1}{f_{aer}}\left(\frac{k_b T}{h_p}\right) A_{death}(9)\, e^{-\frac{E_9}{R_g T}} \frac{H\gamma_1}{[O_2]C_s}$$

$$+ \frac{1}{f_{anaer}}\left(\frac{k_b T}{h_p}\right) A_{death}(9)\, e^{-\frac{E_9}{R_g T}} \frac{H\gamma_1}{C_{NO_3} C_s} \tag{7.17}$$

where C_{NO_3} is the NO_3 concentration in soil solution (mole NO_3/liter pore water); $E_7 = 15.19 + 12.3U$, $E_8 = 12.64 + 12.3U$, $E_9 = 12.83 + 12.3U$, and $A_{death}(7) = 5.0 \times 10^{-35}$, $A_{death}(8) = 4.77 \times 10^{-40}$, $A_{death}(9) = 3.4 \times 10^{-33}$. Based on C:N ratios of microbial biomass, dead biomass was partitioned into the fast soil humus pool. Live microbial biomass is converted to microbial population such as $P_{het} = 950\, C_7$; $P_{aut} = 9500\, C_8$; and $P_{ana} = 9500\, C_9$.

Ammonia volatilization is modeled based on partial pressure gradient of NH_3 in the soil (P_{NH_3}, atm) and air (P'_{NH_3}, 2.45×10^{-8} atm):

$$r_v = -K_v\, T_f \left(P_{NH_3} - P'_{NH_3}\right) C_{NH_4} \tag{7.18}$$

where K_v is a volatilization constant and is affected by wind speed (W, km/d) and soil depth (Z,cm):

$$K_v = k_{v0}\, \log(W)\, e^{-0.25Z} \tag{7.19}$$

where k_{v0} is a volatilization constant (= 4.0×10^3). T_f is a temperature factor and can be expressed as:

$$T_f = T_{k1}\, e^{\frac{-6.0}{T_{k2}(T+273)}} \tag{7.20}$$

where T is soil temperature in °C, and T_{k1} and T_{k2} are constant for the temperature factor with values of 2.9447×10^4 and 1.99×10^{-3}, respectively. P_{NH_3} is calculated from:

$$P_{NH_3} = \frac{E_k C_{NH_4}}{[H]} \tag{7.21}$$

where E_k is an equilibrium constant between NH_4 and NH_3 (= 8.79×10^{-12}). Some of the model parameters were derived from Crane et al. (1981) and Hoff et al. (1981).

Applied urea is hydrolyzed to NH_4^+ through the activity of urease. The hydrolysis rate is simulated as a first-order process:

$$r_{urea} = -k_{urea} C_{urea} \tag{7.22}$$

where r_{urea} and C_{urea} are the urea hydrolysis rate (moles urea/LPW/d) and concentration (moles urea/LPW], respectively. The rate constant k_{urea} is calculated from:

$$k_{urea} = f_{aer} \left(\frac{k_b T}{h_p} \right) A_u \, e^{-\frac{E_u}{R_g T}} \tag{7.23}$$

where A_u is the rate coefficient for urea hydrolysis [s/d] (= 2.5×10^{-4}), E_u is the activation energy in kcal/mole (= 12.6).

Plant Growth and Nitrogen Uptake

A generic plant growth model was developed by Hanson (2000). It simulates both plant growth and plant population development. Seven phenological growth stages (dormant, germinating, emergent, four-leaf, vegetative, reproductive, and senescent) are used in the model to monitor the progress of plant development. The model assumes a minimum time spent between the phenological stages under optimum conditions. Environmental factors may delay or speed the rate of plant development. A plant in any stage can remain alive in the current stage, advance to the next stage, or die, depending on environmental stresses. A modified Leslie probability matrix is used to calculate plant population, and its phenological stage is determined by the dominant class of all the plants. Plants may uptake nitrogen in the form of NO_3 or NH_4. The model assumes that plants extract each nitrogen form in proportion to their percentage in the soil without giving preference to NO_3 or NH_4. In addition, plants meet nitrogen demands first by the concentration of dissolved nitrogen in the transpiration water stream. If such a passive uptake cannot meet the nitrogen demand, an active uptake mechanism is invoked using the Michaelis-Menten equation (Hanson, 2000):

$$U_{ac} = \frac{\mu_1 N_c}{\mu_2 + N_c} \tag{7.24}$$

where U_{ac} is the amount of active uptake nitrogen (g/plant/day), μ_1 is the maximum nitrogen uptake rate (g/plant/day), μ_2 is the Michaelis-Menten constant, and N_c is the total nitrogen concentration in the soil layer (g). Plant demand for nitrogen is crop specific and depends on the growth stage as well.

Nitrogen Inputs and Losses

Nitrogen inputs into the agricultural system include crop residue, manure application, fertilizer, nitrogen in rain and irrigation water, N fixation, and crop root system. Crop residue (a combination of senescent plant material falling from the growing crop plus above-ground stubble and dead plant material deposited at harvest) is partitioned into two residue pools as described above through tillage practices or biological means. Manure is treated as residue and partitioned between the two residue pools (Rojas and Ahuja, 2000). Conservation of mass is preserved for nitrogen in all the transformations. Carbon dioxide is used as a sink/source for carbon. Ammonium contained in manure is added into the soil ammonium pools directly. Applied inorganic nitrogen through fertilization and water is added into its corresponding pools. Crop roots are partitioned into the two residue pools directly, based on their C:N ratios.

In addition to nitrogen losses through volatilization and denitrification, it can also be lost to runoff and groundwater, and harvested plant parts. Inorganic nitrogen in the top 2 cm of the soil profile can be extracted by rainwater and carried away in runoff water (Ahuja et al., 2000b). Nitrate in the soil is subject to leaching in the soil profile along with water movement. Ammonium, on the other hand, is assumed to be immobile in the soil profile.

A CASE STUDY FOR MODEL SENSITIVITY AND MANAGEMENT SCENARIOS

The example scenario was from a field experiment designed to study residual effects of manure on irrigated corn silage (maize silage) production in eastern Colorado (Ma et al., 1998a). The field had a history over the past decade of receiving beef manure (44.8 mg/ha) as fertilizer every fall after corn silage (maize silage) was harvested. No inorganic fertilizer was applied. The experimental plots were on a Vona sandy loam soil (coarse-loamy, mixed, mesic, Ustollic Haplargid). The water table was approximately 8 m below ground level. The field was irrigated in alternate furrows with ditch water containing 1.3 ppm NO_3-N. Each irrigation event lasted 12 hours, with a total application quantity of 20 cm. The farmer irrigated infrequently — usually only four to six times during the months of July and August (Table 7.1).

The experiment was established in the fall (autumn) of 1993 and completed in the fall of 1996. The whole field received approximately 44.8 mg/ha of beef manure (on dry weight basis) as usual in the fall of 1993 (mid October). Manure treatments were started in the fall of 1994 and 1995, with the eastern half of the field receiving 44.8 mg/ha of manure, whereas the western half of the field received no manure, to

Table 7.1 Management Practices for 1993 to 1996 from a Field Experiment Designed to Study Residual Effects of Manure on Irrigated Corn Production in Eastern Colorado

Management Practice	Timing	Method	Specific Information
Planting and harvesting	April 15, 1994 and Sept. 10, 1994		Corn always planted at 76-cm row spacing and 8000 to 8700 seeds/ha
	April 22, 1995 and Sept. 15, 1995		
	April 20, 1996 and Sept. 14, 1996		Harvest always for silage corn
Manure and fertilizer	October 15, 1993	Surface broadcast	No fertilizer was applied
	October 15, 1994		Nitrogen applied with manure 582 kg/ha
	October 15, 1995		
Irrigation	June 14, 25, 1994	Furrow irrigation	20 cm/event
	July 7, 17, 29, 1994		
	Aug. 18, 1994		
	July 13, 1995		
	Aug. 2, 16, 31, 1995		
	May 20, 1996		
	June 29, 1996		
	July 12, 26, 1996		
	Aug. 10, 25, 1996		
Tillage	Oct. 17, 1993	Moldboard plow	15 cm of effective tillage depth
	15 days before planting, 1993	Field cultivator	10 cm of effective tillage depth
	2 days before planting, 1993	Field cultivator	
	May 21, 1994	Field cultivator	
	Oct. 17, 1994	Moldboard plow	
	15 days before planting, 1994	Field cultivator	
	2 days before planting, 1994	Field cultivator	
	June 13, 1995	Field cultivator	
	July 2, 1995	Field cultivator	
	Oct. 17, 1995	Moldboard plow	
	15 days before planting, 1995	Field cultivator	
	2 days before planting, 1995	Field cultivator	
	June 13, 1996	Field cultivator	
	July 2, 1996	Field cultivator	

From Ma, L., J.C. Ascough II, L.R. Ahuja, M.J. Shaffer, J.D. Hanson, and K.W. Rojas. 2000b. Root Zone Water Quality Model sensitivity analysis using Monte Carlo simulation. *Trans. ASAE,* 43:883-895. With permission.

compare the residual effects of long-term manure application in northeastern Colorado. Applied manure was incorporated into the soil after 1 to 2 days with a moldboard plow. A sample of the manure was collected and analyzed for total N, NH_4-N, NO_3-N, and moisture content. Total manure-N applied cach year was 582 kg N/ha.

Model calibration for the data set was reported by Ma et al. (1998a), and a sensitivity analysis for selected model input variables using Monte Carlo simulation was presented by Ma et al. (2000b). To calibrate RZWQM for the study site, Ma et al. (1998a) used the fact that there was no silage yield reduction in 1995 when manure application was stopped on the west side of the field in the fall of 1994. This information was important primarily for calibrating inter-pool mass transfer coefficients (R_{14}, R_{23}, R_{34}, and R_{45}; Figure 7.1). In addition, several years of simulation runs were required to stabilize organic matter pools and microbial pools in order to simulate management effects without influence of initial conditions. As shown in Figure 7.3, it takes 6 years to initialize the fast humus pool, and 9 years to stabilize the intermediate humus pool. It takes more than 600 years to initialize the slow humus pool. However, because the slow humus pool is stable in soil and makes less contribution to short-term C/N dynamics, stabilization of the slow humus pool is not important (Ma et al., 1998a). Figure 7.4 shows the effects of initial condition on three microbial pools. Without appropriate initialization procedures, simulation results can be misleading, and simulated management effects contain effects of initial conditions.

RZWQM was calibrated based on 1994 and 1995 data and used to predict 1996 silage yield and nitrogen uptake. Figure 7.5 shows simulated silage yield and N uptake for 1994, 1995, and 1996. Generally, the model was able to predict 1996 crop production based on information from 1994 and 1995. Simulation results were further improved with a modified crop growth model (Ma et al., 2000c). Simulated total soil nitrate-N was also adequate (Figure 7.6). Simulated manure-N mineralization rates were 22% in the first year and 18% in the second year, which are as close to the ranges (20 to 35% in the first year and 10 to 15% in the second year) given by Schepers and Mosier (1991) for beef manure with 13 g/kg nitrogen content as those used in this study. The calibrated model was further used to evaluate alternative management practices of water and manure applications, such as reducing water and manure application rate and decreasing manure application frequency (Ma et al., 1998a). It was found that reducing water application rate by half increased silage yield slightly and N uptake significantly, with a 30 to 48% decrease in nitrogen losses. With a 50% cutback in manure application, silage yield was reduced by 13%, whereas denitrification loss of nitrogen was reduced by 63%, volatilization by 68%, and leaching by 46 to 58%. Applying manure every other year reduced silage yield by 10% with a return of 58%, 57%, and 53 to 66% decrease in denitrification, volatilization, and leaching losses, respectively (Table 7.2).

Ma et al. (2000b) conducted an extensive sensitivity study with the above calibrated model and demonstrated the importance of key model parameters on selected model outputs (silage yield, N uptake, and nitrate leaching). Table 7.3 shows the tested parameters, their ranges, and assumed distribution within the ranges. A Latin Hypercube Sampling (LHS) method was used to generate model parameters for model simulations. Figure 7.7 shows simulated distributions of N uptake, silage yield, and N leaching from variations in C/N parameters shown in Table 7.3, and Figure 7.8 shows those from sampled plant growth parameters. Wider distributions of simulation results were obtained with variations in plant growth parameters than in C/N parameters within the given sampling ranges. These sensitivity analyses

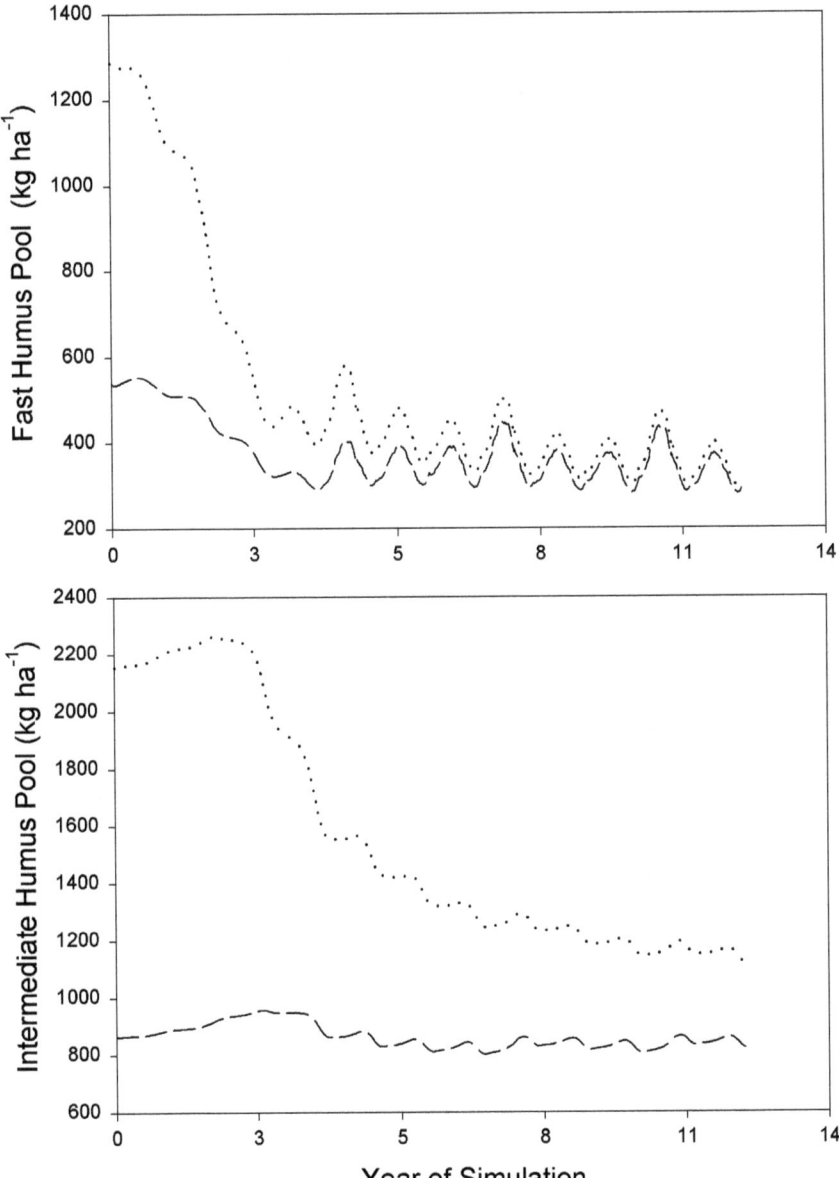

Figure 7.3 Stability analysis of soil organic matter (OM) pools as affected by initial conditions. (From Ma, L., M.J. Shaffer, J.K. Boyd, R. Waskom, L.R. Ahuja, K.W. Rojas, and C. Xu. 1998a. Manure management in an irrigated silage corn field: experiment and modeling. *Soil Sci. Soc. Am. J.,* 62:1006-1017. With permission.)

provide insight into N fates as affected by different model parameters. Generally, nitrogen fates are more sensitive to manure application rate, followed by specific leaf weight, photorespiration, death rate of heterotrophs, and denitrification rate and decay constant of the fast residue pool.

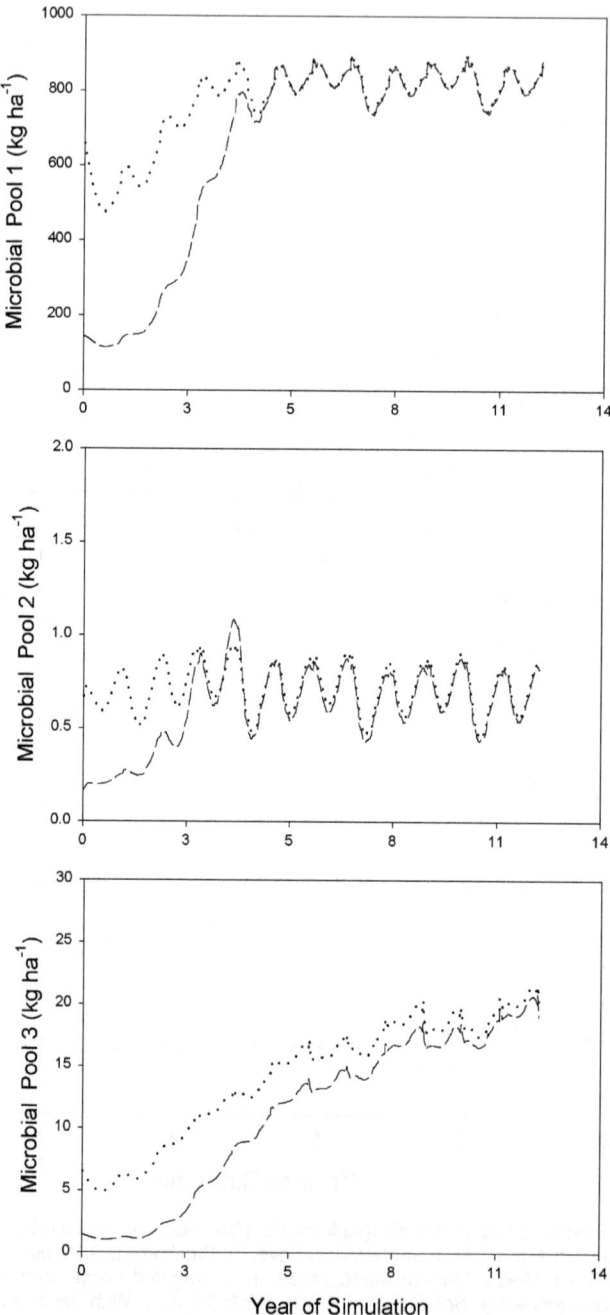

Figure 7.4   Stability analysis of soil microbial pools as affected by initial conditions. (From Ma, L., M.J. Shaffer, J.K. Boyd, R. Waskom, L.R. Ahuja, K.W. Rojas, and C. Xu. 1998a. Manure management in an irrigated silage corn field: experiment and modeling. *Soil Sci. Soc. Am. J.,* 62:1006-1017. With permission.)

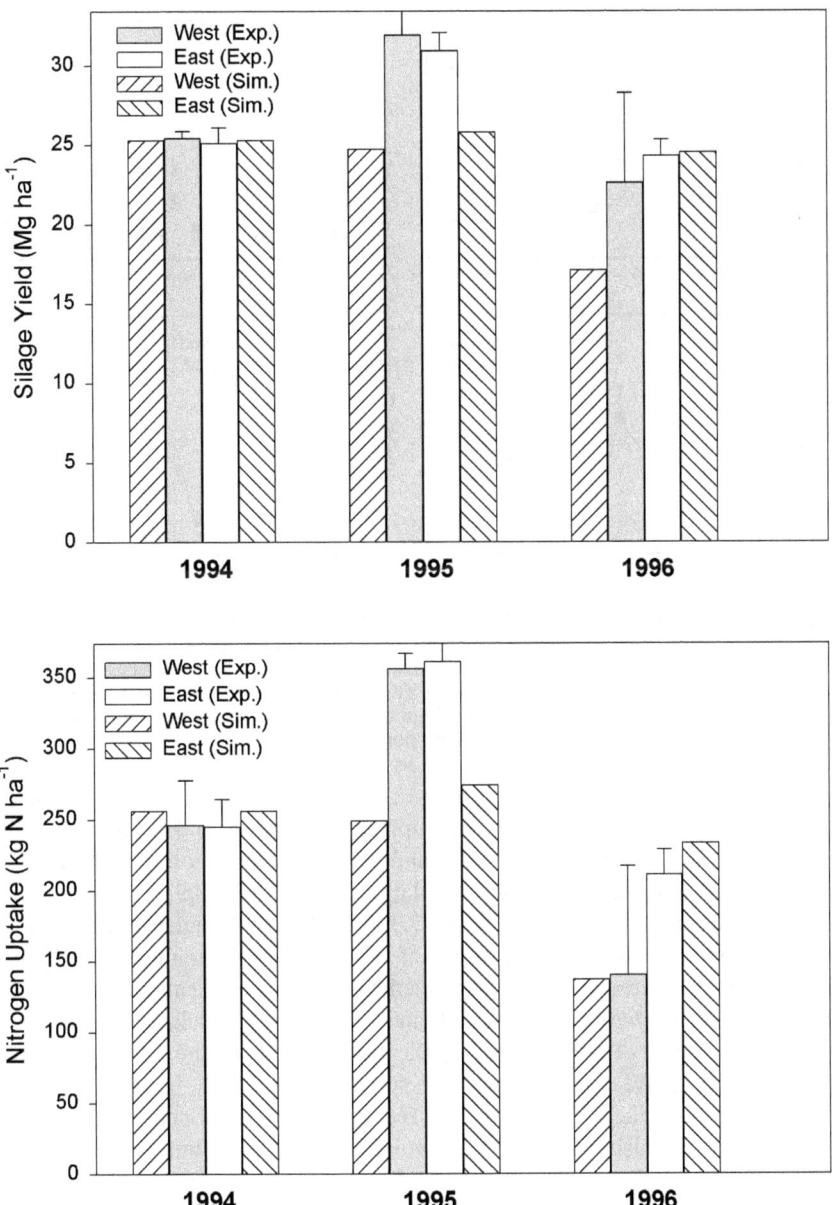

Figure 7.5 Measured and RZWQM simulated silage yield (top) and total plant nitrogen uptake (bottom) for 1994, 1995, and 1996 growing seasons. East site received manure applications in the fall of 1993, 1994, and 1995; west site received manure application in the fall 1993 only. (From Ma, L., M.J. Shaffer, J.K. Boyd, R. Waskom, L.R. Ahuja, K.W. Rojas, and C. Xu. 1998a. Manure management in an irrigated silage corn field: experiment and modeling. *Soil Sci. Soc. Am. J.,* 62:1006-1017. With permission.)

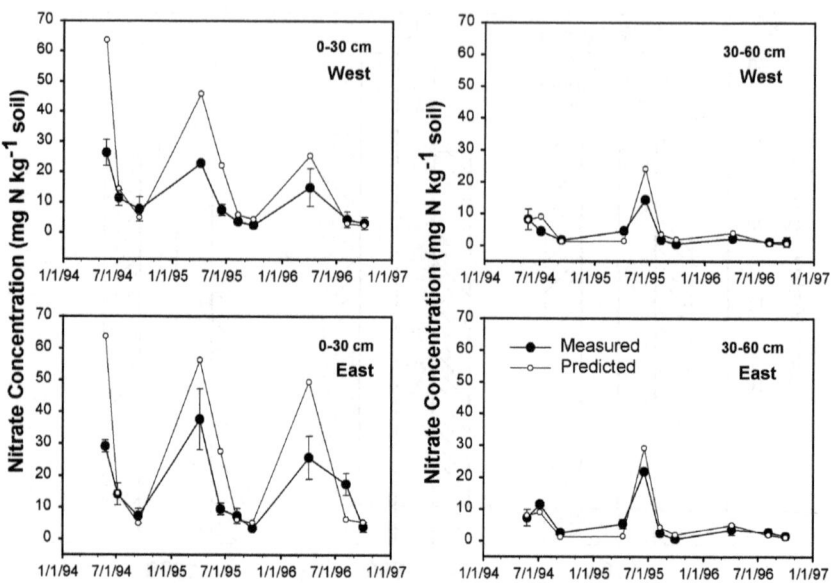

Figure 7.6 Measured and RZWQM predicted nitrate concentrations in the top 60 cm soil profile. East site received manure applications in the fall of 1993, 1994, and 1995; west site received manure application in the fall of 1993 only. (From Ma, L., M.J. Shaffer, J.K. Boyd, R. Waskom, L.R. Ahuja, K.W. Rojas, and C. Xu. 1998a. Manure management in an irrigated silage corn field: experiment and modeling. *Soil Sci. Soc. Am. J.*, 62:1006-1017. With permission.)

The calibrated RZWQM was further extended to simulate the effects of various agricultural management practices on N uptake, silage yield, and N leaching, including rate, timing, and method of water and fertilizer application, type of fertilizer, tillage, and planting dates. Selected simulation results for N application rate, timing, and planting dates are shown in Figures 7.9 through 7.12. Simulated plant N uptake and leaching responses to three types of N fertilizer (ammonium [NH_4], nitrate [NO_3], and urea) rates and application methods (surface application, incorporation, and injection) are shown in Figures 7.9 and 7.10. Plant N uptake changed dramatically with the type of fertilizer and application method (Figure 7.9). Considerably lower plant N uptake was predicted with surface-applied NH_4, because about 93 to 99% of surface-applied NH_4 was volatilized within a week of application. Surface-applied urea was also subject to high volatilization losses, but less so than NH_4 (Figure 7.9c). Injection or incorporation of NH_4 and urea increased plant N uptake compared to surface application. More plant N was taken up when NH_4 was incorporated into the top 10 cm of surface soil profile as compared to NH_4 injection to a depth of 25 ± 5 cm. No difference in plant N uptake was observed between incorporation and injection for NO_3 and urea fertilizers. More N leaching was simulated for NO_3 application than for urea application. Figure 7.10 shows that the responses of NO_3-N leaching to fertilizer application methods were opposite to those of plant N uptake (Figure 7.9), except for surface-applied NH_4 and urea (where NO_3-N leaching was also low due to high volatilization losses). When N was applied at

Table 7.2 Average Yearly Summary for Manure and Water Management Based on RZWQM

Management	Manure	Water Applied (cm/event)	Silage Yield (mg/ha/yr)	Nitrogen Uptake	N-seepage	Denitrification	Volatilization
				(kg N/ha/yr)			
1	Full rate every year	20	25.2	254	196	130	11.1
2	Half rate every year	20	21.9	182	106	48	3.5
3	Full rate every 2 years	20	22.8	192	91	57	4.7
4	Full rate every 3 years	20	19.0	153	83	34	3.2
5	Full rate, split application	20	25.1	250	188	140	7.1
6	Full rate every year	10	25.8	307	116	141	11.1
7	Half rate every year	10	22.7	228	49	50	3.5
8	Full rate every 2 years	10	22.9	232	39	59	4.7
9	Full rate every 3 years	10	20.1	186	40	35	3.2
10	Full rate, split application	10	25.6	302	109	150	7.1

Note: Full manure application rate was 44.8 mg/ha and half rate was 22.4 mg/ha. Manure was applied on October 15 for one application and October 15/April 1 for split application (22.4 mg/ha each).

From Ma, L., M.J. Shaffer, J.K. Boyd, R. Waskom, L.R. Ahuja, K.W. Rojas, and C. Xu. 1998a. Manure management in an irrigated silage corn field: experiment and modeling. *Soil Sci. Soc. Am. J.*, 62:1006–1017. With permission.

Table 7.3 Baseline Values, Testing Ranges, and Probability Distributions of Model Input Parameters Selected for RZWQM Sensitivity Analysis

Simulated Process Group	Model Input Parameter	Unit	Base Value	Testing Range	Distribution
Organic matter/N cycling	A_{nit}: Nitrification rate constant	sec/day/org.	1.0×10^{-9}	0.1×10^{-9}–10×10^{-9}	Log normal
	A_{den}: Denitrification rate constant	sec/day/org.	1.0×10^{-13}	0.1×10^{-13}–10×10^{-13}	Log normal
	A_1: Decay rate constant for the SRP	sec/day	1.67×10^{-7}	0.167×10^{-7}–16.7×10^{-7}	Log normal
	A_2: Decay rate constant for the FRP	sec/day	8.14×10^{-6}	0.814×10^{-6}–81.4×10^{-6}	Log normal
	A_3: Decay rate constant for the FHP	sec/day	2.5×10^{-7}	0.25×10^{-7}–25×10^{-7}	Log normal
	A_4: Decay rate constant for the IHP	sec/day	5.0×10^{-8}	0.5×10^{-8}–50×10^{-8}	Log normal
	A_5: Decay rate constant for the SHP	sec/day	4.5×10^{-10}	0.45×10^{-10}–45×10^{-10}	Log normal
	A_{death} (7): Death rate constant for heterotrophs	sec/day	5.0×10^{-35}	0.5×10^{-35}–50×10^{-35}	Log normal
	A_{death}(8): Death rate constant for nitrifiers	sec/day	4.77×10^{-40}	0.477×10^{-40}–47.7×10^{-40}	Log normal
	A_{death}(9): Death rate constant for denitrifiers	sec/day	3.4×10^{-33}	0.34×10^{-33}–34×10^{-33}	Log normal
Plant growth	μ_1: Maximum active daily N uptake rate	g/plant/day	0.5	0.0–3.3	Normal
	R_1: Photorespiration rate	percentage	0.08	0.0–0.525	Normal
	SLW: Specific leaf weight	g/LAI	9.0	0.0–27.3	Normal
	A_p: Photosynthesis reduction factor at propagules	percentage	0.9	0.708–1.0	Normal
	A_s: Photosynthesis reduction factor at seed stage	percentage	0.8	0.48–1.0	Normal

Modified from Ma, L., J.C. Ascough II, L.R. Ahuja, M.J. Shaffer, J.D. Hanson, and K.W. Rojas. 2000b. Root Zone Water Quality Model sensitivity analysis using Monte Carlo simulation. *Trans. ASAE*, 43:883–895. With permission.

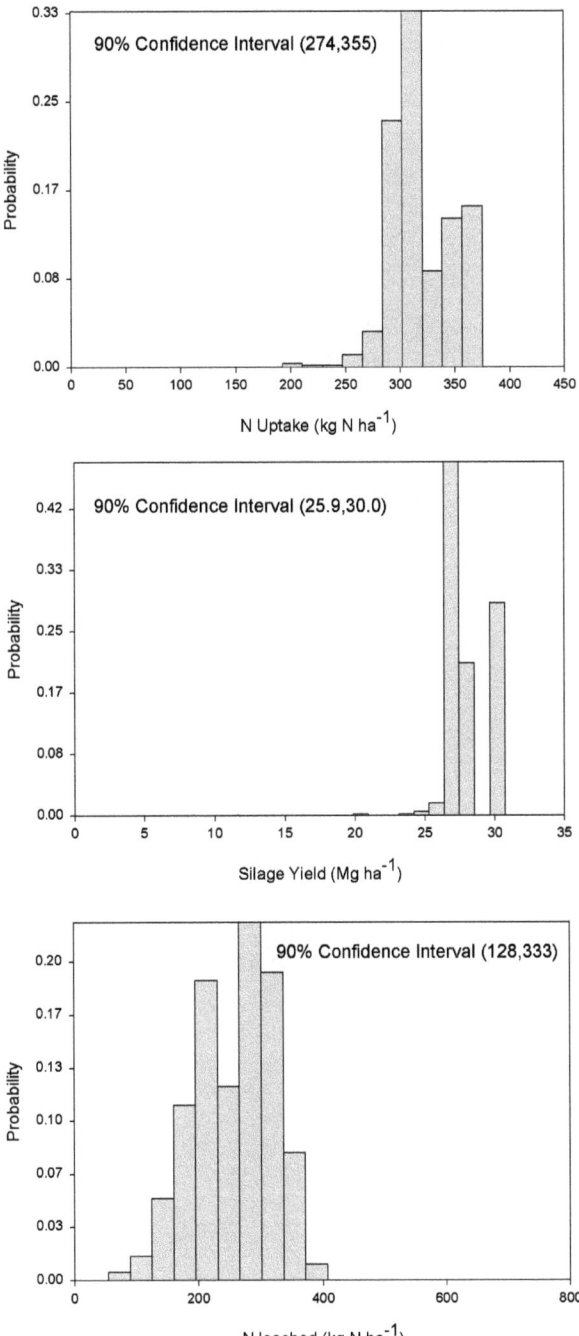

Figure 7.7   Probability distributions of simulated plant N uptake, silage yield, and NO₃-N leaching beyond the root zone due to variation in organic matter/N cycling parameters. (From Ma, L., J.C. Ascough II, L.R. Ahuja, M.J. Shaffer, J.D. Hanson, and K.W. Rojas. 2000b. Root Zone Water Quality Model sensitivity analysis using Monte Carlo simulation. *Trans. ASAE,* 43:883-895. With permission.)

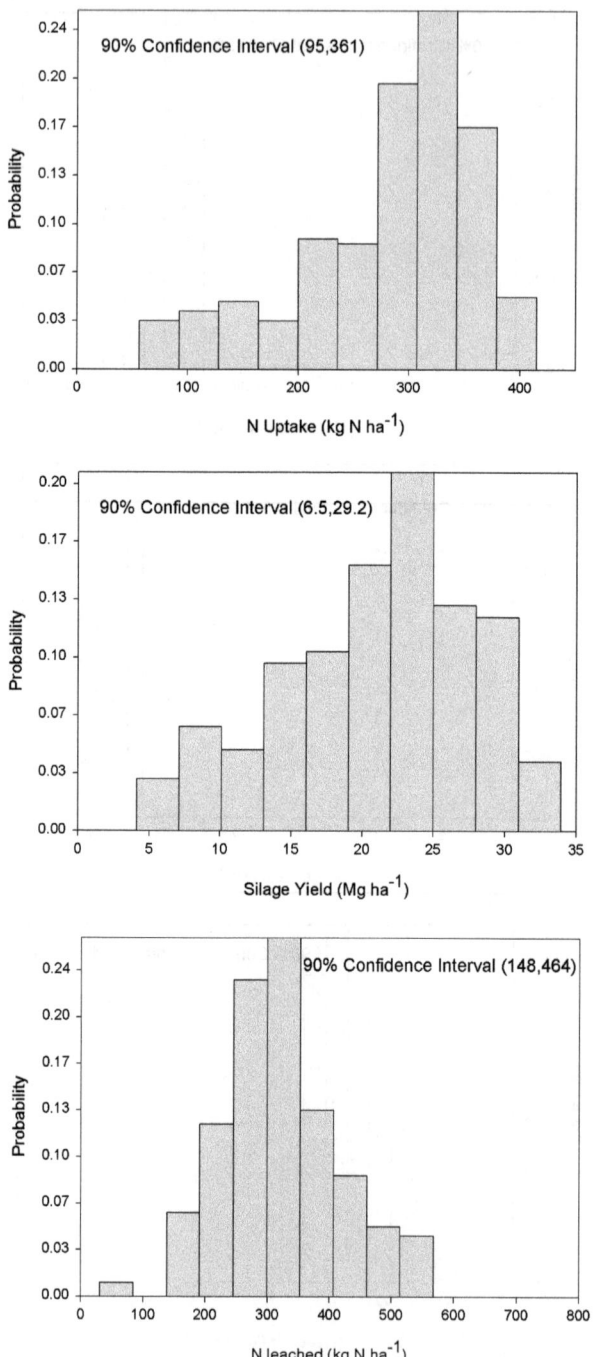

Figure 7.8   Probability distributions of simulated plant N uptake, silage yield, and NO₃-N leaching beyond the root zone due to variation in plant growth parameters. (From Ma, L., J.C. Ascough II, L.R. Ahuja, M.J. Shaffer, J.D. Hanson, and K.W. Rojas. 2000b. Root Zone Water Quality Model sensitivity analysis using Monte Carlo simulation. *Trans. ASAE*, 43:883-895. With permission.)

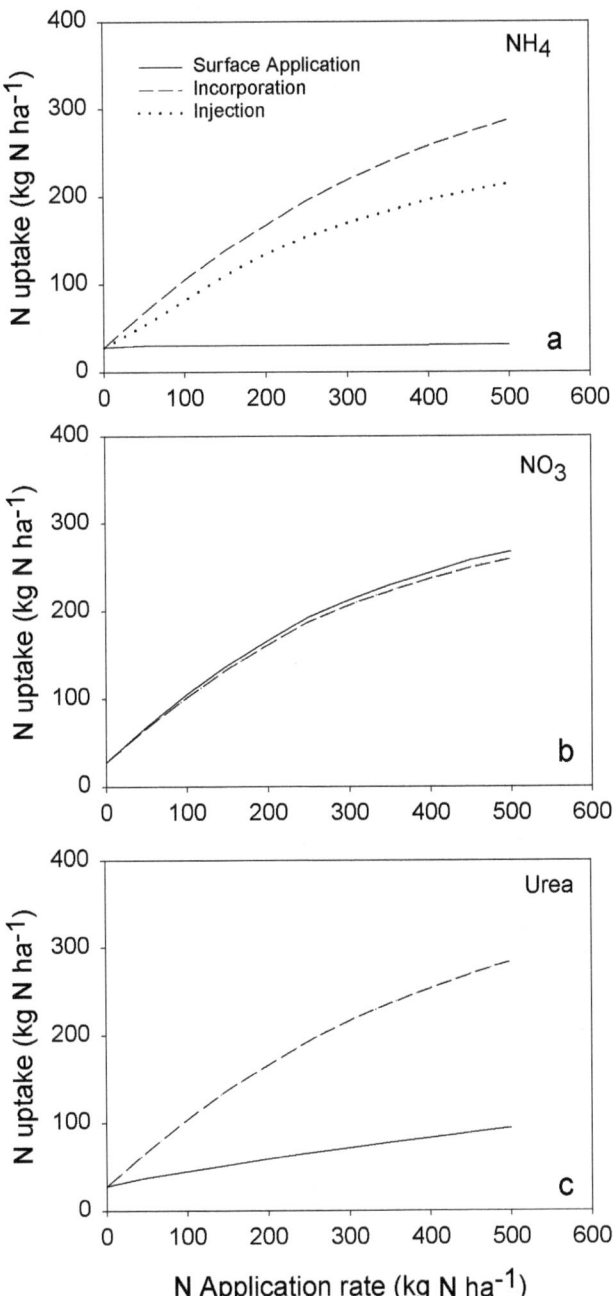

Figure 7.9 RZWQM simulated average yearly plant N uptake output responses to N fertilizer application rates and methods for NH_4 (a), NO_3 (b), and urea (c). Nitrogen was applied 1 day before planting. The model was calibrated by Ma et al. (1998a).

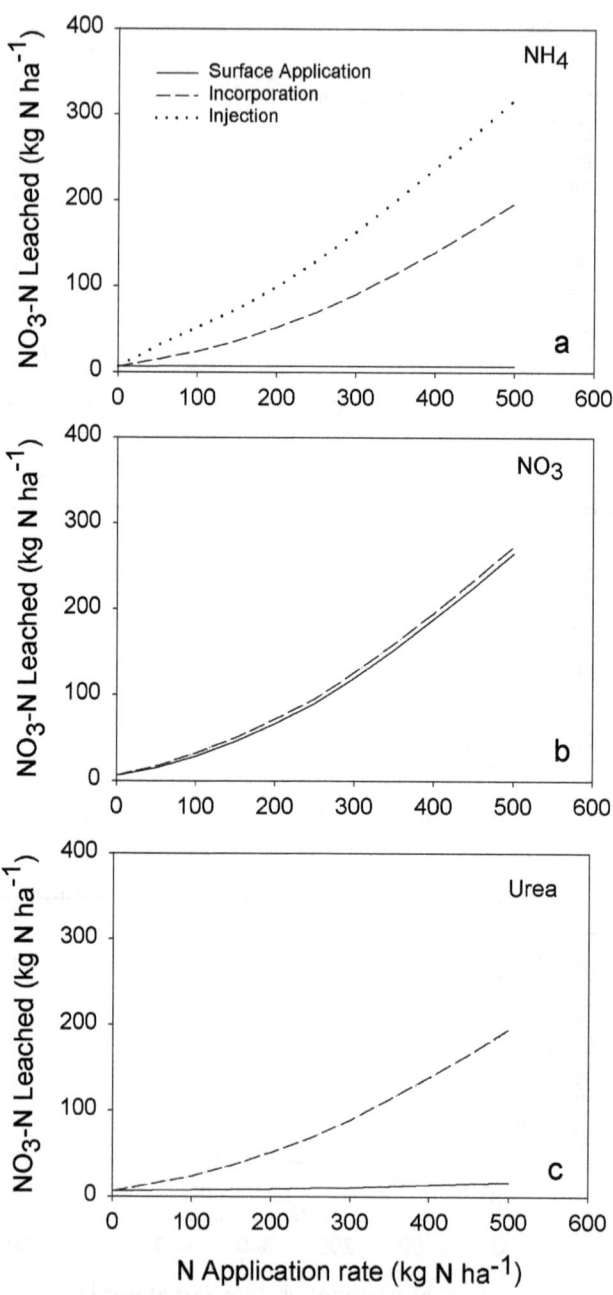

Figure 7.10 RZWQM simulated average yearly NO₃-N leaching output responses to N application rates and methods for NH₄ (a), NO₃ (b), and urea (c). Nitrogen was applied
1 day before planting. The model was calibrated by Ma et al. (1998a).

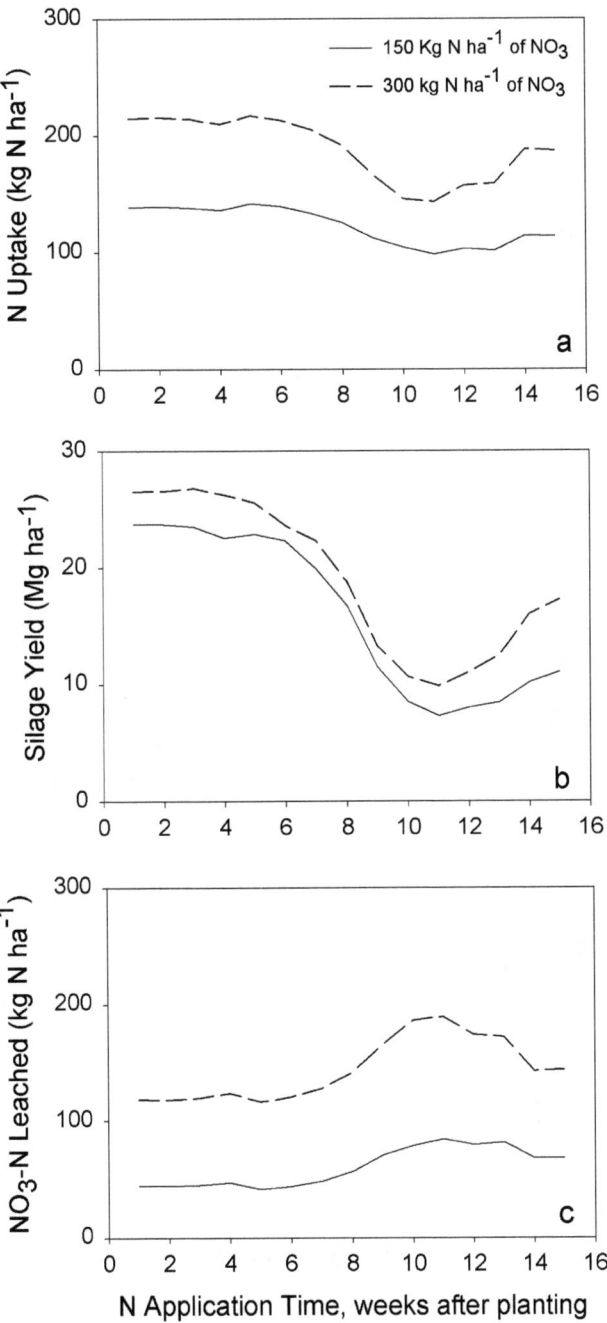

Figure 7.11   RZWQM simulated average yearly plant N uptake (a), silage yield (b), and NO_3-N leaching (c) in responding to N application timing. NO_3-N was surface applied once per growing season at a rate of 150 or 300 kg N/ha. The model was calibrated by Ma et al. (1998a).

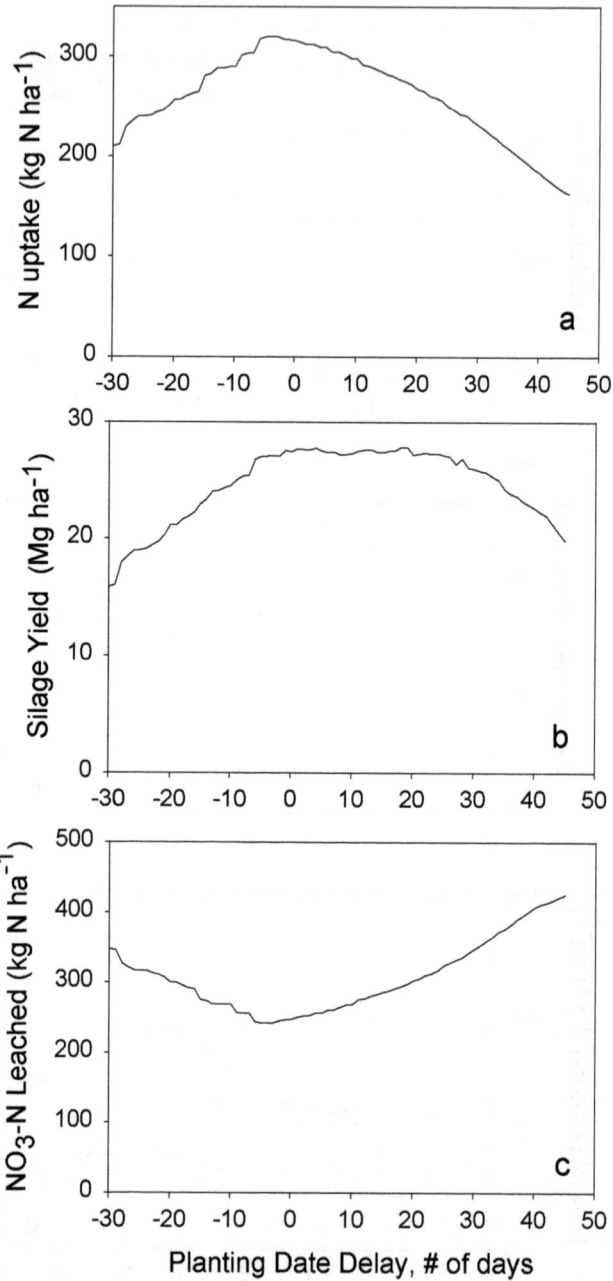

Figure 7.12 RZWQM simulated average yearly plant N uptake (a), silage yield (b), and NO_3-N leaching (c) in responding to planting date. The model was calibrated by Ma et al. (1998a).

a depth of 25 cm (injection), it was subject to more leaching than N incorporated into the top 10 cm of surface soil due to the shorter travel distance to the bottom of the root zone. Because more plant roots were present near the soil surface at the early stage of plant growth, injected N may not be available for plant uptake during that time.

RZWQM output responses to fertilizer application (surface NO_3) timing at different growth stages are shown in Figure 7.11. Early N application increased silage yield and plant N uptake, whereas NO_3-N leaching was lower when N was applied at the early stage of plant growth. Significant reduction in silage yield and plant N uptake was simulated (Figures 7.11a and b) when N was applied 7 to 11 weeks after planting. Thus, early fertilization was especially beneficial to silage yield. An N application rate of 300 kg N/ha increased plant N uptake and silage yield, as well as NO_3-N leaching, as compared to the 150 kg N/ha rate (Figure 7.11).

When to plant is a critical agricultural management decision, especially under persistent unfavorable weather conditions. Figure 7.12 shows RZWQM output responses to advances or delays in planting date. Delaying the planting date decreased plant N uptake and silage yield (Figure 7.12a and b). NO_3-N leaching increased slowly with planting date delay (Figure 7.12c), which was attributed to less plant N uptake and more N available for leaching. Early planting also decreased plant N uptake and silage yield (Figures 7.12a and b) because of lower soil temperatures at the early plant growth stages. Nitrate leaching increased as planting date advanced, due to less competition from plant N uptake.

MODEL APPLICATIONS IN THE LITERATURE

RZWQM has been used to simulate nitrogen management effects on crop production and water quality under various agronomic conditions. Management practices related to nitrogen management in RZWQM are manure application, fertilization, irrigation, tile drainage, and crop rotations. Simulated nitrogen fates that have been compared against experimental measurements are nitrate in the soil profile, nitrate leaching, nitrate in runoff, and plant N uptake.

Manure Management

Thus far, there have been several applications of RZWQM for manure management reported in the literature. The study reported by Ma et al. (1998a) was conducted in northeastern Colorado. It was designed to examine the residual effects of manure (beef manure) after a decade of heavy application. Plant N uptake and soil nitrate-N were measured and used to calibrate nitrogen dynamics in the model (Figures 7.5 and 7.6). The calibrated model was then used to predict nitrogen transformation in an Arkansas study (Ma et al., 1998b). The objective of the Arkansas study was to compare manure effects on tall fescue growth. Plant N uptake and nitrate-N in soil, runoff, and soil water were measured. A third study on manure management in Iowa was reported by Bakhsh et al. (1999) and Kumar et al. (1998), where manure effects on nitrate transport to subsurface drainage water were simulated. Swine manure was applied every other year based on the N recommendation for the soil and crop conditions. Nitrate-N in subsurface drainage water samples was measured.

The Colorado study on manure management was previously summarized, and the calibrated model was used in the Arkansas study (Ma et al., 1998b). Soil carbon/nitrogen parameters used in the Arkansas study were the same as those used in Colorado (Ma et al., 1998a) except using local soil, weather, and crop conditions. As shown in Figure 7.13, simulated soil nitrate-N in the top 90 cm soil profile responded well to manure application and was in agreement with experimental observations. Simulated nitrate-N in runoff was 0.04 kg N/ha for the control plots and 3.87 kg N/ha for the manured plots, which are comparable to experimental measured values (0.28 kg N/ha for control and 1.39 kg N/ha for manure plots). Volatilization loss of N due to manure application was about 9% of total N applied and is within the range (1.7 to 12.8%) reported by Scott et al. (1995). Denitrification after manure application accounted for 6% of the total N applied in broiler litter and was close to the 10% denitrification loss for animal waste amended silt loam soils (Sims and Wolf, 1994). The estimated N mineralization rate of broiler litter in the first year was 70%, which is comparable to the 75% mineralization rate reported by Schepers and Mosier (1991).

The Iowa study concentrated on nitrate-N loss to subsurface drainage water, and RZWQM was used to assess the environmental impacts of swine manure application in tile drained soils (Bakhsh et al., 1999). This was a corn–soybean rotation field with corn (maize) in 1993 and 1995, and soybean in 1994 and 1996. Swine manure was applied in 1993 and 1995 only for the corn phase of crop rotation. The model was calibrated for tile flow first (1993) and then for nitrate-N loss in 1993 (corn) and 1994 (soybean). The calibrated model was used to predict nitrate losses in 1995 and 1996. As shown in Table 7.4, RZWQM correctly predicted nitrate-N losses to subsurface drain in response to manure application on the three experimental plots. Average nitrate-N concentrations were also well-predicted (Table 7.4). Similar results were reported on other experimental sites as well (Kumar et al., 1998).

Fertilizer Application

One of the key components of best management practices is fertilizer application. It is essential for an agricultural system model to correctly simulate fertilizer management. Although all field experiments involved some type of fertilizer application, true testing of RZWQM for various fertilizer treatments was carried out only in Iowa (Bakhsh et al., 2000) and Nebraska (Martin and Watts, 1999) in the United States, and Portugal (Cameira et al., 1998). Two sensitivity analyses of RZWQM responses to fertilizer and manure applications were presented by Azevedo et al. (1997) and Ma et al. (2000b).

In the Iowa study by Bakhsh et al. (2000), the cropping system is a corn–soybean rotation with corn in 1996 and 1998 and soybean in 1997 and 1999. Fertilizer was applied as anhydrous ammonia in 1996 and urea ammonium nitrate (UAN) in 1998.

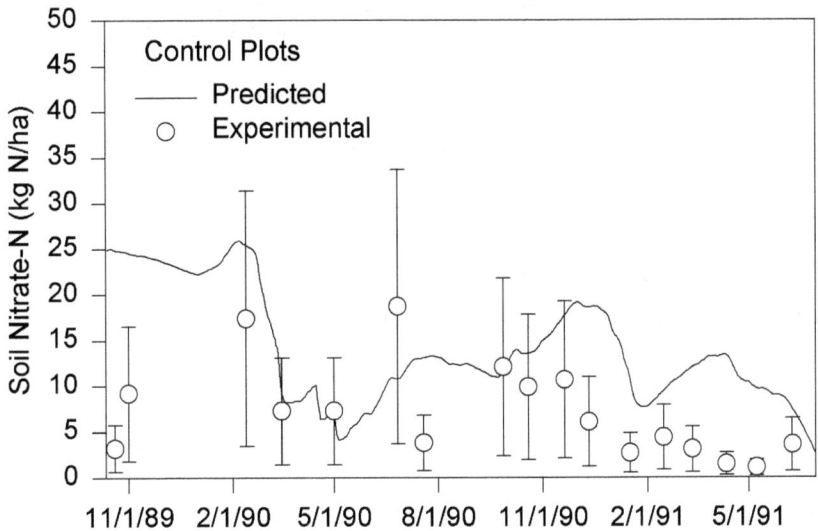

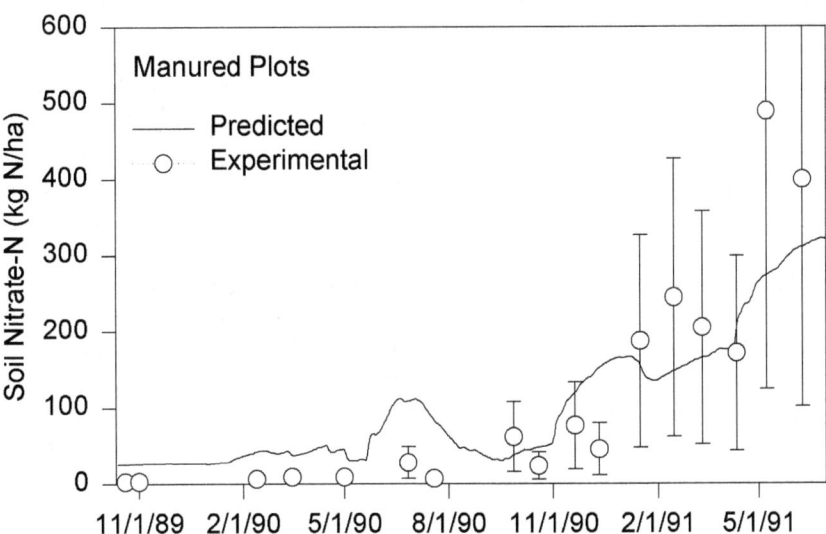

Figure 7.13 Measured and RZWQM predicted soil nitrate in the top 90 cm soil profile in the control (top) and manured (bottom) plots. (From Ma, L., H.D. Scott, M.J. Shaffer, and L.R. Ahuja. 1998b. RZWQM simulations of water and nitrate movement in a manured tall fescue field. *Soil Sci.*, 163:259-270. With permission.)

Three nitrogen treatments were 67, 135, 202 kg N/ha in 1996 and 57, 115, and 172 kg N/ha in 1998. No fertilizer was applied in 1997 and 1999 when soybean was planted. RZWQM was calibrated with data collected in 1996 under the 67 kg N/ha nitrogen treatment, including corn yield, tile drainage, and nitrate-N loss in tile flow.

Table 7.4   Measured and Simulated Nitrate-N Losses (kg/ha) and Average Nitrate-N Concentrations in Subsurface Drain Flow in an Iowa Study

	Plot No. 11		Plot No. 23		Plot No. 27	
Year	Measured	Predicted	Measured	Predicted	Measured	Predicted
			Nitrate Losses (kg/ha)			
1993	49.48	47.06	49.93	46.17	13.42	14.56
1994	6.37	6.23	7.14	7.57	1.97	2.31
1995	13.43	11.18	14.58	13.28	4.24	3.90
1996	7.28	6.02	10.76	10.53	3.13	2.83
Total	76.56	70.49	82.41	77.55	22.76	23.60
			Average Nitrate Concentration (mg/L)			
1993	12.9	12.4	12.1	11.3	9.3	7.2
1994	10.3	11.1	10.1	10.3	7.7	7.5
1995	17.3	17.8	12.6	12.1	14.4	14.7
1996	16.7	17.3	21.4	22.7	16.5	17.2
Average	14.3	14.6	14.0	14.1	11.9	11.6

Modified from Bakhsh, A., R.S. Kanwar, and L.R. Ahuja. 1999. Simulating the effect of swine manure application on NO3-N transport to subsurface drainage water. *Trans. ASAE,* 42:657-664. With permission.

The model was also calibrated for soybean yield in 1997. As shown in Figure 7.14, the model adequately predicted corn yield responses to nitrogen application rates in both 1996 and 1998. Responses of nitrate-N loss (to tile drainage) to nitrogen rates were well predicted in 1998, but not in 1996. The reason for the lack of responses in terms of nitrate loss in 1996 might be inadequate initial conditions for 1996, the first year of model simulation (Bakhsh et al., 2000). The model was better initialized for the year 1998 because more information was available in 1996 and 1997. In a Portuguese study, Cameira et al. (1998) compared soil nitrate amount using three different fertilization methods in a corn field; a single broadcast application of 150 kg N/ha as urea, a single fertigation of UAN (50-25-25), and multiple UAN fertigation. As shown in Figure 7.15, RZWQM provided good simulations of the total soil nitrate dynamics in the top 120 cm soil profile. However, in a Nebraska study, Martin and Watts (1999) tested the responses of simulated N uptake and residue soil nitrate to five fertilizer application rates (0 to 202 kg N/ha). They found that the model overpredicted soil residual N and underpredicted plant N uptake. This is a typical example of how plant growth affects soil N dynamics, and the data should now be reevaluated with the improved plant growth component in the newly released version of RZWQM (Ma et al., 2000c).

Sensitivity analyses of nitrogen management were conducted by Azevedo et al. (1997). Using the model calibrated for a corn field with subsurface drainage, Azevedo et al. (1997) simulated nitrogen application rate and method on corn yield and nitrate loss to tile drains under both moldboard plow and no-till systems in a 15-year simulation. Single nitrogen application was 10 days before planting at rates of 50,

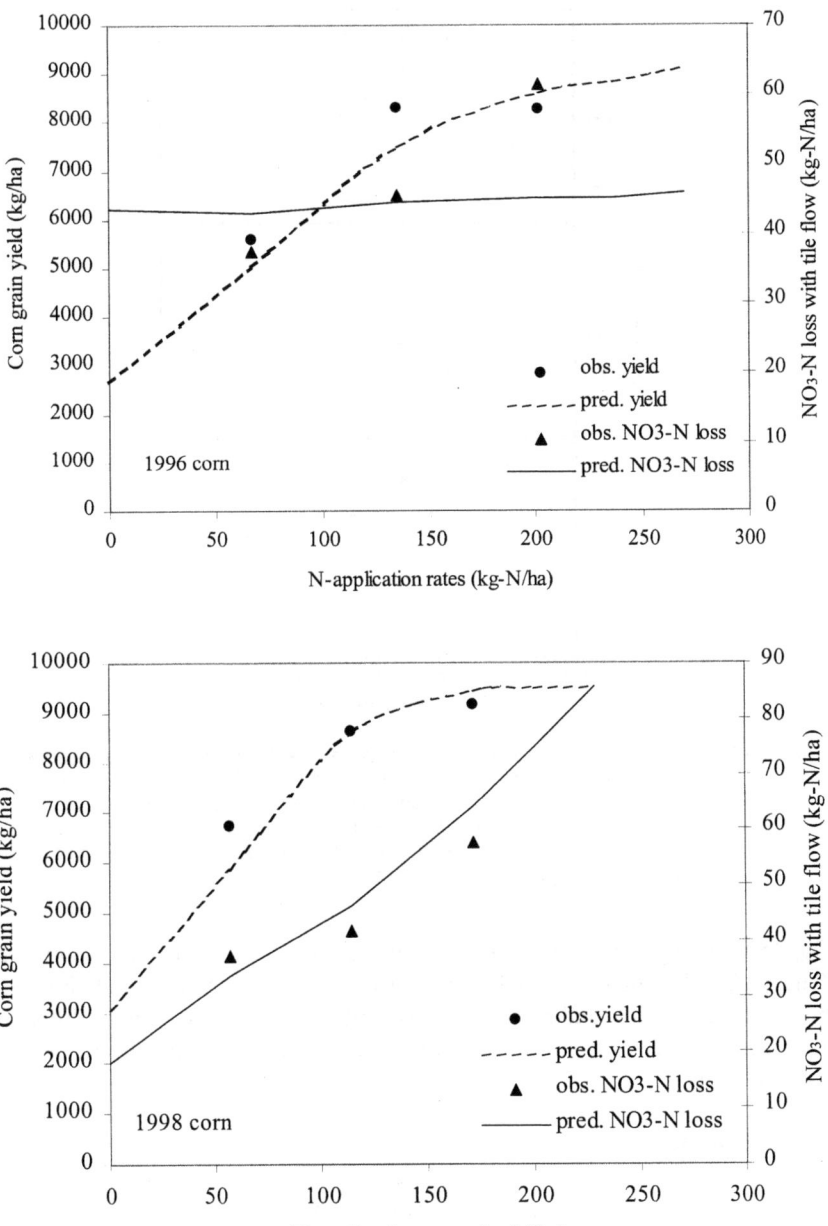

Figure 7.14 RZWQM simulated responses of corn yield and nitrate-N losses to N application rates. (From Bakhsh, A., R.S. Kanwar, D.B. Jaynes, T. S. Colvin, and L.R. Ahuja. 2000. Predicting effects of variable nitrogen application rates on corn yields and NO₃-N losses with subsurface drain water. *Trans ASAE.*, (in press). With permission.)

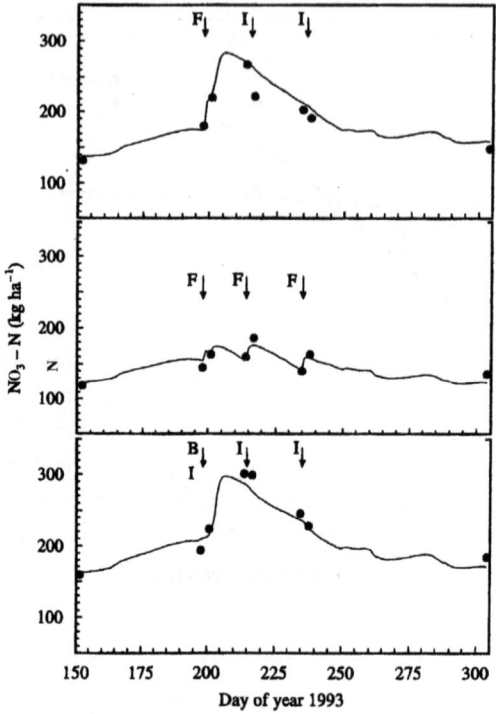

Figure 7.15 Total soil profile (120 cm) nitrate-N during the growing season for treatments of single fertigation, multiple fertigation, and broadcast application of fertilizer (F: fertigation, I: irrigation, B:broadcast). (From Cameira, M.R., P.L. Sousa, H.J. Farahani, L.R. Ahuja, and L.S. Pereira. 1998. Evaluation of the RZWQM for the simulation of water and nitrate movement in level-basin, fertigated maize. *J. Agric. Engng. Res.*, 69:331-341. With permission.)

100, 150, 200 kg N/ha, and split applications were 10 days before planting and 20 days after planting with a total amount of 150 and 200 kg N/ha. Simulation results showed that grain yield and N loss increased with application rates, but no differences were found between single and split applications under both moldboard plow and no-till systems.

Tillage Effects

The major effects of tillage on nitrogen dynamics are residue incorporation and soil aeration. Surface broadcast manure is generally not available to plants where no tillage practice is performed because mixing by biological means is inadequate. In a tillage study over 15 years (1978 to 1992), Karlen et al. (1998) compared four tillage practices (moldboard plow, chisel, ridge tillage, and no-till) with respect to their effects on corn yield, N uptake, soil nitrogen balance, and N losses to tile flow. Measured average corn yields over the 15 years were significantly different among tillage treatments, as were average N uptakes. Nitrogen change in the top 5 cm of soil was also significant among the various tillage practices. However, N losses to tile drainage were not significantly different among tillage treatments. RZWQM predicted that corn yields were not significantly different although the simulated trend agreed with the observed. Large discrepancies existed for simulated N pools in the soil. However, Singh and Kanwar (1995), using a set of data from 1990 to 1992, found that the RZWQM model correctly simulated nitrate losses to tile drainage in all three years under all four tillage treatments. Similar results were obtained by Kumar et al. (1999). Azevedo et al. (1997) used the calibrated RZWQM for the above experimental site to simulate nitrate loss to tile drainage under moldboard plow and no-till systems, and found more nitrate loss under no-till systems due to more continuous macropore flow under no-till than under tilled conditions.

Other Agricultural Management Practices

Irrigation affects nitrate availability in the soil for both leaching and plant uptake. In a Colorado study, Ma et al. (1998a) showed that nitrogen losses were considerably reduced and plant N uptake was increased by reducing irrigation rates to 10 cm per event rather than 20 cm per event (Table 7.2). Similarly, Buchleiter et al. (1995) calibrated the RZWQM for a center pivot irrigated corn system and then used the model to predict nitrate leaching under various irrigation regimes. They found that increasing water irrigation by 40% resulted in an increase of 110 kg N/ha of nitrate leaching and a reduction of 50% in corn biomass. Tile drainage is a commonly used practice in the Midwest and has profound effects on soil nitrogen dynamics (Singh and Kanwar, 1995). The RZWQM correctly simulated nitrate losses to tile flow in several studies in Nashua, Iowa (Singh and Kanwar, 1995; Kumar et al., 1998; 1999; Bakhsh et al., 1999).

SUMMARY AND CONCLUSIONS

The Root Zone Water Quality Model (RZWQM) has been applied to a variety of studies under different soil and climate conditions (Table 7.5). However, all the applications were based on limited data and, therefore, the model was seldom calibrated for all the nitrogen components and processes. Experimental data commonly available include soil residual nitrate content and plant N uptake, which may reflect our emphasis on and interest in soil quality and N use efficiency in the agricultural community. Predictability of the RZWQM for soil N fates depends not only on correct parameterization of the C/N processes, but also on soil water simulation (Martin and Watts, 1999; Singh and Kanwar, 1995); plant growth (Nokes et al., 1996; Martin and Watts, 1999); and initial conditions (Bakhsh et al., 2000; Ma et al., 1998a). Simulation errors on an individual component of an agricultural system will be propagated and reflected in the simulation results of other components. In addition, the RZWQM is a one-dimensional point scale model, and it may be only used for a representative point in a field. Therefore, it is important to take into account the field variability when comparing simulation results to field measurements.

Table 7.5 Field Studies where RZWQM Was Used to Simulate C/N Dynamics

Authors	Exp. Site/Year	Soil/Crop	Tillage	Management Practices Fertilizer	Irrigation	Pesticide	Exp. Measurements
Bakhsh et al., 1999	Iowa 1993–1996	Loam/corn, soybean	Chisel plow	Swine manure	N/A	N/A	Tile flow, nitrate in tile flow
Bakhsh et al., 2000	Iowa 1996–1999	Clay/corn, soybean	Moldboard plow, chisel plow	UAN, anhydrous ammonia	N/A	N/A	Tile flow, nitrate in tile flow, crop yield
Borah and Kalita, 1998	Kansas 1995–1997	Silty clay loam, sandy loam/corn	N/A	UAN	N/A	Atrazine	Nitrate and atrazine in soil water samples (suction lysimeters)
Cameira et al., 1998	Portugal 1993	Silty loam/corn	Disk harrow, rotary tiller	Urea, UAN	Flood	N/A	Soil water content, water table, soil nitrate-N
Cameira 1999	Portugal 1996–1998	Sand,silty loam/corn	N/A	Urea, UAN	Flood, sprinkler	N/A	Soil water content, soil water pressure, water uptake, ET, LAI, yield, biomass, soil N, N uptake, N leaching, plant height, rooting depth
Farahani et al., 1999	Colorado 1991	Clay loam, loam/corn	No-till	UAN	N/A	N/A	Soil water content, yield
Farahani et al., 1999	Colorado 1972–1973	Loamy sand/corn	Disking, chisel plow, cultivator	NH_4NO_3 Anhydrous NH_3	Sprinkler	N/A	Soil water content, soil nitrate yield, biomass, N uptake
Ghidey et al., 1999	Missouri 1992–1994	Silt loam/corn, soybean	Field cultivator, no-till	UAN	N/A	Atrazine, alachlor	Soil water content, above-ground biomass, yield, N uptake, soil nitrate, atrazine and alachlor concentrations in soil profile and runoff
Ghidey et al., 1999	Missouri 1983, 1985 1990, 1993	Silt loam/corn, soybean fallow	Moldboard plow disking, field cultivator, no-till	NH_4, NO_3	N/A	N/A	Surface runoff

Reference	Location/Years	Soil/Crop	Tillage	Fertilizer	Irrigation	Pesticide	Model outputs
Jaynes and Miller, 1999	Iowa 1992–1994	Loam/corn, soybean	Disking, no-till	MAP, anhydrous NH_3	N/A	Atrazine, metribuzin	Soil water content, nitrate-N and pesticide concentrations in soils and deep seepage (drainage), ET
Karlen et al., 1998	Iowa 1977–1992	Loam/corn, soybean	Moldboard plow ridge till, chisel plowing, no-till	Anhydrous NH_3	N/A	Atrazine, alachlor	Plant N uptake, soil OC, C:N ratio, nitrate-N loss in tile drainage
Landa et al., 1999	Ohio 1991–1993	Silt loam/corn, soybean	Chisel plow, disking	Liquid 28 anhydrous NH_3	N/A	Atrazine, alachlor, metribuzin	Soil water content, soil nitrate and above-ground biomass, and grain yield
Ma et al., 1998a	Colorado 1993–1996	Sandy loam/ silage corn	Moldboard plow, field cultivator	Beef manure	Alternative furrow Flood	N/A	Soil water content, soil N, N uptake, silage yield
Ma et al., 1998b	Arkansas 1989	Silt loam/tall fescue	No-till	Broiler litter	None		Soil water content, soil water pressure head, soil N, N uptake, soil temperature, biomass, runoff, N in runoff and suction lysimeter waters
Martin and Watts, 1999	Nebraska 1992–1994	Silt loam/corn	Disking	NH_4NO_3	Sprinkler	Atrazine, metolachlor	Soil water content, nitrate and ammonium-N in soil, LAI, biomass, N uptake, yield
Singh and Kanwar, 1995	Iowa 1990–1992	Loam/corn	Moldboard plow, chisel plow, no-till, ridge-till	N/A	N/A	N/A	Tile drainage flow, nitrate in tile flow and soil profile
Walker 1996	Illinois 1992–1993	Silty clay loam/ corn, soybean	Conventional till, reduced till, no-till	N/A	N/A	N/A	Tile flow, nitrate in tile flow

Note: UAN: urea ammonium nitrate; N/A: not available; LAI: leaf area index; ET: evapotranspiration; MAP: monoammonium phosphate.
Modified from Ma, L., L.R. Ahuja, J.C. Ascough II, M.J. Shaffer, K.W. Rojas, R.W. Malone, and M.R. Cameira. 2000a. Integrating system modeling with field research in agriculture: applications of the Root Zone Water Quality Model (RZWQM). *Adv. Agron.,* 71:233-292.

Generally, the RZWQM has the capability of simulating soil N dynamics such as manure decomposition (Ma et al., 1998a; b), nitrate loss to tile flow (Singh and Kanwar, 1995), residual soil nitrate (Ma et al., 1998a; b; Cameira et al., 1998), and plant N uptake (Ma et al., 1998a). In cases where quantitative agreement between experimental and simulation results was not achieved, the RZWQM simulated the correct responses to various management practices (Azevedo et al., 1997; Karlen et al., 1998). Such a qualitative agreement is important for the purpose of technology transfer and decision-making. With further applications of the RZWQM and a better understanding of carbon/nitrogen dynamics in the soil and their relationship to other components of agricultural systems, greater use of agricultural system models, such as RZWQM, in addressing real-world problems is anticipated.

REFERENCES

Ahuja, L.R., K.W. Rojas, J.D. Hanson, M.J. Shaffer, and L. Ma (Eds.). 2000a. *The Root Zone Water Quality Model.* Water Resources Publications LLC. Highlands Ranch, CO, 372pp.

Ahuja, L.R., K.E. Johnsen, and K.W. Rojas. 2000b. Water and chemical transport in soil matrix and macropores. In L.R. Ahuja, K.W. Rojas, J.D. Hanson, M.J. Shaffer, and L. Ma (Eds.). *The Root Zone Water Quality Model.* Water Resources Publications LLC. Highlands Ranch, CO, 13-50.

Azevedo, A.S., P. Singh, R.S. Kanwar, and L.R. Ahuja. 1997. Simulating nitrogen management effects on subsurface drainage water quality. *Agric. Systems.,* 55:481-501.

Bakhsh, A., R.S. Kanwar, and L.R. Ahuja. 1999. Simulating the effect of swine manure application on NO3-N transport to subsurface drainage water. *Trans. ASAE,* 42:657-664.

Bakhsh, A., R.S. Kanwar, D.B. Jaynes, T. S. Colvin, and L.R. Ahuja. 2000. Predicting effects of variable nitrogen application rates on corn yields and NO_3-N losses with subsurface drain water. *Trans ASAE.,* (in press).

Berghuijs van Dijk, J.T., P.E. Rijtema, and C.W.J. Roest. 1985. ANIMO Agricultural Nitrogen Model. NOTA 1671, Institute for Land and Water Management Research, Wageningen.

Borah, M.J., and P.K. Kalita. 1998. Evaluating RZWQM and LEACHM for agricultural chemical transport in two Kansas soils. Paper presented at the July 12-16, *1998 ASAE Annual International Meeting,* Orlando, FL, Paper No. 982215. ASAE, St. Joseph, MI.

Bradbury, N.J., A.P. Whitmore, P.B.S. Hart, D.S. Jenkinson. 1993. Modeling the fate of nitrogen in crop and soil in the years following application of ^{15}N-labeled fertilizer to winter wheat. *J. Agric. Sci., Cambridge,* 121:363-379.

Buchleiter, G.W., H.J. Farahani, and L.R. Ahuja. 1995 Model evaluation of groundwater contamination under center pivot irrigated corn in eastern Colorado. In C. Heatwole (Ed.). *Proceedings of the International Symposium on Water Quality Modeling.* Orlando, FL, 41-50.

Cameira, M.R. 1999. Water and Nitrogen Balance in Irrigated Corn in the Sorraia Valley. Discussion of the Transfer Processes and Application of the Model RZWQM98. Ph.D. thesis. Technical University of Lisbon, Agronomy Institute, Portugal.

Cameira, M.R., P.L. Sousa, H.J. Farahani, L.R. Ahuja, and L.S. Pereira. 1998. Evaluation of the RZWQM for the simulation of water and nitrate movement in level-basin, fertigated maize. *J. Agric. Engng. Res.,* 69:331-341.

Crane, S.R., P.W. Westerman, and M.R. Overcash. 1981. Short-term chemical transformations following land application of poultry manure. *Trans. ASAE,* 24:382-390.

Eckersten, H., and P.-E. Jansson. 1991. Modeling water flow, nitrogen uptake and production for wheat. *Fert. Res.,* 27:313-329.

Farahani, H.J., G.W. Buchleiter, L.R. Ahuja, and L.A. Sherrod. 1999. Model evaluation of dryland and irrigated cropping systems in Colorado. *Agron. J.,* 91:212-219.

Ghidey, F., E.E. Alberts, and N.R. Kitchen. 1999. Evaluation of RZWQM using field measured data from the Missouri MSEA. *Agron. J.,* 91:183-192.

Grant, R.F. 1995a. Dynamics of energy, water, carbon and nitrogen in agricultural ecosystems: simulation and experimental validation. *Ecol. Modelling,* 81:169-181.

Grant, R.F. 1995b. Salinity, water use and yield of maize: testing of the mathematical model *ecosys. Plant Soil,* 172:309-322.

Grant, R.F. 2001. A review of the Canadian ecosystem model — *ecosys.* In M.J. Shaffer, L. Ma, and S. Hansen (Eds.). *Modeling Carbon and Nitrogen Dynamics for Soil Management.* Lewis Publishers, Boca Raton, FL, 173-264.

Hanks, J., and J.T. Ritchie. 1991. *Modeling Plant and Soil Systems.* American Society of Agronomy, Inc., Madison, WI, 545pp.

Hansen, S., H.E. Jensen, N.E. Nielsen, and H. Svendsen. 1991. Simulation of nitrogen dynamics and biomass production in winter wheat using the Danish simulation model DAISY. *Fert. Res.,* 27:245.

Hanson, J.D. 2000. Generic crop production model for the root zone water quality model. In L.R. Ahuja, K.W. Rojas, J.D. Hanson, M.J. Shaffer, and L. Ma (Eds.). *The Root Zone Water Quality Model.* Water Resources Publications LLC, Highlands Ranch, CO, 81-118.

Hoff, J.D., D.W. Nelson, and A.L. Sutton. 1981. Ammonia Volatilization from liquid swine manure applied to cropland. *J. Environ. Qual.,* 10:90-95.

Jaynes, D.B., and J.G. Miller. 1999. Evaluation of RZWQM using field measured data from Iowa MSEA. *Agron. J.,* 91:192-200.

Johnsson, H., L. Bergstrom, P.-E. Janson, and K. Paustian. 1987. Simulated nitrogen dynamics and losses in a layered agricultural soil. *Agriculture, Ecosystems and Environment,* 18:333-356.

Karlen, D.L., A. Kumar, R.S. Kanwar, C.A. Cambardella, and T.S. Colvin. 1998. Tillage system effects on 15-year carbon-based and simulated N budgets in a tile-drained Iowa field. *Soil Till. Res.,* 48:155-165.

Kumar, A., R.S. Kanwar, and L.R. Ahuja. 1998. RZWQM simulation of nitrate concentrations in subsurface drainage from manured plots. *Trans. ASAE,* 41:587-597.

Kumar, A., R.S. Kanwar, P. Singh, and L.R. Ahuja. 1999. Simulating water and NO_3-N movement in the vadose zone by using RZWQM for Iowa soils. *Soil and Tillage Res.,* 50:223-236.

Landa, F.M., N.R. Fausey, S.E. Nokes, and J.D. Hanson. 1999. Evaluation of the root zone water quality model (RZWQM3.2) at the Ohio MSEA. *Agron. J.,* 91:220-227.

Leonard, R.A., W.G. Knisel, and D.S. Still. 1987. GLEAMS: Groundwater loading effects of agricultural management systems. *Trans. ASAE,* 30:1403-1418.

Linn, D.M., and J.W. Doran. 1984. Effect of water -filled pore space on carbon dioxide and nitrous oxide production in tilled and non-tilled soils. *Soil Sci. Soc. Am. J.,* 48:1267-1272.

Ma, L., M.J. Shaffer, J.K. Boyd, R. Waskom, L.R. Ahuja, K.W. Rojas, and C. Xu. 1998a. Manure management in an irrigated silage corn field: experiment and modeling. *Soil Sci. Soc. Am. J.,* 62:1006-1017.

Ma, L., H.D. Scott, M.J. Shaffer, and L.R. Ahuja. 1998b. RZWQM simulations of water and nitrate movement in a manured tall fescue field. *Soil Sci.,* 163:259-270.

Ma, L., L.R. Ahuja, J.C. Ascough II, M.J. Shaffer, K.W. Rojas, R.W. Malone, and M.R. Cameira. 2000a. Integrating system modeling with field research in agriculture: applications of the Root Zone Water Quality Model (RZWQM). *Adv. Agron.,* 71:233-292.

Ma, L., J.C. Ascough II, L.R. Ahuja, M.J. Shaffer, J.D. Hanson, and K.W. Rojas. 2000b. Root Zone Water Quality Model sensitivity analysis using Monte Carlo simulation. *Trans. ASAE*, 43:883-895.

Ma, L., D.C. Nielsen, L.R. Ahuja, K.W. Rojas, J.D. Hanson, and J.G. Benjamin, 2000c. Modeling corn responses to water stress under various irrigation levels in the Great Plains. *Trans. ASAE*, (in review).

Ma, L., and M.J. Shaffer. 2001. A review of carbon and nitrogen processes in nine U.S. soil nitrogen dynamics models. In M.J. Shaffer, L. Ma, and S. Hansen (Eds.). *Modeling Carbon and Nitrogen Dynamics for Soil Management*. Lewis Publishers, Boca Raton, FL, 55-102.

Martin, D.L., and D.G. Watts. 1999. Application of the root zone water quality model in central Nebraska. *Agron. J.*, 91:201-211.

McGechan, M.B., and L. Wu. 2001. A review of carbon and nitrogen processes in European soil nitrogen dynamics models. In M.J. Shaffer, L. Ma, and S. Hansen (Eds.). *Modeling Carbon and Nitrogen Dynamics for Soil Management*. Lewis Publishers, Boca Raton, FL, 103-171.

Nokes, S.E., F.M. Landa, and J.D. Hanson. 1996. Evaluation of the crop component of the root zone water quality model for corn in Ohio. *Trans. ASAE*, 39:1177-1184.

Parton, W.J., D.S. Ojima, C.V. Cole, and D.S. Schimel. 1994. A general model for soil organic matter dynamics: sensitivity to litter chemistry, texture and management. In R.B. Bryant, and R.W. Arnold (Eds.). *Quantitative Modeling of Soil Forming Processes*. SSSA Special Publication No. 39, Madison, WI, 147-167.

Rojas, K.W., and L.R. Ahuja. 2000. Management practices. In L.R. Ahuja, K.W. Rojas, J.D. Hanson, M.J. Shaffer, and L. Ma (Eds.). *The Root Zone Water Quality Model*. Water Resources Publications LLC, Highlands Ranch, CO, 245-280.

Schepers, J.S., and A.R. Mosier. 1991. Accounting for nitrogen in nonequilibrium soil-crop systems. In R.F. Follett, D.R. Keeney, and R.M. Cruse (Eds.). *Managing Nitrogen for Groundwater Quality and Farm Profitability*. Soil Society of America, Inc. Madison, WI.

Scott, H.D., A. Mauromoustakos, and J.T. Gilmour. 1995. Fate of inorganic nitrogen and phosphorus in broiler litter applied to tall fescue. *Arkansas Agri. Exp. Stn. Bull.* 947.

Shaffer, M.J., and W.E. Larson. (Eds.) 1987. NTRM, A Soil-Crop Simulation Model for Nitrogen, Tillage, and Crop-Residue Management. U.S. Department of Agriculture Conservation Research Report 34-1. 103pp.

Shaffer, M.J., A.D. Halvorson, and F.J. Pierce. 1991. Nitrate leaching and economic analysis package (NLEAP): model description and application. In R.F. Follett, D.R. Keeney, and R.M. Cruse. (Eds.). *Managing Nitrogen for Ground Water Quality and Farm Profitability*. Soil Science Society of America, Madison, WI, 285–322.

Shaffer, M.J., K.W. Rojas, D.G. DeCoursey, and C.S. Hebson. 2000a. Nutrient chemistry processes (OMNI). In L.R. Ahuja, K.W. Rojas, J.D. Hanson, M.J. Shaffer, and L. Ma (Eds.). *The Root Zone Water Quality Model*. Water Resources Publications LLC, Highlands Ranch, CO, 119-144.

Shaffer, M.J., P.N.S. Bartling, and J.C. Ascough, II. 2000b. Object-oriented simulation of integrated whole farms: GPFARM framework. *Computers and Electronics in Agriculture*, 28:29-49.

Sims, J.T., and D.C. Wolf. 1994. Poultry waste management: agricultural and environmental issues. *Adv. Agron.*, 52:2-83.

Singh, P., and R.S. Kanwar. 1995. Simulating NO_3-N transport to subsurface drain flows as affected by tillage under continuous corn using modified RZWQM. *Trans. ASAE*, 38:499-506.

Walker, S.E. 1996. Modeling Nitrate in Tile-Drained Watersheds of East-central Illinois. Ph.D. thesis. University of Illinois at Urbana-Champaign. 164pp.

Williams, J.R. 1995. The EPIC model. In V.P. Singh (Ed.). *Computer Models of Watershed Hydrology.* Water Resources Publications, Highlands Ranch, CO, 909-1000.

Wu, L., W. Chen, J.M. Baker, and J.A. Lamb. 1999. Evaluation of RZWQM field measured data from a sandy soil. *Agro. J.,* 91:177-182.

CHAPTER 8

Simulated Interaction of Carbon Dynamics and Nitrogen Trace Gas Fluxes Using the DAYCENT Model[1]

S.J. Del Grosso, W.J. Parton, A.R. Mosier, M.D. Hartman, J. Brenner, D.S. Ojima, and D.S. Schimel

CONTENTS

INTRODUCTION

Cycling of carbon (C) and nitrogen (N) among living and non-living systems interacts with climate to maintain the environmental conditions that support life on

[1] The research for this chapter was supported by the following grants; NASA-EOS NAGW 2662, NSF-LTER BSR9011659, DOE NIGEC LWT62-123-06516 and EPA Regional Assessment R824939-01-0.

Earth. Industrialization and human population growth have profoundly altered C and N flows. The human-induced input of reactive N into the global biosphere increased to approximately 150 Tg N (1 Tg = 10^{12} g) in 1996 and is expected to continue to increase for the foreseeable future (Bumb and Baanante, 1996; Smil, 1999). Of this 150 Tg N, about 85 Tg is from synthetic fertilizer, ~35 Tg from biological N fixation in crops, and ~25 Tg from fossil fuel combustion. This input of newly fixed N into the biosphere exceeds the N fixed annually in natural ecosystems (Smil, 1999; Vitousek et al., 1997; Galloway et al., 1995). Atmospheric CO_2 concentration has been increasing since the industrial revolution because the sum of CO_2 emissions from fossil fuel combustion and respiration in natural and agricultural systems exceeds the uptake of CO_2 by photosynthesis and by transport to deep ocean water. Changes in the pool sizes of some important compounds of C and N have been reliably measured; for example, the current and historical atmospheric concentrations of the long-lived trace gases CO_2, CH_4, and N_2O have been well documented (Prather et al., 1995). However, the absolute and relative contributions of the terrestrial and aquatic systems that act as sources and sinks of C and N are somewhat uncertain (Matson and Harris, 1995). Complete accounting of C and N flows through ecosystems by direct measurement would require continuous, spatially intensive monitoring at various scales, and thus, is not feasible. Models test our understanding of the controls on biogeochemical processes and are necessary to scale up results of plot size measurements and calculate the contributions of various natural and managed ecosystems to global C and N budgets.

Ecosystem models can be classified along a spectrum that includes at one extreme complex models that explicitly simulate the biological, chemical, and physical processes that control C and N flows. For example, the model *ecosys* (Grant et al., 1993) simulates microbial growth, aqueous and gaseous transport of metabolic reactants and products, as well as other small-scale processes to model large-scale ecosystem characteristics such as N_2O emissions (Grant and Pattey, 1999) from soils. At the other extreme, simple, empirical models correlate ecosystem-scale processes with parameters that are frequently measured in the field. For example, the Miami model (Lieth, 1975) has been used to calculate global net primary productivity (NPP) from average annual precipitation and temperature values (Alexandrov et al., 1999). Complex models require detailed parameterization and intensive computation, and the usefulness of highly mechanistic models for ecosystem managers has been questioned (Nuttle, 2000). Simpler models are also of limited use to managers because such models are likely to be overly generalized and cannot represent the heterogeneity characteristic of most real-world systems. DAYCENT (Kelly et al., 2000; Parton et al., 1998) is a terrestrial ecosystem model used to simulate exchanges of C, N, and trace gases among the atmosphere, soils, and vegetation. DAYCENT is of intermediate complexity; important processes are represented mechanistically but the model makes use of empirically derived equations, and the required input parameters are often available for many regions.

First, we describe the DAYCENT model in detail. Then, observed and simulated values of soil water content, temperature, mineral N, and N gas emissions from native and agricultural soils in Colorado are compared to demonstrate the validity of DAYCENT. Finally, an application of the model is presented to calculate net greenhouse gas emissions associated with alternative crop management practices for a typical midwestern U.S. soil and a typical U.S. Great Plains soil. Sequestration of C in soil organic matter has been suggested as a means to compensate for emissions of greenhouse gases associated with human activities (Bruce et al., 1999; Lal et al., 1998). Although addition of fertilizer to soils often results in an increase in soil C, objectivity requires that the greenhouse gas emissions associated with N fertilizer production and application be included in a total greenhouse gas accounting. N fertilizer manufacture requires energy consumption resulting in CO_2 emissions, and application of N fertilizer typically results in elevated emissions of nitrous oxide (N_2O), an important greenhouse gas (Granli and Bockman, 1994). Agricultural soils amended with N fertilizer can also contribute to nitrate (NO_3^-) leaching and reduce the quality of water that supplies underground aquifers, waterways, and estuaries. DAYCENT simulations were used to explore how different land management strategies effect soil C and N levels, N_2O emissions, net C storage, NO_3^- leaching, and crop yields associated with a Great Plains soil that has been used for winter wheat/fallow rotations for at least 60 years and a Midwestern soil that has been used for corn/winter wheat/pasture rotations for over 100 years.

DAYCENT MODEL DESCRIPTION

DAYCENT (Parton et al., 1998; Kelly et al., 2000) is the daily time step version of the CENTURY ecosystem model. CENTURY (Parton et al., 1994) operates at a monthly time step because this degree of resolution is adequate for simulation of medium- to long-term (10 to >100 years) changes in soil organic matter (SOM), plant productivity, and other ecosystem parameters in response to changes in climate, land use, and atmospheric CO_2 concentration. However, simulation of trace gas fluxes through soils requires finer time-scale resolution because a large proportion of total gas fluxes is often the result of short-term rainfall, snow melt, or irrigation events (Frolking et al., 1998; Martin et al., 1998), and the processes that result in trace gas emissions often respond nonlinearly to changes in soil water levels. DAYCENT and CENTURY both simulate exchanges of carbon and the nutrients nitrogen (N), potassium (P), and sulfur (S) among the atmosphere, soil, and plants, and use identical files to simulate plant growth and events such as fire, grazing, cultivation, harvest, and organic matter or fertilizer additions. In addition to modeling decomposition, nutrient flows, soil water, and soil temperature on a finer time scale than CENTURY, DAYCENT also has increased spatial resolution for soil layers.

DAYCENT includes sub-models for plant productivity, decomposition of dead plant material and SOM, soil water and temperature dynamics, and trace gas fluxes (Figure 8.1). Plant growth is limited by temperature, water, and nutrient availability. Carbon and nutrients are allocated among leaf, woody, and root biomass based on vegetation type. Transfer of C and nutrients from dead plant material to the soil organic matter and available nutrients pools is controlled by the lignin concentration and C:N ratio of the material, abiotic temperature/soil water decomposition factors,

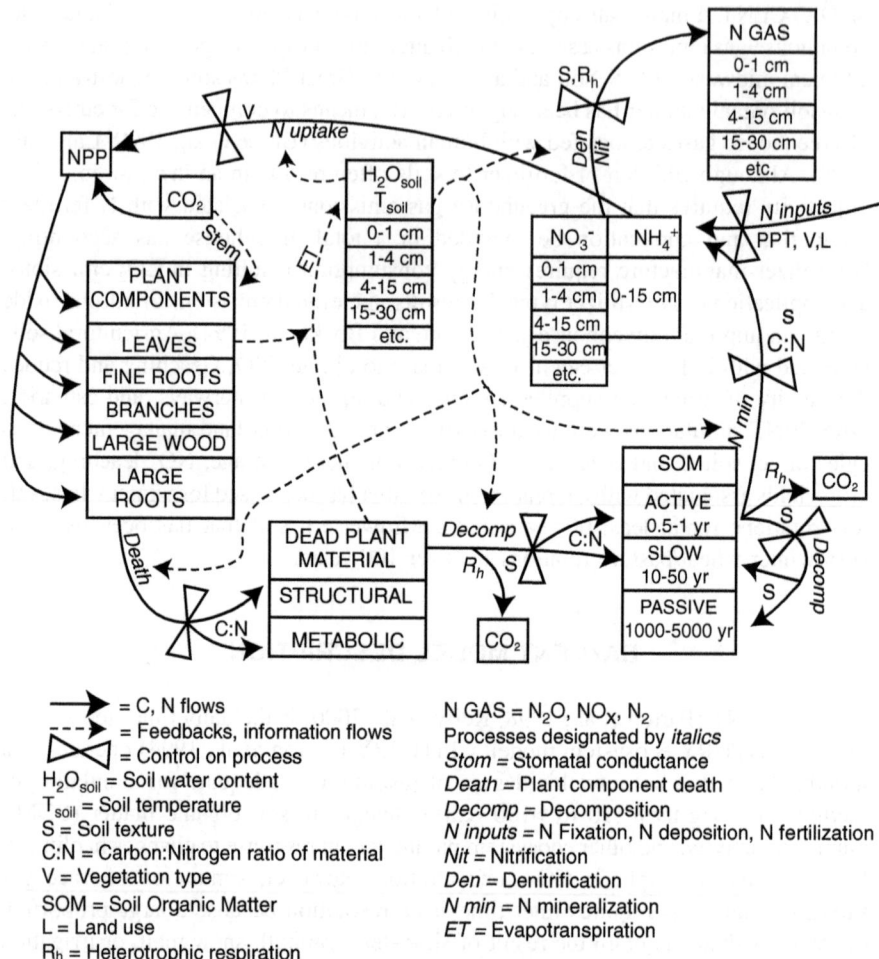

Figure 8.1 Conceptual diagram of the DAYCENT ecosystem model.

and soil physical properties related to texture. Detrital material with low C:N ratios and low proportions of lignin (metabolic) goes to the active SOM pool. The active SOM pool has a rapid turnover time (0.5 to 1 yr) and includes microbial biomass and the highly labile by-products of microbial metabolism. Structural detritus, characterized by high C:N ratios and high lignin contents, flows to the slow SOM pool. The slow SOM pool has intermediate turnover rates (10 to 50 yrs) and includes the microbial by-products that are moderately resistant to further decomposition. Products of SOM decomposition that are extremely resistant to further breakdown make up the passive SOM pool, which has very slow turnover (1000 to 5000 yr). A lower proportion of decomposing SOM is respired as CO_2 and more organic matter is retained in stable forms due to chemical and physical protection as soils become finer textured. The available nutrient pool (NO_3^-, NH_4^+, P, S) is supplied by decomposition of SOM, biological N fixation, and external nutrient additions such as fertilization and N deposition. The rate of nutrient supply from decomposition is determined by the proportions of SOM in the respective pools, and soil water, temperature, and texture. NO_3^- is distributed throughout the soil profile while NH_4^+ is modeled for the 0–15 cm layer only. Available nutrients are distributed among soil layers by assuming that the concentrations of mineral N and SOM are highest near the soil surface and drop exponentially with depth. NO_3^- and NH_4^+ are available for both plant growth and for biochemical processes that result in N transformations (nitrification and denitrification) and N gas emissions. Significant amounts of NO_3^- can be lost from soils via leaching when water flow through the soil profile is sufficiently high. A detailed description of the SOM model used in DAYCENT can be found in Parton et al. (1993) and Parton et al. (1994).

Plant Growth Sub-model

The DAYCENT plant production sub-model (Metherell et al., 1993) can simulate the growth of various crops, grasses, and trees. However, only one type of grass (or crop) and one type of tree can be simulated at a time along with competition between the grass/crop and trees. Different crop, grass, and forest systems are distinguished by varying the parameters that control maximum growth rate, C allocation among plant parts, and the C:N ratios of plant parts. The model also simulates disturbance events such as burning, grazing, and plowing, as well as addition of water and nutrients via irrigation, fertilization, organic matter application, N deposition, and N fixation. For a given grass/crop, forest, or savanna (tree and grass) system, plant production is controlled by a maximum plant growth parameter, nutrient availability, and 0–1 multipliers that reflect shading, water, and temperature stress. Parameters in the equations that account for shading, water, and temperature limitation, maximum plant growth rate, ranges of C:N ratios for plant compartments, etc., can be adjusted to reflect the physiological properties of various vegetation types and particular species of grasses, crops, or trees.

The grass/crop sub-model (Figure 8.2) divides plant biomass among grain, shoot, and root compartments, while the forest sub-model divides biomass among leaves, fine branches, large wood, fine roots, and coarse roots. The allocation of net primary productivity (NPP) among plant compartments is a function of season, the relative sizes of the compartments, tree age, precipitation, and nutrient availability. The proportion of NPP allocated to roots increases as precipitation decreases and can be made a function of time since planting for crops. The N, P, and S concentrations of biomass in the plant compartments are functions of nutrient availability and the component-specific C to element ratio ranges for these nutrients. Plant material death rate is a function of soil water content and a plant component-specific death rate. Coniferous and deciduous tree leaves have a death rate that changes monthly, while deciduous tree leaves also have a much higher death rate during the senescence month. Biomass can be removed or transferred to the litter pool by disturbance events such as harvesting, grazing, plowing, burning, clear cutting, etc. Disturbance events affect both the quantity and nutrient concentration of litter that supplies the SOM pool. For example, fire volatilizes some C and nutrients from the live biomass and litter pools, but the ash provides a nutrient-rich soil input.

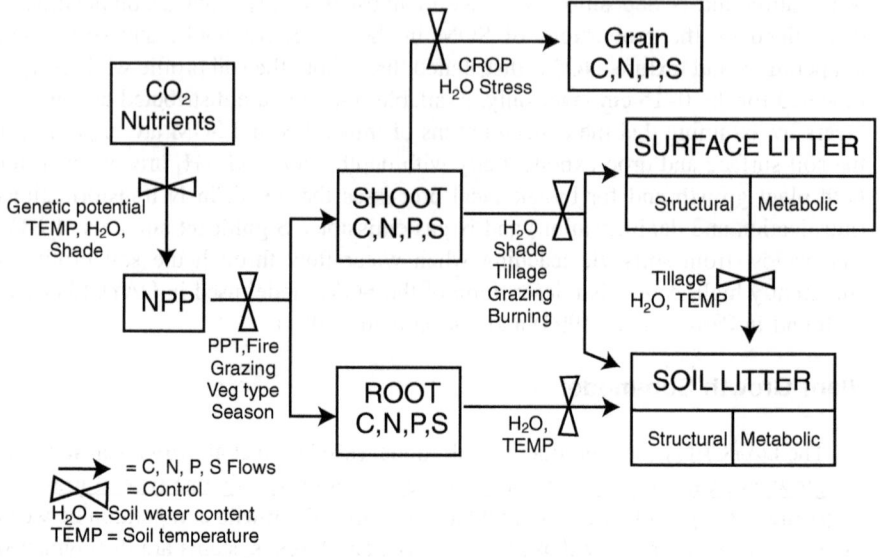

Figure 8.2   The grassland/crop growth sub-model of DAYCENT.

Land Surface Sub-model

The land surface sub-model of DAYCENT (Parton et al., 1998) simulates water flow through the plant canopy, litter, and soil (Figure 8.3), as well as soil temperature. Water content and temperature are simulated for each soil layer. Different types of soils can be represented by varying the number of soil layers simulated, the thickness of each layer, and the depth of the soil profile. An example of a 12-layer structure is: 0–1 cm, 1–4 cm, 4–15 cm, 15–30 cm, 30–45 cm, ..., 135–150 cm. Model inputs include daily precipitation, maximum and minimum air temperature values, soil bulk density (BD), field capacity (FC), saturated hydraulic conductivity (K_{sat}), and the proportion of plant roots for each layer. Precipitation that is intercepted by vegetation and litter is evaporated at the potential evapotranspiration rate (PET). The amount of precipitation intercepted is a function of the total amount of precipitation, the above-ground biomass, and the litter mass. When the average daily air temperature is below freezing, precipitation is assumed to fall as snow and is accumulated in the snowpack. Precipitation that is not intercepted and water inputs from snow melting or irrigation infiltrate the soil or run off of the soil surface. Soil saturated flow and surface runoff are simulated first; then water is evaporated and distributed throughout the soil profile using an unsaturated flow algorithm. PET is calculated using the equation of Penman (1948), and the actual evapotranspiration rate (AET) is limited by soil water potential in the root zone. As biomass decreases, transpiration is assumed to decrease and evaporation increases (Parton, 1978).

Soil water input events initiate a 4-hour infiltration/saturated flow period that is unidirectional downward, and each layer is filled before water flows to the next layer. If the water input rate is greater than K_{sat} (cm/s), the difference goes to surface

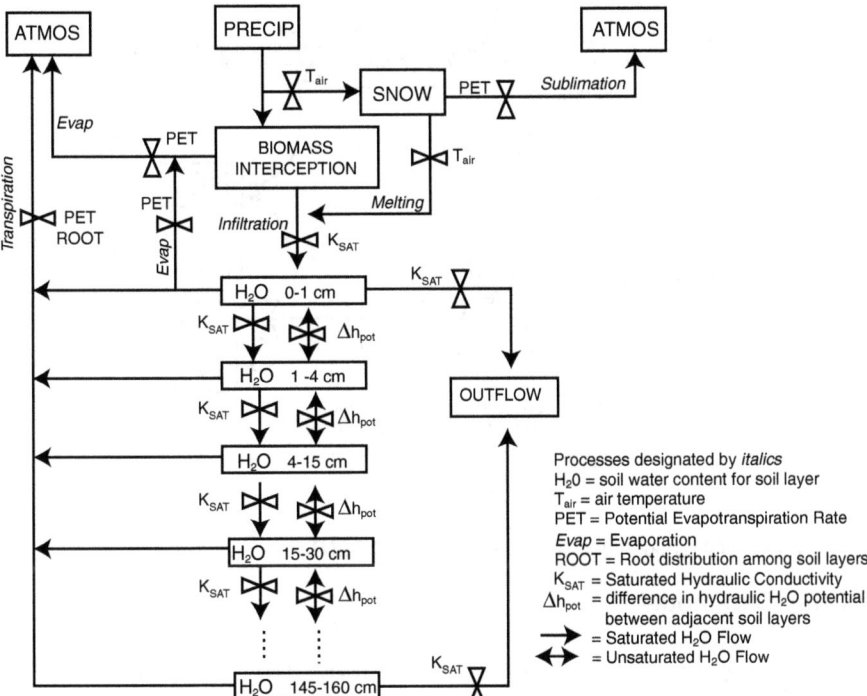

Figure 8.3   The water flow sub-model of DAYCENT.

runoff. The K_{sat} of frozen soil is greatly reduced. After the 4-hour input event ends, water in excess of field capacity is drained into the layer below it. Water that exits the bottom layer of the soil profile goes to outflow and can include leached NO_3^-.

Unsaturated, bidirectional flow is calculated using Darcy's law as a function of the hydraulic conductivity of the soil layers and the difference in hydraulic potential calculated at the centers of adjacent soil layers. The gravitational head and matric potential are summed to obtain the hydraulic potential for each layer. Flux out of the top layer depends on PET, the water content of the top layer, and the minimum water content designated for the top layer. The calculated water fluxes are adjusted if the flow would dry a layer below its minimum allowable water content.

The temperature sub-model (Eitzinger et al., in press; Parton, 1984) calculates thermal diffusivity and daily maximum/minimum temperature for each soil layer. Inputs include daily maximum/minimum air temperatures at 2 meters, plant biomass, snow cover, soil moisture, soil texture, and day length. The upper boundary for the one-dimensional heat flow equation is the soil surface temperature, and the lower boundary is a sine function of the Julian date and the average annual temperature at the bottom of the soil profile. Soil surface temperature is a function of the maximum/minimum air temperature, snow cover, plant biomass, and litter. As snow cover, plant biomass, or litter increases, maximum and minimum soil temperature values are less responsive to changes in air temperature.

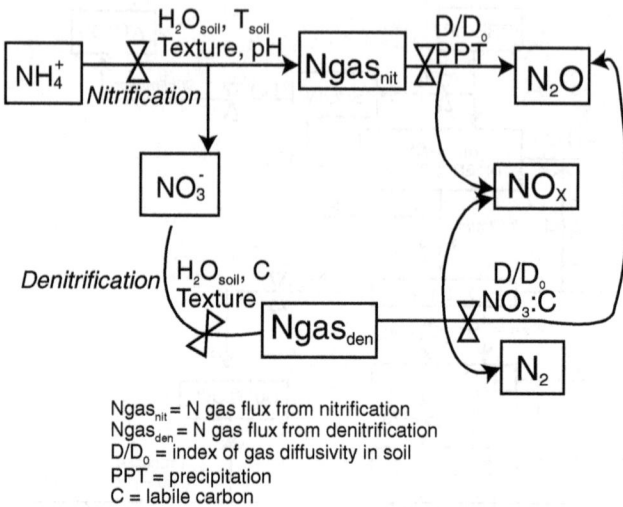

Ngas_nit = N gas flux from nitrification
Ngas_den = N gas flux from denitrification
D/D_0 = index of gas diffusivity in soil
PPT = precipitation
C = labile carbon

Figure 8.4   The nitrogen gas flux sub-model of DAYCENT.

N Gas Sub-model

The N gas sub-model (Parton et al., in press; Del Grosso et al., 2000) of DAYCENT simulates N_2O, NO_x, and N_2 gas emissions from soils resulting from nitrification and denitrification. Nitrification is an aerobic process in which NH_4^+ is oxidized to NO_3^-, with N_2O and NO_x being released as by-products during the intermediate steps (Figure 8.4). DAYCENT assumes that releases of N gases from soils due to nitrification are proportional to nitrification rates and that nitrification rates are controlled by soil NH_4^+ concentration, water content, temperature, pH, and texture (Parton et al., 1996; Parton et al., in press). For a given soil NH_4^+ concentration, nitrification increases with soil temperature (T_{soil}) until T_{soil} reaches the average high temperature for the warmest month of the year. Nitrification is assumed to be not limited by temperature when T_{soil} exceeds the site-specific average high temperature for the warmest month of the year. Nitrification is limited by moisture stress on microbial activity when WFPS (water-filled pore space) is low and by O_2 availability when WFPS is high. Peak nitrification rates are assumed to occur when soil water content is ~50% WFPS, with finer textured soils having a slightly higher optimum WFPS. Nitrification is not limited when pH is greater than 7, but decreases exponentially as pH falls below 7 due to acidity.

Denitrification is a biochemical process in which heterotrophic microbes reduce NO_3^- to NO_x, N_2O, and N_2 under anaerobic conditions. The denitrification sub-model (Del Grosso et al., 2000; Parton et al., 1996) first calculates total N gas flux from denitrification ($N_2 + N_2O$) and then uses a $N_2:N_2O$ ratio function to infer N_2O and N_2 emissions (Figure 8.4). Denitrification is controlled by labile C availability (e^- donor), soil NO_3^- concentration (e^- acceptor), and O_2 availability (competing e^- acceptor). Modeled soil heterotrophic respiration is used as a proxy for labile C availability. O_2 availability is a function of soil WFPS and O_2 demand. O_2 demand is a function of simulated heterotrophic CO_2 respiration and soil gas diffusivity. Gas diffusivity is calculated as a function of soil WFPS, bulk density, and field capacity according to equations presented by Potter et al. (1996). The model predicts that O_2 is readily available, and little denitrification can occur, in coarse-textured soils with high gas diffusivity unless WFPS exceeds ~80% WFPS. However, fine-textured soils with low gas diffusivity are predicted to contain a substantial proportion of aggregates that can become anaerobic when O_2 demand is high and facilitate denitrification at WFPS values as low as 60%. The ratio of N_2/N_2O gases emitted from soil due to denitrification is a function of soil gas diffusivity and the ratio of e^- acceptor (NO_3^-) to e^- donor (labile C). The probability that N_2O produced from denitrification is further reduced to N_2 before diffusing from the soil surface increases as gas diffusivity decreases because residence time in the soil increases. The $N_2:N_2O$ ratio is high when labile C is in excess compared to NO_3^- because N_2O can act as an alternative e^- acceptor when NO_3^- is in short supply.

On a daily time step, simulated values of soil NH_4^+, NO_3^-, CO_2, water content, and temperature are used to calculate N_2O and N_2 emissions from nitrification and denitrification for each soil layer; then the N gas values are summed to yield simulated N_2O and N_2 gas emissions for the soil profile. NO_x emissions are a function of total N_2O emissions from nitrification and denitrification, a $NO_x:N_2O$ ratio function, and a pulse multiplier (Parton et al., in press). The majority of NO_x emissions from soil are assumed to be from nitrification because NO_x is highly reactive under the reducing conditions that facilitate denitrification (Conrad, 1996). The $NO_x:N_2O$ ratio is low (~1) when gas diffusivity, and hence O_2 availability, is low because denitrification is the dominant process under such conditions. The ratio increases to a maximum of ~20 as gas diffusivity increases because nitrification dominates when soils are well-aerated. The modeled total N_2O emission rate is multiplied by the ratio function to obtain a base NO_x emission rate. The base NO_x emission rate can be modified by a pulse multiplier. Large pulses of NO_x are often initiated when precipitation falls on soils that were previously dry, independent of other controlling factors (Smart et al., 1999; Martin et al., 1998; Hutchinson et al., 1993). The pulses are thought to be related to substrate accumulation and activation of water-stressed bacteria upon wetting (Davidson, 1992). To account for these pulses, the model incorporates the pulse multiplier sub-model described by Yienger and Levy (1995). The magnitude of the multiplier is proportional to the amount of precipitation and the number of days since the precipitation event, with a maximum multiplier of 15. N balance is verified on a daily basis, and calculated N gas emission rates are revised downward if there is not enough NO_3^- and NH_4^+ available to supply the calculated N gas emission for that day.

MODEL VALIDATION

The CENTURY model has been used to successfully simulate SOM and N cycling in various natural and managed systems (Kelly et al., 1997; Parton and Rasmussen, 1994; Paustian et al., 1992). The ability of DAYCENT to simulate soil water content, mineral N, and N_2O and CO_2 emissions from field sites in Scotland,

Germany, and Colorado was demonstrated and compared to alternative models by Frolking et al. (1998). Simulated values of soil water content, temperature, and N_2O and NO_x emissions were shown to agree favorably with observed data from rangeland soils of varying texture and fertility levels (Parton et al., in press). This chapter section presents results of model simulations of soil water content, temperature, mineral N concentration, and N_2O emissions from rangeland (Mosier et al., 1996) and from fertilized, irrigated corn and barley crops in northeastern Colorado (Mosier et al., 1986). Mosier et al. (1996) collected weekly soil water, temperature, and N gas flux data from soils of different textures and fertility levels at the Central Plains Experimental Range (CPER). The CPER has an average annual precipitation of 35 cm, and texture ranges from sandy loam to clay loam. Mosier et al. (1986) collected soil water, temperature, mineral N, and N_2O gas flux data from a field used for corn cropping in 1982 and barley cropping in 1983. The soil used for the corn and barley cropping is a moderately well-drained Nunn clay. It was amended with 20 g $N/m^2/yr$ of ammonium sulfate $[(NH_4)_2SO_4]$ and irrigated during the growing season (Mosier et al., 1986). Measured values of soil C, mineral N, texture, and bulk density (Mosier et al., 1996) were used to initialize the SOM and nutrient pools and to parameterize soil physical properties (field capacity, wilting point, K_{sat}). No model coefficients or equations were adjusted for the validation simulations. First consider how well DAYCENT simulated the drivers of the N gas sub-model and then compare observed and simulated N_2O emissions.

Figures 8.5a and b show the observed and simulated time series for soil water content and temperature for a representative year of measurements for a native sandy loam soil at the CPER. Temperature was simulated well during all seasons but the model failed to simulate the variability in soil water observed during the spring and winter months. The poor model performance during the non-growing season is related to the high spatial heterogeneity of measured soil water content from snow drifting and melting. The model correctly simulated the trends in soil water content for the cropped soil (Figure 8.5c) but did not dry to the extent indicated by the data. Similarly, the model simulated the general trends in soil temperature (Figure 8.5d), but some data points were greatly over- or underestimated. Figure 8.6a shows that the model correctly simulated high NH_4^+ levels after fertilization events for the cropped soil and the observed depletion of NH_4^+ from plant uptake and nitrification. The measured soil NO_3^- concentrations were quite variable, but the model exhibited the general pattern of the data, high levels in the spring, and decreasing NO_3^- during the growing season (Figure 8.6b). Figures 8.6c and d show the observed and simulated N_2O gas emissions from the native sandy loam pasture at the CPER for a representative year and from the corn and barley crops. Although the day-to-day variability is high, DAYCENT accurately captured the monthly and seasonal trends in N_2O emissions from the native soil, although winter season fluxes tended to be underestimated. The model represented the daily variability and seasonal patterns of N_2O emissions from the corn crop (1982) reasonably well. However, N_2O emissions were greatly overestimated for the barley (1983). The overestimation of N_2O emissions in May and June is due to denitrification events simulated by DAYCENT. Mosier et al. (1986) pointed out that the spring of 1983 was unusually wet and soil water content was at or near field capacity most of the time, but the expected N_2O

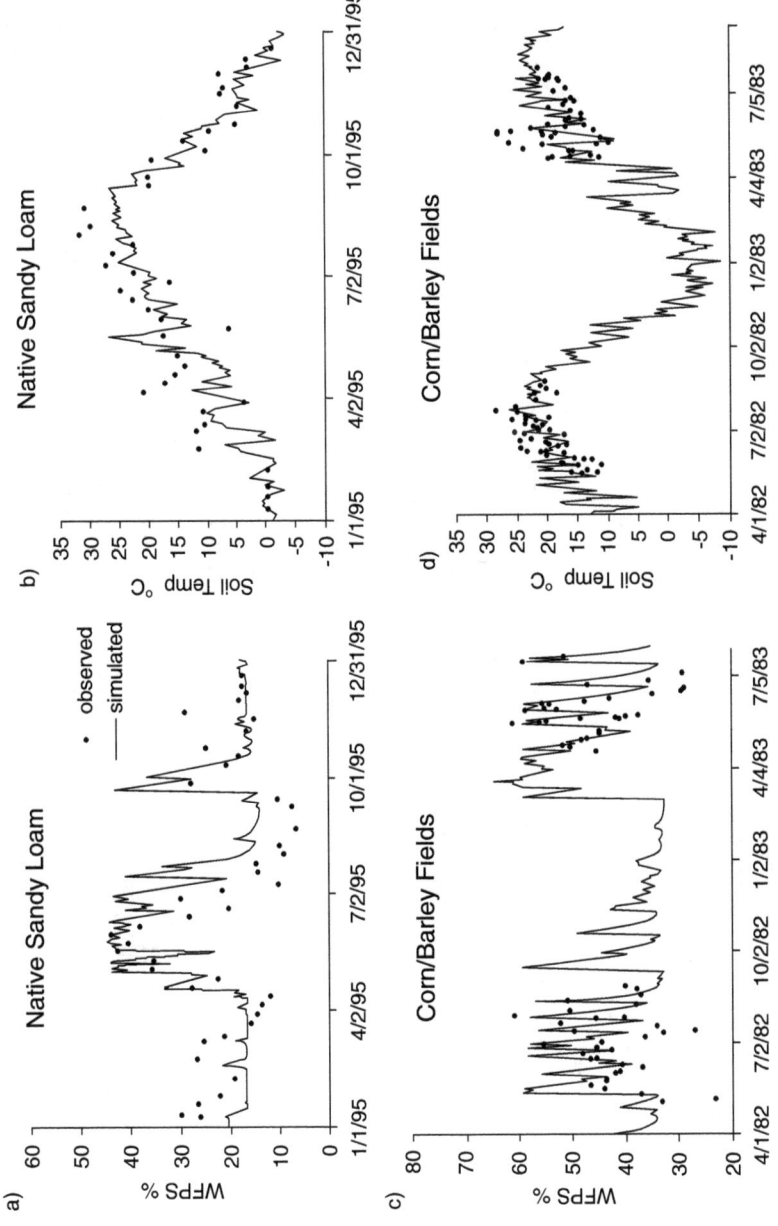

Figure 8.5   Observed and simulated values of soil water content and temperature for a native sandy loam grassland soil (a, b) and for a fertilized, irrigated field cropped with corn in 1982 and barley in 1983 (c, d).

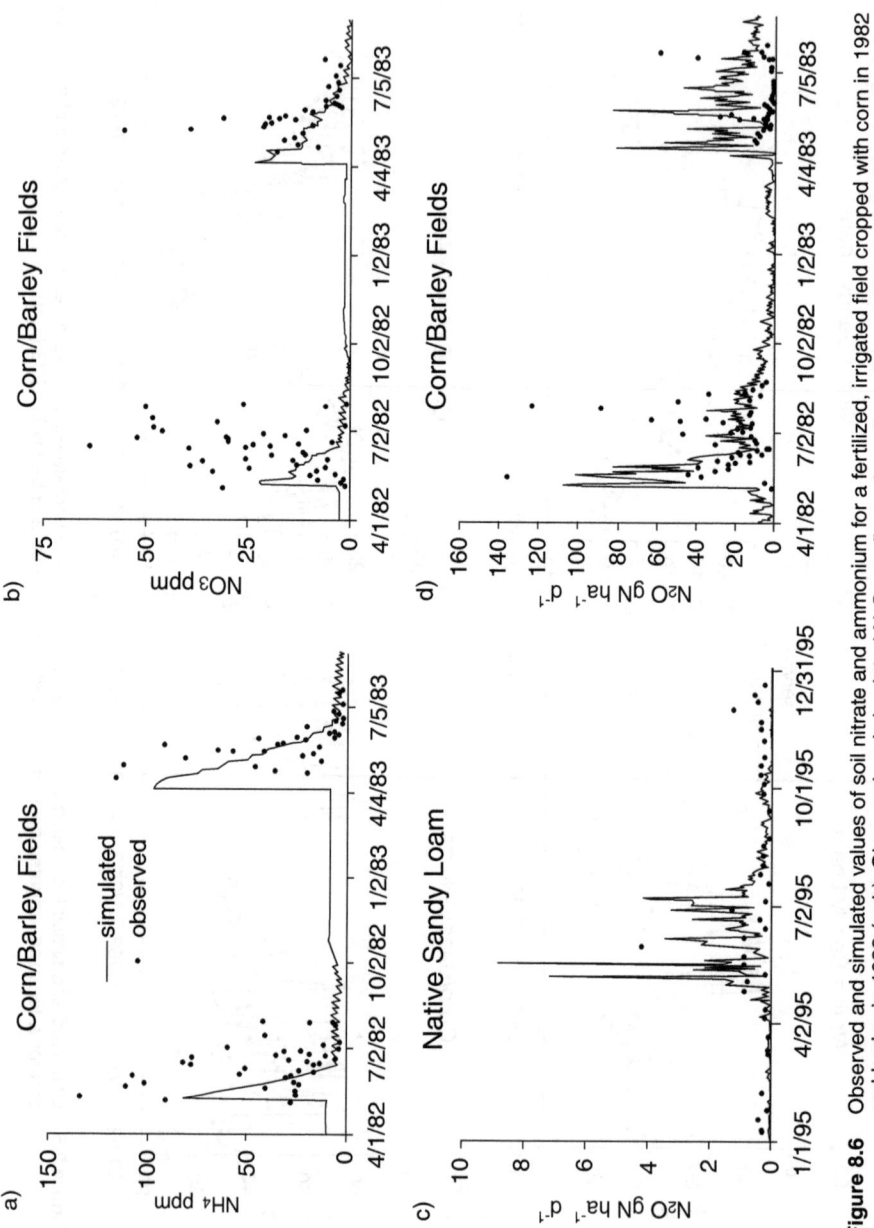

Figure 8.6 Observed and simulated values of soil nitrate and ammonium for a fertilized, irrigated field cropped with corn in 1982 and barley in 1983 (a, b). Observed and simulated N$_2$O gas fluxes from the native sandy loam grassland soil (c) and from the cropped soil (d).

emissions from denitrification (an anaerobic process) were not observed. The authors speculate that C availability may have been limiting denitrification in the spring and that the model overestimated C availability.

These results show that DAYCENT generally simulates the inputs used to calculate N gas emissions reasonably well but that high spatial or temporal variation in the data can contribute to a poor comparison between observed and simulated values. Comparisons with observed data from the CPER showed that soil WFPS was modeled significantly better in the growing season compared to the winter with r^2 values of 0.66 and 0.28, respectively, for aggregated multiyear data from rangeland soils of varying texture and fertility levels. Variability in N gas sub-model inputs and high coefficients of variation for measured N_2O flux rates contribute to the lack of fit on a daily scale between observed and simulated N_2O emission rates (Figures 8.6c and d). Although the model often miss-times large N_2O flux events, on average, the model mimicked the seasonal trends in N_2O flux well, except for the barley crop.

A more extensive validation of DAYCENT (Parton et al., in press) showed that seasonal N gas fluxes were fairly well represented by the model with observed vs. simulated r^2 values of 0.29 and 0.43 for monthly N_2O and NO_x emissions, respectively, for the combined data from five soils of different texture and fertility levels. Frolking et al. (1998) compared the ability of DAYCENT and three other models to simulate the observed soil water content and N_2O emissions from soils in Colorado, Scotland, and Germany. The Colorado site is a dry shortgrass steppe, the Scotland site is a fairly wet ryegrass pasture, and the German sites are perennially cropped. Average yearly N_2O emissions varied by a factor of ~100 among these sites. DAYCENT correctly simulated the high N_2O emission rates from the agricultural soils, the intermediate N_2O emission rates from the pasture, and the very low N_2O emission rates from the shortgrass steppe. DAYCENT also simulated reasonably well the daily variations in N_2O emission rates and soil water content observed at the respective sites (Frolking et al., 1998).

MODEL APPLICATION

The authors used DAYCENT to compare the trade-offs involving net greenhouse gas emissions, NO_3^- leaching, and crop yields associated with different land use scenarios in a Great Plains soil used for winter wheat/fallow rotations and midwestern soil used for corn/winter wheat/pasture rotations. For the Great Plains soil, alternatives to conventional winter wheat/fallow cropping that have been suggested to increase SOM include no-till winter wheat cropping, perennial cropping of corn, corn/soybeans and silage, and reversion to rangeland. The management alternatives considered for the midwestern soil included continuous corn and corn/soybean rotations, each under conventional or no-till cultivation.

To compare the net greenhouse gas effect of the management options, the authors simulated SOM levels and inferred the flux of CO_2 between the atmosphere and the soil from changes in system C values. System C (the sum of C in soil organic matter, surface organic matter, and surface litter) is used because the effects of management options should consider residue C on the soil surface, as well as soil C (Peterson et al., 1998). The N_2O emissions were also simulated for each option, and a yearly CO_2 equivalent for N_2O gas emissions was calculated by assuming that each N_2O molecule has 310 times the global warming potential of a CO_2 molecule (Prather et al., 1995). The greenhouse gas emissions of N fertilizer production were calculated by assuming that each gram of N fertilizer results in the release of 0.8 g CO_2-C due to fossil fuel combustion (Schlesinger, 1999). The preceding three quantities were summed to obtain a measure of the net greenhouse contribution associated with each land use practice. Crop yields and NO_3^- leaching associated with the management alternatives were also compared. The authors present an analysis for the relatively dry Great Plains soil similar to that performed by Del Grosso et al. (2000) using an earlier version of DAYCENT, followed by the results of the simulations for the wetter midwestern soil.

Simulation of Conventional Winter Wheat/Fallow Rotations and Six Alternatives

The crop/fallow system is used to grow winter wheat in the Great Plains because precipitation ranges from 30 to 60 cm per year and this region often experiences drought conditions. To conserve soil water, the land is left to lie fallow for a season after each cropping period. This practice results in a significant decrease (~50%) in SOM after 30 to 50 years (Peterson et al., 1998). Measured texture and bulk density values for a sandy loam soil from the Central Plains Experimental Range (CPER) in northeastern Colorado (Mosier et al., 1996) were used to parameterize soil physical properties for the simulations. DAYCENT was run for 100 years of winter wheat/fallow rotations (1880 to 1980) using a weather file from the CPER to initialize the SOM and nutrient pools. The average annual precipitation for this weather file is 35 cm and the average annual temperature is 9.5°C. The model was then run for another 100 years (1981 to 2080) of conventional till winter wheat/fallow rotations (wwct) and six alternative management scenarios (Table 8.1): no-till winter wheat/fallow rotations (wwnt), perennial corn cropping (corn), corn/alfalfa rotations (corn/alf), range grass (rg), fertilized range grass (rg + N), and continuous silage cropping (silage). Fertilizer was added in the form of ammonium nitrate. The corn, corn/alf, and silage alternatives included irrigation during the growing season (May to October). The winter wheat was fertilized during the planting month (September) and in April of the following year before harvest in July. The simulated fallow period was from August of the harvest year through August of the following year. The simulated conventional tillage winter wheat/fallow soil had a 1995 SOM value of ~1.3 kg C/m^2 soil, which agrees closely with measured values (1.4 to 1.5 kg C/m^2 soil) for a similar soil used for winter wheat/fallow rotations for many years (Mosier et al., 1997). Yearly values of soil C, NO_3^- leached, N_2O gas emitted, and aboveground NPP were compiled for winter wheat/fallow cropping and each alternative scenario.

Figure 8.7a shows that winter wheat/fallow rotations maintain soil C at fairly low levels. Alternative scenarios that included perennial cropping, N addition, and irrigation (corn, corn/alfalfa, and silage) showed significant increases in SOM. Reversion

Table 8.1   Land Management Practices Simulated by DAYCENT

Crop/Vegetation	N Addition (g N/m²/yr)	Month(s) of N Addition	Irrigation	Tillage
Great Plains Soil				
Winter wheat/fallow (wwct)	2.5	9,4	No	Conventional
Winter wheat/fallow (wwnt)	2.5	9,4	No	No-till
Corn	10	5,7	Yes	Conventional
Corn/alfalfa (corn/alf)	6.3	5,7	Yes	Conventional
Silage	6.25^m	12	Yes	Conventional
Range grass (rg)^g	0		No	None
Range grass (rg+N) ^{g}	1	6	No	None
Midwestern Soil				
Corn (cct)	15	5,6	No	Conventional
Corn (cnt)	15	5,6	No	No-till
Corn/soybean (cbct)	7.5	5,6	No	Conventional
Corn/soybean (cbnt)	7.5	5,6	No	No-till

Note: wwct = winter wheat/fallow rotations, conventional till

wwnt = winter wheat/fallow rotations, no-till

corn = perennial corn cropping, conventional till

corn/alf = corn/alfalfa rotations, conventional till

silage = perennial silage cropping, conventional till

rg = native range grass

rg + N = native range grass amended with fertilizer

cct = perennial corn, conventional till

ccnt = perennial corn, no-till

cbct = corn/soybean rotations, conventional till

cbnt = corn/soybean rotations, no-till

^{g} = grazed from April-November

^{m} = N from manure, all others from ammonium nitrate fertilizer

to native rangeland or conversion to no-till wheat/fallow rotations resulted in a small increase of SOM compared to the standard winter wheat/fallow conventional till practice. However, conversion to rangeland augmented with a small amount of N fertilization resulted in sustained increases of SOM. The scenarios that included irrigation and N fertilization (corn and corn/alfalfa) emitted much more N$_2$O than the other alternatives (Figure 8.7b). Corn/alfalfa rotations had the highest N$_2$O emissions, with the corn years showing higher emissions than the alfalfa years. Perennial silage cropping augmented with manure resulted in only slightly higher N$_2$O emissions than the conventional winter wheat/fallow system. No-till winter/wheat fallow rotations showed similar N$_2$O emissions as the conventional till winter wheat/fallow cropping, with the majority of the N$_2$O emitted during the fallow season independent of the cultivation method. Reversion to rangeland, both with and without N addition, resulted in lower N$_2$O emissions compared to the baseline winter wheat/fallow system. Across management alternatives, from 1 to 3% of the applied N fertilizer was lost as N$_2$O gas. The alternatives that did not involve irrigation (wwct, wwnt, rg, rg + N) showed

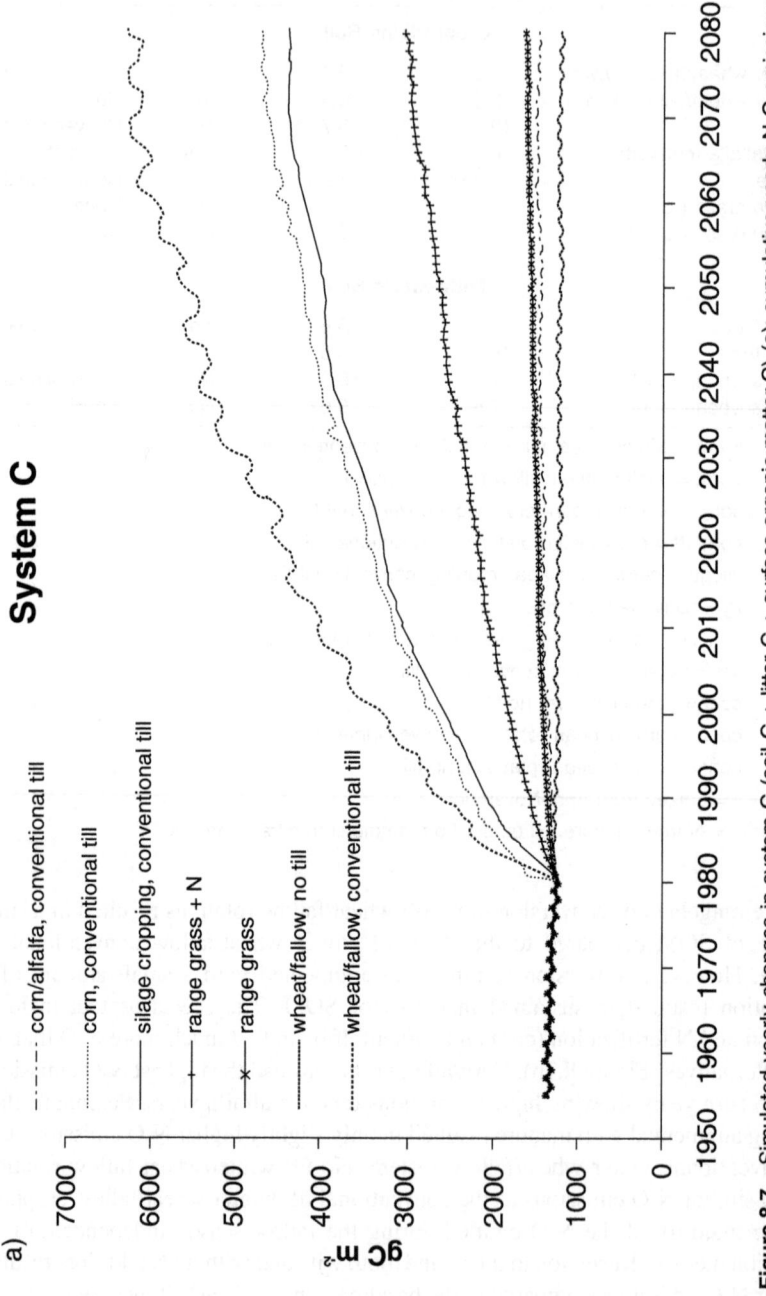

Figure 8.7 Simulated yearly changes in system C (soil C + litter C + surface organic matter C) (a); cumulative yearly N_2O emissions (b); and cumulative yearly NO_3^- leaching (c) for a Great Plains soil under conventional till winter wheat/fallow rotations and six alternative land use scenarios.

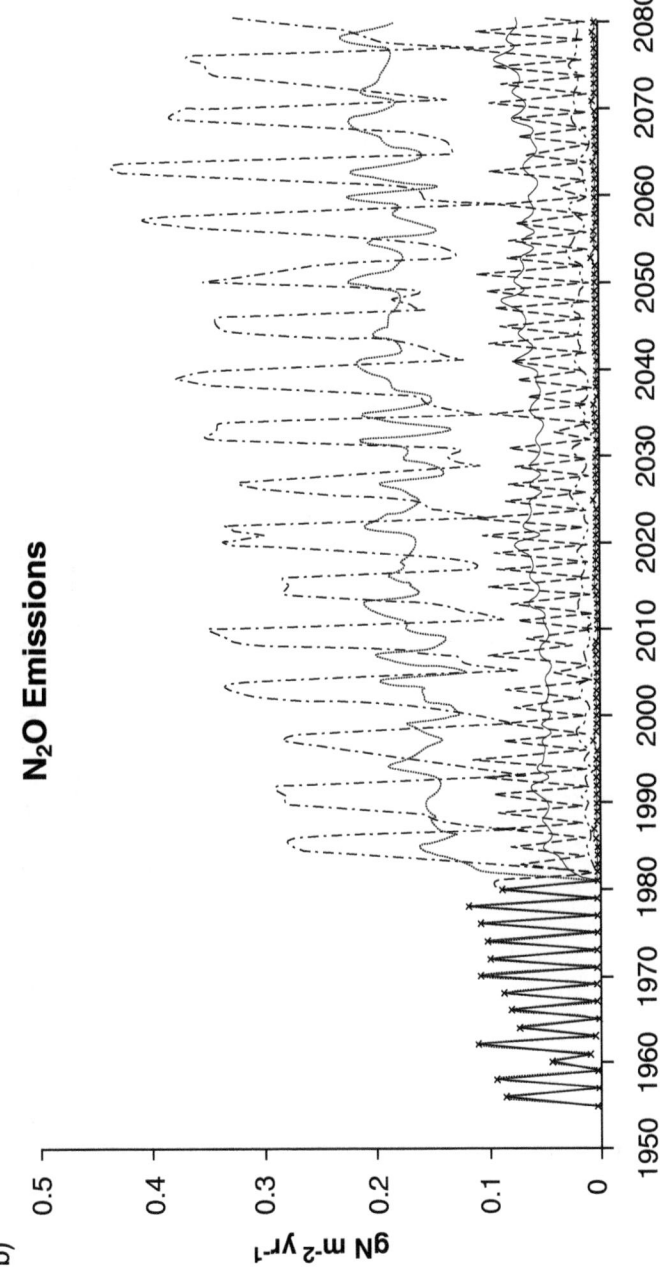

Figure 8.7 (continued).

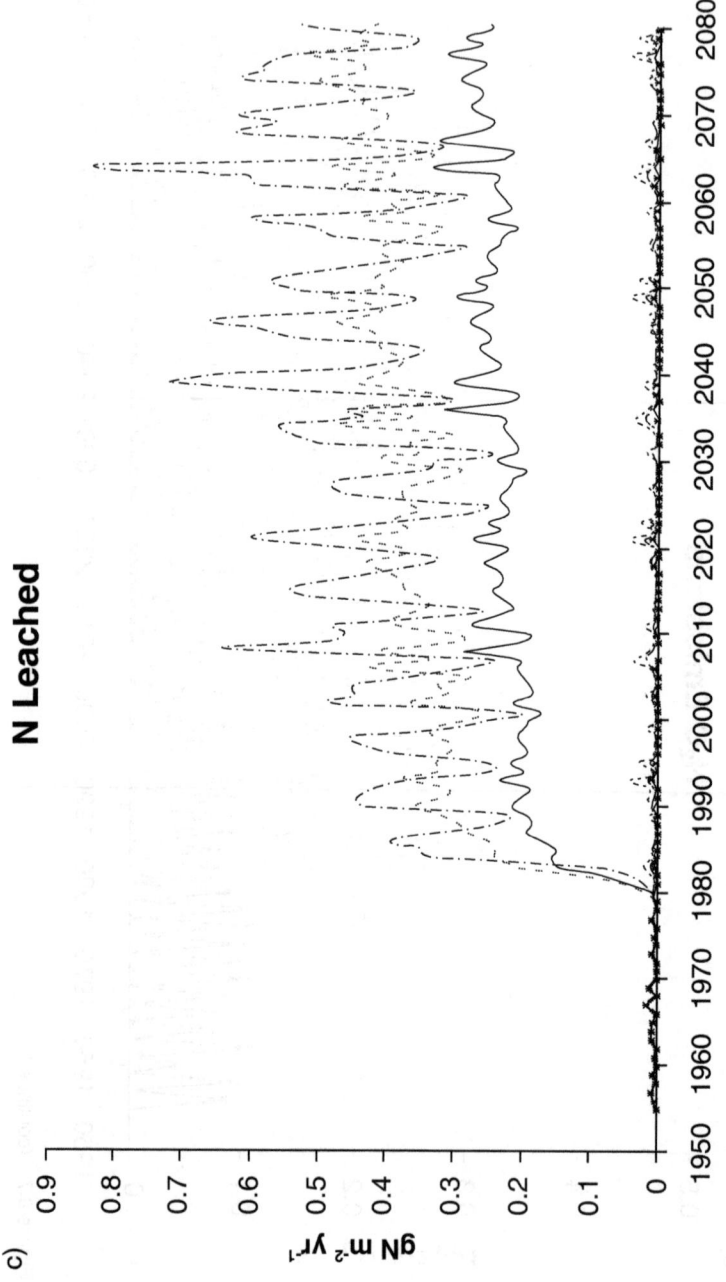

Figure 8.7   (continued).

no significant amounts of NO_3^- leaching (Figure 8.7c). However, 3 to 6% of the applied N fertilizer was leached from the irrigated corn, corn/alfalfa, and silage crops.

To summarize the results and compare scenarios, the output for system C (SOM C + surface organic matter C + litter C), N_2O emissions, NO_3^- leaching, and crop yields were aggregated into 25-year periods. Figure 8.8a shows the change in system C for each scenario from the perspective of the atmosphere. Thus, conversion from

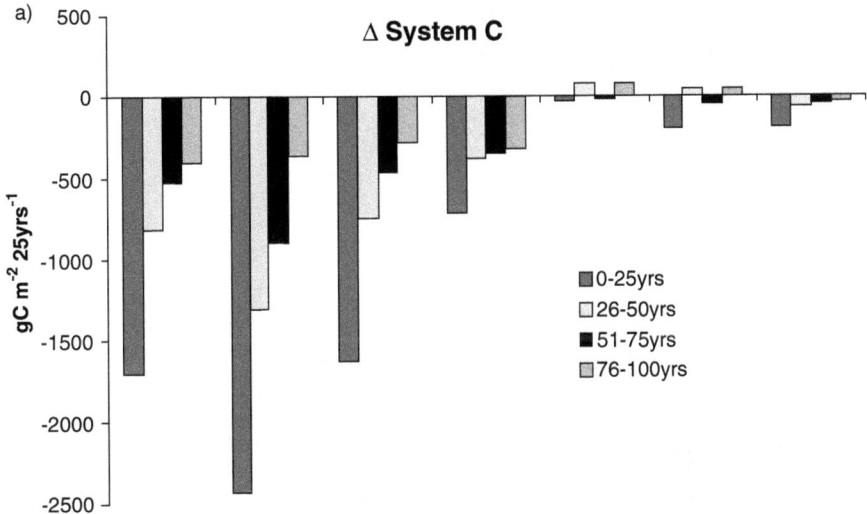

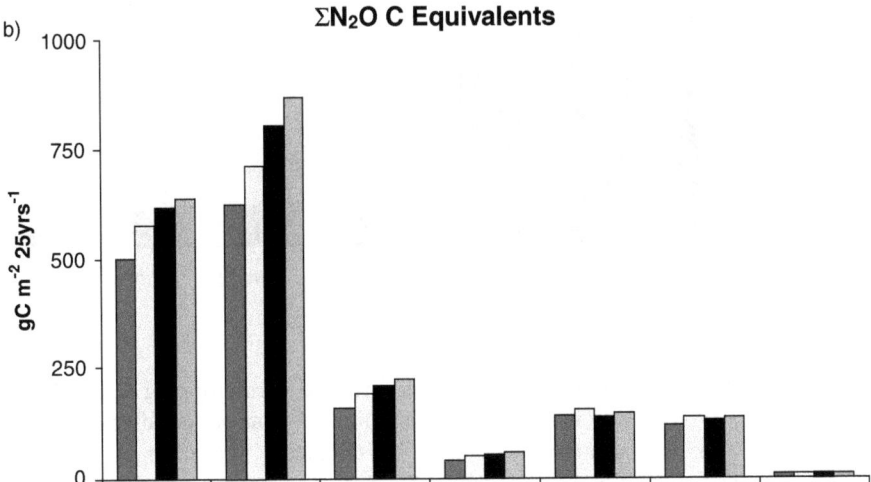

Figure 8.8   Summaries of DAYCENT simulations of changes in system C (a); cumulative CO_2-C equivalents of N_2O gas emissions (b); net C (c); and cumulative above-ground plant productivity (d) for 25-year periods of conventional tillage winter wheat/fallow and six alternative land uses.

c)

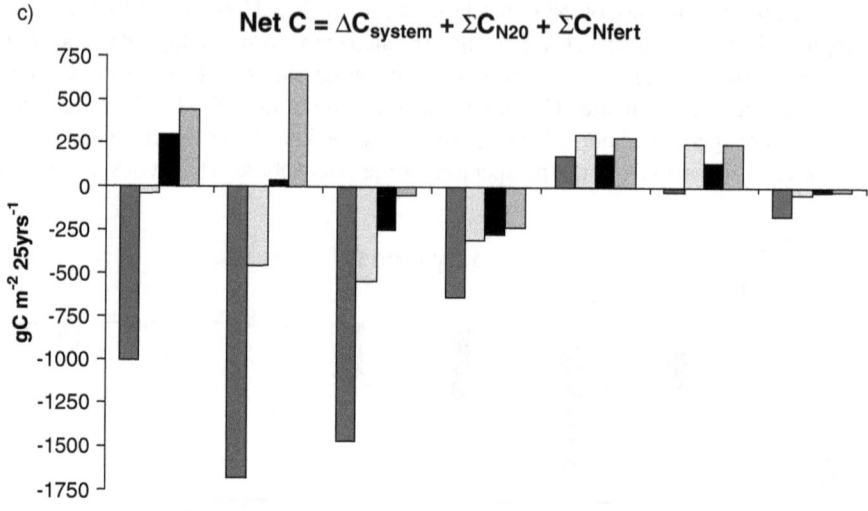

d)

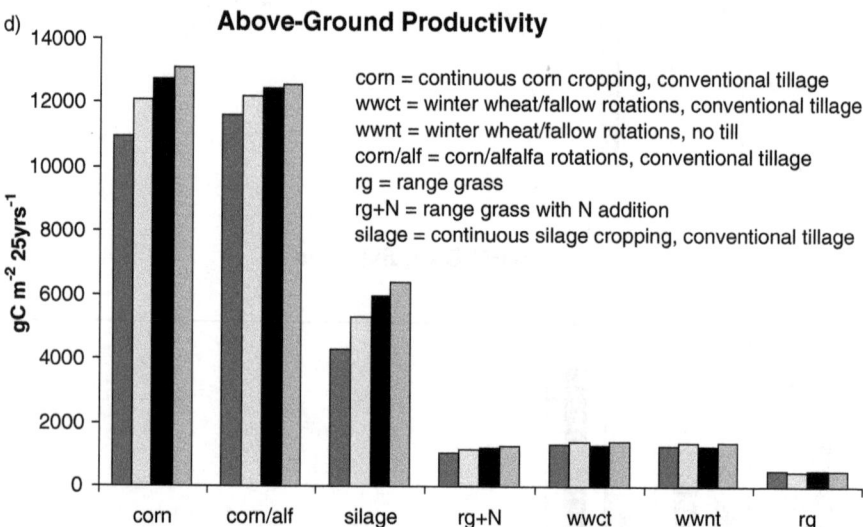

Figure 8.8   (continued).

winter wheat/fallow rotations to continuous, irrigated corn cropping resulted in an increase of ~1700 g C/m² in SOM during the first 25-year period after conversion with the C supplied by atmospheric CO_2. This indicates that C inputs from root senescence and surface litter supplied the SOM pool at a faster rate than decomposition depleted SOM by respiring it as CO_2. Conventional tillage winter wheat/fallow cropping showed no significant change in SOM because that system had been in place for a sufficient time period such that SOM had reached an equilibrium. The no-tillage management option for winter wheat/fallow copping sequestered only small amounts of CO_2-C in SOM. However, conversion to perennial cropping and increased N or organic matter addition (the corn, corn/alfalfa, and silage options) resulted in large increases in SOM, particularly during the first 25-year period following conversion. Range grass amended with N also showed significant increases in SOM.

Figure 8.8b shows that the corn and corn/alfalfa alternatives, which stored large amounts of CO_2-C in the soil, also emitted comparatively large amounts of N_2O as a result of N fertilizer application. However, the silage treatment that included manure addition also stored a large amount of C in SOM but emitted only slightly more N_2O than the wheat/fallow system. Range grass amended with N emitted more N_2O than the non-fertilized range grass, but still significantly less N_2O than the baseline winter wheat/fallow system. For the alternatives that sequestered C on the soil, N_2O emissions tended to increase with time as SOM levels began to stabilize and soil N levels increased.

Figure 8.8c shows net C sequestration for each scenario by accounting for flows of CO_2 between the atmosphere and the soil (Figure 8.8a), the CO_2 equivalents of N_2O emissions (Figure 8.8b), and the CO_2 costs of N fertilizer production. Winter wheat/fallow rotations showed a net input of greenhouse gases to the atmosphere under both conventional and no-tillage cultivation, except that the no-till option stored a small amount of net C during the first 25-year period. The systems that use high N fertilizer inputs (corn and corn/alfalfa) showed large amounts of net C sequestration during the first 25 years after conversion, but increases in SOM quickly diminished while N_2O emissions increased so that after 50 to 75 years these systems became net emitters of greenhouse gases. Conversion to silage showed net C uptake during the entire 100-year simulation. Reversion to rangeland accompanied with the addition of a small amount of N (1 g N/m^2) also resulted in long-term net C sequestration. Figure 8.8d shows the cumulative above-ground NPP for the land use options. As expected, plant growth increased with N inputs and irrigation. Yields for winter wheat were not significantly affected by cultivation method. Addition of N to range grass resulted in an increase in NPP by about a factor of 2 compared to non-fertilized range grass. This increase in NPP accounts for the higher SOM simulated for the rg + N treatment (Figure 8.7a) and indicates that these soils are strongly N limited. Although N addition to the range grass resulted in higher N_2O emissions than the non-fertilized range grass, the increase in soil C storage more than compensates for this from the net greenhouse gas perspective.

Simulations of Conventional Till and No-till Corn and Corn/Soybean Cropping

Corn and soybeans are major crops in the midwestern region of the United States. To compare the effects of tillage practices on soil C levels, N_2O emissions, crop yields, and NO_3^- leaching, the authors first performed a 100-year simulation to derive initial conditions for soil C and nutrient levels. Using a weather file for Lafayette, Indiana (mean annual precipitation = 94 cm, mean annual temperature = 10.2°C), the authors simulated traditional land use for a loam soil from this region. Corn/winter wheat/grass clover pasture rotations were simulated from 1880 to 1950. In 1951, the rotation was changed to corn/soybean/winter wheat/grass clover pasture. Chemical fertilizer was added beginning in 1951 and crop parameters were altered to reflect improved varieties of corn that were introduced. Beginning in 1980, the cropping scheme was changed to one of four alternatives: conventional till perennial corn cropping (cct), no-till perennial corn (cnt), conventional till corn/soybean rotations (cbct), and no-till corn/soybean rotations (cbnt) as shown in Table 8.1. Fertilizer (ammonium nitrate) was added at a rate of 15 g N/m^2 in a split application (7.5 g N/m in May and 7.5 g N/m^2 in June) each year that corn was grown. Simulations were run until 2010 and output files were compiled for SOM, N_2O emissions, grain yields, and NO_3^- leaching.

DAYCENT simulations showed SOM decreasing from 1880 levels and reaching a minimum value of ~4000 g C/m^2 in 1950. Soil C then increased to ~5000 g C/m^2 in 1980 as a result of N fertilization. This is similar to the historical soil C data for some soils in the U.S. corn belt (Lal et al., 1998; Higgins et al., 1998). Figure 8.9a shows the changes in the base soil C level (1980 value) simulated for corn cropping under conventional tillage (cct), no-till corn cropping (cnt), corn/soybean rotations under conventional tillage (cbct), and no-till corn/soybean rotations (cbnt). Continuous corn cropping tends to store more C in the soil than corn/soybean rotations, and no-till cultivation results in higher C storage for both cropping alternatives. No-till cultivation reduces N_2O emissions slightly and NO_3^- leaching significantly (Figures 8.9b and c). Less N is leached under no-till because more N is retained in SOM that is not susceptible to leaching. Model results suggest that no-till has benefits from both net greenhouse gas emission and NO_3 leaching perspectives.

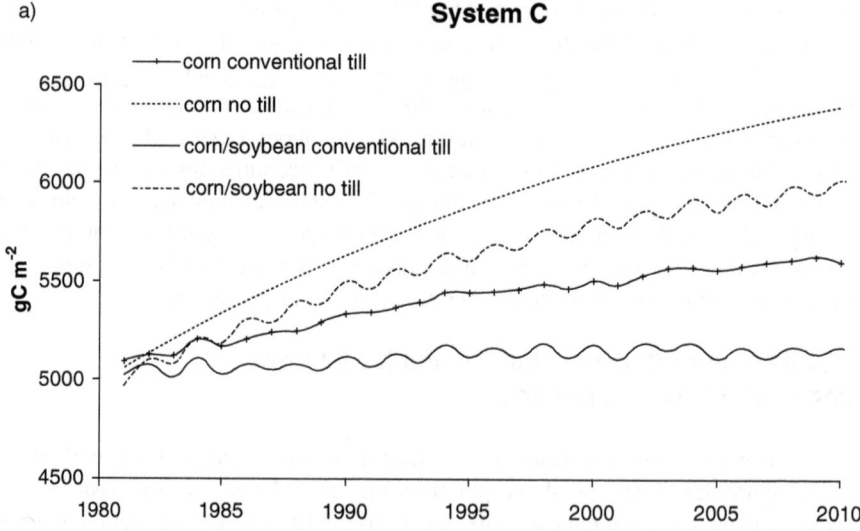

Figure 8.9 Simulated yearly changes in system C (soil C + litter C + surface organic matter C) (a); cumulative yearly N_2O emissions (b); and cumulative yearly NO_3^- leaching (c) for a midwestern soil under conventional till and no-till perennial corn and corn/soybean cropping.

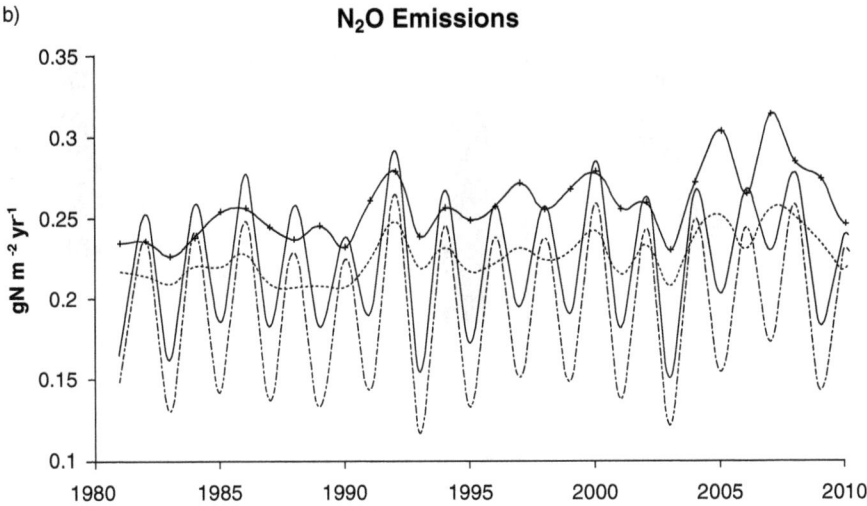

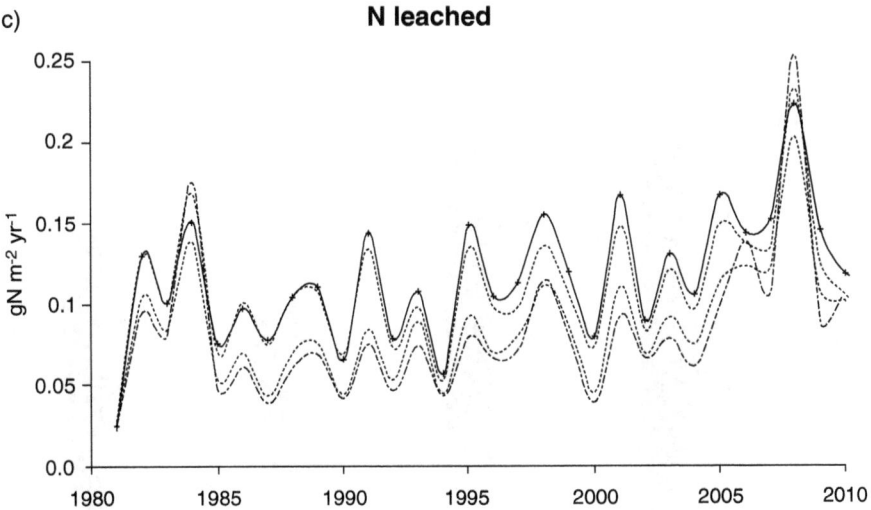

Figure 8.9   (continued).

To summarize and compare the results, the authors aggregated changes in SOM, total C equivalents for N_2O emissions, and grain yields for 10-year periods. Figure 8.10a shows that C sequestration in soil diminishes with time as SOM approaches an equilibrium and that no-till enhances soil C storage. Soil N_2O emissions show less dramatic differences among the treatments but no-till tends to decrease N_2O emissions (Figure 8.10b). Corn/soybean rotations emitted less N_2O than continuous corn cropping because N fertilizer was not added during the soybean year. Net C (Figure 8.10c) was calculated by summing the change in system C (Figure 8.10a), the C equivalents of the aggregated N_2O emissions (Figure 8.10b),

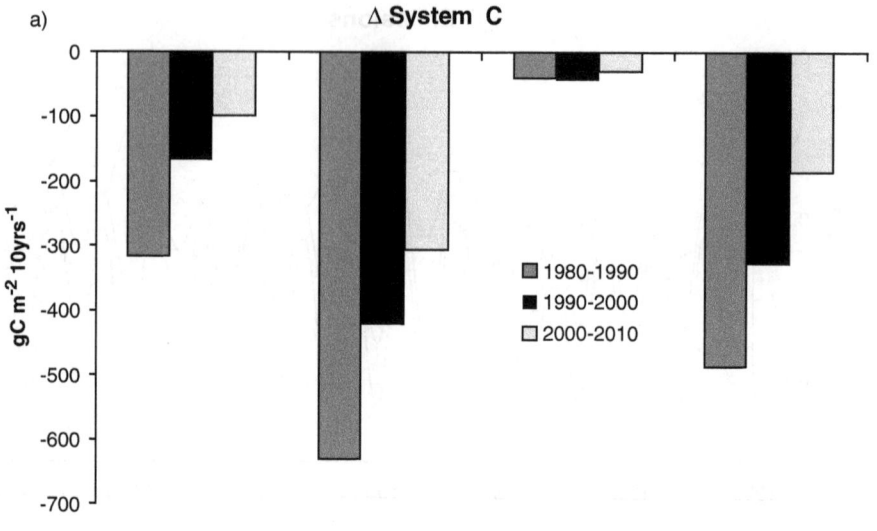

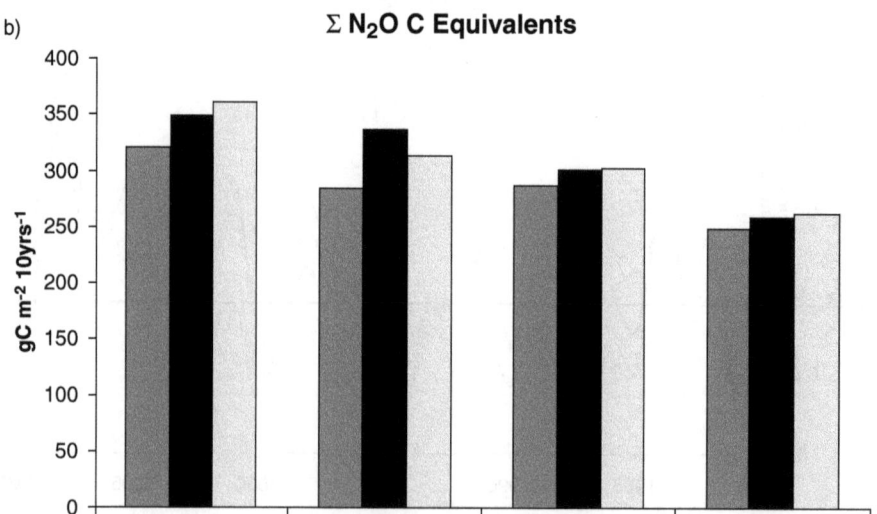

Figure 8.10 Summaries of DAYCENT simulations of changes in system C (a); cumulative CO_2-C equivalents of N_2O gas emissions (b); net C (c); and cumulative above-ground plant productivity (d) for 10-year periods of conventional till and no-till perennial corn and corn/soybean cropping.

and the CO_2 costs associated with N fertilizer production. Both no-till treatments (cnt and cbnt) showed net C sequestration for the first 10-year period, were essentially C neutral during the second 10-year period, and became net C emitters during the third 10-year period. The crops that were conventionally tilled were net emitters of C during each 10-year period. Tillage practice did not significantly affect grain

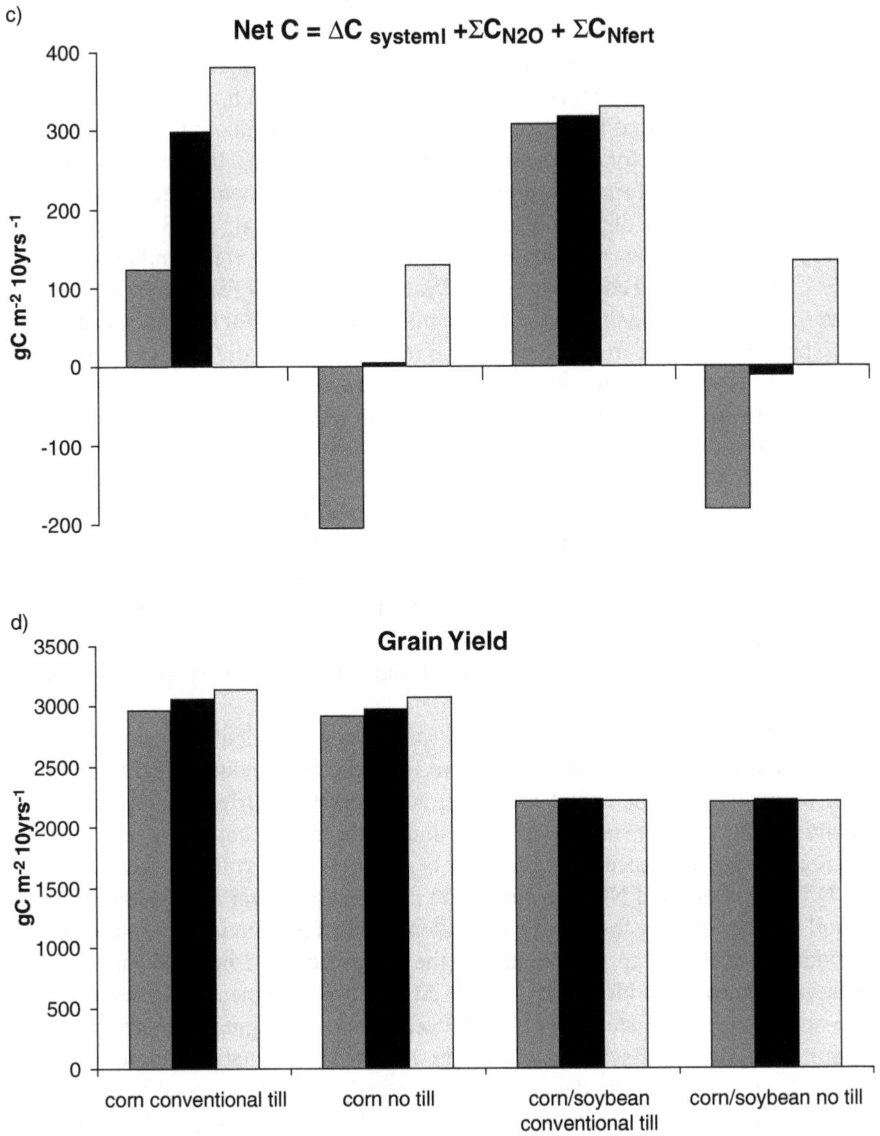

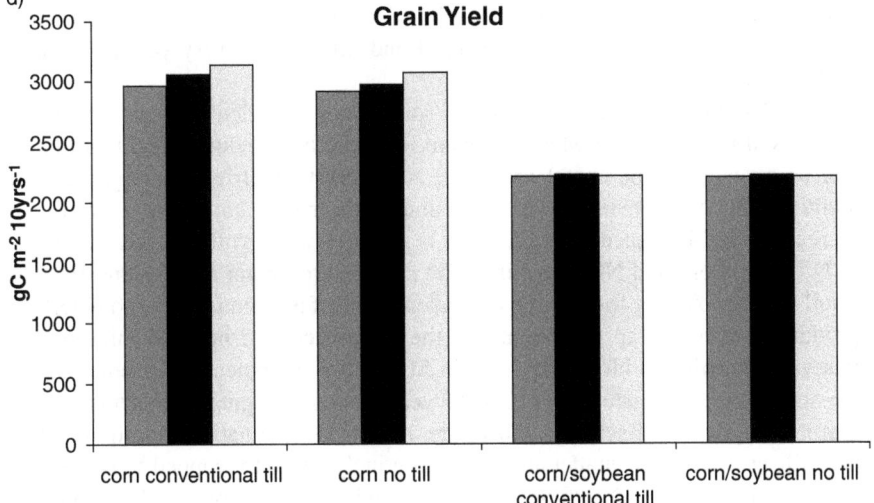

Figure 8.10 (continued).

yield (Figure 8.10d). Corn has a higher growth rate than soybeans and produces a larger grain yield than soybeans (Figure 8.10d). Higher NPP for corn implies more C inputs to the soil, so SOM values for a given tillage practice are higher for perennial corn cropping compared to corn soybean rotations (Figure 8.10a). To summarize, DAYCENT simulations suggest that conversion to no-till in the U.S. corn belt would have a minor effect on N_2O emissions but would significantly increase soil C and decrease NO_3^- leaching while maintaining high crop yields.

DISCUSSION

The authors used a well-validated ecosystem model to compare the effects of land management on SOM, N_2O emissions, plant production, and NO_3^- leaching for a Great Plains soil that has been used for winter wheat fallow rotations and for a midwestern soil used for corn/winter wheat/pasture rotations. In both cases, 100-year simulations were performed to initialize SOM and soil nutrient levels. Model results support observations of previous researchers (Higgins et al., 1998; Paustian et al., 1997) that crop fallow rotations and conventional tillage can significantly reduce SOM. The established equilibrium soil C value for the Great Plains soil (1.3 kg C/m^2) was somewhat less than the value for the midwestern soil (5.0 kg C/m^2). Both values are supported by data from the respective regions, and the differences in base SOM levels are related to soil texture, precipitation, and historical cropping practices. The midwestern soil chosen for simulations was finer textured, so SOM was more likely to become stabilized than SOM in the coarser-textured Great Plains soil used for simulations. The midwestern soil also receives ~2.5 times the precipitation of the Great Plains soil. Higher precipitation implies more plant growth and a higher rate of input to the SOM pool. The midwestern soil was historically cropped perennially, whereas the Great Plains soil was used for crop/fallow rotations. During the fallow season, inputs from plant growth to the SOM and litter pools are highly reduced, but decomposition continues and thus SOM decreases during the fallow year. Perennial cropping supplies inputs to the SOM and litter pools every year and leads to higher SOM levels.

Results from both regions show that some types of agriculture can dramatically reduce soil C levels from what they were in the native condition, and that this loss can be reversed by perennial cropping, N fertilization, irrigation, organic matter additions, no-till cultivation, and reversion to the native condition. However, the increase in SOM induced by N additions is partially offset by the associated increase in N_2O emissions and NO_3^- leaching. The potential to obtain net C sequestration in a soil after accounting for the CO_2 equivalent costs of N_2O emissions and N fertilizer production is related to the extent that the soil is depleted in SOM initially. Soils that are somewhat to highly depleted in SOM have the potential to sequester C in the short to intermediate term (10 to 50 years) upon changing management. Given that most agricultural systems will not be returned to the native condition and that N fertilizer and organic matter are already routinely added to cropped soils, perennial cropping rather than crop/fallow rotations and no-till cultivation have the most potential to sequester C in these systems.

Crop/fallow rotations are disadvantageous for two reasons. During the fallow season, SOM decreases and N_2O emissions are high. SOM decreases because inputs to SOM are greatly reduced but decomposition continues. N_2O emissions are high because soil water is higher and nitrifying microbes face little competition from plant growth for available NH_4^+ during the fallow season. Another factor that contributes to higher N_2O emissions during the fallow season with winter wheat cropping is that the cropping season includes the winter months when N_2O emissions tend to be low. Although the crop/fallow system is necessary for extremely dry regions, the range of precipitation levels that can support continuous cropping is expanded upon conversion to no-till cultivation because no-till conserves soil moisture. Under no-till, soils take more time to warm up during the spring, and soil aggregate structure is maintained so less water is evaporated. Great Plains soils that are highly depleted of SOM related to crop/fallow rotations could store large amounts of C upon conversion to intensive, perennial cropping provided that water for irrigation is made available.

The effect of the no-till alternative was much stronger in the midwestern soil compared to the Great Plains soil. Conventional tillage results in decreased SOM because the aggregate structure of the soil is altered, aggregate turnover is increased, and aggregate formation is decreased so that SOM is less protected from decomposition (Six et al., 1999). The midwestern soil was perennially cropped, and the increase in SOM decomposition rate induced by plowing resulted in lower SOM compared to the no-till practice. The Great Plains soil was not sensitive to cultivation style because of the crop/fallow rotation schedule used in this region. Plowing during the cropping year increases decomposition compared to no-till for that year. But during the fallow year, there is very little input to the SOM and litter pools, so that labile organic matter that was added to the system the previous year has sufficient time to decompose before the next cropping season, regardless of tillage practice. Similar to observations reported by Robertson et al. (2000), DAYCENT simulations suggest that no-till has a small effect on N_2O emissions but a significant effect on SOM levels. To summarize, continuous cropping and N additions increase soil C because inputs to SOM are increased to a greater extent than decomposition. No-till cultivation is not likely to lead to increased soil C sequestration under two-year crop/fallow rotation schedules, but will increase soil C for perennially cropped systems and for crop/fallow systems if the increase in soil moisture as a result of no-till allows the proportion of fallow years in the rotation schedule to be reduced.

NO_3^- leaching from the irrigated corn and corn/alfalfa crops was higher for the Great Plains soil than for similar cropping in the midwestern soil. This can be explained by soil textural differences. The Great Plains soil simulated was a sandy loam and hence had a relatively high saturated hydraulic conductivity (K_{sat}) parameter. This facilitates movement of water through the soil profile and NO_3^- leaching. The simulated midwestern soil was much finer textured, K_{sat} was lower, and less water was lost from the bottom of the profile so NO_3^- leaching was reduced. Although annual precipitation for the simulated midwestern soil was ~2.5 times that of the Great Plains soil, enough water was added from irrigation of the corn and corn/soybean Great Plains scenarios such that total water inputs for these options were roughly equivalent to those of the midwestern soil. Model results suggest that no-till can reduce NO_3^- leaching in some soils, but the authors are not aware of any experimental evidence to support this assertion. The increase in soil water content associated with no-till would tend to increase leaching, but the increased N immobilization as a result of no-till more than compensates for this in the simulated Great Plains soil.

DAYCENT simulations suggest that soils that are depleted in SOM can temporarily compensate for greenhouse gas emissions by changing land management. Although net carbon sequestration will not continue for more than 10 to 50 years, from the perspective of the net greenhouse gas emissions emitted by the conventional tillage crop/fallow and perennial cropping systems, the benefits of no-till extend at least 100 years into the future. No-till should not reduce crop yields and may reduce NO_3^- leaching, thus providing additional societal benefits.

REFERENCES

Alexandrov, G.A., T. Oikawa, and G. Esser. 1999. Estimating terrestrial NPP: what the data say and how they may be interpreted? *Ecol. Modelling,* 117:361-269.

Bruce, J.B., M. Frome, E. Haites, H. Janzen, R. Lal, and K. Paustian. 1999. Carbon sequestration in soils. *J. Soil Water Conserv.,* 54:382-389.

Bumb, B.L., and C.A. Baanante. 1996. The role of fertilizer in sustaining food security and protecting the environment to 2020. *Food, Agriculture, and the Environment Discussion Paper 17.* International Food Policy Research Institute. Washington, D.C., 54 pp.

Conrad, R. 1996. Soil microorganisms as controllers of atmospheric trace gases (H_2, CO, CH_4, OCS, N_2O, and NO). *Microbiol. Rev.,* 60:609-640.

Davidson, E.A. 1992. Sources of nitric oxide and nitrous oxide following the wetting of dry soil. *Soil Sci. Soc. Am. J.,* 56:95-102.

Del Grosso, S.J., W.J. Parton, A.R. Mosier, D.S. Ojima, and M.D. Hartman, 2000. Interaction of soil carbon sequestration and N_2O flux with different land use practices. In J. van Ham et al. (Eds.). *Non-CO_2 Greenhouse Gases: Scientific Understanding, Control and Implementation,* Kluwer Academic, The Netherlands, 303-311.

Del Grosso, S.J., W.J. Parton, A.R. Mosier, D.S. Ojima, A.E. Kulmala, and S. Phongpan, 2000. General model for N_2O and N_2 gas emissions from soils due to denitrification. *Global Biogeochem. Cyc.* 14:1045-1060.

Eitzinger, J., W.J. Parton, and M.D. Hartman. In press. Improvement and validation of a daily soil temperature sub-model for freezing/thawing periods, *Soil Sci.*

Frolking, S.E., A.R. Mosier, D.S. Ojima, C. Li., W.J. Parton, C.S. Potter, E. Priesack, R. Stenger, C. Haberbosch, P. Dörsch, H. Flessa, and K.A. Smith. 1998. Comparison of N_2O emissions from soils at three temperate agricultural sites: simulations of year round measurements by four models. *Nutrient Cycling in Agroecosystems,* 52:77-105.

Galloway, J.N., W.H. Schlesinger, H. Levy II, A. Michaels, and J.L. Schnoor. 1995. Nitrogen fixation: anthropogenic enhancement-environmental response. *Global Biogeochem. Cyc.,* 9:235-252.

Granli, T., and O.C. Bockman. 1994. Nitrous oxide from agriculture. *Nor. J. Agric. Sci.,* (Suppl. 12):1-128.

Grant R F., N.G. Juma, and W.B. McGill. 1993. Simulation of carbon and nitrogen transformations in soils. I. Mineralization. *Soil Biol. Biochem.,* 25:1317-1329.

Grant, R.F., and E. Pattey. 1999. Mathematical modeling of nitrous oxide emissions from an agricultural field during spring thaw. *Global Biogeochem. Cyc.,* 13:679-694.

Higgins, D.R., G.A. Buyanovsky, G.H. Wagner, J.R. Brown, R.G. Darmody, T.R Peck, G.W. Lesoing, M.B. Vanotti, and L.G. Bundy. 1998. Soil organic C in the tallgrass prairie-derived region of the corn belt: effects of long-term crop management. *Soil Tillage & Research,* 47:219-234.

Hutchinson, G.L., G.P. Livingston, and E.A. Brams. 1993. Nitric and nitrous oxide evolution from managed subtropical grassland. In R.S. Oremland (Ed.). *Biogeochemistry of Global Change: Radiatively Active Trace Gases,* Chapman & Hall, New York, 290-316.

Kelly, R.H., W.J. Parton, M.D. Hartman, L.K. Stretch, D.S. Ojima, and D.S. Schimel. 2000. Intra- and interannual variability of ecosystem processes in shortgrass steppe: new model, verification, simulations. *J. Geophys. Res.: Atmospheres,* 105:93-100.

Kelly, R.H., W.J. Parton, G.L. Crocker, P.R. Grace, J. Klir, M. Korschens, P.R. Poulton, and D.D. Richter. 1997. Simulating trends in soil organic carbon using the century model. *Geoderma,* 81:75-90.

Lal, R., J.M. Kimble, R.F. Follett, and C.V. Cole. 1998. *The Potential of U.S. Cropland to Sequester Carbon and Mitigate the Greenhouse Effect.* Ann Arbor Press, Chelsea, MI.

Lieth, H. 1975. Modeling the primary productivity of the world. In H. Lieth and R. Whittaker (Eds.). *Primary Productivity of the Biosphere,* Springer-Verlag, New York.

Martin, R.E., M.C. Scholes, A.R. Mosier, D.S. Ojima, E.A. Holland, and W.J. Parton. 1998. Controls on annual emissions of nitric oxide from soils of the Colorado shortgrass steppe. *Global Biogeochem. Cyc.,* 12:81-91.

Matson, P.A., and R.C. Harriss. 1995. Trace gas exchange in an ecosystem context: multiple approaches to measurement and analysis. In P.A. Matson and R.C. Harriss (Eds.). *Biogenic Trace Gases: Measuring Emissions from Soil and Water,* Blackwell Scientific, Malden, MA, 1-13.

Metherell, A.K., L.A. Harding, C.V. Cole, and W.J. Parton. 1993. CENTURY Soil Organic Matter Model Environment. Technical documentation, Agroecosystem version 4.0, Great Plains System Research Unit Technical Report No. 4, USDA-ARS, Fort Collins, CO.

Mosier, A.R., W.D. Guenzi, and E.E. Schweizer. 1986. Soil losses of dinitrogen and nitrous oxide from irrigated crops in northeastern Colorado. *Soil Sci. Soc. Am. J.,* 50:344-348.

Mosier, A.R., W.J. Parton, D.W. Valentine, D.S. Ojima, D.S. Schimel, and J.A. Delgado. 1996. CH_4 and N_2O fluxes in the Colorado shortgrass steppe. 1. Impact of landscape and nitrogen addition. *Global Biogeochem. Cyc.,* 10:387-399.

Mosier, A.R., W.J. Parton, D.W. Valentine, D.S. Ojima, D.S. Schimel, and O. Hienemeyer. 1997. CH_4 and N_2O fluxes in the Colorado shortgrass steppe. 2. Long-term impact of land use change. *Global Biogeochem. Cyc.,* 11:29-42.

Nuttle, W.K. 2000. Ecosystem managers can learn from past successes. *EOS,* 81:278.

Parton, W.J. 1978. Abiotic section of ELM. In G.S. Innis (Ed.). *Grassland Simulation Model, Ecological Studies Analysis and Synthesis,* Vol. 45, 31-53.

Parton, W.J. 1984. Predicting soil temperatures in a shortgrass steppe. *Soil Sci.,* 138:93-101.

Parton, W.J., and P.E. Rasmussen. 1994. Long-term effects of crop management in wheat/fallow. II. CENTURY model simulations, *Soil Sci. Soc. Am. J.,* 58:530-536.

Parton, W.J., J.M.O. Scurlock, D.S. Ojima, T.G. Gilmanov, R.J. Scholes, D.S. Schimel, T. Kirchner, J.C. Menaut, T. Seastedt, E. Garcia Moya, Apinan Kamnalrut, and J.L. Kinyamario. 1993. Observations and modeling of biomass and soil organic matter dynamics for the grassland biome worldwide. *Global Biogeochem. Cyc.,* 7:785-809.

Parton, W.J., D.S. Ojima, C.V. Cole, and D.S. Schimel. 1994. A general model for soil organic matter dynamics: Sensitivity to litter chemistry, texture and management. In *Quantitative Modeling of Soil Forming Processes,* SSSA, Spec. Pub. 39, Madison, WI, 147-167.

Parton, W.J., A.R. Mosier, D.S. Ojima, D.W. Valentine, D.S. Schimel, K. Weier, and K.E. Kulmala. 1996. Generalized model for N_2 and N_2O production from nitrification and denitrification. *Global Biogeochem. Cyc.,* 10:401-412.

Parton, W.J., M. Hartman, D.S. Ojima and D.S. Schimel. 1998. DAYCENT: its land surface sub-model: description and testing. *Global Planetary Change,* 19:35-48.

Parton, W.J., E.A. Holland, S.J. Del Grosso, M.D. Hartman, R.E. Martin, A.R. Mosier, D.S. Ojima, and D.S. Schimel. In press. Generalized model for NO_x and N_2O emissions from soils, *J. Geophys. Res.*

Paustian, K., W.J. Parton, and J. Persson. 1992. Modeling soil organic matter in organic amended and nitrogen-fertilized long term plots. *Soil Sci. Soc. Am. J.,* 56:476-488.

Paustian, K., O. Andren, H.H. Janzen, R. Lal, P. Smith, G. Tian, H. Tiessen, M. Van Noordwijk, and P.L. Woomer. 1997. Agricultural soils as a sink to mitigate CO$_2$ emissions. *Soil Use & Management,* 13:230-244.

Penman, H.L. 1948. Natural evaporation from open water, bare soil and grass. *Proc. R. Soc. London, Ser. A.,* 193:120-145.

Peterson, G.A., A.D. Halvorson, J.L. Havlin, O.R. Jones, D.J. Lyon, and D.L. Tanaka. 1998. Reduced tillage and increasing cropping intensity in the Great Plains conserves soil C. *Soil Tillage & Research,* 47:207-218.

Potter, C.S., E.A. Davidson, and L.V. Verchot. 1996. Estimation of global biogeochemical controls and seasonality in soil methane consumption. *Chemosphere,* 32:2219-2245.

Prather, M, R. Derwent, D. Ehhalt, P. Fraser, E. Sanhueza, X. Zhou, 1995. Other trace gases and atmospheric chemistry. In J.T. Houghton, L.G. Meira Filho, J.B.H. Lee, B.A. Callander, E. Haites, N. Harris, and K. Maskell (Eds.). *Climate Change 1994. Radiative Forcing of Climate Change and an Evaluation of the IPCC IS92 Emission Scenarios.* published for the Intergovernmental Panel on Climate Change, Cambridge University Press, chap. 2, 77-126.

Robertson, G.P., E.A. Paul, and R.R. Harwood. 2000. Greenhouse gases in intensive agriculture: contributions of individual gases to the radiative forcing of the atmosphere. *Science,* 289:1922-1925.

Schlesinger, W.H. 1999. Carbon sequestration in soils. *Science,* 284:2095.

Six, J., E.T. Elliot, and K. Paustian. 1999. Aggregate and soil organic matter dynamics under conventional and no-tillage systems. *Soil Sci. Soc. Am. J.,* 63:1350-1358.

Smart, D.R., J.M. Stark, and V. Diego. 1999. Resource limitation to nitric oxide emissions from a sagebrush-steppe ecosystem. *Biogeochemistry,* 47:63-86.

Smil, V. 1999. Nitrogen in crop production: An account of global flows. *Global Biogeochem. Cyc.,* 13:647-662.

Vitousek, P.M., J. Aber, R.W. Howarth, G.E. Likens, P.A. Matson, D.W. Schindler, W.H. Schlesinger, and D.G. Tilman. 1997. Human alteration of the global nitrogen cycle: causes and consequences. *Issues in Ecology,* 1:1-15.

Yienger, J.J., and H. Levy. 1995. Empirical model of global soil biogenic NO$_x$ emissions. *J. Geophys. Res.,* 100:11447-11464.

CHAPTER **9**

NLEAP Water Quality Applications in Bulgaria

Dimitar Stoichev, Milena Kercheva, and Dimitranka Stoicheva

CONTENTS

INTRODUCTION

Fresh groundwater resources in Bulgaria total approximately 5.5 billion m^3, with 0.8 to 1.0 billion m^3 used annually (Yordanova and Donchev, 1997). The main characteristics of freshwater resources in the country are their uneven distribution across the landscape and high annual variation in amount. According to officially published data (Yordanova and Donchev, 1997), about 24% of the country's population is under a constant or temporary water supply regime.

More than 70% of the drinking water in the centralized water supply system in Bulgaria is from groundwater sources (Technical Report No. 27, 1988). The main aquifers consist of alluvial and deluvial quaternary deposits in the plains of the Danube and Maritsa Rivers and their tributaries. The common depth to groundwater level is 5 to 10 m (Yordanova and Donchev, 1997). The nitrate pollution of drinking water arose as a significant ecological problem in Bulgaria at the beginning of the 1980s. An extract of generalized information for a thousand settlements with centralized water supply systems that supply about 90% of the country's population indicates that drinking water nitrate content ranges across very broad limits. About 5.2% of monitored sources for the period 1971 to 1975 supplied waters with nitrate contents of 45 to 100 ppm, and 1% were classified as highly polluted, as the nitrates in their water exceeded 100 ppm. It has been established that the polluted groundwater is situated mainly in the zone of Vertisols, Luvisols, and Fluvisols in southern Bulgaria. The spots with the highest nitrate groundwater pollution were registered in the zone of Vertisols around the town of Stara Zagora (Drinking Water Quality in the Country, 1979).

According to the data of the World Health Organization, 4% of the total population of Bulgaria (8.8 million in 1985) has been exposed to nitrate levels of 50 ppm or above (Technical Report No. 27, 1988). Later approximate estimations (Gitsova et al., 1991) showed that contaminated sources of drinking water reached 20% despite the higher value (50 ppm NO_3) of MPCL (maximum permissible contaminant level) adopted in 1983.

The situation in rural territories where home wells are significant sources for drinking and irrigation water is extremely alarming and complicated. There are many potential sources for nitrogen contamination there, such as big dairy farms, fertilized arable lands, septic holes, and intensive manuring and irrigation in home vegetable gardens.

Assessments of agriculture as the main non-point potential source of surface and groundwater nitrate contamination depend on the available information and could be accomplished using a simplified nitrogen balance, experimental investigations, and simulation models. The simplified nitrogen balance gives the potential nitrogen surplus on a country level. The experimental studies concerning nitrogen leaching in Bulgaria based on long-term controlled field trials in different agroecological regions and on the watershed pilot project in a real-life situation are reviewed as a valuable information source for implementation of the simulation model NLEAP (Shaffer et al., 1991; Hansen et al., 1995). Some of the results achieved in Bulgaria and the remaining problems are discussed in this chapter.

NITROGEN BALANCE ON A COUNTRY LEVEL

The area of agricultural land in Bulgaria is 6.2×10^6 ha (56% of 111,000 km^2 total territory), woodlands account for 3.9×10^6 ha (35%), and the other 1.1×10^6 ha (9%) are used for infrastructure and communal purposes. Arable and permanent croplands amount to 4.8×10^6 ha (about 0.58 ha per capita), 89.6% of which is cropland, 6.3% grassland, and 4.2% vineyards and orchards (Statistical Yearbook,

1960–1998). About 0.62×10^6 ha are technically equipped for irrigation, but only 5 to 10% of them were actually irrigated last year (Petkov et al., 2000).

The nitrogen balance in this study is performed assuming the entire country as a unified system. Nitrogen supplied to agriculture by commercial fertilizers, manure from all animals bred in the country, and from total annual precipitation is considered as an input into the system, and the nitrogen uptake by gross national plant production is defined as an output. This approach is classified as the Soil Surface Balance (Oenema, 1999; Vermeulen et al., 1998; Fotyma and Fotyma, 1999) and does not include the possible soil N transformations in the nitrogen mass balance. The positive difference between the input and the output is defined as a net nitrogen surplus. From an ecological point of view, this amount of residual soil nitrogen is considered a potential source of surface and groundwater nitrate pollution.

The data needed for the calculation of N balance on a country level is taken from Statistical Yearbooks (1960–1998), published information for the annual amount and quality of manure from different livestock types in Bulgaria (Petrov et al., 1983), nitrogen uptake by statistically monitored crops (Nikolova et al., 1995), and the authors' own long-term monitoring data for the input of nitrogen by precipitation in the country's agricultural area (Stoichev, 1997).

The annual nitrogen inputs calculated over the period 1960 to 1998 across the entire country for precipitation vary from 68×10^3 to 133×10^3 tons; for manure from 50×10^3 to 57×10^3 tons; and for commercial fertilizers from 53×10^3 to 550×10^3 tons.

The generalized estimations for the annual yield production of 40 crops statistically registered in the country show that the major nitrogen output is by cereals (60 to 80% from the total agricultural nitrogen output) and by forage crops (13 to 24%). The portion of the technical crops in the nitrogen output is approximately 6 to 13%; of permanent crops 3 to 6%; and of vegetables 2 to 4%. The average yields of the two main cereal crops (wheat and corn) are presented to illustrate the relationship between total nitrogen input and yield production. As can be seen from Figure 9.1, the significant increase in nitrogen input, especially from 1970 to 1990, is not followed by a corresponding increase in the yields; and as a result, a significant surplus of nitrogen is created. The amount of nitrogen not included in the biological cycle of nutrients for the whole monitored period varies from 52 to 490×10^3 tons, which is 8 to 79 kg/ha for the total area of the country's agricultural lands. Total annual surplus of nitrogen reaches 21 to 65% from the total nitrogen input to the system. For the period 1970 to 1990, the nitrogen imported to the system is about twice the estimated output by plant production. During the past 10 years, a drastic decrease in fertilizer application has been observed in the country, but the nitrogen balance still remains positive.

During the study period as a whole (Figure 9.1), the yields from the main cultivated crops and their respective total nitrogen outputs do not correspond to the trend of the total nitrogen input. Under these conditions, a positive nitrogen balance has been maintained. Nevertheless, this fact is not reason to charge the whole of agriculture as a source of water nitrate pollution without determining the vulnerable components of the biological cycle of nitrogen in the plant-soil-groundwater system. It is important to evaluate the factors that lead to nitrate accumulation in the root zone and its movement to the surface and groundwater.

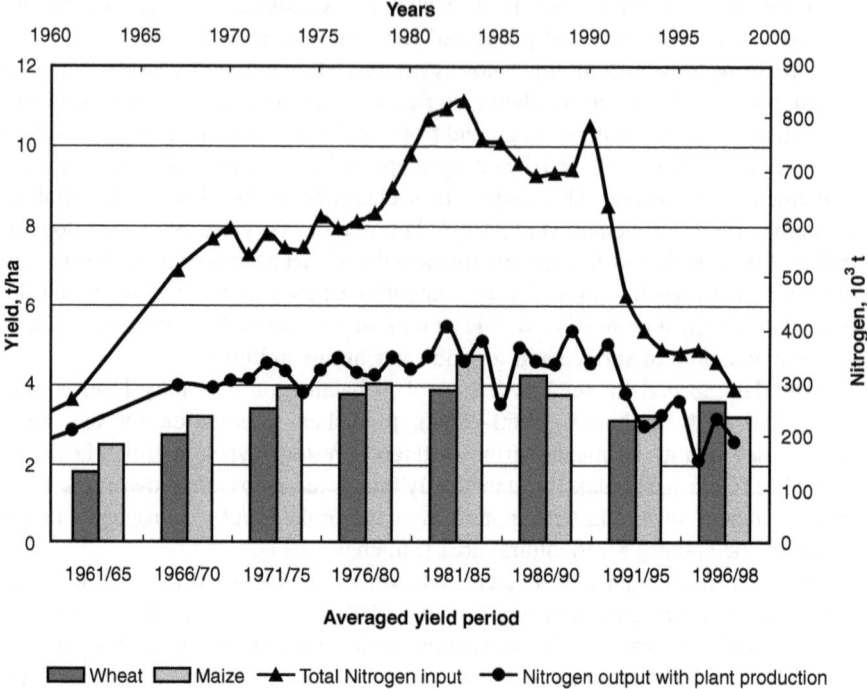

Figure 9.1   Total nitrogen balance for agricultural area of Bulgaria and average yields of main cereal crops.

When considering animal waste as a source of nitrogen input, it should be remembered that before intensive use of fertilizers, manure was applied to arable lands without any evidence of surface or groundwater pollution. This is confirmed by the fact that only 1% of the monitored sources of drinking water for the period 1961 to 1965 exceeded the maximum permissible contaminant level (MPCL) of 30 ppm NO_3 adopted at that time (Drinking Water Quality in the Country, 1968). Animal wastes became an ecological problem for the country as a result of intensification of agriculture and animal breeding after the 1960s. At that time, farm animals were collected in huge production units, and the arable lands were concentrated in large plant production cooperatives. Commercial fertilizer application was stimulated by government subsidies. For a long time, their price was lower than the expenses needed for the transformation of animal wastes into manure, its collection, storage, and proper application in agriculture. This fact gave rise to changing the animal waste from a source of nutrients into a potential source of pollution. According to the authors' estimations, during the period with the highest surplus (1981 to 1985) (Figure 9.1) approximately 10% of the manure produced in the country was applied to agriculture. The total amount of the main nutrients in manure produced at that time consisted of about 70% of the nitrogen (N), 110% of the phosphorus (P), and 240% of the potassium (K) applied by commercial fertilizers. However, the nutrients (NPK) turned back into the agricultural land by the applied manure constituted approximately 8 to 26% of nutrients supplied by the commercial fertilizers and restored only 10 to 48% of total plant production uptake.

It should be taken into account that the N-balance approach used is applicable only for an approximate estimation of the trend of N surplus over a long-term period in Bulgaria. The decrease of nitrogen surplus in the past 10 years is based on the economic restrictions and could not be considered as conscious agroecological management. Obviously, the fate of the residual nitrogen will strongly depend on water balance, soil properties, topography, geology, and land management of the concrete watershed, farm area, and separate field.

EXPERIMENTAL STUDIES ON NITROGEN LEACHING

The investigations on nitrogen leaching in Bulgaria (Atanassov et al., 1974; Stoichev et al., 1985; 1996; Stoichev and Stoicheva, 1987) were conducted as a part of long-term (1972 to 1990) irrigated corn experiments at the field stations of the N. Poushkarov Institute of Soil Science and Agroecology in the framework of the "yield prediction" complex program. The main goals of this program were to obtain information concerning parameters of carbon, water, and nutrient balances for the main agroecological regions in Bulgaria.

The experiments involved four treatments with different nitrogen and phosphorus rates of fertilizer application compared with non-treated plots. The experimental design made it possible to include all existing situations of the biological nitrogen cycle in the ecosystems. In the non-treated plots (N_0P_0), the yield was controlled by natural nitrogen sources (rainfall, irrigation, fixation from the air, and organic matter mineralization). In this treatment, a situation typical for the natural ecosystems was simulated, where one-way nitrogen uptake was stimulated by harvesting and a constant negative nitrogen balance was maintained. In the optimal treatment, the rate was calculated for full compensation of the nitrogen uptake by crop production. For the remaining three rates of fertilization, different levels of deficit and surplus N were maintained in the soil-crop system.

Later (1994 to 1997), during the joint USDA-Bulgaria project "Agro-environmental Water Quality Program in the Yantra River Basin," monitoring and educational programs were initiated in a region with significant groundwater nitrate contamination (Stoichev et al., 1997; Stoichev et al., 1999). The village of Parvomaitsi (lat = 43°10′N, lg = 25°36′E, alt. = 87 m) is situated in the central part of the Yantra River basin on a well-defined small watershed with climate, soil, municipal economy activities, and habits representative of this part of the country. The total area of the village is about 3000 ha, including 2262 ha of arable land, 213 ha of common and pastures, and 257 ha of woods. The village itself occupies 267 ha.

The pre-project suggestion of the local authorities and inhabitants was that the main sources of groundwater nitrate pollution were arable lands above the village and the big dairy cooperative farm. That is why a monitoring profile line was selected in the northwestern part of the village passing through arable lands, the territory of the large dairy farm, and the households in the village. Four field sites with crop rotations of winter wheat, maize, sunflower, and alfalfa; a peach orchard; a pasture;

Table 9.1 Average (1977–1989) Nitrogen Content in the Lysimetric Water (mg/L) at 50 and 100 cm Soil Depths under Different Treatments of Irrigated Corn Experiments

Location	Soil Type (FAO)	Soil Texture Class[a]	V_1 (N_0)		V_5 ($N_{111-137}$[b])		V_3 ($N_{222-276}$[b])	
			50 cm	100 cm	50 cm	100 cm	50 cm	100 cm
Gorni Dabnik	Haplic Chernozem	SiC to SiCL	7.1	7.1	16.6	10.7	25.6	14.3
Bejanovo	Phaeozem	SiC to C	14.8	6.5	11.4	9.7	20.1	19.6
Sredets	Vertisol	C	13.4	10.7	12.0	9.4	39.6	20.7
Sadievo	Chromic Luvisol	CL	14.8	23.8	25.1	34.2	28.1	50.4
Tsalapitsa	Fluvisol	SL to SCL	10.6	10.1	15.7	15.3	20.6	26.1
Gorni Lozen	Chromic Luvisol	CL	13.4	10.7	12.0	9.4	39.6	20.7

[a] SiC = silty clay; C = clay; SiCL = silty clay loam; SL = sandy loam; SCL = sandy clay loam.
[b] Average annual nitrogen fertilizer rate (a.s. in kg/ha).

two home gardens with intensive vegetable growth situated on Haplic Chernozems; and two other home gardens situated on Fluvisols were monitored for soil chemical composition, soil water content, and agricultural activities (Final Report, 1997). The joint project was a contribution to the large Danube River program.

Nitrogen in the Drainage Flow

Drainage-flow collecting lysimeters (0.11 m²) were used to study the nitrogen leaching at 50 and 100 cm soil depth. The lysimetric devices represent a modification (Stoichev, 1974) of an Ebermaier–Shilova type of lysimeter. The main difference between this type of lysimeter and lysimeters with disturbed soil and monolith field lysimeters is that the contact plate was cut into the soil profile without disturbing the overlying soil layers.

As a whole, the drainage flow in the root zone of the studied soil units (Table 9.1) is enriched by nitrogen compared to MPCL (11.3 ppm NO_3-N) for drinking water, even in the treatments with no fertilizer. The application of fertilizer rate ($N_{111-137}$) for compensation of about 50% of the uptake by corn yield production increases the nitrogen content in the lysimetric water from Chromic Luvisols and Fluvisols. When the rate of the applied fertilizer is doubled, a significant increase in nitrogen (14.3 to 50.4 ppm) in the active root zone leachate is observed for all studied soils. Nevertheless, the elevated nitrogen content in the lysimetric water and formation of drainage flow out of the 1-m layer is only a potential source of groundwater contamination. From a theoretical point of view, the nitrates leached out of the soil rooting zone could be distributed in three main directions:

1. Nitrates could go back into the active root zone through upward capillary movement.
2. Nitrates could be lost from the topsoil as a result of various microbiological transformations.
3. Some part of the total amount of N leached from soil could be transported through the unsaturated zone to shallow groundwater as a result of deep percolation.

Nitrogen in Deep Vadose Zone

To study the fate of nitrogen leached below the 1-m soil depth in the years of the investigations, deep drillings were completed at some of the experimental stations. The analyses of soil samples and geological materials indicate that in Fluvisol (Tsalapitsa experimental station), there is a certain amount of nitrogen available to move, and deep nitrogen leaching exists even in the case of no fertilizer applications. Nitrogen input in excess of the uptake by corn production has led to considerable enrichment with nitrogen of the entire soil profile (0 to 5 m) and groundwater as well (Stoichev et al., 1980).

The experiments carried out on a Vertisol (Sredets experimental station) show that among the tested treatments (N_0, N_{110}, N_{220}, N_{276}), only the maximum rate results in a statistically proved nitrogen accumulation in the 0- to 0.6-m layer. Detectable nitrogen is found in the 2.0-m layer and there is no direct relation between the fertilizer rate and contaminated shallow (3.5 to 4.6 m) groundwater (Stoichev et al., 1983). Almost the same results were found in the field trial carried out on Luvisol (Gorni Lozen experimental station) (Mateva et al., 1982).

The deep drilling completed during the joint Bulgaria-USDA pilot project in Haplic Chernozems and Fluvisols in the village of Parvomaitsi showed that, independent of land use type, NO_3-N was detected in all samples from the soil surface to the groundwater table and varied from 0.2 to 33.0 mg/kg (Stoichev et al., 1998). The area of the large dairy farm could be classified as a point source of contamination because about 400 kg/ha NO_3-N was accumulated in the 2-m layer. The entire geological profile from the soil surface to the groundwater level contained approximately 1200 kg/ha residual nitrogen. Fluvisols on the Yantra River terrace are more vulnerable because of their narrow soil and geological profile, low water storage capacity, high water permeability, and shallow groundwater. This area is subject to intensive anthropogenic loading, thus, it is very important to be under more precise ecological control than Haplic Chernozems covering the second river terrace.

Nitrogen in Groundwater

During the period (1995 to 1997) of the pilot project on the Yantra River watershed, about 24 home wells and seven reference wells specially built in May 1995 in the arable and pasturelands in the village of Parvomaitsi were monitored for groundwater quality. A total of 1193 wells in the village of Parvomaitsi are used as a source of drinking water, communal, and irrigation needs, which means that almost every family has its own well. The NO_3-N content in the groundwater of the first river terrace in the zone of medium coarse-textured Fluvisols was 4 to 7 times above the maximum permissible contaminant level (11.3 ppm NO_3-N) and had the greatest seasonal variability (Figures 9.2 and 9.3, Table 9.2). In home wells of the second river terrace in the zone of medium-textured Haplic Chernozems, NO_3-N was 2 to 4 times above MPCL and still had significant seasonal variability. In the highest part of the region under the arable lands, the content and variability of NO_3-N in the groundwater was small and seldom exceeded the MPCL (Table 9.2,

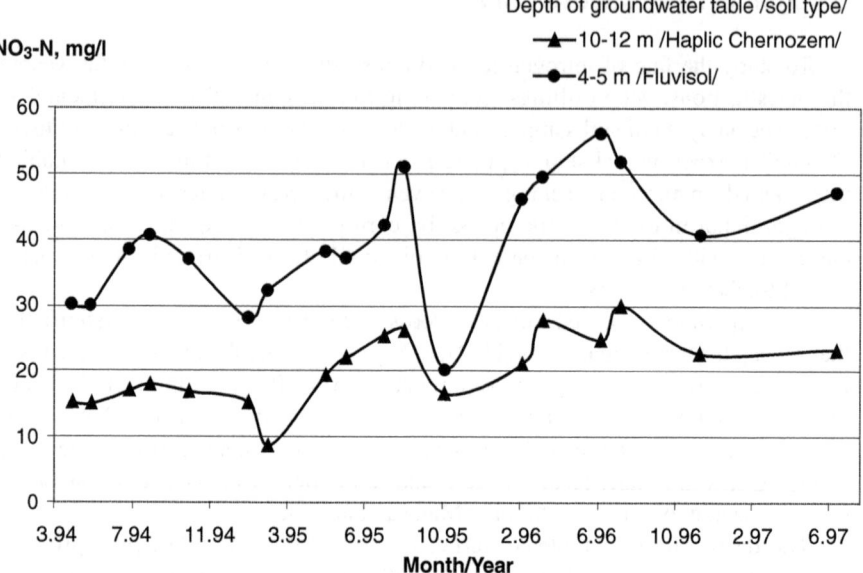

Figure 9.2   Dynamics of NO₃-N of groundwater in home wells in the region of Fluvisol and Haplic Chernozem, Parvomaitsi (northern Bulgaria).

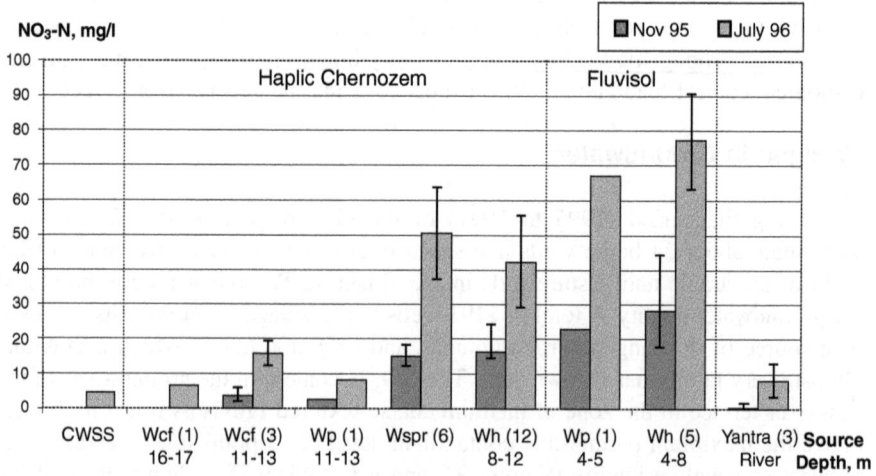

Figure 9.3   NO₃-N content in central water supply system (CWSS), in different sources of groundwater (W), and in the Yantra River in November 1995 and July 1996 in Parvomaitsi. Wcf = reference wells in crop fields; Wp = reference wells in pasture; Wspr = springs; Wh = home wells; (number of sources studied).

Table 9.2 Statistical Parameters of NO₃-N (mg/L) in Central Water Supply System (CWSS), in Groundwater in Some Reference and Home Wells, and in the Yantra River During the Project Period 1994 to 1997 in Parvomaitsi, Northern Bulgaria

Soil Type Source/Land Use	Groundwater Depth (m)	n	Ave.	Min.	Max.	Std. Dev.
CWSS		14	1.3	0.0	4.5	1.2
Yantra River		7	3.1	0.0	5.8	2.1
Haplic Chernozem						
Crop field	16–20	6	4.3	0.0	6.5	2.5
Crop field	12–13	8	9.8	3.3	12.5	3.1
Dairy farm (rw)	11–12	8	14.0	6.1	18.8	4.2
Pasture	11–12	8	8.2	2.3	12.6	3.8
Home well	11–12	18	20.1	8.6	29.6	5.4
Fluvisol						
Riverbank	4–5	7	43.5	22.8	67.0	18.3
Home well	4–5	18	39.7	20.0	56.0	9.4

Table 9.3 Statistical Parameters of NO₃-N (mg/L) in Central Water Supply System (CWSS), in Groundwater in Some Reference and Home Wells, and in the Maritsa River During the Period 1994 to 1996 in Tsalapitsa, Southern Bulgaria

Soil Type Source/Land Use	Groundwater Depth (m)	n	Ave.	Min.	Max.	Std. Dev.
CWSS		23	7.1	4.9	10.5	1.8
Maritsa River		23	3.3	0.0	6.7	2.1
Fluvisol						
Crop field	3–5	23	17.2	9.5	23.5	3.9
Dairy farm	>50	23	12.5	2.8	49.3	13.6
Home well	4–8	23	62.6	46.5	115.2	18.5

Figure 9.3). The space variability of NO₃-N content was also highest in the shallow (2 to 5 m) groundwater (Figure 9.3).

The impacts of moderately coarse-textured Fluvisols and land use activities on groundwater nitrate contamination were studied during the same period, also around the experimental station in the village of Tsalapitsa, situated on the Maritsa River watershed, in southern Bulgaria. The central water system in this village was supplied by deep groundwater, which explained the higher nitrate contents compared to CWSS (central water supply system) in Parvomaitsi, which is supplied by Yovkovtzi Dam Lake (Tables 9.2 and 9.3). The groundwater data from shallow (4 to 5 m) and deep (second aquifer layer on the depth >50 m) wells in the village of Tsalapitsa show (Figure 9.4, Table 9.3) a significant vulnerability to nitrogen loading of all

NO₃-N, mg/l

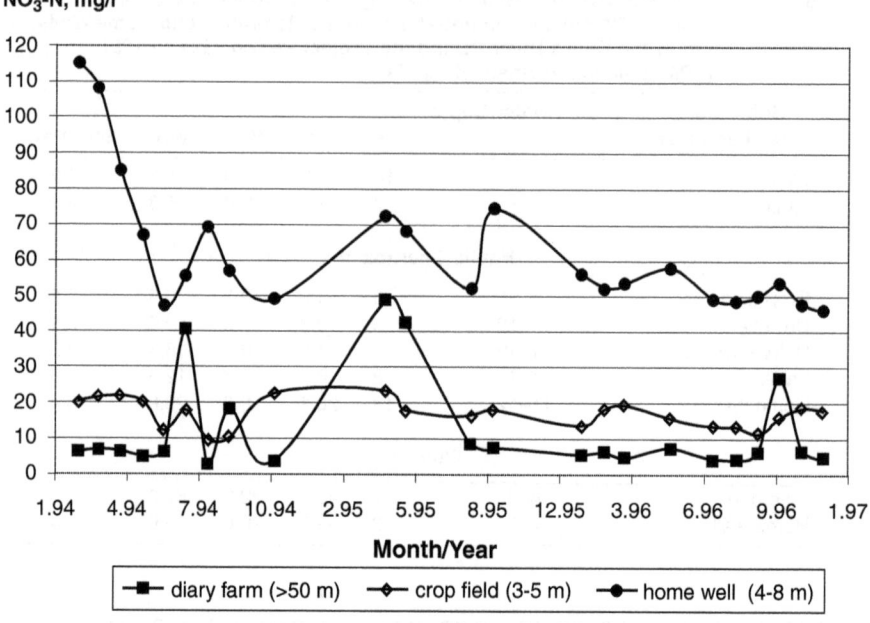

Month/Year

— diary farm (>50 m)   —◆— crop field (3-5 m)   —●— home well (4-8 m)

Figure 9.4 Dynamics of NO$_3$-N in shallow (3–8 m) and deep (>50 m) groundwaters under different land use in the region of Fluvisol in Tsalapitsa, southern Bulgaria.

land use practices. Although the nitrate contamination of the groundwater was higher than the contamination in Parvomaitsi, both the Maritsa and Yantra Rivers had the same content of NO$_3$-N.

NLEAP SIMULATIONS

The information obtained during the pilot project in Parvomaitsi on Haplic Chernozems and Fluvisols, along with the information gathered from the long-term field experiments with continuously irrigated corn on Fluvisols in Tsalapitsa, were used to evaluate the influence of climate, soil properties, and land use on nitrogen leaching predicted by NLEAP (Nitrate Leaching and Economic Analysis Package) (Shaffer et al., 1991). The simulated case studies were:

1. Long-term irrigated corn experiments
2. Crop rotations under rain-fed conditions
3. Tomato growing under intensive irrigation and manuring in household gardens

NLEAP Simulation of Long-term Irrigated Corn Experiments

Sequential runs with the event-by-event time scale of the NLEAP model were performed using the information obtained during the long-term (1972 to 1990) continuous experiments with irrigated corn at three fertilization rates at the experimental

Table 9.4 Soil Texture and Soil Properties of Fluvisol in the Village of Tsalapitsa Used as Input Data for NLEAP Simulations

Parameter	0–30 cm	30–60 cm	60–90 cm
Sand (%)	57	53	58
Clay (%)	20	26	21
Organic matter (%)	0.70	0.55	0.42
pH in H_2O	6.0	6.4	6.5
CEC (meq/100 g)	20.6	23.6	23.8
BD (mg/m³)	1.55	1.50	1.47
PAWHC (v/v)	0.155	0.150	0.147
WP (v/v)	0.140	0.180	0.147

station of the N. Poushkarov Institute in the village of Tsalapitsa (lg = 24°35′E, lat = 42°14′N, alt. = 180 m). The field study was conducted on a sandy clay loam Fluvisol. The soil properties used as input data are presented in Table 9.4.

The daily climate data for precipitation, air temperature, relative humidity, wind velocity, and cloudiness were obtained from the nearest MTO station in Plovdiv (Meteorological annual references, 1972 to 1984) and from the MTO station of the N. Poushkarov Institute in Tsalapitsa for the period 1985 to 1990. The potential evapotranspiration (PET) was calculated by a grass referenced Penman equation (Doorenbos and Pruitt, 1984). The data were loaded into the NLEAP user database for the period 1972 to 1990.

The annual (October to September) precipitation for average, wet, and dry years calculated over a 35-year period for the region of Tsalapitsa is 541, 700, and 343 mm, respectively, and the precipitation sums for the growing periods in these years are 305, 384, and 201 mm, respectively. The average amounts of NH_4-N and NO_3-N in precipitation are 1.9 and 2.1 ppm, respectively (Stoichev, 1997).

During each sequential run of the NLEAP model, the actual information was used for the agrotechnics; irrigation and fertilization management; yields (Stoyanov and Donov, 1996a); and monthly values of crop coefficients calculated on the basis of measured duration of corn development stages. The late hybrids of corn (FAO group 600) were grown in 200 m² plots under three treatments of fertilization: optimal (V_3), ecologically based (V_5), and without fertilization (V_1). The optimal and ecologically based fertilization rates were established corresponding to 232 kg/ha N (a.s.) and 116 kg/ha N (a.s.). It should be mentioned that during the first 3 years, the rates for V_3 and V_5 were much higher (i.e., 512 to 443 and 256 to 221 kg/ha N (a.s.), respectively). The ammonium nitrate fertilizer was applied broadcast, two thirds of the rate before seeding and the remaining one third was incorporated at the time of the 10 to 11 leaf phenological phase. Sprinkler irrigation was scheduled to ensure a non-deficit water regime for crop growth. The average grain yields for non-fertilized, ecologically based, and optimal fertilized treatments were 3765, 9663, and 10618 kg/ha, respectively. The greatest yield variation (44%) was in the non-fertilized treatment, and in the other treatments it was about 18%. Maximum obtained yields — 9230, 12050, and 13450 kg/ha — for the V_1, V_5, and V_3 treatments were loaded into the NLEAP parameter file Region.idx. Nitrogen uptake by the corn grown in these treatments was 1.9, 1.8, and 2.3 kg for 100-kg grain and corresponding by-products (Stoyanov and Donov, 1996b). The primary soil tillage — moldboard plow — was usually performed in November and was adopted as a time for corn residue incorporation. The root portions were estimated (Qawasmi, 1982) to be 31, 16, and 14% of the yield for V_1, V_5, and V_3 treatments, respectively, and the top residue was 15% of the yields for all treatments. These lower values of corn residue compared to other published data (Peterson and Power, 1991) reflect the tradition in Bulgarian agriculture of removing the crop residue from the field for later use in animal husbandry.

Validation tests of the NLEAP model simulations for this case study were conducted using data for measured soil water contents and residual soil nitrogen contents, as well as using leached water and nitrogen data from the Ebermayer lysimeters at the 1.0-m depth. Good agreement ($R^2 = 75\%$) between measured and predicted values of leached nitrogen (NL) was found in the non-fertilized treatment, although the predicted leachate volume was higher than measured. In the fertilized treatments, the predicted values of leached nitrogen were higher than measured. Taking into account the large variability and the irregularity of the measured data, the simulated results could be accepted as good enough for NLEAP utilization for assessment of the climate and management impacts on leached nitrogen in this case.

The measured (precipitation, irrigation, potential evapotranspiration) and NLEAP simulated components of the water balance (leachate volume LP, real evapotranspiration) for the fallow state of the field (Figure 9.5) and for the irrigated corn (V_5 treatment) growing season (Figure 9.6) show that climate conditions in southern Bulgaria are favorable for water percolation during both seasons. The applied, scheduled sprinkler irrigation (96 to 400 mm) and the precipitation (98 to 444 mm) fully compensated for the real evapotranspiration sums (432 to 605 mm) for the vegetation period. The maximum evapotranspiration calculated by the model is 6.1 mm/day, which was lower than reported for the same location data (6.5 to 7.0 mm) obtained by the water balance method using soil moisture data (Varlev and Popova, 1999; Popova and Feyen, 1996). This could be explained by the model default crop coefficient used for the mid-season of 1.0, while the cited studies obtained values of 1.05 to 1.14. Average percolated water for the 19-year period in the V_5 treatment was 33.7 mm for May to September, and 34.2 mm for the October to April period with high variation (39 and 23%, respectively) in these years. Although equal, these quantities were unevenly distributed through the months. The monthly maxima of the percolated water below the corn root zone were in August (16 mm) when the applied irrigation exceeded the crop demands, and in February (11 mm) when the autumn-winter precipitation refilled the soil water holding capacity. Percolation was negligible in September and in October.

As expected, the simulated nitrogen available for leaching (NAL) and leached nitrogen (NL) in the three treatments were ordered according to the applied fertilizer. Quantitative estimates on an annual basis (Table 9.5) revealed the assessment of the V_5 treatment to be ecologically sound. The nitrogen losses through leaching according to the simulation results for the V_5 and V_3 treatments were 17 and 26% of the applied fertilizer rates, respectively. The average results of 19 years of sequential NLEAP model runs showed that July and August were the most susceptible to leaching, which suggested that irrigation rates were too high for this type of soil

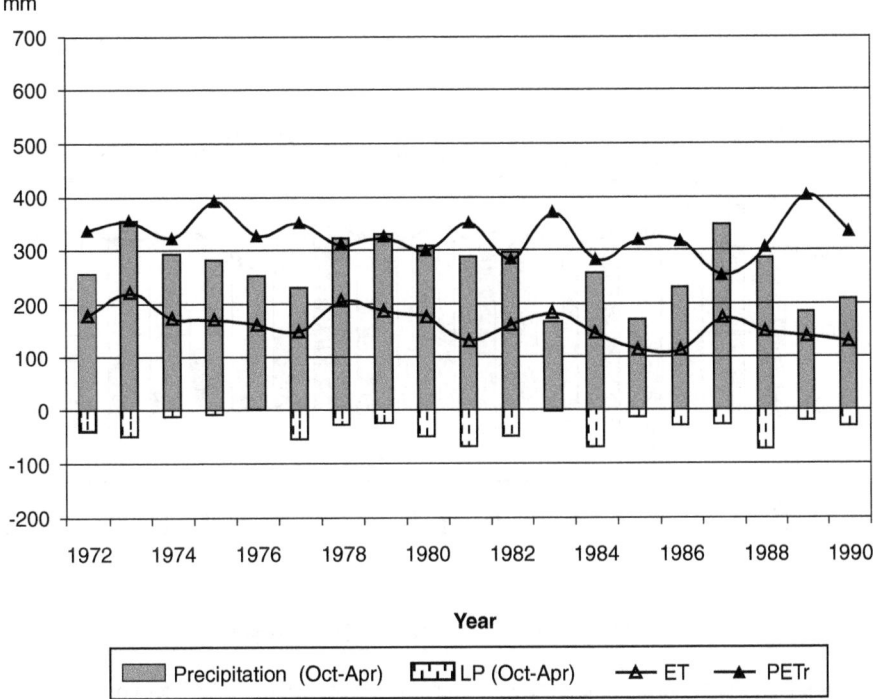

Figure 9.5 Precipitation and simulated LP and ET by NLEAP for the fallow state period of the field (October–April) of V_5 treatment in Tsalapitsa.

(Figure 9.7 and 9.8). The recommended irrigation rates should be lower (especially in August) but more frequently applied.

NLEAP Simulation of Crop Rotations under Rain-fed Conditions

Data from the "Agro-environmental Water Quality Program in the Yantra River Basin" project was used to simulate the crop rotations under rain-fed conditions on a silty clay Haplic Chernozem. The dominant clay minerals in this soil are montmorillonite and illite. The fine texture, high bulk density, and low content of drainage-aeration pores indicated the B horizon to be a layer with low hydraulic conductivity that can influence the water movement and nitrate leaching of the soil profile. The soil cracks, which developed during dry periods, may lead to a higher permeability of this soil. The Haplic Chernozem could be classified as moderately well-drained (Soil Survey Manual, 1993). The soil properties used as input data in NLEAP are presented in Table 9.6.

Climate parameters were observed using a Campbell weather station CR-10 situated at the airport of Gorna Oryahovitsa (2 km from Parvomaitsi) and equipped with sensors for air (at 2 m height) and soil (at 5 and 15 cm depth) temperatures, relative humidity, wind velocity and direction (at 2 m), incoming solar radiation (LI-200X pyranometer), and a rain gage. Additionally, an atmometer, min-max air

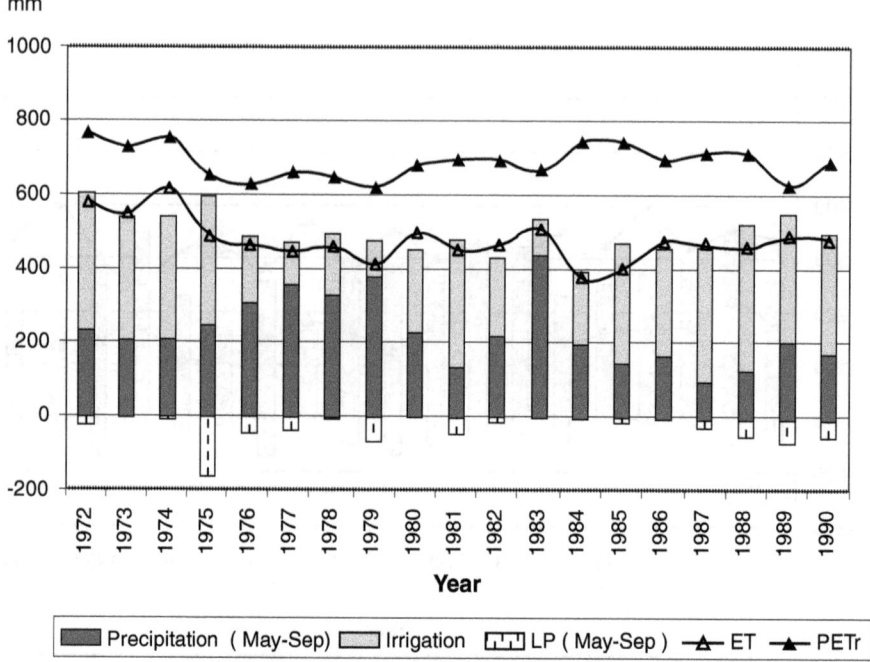

Figure 9.6 Precipitation, irrigation, and simulated LP and ET by NLEAP for corn vegetation period (May–September) of V_5 treatment in Tsalapitsa.

Table 9.5 The Annual Results for Leached Nitrogen (NL) in Non-fertilized (V_1), Ecologically Oriented (V_5), and Optimally Fertilized (V_3) Treatments over 19 Years in Tsalapitsa

Treatment	NL (kg/ha)			Frequency (%) of Three Levels of NL		
	Ave.	Min.	Max.	>80 kg/ha	80-40 kg/ha	<40 kg/ha
V_1	7	0	15	0	0	100
V_5	27	0	121	5	5	80
V_3	83	0	328	32	47	21

thermometers, and a rain gage were set in one of the monitored household gardens. The monthly sums of potential evapotranspiration (PET) calculated by a grass referenced Penman equation (Doorenbos and Pruitt, 1984) and precipitation, mean monthly air temperature, and daily precipitation data for the period November 1994 to October 1997 were loaded into the NLEAP user data file.

The study period (October 1994 to December 1997) was characterized by great variability in precipitation. The data in Table 9.7 show that the climate was wet in 1995 and 1997 and dry in 1996. The precipitation distribution during 1995 is typical for a European moderate continental climate with maximum precipitation in June. It should be taken into account that the wet autumn-winter period of 1994 to 1995 occurred after a very dry 3-year period.

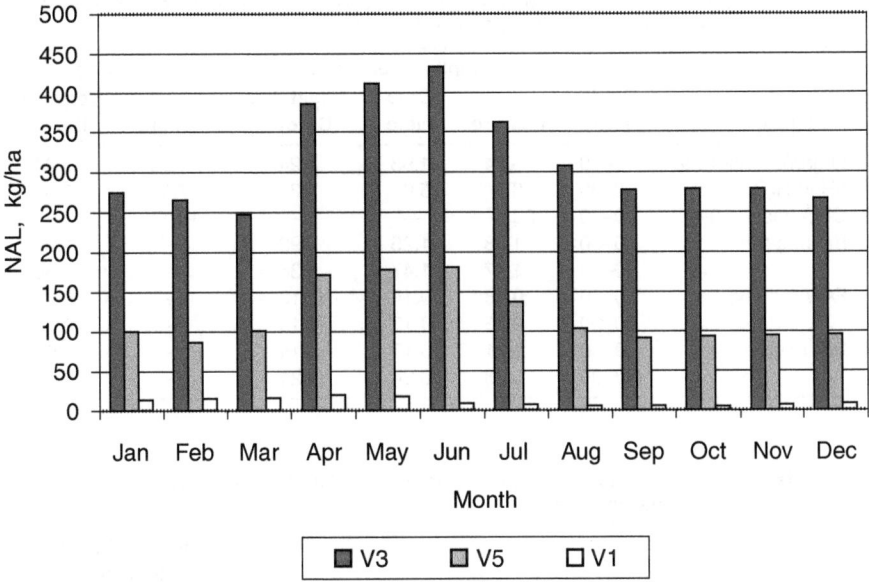

Figure 9.7   Average NAL over 19-year period in the root zone (0–0.9 m) simulated by NLEAP in non-fertilized (V_1), ecologically based (V_5), and optimally fertilized (V_3) treatments of irrigated corn in Tsalapitsa, southern Bulgaria.

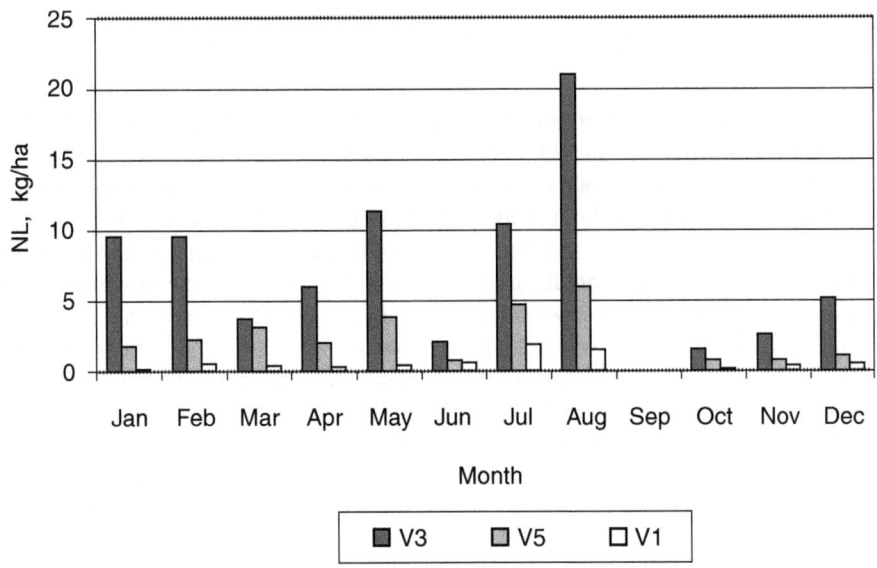

Figure 9.8   Average NL over 19-year period simulated by NLEAP in non-fertilized (V_1), ecologically based (V_5), and optimally fertilized (V_3) treatments of irrigated corn in Tsalapitsa, southern Bulgaria.

Table 9.6 Soil Properties of Studied Land Use Variants in the Village of Parvomaitsi Used as Input Data for NLEAP Simulations

| Parameter | Depth (cm) | Haplic Chernozem | | | Fluvisol | |
		Pasture	Crop Field	Vegetable Garden	Pasture	Vegetable Garden
Organic matter (%)	0–30	3.24	1.93	5.24	1.50	3.24
pH in H_2O	0–30	6.7	5.0	7.3	7.5	7.6
CEC (meq/100 g)	0–30	35	—	—	16.1	—
BD (mg/m³)	0–30	1.38	1.25	0.93	—	1.09
	30–90	1.47	1.47	1.36	—	1.36
PAWHC (v/v)	0–30	0.18	0.14	0.18	—	0.19
	30–90	0.15	0.13	0.15	—	0.13
WP (v/v)	0–30	0.21	0.17	0.14	—	0.08
	30–90	0.23	0.24	0.21	—	0.13

Table 9.7 Seasonal and Annual Precipitation (mm) in the Region of Parvomaitsi

Year	October–April	May–September	Annual	Probability of Exceedance (%)
Average	331	258	589	50
Wet	395	319	714	11
Dry	241	194	435	87
1994–1995	341	335	676	22
1995–1996	244	241	485	78
1996–1997	331	374	705	15

Note: The probability of exceedance of annual precipitation is calculated over a 45-year period.

The average concentrations of NH_4-N and NO_3-N in precipitation during the studied period were 5.2 and 3.8 ppm, respectively.

Two variants of crop rotation under rain-fed conditions during the project period were chosen for sequential runs with the NLEAP model. The runs started on November 1, 1994, which corresponded to the time of autumn soil sampling and the usual time for primary tillage (moldboard plow), as well as with the time of winter wheat sowing. The corn-sunflower-winter wheat rotation was monitored on a 31-ha field, and the sunflower-winter wheat-sunflower rotation was monitored on a 48-ha field. The reduced fertilization (41 to 51 kg/ha N [a.s.]) was applied (Figures 9.9 and 9.10) broadcast in the spring of 1995 and 1997. While the yields in the corn-sunflower-winter wheat rotation were quite good for this type of agrotechnics (5280, 1700, and 5640 kg/ha, respectively), in the sunflower-winter wheat-sunflower rotation they were quite low (1570, 2500, and 1250 kg/ha, respectively). Nitrogen uptake by sunflower was set to 5 kg for 100-kg production.

The predicted values of residual soil NO_3-N content in the root zone (0 to 0.9 m) were quite close to the measured ones ($r^2 = 92\%$), taking into account the space variability of the measured NO_3-N (Figures 9.9 and 9.10). In the regression analyses, the extremely high residual soil NO_3-N values measured in December 19, 1996,

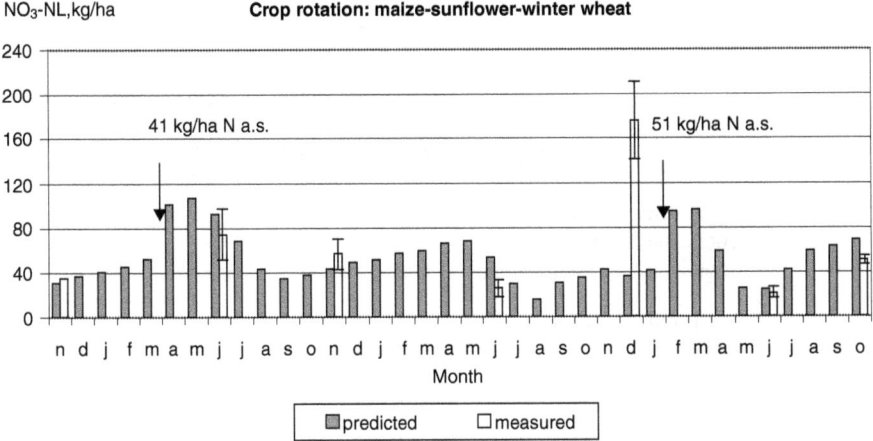

Figure 9.9 Predicted NLEAP and measured residual soil NO_3-N in the root zone (0–0.9 m) of Haplic Chernozem during the maize-sunflower-winter wheat rotation in Parvomaitsi. (Simulation began on Nov. 1, 1994.)

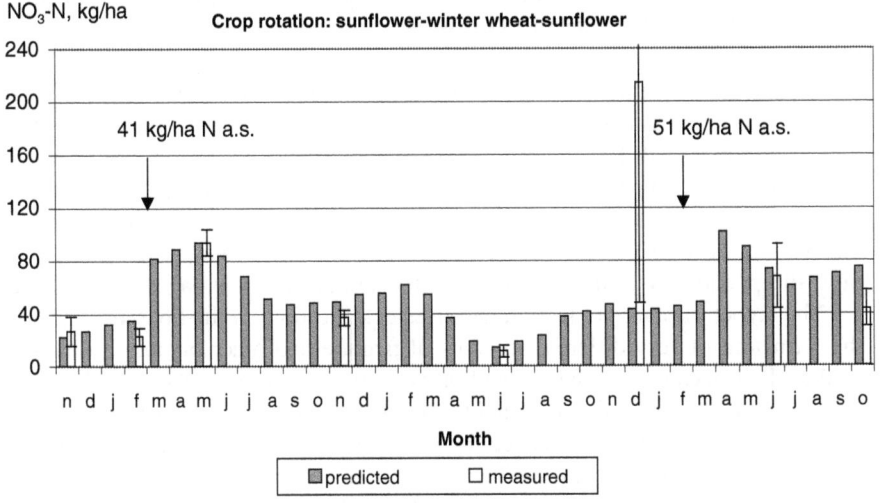

Figure 9.10 Predicted NLEAP and measured residual soil NO_3-N in the root zone (0–0.9 m) of Haplic Chernozem during the sunflower-winter wheat-sunflower rotation in Parvomaitsi. (Simulation began on Nov. 1, 1994.)

were excluded. Obviously, the model missed this peak, possibly the result of wet conditions after moldboard plowing.

The predicted nitrogen leaching (Table 9.8) reflected the climate impact and specificity of the crop rotation. After prolonged drought, the leaching in the wet 1994/1995 year and in the dry 1995/1996 year in both fields was small. The low yield of winter wheat in 1996 due to climatic (cold winter and dry summer) and agrotechnical (without fertilization) conditions led to later accumulation of soil

Table 9.8 NLEAP Simulated Leachate Water Volume (LP) and Leached Nitrogen (NL) below Root Zone (0–0.91 m) in Two Fields (1, 2) with Different Crop Rotation in Parvomaitsi

	Crop	November–April		May–October	
		LP (mm)	NL (kg/ha)	LP (mm)	NL (kg/ha)
1					
1994–1995	Maize	25	3	20	3
1995–1996	Sunflower	15	2	0	0
1996–1997	Winter wheat	173	21	20	2
2					
1994–1995	Sunflower	23	2	20	1
1995–1996	Winter wheat	20	3	33	3
1996–1997	Sunflower	185	25	66	15

residual nitrogen leaching by the heavy rains in September and in November/December. As a whole, the soil properties of the Haplic Chernozem are not risky for significant nitrogen leaching during crop rotations under rain-fed conditions and reduced fertilization.

NLEAP Simulations of Tomatoes Growing under Intensive Irrigation and Manuring in Household Gardens

Typical for the village of Parvomaitsi is that every house (about 1200 in total) is surrounded with a small yard (0.03 to 0.7 ha), separated into a garden and a small domestic animal shed. The garden plots are used for producing a variety of vegetables such as tomatoes, peppers, cucumbers, aubergine, cabbage, and spices, mainly for households needs. In the households surveyed, almost all kinds of animals are bred that are typical for Bulgarian villages, for example, poultry (in almost every house), pigs, sheep, and cows. The distribution of the animals in the village is not uniform.

The layout of plant production and livestock breeding in the households causes the formation of very different amounts of manure and other organic residues, as well as their unequal distribution within the framework of household yards.

The limited area of the households causes a significant density of home wells within the village and short distances between them. The same situation occurs with the distribution of septic tanks and animal sheds. According to the survey, approximately 50% of household septic tanks and farm buildings were situated no farther than 25 to 30 m from their own well and their neighbors' home wells. The septic tanks in 20% of the cases were situated about 10 m from the wells. Additionally, traditional livestock residues and organic home wastes were collected in piles in the farmyard and remained there 1 to 6 months. The collected manure is primarily applied to the home gardens and only a small portion of it is broadcast on the arable lands.

Taking into account the above considerations, the land influenced by manure application is about 133 ha. In the case of uniform distribution, the manure loading of the household gardens could be 27 t/ha and the NPK inputs 184, 136, and 142 kg/ha, respectively.

Table 9.9   Residual NO$_3$-N (kg/ha) in the Soil of
the Root Zone (0–0.90 m) in Manured
Vegetable Plots of Monitored Household
Gardens (Oct. 1994 to Nov. 1997)

Monitored Plots	Ave.	Min.	Max.
Garden 1 (Haplic Chernozem)	260	80	640
Garden 2 (Haplic Chernozem)	453	180	1211
Garden 1 (Fluvisol)	290	34	700

The use of composting technologies as an ecological approach for management of animal and home organic residues was not practiced by the people of the pilot area. Usually, householders burned their home plant residues, and the families living near the river used the riverbank as a waste disposal site. As a rule, the kitchen residues were used as animal feed, but the families without animals threw these residues into waste containers and mixed them with other home wastes.

The groundwater data (Figures 9.2 and 9.3) confirmed that agricultural practices in household gardens were the main factors of nitrate contamination. Although it was not possible to determine the exact rate of manuring, the high values of residual NO$_3$-N measured after its application in autumn suggest high rates, especially in one of the monitored gardens (Table 9.9).

The physical properties in the top layer of both monitored soils (Haplic Chernozem and Fluvisol) changed significantly (Table 9.6) in household gardens as a result of manuring.

The NLEAP model was validated with the data for 1997 using tomato as the most common vegetable grown in the household under irrigated conditions with a starting date after manure application (Stoichev et al., 1998). It was established that for residual NO$_3$-N and water content of the entire soil root zone (0 to 0.9 m) predicted by NLEAP, values are in the range of the measured ones.

In such complicated cases as the household vegetable gardens described above, the application of a modeling approach can outline different management scenarios and help determine the best one for recommendation. For this purpose, the NLEAP model was run using four variants of manure application for the period 1995 to 1997 (Table 9.10), assuming a yield of tomato at about 60 t/ha. The frequent irrigation applied in household gardens is a significant source of nitrogen (44 to 84 kg/ha) when the NO$_3$-N concentration in the groundwater is 20 ppm. The most risky variant is every-year manuring; the amount of 27 t/ha manure is high even for applications once every 3 years. The effect of time of application depended on climate conditions over the 3 years. On the basis of these results, the householders were advised to reduce the nitrogen loading in their vegetable gardens through the use of only irrigation water as a source of nitrogen.

CONCLUSION

The fate of nitrogen was traced by nitrogen balance calculation on a country level and by experimental studies and model simulation of nitrogen leaching under

Table 9.10 NLEAP Predicted Values of Annual NO$_3$-N Available to Leach (NAL), NO$_3$-N Leached (NL), Leachate Volume (LP), and Leaching Depth Below the Root Zone (D) at Different Variants of Manure Application to Tomato Grown in Household Garden under Irrigation with Groundwater

Variant	Parameters	1995	1996	1997
Every-autumn manuring with 27 t/ha	NAL (kg/ha)	385	676	909
	NL (kg/ha)	19	144	453
Manuring (27 t/ha) applied in the autumn of the first year	NAL (kg/ha)	385	383	289
	NL (kg/ha)	19	101	196
Manuring (27 t/ha) applied in the spring of the first year	NAL (kg/ha)	362	384	363
	NL (kg/ha)	16	57	221
Without manuring (N input by irrigation water)	NAL (kg/ha)	78	73	63
	NL (kg/ha)	16	12	40
For all variants	LP (mm)	76	127	330
	D (m)	0.5	0.8	2.1

specific land use. These approaches elucidate different elements of the reason-consequences links on the dynamic behavior of the nitrogen in the plant-soil-ground-water system. The information obtained during long-term investigations was linked and assessed using the simulation model NLEAP.

The NLEAP model was successfully applied for describing long-term irrigated corn experiments located on the Maritsa River watershed, as well as for typical Bulgarian agricultural crop rotations and tomato production in household gardens realized during the pilot project on the Yantra River watershed. The simulated scenarios clarified the impact of different driving forces on nitrogen leaching and hence on groundwater contamination registered in both watersheds. The conclusions concerning rates and timing of mineral fertilization, manuring, and irrigation can be used as recommendations for improving agricultural management.

The results obtained are encouraging for adoption of simulation modeling as an essential part of scientific projects for solving different agroecological problems. Thorough planning of the monitoring and information on local conditions is needed to supply the required model databases.

REFERENCES

Atanassov, I., Stoichev, D., and Glogov, L. 1974. Migration of some elements with lysimetric waters in the experimental field plots with irrigated maize, *Agric. Sci.*, XIII(2):103-108.

Doorenbos, J. and Pruitt, W.O. 1984. Crop-water requirements. *Irrigation and Drainage,* paper 33, FAO, Rome.

Drinking Water Quality in the Country. 1979. Medicine and Sport, Sofia, Bulgaria.

Drinking Water Quality. 1968. Medicine and Sport, Sofia, Bulgaria.

Final Report, 1997. Agro-Environmental Water Quality Program in the Yantra River Basin, N. Poushkarov Institute of Soil Science and Agroecology, Sofia, Bulgaria.

Fotyma, M. and Fotyma, E. 1999. Balance of nitrogen in soil-fertilizer crop system, in *Conference Proc. Nitrogen Cycle and Balance in Polish Agriculture*, Publisher Institute for Land Reclamation and Grassland Farming, Falenty, Poland, 90-95.

Gitsova, S., Matev, B., and Stoichev, D. 1991. On some aspects of water quality management and water pollution problems in Bulgaria. *Report in Water Quality Workshop*, Poznan, Poland.

Hansen, S., Jensen, H.E., and Shaffer, M.J. 1995. Developments in modeling nitrogen transformation in soil. In P.E. Bacon (Ed.). *Nitrogen Fertilization in the Environment*, Marcel Dekker, New York, 83-107.

Mateva, K., Stoichev, D., and Ahchiski, P. 1982. Studying of nitrogen movement under intensive agriculture at different soils. *Symposium Der Technische Fertschritt in der Wasserversorgung und in der Wasseraufbereittung Band.* 1, Varna, 28-40. *Meteorological Annual References.* 1972–1984. Sofia, Bulgaria.

Nikolova, M., Andres, E., and Glas, K. 1995. *Potassium — Nutrient Element for Yield and Quantity.* International Potash Institute, Basel, Switzerland, 15.

Oenema, O. 1999. Nitrogen cycling and losses in agricultural systems; Identification of sustainability indicators. In *Conference Proc. Nitrogen Cycle and Balance in Polish Agriculture*, Publisher Institute for Land Reclamation and Grassland Farming, Falenty, Poland, 25-43.

Peterson, G.A. and Power, J.F. 1991. Soil, crop, and water management. In R.F. Folett et al. (Eds.). *Managing Nitrogen for Groundwater Quality and Farm Profitability.* Soil Sci. Soc. Am., Madison, WI, chap. 9, 189-198.

Petkov P., Ivanov, S., Jivkov, J., Popov, B., Georgiev, D., Popov, I., and Kiosev, G. 2000. A strategy for the development of irrigation in the free market economy in Bulgaria. *Vodno delo*, 1/2:29-38.

Petrov, P., Bojilov, A., Vankov, I., Mladenov, I., Petrov, G., Marinova, S., and Bildirev, N. 1983. *Slurry — Management and Use in Agriculture.* Zemizdat, Sofia, Bulgaria.

Popova, Z. and Feyen, J. 1996. Adjustment of WAVE model Kc-factors for Bulgarian weather conditions. In *Selected Papers of the Workshop "Crop-Water-Environment Models" During the XVI ICID Congress.* Cairo, 176-186.

Qawasmi, Y. 1982. Influence of Fertilization on Nutrient and Water Migration at Maize Growing Under Leached Cinnamonic Soil, Ph.D. thesis, Sofia, Bulgaria.

Shaffer, M.J., Halvorson, A.D., and Pierce, F.J., 1991. Nitrate Leaching and Economic Analysis Package (NLEAP): model description and application. In R.F. Follett et al. (Eds.). *Managing Nitrogen for Groundwater Quality and Farm Profitability.* Soil Sci. Soc. Am., Madison, WI, chap. 13, 285-322.

Soil Survey Manual, 1993. *USDA Handbook*, No. 18.

Statistical Yearbook, 1960-1998. National Statistical Institute, Sofia, Bulgaria.

Stoichev, D., Kercheva, M., and Stoicheva, D. 1998. Use of NLEAP model for estimation of nitrates fate in leached Chernozem. In *Congres Mondial de Sci. du sol*, 16e, Montpellier, France, Resumes, Vol. 1, Symp. N 8, 167.

Stoichev, D., Stoicheva, D., Angelov, G., and Dimitrova, K. 1996. Influence of long-term fertilization on nitrogen leaching. In Van Cleemput et al. (Eds.). *Progress in Nitrogen Studies*, Kluwer Academic Publishers, The Netherlands, 689-693.

Stoichev, D. and Stoicheva, D. 1987. Nitrate content of liquid phase of soil under intensive agriculture. In *Proc. 5th International Symposium of CIEC*, Vermes, L., Mihalefy, A., and Nemeth, T. (Eds.). MTESZ, Balatonfured, Hungary, 284-291.

Stoichev, D. 1974. A device to obtain lysimetric water. *Soil Sci. Agrochem.,* 9:13-18.

Stoichev, D. 1997. Ecological Aspects of Anthropogenic Loading of Soils. Dissertation, Sofia, Bulgaria.

Stoichev, D., Atanassov, I., and Qawasmi, Y. 1985. Effect of fertilizing of the chemical composition of lysimetric water from leached cinnamonic forest soil. *Soil Sci. Agrochem. Plant Proc.,* 20, 42-47.

Stoichev, D., Atanassov, I., and Glogov, L. 1980. Vertical movement of nitrogen in alluvial-meadow soil, *Soil Sci. Agrochem.*, XV(4):10-17.

Stoichev, D., Shaffer, M., Starr J., Lemunyon, J., Stoicheva, D., Kercheva, M., and Koleva, V. 1999. Cooperation between USDA and Bulgaria in Agro-Environmental Water Quality Program. *ISCO'99 Conference,* Indiana, USA.

Stoichev, D., Stoichev, D., Kercheva, M., Lazarov, A., Marinova, S., Angelov, G., Alexandrova, P., Filcheva, E., Mitova, T., Tzvetkova, E., Stoimenova, M., and Koleva, V. 1997. Assessment of the agricultural pollution sources in the Yantra River Basin. *Chimia i Industria,* 68(1-3):51-53.

Stoichev, D., Stoicheva, D., Kercheva, M., Lazarov, A., and Dervenkov, K. 1998. Distribution of nitrogen in the intermediate vadose zone of a watershed. *Bulg. J. Agric. Sci.,* 4:575-582.

Stoichev, D., Stoicheva, D., Mateva, K., and Ahchiiski, P. 1983. Distribution of nitrates in the profile of a leached Smolnitza and their relationship to underground water. *Soil Sci. Agrochem.,* 18:30-40.

Stoyanov P. and Donov D. 1996a. Sustainable effectiveness of nitrogen fertilization on various soils in maize grown as a monoculture. I. Yields and nitrogen uptake with the above-ground biomass, *Soil Science, Agrochemistry and Ecology, Sofia,* XXXI(III):144-146.

Stoyanov P. and Donov D. 1996b. Sustainable effectiveness of nitrogen fertilization on various soils in maize grown as a monoculture. II. Nitrogen economy of maize crops. *Soil Science, Agrochemistry and Ecology, Sofia,* XXXI(III):147-149.

Technical Report No. 27, 1988. *Nitrate and Drinking Water,* Copyright (European Chemical Industry, Ecology and Toxicology Centre), ISSN 0773-8072-27.

Varlev, I. and Popova, Z. 1999. *Water-Evapotranspiration-Yield.* Sofia, 67-77.

Vermeulen, S.E., Steen L., and Schnug H. 1998. Nutrient balances at the farm level. *Bibliotheca Fragmenta Agronomica,* 3:108-123.

Yordanova, M. and Donchev, D. (Eds.). 1997. *Geography of Bulgaria,* "Prof. Marin Drinov" Academic Publishing House, Sofia, Bulgaria.

CHAPTER 10

Use of Simulations for Evaluation of Best Management Practices on Irrigated Cropping Systems

Jorge A. Delgado

CONTENTS

INTRODUCTION

Optimum yields cannot be achieved without an adequate supply of nitrogen (N) to the root zone and, in general, natural sources of N in irrigated agroecosystems are not adequate for maximum crop production or the economical sustainability of these systems. Newbould (1989) reported that N use efficiency (NUE) is usually lower than 50%. The USEPA (1989) reports that drinking water with more than 10 mg/L of NO_3^--N is considered unfit for human consumption. This problem of NO_3^--N concentrations that are higher than the established USEPA limit has been reported for many parts of North America as well as in other areas (e.g., Milburn et al., 1990; Follett et al., 1991; McCracken et al., 1994; Owens and Edwards, 1994). It is important to use recommended best management practices (BMPs) to minimize losses of N during the growing season and to minimize potential additional losses of residual soil NO_3^--N after harvest. There is a need to continue developing and evaluating BMPs that contribute to maintaining an adequate supply of N in the root zone of irrigated crops while maintaining minimal N losses to reduce potential off-site impacts of N, including the potential leaching losses of NO_3^--N from the root zone that can potentially impact underground water.

For some crops, quality is also an important economic factor in management and harvest. For these irrigated cropping systems where quality becomes an economic driving factor, N management becomes more complicated. Not only is there need to maintain an adequate supply of N during the growing season, but there is also the need to deliver it during periods of maximum demand and then reduce its availability during those periods when high N uptake can potentially affect the product quality and reduce economic returns. It is difficult to achieve the goal of maximizing the optimum production with the best product quality and the highest NUE. This is particularly true in irrigated cropping systems in which water demands can quickly fluctuate due to changes in weather and crop growth stages, thus increasing the potential of NO_3^--N leaching due to irrigation and/or local thunderstorms.

Maximizing NUE while increasing yields and product quality is also more difficult for those management areas (e.g., center irrigated pivot) where spatial variability in soil type may be present. This variability in soil type will affect the soil water holding capacities and may affect the residual soil NO_3^--N due to management (Delgado, 1999; Delgado and Duke, 2000; Delgado et al., 2001). Figure 10.1 depicts the spatial variability of residual soil NO_3^--N after potato harvest and soil texture (% sand) and organic matter. To maximize the potential uptake of NO_3^--N by barley following potato, and to reduce lodging and improve grain quality, N management will have to consider this spatial variability of texture, residual NO_3^--N, and yield potentials when making final fertilizer recommendations.

When product quality is an important factor in maximizing economic returns, management practices that maximize the production and quality are needed. For example, excessive N applications can affect the quality of grains, tubers, fruits, and other cropping systems. This is very important for small grain crops such as barley (*Hordeum vulgare* L.), especially when managed for malting, because higher than needed N levels can increase the levels of proteins, thereby reducing its quality

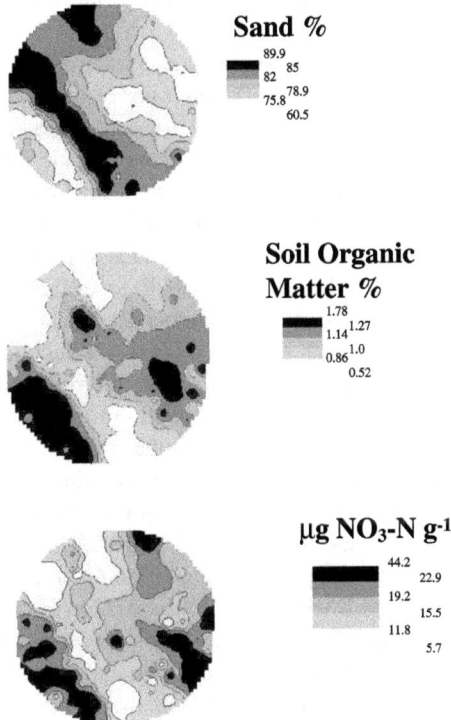

Figure 10.1 Spatial variability of sand (%), soil organic matter (%), and nitrate (μg NO₃⁻-N/gr) in a center irrigated pivot after potato harvest.

(Zubriski et al., 1970; Bishop and MacEachern, 1971). A tuber crop, such as potato (*Solanum tuberosum* L.) can also have its quality affected by excessive application of N fertilizer (Laughlin, 1971; Paintier et al., 1977; Westermann and Kleinkopf, 1985; Westermann et al., 1988; Errebhi et al., 1998). Sugarbeets can also be affected and the sugar content reduced by excessive application of N that can stimulate vegetative growth later in the growing season and potentially reduce the storage of sugars (Hills and Ulrich, 1971; Cole et al., 1976; Roberts et al., 1981; Carter and Traveller, 1981; Hill, 1984). Fruit crop quality can also be affected by high N rates (Locascio et al., 1984). Nitrogen inputs can affect grain, fruit, and tuber qualities such as size, color, as well as sugar and protein contents and other physiological characteristics that can affect the culinary properties considered by farmers, producers, distributors, and consumers.

For rotations that include vegetable and small grain crops such as malting barley, the producer has the dilemma of applying high N rates that can optimize yields but can also maximize quality. It is important, then, to evaluate how recommended BMPs optimize yields and product quality and their impact on NUE. By improving the

BMPs that increase the NUE while maximizing yield and quality, greater net returns to farmers, producers, and distributors will be achieved, along with a better-quality product for the consumer.

Improvements in NUE across irrigated systems can contribute to savings of millions and billions of dollars worldwide. For example, Raun and Johnson (1999) reported that if NUE is increased by 1% for cereal production alone, $234 million can be saved worldwide. Their estimate of worldwide NUE for cereals was 33%. Their conservative estimate also concluded that there was potential to save $4.7 billion worldwide by increasing the NUE on the cereal crop. Because it is difficult to extrapolate research results to a wide variety of worldwide scenarios, it is necessary to model N budgets and NUE to evaluate how BMPs can protect environmental quality across different agroecosystems and regions.

New computer models can serve as tools to conduct quick evaluations of soil N management practices and can be used across different agroecosystems worldwide. However, there is limited information regarding modeling of NUE across different cropping systems, as well as a limited number of manuscripts with extensive simulations that have been conducted to calibrate/validate (Delgado et al., 2000) the evaluation of NUE, N losses, N budgets, and the balance between the NO_3^--N that is leaving the rooting zone and the NO_3^--N that is being pumped back with the irrigation water into the cropping systems of commercial operations. For example, to compare the effects of BMPs under a lettuce (*Lactuca sativa* L.) crop with winter wheat (*Triticum aestivum* L.), one must consider the soil depth or a similar soil profile (Delgado et al., 1998a; b). This comparison of similar soil depths is required because the NO_3^--N that may be leached from the lettuce root zone can be scavenged with a deeper rooting crop such as wheat and rye (*Secale cereale* L.) (Delgado, 1998; Delgado et al., 1998a). Validation of these models across different commercial fields and how the root zone of different agroecosystems interacts with N budgets and NUE is limited.

One such computer software package is the Nitrate Leaching and Economic Analysis Package (NLEAP) model (Shaffer et al., 1991). NLEAP can be used to simulate a wide variety of BMPs at a specific farmer's field and how they affect N budgets. A new version of NLEAP (1.20) is capable of evaluating not only N budgets on the root zone of lettuce, potato, barley, wheat, canola (*Brassica napus* L.) and winter cover rye, but also the N budgets below the root zone of these crops to a similar soil depth for these systems, which was developed by Shaffer et al. (1998). For more information on version 1.20, see Delgado et al. (1998a) and Delgado (1999).

Khakural and Robert (1993) and Beckie et al. (1994) conducted NLEAP simulations in the northern United States and Canada, respectively, and found that the previous NLEAP version 1.10 was able to simulate residual soil NO_3^--N, NO_3^--N leaching, and N and water budgets. They reported that NLEAP simulations of residual soil NO_3^--N and soil water content in the root zone were as accurate as the simulations conducted by other computer models such as the Erosion/Productivity Impact Calculator (EPIC) (Williams, 1982), the Crop Estimation through Resource and Environment Synthesis (CERES) (Ritchie et al., 1985), Nitrogen Tillage, and Residue Management (NTRM) (Shaffer and Larson, 1987), and LEACHM-N

(Wagenet and Hutson, 1989). Khakural and Robert (1993) found that NO_3^--N leaching simulations were similar to those of the LEACHM-N model and were correlated to measure NO_3^--N leaching values.

Several other authors have conducted NLEAP simulations with the NLEAP 1.10 version, which is only capable of conducting simulations for the maximum rooting zone of the modeled crop (Shaffer et al., 1991; Follett et al., 1994; Wylie et al., 1994; Shaffer et al., 1995). To evaluate cropping systems that include shallow (potato and vegetables) and deeper (small grains) rooted crops and how they interact with N dynamics in these systems, a model that can conduct simulations for multiple horizons such as the rooting zone of the modeled crop and below the root zone of the modeled crop to a similar depth for all the crops in the rotation is needed. The primary objective of this chapter is to present a unique, extensive, first simulation of seven cropping systems that evaluate N budgets, NO_3^--N dynamics, and water budgets in the root zone, and NO_3^--N leaching from the root zone and from a similar soil depth for all crops on commercial farming systems. This is a unique presentation of how to use the new NLEAP 1.20 version to model NUE and determine how the root zone of these crops interacts with environmental quality across different agro-ecosystems and NO_3^--N dynamics and losses from different soil horizons.

MODEL INPUTS

To conduct the simulations, the author entered the following information into the NLEAP model: crop planting and harvesting dates; N-, water-, cultural-management inputs and timing, soil and climate information; expected yield; and all N additions (e.g., initial NO_3^--N content of the soil, amount and type of N fertilizer added, amount of N in the irrigation water, crop residue mass and its N content).

Field Sites, Climate, and Management Practices

BMPs were monitored on 38 fields located across various farms in the San Luis Valley (SLV) of south-central Colorado (Figure 10.2). These studies were conducted from 1992 to 1999, during which time all needed information was collected regarding management, yields, soil, and plant samples. The SLV has an area of approximately 8288 km^2 (Edelmann and Buckles, 1984); the mean altitude is approximately 2341 m; and the mean average annual precipitation is 180 mm.

At each site, recommended BMPs were applied. The N fertilizer rates were based on laboratory analysis of soil, plant tissue, and irrigation. Farmers used split applications of N fertilizer into preplant, sidedressing, and fertigation applications for sensitive crops such as potato and lettuce. All N fertilizer applications at preplant and side dressing were banded. There were no recommendations for fall N fertilizer application of spring planted crops. The recommended BMPs for this region were revised and published in 1999 by the San Luis Valley Water Quality Demonstration Project (SLVWQDP) (Ristau, 1999). This publication was put together and revised by a team composed of the SLVWQDP, USDA-NRCS-Alamosa Regional Area,

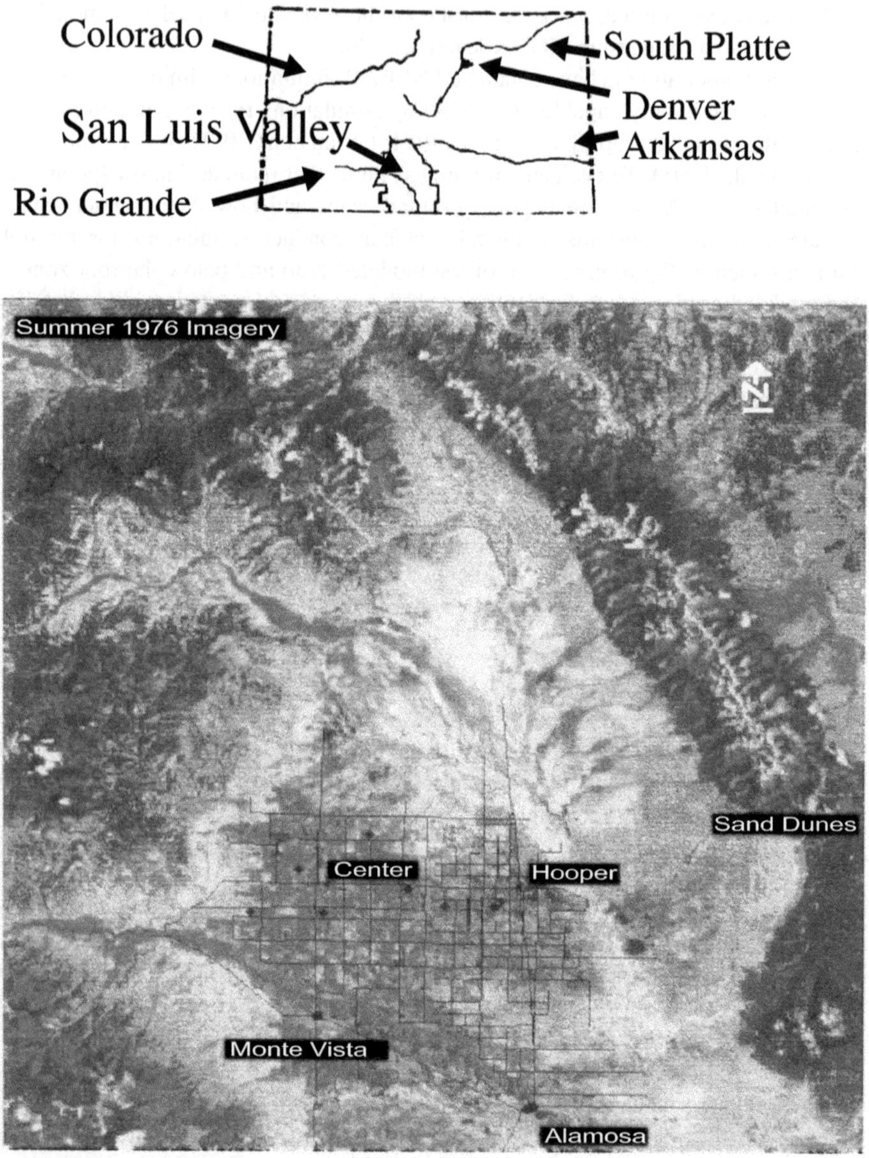

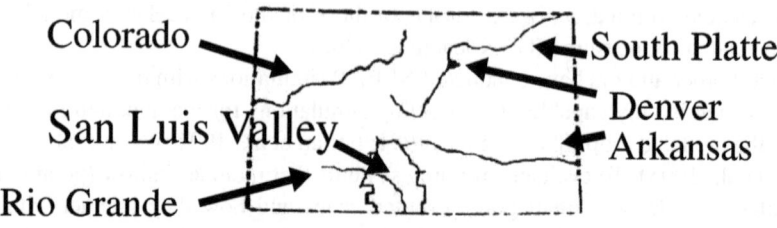

Figure 10.2   Major bodies of water in Colorado, such as the Arkansas, Colorado, Rio Grande, and South Platte Rivers and the San Luis Valley study area localized in south-central Colorado, are labeled in the map. The close-up is a Landsat image of the valley where commercial farms were simulated around the San Luis Valley, from Center to Hooper, and Monte Vista to Alamosa, and also around Blanca, which is located east of Alamosa, but not shown on the map.

USDA-ARS-Soil Plant Nutrient Research Unit, Colorado State University, CSU-Cooperative Extension, consultants, and farmers. For more specific information on the BMPs used in this region, refer to Best Management Practices for Nutrient and Irrigation Management in the SLV (Ristau, 1999).

At each commercial farm site, rain and/or snow was measured during the growing season. Additional needed climatic data from the nearest weather station in the SLV (Center, CO) was used to conduct the simulations. During the growing season, potential evapotranspiration (E_{tp}) using the modified Jensen-Haise (JH) estimates was calculated (Follett et al., 1973; Jensen et al., 1990). At each farm site, center-pivot sprinklers were evaluated for efficiency and irrigation, and water samples were collected three times and analyzed for NO_3^--N content during the growing season. All other information necessary for conducting simulations (e.g., irrigation amounts and scheduling, N fertilizer application, planting, harvesting, cultivation, and other agricultural management practices) were collected at each field site (Shaffer et al., 1991).

At some of these commercial fields, four to sixteen 20.9-m^2 plots were located under a 54.7-ha center-pivot irrigation sprinkler system. Transponders were placed permanently in each field so that soils could be sampled at the same site during the spring and fall. Additionally, complete fields were monitored using the entire center-pivot irrigation sprinkler (54.7 ha) and commercial farmer's yields (truck loads). In addition, plant and soil samples were collected randomly to monitor N status and soil NO_3^--N at these large-scale commercial fields.

Soil Physical and Chemical Characteristics

Soil samples were collected through these commercial fields during the spring before planting and the fall after harvesting. At each commercial operation, for whole field sites, 20 randomly located soil cores were composited for the initial and final soil samples. At the sites where small plots were used to intensively monitor the commercial operations, samples were collected in each of the 20.9-m^2 plots in the spring before planting and in the fall after harvesting. All soil samples were collected in 0.3-m intervals to 1.5 m. Because the simulations were conducted on a 0- to 0.91-m profile, the data presented here will only describe the effects of NO_3^--N dynamics on this 0- to 0.91-m soil profile depth. At each site, initial soil water content in the profile was measured. The soil organic matter content, pH, and CEC (cation exchange capacity) of the surface soil (0–0.31 m) were measured. Bulk densities were estimated from texture as described by USDA-SCS (1988), and available soil water at harvesting in the root zone was measured at selected sites.

After sampling, soils were air-dried and sieved through a 2-mm sieve. Because most of the soils at these commercial sites were coarse in texture, the percentage of coarse fragment by volume was calculated as described by Delgado et al. (1998b). Sieved samples were extracted with 2N KCl and the NO_3^--N and NH_4^+-N were colorimetrically determined by automated flow injection analysis. The soils used in these studies included (1) Gunbarrel, mixed, frigid Typic Psammaquents; (2) Kerber, coarse-loamy, mixed, frigid Aquic Natrargids; (3) McGinty, coarse-loamy, mixed, frigid Typic Calciorthids; (4) Mosca, coarse-loamy, mixed, frigid Typic Natrargids; (5) Norte, loamy-skeletal, mixed (calcareous), frigid Aquic Ustorthents; and (6) San Luis, fine-loamy over sandy or sandy-skeletal, mixed, frigid Aquic Natrargids. Additional chemical and physical properties of these soil types are described in the USDA-SCS survey of the region (USDA-SCS, 1973).

Cropping System Information

A crop database with mean rooting depths, N crop indices for whole plants, and other variables was developed as a "region.idx" file that was used to conduct the simulations (Delgado et al., 1998a; b). Plant samples for small grain, potato, and lettuce were collected by harvesting samples from each plot. Potato vines, roots, and tubers were collected prior to farmers harvesting their fields. Plant samples were dried at 55°C, ground, and analyzed for total C and N content by automated combustion using a Carlo Erba automated C/N analyzer. The mean root depth was measured for all crops by digging a hole at each site and measuring root depth. The mean root depth was measured for barley (0.61 m), canola (0.76 m), lettuce (0.37 m), potato (0.40), spring wheat (0.91m), winter wheat (0.91m), and winter cover rye (0.91m).

NLEAP 1.20 SIMULATION OUTPUTS AND STATISTICAL ANALYSIS

Version 1.20 was used to simulate the residual soil NO_3^--N and available soil water (Figures 10.3 and 10.4; $p < 0.001$). The computer-simulated outputs for soil NO_3^--N in the root zone, bottom of the root zone to the 0.91-m depth, and the entire soil profile (0 to 0.91 m) were compared to the observed residual NO_3^--N values (Figures 10.3b and 10.4; $p < 0.001$). Using these computer simulations of the NO_3^--N leached from the root zone of each crop and from a similar soil depth for the entire system (0 to 0.91 m), a NO_3^--N balance between inputs and outputs was conducted. The simulated amount of NO_3^--N leaching was subtracted from the amount of NO_3^--N that was pumped from the groundwater with irrigation and identified as NO_3^--N mining potential. The NO_3^--N mining potential was calculated as follows: (1) NO_3^--N mining for the root zone = NO_3^--N in the groundwater added to the field minus NO_3^--N leached from the root zone; and (2) NO_3^--N mining for the soil profile = NO_3^--N in the groundwater added to the field minus NO_3^--N leached from the 0.91-m soil profile.

Simulated outputs were also used to estimate NUE for each agroecosystem. The system NUE was calculated as NUE = [(Total N uptake by crop/Total N available in the 0–0.91-m soil profile) × 100]. Total N available included all N inputs such as the initial NO_3^--N in the 0–0.91-m soil profile, added fertilizer, added fertilizer in irrigation, background N in water, and simulated soil and crop residue mineralized N. By including mineralization and crop residue N, the effects of BMPs on the agroecosystem NUE are being evaluated.

Statistical analyses were performed using the SAS analysis of variance GLM procedure (SAS Institute Inc., 1988) to test for differences between cropping

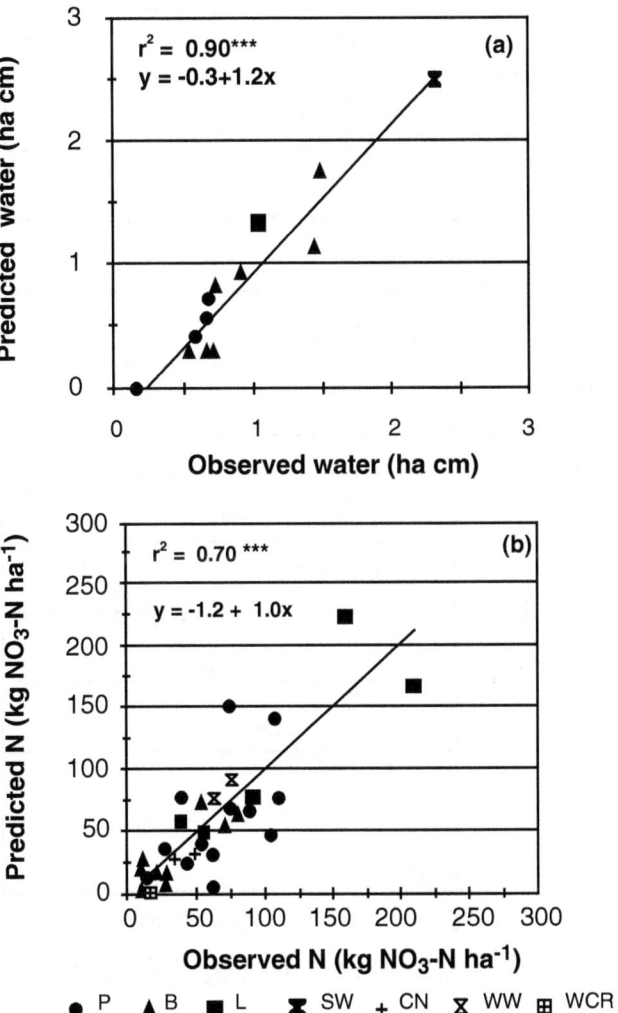

Figure 10.3 Observed and NLEAP-simulated available water in the root zone (a) and residual soil NO_3^--N in the root zone (b). Observed and simulated data for potato (P), barley (B), lettuce (L), spring wheat (SW), canola (CN), winter wheat (WW), and winter cover rye (WCR) grown in the San Luis Valley (*** = r^2 significant at $p < 0.001$).

systems. The shallow-rooted crops such as potato and lettuce were also compared to the deeper-rooted crops such as spring wheat, canola, malting barley, winter cover rye, and winter cover wheat. Differences between shallow- and deeper-rooted agro-ecosystems in simulated NO_3^--N leaching, and NUE were conducted using the SAS analysis of variance GLM procedure (SAS Institute Inc., 1988). Correlations were made between predicted and observed available soil water using SAS REG (SAS Institute Inc., 1988). The SAS REG procedure was also used for correlation between predicted and observed residual soil NO_3^--N. The intercept (b_0) and slope (b_1) were tested with SAS REG for differences from 0 and 1, respectively.

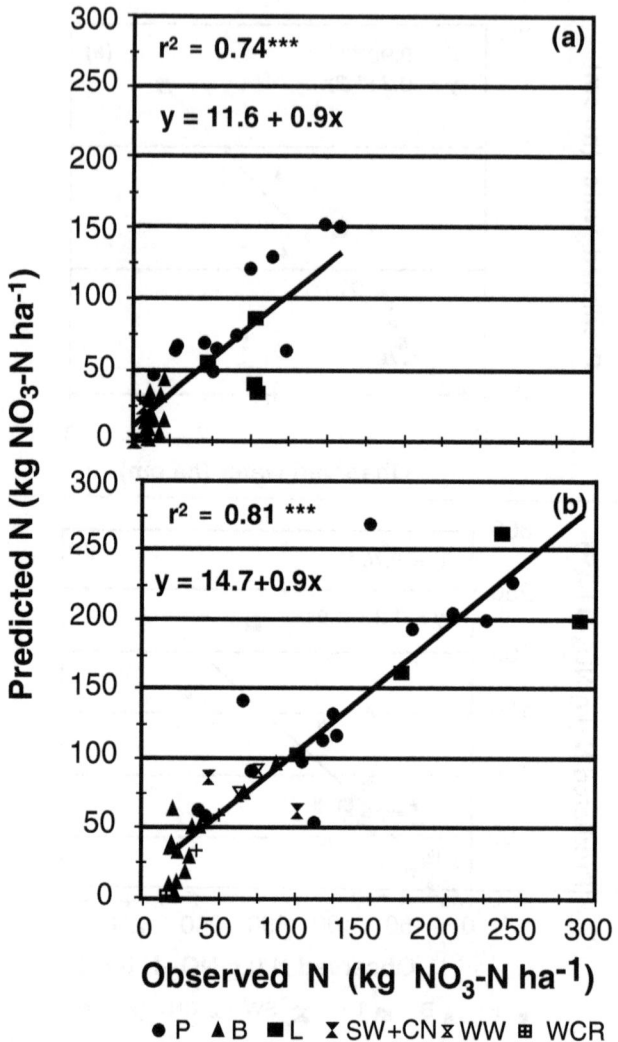

Figure 10.4 Observed and NLEAP-simulated residual soil NO_3^--N from the bottom of the root depth (BRD) to the soil depth (BRD to 0.9 m) (a); and for the entire soil depth (0.0 to 0.9 m) (b). Observed and simulated data for potato (P), barley (B), lettuce (L), spring wheat (SW), canola (CN), winter wheat (WW), and winter cover rye (WCR) grown in the San Luis Valley (*** = r^2 significant at $p < 0.001$).

OBSERVED NO_3^--N DYNAMICS

The N inputs for the lettuce and potato cropping systems were, on average, higher than those inputs added to the small grains (Table 10.1). The mean total crop N uptakes were 127, 174, 150, 195, 293, 171, and 207 kg N/ha for lettuce, potato, barley, canola, spring wheat, winter cover rye, and winter wheat, respectively. The

Table 10.1 **Mean Fertilizer Application, and Initial and Residual Soil NO_3^--N (RSN) from Lettuce, Potato, Barley, Canola, Spring Wheat (SW), Winter Cover Rye (WCR), and Winter Wheat (WW) Fields Grown under Commercial Operations**

Crop	Total N Fertilizer Applied[a] (kg N/ha)	Soil NO_3^--N at Planting	Residual Soil NO_3^--N at Harvest	
			Root Zone	(Soil Depth 0–0.9 m)
		(kg NO_3^--N/ha)		
Lettuce	297 a	21 c	148 a	222 a
Potato	210 b	38 c	65 b	123 b
Barley	42 de	79 b	20 d	31 cef
Canola	106 cd	129 a	40 cd	44 cde
SW	234 ab	73 b	59 cd	73 cd
WCR	0 e	131 a	15 d	15 f
WW	202 bc	80 b	69 b	69 c

[a] Total N fertilizer includes all dry and liquid N applications at pre-planting, planting, side-dress, and through fertigations during the season. Within a column, different letters are significantly different at $p < 0.05$.

297 and 210 kg N fertilizer/ha applied to the lettuce and potato, respectively, were higher than the amounts applied to canola and barley and higher than the non-fertilized winter cover rye. However, the 202 kg N/ha applied to the winter wheat was not significantly different from the amount applied to potato.

The mean initial soil NO_3^--N for the 0–0.91-m soil profile was higher for the barley, wheat, canola, and rye than the initial content in the potato and lettuce fields. Although these commercial sites were in a vegetable-small grain rotation, none of the initial soil NO_3^--N values for the small grains was collected on a lettuce-small grain rotation; thus, the initial average of 100 kg NO_3^--N/ha for small grains only reflects an initial soil NO_3^--N under a potato-small grain rotation. The 123 kg of residual NO_3^--N/ha after potato harvest is in agreement with the initial 100 kg NO_3^--N/ha observed for small grains. The mean residual soil NO_3^--N of 46 kg NO_3^--N/ha after small grain harvest is in agreement with the 30 kg NO_3^--N/ha initial for potato. After lettuce harvest residual soil NO_3^--N was significantly higher (at 222 kg NO_3^--N).

These results show that it is important to collect soil samples before small grain planting, and especially a potato- or lettuce-small grain rotation, to account for the potential available soil NO_3^--N. In particular, a significant amount of NO_3^--N can be available for deeper-rooted crops such as spring wheat, malting barley, and winter wheat that can potentially scavenge soil NO_3^--N at the 0.3 to 0.6 m and or 0.3 to 0.91 m lower depths. For the shallow varieties of small grains, deeper initial cores should be collected at 0.3 to 0.6 m, or at the very least, consideration should be given to measuring the content on the 0.30 to 0.45 m for those small grain varieties that are shallow rooted.

Significant savings in N input can potentially be gained by accounting for the initial soil NO_3^--N for small grains in a vegetable-small grain rotation. Assuming an average 100 kg NO_3^--N/ha across 29,039 and 2430 hectares of potatoes and vegetables in this region, respectively, one could save approximately 3.15×10^6 kg N fertilizer input. If a value of $0.66 per kg of N is assumed, this will equal a potential savings of \$2.1 million for this region in N input without accounting for other potential savings (e.g., cost of application, storage, transport, etc.). To get an idea of the magnitude of importance of accounting for the initial N fertilizer at planting of the small grain systems, it is potentially equivalent to the N fertilizer needed for 274 center irrigated potato pivots, or 52% of the area of potatoes planted in this region.

The NO_3^--N changes for these systems is very dynamic between planting and harvesting with a significant change in initial and final soil NO_3^--N content. Small grains significantly reduce the residual soil NO_3^--N that is available for leaching in these systems, serving as scavengers and contributing to the protection of the water resources in the region. By serving as scavenger crops and reducing the NO_3^--N content after harvesting, there will be less NO_3^--N available to leach during the non-growing season due to local thunderstorms and/or snow precipitation and melting. On the other hand, the increase in residual soil NO_3^--N under potato and vegetable systems reflects that these crops increase the potential for NO_3^--N leaching because they leave significantly higher residual soil NO_3^--N that is available to leach after harvesting. Winter cover crops can contribute to scavenging residual soil NO_3^--N from these systems and conserve water quality (Delgado, 1998).

The observed residual soil NO_3^--N below the root zone of the potato and lettuce systems has an average of 66 kg NO_3^--N/ha higher than the 6 kg NO_3^--N/ha observed in the small grains (Figure 10.4a; $p < 0.001$). This shows a greater movement of NO_3^--N below the root zone of the shallow-rooted potato and lettuce crops than with the deeper-rooted small grains such as spring or winter wheat, and/or barley, canola, and rye ($p < 0.001$).

SIMULATION OF WATER BUDGETS AND NO_3^--N DYNAMICS

Simulation of Water Budgets

The water balance is presented in Table 10.2. It is important to collect precipitation data at each site, especially for the 8288-km^2. There were differences in precipitation reflecting the spatial variability of local thunderstorms at each individual site. The irrigation and simulated evapotranspiration rates were different among crops and were higher for wheat (spring or winter grown) than for barley or canola. The lowest evapotranspiration rates were for potato and lettuce. The lower irrigation for the winter cover rye reflects its use as a scavenger crop that was killed and incorporated into the soil early during spring. The lower evapotranspiration rate for winter cover rye also reflects that it was grown during the late fall, winter, and early spring.

There were no great differences in leachate from the 0–0.91-m soil profiles between the small grains (not including winter cover crops) and the potato and lettuce; this shows that the higher N inputs for the vegetable should be the driving force for the higher rate of NO_3^--N losses from these systems. However, the deeper-rooted winter cover wheat and spring wheat had a lower mean leachate. Taking into account the higher irrigation water inputs to the small grain and normalizing the leachate as a percentage of the irrigation, the average 18.5% loss of the applied

Table 10.2   Total Irrigation, Precipitation, and Simulated Evapotranspiration and Water Leachate in the System from Lettuce, Potato, Barley, Canola, Spring Wheat (SW), Winter Cover Rye (WCR), and Winter Wheat (WW) Fields Grown under Commercial Operations

Crop[a]	Irrigation (ha·cm)	Precipitation (ha·cm)	Evapotranspiration (ha·cm)	Leachate[b] (ha·cm)
Lettuce	12.5 d	3.5 c	12.8 d	2.6 a
Potato	14.8 c	5.6 ab	18.1 c	2.4 a
Barley	17.6 b	5.7 ab	19.9 b	2.9 a
Canola	13.1 cd	6.5 a	18.8 bc	1.4 ab
SW	21.1 a	4.5 bc	24.0 a	1.9 ab
WCR	4.9 e	6.4 a	10.4 e	0.7 b
WCW	24.0 a	3.5 c	25.0 a	1.2 b

[a] Within a column, different letters are significantly different at $p < 0.05$.
[b] Leachate from the 0–0.9-m profile.

irrigation to potato and lettuce was higher than the average loss of 10.3% for the small grain (without including the winter cover rye) ($p < 0.05$). If we further normalize and use simulated evapotranspiration rates to account for variability in precipitation, the 16.7% of the leachate loss of all the water inputs for the potato and lettuce was greater than the 8.7% of the small grains (without winter cover rye)($p < 0.05$). These data show that the small grains are more effective users of irrigation and rainwater resources than the lettuce and potato crop grown in this region.

Simulation of Available Soil Water and NO_3^--N Dynamics

The new NLEAP version 1.20 simulated the different BMP effects for all these different agroecosystems and their effects on the water and N budgets (Figures 10.3 and 10.4; $p < 0.001$). The effects of BMPs on available soil water for the root zone of these shallow- and deeper-rooted agroecosystems were simulated (Figure 10.3a; $p < 0.001$). The model also simulated the effects of BMPs on NO_3^--N dynamics of these shallow- and deeper-rooted cropping systems and the movement between the root zone and below the root zone of these crops (Figures 10.3 and 10.4; $p < 0.001$).

The model simulated the NO_3^--N dynamics and changes in concentrations such as the mean increase from 30 to 100 kg NO_3^--N/ha for the potato and its partition between the root zone and below the root zone. Similarly, it also simulated the increase from 21 to 222 kg NO_3^--N/ha for the lettuce crop, as well as the capability of the small grains to serve as scavenger crops and the reduction from 100 to 46 kg NO_3^--N/ha that is potentially available to leach. For all these simulations, the b_0 and b_1 were not significant from 0 and 1, respectively ($p < 0.001$). Because the model shows a potential to simulate these increases and decreases in NO_3^--N dynamics across cropping systems and the transfer between soil horizons, it was then used as a tool to simulate the N budgets and to evaluate the NUE and effects of BMPs on NO_3^--N leaching.

The lowest NUE (31%) was for lettuce, the most heavily fertilized crop (297 kg N/ha). Although NLEAP NUE simulations account for all the N available in the systems, in relative terms approximately 93 kg N/ha of the N fertilizer applied was recovered by the crop, leaving a significant 204 kg NO_3^--N/ha available for leaching. The NUE for potato and winter wheat was higher than that of lettuce, by about 50%. These two crops were also heavily fertilized with about 206 kg N/ha but had a 20% higher NUE. One of the factors that also contributes to lowering the NUE of the winter wheat, in comparison to the potato, may have been the higher background of 95 kg NO_3^--N/ha accounted for in the irrigation water. The other small grains had NUE of about 70%, which will contribute to leaving low amounts of NO_3^--N available to leach. These simulations clearly show a pattern where the NUE is lower with the higher N rate applications and shallow-rooting crops such as potato and lettuce when compared to the other deeper-rooted crops such as the small grains.

SYSTEM SIMULATION FOR EVALUATION
OF NO_3^--N MINING POTENTIAL

Simulation of Cropping Systems NO_3^--N Leaching and Mining and Their NUE

Because NLEAP 1.20 simulated the residual soil NO_3^--N in the root zone and below the root zone as well as the changes between the initial and final soil NO_3^--N concentrations and the partition between soil horizons, it was assumed that it was also capable of simulating NO_3^--N leaching from the root zone and below the root zone of these cropping systems (Figure 10.4; Table 10.3). NLEAP version 1.20 simulated N and water budgets, NO_3^--N leaching from the root zone, and the transport between the root zone and below the root zone. These results agree with those presented by Khakural and Robert (1993), who found that NLEAP 1.10 was capable of simulating these mechanistic losses.

The simulated NO_3^--N losses due to NO_3^--N leaching from the shallow-rooted crops were 3 times higher than for the deeper-rooted small grain cropping systems (Table 10.3). The NO_3^--N losses from the 0.91-m soil depth were 2 times higher in the shallow-rooted crops of potato and lettuce than in the deeper-rooted small grains. Taking into account the balance of NO_3^--N between the amount put into the field with irrigation and NO_3^--N leaching (defined as NO_3^--N mining), it was found that the shallow-rooted crops were losing 47 and 9 kg NO_3^--N/ha from the root zone and the 0.91-m soil profile depth, respectively, while the small grains were mining 33 kg NO_3^--N/ha at this similar 0.91-m soil profile depth (Table 10.3).

The NO_3^--N leaching from the root zone (75 kg NO_3^--N/ha) was significantly higher for potato and lettuce than for other crops. However, when compared with the amount leached from the 0.91-m soil depth, it was lower (39 kg NO_3^--N/ha). The small grains mean leachate of 21 NO_3^--N/ha from the 0.91-m depth was lower than that of the potato and lettuce (39 NO_3^--N/ha) ($p < 0.05$).

Commercial sites can be compared with respect to their mining potential from the root zone and from a similar soil depth (0 to 0.91 m). By doing this, the effect of BMPs is normalized, to the management practices, on their net effect in NO_3^--N losses. However, the effect of higher or lower NO_3^--N background content

Table 10.3 Background NO_3^--N in Irrigation, Mean Simulated NO_3^--N Leaching from Root Zone and Soil Depth (0–0.9 m). Simulated NO_3^--N Mining for the Root Zone and Soil Depth (0–0.9 m), and N Use Efficiency for the System from Lettuce, Potato, Barley, Canola, Spring Wheat (SW), Winter Cover Rye (WCR), and Winter Wheat (WW) Fields Grown under Commercial Operations

Crop[a]	Background NO_3^--N in Irrigation	NO_3^--N Leached from Root Zone	NO_3^--N Leached from Soil Profile (0–0.91 m)	Mining[b] of NO_3^--N Root Zone	Mining[b] of NO_3^--N Soil Profile (0–0.91 m)	N Use Efficiency in the System[c]
	(kg NO_3^--N/ha)					(%)
Lettuce	27 c	64 a	39 a	−41 c	−12 c (c)[*.20]	31 c (d)[*.10]
Potato	33 c	86 a	38 a	−53 c	−5 c (c)	52 bc (c)
Barley	51 b	38 b	38 a	13 b	13 bc (b)	68 b (b)
Canola	34 c	26 bc	21 ab	8 b	13 bc (b)	67 b (b)
SW	52 b	38 b	31 ab	14 b	21 bc (b)	70 ab (b)
WCR[d]	37 c	0 c	0 b	37 b	37 b (b)	89 a (a)
WW	95 a	14 c	14 ab	79 a	79 a (a)	50 bc (c)

[a] Within a column, different letters are significantly different at $p < 0.05$. Letters in parentheses $p < 0.20$ and $p < 0.10$.
[b] NO_3^--N mining for the root zone = NO_3^--N in the groundwater added to the field − NO_3^--N leached from the root zone; and b) NO_3^--N mining for the soil profile (0–0.9 m) = NO_3^--N in the groundwater added to the field − NO_3^--N leached from the soil profile (0 to 0.91 m).
[c] Nitrogen Use Efficiency in the system NUE_{sys} = ((total N uptake by crop/total N available in the 0–0.9-m soil profile) × 100).
[d] Background includes 27 kg NO_3-N/ha background applied during the lettuce growing season to the lettuce crop.

can potentially affect the NUE response. One cannot assume that the uptake efficiency will be similar for the N fertilizer that is applied at planting, uphill, fertigations, and/or irrigation background. The lower efficiency of recovery will be from the N fertilizer applied at planting (Delgado et al., 1998c; Shoji et al., 2001). The efficiency of the NO_3^--N applied through fertigations will depend on the rooting depth of the crop used; because the NO_3^--N is in solution, it will be susceptible to leach-out from the root zone.

The net losses from the potato and lettuce root zone averaged 47 kg NO_3^--N/ha, while the small grain systems were capable of mining 33 kg NO_3^--N/ha from the underground water. The net losses from a similar soil depth of 0.91 m were 12 and 5 kg NO_3^--N/ha for the lettuce and potato, respectively. This simulation shows that by applying the recommended BMPs and by rotating the small grains, NO_3^--N can be mined in this region and water quality can be protected and conserved. If one accounts for the area planted in potato, vegetables, and small grains (using a mean mining of 4 kg NO_3^--N/ha), there will be a total potential for mining 0.25×10^6 kg NO_3^--N. These BMPs can potentially save another $0.17 million, or enough fertilizer for 22 circles of potato or 4.2% of the region-planted area of potato.

It cannot be assumed that 100% of all the NO_3^--N/ha that leaves the root zone will eventually make it to underground water resources. This NO_3^--N can potentially be immobilized or denitrified by microorganisms below the root zone. These simulations show that there is a significant dynamic between the NO_3^--N outputs (mining by small grains) from the aquifer and the potential NO_3^--N/ha of yearly inputs such as the potential leaching from potato and lettuce fields that may reach the aquifer. The BMPs that were put together and revised by a team composed of the SLVWQDP, USDA-NRCS-Alamosa Regional Area, USDA-ARS-Soil Plant Nutrient Research Unit, Colorado State University, CSU-Cooperative Extension, consultants, and farmers (Ristau, 1999) are implemented across the entire region. Simulations of these BMPs suggest that they are contributing to reducing the transport of NO_3^--N to the underground water resources and potentially mining NO_3^--N.

Evaluation of Active Rooting Zones

The NO_3^--N leaching was correlated significantly with the mean rooting depth and decreased as root depth increased (Figure 10.5). For these coarser sandy soils, the deeper-rooted crop provided a significant advantage that increased the availability of soil water and reduced the frequency of irrigation between applications. Additionally, the deeper rooting systems have a higher potential to scavenge and uptake NO_3^--N from deeper areas in the soil profile and will be less vulnerable to leaching events due to local thunderstorms.

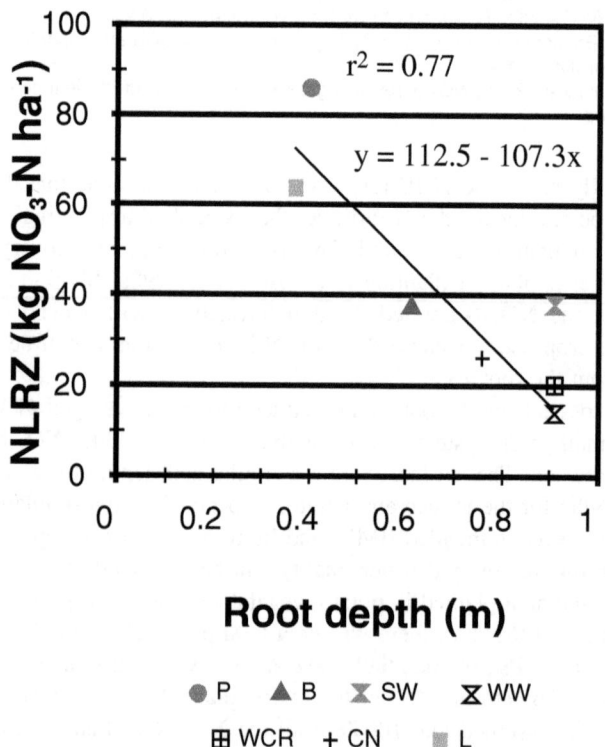

Figure 10.5   Correlation between the simulated nitrogen leached below the root zone (NLRZ) and the mean rooting depth of potato (P), barley (B), spring wheat (SW), winter wheat (WW), winter cover rye (WCR), canola (CN), and lettuce (L) grown in commercial farm operations of the San Luis Valley (* = r^2 significant at $p < 0.05$).

For shallow-rooted crops grown on sandy soils, there is not much available water in the top 35 cm of sand, loamy sand, or sandy loam textured soils. This is especially accentuated at some of the sites where water with a higher content of salts is used for irrigation. This high salinity lowers the osmotic potential around the root zone, thus reducing the available water even further for these sandy, coarser soils. This increases the need for higher frequency of irrigation between irrigation events and increases the potential for NO_3^--N leaching, especially when shallow-rooted crops are grown.

NUE, NO_3^--N leaching, and NO_3^--N mining were significantly correlated with mean root depth (Figures 10.5, 10.6, 10.7). The inclusion of small grains such as winter cover rye, winter wheat, and barley with the traditional barley-potato rotation is a significant improvement in management that has the potential to increase the system's NUE over rotations that include multiple years of potato without grain, or lettuce-potato without grain or winter cover rye. When developing and implementing BMPs that conserve water quality, the active rooting depth should be considered for these irrigated cropping systems that are grown in sandy and sandy coarse soils. These results show that with BMPs, not all the NO_3^--N that is leaving the root zone

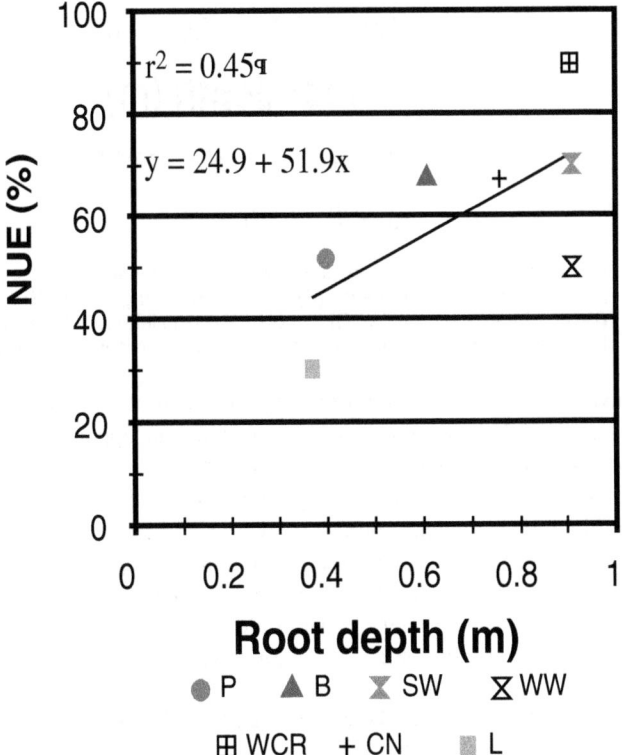

Figure 10.6 Correlation between the simulated nitrogen use efficiency (NUE) and the mean rooting depth of potato (P), barley (B), spring wheat (SW), winter wheat (WW), winter cover rye (WCR), canola (CN), and lettuce (L) grown in commercial farm operations of the San Luis Valley ($\P = r^2$ significant at $p < 0.10$).

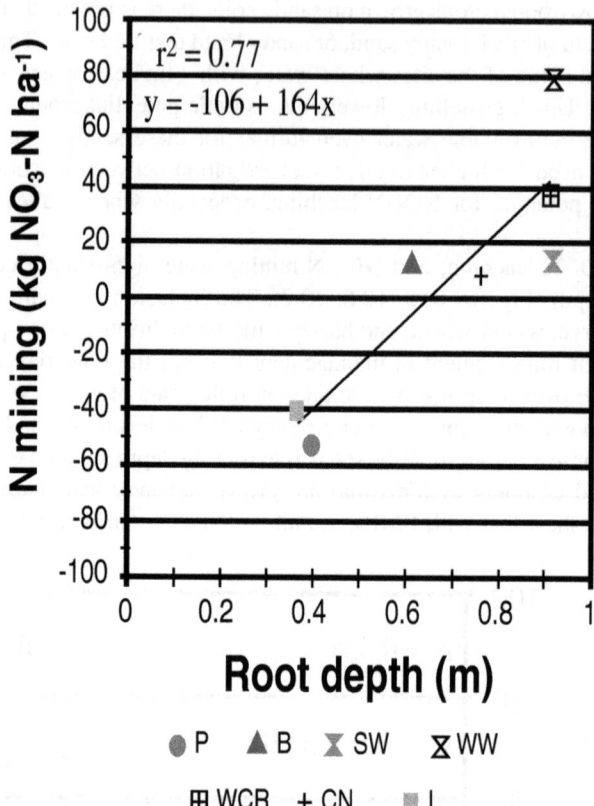

Figure 10.7 Correlation between the simulated root zone nitrogen mining potential and the mean rooting depth of potato (P), barley (B), spring wheat (SW), winter wheat (WW), winter cover rye (WCR), canola (CN), and lettuce (L) grown in commercial farm operations of the San Luis Valley. (* = r^2 significant at $p < 0.01$).

of potato leaves the root zone of the following, deeper-rooted small grain crop. This keeps losses of NO_3^--N to a minimum and provides an opportunity for the next, deeper-rooted crop to scavenge residual soil NO_3^--N.

Different morphological and physiological characteristics affect the NUE recovery and assimilation. It appears there is potential to use these physiological and morphological properties to decrease the N use index (N uptake needed to produce one unit of yield), increase the NUE, and reduce the NO_3^--N losses from these systems. Varieties with a lower N use index, a higher NUE, and a higher NO_3^--N mining potential will contribute to conserving water quality.

Selection of Crop Rotations That Increase NUE and Maximize Water Quality Conservation

The simulations of these crop rotations show that by inclusion of small grains into the rotation, these systems that are using BMPs are in balance or mining underground NO_3^--N. These results are in agreement with other researchers who also

Table 10.4   Nitrogen Use Efficiency, Simulated NO_3^--N Mining for the Root Zone and Soil Profile (0–0.91 m), of Different Rotations that Include Lettuce (L), Potato (P), Barley (B), Canola (CN), Winter Cover Rye (WCR), and Winter Wheat (WW) That Were Grown at Commercial Operations

| | | Mining[b] of NO_3^--N | |
| | N Use Efficiency in the Rotation[a] | Root Zone | Soil Profile (0–0.91 m) |
Rotation	(%)	(kg NO_3^--N/ha)	
L-L-L	31.0	−41.0	−12.0
L-L-P	38.0	−45.0	−9.7
L-P	41.5	−47.0	−8.5
P-P	52.0	−53.0	−5.0
L-L-B	43.3	−23.0	−3.7
L-P-B	50.3	−27.0	−1.3
L-B	49.5	−14.0	0.5
P-P-B	57.3	−31.0	1.0
L-WCR-P-B	60.0	−23.7	2.0
P-B	60.0	−20.0	4.0
P-CN	59.5	−22.5	4.0
L-WCR-P-WW	54.0	−1.7	24.0
P-WW	51.0	13.0	37.0

[a] Nitrogen Use Efficiency in the rotation NUE_{rot} = ((total N uptake by crop/total N available in the 0–0.9 m soil profile) × 100).
[b] NO_3^--N mining for the root zone = NO_3^--N in the groundwater added to the field − NO_3^--N leached from the root zone; and (b) NO_3^--N mining for the soil depth (0–0.91 m) = NO_3^--N in the groundwater added to the field − NO_3^--N leached from the soil depth (0–0.91 m).

found that NO_3^--N leaching losses below the root zone in irrigated sandy soils can be kept to a minimum with proper BMPs (Smika et al., 1977; Hergert, 1986; Westermann et al., 1988; Schepers et al., 1995; Thompson and Doerge, 1996a; b).Our results show that small grains are key in protecting water quality in these irrigated systems and that potato-barley or lettuce-winter cover rye rotations can contribute to significant scavenging of NO_3^--N from the underground water.

The lettuce and lettuce-potato cropping system had the lowest NUE (Table 10.4). The NUE_{rot} increased as more small grains were incorporated into the rotation (Table 10.4). The most NO_3^--N leaching-sensitive systems were the shallow-rooted systems such as the lettuce and potato rotation. By including small grains, NO_3^--N leaching is significantly reduced and potential mining increases.

The presentation and evaluation of these cropping systems are general evaluations based on the studies conducted at these commercial fields. These evaluations were based on plot and whole center-pivot simulations (Delgado et al., 2000). At some of the simulated sites, there will be the need for more precise simulations by zones. At these sites, precision farming, zone management, and the use of NLEAP simulations for evaluation of BMPs (Delgado, 1999) will significantly contribute to improving NUE when significant spatial variability in soil parameters is observed across the field.

What Happens When Significant Spatial Variability Is Found?

On average, the residual (0 to 0.91 m) soil NO_3^--N for irrigated sandy loam fields was 42, 51, and 136 kg NO_3^--N/ha for barley, canola, and potato, respectively. This was higher than the 20, 44, and 109 observed for barley, canola, and potato grown in the loamy sand, irrigated fields. The leaching of 32, 39, and 91 kg NO_3^--N/ha from the root zone of barley, canola, and potato, respectively, was higher for the loamy sand fields than the 29, 13, and 72 kg NO_3^--N/ha simulated for the sandy loam fields, respectively. These data show that loamy sand soils are more susceptible and have a higher leaching potential, especially for shallow-rooted crops.

NLEAP can potentially be used for evaluation of management zones in precision farming (Delgado, 1999). Data from Delgado (1999) is adapted and presented in Figure 10.8. These data were collected in intensively monitored plot studies located at commercial operations. For the sandy loam areas of the field, with all other management practices similar, the residual soil NO_3^--N was higher ($p < 0.05$). NLEAP was capable of simulating the effects of soil type on these sandy loam and loamy sand areas (Figures 10.9 and 10.10).

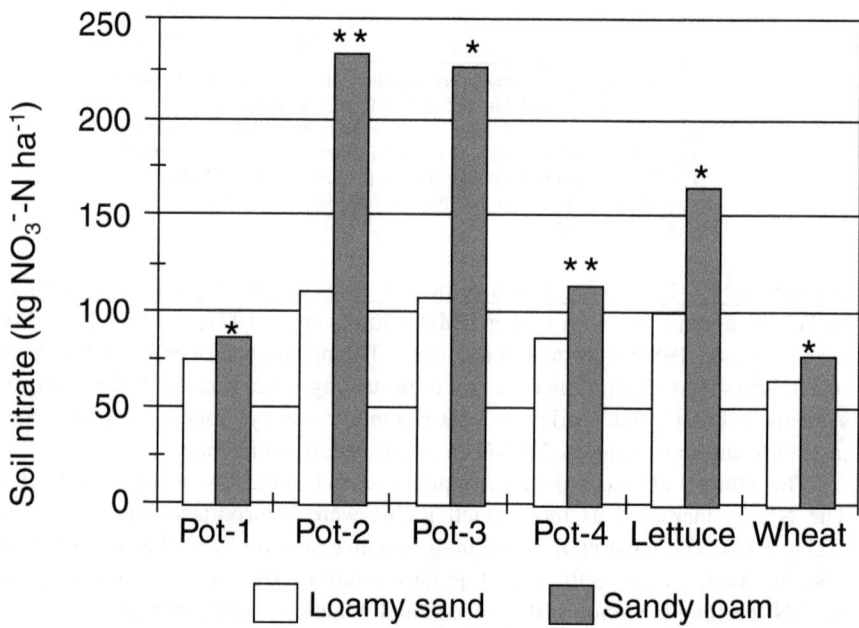

Figure 10.8   Observed residual soil NO_3^--N for the 0 to 0.91-m soil depth after harvesting of four fields of potato (Pot), lettuce, and winter wheat grown in loamy sand and sandy loams of commercial farm operations in the San Luis Valley. These data were collected in different soil type areas under an irrigated center pivot. The entire area was managed identically with the same N inputs and irrigation and management practices; the main differences were the soil types, texture, and the difference in chemical and physical characteristics between soil types. Differences in residual soil NO_3^--N were * significant at $p < 0.05$, and ** $p < 0.01$. The Pot-1 (site one) was significant at $p < 0.07$.

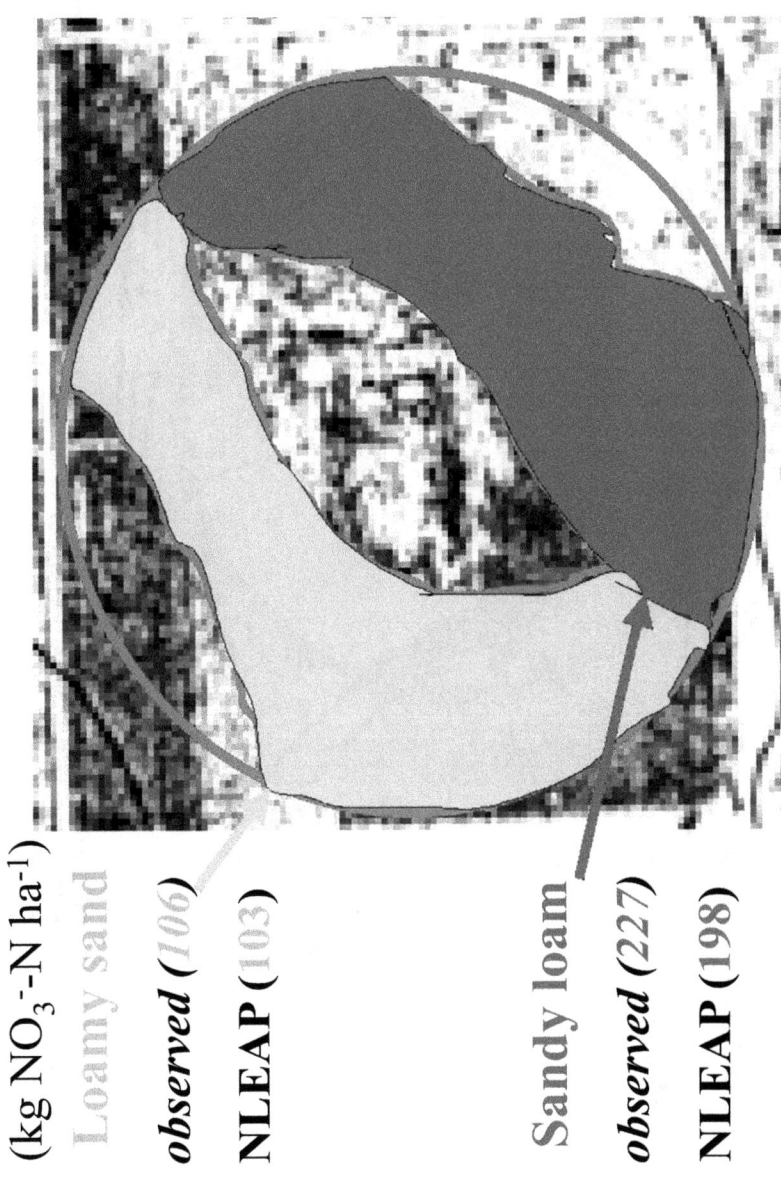

(kg NO$_3$-N ha^{-1})

Loamy sand

observed (106)

NLEAP (103)

Sandy loam

observed (227)

NLEAP (198)

Figure 10.9   Observed residual soil NO$_3$-N for the 0 to 0.91-m soil depth after harvesting of potato fields grown in loamy sand and sandy loams of commercial farm operations of the San Luis Valley. The simulated NLEAP values were not significantly different from those observed. Each site was managed uniformly through the whole circle with the recommended BMPs for lettuce. The main differences were the soil type, texture, and the difference in chemical and physical characteristics between soil types. Four plots were established and sampled at each site-soil type and were simulated.

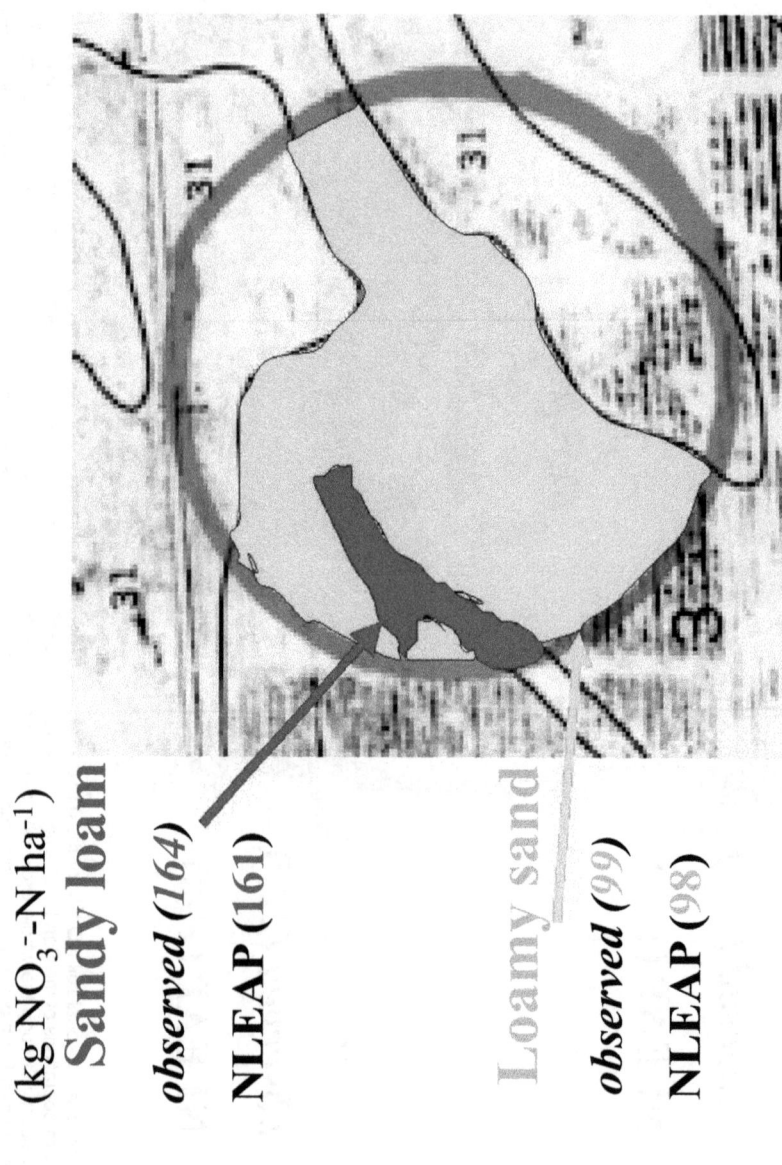

(kg NO$_3$-N ha^{-1})

Sandy loam

observed (164)

NLEAP (161)

Loamy sand

observed (99)

NLEAP (98)

Figure 10.10 Observed residual soil NO$_3$-N for the 0 to 0.91-m soil depth after harvesting of lettuce fields grown in loamy sand and sandy loams of commercial farm operations in the San Luis Valley. The simulated NLEAP values were not significantly different from those observed. Each site was managed uniformly through the whole circle with the recommended BMP's for potato. The main differences were the soil type, texture, and the difference in chemical and physical characteristics between soil types. Four plots were established and sampled at each site-soil type and were simulated.

The data show that soil type can affect the residual soil NO_3^--N in the system as well as the potential losses. When there are significant differences among soil type in a center pivot, the losses from the loamy sand can be higher than from the sandy loam. NLEAP was shown to have the potential to be used as a precision farming tool to simulate the NO_3^--N dynamics in these cropping systems and to simulate the effects of the soil type in the residual soil NO_3^--N (Delgado, 1999).

These BMPs didn't affect the product quality at these sites; however, when the spatial variability of the site was higher due to soil type, residual soil NO_3^--N, N uptake, and product quality were also different (e.g., lodging of small grains due to higher residual soil NO_3^--N was observed at the sandy loam areas). If precision farming techniques or management zones are not used at these specific sites with higher variability, it will be more difficult to improve the management at these sites, which can then contribute to lower NUE and higher NO_3^--N leaching. NLEAP can then be used to identify these areas and help manage them better (Delgado, 1999).

SUMMARY AND CONCLUSION

Irrigated cropping systems need N inputs to maintain economically optimum yields. Because NO_3^--N is a readily mobile form of N that can leach out of the root zone due to rain and/or irrigation events, it is important to use BMPs that maximize NUE and recovery; otherwise, NO_3^--N leaching can potentially create environmental problems, especially on sandier coarse-textured soils. In general, BMPs are developed from research studies conducted on small plots that are further tested at field scale. Because it is impossible to evaluate every possible combination of BMPs and commercial field scenarios and their interactions with weather, rotations, and other factors across different farmers' fields, computer model simulations offer the potential solution of evaluating the effects of these BMPs under commercial operations across different regions of the United States and other countries. The data needed for conducting these evaluations were collected on 38 commercial fields across south-central Colorado and were used to conduct over 200 simulations to evaluate the effects of management on the N dynamics and NUE, N losses, budgets, and balance between the NO_3^--N that is leaving the rooting zone and the NO_3^--N that is being pumped back with the irrigation water into these seven cropping systems.

The results reveal that NUE, NO_3^--N leaching, and N mining was significantly correlated with root depth ($p < 0.05$). When small grains such as winter cover rye, winter wheat, and barley are included in a rotation with lettuce and potato, the system's NUE is significantly increased. Additionally, with the inclusion and rotation of vegetables with small grains, the system becomes more tied and closed and reduces the NO_3^--N leaching losses, thus increasing the potential to recover NO_3^--N from the underground water. These winter cover crops and small grains also reduce the losses of nutrients, soil fine particles, organic matter, and protect soil quality (Delgado et al., 1999; Dabney et al., 2001). These simulations suggest that with adequate implementation of BMPs and the use of crop rotations, these small grain and cover crop systems will contribute to the mining of NO_3^--N from the underground aquifer, protecting environmental quality, and increasing the NUE of the agroecosystems in this region.

There is potential to further increase NUE. Breeding programs are needed that can contribute to increasing the water and NUE of these systems. In addition to increasing N and water use efficiency, and lowering use per unit of product, breeders should develop new potato and lettuce varieties that have deeper-rooted systems and higher yields that can potentially contribute to the increase of N and water use efficiency for these shallow-rooted crops, especially on sites with salinity problems. Research into the effects of other physiological and morphological parameters is needed. There is potential to develop new varieties with a lower N use index, a higher NUE, and a higher NO_3^--N mining potential that will contribute to conserving water quality.

The author's simulations of these N budgets show that for this region, when the recommended BMPs are applied in commercial operations, the system can potentially be in balance if small grains are rotated with vegetables. These simulations suggest that BMPs for this region can save up to $2.27 million in N cycling and recovery a year. These recommended BMPs have been published by Ristau (1999), are available for the farmers of this region, and are being applied in their commercial operations.

This is one of the few simulation studies conducted on a similar soil depth for shallow- and deeper-rooted crops to evaluate the N budgets and NO_3^--N dynamics of system inputs and outputs. This was possible with the new NLEAP version 1.20. Without this model simulation, it would not have been possible to compare the lettuce and winter wheat simulations on a similar basis. There is the potential application of precision farming techniques, management zones, and of more intense NLEAP simulations to further increase the NUE of these systems and further improve the N management at sites with gravel bars, high soil type variability, and with salty spot areas. NLEAP is a computer tool that can be used by scientists, agronomists, extension personnel, consultants, and farmers to evaluate BMP effects on N dynamics of irrigated cropping systems across the United States and in other countries.

REFERENCES

Beckie, H.J., A.P. Moulin, C.A. Campbell, and S.A. Brandt. 1994. Testing effectiveness of four simulation models for estimating nitrates and water in two soils. *Can. J. Soil Sci.,* 74:135-143.

Bishop, R.F., and C.R. MacEachern. 1971. Response of spring wheat and barley to nitrogen, phosphorous and potassium. *Can. J. Soil Sci.,* 51:1-11.

Carter, J.N., and D.J. Traveller. 1981. Effect of time and amount of nitrogen uptake on sugar beet growth and yield. *Agron. J.,* 73:665-671.

Cole, D.F., A.D. Halvorson, G.P. Hartman, J.E. Etchevers, and J.T. Morgan. 1976. Effect of nitrogen and phosphorous on percentage of crown tissue and quality of sugar beets. *N.D. Farm Res.,* 33(5):26-28.

Dabney, S.M., J.A. Delgado, and D.W. Reeves. 2001. Using winter cover crops to improve soil and water quality. *J. Comm. Soil Sci. Plant Anat.,* in press.

Delgado, J.A. 1998. Sequential NLEAP simulations to examine effect of early and late planted winter cover crops on nitrogen dynamics. *J. Soil Water Conserv.*, 53:241-244.

Delgado, J.A., M.J. Shaffer, and M.K. Brodahl. 1998a. New NLEAP for shallow and deep rooted crop rotations. *J. Soil Water Conserv.*, 53:338-340.

Delgado, J.A., R.F. Follett, J.L. Sharkoff, M.K. Brodahl, and M.J. Shaffer. 1998b. NLEAP facts about nitrogen management. *J. Soil Water Conserv.*, 53:332-337.

Delgado, J.A., A.R. Mosier, A. Kunugi, and L. Kawanabe. 1998c. Potential use of controlled release fertilizers for conservation of water quality in the San Luis Valley of south central Colorado. In A.J. Schlegel (Ed.). *Proc. Great Plains Soil Fertility Conf.*, Denver, DO, Kansas State University. Manhattan, KS, 151-156.

Delgado, J.A. 1999. NLEAP simulation of soil type effects on residual soil NO_3-N in the San Luis Valley and potential use for precision agriculture. In P.C. Robert, R.H. Rust, and W.E. Larson (Eds.). *Proc. 4th Int. Conf. on Precision Agriculture*. ASA, Madison, WI. 1367-1378.

Delgado, J.A., R.T. Sparks, R.F. Follett, J.L. Sharkoff, and R.R. Riggenbach. 1999. Use of winter cover crops to conserve soil and water quality in the San Luis Valley of south central Colorado. In R. Lal (Ed.). *Soil Quality and Soil Erosion*. CRC Press, Boca Raton, FL, 125-142.

Delgado, J.A., and H. Duke. 2000. Potential use of precision farming to improve nutrient management of an irrigated potato-barley rotation. In *Proc. 5th Int. Conf. Precision Agriculture*. Precision Agriculture Center Department of Soil Water and Climate University of Minnesota, St. Paul, MN, in press.

Delgado, J.A., R.F. Follett, and M.J. Shaffer. 2000. Simulation of NO_3^--N dynamics for cropping systems with different rooting depths. *J. Soil Sci. Soc. Am.*, 64:1050-1054.

Delgado, J.A., R.J. Ristau, M.A. Dillon, R.F. Follett, M.J. Shaffer, H.R. Duke, R.R. Riggenbach, R.T. Sparks, A. Thompson, L.K. Kawanabe, A. Stuebe, A. Kunugi, and K. Thompson. 2001. Use of innovative tools to increase nitrogen use efficiency and protect water quality in crop rotations. In Delgado, J.A. (Ed.). Special Issue, Potential Use of Innovative Nutrient Management Alternatives to Increase Nutrient Use Efficiency, Reduce Losses, and Protect Soil and Water Quality, *J. Comm. Soil Sci. Plant Anat.*, in press

Edelmann, P., and D.R. Buckles. 1984. Quality of Ground Water in Agricultural Areas of the San Luis Valley, South Central Colorado. Water-Resources Investigations Report 83-4281. US. Geological Survey.

Errebhi, M., C.J. Rosen, S.C. Gupta, and D.E. Birong. 1998. Potato yield response and nitrate leaching as influenced by nitrogen management. *Agronomy J.*, 90:10-15.

Follett, R.F., G.A. Reichman, E.J. Doering, and L.C. Benz. 1973. A nomograph for estimating evapotranspiration. *J. Soil Water Conserv. Soc.*, 28(2):90-92.

Follett, R.F., D.R. Keeney, and R.M. Cruse (Eds.). 1991. *Managing Nitrogen for Groundwater Quality and Farm Profitability*. SSSA, Madison, WI.

Follett, R.F., M.J. Shaffer, M.K. Brodahl, and G.A. Reichman. 1994. NLEAP simulation of residual soil nitrate for irrigated and non-irrigated corn. *J. Soil Water Conserv.*, 49:375-382.

Hergert, G.W. 1986. Nitrate leaching through sandy soil as affected by sprinkler irrigation management. *J. Environ. Qual.*, 15:272-278.

Hill, W.A. 1984. Effect of nitrogen nutrition on quality of three important root/tuber crops. In Hauck, R.D. (Ed.). *Nitrogen in Crop Production*. ASA/CSSA/SSSA, Madison, WI, 627-641.

Hills, F.T., and A. Ulrich. 1971. Nitrogen nutrition. In Johnson, R.T. et al. (Ed.). *Advances in Sugarbeet Production: Principals and Practices*. Iowa State University Press, Ames, IA, 111-115.

Jensen, M.E., R.D. Burman, and R.G. Allen (Eds.). 1990. In *Evapotranspiration and Irrigation Water Requirements*. ASCE Manuals and Reports on Engineering Practice No. 70. American Society of Civil Engineers, New York.

Khakural, B.R., and P.C. Robert. 1993. Soil nitrate leaching potential indices: using a simulation model as a screening system. *J. Environ. Qual.*, 22:839-845.

Laughlin, W.M. 1971. Production and chemical composition of potatoes related to placement and rate of nitrogen. *Am. Potato J.*, 48:1-15.

Locascio, S.J., W.J. Wiltbank, D.D. Gull, and D.N. Maynard. 1984. Fruit and vegetable quality as affected by nitrogen nutrition. In Hauck, R.D. (Ed.). Nitrogen in crop production. ASA/CSSA/SSSA, Madison, WI, 617-626.

McCracken, D.V., M.S. Smith, J.H. Grove, C.T. MacKown, and R.L. Blevins. 1994. Nitrate leaching as influenced by cover cropping and nitrogen source. *Soil Sci. Soc. Am. J.*, 58:1476-1483.

Milburn, P., J. E. Richards, C. Gartley, T. Pollock, H. O = Neill, and H. Bailey. 1990. Nitrate leaching from systematically tilled potato fields in New Brunswick, Canada. *J. Environ. Qual.*, 19:448-454.

Newbould, P. 1989. The use of nitrogen fertilizer in agriculture. Where do we go practically and ecologically? *Plant Soil*, 115:297-311.

Owens, L.B., and W.M. Edwards. 1994. Groundwater nitrate levels under fertilized grass and grass legume pastures. *J. Environ. Qual.*, 23:752-758.

Painter, C.G., R.E. Ohms, and A. Walz. 1977. The effect of planting date, seed spacing, nitrogen rate and harvest date on yield and quality of potatoes in southwestern Idaho. *Univ. of Idaho Agric. Exp. Stn. Bull.*, No. 571.

Raun, W.R., and G.V. Johnson. 1999. Improving nitrogen use efficiency for cereal production. *Agron. J.*, 91:357-363.

Ristau R.J. (Ed.). 1999. *Best Management Practices for Nutrient and Irrigation Management in the San Luis Valley*. Colorado State University Cooperative Extension, Fort Collins, CO.

Ritchie, J.T., D.C. Godwin, and S. Otter-Nacke. 1985. *CERES Wheat. A Simulation Model of Wheat Growth and Development*. Texas A&M University Press, College Station, TX.

Roberts, S., W.H. Weaver, and A.W. Richards. 1981. Sugar beets response to incremental application of nitrogen with high frequency sprinkler irrigation. *Soil Sci. Soc. Am. J.*, 45:448-449.

SAS Institute Inc. (Statistical Analysis System). 1988. *SAS/STAT Users Guide. Ver 6.03*, 3rd ed. SAS Inst., Cary, NC.

Schepers, J.S., G.E. Varvel, and D.G. Watts. 1995. Nitrogen and water management strategies to reduce nitrate leaching under irrigated maize. *J. Contaminant Hydrol.*, 20:227-239.

Shaffer, M.J., A.D. Halvorson, and F.J. Pierce. 1991. Nitrate leaching and economic analysis package (NLEAP): model description and application. In R.F. Follett, D.R. Keeney, and R.M. Cruse (Eds.), *Managing Nitrogen for Groundwater Quality and Farm Profitability*. Soil Science Society America, Madison, WI, chap. 13, 285-322.

Shaffer, M.J., B.K. Wylie, and M.D. Hall. 1995. Identification and mitigation of nitrate leaching hot spots using NLEAP-GIS technology. *J. Contaminant Hydrol.*, 20:253-263.

Shaffer, M.J., and W.E. Larson. 1987. NTRM, A Soil-Crop Simulation Model for Nitrogen, Tillage, and Crop Residue Management. USDA-ARS Conserv. Res. Rep. 34-1.

Shaffer, M.J., M.K. Brodhal, and J.A. Delgado. 1998. *NLEAP Software Version 1.20*. USDA-ARS-GPRSU, Fort Collins, CO.

Shoji, S., J.A. Delgado, A. Mosier, and Y. Miura 2001. Use of controlled release fertilizers and nitrification inhibitors to increase nitrogen use efficiency and to conserve air and water quality. In J.A. Delgado (Ed.). *J. Comm. Soil Sci. Plant Anat.*, in press.

Smika, D.E., D.F. Heermann, H.R. Duke, and A.R. Batchelder. 1977. Nitrate-N percolation through irrigated sandy soil as affected by water management. *Agron. J.,* 69:623-626.

Thompson, T.L., and T.A. Doerge. 1996a. Nitrogen and water interactions in subsurface trickle-irrigated leaf lettuce. I. Plant response. *J. Soil Sci. Soc. Am.,* 60:163-168.

Thompson, T.L., and T.A. Doerge. 1996b. Nitrogen and water interactions in subsurface trickle-irrigated leaf lettuce. II. Agronomic, economic, and environmental outcomes. *J. Soil Sci. Soc. Am.,* 60:168-173.

USDA-SCS (United States Department of Agriculture-Soil Conservation Service) 1973. *Soil Survey of Alamosa Area, Colorado.* USDA-SCS, Washington, D.C.

USDA-SCS (United States Department of Agriculture-Soil Conservation Service). *1988. National Agronomy Manual,* 2nd ed. Washington, D.C.

USEPA (United States Environmental Protection Agency). 1989. *Fed. Reg.* 54 FR 22062, 22 May. USEPA, Washington, D.C.

Wagenet, R.J., and J.L. Hutson. 1989. LEACHM: Leaching Estimation and Chemistry Model — A Process Based Model of Water and Solute Movement, Transformations, Plant Uptake and Chemical Reactions in the Unsaturated Zone. Continuum Vol. 2. Water Resour. Inst., Cornell University, Ithaca, NY.

Westermann, D.T., and G.E. Kleinkopf. 1985. Nitrogen requirements of potatoes. *Agron. J.,* 77:616-621.

Westermann, D.T., G.E. Kleinkopf, and L.K. Porter. 1988. Nitrogen fertilizer efficiencies on potatoes. *Am. Potato J.,* 65:377-386.

Williams, J.R. 1982. EPIC — A model for assessing the effects of erosion on soil productivity. *Proc. Third Int. Conf. on State of the Art in Ecol. Modeling.* Colorado State Univ., May 24-28, 1982.

Wylie, B.K., M.J. Shaffer, M.K. Brodahl, D. Dubois, and D.G. Wagner. 1994. Predicting spatial distributions of nitrate leaching in northeastern Colorado. *J. Soil Water Conserv.,* 49:346-351.

Zubriski, J.C., E.H. Vasey, and E.B. Norum. 1970. *Agron. J.,* 62:216-219.

CHAPTER 11

NLOS (NLEAP On STELLA®) — A Nitrogen Cycling Model with a Graphical Interface: Implications for Model Developers and Users

Shabtai Bittman, Derek E. Hunt, and M.J. Shaffer

CONTENTS

INTRODUCTION

Simulation models of the nitrogen (N) and carbon dynamics in the soil are needed by agrologists to help optimize system efficiency and reduce nutrient losses to the environment. Everyone concerned with the challenge of managing inputs of nutrients on the land, from farmer to scientist, mentally models the system based on their knowledge and experience to make best management decisions. Unfortunately, mental models tend to be poorly defined, are difficult to communicate, and are not easily manipulated. Gone are the days when farmers could mentally integrate the new research information into their production system (Miller and Kay, 1993). Indeed, no one person can have the understanding to integrate the vast body of knowledge to deal with the economic and environmental constraints of farming systems. Simulation models integrate the many system components at once in a systematic and quantitative way. The goal of such soil models is to explain the combined activities and processes in the soil and ultimately to enable predictions.

Most soil-N models are either empirical (also called functional) or mechanistic. The best-known empirical models are soil test correlations. In these, vast numbers of field trials are combined to predict the optimum N inputs, usually based on measurements of soil nitrate, for achieving maximum crop yield or maximum economic yield, and in some cases taking into account possible environmental outcomes. The N input recommendations are specific to crop, soil type, and may be modified according to expected moisture conditions. Empirical models are usually well accepted by practitioners because they are based on extensive field data and experience. It is noteworthy that soil test correlation results are often adjusted according to the experience of the expert recommendation committee and further modified by practitioners in the field (probably invoking their mental models). While N recommendations based on soil nitrate tests are effective for annual crops, they generally cannot be used for permanent grassland because little nitrate accumulates in the soil due to continual crop uptake. Here, predicting N requirements requires an understanding of soil processes, especially mineralization and immobilization of N, which are best described by mechanistic models.

While empirical models are effective, they are not robust. They cannot be easily transferred from one location to another, from one crop to another, nor can they easily accommodate new considerations or new information that was not included in the model development. Mechanistic models are more robust than empirical models because they describe the actual physical and biological processes in the soil-crop system.

Despite the clear need for an understanding of the soil carbon and N system, mechanistic models are often not well trusted by extension agents working with farmers and by many field-oriented scientists. To the user, these models appear to be black boxes, and the user cannot easily find out how the results were obtained (Young, 1994). Even the copious notation and acronyms are rather opaque and not consistent across models.

Because of their complexity, mechanistic models require much input data and are difficult to validate (Mary et al., 1999). Although modelers expend significant effort to obtain data to prove that their model is "valid," models can probably never be truly validated (Molina and Smith, 1998). In fact, the term "validate" may be misleading. It does not help that the applicability of a model is limited by those processes the modeler has chosen to include (Diekkruger et al., 1995). Young (1994) reflected that models tend to be used primarily by their authors, perhaps explaining the existence of a rather large number of similar models. Ironically, the body of knowledge that could be used to assess and validate models lies not with the modelers, but in the collective minds of the potential users — scientists, extension agents, and producers.

This chapter discusses how a graphical interface can help overcome some of the obstacles and increase acceptance of soil N models. Also presented is the translation of FORTRAN-based model Nitrogen Leaching and Economic Analysis Package (NLEAP) (Shaffer et al., 1991) to the graphic-based STELLA® II platform (to create NLOS (NLEAP On STELLA®).

EXPANDING USE OF N MODELS FOR PUBLIC POLICY

Although many scientists and producers mistrust models because they oversimplify (Molina and Smith, 1998), the use of nutrient flow and related models by regulators and policy-makers has been rapidly expanding for several years (Addiscott, 1993; Vinten et al., 1996). For example, Romstad et al. (1997) and Vatn et al. (1999) adapted several biophysical models, including the model for soil N dynamics from Sweden called SOILN, and a socioeconomic component called ECMOD to produce a biophysical-economic model called ECECMOD. This model was used in Norway to test the outcome of several policy scenarios such as tax levies on the use of N on grassland and their impact on the environment. A similar analysis in the United States, using the Erosion-Productivity Impact Calculator (EPIC) model, determined that a 69% surtax on N fertilizer would reduce leaching of N nationwide by an average of 19 kg/ha (House et al., 1999). EPIC has been linked with an economic model called CRAM to analyze economic and environmental impacts of policies such as crop rotations on the Canadian prairies (Izaurralde et al., 1997).

The EPIC model was originally developed by the USDA Agricultural Research Service in Temple, Texas, to assess the effect of soil erosion on soil productivity (Williams et al., 1984). The model was later expanded and refined to allow simulation of many processes important in agricultural management (Sharpley and Williams, 1990). EPIC has been widely employed both at the local level for farm planning and on a regional scale to assess environmental and agricultural policy. In 1997–1998, the USDA Natural Resource Conservation Service (NRCS) tested the EPIC model on irrigated agriculture in Idaho where nutrient loading to the lower Boise River had been identified as a water quality concern. In an unpublished report, the NRCS noted several difficulties with using EPIC in this resource setting (personal communication, T.L. Nelson, USDA-NRCS, Portland, OR):

1. There was incomplete documentation of physical processes being modeled.
2. Interactions between data inputs were unknown.
3. Plant growth and hydrologic (irrigation) processes in the model were not representative of southwestern Idaho.
4. Changes in the model and even adjustments to input parameters required direct assistance from the model developers.

The NRCS concluded that to provide the best information to farmers and ranchers, computerized models must either be applicable to a wide range of resource conditions or be easily adaptable to local conditions.

There is concern that the usage of biophysical models over the past few years has increased more rapidly than the fundamental understanding of the modeling process. Lee et al. (1996) reviewed several reports in which groundwater modeling was used to choose the best actions to remediate groundwater contamination. Mistakes in model use, found in every report, included misunderstanding the models, improper application of boundary conditions, poor estimation of input data, lack of validation or calibration, and misinterpretation of results. Of these, "inappropriate input parameters" was rated as the worst problem. This study shows an urgent need for increased education in the use of models.

Addiscott (1993) cautioned about the importance of using the models correctly, particularly where public policy is at stake. Molina and Smith (1998) indicated the importance of providing information on how widely applicable and reliable a model is, especially when the model may influence policy. Diekkruger et al. (1995) concluded that model outputs are significantly affected by the knowledge of the users and emphasized the need for good training for model users. Graphical interfaces should make dynamic models easier to understand and may encourage more people to make use of these models. No less important, the graphical interface serves well for education in systems and modeling (see below).

CONCERNS ABOUT MODELS

Is the Model Valid?

The aim of making mechanistic models robust is that they should be generally applicable and require little local calibration (Ritchie, 1983). In fact, the complexity of mechanistic models is required to make them widely applicable.

Most model developers provide experimental data in support of their model. Sometimes, models are presented despite the absence of validating data (Pang et al., 1999), but these models are less likely to be trusted. It is much better for researchers to extensively test their models to gain confidence in them and to demonstrate to others their reliability and usefulness (Qureshi et al., 1999). A model that consistently performs well may be accepted, but its validity is demonstrated only for use in particular situations (Molina and Smith, 1998). The uneven performance of models under different soil conditions was demonstrated by Vinten et al. (1996). They evaluated two N models, one based on aggregated assembly and the other on simple soil structure, for simulating water flow in a clay loam and sandy loam soil. Both models accurately simulated total nitrate leaching and denitrification in the clay loam, but neither model accurately simulated the disappearance of nitrate on the sandy loam soil. Moreover, for the sandy loam, only the aggregate assembly model simulated denitrification. Conversely, Whitmore (1995) noted a close fit between simulated and measured values of soil mineral N for a sandy but not a loam soil. Whitmore (1995) and others have pointed out that the disagreement between simulated and measured soil N values may be, in part, due to problems in getting accurate samples in the field. Diekkruger et al. (1995) concluded that it is more important to improve field measurement techniques than to develop new models. These studies show that validation in one environment does not mean that the model can be used without calibration and validation in a new environment. Validation always involves parameters that are relatively easily measured in the field. That outputs agree with field measurements gives no indication whether the internal processes within the model, such as mineralization or denitrification, have been well simulated. Sometimes, field workers have experience and knowledge that can help them assess some of these internal model processes. For example, under high rainfall conditions in south coastal British Columbia, there is evidence of unexpected patterns of uptake of applied N by a perennial grass crop, probably due to high rates of immobilization (Bittman and Kowalenko, 1998). Similarly, local field workers often have extensive experience with nitrate leaching events from measurement of water quality in wells and with runoff events from monitoring fish kills.

It may not be possible to really validate a model in the same way that a hypothesis cannot be proved (the null hypothesis is disproved) (Addiscott, 1993). Validation is more a matter of degree than a process with a clearly defined endpoint (Qureshi et al., 1999). The process continues until sufficient confidence is built up and confidence is gained as the model is improved; confidence may be built up more rapidly if there are more model testers. Because objectives vary, general rules for validation are impractical; in the end, model validity is in the eyes of the user.

The validation provided with some models is sometimes called into question because the same data set used for calibrating the model is also used for validating it (Molina and Smith, 1998). The reason for this approach is usually a dearth of useful field data. As a solution to scarcity of appropriate data for model calibration and evaluation, researchers have proposed that all field experiments should involve recording a minimum data set (Hunt, 1998). To facilitate this, they have proposed that a set of standards be used. However, more and more scientific journal articles tend to include only essential data and often display the data in graphical form, which is more difficult to apply to models than tabulated data (Kowalenko, 1997). Few papers furnish additional data in an appendix. It would appear that researchers are more likely to record additional data if they understand their importance and if they required the data for their own model. When data for validation are lacking, using the opinions of experts to gain confidence in a model, called *subjective face validity approach*, was suggested for multicriteria analysis projects by Qureshi et al. (1999). This approach helps to engage large numbers of knowledgeable specialists in the process of model validation. Engaging a larger population of model users is imperative because the majority of data and knowledge for model inputs and evaluation resides with possible model users.

Are Dependable Input Parameters Available?

Modelers often try to make their models very robust so that they can be applied under new circumstances without the need for recalibration (Ritchie, 1983). Robust models often have the drawback of being very complex and requiring a great deal of input data that are often unavailable. Accurate input data are essential to producing valid simulation output (Shayya et al., 1993).

Typical soil N models require inputs for extractable soil nitrate and ammonium concentrations, organic carbon, bulk density, and pH for each of the model soil layers. In addition, the models require inputs of crop residue with C:N ratios, and information on N inputs, including amount, time, form, etc. (Godwin and Jones, 1991). Ritchie and Dent (1994) provided lists of basic and desirable input data for running soil-crop models. They suggested that the need for detailed simulation systems that require detailed data inputs must be balanced against the availability of needed data. Inadequate input data, often soil hydraulic conductivity, was cited as a problem for running some of the models that were compared in a workshop (Diekkruger et al., 1995). It has been shown that simple functional models may perform better than complex mechanistic models because the former require fewer hard-to-measure input parameters (de Willigen, 1991). Simple models were developed for simulating N mineralization and leaching (Mary et al., 1999) and soil erosion (Greer et al., 1993) to reduce the need for input data. The NLEAP model uses archived USDA soil survey data and standard weather data as inputs. The user is required to enter initial soil conditions.

Ritchie and Dent (1994) recognized the knowledge of local experts and suggested that they be consulted for information on initial soil conditions (residue amounts, approximate water status, ammonium and nitrate concentrations) when simulating soil processes for strategic planning purposes.

Proliferation of Models (and Notation): Which Model to Use?

Molina and Smith (1998) reported that of the 200 agroecosystem models registered in the CAMASE register, 98 include components that simulate soil processes. There are at least 27 operational models for soil organic matter (Smith et al., 1996, cited in Molina and Smith, 1998; see http://yacorba.res.bbsrc.ac.uk/cgi-bin/somnet).

The number of published models suggests that it is easier to develop new models than go to the trouble and expense of first verifying and validating existing models with field experimentation and then modifying these models. It has been suggested that it is often easier to obtain research funding for model development than for conducting experiments (Diekkruger et al., 1995).

Several studies have been conducted to compare the performance of related models against a single set of field data. Such comparisons are necessarily based on easily measured parameters. In the de Willigen (1991) study, simulations were more accurate for above- than below-ground variables and none of the models simulated the mid-season loss of mineral N, probably due to ineffective simulation of microbiological processes in the soil. The simple models performed as well as the complex ones under the trial conditions but could not predict denitrification. The study of 19 models by Diekkruger et al. (1995) concluded that the simple models run by experts produced better results than complex models run by non-experts. They stressed that even one poorly estimated process can lead to general error and that while the overall results may fit experimental data, it does not mean that each process has been well described. Importantly, they observed that whether or not most of the models in their study were validated is ultimately a matter of viewpoint. In their evaluation of 24 models, Molina and Smith (1998) were surprised that models with a different number of organic matter pools, having different half-lives, and organized differently, gave simulation kinetics that closely followed measured data. They suggested that good fit does not prove good simulation of processes. Similarly, Jabro et al. (1998) reported that five models having widely different methods of simulating water content and flow through the soil profile (ranging from the Richards equation without preferential flow to the field capacity approach with preferential flow) nevertheless produced similar values for drainage flux below 1.2 m depth. In a comparison of four models for estimating nitrates and water in two prairie soils in Canada, the authors concluded that, overall, all the models performed well (Beckie et al., 1995), although enough differences were found in particular cases to leave some doubt for users requiring precision. The key to the performance of the models in these studies was calibration.

Of course, users may find it very difficult to choose the most suitable model, and it is probably safe to say that the most familiar model is often chosen. Wu and McGechan (1998) made a detailed comparison of the algorithms, equations, and parameters used in four well-known European N dynamic models (SUNDIAL from the United Kingdom, SOILN from Sweden, ANIMO from The Netherlands, and DAISY from Denmark). The authors found differences in both complexity and treatment of individual processes. SOILN had the most detailed plant growth model, and ANIMO had the most detailed simulation of animal slurries and the most mechanistic simulation of denitrification, but generally there were many more similarities than differences among these models. The authors do not explain why a different model for N dynamics was created in each country. The user is left to choose between a model that considers volatilization from fertilizer (SUNDIAL) or from slurry (ANIMO). The information in this study may help users choose which model is most suitable, but such a decision requires knowledge of these and other models. Model users would be well-served if they could select and combine the most suitable elements from any of the existing models.

ADVANTAGES OF A GRAPHICAL INTERFACE

The principal advantage of a graphical interface for soil N models is that symbols take the place of programming code. This has several consequences. Foremost is the fact that, to the user, the graphical model is transparent rather than a black box (Young, 1994). Dynamic, mechanistic soil N models are inherently very complex because of the many processes involved, but the graphical format enables non-programmers to more easily understand the structure and assumptions of the model. As discussed above, users must understand models to use them correctly. The user can reflect on the model construction and assess if the modeler has included all essential system elements (Diekkruger et al., 1995). In part, understanding of the model is gained from being able to access output data from any part of the model (see below).

When they understand the model, users can compose changes to the model using the graphical interface instead of the programming language. Useful sub-models may be translated from other published models and included or interchanged (see above).

The graphical model can be changed without obtaining the source code (which is not always available) or requiring explanations from the modeler (who may not be available). In older models written in procedural languages like FORTRAN and C++, to add new processes or pools or even make small changes usually requires extensive modification of source code, which might introduce new errors into the code that will need to be debugged. Adding soil processes that are of current interest (e.g., carbon sequestration) to an existing model may be more costly than writing an entirely new model because redesigning code in procedural languages is expensive. Worthwhile features may be lost from older models if they become obsolete through disuse.

Another important outcome of the graphical interface is that it places some of the responsibility of model validation in the hands of the user. Given that models can never be validated for every location and circumstance, the only possible solution is that users carry out the validation for their own needs and environment. With conventional models, the user is forced to decide whether to keep or discard an entire model. The graphical interface enables users to observe any flow or transformation for each day of the simulation (see "Model Output" below). Also, users can access and manipulate any of the equations or parameters in the model; they can add to or modify any algorithm in the model. Users can compare model performance with their own data as well as personal experience. In this way, the vast knowledge of non-modelers can gradually be incorporated into future versions. Conversely, the modeler can perhaps lessen the onerous and largely unappreciated task of validating the model.

TEACHING THE NITROGEN CYCLE WITH NLOS

In the Altheimer Lecture Series at Fayetteville, Arkansas, Ritchie (1983) underlined the need to strengthen interdisciplinary teaching programs in climate-plant-soil relationships. Advanced degrees have traditionally been oriented to specialty disciplines because institutional organizations and reward systems tend to encourage the reductionist approach (MacRae et al., 1989). Boyer (1990) pointed out the need to strengthen the "scholarship of integration" in both graduate and undergraduate programs. At the 1st International Crop Science Congress, Miller and Kay (1993) stated that although many different institutional arrangements have been used to foster multidisciplinary research programs, none has been particularly successful.

Even now, the use of models has largely been restricted to research and graduate-level education (Gunn et al., 1999). These writers found that models with easy-to-use interfaces, databases, and supporting material were an effective educational tool for early undergraduate education. In fact, studies on natural resource utilization and water quality protection provide a means to educate students about agricultural systems and their many interactions. Torbert et al. (1994) developed a simple model of the N cycle on the STELLA® II graphical platform for use as an educational tool in university. Their model served as a general introduction to students on the fundamental components of the N cycle. For advanced students, the model provided direct access to the model equations and information about sources of the equations and the possibility to carry out research by extending certain components and exploring alternative algorithms.

This organization is consistent with the layered structure of STELLA®. STELLA® was designed primarily to teach model building from the ground up to a diverse clientele (Richmond, 1994). However, many students understand models better by starting with a complete working model, particularly in an area of interest. The STELLA® authoring version was designed to facilitate teaching model operation and manipulation prior to model construction. In the past, STELLA® was useful primarily for teaching the concepts of modeling but STELLA®-based models could not be used for large models and data sets (Lee, 1993); however, it is evident that current versions and computers can easily handle mechanistic models of soil N.

Because of the complexity of the transformations and flows of N in the soil, the (100-page) NLOS model is not immediately clear to the user. As described below, NLOS can be used at three levels. Students at the introductory level learn from examining the components of the model with the high-level map, manipulating the inputs, and studying outputs. At this level, NLOS is essentially similar to NLEAP except that more of the activities within the model can be observed (e.g., movement of water between the layers or carbon transformations) and the output is in daily (not monthly) time steps. Students taking advanced courses in soils can explore the mechanics of the model, the stocks, flows, converters, and connectors, and learn about some of the assumptions. They can game with the model and test sensitivity. These students will gain understanding about the behavior of systems in general and the soil N system in particular. At the graduate level, students can adapt NLOS to suit specific needs, such as looking at new crops, incorporating preferential flow, considering fluctuating water tables, etc.

Knowledge of the soil N system through simulation models is no longer the domain of modelers alone but essential for everyone involved in managing nutrient resources on farms and in the environment. This knowledge is required to generate better agricultural practices and environmental policies. When many field workers and specialists are trained in soil N simulations, the quality and validity of models will greatly improve and more workers will be inclined to make use of the resource. Also, there will be fewer cases of model misuse.

BRIEF DESCRIPTION OF THE GRAPHICAL INTERFACE STELLA® II

The authors have translated NLEAP to the graphical platform called STELLA® II 5.1.1 (High Performance Systems, Inc., Hanover, New Hampshire) to produce the model called NLOS (NLEAP On STELLA®). A complete description of STELLA® II can be found in Peterson and Richmond (1994). STELLA® II, which runs almost identically in a Windows® or Apple® environment, is a multi-level hierarchical platform for constructing and manipulating models. There are two main layers: (1) the high-level mapping and input/output layer, and (2) the model construction layer.

The screen display in the model construction layer uses diagram icons to schematically represent the process being modeled; simply drawing the model diagram on the screen creates the model. The model is drawn with five main icons as described below. Each icon is identified by an assigned name that is written nearby.

1. **Stocks (rectangles).** Stocks represent quantity. They can accumulate quantity or be depleted of quantity. Stocks in NLOS include lb. NH_4-N/acre, or in. soil water in the top 12 in. of soil. Stocks may be stacked allowing, for example, carbon from each of several manure applications to be tracked separately.
2. **Flows (pipes).** Flows represent activities or movement of quantities. Flows fill and drain stocks; flows can also move quantities into and out of the system. For example, a soil process such as denitrification is an activity, and flow of water between soil layers is a movement.
3. **Converters (circles).** Converters represent either information or material quantities. They are used to modify stocks and flows. For example, a converter is used to establish the initial values of soil NO_3-N or to define the rate constant for nitrification of NH_4-N to NO_3-N.
4. **Connectors (arrows).** Connectors link stocks, flows, and converters to indicate which components influence other components. For example, the mineralization rate constant is connected to the mineralization flow to indicate that the rate of mineralization is influenced by the mineralization rate constant.
5. **Clouds (depicted as clouds).** Clouds represent sources and sinks outside the model system. For example, the atmosphere is a sink for denitrified N.

When a stock or converter is needed elsewhere in the model, the icon is "ghosted" (dotted lines) to show that the information occurs elsewhere. This simplifies the diagram by reducing the number of interconnecting lines and pipes.

Clicking on an icon brings up a menu box that contains the controlling information, in the form of an equation or data, which is inputted as a table or as a two-dimensional graph. The menu box accommodates explanations that form an integral documentation. The stock and converter names used in the diagram are the same as the variable names in the equations.

To the user, STELLA®-based models have three layers. The upper layer contains the inputs and outputs. For the NLOS model, the user can specify soil attributes, feed in weather data, input N as fertilizer or manure on particular dates, choose a crop, etc. Management information is entered into input windows (numerically or with slider bars). The output for any variable in the model can be produced as tables or graphs. Several variables can be plotted on individual graphs. The outputs can be printed or downloaded to data handling software such as Microsoft Excel or Microsoft Access®. In STELLA® Version 5.1.1, the input and output data can be dynamically exchanged with either spreadsheet or database files (via DDE in Windows and Publish and Subscribe in Macintosh®). This greatly simplifies running NLOS with different sets of weather or soil data. Dynamic data exchange allows the model output to be linked with other models that require these data as input.

The second layer of the model is the schematic representing the flow of material (NO_3, NH_4, H_2O) and information (rate controls such as the environmental stress factors) through the system. This is the diagram of the "soil N cycle" according to NLEAP. Navigation through this layer is done by scrolling or by using the search menu to find specific terms in the model. The layer appears as sheets that can be printed directly from the screen. To view the entire model at once, all pages would have to be printed and attached. The current graphical arrangement of NLOS (which essentially represents the soil N cycle with all the associated input data, parameter values, equations, algorithms, and outputs) occupies approximately 100 pages.

Every icon on the modeling layer can be double-clicked to open a dialog box that contains the equation (or constant) associated with that icon and its dependent variables. Rates can be controlled by numerical or graphical data rather than by an equation. Comments can also be added to the dialog box to give documentation relating to that constant, equation, or graphical input.

STELLA® II looks after the "bookkeeping" while the modeler designs the model graphically (with stocks, flows, converters, and connectors) and specifies constants or equations (using dialog boxes). This bookkeeping functionality generates the equation structure that is needed to define and simulate the system. Every time a stock and associated flows are added to the model, STELLA® II creates a generic equation (in differential equation format). As the equations and constants are added to the dialog boxes for stocks and flows, STELLA® II fills these details into the generic equation.

All the equations, constants, and comments exist together in the third layer of the STELLA® II modeling environment and are accessed by a simple navigation arrow on the modeling layer. This third layer is organized by sub-model structure defined on the mapping layer and icons are organized by icon type (stocks, flows, and converters) and then alphabetically within each icon type. Any icon can be found by searching for its name. Comments added to the dialog boxes in the mapping layer are associated with the icon and can be viewed in the third layer.

The large number of equations in NLOS is due mainly to identical processes operating in the different soil layers, flows between the layers, and the repetition of processes for multiple applications of fertilizer, amendments, crop residue, manure, etc.

TRANSLATING NLEAP TO STELLA® II TO CREATE NLOS

The process of translating NLEAP (http://www.gpsr.colostate.edu) to NLOS has proven to be instructive in relation to model use where the modelers and the model users are not co-located on site. The source code for NLEAP is public domain and freely available thanks to Dr. M.J. Shaffer of USDA-ARS. Some models are not in the public domain and may not be freely accessed by all potential users in the scientific community (Diekkruger et al., 1995). Can a user be certain that a model actually carries out the calculation described in the documentation?

Because there is no software available to directly transfer FORTRAN 77 code (which is the driving code of NLEAP) into the STELLA® software environment, the entire process had to be done manually. The authors began to reproduce NLEAP in STELLA® by reviewing the NLEAP model description (Shaffer et al., 1991) and running NLEAP in the DOS environment. This enabled identification of the structure of NLEAP and the overall layout of its components, and the development of a conceptual model of the N cycle in NLEAP. The cycle was reproduced by placing the stocks, flows, and controllers into the "modeling" layer of STELLA®. Using these features, the NLEAP N cycle was re-created without equations or numbers. For example, a part of the N cycle in NLEAP consists of the top soil zone which contains NH_4-N and NO_3-N pools. The second soil layer may contain only a pool of NO_3-N. Conceptually, the STELLA® model contains three stocks (where L represents the soil layer):

NH_4 L1 NO_3 L1 NO_3 L2

N flows associated with these stocks include:

- N additions from precipitation, irrigation, manure, and fertilizer
- N removal by runoff, crop uptake, volatilization, denitrification, and leaching
- N transformations by mineralization, nitrification, and immobilization

Figure 11.1 shows the three stocks with associated flows.

Flows can enter or leave the system (shown as "clouds" in STELLA®). The scope of the system can be expanded by turning the "clouds" into new stocks. In this example, stocks representing crop N-uptake, N in soil organic matter, and applied manure-N were added to the model (Figure 11.2).

After the overall N cycle was constructed in the modeling layer, the details regarding the content of each stock and flow were filled in. For example, NH_4-N nitrification in NLEAP is calculated as (Shaffer et al., 1991):

$$Nn = kn\,(TFAC)\,(WFAC)\,(ITIME) \tag{11.1}$$

where Nn   = Nitrification (lb. N/acre/day)
                      (Nn $\leq$ NH_4-N content of the top soil layer)
        kn   = Zero-order rate coefficient for nitrification (lb. N/acre/day)
        TFAC = Temperature stress factor (0 to 1)
        WFAC = Water stress factor (0 to 1)
        ITIME = Time step (days)

In NLOS, the nitrification flow is linked to controllers and connectors for kn, TFAC, and WFAC as represented in Figure 11.3. TFAC and WFAC are, in turn, defined in their own dialog boxes by temperature (Temp) and water-filled pore space (WFP).

Double-clicking on the nitrification flow icon opens a dialog box (Figure 11.4) where the equation for nitrification is entered. All the dependent variables connected to the flow icon (kn, TFAC, and WFAC) must be entered into the equation (defined) before the model can run. The ITIME term in the NLEAP equation is dealt with in STELLA® by defining the time step (days) for the entire simulation model.

Understanding unfamiliar source code was difficult, so some direct interaction with the model developers proved necessary. Because model developers know the source code best, it was fortunate that they were willing to provide help and support

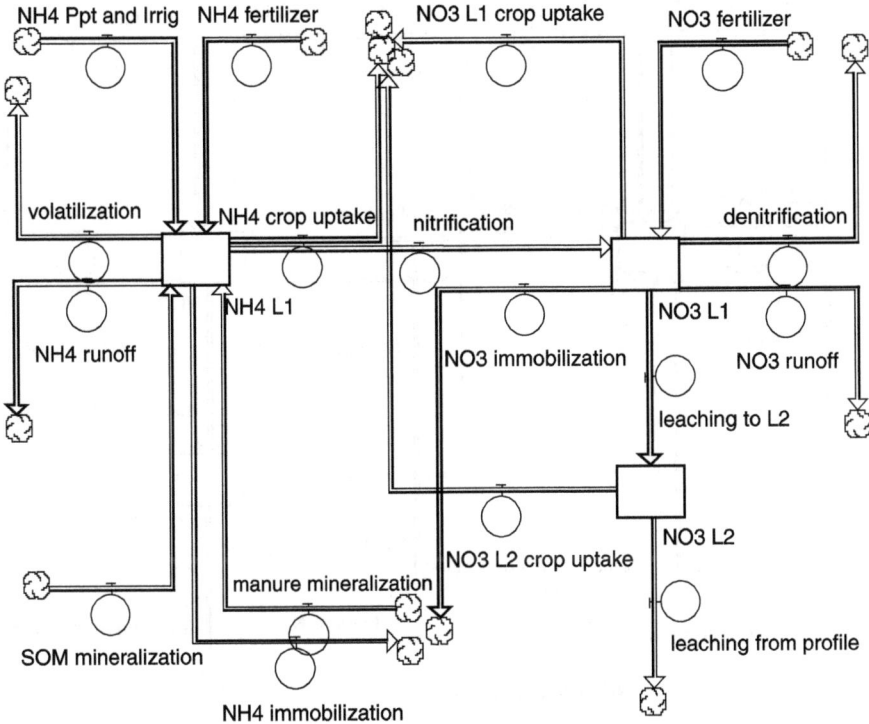

Figure 11.1   A STELLA® II model showing pools (stocks) containing NO_3-N in soil layer 1 (NO3 L1) and soil layer 2 (NO3 L2) and NH_4-N pool in soil layer 1 (NH4 L1). N flows associated with these stocks include N additions from precipitation, irrigation, manure, and fertilizer; N removal by runoff, crop uptake, volatilization, denitrification, and leaching; and N transformations by mineralization, nitrification, and immobilization. Clouds represent sources or sinks outside the system model. No connectors are shown, for simplicity.

when needed. There is a concern that knowledge of programming fades with the passage of time after any model is complete, especially if further development of the model has ceased and no funding is available to provide support for users. Ironically, the models most likely to maintain ongoing support are those that can generate revenue while their commercial success may rely on restricting access to the source code.

The strength of model building with STELLA® is the relative ease with which it can be done. Thanks to the convenience of the interface, approximately 75% of the model translation was done in about 25% of the time. This involved modeling the basic algorithms that were documented in Shaffer et al. (1991). Correctly incorporating every detail of NLEAP into NLOS required several discussions and even some on-site visits to sort out a few nagging questions. Perhaps the frustration of figuring out the models of others has stimulated many workers to develop their own versions, albeit following many similar strategies. But without fully understanding the model, users are condemned to trusting the modeler and working in partial darkness.

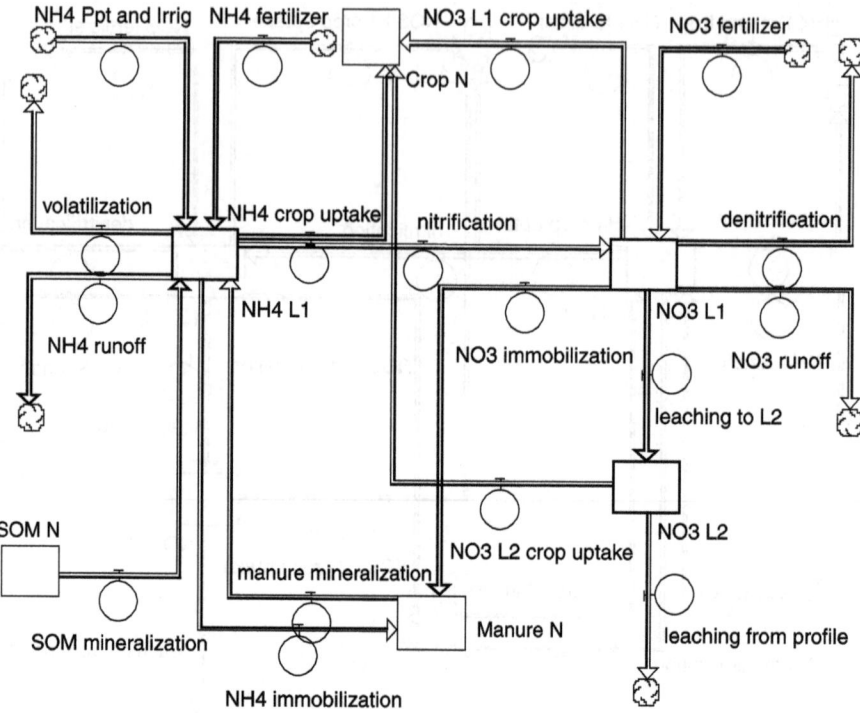

Figure 11.2 STELLA® II model shown in Figure 11.1 whose scope has been increased by changing some of the clouds (external sources and sinks) to stocks, including crop N-uptake (Crop N), N in soil organic matter (SOM N), and applied manure-N (Manure N).

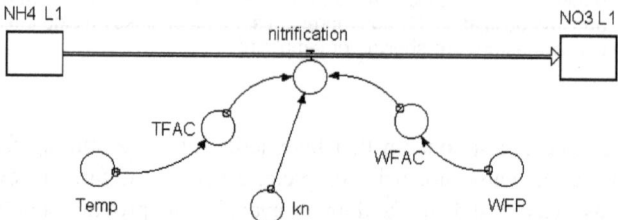

Figure 11.3 STELLA® II model of the NLEAP nitrification equation: Nn = kn (TFAC) (WFAC) (ITIME), where Nn is nitrification, kn is the zero-order rate coefficient for nitrification, TFAC is the temperature stress factor (0 to 1) calculated from temperature (Temp), WFAC is the water stress factor (0 to 1) calculated from soil water (WFP), and ITIME is the time step (days). The rate controllers (converters) are shown as circles, which are linked to the nitrification flow by connectors (arrows).

COMPARING THE OUTPUT OF NLOS AND NLEAP

Table 11.1 summarizes the annual outputs obtained from NLEAP and NLOS for different trial runs; the same parameters and input variables were used for both models for each run. For these simulations, soil data were based on the NLEAP database

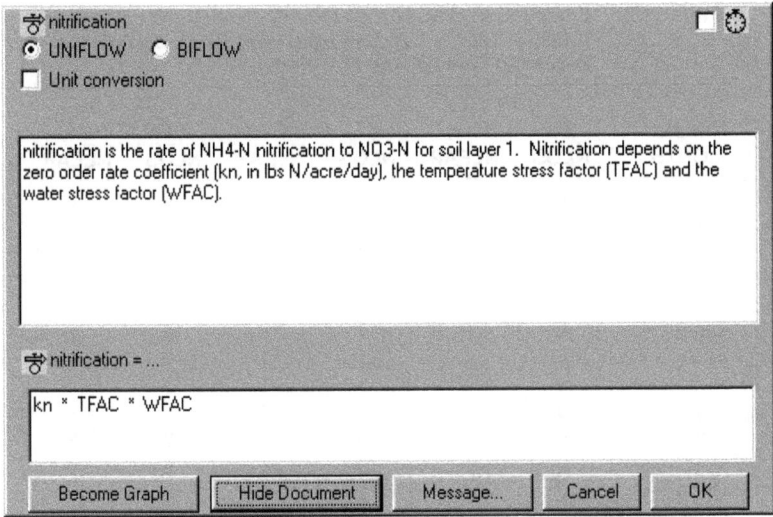

Figure 11.4   Example of a STELLA® II dialog box from the NLOS model (accessed by double-clicking the icon) that shows the nitrification equation and comment on the equation.

Table 11.1   **Comparison of Outputs from Simulation Runs with Three Rates of N Fertilizer Application on Grain Corn Using Computer Model NLEAP and Its STELLA® Version, NLOS**

Fertilizer Input	RUN 1 (100 lb. N/acre)		RUN 2 (200 lb. N/acre)		RUN 3 (300 lb. N/acre)	
	NLOS	NLEAP	NLOS	NLEAP	NLOS	NLEAP
Water summary (in.)						
Runoff	0.1	0.0	0.1	0.0	0.1	0.0
Infiltration	30.6	30.7	30.6	30.7	30.6	30.7
Evapotranspiration	10.6	10.7	10.6	10.7	10.6	10.7
Leached	18	18	18	18	18	18
Nitrogen summary (lb./acre)						
NO_3 leached	25	24	47	46	76	76
Soil organic matter mineralization	5.8	5.7	5.8	5.7	5.8	5.7
Crop residue	0.0	0.0	0.0	0.0	0.0	0.0
Manure/waste mineralization	0.0	0.0	0.0	0.0	0.0	0.0
Nitrification	53.0	52.8	93.4	93.3	131.3	130.6
Crop uptake	112	110	145	144	145	144
Runoff	0.0	0.0	0.1	0.0	0.2	0.0
Volatilization of NH_3-N	0.0	0.0	0.0	0.0	0.0	0.0
Denitrification	18.9	20.0	38.4	40.8	64.7	67.9
Residual soil NO_3	0.7	0.7	27.3	25.3	71.4	68.4

Note: Values are annual totals. Soil data were based on NLEAP database information for Adair soil in Iowa. Annual rainfall was 30.7 in. Ammonium-nitrate fertilizer was applied and incorporated on April 29 (at three rates). Disking was done on April 30. Grain corn was planted May 1 and harvested September 30; target yield was set at 120 bushels/acre. There was no input of crop residue or waste. No irrigation was applied. No N was added from precipitation. At start of simulation, soil water content was 21.0 in. and soil nitrate content was 51.2 lb. N/acre.

Table 11.2   Comparison of NLEAP and NLOS Simulations of Monthly Totals of Nitrate Leached from Iowa Grain Corn Crops Receiving Three Rates of N Fertilizer

N Input	RUN 1 (100 lb./acre)		RUN 2 (200 lb./acre)		RUN 3 (300 lb./acre)	
MONTH	NLOS	NLEAP	NLOS	NLEAP	NLOS	NLEAP
January	0	0	0	0	0	0
February	1	1	1	1	1	1
March	4	5	4	5	4	5
April	4	3	4	3	4	3
May	8	8	9	10	11	12
June	8	7	11	10	13	12
July	1	0	1	1	2	2
August	0	0	1	1	1	1
September	0	0	2	3	6	8
October	0	0	5	4	13	12
November	0	0	3	4	9	10
December	0	0	5	4	12	10

Note: See Table 11.1 for input information.

information for Adair soil in Iowa. Annual rainfall was 30.7 in. Ammonium-nitrate fertilizer was applied and incorporated on April 29 (at three rates). Disking was done on April 30. Grain corn was planted May 1 and harvested September 30; target yield was set at 120 bushels per acre. There was no input of crop residue or waste. No irrigation was applied. No N was added from precipitation. At the start of the simulations, soil water content was 21.0 in. and soil contained 51.2 lb. nitrate-N per acre.

Table 11.2 shows the monthly totals for nitrate leaching generated by NLEAP and NLOS. Input data in Table 11.2 are the same as for comparable runs in Table 11.1.

Small differences can be seen between the two models in some of the output variables, while others are equal. The discrepancies in output between NLEAP and NLOS may be due to subtle differences in methods of calculation. For example, the two models use different time steps for performing calculations. In NLEAP Version 1.13, calculations are triggered by events such as irrigation, rainfall, application of any source of N, and tillage. In contrast, NLOS recalculates daily. Calculations are based on average conditions during the period in question. For example, precipitation each day from June 1 to 11 would result in daily calculations in both models. However, if no calculation-triggering event occurred from June 1 to 11, NLEAP would average all calculations over the average conditions during the 10-day period, whereas NLOS would use average conditions for a single day. It is possible that rounding errors also contributed to the differences in results. Finally, one cannot dismiss the possibility that NLOS does not perfectly represent the algorithms in NLEAP, although such differences would likely be small. That it has proved so difficult to produce exact results demonstrates the challenge of fully understanding any model.

Most of the output variables provided by NLEAP are shown in Table 11.1. These variables are useful for model users whose primary interest is the N balance and nitrate leaching. For some users, the output may be restrictive because it does not enable detailed scrutiny of, for example, timing of events and movement of N within the soil layers. Timing of leaching associated with specific events such as rainfalls, irrigation, or nutrient applications would be more apparent in NLOS than in NLEAP.

In NLOS, any stock and flow, including inputs, within the model can be selected for graphical or tabular output in daily time steps. Several graphs can be generated, and each graph can have several parameters plotted for convenient comparisons. The tabulated outputs can be downloaded to database or spreadsheet files for further processing (e.g., linking with other models).

The usefulness and applicability of all models are inherently limited by those processes and situations that the modeler has chosen to include (Diekkruger et al., 1995). The NLEAP water flux sub-model is based on the "field capacity" approach. This approach allows neither for upward water flow nor for preferential water flow, which is important especially in soils with a high content of expanding clay (although an adjustment is made by set factors according to soil management). NLEAP has limited simulation capability for several aspects of managing liquid manure-N relating to source, pH, moisture content, and method of application. Crop choices include grain corn and alfalfa, but not forage corn or pasture. NLEAP does not grow crops; rather, the user specifies a final yield and NLEAP provides a growth curve particular to the crop specified. These are some of the aspects that NLEAP users might want to have enhanced.

LIMITATIONS OF THE STELLA®-BASED APPROACH

The runtime per simulation is determined by the amount of graphical output being generated. On a Pentium III 466 MHz with 64 Mbytes of RAM, the basic runs of NLOS took about 15 sec. When NLOS generated extensive output charts and tables (over 40 tables and graphs, 400 variables), runs took up to 45 sec. By comparison, NLEAP simulations ran almost instantaneously.

Although it has the same logical structure and algorithms as NLEAP, NLOS has some additional capabilities that contribute to the greater runtime. NLOS conducts more calculations than NLEAP because it uses daily not event-based time steps. NLOS allows each application of manure, fertilizer, or amendment to be treated as a separate pool for the entire simulation. NLOS also allows the user to either choose standard fertilizers or customize blends. NLOS also has several user-input tables allowing the user to enter only the standard inputs (as in NLEAP) or to enter input data from associated databases that can be displayed. In NLEAP, these data come from the soil database and the REGION.IDX file. Unlike NLEAP, NLOS displays internal coefficients and rate constants and allows them to be modified by the user.

While the full NLOS model runs efficiently using standard PC computers, there is a concern that adding significantly to the model will again necessitate additional computing power. Because of hardware limitations, expansion of the NLOS model must be done judiciously. To reduce runtimes, as new sub-models are added or existing sub-models elaborated, unused portions or components of the model may be easily turned off. For example, typically where manure is being added, the contribution of crop residues might be relatively insignificant, particularly in forage-based systems where only roots and short stubble remain after harvesting.

The STELLA® II software also has limitations in terms of its computational capability beyond differential equations. An example is the need to solve a set of simultaneous equations for linear optimization as part of a computer simulation. This would occur for a whole farm/livestock herd simulation where a diet must be optimized based on rations available. STELLA® II for Research ver. 5.1.1 is also limited by its inability to change time steps during a simulation. This feature is required in soil models where heavy rainfall and rapid infiltration necessitates that time steps be reduced from day to hour, or even to minutes. Future versions of STELLA® may overcome these shortcomings.

CONCLUSION

Thanks to advances in computer hardware in recent years and the sophistication of the modeling platforms (e.g., STELLA® II), graphical versions of dynamic soil-N models can be run efficiently, even while generating substantial graphical and tabular output. Therefore, the graphical interface is now more than a teaching tool; it is also a convenient platform for routinely running the model. There are several advantages to developing and running soil N models on a graphical platform: (1) there is no source code; (2) the entire model is visible — not a black box; and (3) model construction is rapid and requires little technical programming skill. In effect, the model's exterior is made more transparent so that the internal functioning can be more easily seen.

The graphical interface makes models accessible to more people involved with nutrient management issues. Non-modelers can more easily become involved in the modeling process, effecting changes and improvements according to their needs and understanding of the system. This approach involves the user in the validation process and encourages users to obtain the required input parameters, and perhaps even to develop better techniques for measuring the parameters. Knowledge of the N cycle as a system is necessary for developing advanced techniques and policies for managing nutrients.

While involvement with graphical models does not require knowledge of programming, it does require an understanding of systems, modeling, and simulation. Because of the complexity of the processes, the NLOS model may be difficult to grasp without instruction. Fortunately, NLOS itself is a useful tool for teaching the N cycle specifically, and "system thinking" in general. Students learn first by operating the model, then by manipulating it, and finally by modifying and enhancing it. Having resource managers skilled in graphical modeling will help to prevent the misuse of models, which is a growing concern as policy-makers rely increasingly on available models.

The graphical interface makes models transparent, and hence accessible, to those outside the circle of model developers. Knowledge of the model does not depend on the availability of these individuals. Using this common platform, modelers with different specialties can more easily combine their work, and modelers from different parts of the world can compare their creations by interchanging component parts. In the end, transparency is at the heart of all scientific progress.

REFERENCES

Addiscott, T.M. 1993. Simulation modelling and soil behavior. *Geoderma,* 60:15-40.

Beckie, H.J., A.P.Moulin, C.A. Campbell, and S.A. Brandt. 1995. Testing effectiveness of four simulation models for estimating nitrates and water in two soils. *Can. J. Soil Sci.,* 75:135-143.

Bittman, S., and C.G. Kowalenko. 1998. Whole-season grass response to and recovery of nitrogen applied at various rates and distributions in a high rainfall environment. *Can. J. Plant Sci.,* 78:445-551.

Boyer, E.L. 1990. Scholarship Considered: Priorities for the Professoriate. A special report of the Carnegie Foundation for the Advancement of Teaching. Princeton University Press, Lawrenceville, NJ.

de Willigen, P. 1991. Fertilizer research. Nitrogen turnover in the soil-crop system; comparison of fourteen simulation models. *Fert. Res.,* 27:141-149.

Diekkruger, B., D. Sondgerath, K.C. Kersebaum, and C.M. McVoy. 1995. Validity of agro-ecosystem models. A comparison of results of different models applied to the same data set. *Ecol. Modelling,* 81:3-29.

Godwin, D.C., and C.A. Jones. 1991. Nitrogen dynamics in soil plant systems. In J. Hanks and J.T. Ritchie (Eds.). *Modeling Plant and Soil Systems.* American Soc. Agronomy, Madison, WI, 287-321.

Greer, K.J., J.J. Schoenau, D.W. Anderson, and C.R. Hillard. 1993. Simulated productivity lost by erosion (SimPLE): model development, validation, and use. In *Proc. Soils and Crops Workshop,* Saskatoon, Sask.

Gunn, R.L., R.H. Mohtar, and B.A. Engel. 1999. Effectiveness of Internet-Based Soil and Water Quality Modeling in the Undergraduate Education. Paper No. 99-7021. Am. Soc. Agric. Engineers. St. Joseph, MI.

House, R., H. McDowell, M. Peters, and R. Heimlich. 1999. Agricultural sector resources and environmental policy analysis: an economic and biophysical approach. In *1999 Environmental Statistics: Analysing Data for Environmental Policy.* Wiley, Chichester (Novartis Foundation Symposium 220), 243-264.

Hunt, L.A. 1998. Recent attempts to evaluate and apply wheat simulation models, and to simplify storage and exchange of experimental data. In H.-J. Braun et al. (Eds.). *Wheat: Prospects for Global Improvement.* Kluwer Academic, Netherlands, 445-454.

Izaurralde, R.C., P.W. Gassman, A. Bouzaher, J. Tajek, P.G. Laksminarayan, J. Dumanski and J.R. Kiniry. 1997. Application of EPIC within an integrated modelling system to evaluate soil erosion in the Canadian prairies. In D. Rosen, E. Tel-Or, Y. Hadar, and Y. Chen (Eds.). *Modern Agriculture and the Environment.* Kluwer Academic, London, UK, 269-285.

Jabro, J.D., J.D. Toth, and R.H. Fox. 1998. Evaluation and comparison of five simulation models for estimating water drainage fluxes under corn. *J. Environ. Qual.,* 27:1376-1381.

Kowalenko, C.G. 1997. Letter to the editor re: "Is the scientific paper obsolete?" by H. H. Janzen (*Can. J. Soil Sci.,* 1996. Vol. 76 No. 4). *Can. J. Soil Sci.,* 77:331-332.

Lee, J. 1993. A formal approach to hydrological model conceptualization. *J. Hydrol. Sci.,* 38:391-401.

Lee, S.B., V. Ravi, J.R. Williams, and D.S. Burden. 1996. Evaluation of subsurface modeling application at CERCLA/RCRA sites. In J.D. Ritchey and J.O. Rumbaugh (Eds.). *Subsurface Fluid-Flow (Groundwater and Vadose Zone) Modeling,* ASTM STP 1288 American Society for Testing and Materials, Philadelphia, PA, 3-13.

Mary, B., N. Beaudoin, E. Justes, and J.M. Machet. 1999. Calculation of nitrogen mineralization and leaching in fallow soil using a simple dynamic model. *Eur. J. Soil Sci.,* 50:549-566.

MacRae, R.J., S.B. Hill, J. Henning, and G.R. Mehuys. 1989. Agricultural science and sustainable agriculture: a review of existing scientific barriers to sustainable food production and potential solutions. *Biol. Agric. Hort.,* 6:173-219.

Miller, M.H., and B.D. Kay. 1993. New approaches needed for research in sustainable cropping systems. *Intern. Crop Sci.,* 1:15-18.

Molina, J.-A.E., and P. Smith. 1998. Modeling carbon and nitrogen processes in soils. *Adv. Agron.,* 62:253-298.

Pang, H., M. Makarechian, J.A. Basarab, and R.T. Berg. 1999. Application of a dynamic simulation model on the effects of calving season and weaning age on bioeconomic efficiency. *Can. J. Anim. Sci.,* 79:419-424.

Peterson, S., and B. Richmond. 1994. STELLA® II Technical Documentation. High Performance Systems, Hanover, NH.

Qureshi, M.E., S.R. Harrison, and M.K. Wegener. 1999. Validation of multicriteria analysis models. *Agric. Systems,* 62:105-116.

Richmond, B. 1994. Authoring Module. High Performance Systems, Inc., Hanover, NH, 70 pp.

Ritchie J.T. 1983. Integrating weather, management, genetic and soil information into crop-yield models. *Proc. 1983 Ben J. Altheimer Lecture Series,* April 18-20 Fayetteville, Arkansas.

Ritchie, J.T., and J.B. Dent. 1994. Data requirements for agricultural systems research and application. In P. Goldsworthy and F.W.T. Penning de Vries (Eds.). *Opportunities, Use and Transfer of Systems Research Methods in Agriculture to Developing Countries.* Kluwer Academic, The Netherlands, 153-166.

Romstad, E., A. Vatn, L. Bakken, and P. Botterweg. 1997. Economics-ecology modelling — the case of nitrogen. In E. Romstad, E., J. Simonsen, and A. Vatn (Eds.). *Controlling Mineral Emissions in European Agriculture,* CAB International, 225-248.

Shaffer, M.J., A.D. Halvorson, and F.J. Pierce. 1991. Nitrate leaching and economics analysis package (NLEAP): model description and application. In R.F. Follett et al. (Eds.). *Managing Nitrogen for Groundwater Quality and Farm Profitability.* ASA, CSSA, and SSSA, Madison, WI, 285-322.

Sharpley, A.N., and J.R. Williams. 1990. EPIC — Erosion/Productivity Impact Calculator: 1. Model Documentation. U.S. Dept. Agric. Tech. Bull. No. 1768.

Shayya, W.H., R.D. von Bermuth, J.T. Ritchie, and H.L. Person. 1993. A Simulation Model for Land Application of Animal Manure. Paper No. 93-2012. Am. Soc. Agric. Engineers. St. Joseph, MI.

Torbert, H.A., M.G. Huck, and R.G. Hoeft. 1994. Simulation of soil-plant nitrogen interactions for educational purposes. *J. Nat. Resour. Life Sci. Educ.,* 23:35-42.

Vatn, A., L. Bakken, P. Botterweg, and E. Romstad. 1999. ECECMOD: an interdisciplinary modelling system for analyzing nutrient and soil losses from agriculture. *Ecol. Econ.,* 30:189-205.

Vinten, A.J.A., K. Castle, and J.R.M. Arah. 1996. Field evaluation of models of denitrification linked to nitrate leaching for aggregated soil. *Eur. J. Soil Sci.,* 47:305-317.

Whitmore, A.P. 1995. Modelling the mineralization and leaching of nitrogen from crop residues during three successive growing seasons. *Ecol. Modelling,* 81:233-241.

Williams, J.R., C.A. Jones, and P.T. Dyke. 1984. A modeling approach to determining the relationship between erosion and soil productivity. *Trans. ASAE,* 27:129-144.

Wu, L., and M.B. McGechan. 1998. A review of carbon and nitrogen processes in four soil nitrogen dynamics models. *J. Agric. Engng. Res.,* 69:279-305.

Young, A. 1994. Modelling changes in soil properties. In D.J. Greenland and I. Szalbocs (Eds.). *Soil Resilience and Sustainable Land Use.* CAB International, 423-447.

CHAPTER 12

NLEAP Internet Tools for Estimating NO$_3$-N Leaching and N$_2$O Emissions

M.J. Shaffer, K. Lasnik, X. Ou, and R. Flynn

CONTENTS

INTRODUCTION

Recent advances in Internet programming technology have made feasible an extended version of the Nitrate Leaching and Economic Analysis Package (NLEAP) model (Shaffer et al., 1991) that runs in a Web browser from a remote server and can easily be configured to meet user requirements. This tool uses client/server methods allowing the user interface (Java applets) to download and run on a client's computer while the NLEAP model, Geographical Information System (GIS) software, and template databases reside on remote server machines. The package can be run from the NLEAP Java Web page located at http://nleap.usda.gov/nresearch.html (Figure 12.1).

The NLEAP tool windows are programmed in Java to run as client applications in an Internet browser such as Netscape. This allows immediate access to the model over the Internet from a variety of client platforms, including Microsoft Windows,

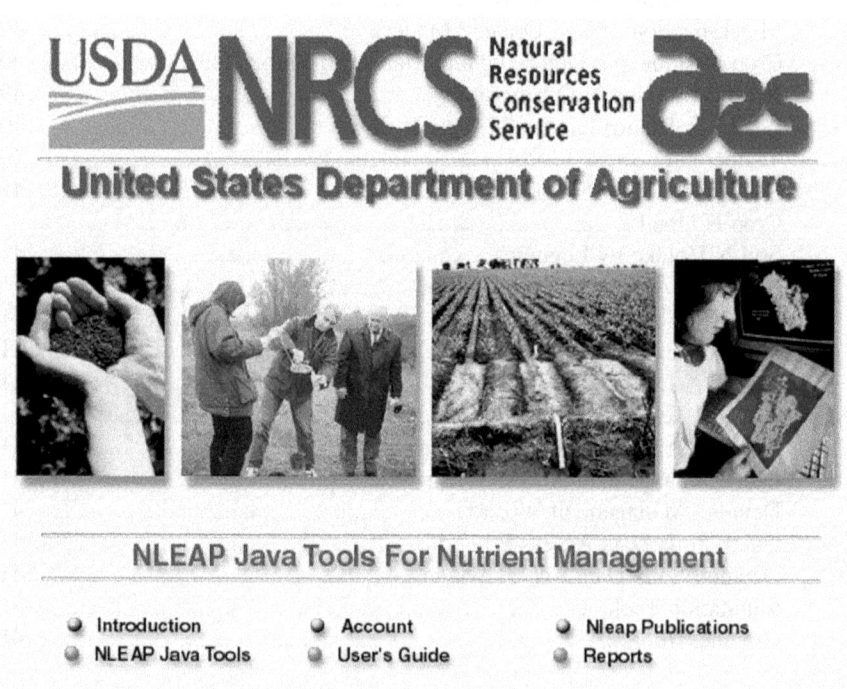

Figure 12.1   NLEAP Java Web site.

UNIX, and Apple. The simulation model is written in FORTRAN/C++ for operation on a server platform located in Fort Collins, Colorado.

Access to compatible NLEAP soils, climate, and management template databases is provided from servers on USDA, Natural Resources Conservation Service (NRCS) Web sites; and the USDA, Agricultural Research Service (ARS) NLEAP-Java Web site in Fort Collins, Colorado. NLEAP user files are derived from template files on the Fort Collins server and the client's own input, and then maintained on the server.

The above configuration structure has a number of advantages. For example, the Java tool windows can be customized for use by individual NRCS field offices and others, but it still retains the same base-window, simulation, GIS, and database servers.

The NLEAP carbon/nitrogen (C/N) cycling process source code (Shaffer et al., 1991) has been modularized and refined for use with a framework driver package written in C++ that was adapted from the Great Plains Framework for Agricultural Resource Management (GPFARM) framework (Shaffer et al., 2000). This combined package served as a starting point for development of the NLEAP simulation server. A rule-based management system has also been added that provides for on-farm event management by user-defined rules rather than by fixed-date events (Shaffer and Brodahl, 1998). Java windows development packages such as Symantec Visual Café were used to assist with development of the NLEAP interface applets. Various databases such as the NRCS Map Unit Interpretation Database (MUIR) and the NRCS climate database are now available or under development, and can be accessed from the NLEAP Java Web site to provide input suitable for use in NLEAP.

A custom GIS package (ESRI, ArcIMS) has been developed that can be run from the NLEAP interface via a custom map server, and provides for spatial visualization and mapping of farms and fields. This tool has been interfaced with the rest of the NLEAP Java package and they work together as a unit. For example, a digitized soil map layer can be sub-sampled as a farm field and the intersecting soil mapping units linked to the NLEAP management and climate databases. When the simulation is run, the user can choose to run this set of multiple scenarios and have the results (e.g., NO_3-N leached, soil residual N, etc.) displayed on the GIS map as a data layer. Other information can also be displayed on the map, including roads and towns, field boundaries, soil properties, and management scenarios.

NLEAP SIMULATION SERVER

An NLEAP model server is available that simulates soil processes in one dimension, starting with the residue cover on the soil surface and continuing down through the crop root zone to the bottom of the soil profile. Processes include infiltration and transport of soil water and nitrates; carbon and nitrogen cycling and transformations on the soil surface and within the soil profile; surface runoff of water, nitrate, and ammonium; nitrate leaching from the root zone; crop uptake of nitrate and ammonium; denitrification losses (including N_2 and N_2O); and ammonia volatilization.

As with previous versions of NLEAP, the user supplies the expected crop yields, and this information is used to distribute crop uptake of water and nitrogen over the

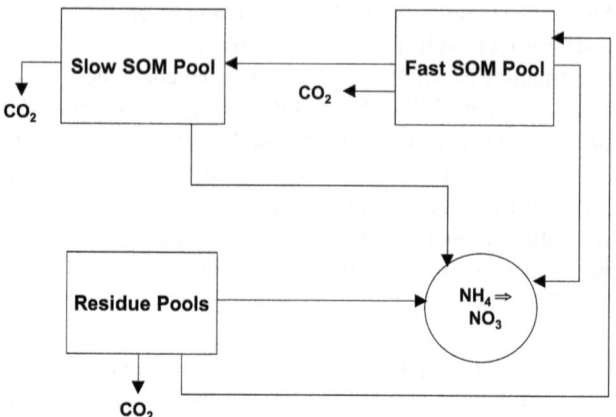

Figure 12.2 NLEAP carbon and nitrogen pools.

growing season. A version of the model with simulated crop growth responses to water and nutrient stress together with yield prediction is under development.

Sub-models for C/N Cycling Processes on the Soil Surface and Within the Soil Profile

A sub-model has been added for C/N cycling on the soil surface. This simulation accounts for decomposition of crop residues, manure, other organics, and inorganic nitrogen fertilizers that are applied to the soil surface. Decay of standing dead crop residues is handled separately from flat-lying residue decay, and an algorithm is included to convert standing to flat-lying residues. The surface sub-model also accounts for denitrification and gaseous losses of NH_3, plus surface runoff of NH_4^+ and NO_3^-.

A similar, related sub-model for residue decomposition and cycling within the soil profile uses most of the base rate equations and computer code, but includes different process rate coefficients and stress functions. With both sub-models, individual applications of organic materials are tracked from the time they enter the soil surface or soil profile until they become soil organic matter (SOM). SOM formed on the soil surface is assumed to be part of the uppermost (Ap) soil horizon. Tillage incorporates surface materials into the soil and infiltration of water moves NO_3^- into the soil.

Mineralization of Soil Organic Matter

Mineralization of SOM is simulated using a two-pool model containing a fast, readily decomposable pool and a slower humus pool (Figure 12.2). Decomposition within each of these pools is done using a first-order rate equation of the form shown in Equation (12.1).

$$\text{NOMR} = k_{omr} \times \text{SOM} \times \text{TFAC} \times \text{WFAC} \times \text{ITIME} \times 0.58/10 \qquad (12.1)$$

where NOMR is the ammonium-N mineralized (kg/ha/time step), k_{omr} is the first-order rate coefficient (fast or slow pool), SOM is soil organic matter (kg/ha), and ITIME is the size of the time step (days). The fraction of carbon in the SOM is 0.58 and the C:N ratio is 10. Factors for temperature stress (TFAC) and water stress (WFAC) are calculated using the relationships described below. Transfer from the fast to slow organic matter pools is accomplished using a transfer coefficient controllable by the user.

Crop Residue and Other Organic Matter Mineralization

Mineralization of crop residues and other organic materials such as manure is computed using the following equations.

$$CRES = fr \times RES \tag{12.2}$$

where RES represents the dry residues (kg/ha), fr is the carbon fraction of the residues, and CRES is the carbon content of the residues (kg/ha).

$$CRESR = k_{resr} \times RADJST \times CRES \times TFAC \times WFAC \times ITIME \tag{12.3}$$

where CRESR is the residue carbon metabolized (kg/ha/time step), k_{resr} is the first-order rate coefficient (per day), and RADJST is the rate adjustment factor depending on the current C:N ratio. RADJST is set equal to 1.0 at a base C:N = 25; to 2.6 at C:N = 9; to 0.29 at C:N = 100; and to 0.57 at C:N = 40. Linear interpolation is used between these points. Transfer of decayed residue material to the fast N_0 pool occurs at a C:N = 6.5 for manure and other organics, at 10 for crop residues starting at <25, and at 12 for crop residues starting at ≥25.

The residue carbon is updated after each time step using

$$CRES = CRES - CRESR \tag{12.4}$$

constrained by CRESR ≤ CRES, and net mineralization-immobilization is determined using

$$NRESR = CRESR \times (1/CN - 0.0333) \tag{12.5}$$

constrained by −NRESR ≤ NAF + NIT1 when NRESR < 0.0, where NRESR is the net residue-N mineralized (kg/ha/time step), CN is the current carbon to nitrogen ratio of the residues, and NAF is the ammonium-N content and NIT1 is the nitrate-N content of the top 30 cm (kg/ha). The N content of the decaying residues is updated after each time step using:

$$NRES = NRES - NRESR \tag{12.6}$$

constrained by NRESR ≤ NRES, and a new value for CN is computed for the next time step as:

$$CN = CRES/NRES \qquad (12.7)$$

where NRES is the N content of the crop residues, manure, or other organic wastes (kg/ha).

The mineralization of manure and other organic wastes is calculated using the same basic equation set as for crop residues given above, with manure or organic wastes substituted for crop residues.

Equations (12.2) through (12.7) assume that (1) crop residues contain a user-supplied percent carbon (manure and other organic wastes are assigned percentages based on separate user supplied analysis); (2) net mineralization/immobilization equals zero at a C:N value of 30; and (3) that the C:N value for soil microbes is 6.0. The corresponding first-order rate coefficients, k_{resr}, k_{manr}, and k_{othr}, have values depending on the material being decomposed and the current C:N values. In general, fresh materials are assigned a higher rate coefficient until a C:N value is reached where most of the faster pool has been decomposed and a lower rate coefficient is required.

In the case of surface standing dead crop residues, a conversion function is used to estimate when standing residues break off and become flat-lying on the ground. This function is driven by decay of the residue base, wind run, and tillage, and can be expressed as:

$$RESMOV = k_{till} \times (1 - RES/SSORIG) \times WINDRUN/250,000 \qquad (12.8)$$

where RESMOV is the daily fraction of the standing residue converted to flat-lying, k_{till} is a tillage coefficient (0.045 with tillage, 0.035 without tillage), RES (kg/ha) is mass of residue contacting the soil, SSORIG (kg/ha) is the mass of original fresh residue contacting the soil, and WINDRUN (km) is the total wind since the residue was fresh.

Nitrification and N$_2$O Emissions

The nitrification of ammonium-N is calculated using

$$N_n = k_n \times TFAC \times WFAC \times ITIME \qquad (12.9)$$

subject to the constraint $N_n \leq NAF$, where k_n is the zero-order rate coefficient for nitrification (kg/ha/time step), TFAC is the temperature stress factor (0–1), WFAC is the soil water stress factor (0–1), ITIME is the length of the time step (days), and NAF is the ammonium-N content of the top 30 cm (kg/ha). The use of nitrification inhibitors is simulated by reducing the magnitude of the rate coefficient, k_n. N$_2$O emissions (NN_{N_2O}) from the nitrification process are computed using the equation

$$NN_{N_2O} = N_n \times alpha \times TFAC \times WFAC \qquad (12.10)$$

where alpha is the maximum fraction of N$_2$O leakage from the nitrification process when temperature and water content are not constraining factors.

Losses to Denitrification (N_2 plus N_2O)

N lost to denitrification (N_{det}) during the time spans ending with precipitation and irrigation events is computed using the equation

$$N_{det} = k_{det} \times NIT1 \times TFAC \times [NWET + WFAC \times (ITIME - NWET)] \quad (12.11)$$

subject to the constraint $N_{det} \leq NIT1$, where N_{det} is nitrate-N denitrified (kg/ha/time step), k_{det} is the rate constant for denitrification, NIT1 is the nitrate-N content of the top 30 cm (kg/ha), and NWET is the number of days with precipitation or irrigation during the time step (for daily time steps, NWET is either 1 or 0). The value assigned to k_{det} is a function of percent soil organic matter, soil drainage class, type of tillage, presence of manure, tile drainage, type of climate, and occurrence of pans (Meisinger and Randall, 1991; Tables 5 through 7). Equation (12.11) has the advantage that maximal denitrification occurs on the wet days, while a separate estimate of deni-trification under dryer soil water conditions is made for other days.

N_2O emissions from denitrification are calculated based on extensions to Equation (12.11), (Xu et al., 1998). Emissions for wet conditions are calculated using

$$NW_{N_2O} = N_w \times alpha_w \quad (12.12)$$

where N_w is total nitrogen denitrified under wet conditions, and $alpha_w$ is the fraction of total denitrification that is N_2O under wet conditions.

For dry soil conditions, N_2O emissions are estimated using

$$ND_{N_2O} = N_d \times alpha_d \times (1 - WFAC) \quad (12.13)$$

where N_d is total nitrogen denitrified under dry conditions and $alpha_d$ is the maximum fraction of total denitrification that is N_2O at 50% water-filled pore space.

Total N_2O emissions (N_{N_2O}) are then calculated as a sum of the components:

$$N_{N_2O} = NN_{N_2O} + NW_{N_2O} + ND_{N_2O} \quad (12.14)$$

N_2 gas emissions are calculated by subtracting N_{N_2O} from N_d.

Temperature Stress Factor

The soil temperature stress factor, TFAC, is computed using an Arrhenius equation of the form

$$TFAC = 1.68E9 \times exp(-13.0/(1.99E - 3 \times (TMOD + 273))) \quad (12.15)$$

where TMOD equals $(T - 32)/1.8$ when $T \leq 86°F$, and TMOD equals $60 - (T - 32)/1.8$ when $T > 86°F$, and T is soil temperature (°F). TFAC has a range of 0.0 to 1.0. This equation was developed using data reported by Gilmour (1984) and Marion and Black (1987). Equation (12.15) approximately doubles the rate for each 18°F increase in soil temperature below a maximum of 86°F, and halves the rate for equivalent increases above the maximum.

The above equations for TFAC apply only to the soil simulation model. TFAC for use on the soil surface is calculated using a modified version of the soil equations.

Soil Water Stress Factor

The soil water factor, WFAC (range 0.0 to 1.0), is computed as a function of percent water-filled pore space (WFP) using curves fitted to data developed by Linn and Doran (1984) and Nommik (1956) for aerobic and anaerobic processes. For aerobic processes such as mineralization and nitrification:

$$\text{WFAC} = 0.0075 \times \text{WFP}, \text{ for NFP} \leq 20 \tag{12.16}$$

$$\text{WFAC} = -0.253 + 0.0203 \times \text{WFP}, \text{ for WFP} \geq 20 \text{ and } < 59 \tag{12.17}$$

$$\text{WFAC} = 41.1 \times \exp(-0.0625 \times \text{WFP}), \text{ for WFP} \geq 59 \tag{12.18}$$

and

$$\text{WFAC} = 0.000304 \times \exp(0.0815 \times \text{WFP}), \tag{12.19}$$

for anaerobic processes such as denitrification

The above equations for WFAC apply only to the soil simulation model. WFAC for use on the soil surface is calculated using a modified version of the soil equations.

Crop N Uptake

Nitrogen taken up by the crop (N_{plt}) is calculated using the following relationships.

$$N_{dmd} = YG \times TNU \times fNU \times ITIME \tag{12.20}$$

where N_{dmd} is N uptake demand (kg/ha/time step), YG is yield goal or maximum yield in appropriate units, TNU is total N uptake (kg/harvest unit), and fNU is fractional N uptake demand at the midpoint of the time step. A normalized curve relating fNU to relative crop growth stage developed by Halvorson (Shaffer et al., 1991) is used to proportion N uptake demand. The N uptake demand is proportioned between the upper and lower soil horizons according to the relative water uptake. N available for uptake in each horizon is computed as follows:

$$\text{Navail}_1 = \text{NAF} + \text{NIT1}, \text{ for the upper horizon} \tag{12.21}$$

$$\text{Navail}_{2\text{or}3} = \text{NIT2 or NIT3, respectively,}$$

$$\text{for the second and third horizons} \tag{12.22}$$

where NIT2 or NIT3 are the nitrate-N contents in the lower horizons (kg/ha). Note that a third horizon has been added as:

$$\text{Navail}_3 = \text{NIT3} \tag{12.23}$$

This three-horizon configuration provides the same capability as that contained in NLEAP version 1.2 reported by Delgado et al. (1998b).

In each case, the uptake demand for each layer is constrained by the nitrogen availability. Therefore, N_{plt} is set equal to the smaller of N_{dmd} or ($\text{Navail}_1 + \text{Navail}_2 + \text{Navail}_3$). Plant uptake of ammonium-N (NPLTA) is calculated from total N uptake in the upper 30 cm according to the fraction of nitrate-N plus ammonium-N that is ammonium-N.

Soil N Uptake by Legumes

Soil nitrogen uptake by legumes is taken as either the nitrogen demand by the crop or the sum of $\text{Navail}_1 + \text{Navail}_2 + \text{Navail}_3$, whichever is smaller. If the nitrogen demand is greater than the nitrogen available in the soil, the plant is assumed to obtain the difference from nitrogen fixation.

N Loss to Ammonia Volatilization

Nitrogen lost to ammonia volatilization (N_{NH_3}) during the same time steps discussed above is calculated using:

$$N_{NH_3} = k_{af} \times \text{NAF} \times \text{TFAC} \times \text{ITIME} \tag{12.24}$$

subject to the constraint $N_{NH_3} \leq \text{NAF}$, where N_{NH_3} is ammonia-N volatilized (kg/ha/time step), k_{af} is the rate constant for ammonia volatilization, and NAF is the ammonium-N content of the top 30 cm (kg/ha). The particular value used for k_{af} is a function of fertilizer application method, occurrence of precipitation, cation exchange capacity of surface soil, and percent residue cover (Meisinger and Randall, 1991, Table 5-6.1). In the case of manure, k_{af} is a function of the type of manure and application method (Meisinger and Randall, 1991, Tables 5-3.1 and 5-3.2).

Water Available for Leaching

Water available for leaching (WAL) is calculated after each precipitation and irrigation event using the three-layer soil model and the following relationships.

$$WAL1 = P_e - ET1 - \left(AWHC1 - S_{t1} \right) \qquad (12.25)$$

constrained by WAL1 $\geq$ 0.0

$$WAL2 = WAL1 - ET2 - \left(AWHC2 - S_{t2} \right) \qquad (12.26)$$

$$WAL = WAL2 - ET3 - \left(AWHC3 - S_{t3} \right), \qquad (12.27)$$

constrained by WAL $\geq$ 0.0, where WAL1 is water available for leaching from the top 30 cm, WAL2 and WAL3 are water available for leaching from the second and third horizons (cm), ET1 and ET2 are potential evapotranspiration associated with the top two horizons (cm/time step), AWHC1 and AWHC2 are the available water holding capacities of the upper horizons (cm), WAL is water available for leaching from the bottom of the soil profile (cm), P_e is effective precipitation (inches), ET2 and ET3 are potential evapotranspiration from the lower horizons (cm), S_{t1} is available water in the top 30 cm at the end of the previous time step (cm), AWHC2 and AWHC3 are the available water holding capacities of the second and third horizons (cm), and S_{t2} and S_{t3} are available water in these two lower horizons at the end of the previous time step.

Potential Evapotranspiration

Potential evapotranspiration is computed using pan evaporation data and appropriate coefficients as follows:

$$ET_p = EV_p \times k_{pan} \times k_{crop} \times ITIME \qquad (12.28)$$

where ET_p is potential evapotranspiration (cm/time step), EV_p is average daily pan evaporation during the time step (cm/day), k_{pan} is the pan coefficient, and k_{crop} is the crop coefficient. ET_p is proportioned between potential evaporation at the soil surface (ET_{ps}), and potential transpiration (ET_{pt}), using normalized curves for each crop. ET_{pt} is then proportioned between the upper and lower soil horizons according to the relative root distributions. Actual surface evaporation for any time step is taken as the minimum value of either ET_{ps} or the soil water available for evaporation. Actual transpiration for each time step, and soil horizon is taken as the minimum value of either the potential transpiration for that layer or the remaining soil water above the permanent wilting point. If one horizon is depleted of water, an attempt is made to extract the water from the other horizon.

Nitrate-N Leached

Nitrate-N leached, NL (kg/ha), during a time step is computed using an exponential relationship (Shaffer et al., 1991).

$$NL1 = NAL1 \times \left(1 - \exp(-1.2 \times WAL1/POR1)\right) \qquad (12.29)$$

$$NAL2 = NAL2 + NL1 \qquad (12.30)$$

$$NL2 = NAL2 \times \left(1 - \exp(-1.2 \times WAL2/POR2)\right) \qquad (12.31)$$

$$NAL = NAL3 + NL2 \qquad (12.32)$$

$$NL = NAL \times \left(1 - \exp(-1.2 \times WAL/POR3)\right) \qquad (12.33)$$

where NL1 and NL2 are nitrate-N leached from the top two horizons (kg/ha); POR1 is the porosity of the top 30 cm (cm); POR2 is the porosity of the second horizon (cm); NAL1, NAL2, and NAL3 are the nitrate-N available for leaching at the start of the time step for each horizon (kg/ha); NAL is nitrate-N available for leaching from the root zone (kg/ha); NL is nitrate-N leached from the bottom of the root zone (kg/ha); and POR3 is the porosity of the lower horizon (cm).

Total nitrate-N leached for any month or year is computed by summing the leaching obtained from each time step during the period of interest.

NLEAP MODEL TESTING

The NLEAP model has been widely used and validated in the United States, Europe, and Canada. Given proper input, the model's predicted values for residual soil nitrates and nitrate leaching rates have been shown to be within reasonable tolerances of actual values over a wide range of circumstances (Shaffer et al., 1991; Shaffer et al., 1995; NCWCD, 1990; 1991; Hoffner and Crookston, 1994; 1995; Crookston and Hoffner, 1992; 1993; Walthall et al., 1996; Beckie et al., 1995; Campbell et al., 1993; Follett et al., 1994; Wylie et al., 1994; 1995; Delgado et al., 2000; Stoichev et al., 2001; Delgado, 2001). As an example, the combined results for 150+ site-years of validation testing of NLEAP under irrigated agriculture in northeastern Colorado and in the San Luis Valley of south-central Colorado are shown in Figure 12.3. Additional details regarding the application of the NLEAP model to field situations and to cases of climatic variability can be found in Delgado et al. (1998a) and Shaffer et al. (1994).

THE NLEAP INTERNET JAVA INTERFACE (BETA VERSION 1.0)

Users must establish an individual account on the Fort Collins NLEAP Java server or use the existing guest account under user name **beta** and with password **xxxxxxxx** (available from the Web site administrator). Currently, all files are maintained on the server, but the capability to also download and store user data files

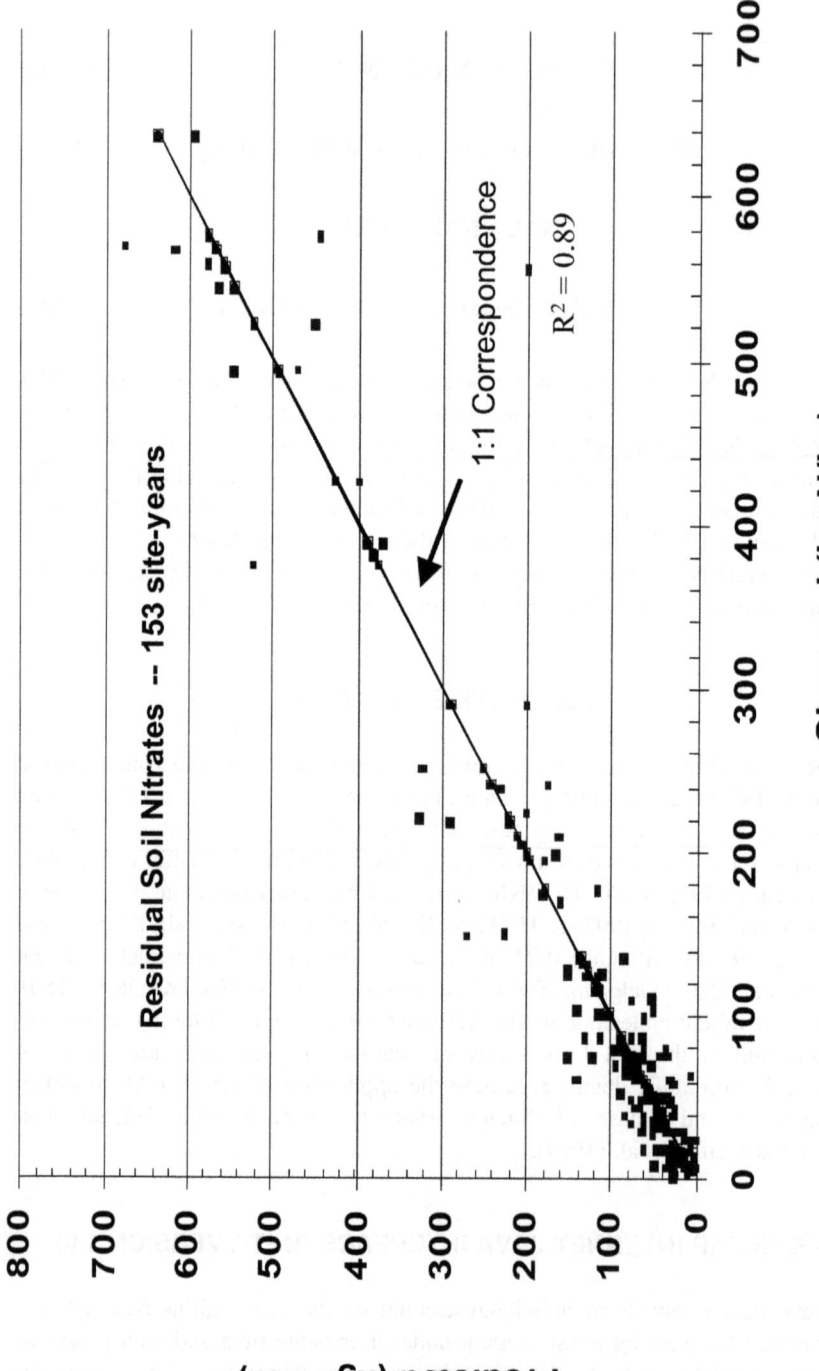

Figure 12.3 NLEAP simulated vs. observed residual soil nitrates.

onto individual machines may be provided in the future. Files created on the beta account on the Fort Collins server are currently being maintained on the site.

The primary Web site page contains a site introduction, user's guide, and account information, along with a list of NLEAP publications and project reports. The account option provides a form for users to request an individual account on the server.

As previously noted, entry to the NLEAP Java applets Web page is through the NLEAP Java Tools button and requires an account name and password. A series of applet tools is available to provide support for the NLEAP model and associated databases and servers. These include farm management and configuration tools, soils data download and formatting tools (for NRCS MUIR and climate databases), an NLEAP simulation tool, a 2-D graphics tool, an ArcIMS GIS tool, a rule-based editor tool, and some additional support tools. These tools are linked together behind the scenes and function as a coordinated unit. The system provides established template databases for a range of typical management scenarios, soil data, climate data, and GIS coverages. Users can easily develop custom databases and map files for their own local areas and conditions.

Once the Java applet Web page is displayed, access to the Quick and Detailed Management, Soils, and Climate Data Wizards are provided from buttons at the top of the screen. Access to tools such as the Simulation, Graphics, and GIS server applets is located along the left side of the page. By clicking the left mouse button over these controls, the appropriate Java applet is downloaded from the server and appears as a floating tool window that can be moved and positioned anywhere on the screen; or it can be iconized at the bottom of the main window for rapid recall.

Quick Management Access Wizard

This tool contains a series of menus that access predefined management scenarios for combinations of local crop rotations, tillage, fertilizer applications, water source rain-fed/irrigated, and irrigation type (Figure 12.4). The exact content of these menus can be pre-set in configuration files by individual users for use on their account. This gives the user maximum flexibility in selecting the types of management scenarios available on a routine basis. Additional refinement of each scenario can be made using the Detailed Management Wizard tool (see below). These refinements can include minor changes in management amounts and types, as well as major modifications such as additional management events or deletion of events. Once a management scenario is selected, a corresponding management code is saved in a file for future reference and use in the NLEAP simulation.

Detailed Management Wizard

This applet (Figure 12.5) provides access to the event database containing the details of all the management events for each scenario over approximately a 10-year period. Several event databases can occur on the user's account and each database can contain many scenarios. The scenario code is used to select a particular management time sequence of interest, while the management type code narrows the selection to a particular management group for that scenario. These groups include

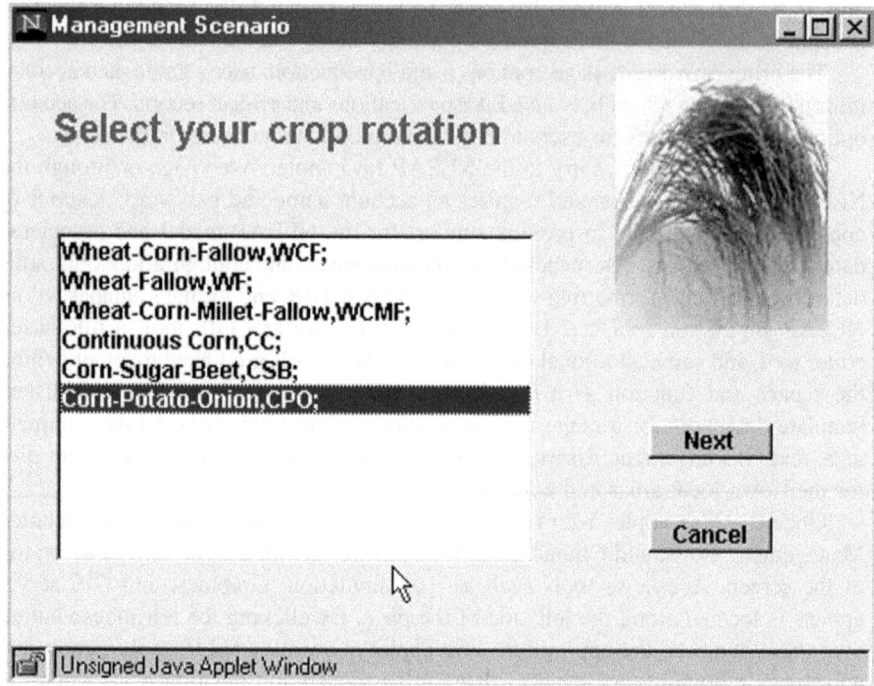

Figure 12.4　Quick Management Access applet.

crop information, fertilizer, manure, tillage, and irrigation. Once a scenario code and management group have been selected, the Next and Previous buttons are used to preview each management event in the database for that category. Management events can be edited or deleted; or new events can be added.

The Management Scenario drop-down menu contains a list of codes for management scenarios currently available in the database. For example, the code shown in Figure 12.5, WF; NT; SA; RF; indicates a scenario for a winter wheat-fallow rotation; no-till tillage; a single application of fertilizer; and rain-fed conditions. The event type NUTC means commercial fertilizer applications, while other available event types include PLNT (planting and crop information), HARV (harvest information), NUTM (manure applications), TILL (tillage), and IRR (irrigation). Once an Event Type and Management Scenario have been selected, the Result View button is used to query the database and display the first record in that group. The Add and Save buttons allow new records to be added and changes in existing records to be saved, respectively.

Quick Soils Data Access Wizard

This applet provides a list of soil types common to the local area or county (Figure 12.6). This provides for rapid selection of the soil type for use in the simulation. The soil selected is automatically saved for use in the simulation. The user has control over soil types included in the Quick Soils Data applet list and can change

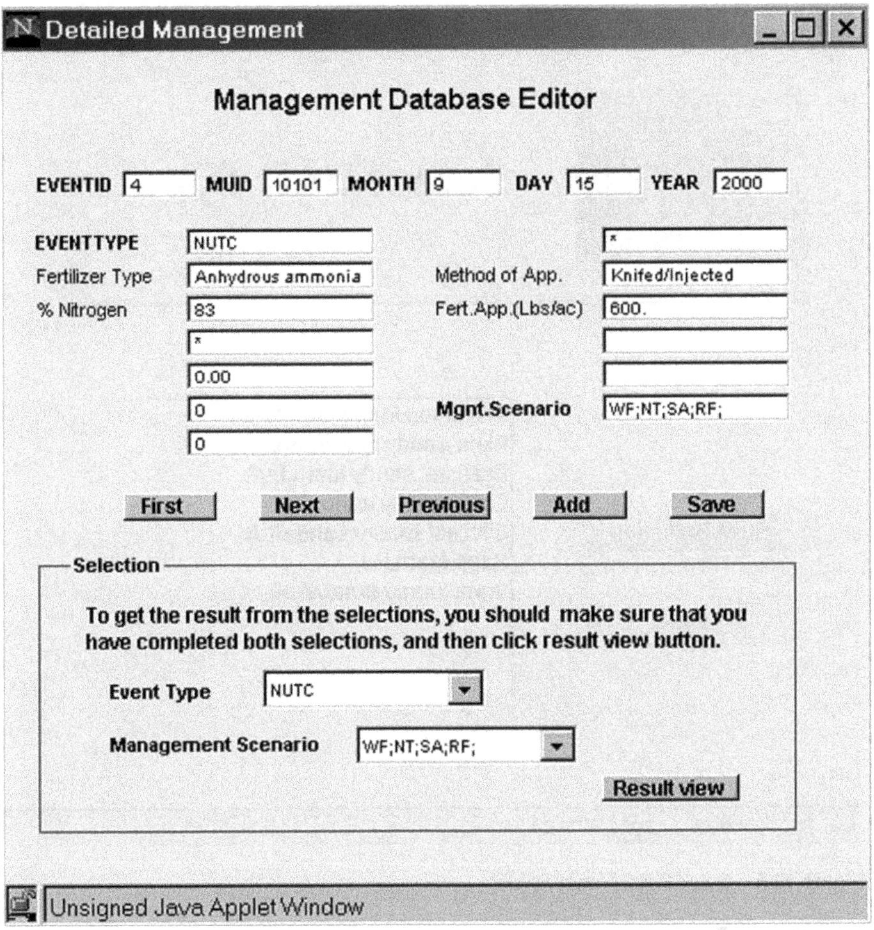

Figure 12.5 Detailed Management applet.

the list's content when needed. In general, the short list is derived from a download of a longer list for a soil survey area (or county) on the MUIR Web site. The short list might also be derived from intersecting a field map with the soils GIS map layer.

Detailed Soils Data Access Wizard

This tool provides access to the soil layer database for the soil survey area (Figure 12.7). Data contained for each soil type and layer in the soil survey area can be viewed and edited. All soil types are made available for selection and use in the simulation, including those not found on the abbreviated soils data list. If local soils data are available, this information can be added to the database. This includes changes to existing records (e.g., updated values for OM%, initial soil NO_3-N and water content, etc.). Saved changes to the form will permanently alter the user's copy of this database, so caution is advised.

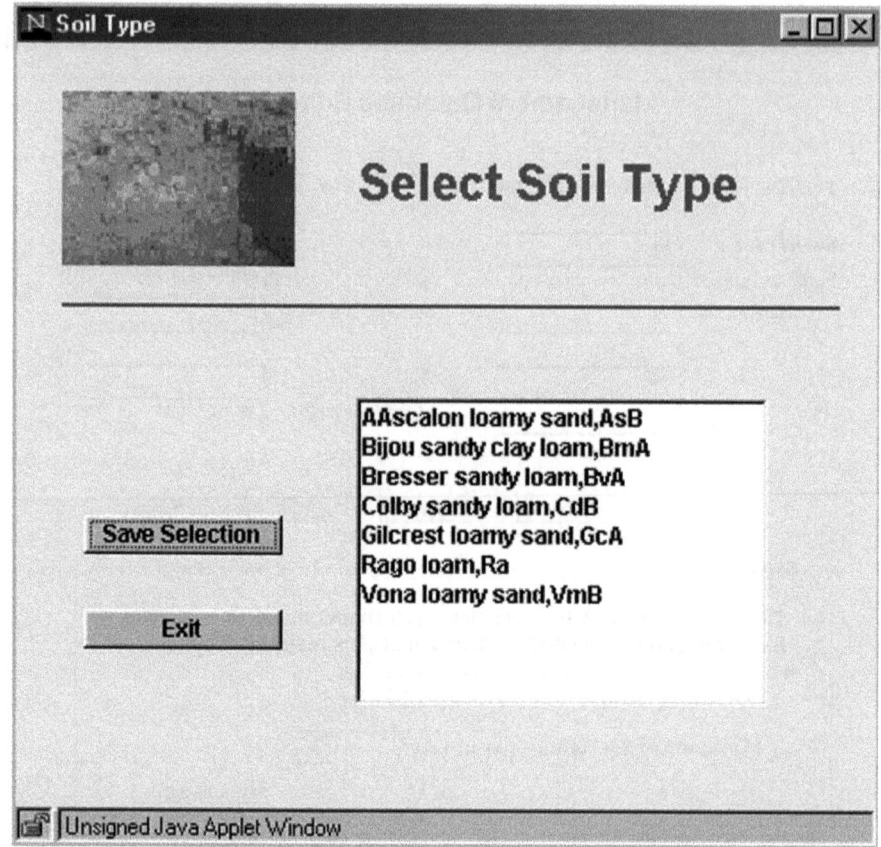

Figure 12.6   Quick Soils Access applet.

Simulation Tool

This applet provides access to the NLEAP simulation server (Figure 12.8). The user must select a management database from the drop-down list and a file name for a previously defined management scenario. Soils data are used according to the previous soil type selection. Once a long-term simulation has been completed, results are saved to a file that the user specifies and status messages are provided as feedback during the simulation both in the error message window of the simulation applet and in an ASCII file named *message.fil* on the user's account.

Graphics Tool

This applet creates and displays bar graphs of selected output from the NLEAP model (Figure 12.9). These include graphs of monthly results for nitrate-N leached, residual nitrate-N in the soil profile, and crop uptake of N similar to that shown in Figure 12.10. The graphs can be saved to files for later reference, or they can be iconized for later recall to the screen during an Internet session.

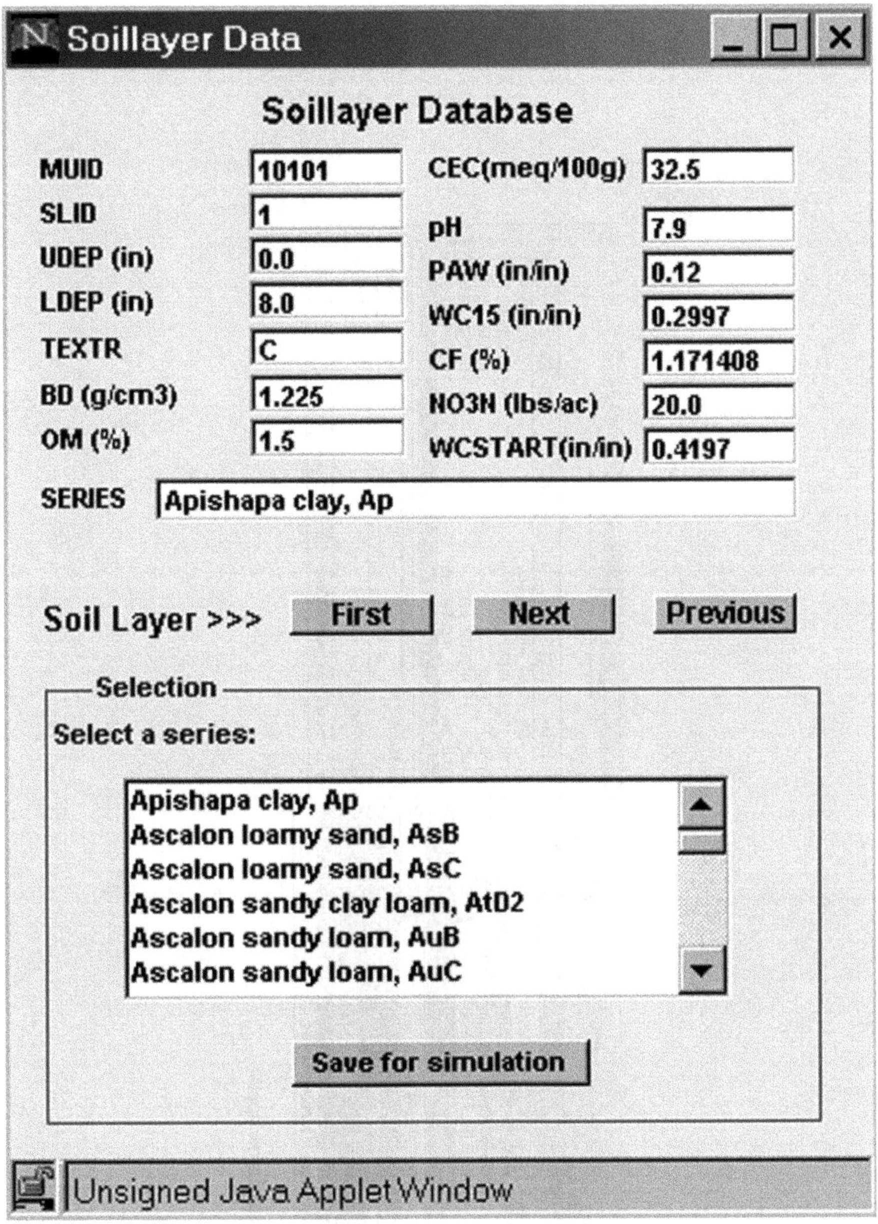

Figure 12.7 Detailed Soil Layer Editor.

GIS Tool

This applet runs the project's ESRI ArcIMS server in Fort Collins (Figure 12.11). The Java interface and server have been customized for use with the NLEAP Java tools package. A linkage mechanism allows the user to assign soil or other management

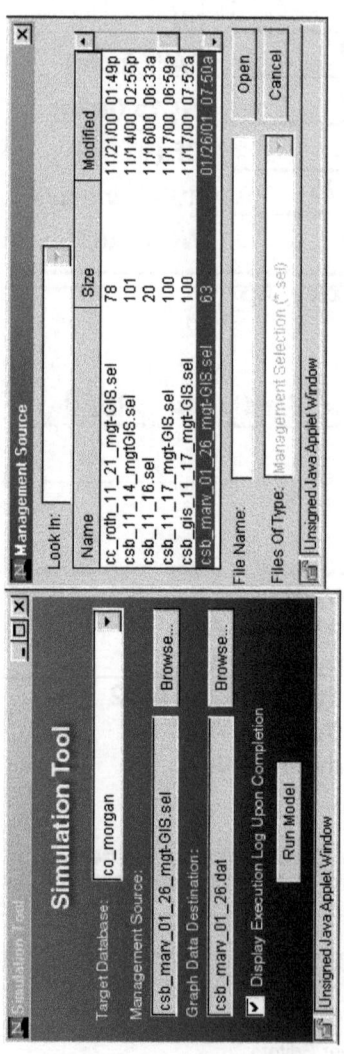

Figure 12.8 NLEAP Simulation applet.

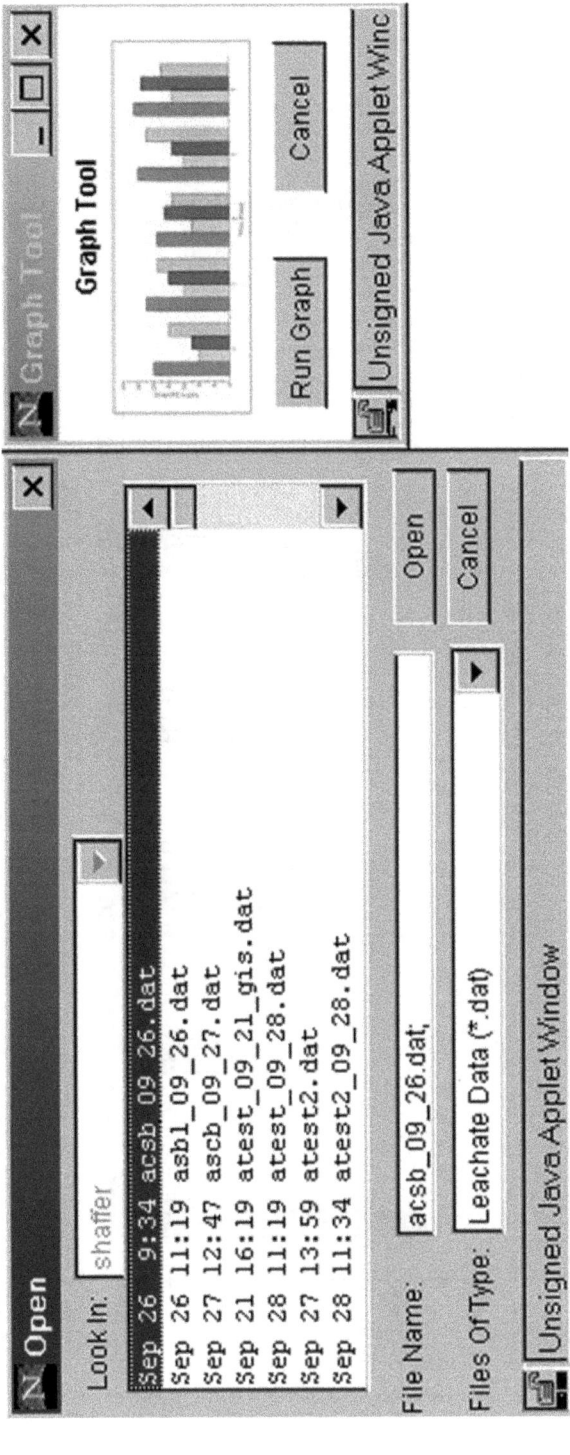

Figure 12.9 Graphics tool applet.

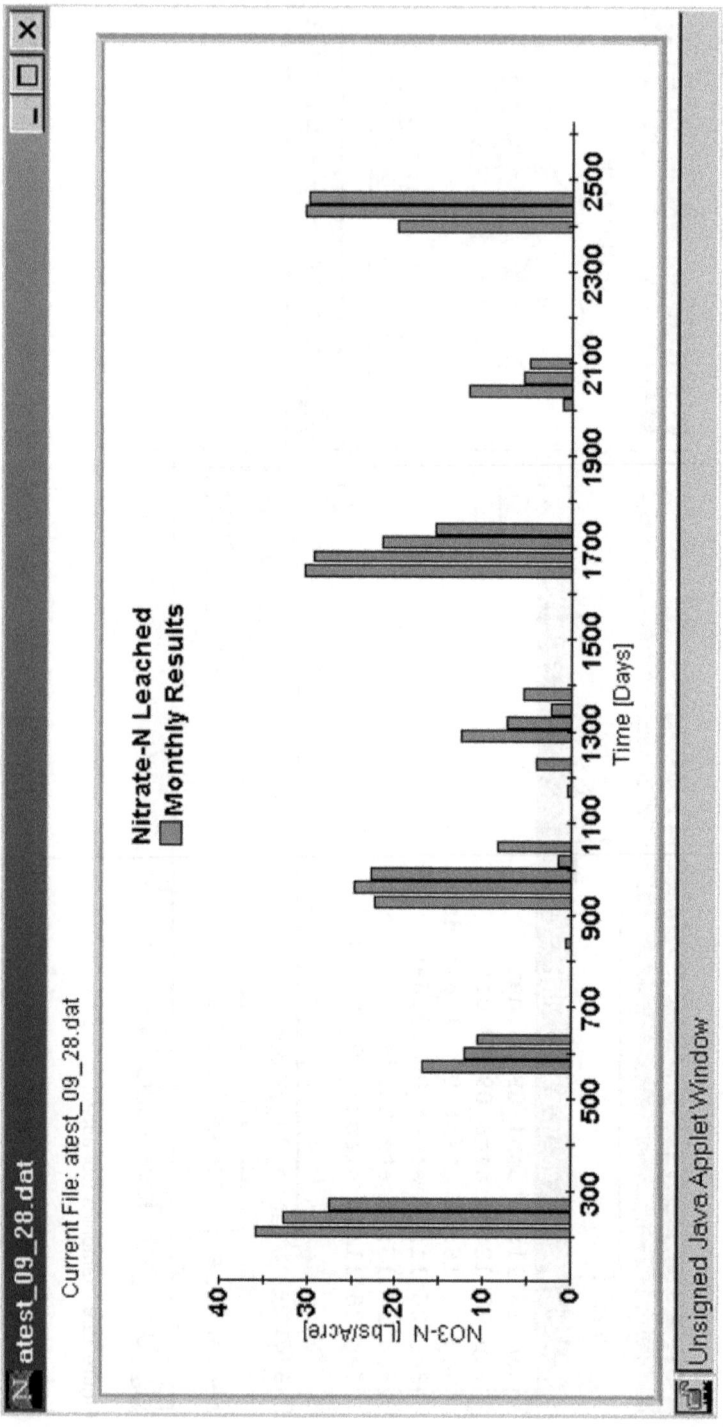

Figure 12.10 Bar graph output from Graphics applet.

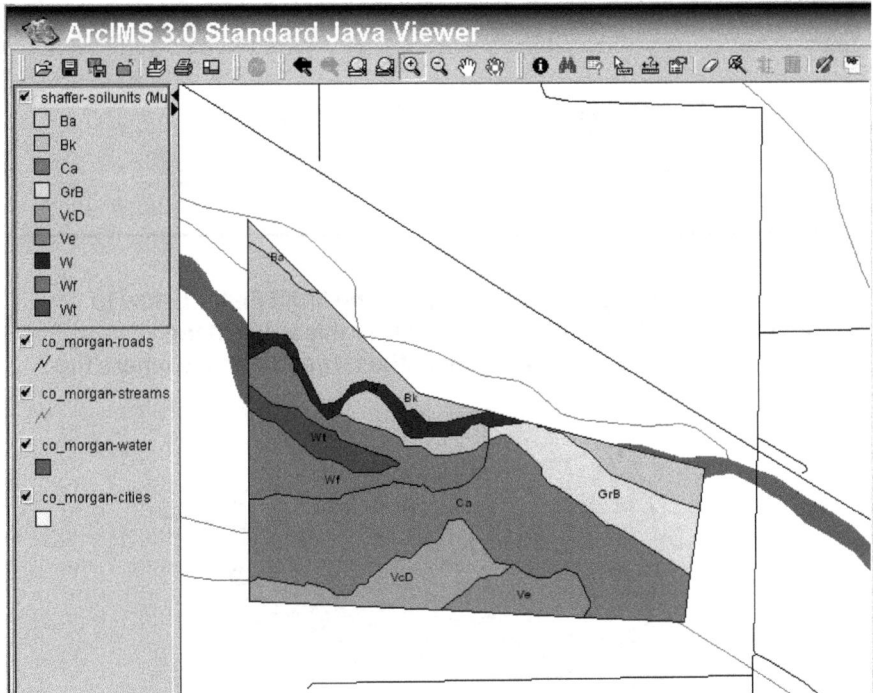

Figure 12.11 GIS Tool applet.

units identified on the GIS map of a farm to management and climate data scenarios stored on the user's account. This means, for example, that map layers specific to each account can be displayed for base location maps, soils and soil layer data, selected management, and nitrate leaching results from the NLEAP model. The user can also add field and management unit boundaries onto the base map and use these to help direct and display the simulation analyses. The GIS tool is designed to allow management by soil type across a farm, sub-farm, or field.

MUIR Database Download Tool

This tool (Figure 12.12) interacts with the NRCS MUIR Web site to automatically download soils data sets for a specified soil survey area, clean and convert them to NLEAP format, and store the results in a specified Microsoft Access NLEAP Java database. The tool cleans the NRCS database by removing soil types with incomplete soil layer records and generating missing values for data fields such as water content at 15 bars and percent coarse fragments. Initial default values also are set for soil nitrate-N and soil water content.

Editor for Rule-based Management Script

NLEAP Java contains a rule-based system for management events that allows some or all on-farm management events to be simulated with the use of rules rather

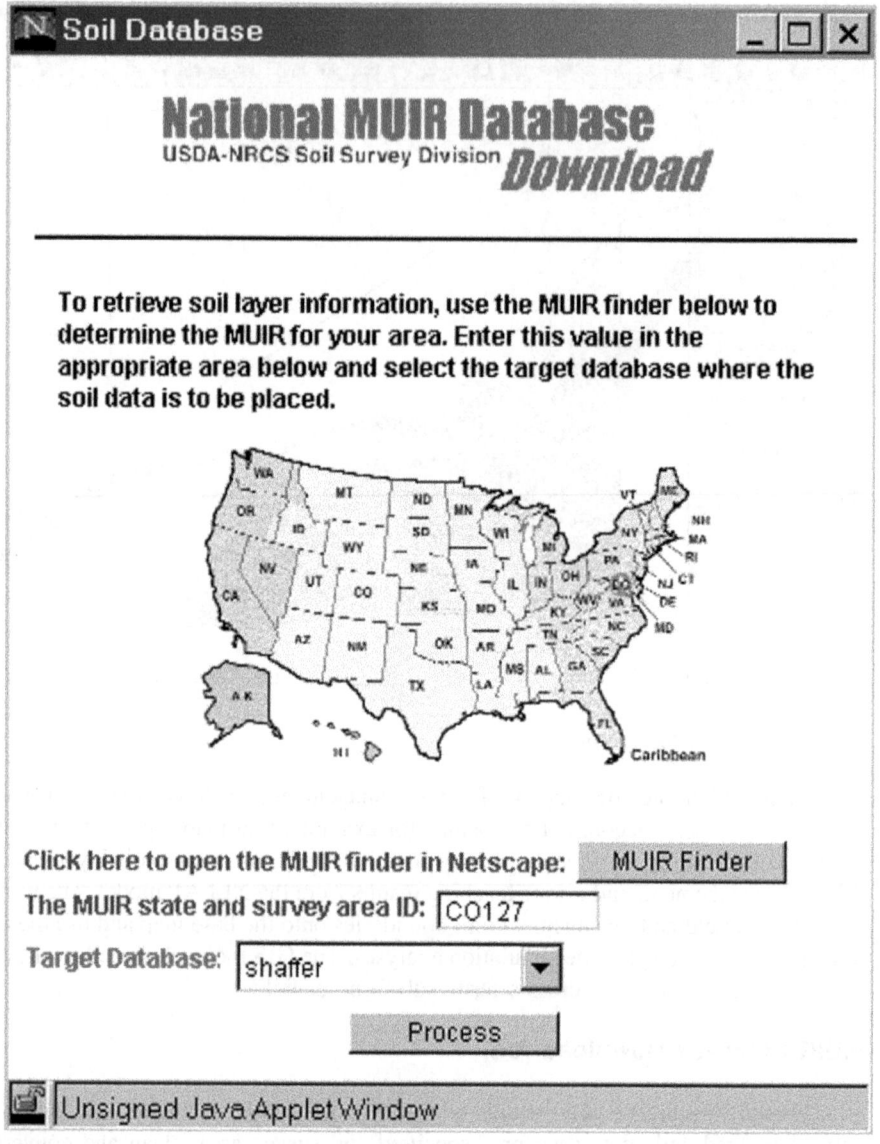

Figure 12.12 MUIR Database Access tool.

than fixed-date management. For example, instead of using a fixed date for planting, a rule can be developed that instructs the model to plant the crop when the soil temperature and soil water content are within specified ranges. The rule set can be simple (e.g., a script for irrigation scheduling), or quite complex (e.g., management instructions for an entire crop rotation). An editor applet is provided to assist the user in writing the appropriate syntax for the rule script (Figure 12.13). The NLEAP model then reads the script during a simulation and uses this information along with input from the model to dynamically schedule management events.

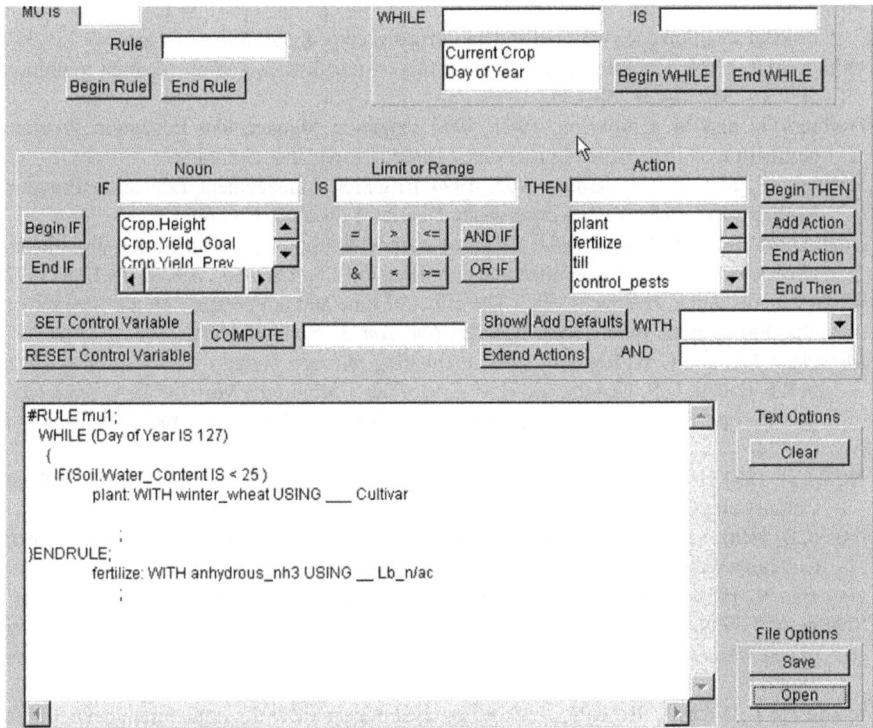

Figure 12.13   Rule Editor applet.

REFERENCES

Beckie, H.J., A.P. Moulin, C.A. Campbell, and S.A. Brandt. 1995. Testing effectiveness of four simulation models for estimating nitrates and water in two soils. *Can. J. Soil Sci.,* 75:135-143.

Campbell, C.A., R.P. Zentner, F. Selles, and O.O. Akinremi. 1993. Nitrate leaching as influenced by fertilization in the brown soil zone. *Can. J. Soil Sci.,* 73:387-397.

Crookston, M. and G. Hoffner. 1993. 1992 Irrigation Management Education Program. Northern Colorado Water Conservancy District, Loveland, Colorado.

Crookston, M. and G. Hoffner. 1992. 1991 Irrigation Management Education Program. Northern Colorado Water Conservancy District, Loveland, Colorado.

Delgado, J. 2001. Use of simulations for evaluation of best management practices on irrigated cropping systems. In M.J. Shaffer, L. Ma, and S. Hansen. (Eds.) *Modeling Carbon and Nitrogen Dynamics for Soil Management.* CRC Press, Boca Raton, FL, 355-381.

Delgado, J.A., R.F. Follett, J.L. Sharkoff, M.K. Brodahl, and M.J. Shaffer. 1998a. NLEAP facts about nitrogen management. *J. Soil Water Conserv.,* 53:332-338.

Delgado, J.A., M.J. Shaffer, and M.K. Brodahl. 1998b. New NLEAP for shallow and deep rooted crop rotations. *J. Soil Water Conserv.,* 53:338-340.

Delgado, J.A., R.F. Follett, and M.J. Shaffer. 2000. Simulation of NO_3^--N dynamics for cropping systems with different rooting depths. *J. Soil Sci. Soc. Am.,* 64:1050-1054.

Follett, R.F., M.J. Shaffer, M.K. Brodahl, and G.A. Reichman. 1994. NLEAP simulation of residual soil nitrate for irrigated and non-irrigated corn. *J. Soil Water Conserv.,* 49:375-382.

Gilmour, J.T. 1984. The effects of soil properties on nitrification and nitrification inhibition. *Soil Sci. Soc. Am. J.,* 48:1262-1266.

Hoffner, G., and M. Crookston. 1994. 1993 Irrigation Management Education Program. Northern Colorado Water Conservancy District, Loveland, Colorado.

Hoffner, G., and M. Crookston. 1995. 1994 Irrigation Management Education Program. Northern Colorado Water Conservancy District, Loveland, Colorado.

Linn, D.M., and J.W. Doran. 1984. Effect of water-filled pore space on carbon dioxide and nitrous oxide production in tilled and non-tilled soils. *Soil Sci. Soc. Am. J.,* 48:1267-1272.

Marion, G.W., and C.H. Black. 1987. The effect of time and temperature on nitrogen mineralization in arctic tundra soils. *Soil Sci. Soc. Am. J.,* 51:1501-1508.

Meisinger, J.J., and G.W. Randall. 1991. Estimating nitrogen budgets for soil-crop systems. In R.F. Follett, D.R. Keeney, and R.M. Cruse (Eds.). *Managing Nitrogen for Groundwater Quality and Farm Profitability,* Soil Science Society of America, Inc., Madison, WI, 85-124.

NCWCD. 1991. 1990 Irrigation Management Education Program. Northern Colorado Water Conservancy District, Loveland, Colorado.

NCWCD. 1990. 1989 Irrigation Management Education Program Summary Report. Northern Colorado Water Conservancy District, Loveland, Colorado.

Nommik, N. 1956. Investigations of denitrification in soil. *Acta Agric. Scand.,* 6:195-228.

Shaffer, M.J., P.N.S. Bartling, and J. Ascough, II. 2000. Object-oriented simulation of integrated whole farms: GPFARM framework. *Computers and Electronics in Agriculture,* 28:29-49.

Shaffer, M.J., and M.K. Brodahl. 1998. Rule-based management for simulation in agricultural decision support systems. *Computers and Electronics in Agriculture,* 21:135-152.

Shaffer, M.J., A.D. Halvorson, and F.J. Pierce. 1991. Nitrate Leaching and Economic Analysis Package (NLEAP): model description and application. In R.F. Follett, D.R. Keeney, and R.M. Cruse (Eds.). *Managing Nitrogen for Groundwater Quality and Farm Profitability,* Soil Science Society of America, Inc., Madison, WI, 285-322.

Shaffer, M.J., B.K. Wylie, R.F. Follett, and P.N.S. Bartling. 1994. Using climate/weather data with the NLEAP model to manage soil nitrogen. *Agricultural and Forest Meteorology,* 69:111-123.

Shaffer, M.J., B.K. Wylie, and M.D. Hall. 1995. Identification and mitigation of nitrate leaching hot spots using NLEAP/GIS technology. *J. Contam. Hydrol.,* 20:253-263.

Stoichev, D., M. Kercheva, and D. Stoicheva. 2001. NLEAP water quality applications in Bulgaria. In M.J. Shaffer, L. Ma, and S. Hansen. (Eds.) *Modeling Carbon and Nitrogen Dynamics for Soil Management.* CRC Press, Boca Raton, FL, 333-354.

Walthall, P.M., W.D. Brady, and R.L. Hutchinson. 1996. Cotton production on the Macon Ridge. *Louisiana Agriculture,* Spring 1996: 5-9.

Wylie, B.K., M.J. Shaffer, M.K. Brodahl, D. Dubois, and D.G. Wagner. 1994. Predicting spatial distributions of nitrate leaching in Northeastern Colorado. *J. Soil Water Conserv.,* 49:346-351.

Wylie, B.K., M.J. Shaffer, and M.D. Hall. 1995. Regional assessment of NLEAP NO_3-N leaching indices. *Water Resources Bull.,* 31(3):399-408.

Xu. C., M.J. Shaffer, and M. Al-kaisi. 1998. Simulating the impact of management practices on nitrous oxide emissions. *Soil Sci. Soc. Am. J.,* 62:736-742.

CHAPTER 13

Modeling the Effects of Manure and Fertilizer Management Options on Soil Carbon and Nitrogen Processes

Malcolm B. McGechan, D.R. Lewis, L. Wu, and I.P. McTaggart

CONTENTS

INTRODUCTION

Concerns about environmental pollution problems associated with both the use of fertilizer and the disposal of farm wastes have focused attention on opportunities for reducing mineral fertilizer requirements by better utilization of the nutrient potential of animal slurry and other manures. Poor utilization and high nutrient losses in current farming practice, resulting from inappropriate technology or procedures, and unsuitable timing of spreading operations, represent both a threat to the environment and an economic loss of a valuable resource.

Nitrogen and phosphorus are the nutrients in manures that cause most environmental concerns, but this chapter concentrates on nitrogenous pollution. Land spread slurry can pollute the environment by various routes. Water pollution takes the form of leached nitrate at field drain level and to deep groundwater, and can also occur due to surface runoff (overland flow) transporting whole slurry with a high biological oxygen demand (BOD) and containing nitrogenous and other components. Air pollution takes the form of volatilized ammonia, which is a major cause of acid rain and eutrification, and of emissions of nitrous oxide, a potent greenhouse gas. Research is being undertaken at various centers to assess the extent of pollution in a number of these categories, and the scope for measures to reduce losses, increase recycling of nutrients, and reduce mineral fertilizer inputs. Leached nitrate resulting from land spread animal slurry (liquid manure) is the main pollution form considered in this chapter, although simple estimates of emissions of the polluting gases ammonia and nitrous oxide are also included. Overland flow of slurry components occurs only on certain fields or soil types that are prone to surface runoff, and this has been the subject of a separate study (Lewis and McGechan, 1999; McGechan and Lewis, 2000). Unlike surface runoff pollution, which can often be avoided by careful management of spreading, nitrate leaching is always present to some extent, but measures can nevertheless be taken to minimize the level of leaching.

The movement of nitrogen as a nutrient and pollutant in agricultural systems has been studied extensively for a number of years (Burt et al., 1993), with particular emphasis in earlier work directed toward efficient use of mineral nitrogen fertilizer to increase agricultural production. Recently, the emphasis has concerned the environmental effects of leakages of nitrogen from the soil, especially nitrate leaching and gaseous nitrogen emissions (Powlson, 1993; Jarvis et al., 1996). Greater understanding of these interlinked processes can potentially lead to better agricultural practices, improving nitrogen use efficiency and minimizing pollution (Goss et al., 1995; Jarvis, 1996).

Overall nitrogen losses can be reduced by a combination of strategic decisions regarding capital purchase of equipment, and of tactical decisions regarding manure applications and reduced fertilizer nitrogen inputs (Jarvis et al., 1996). Incorrect timing of nitrogen applications can reduce nitrogen use efficiency and increase the risk of nitrate leaching. Autumn applications of slurry have been shown to increase nitrogen leaching losses compared with similar applications in the spring (Beckwith et al., 1998). Lack of growth over the winter reduces nitrogen use efficiency after autumn applications and, in addition, can lead to reduced yields (Sieling et al., 1998; Jackson and Smith, 1997). However, autumn and winter applications can only be avoided if there is sufficient storage capacity to hold all the slurry produced during the winter housing period to enable it all to be spread in the spring. Ammonia volatilization can be a major source of nitrogen losses from applied slurry. Injection or rapid incorporation has been shown to greatly reduce losses (Sommer and Hutchings, 1995; Misselbrook et al., 1996), but can sometimes lead to increased denitrification losses (Misselbrook et al., 1996; Ellis et al., 1998).

The aim of an improved manure management system must be to apply both slurry and complementary mineral fertilizer to match the nitrogen requirements of the crop, both in terms of quantity and timing, in a manner that both minimizes losses and maintains adequate crop yields. The object of the study described in this chapter is to use simulation modeling to assess the environmental impacts, as well as the benefits in terms of exploiting recycled nutrients as fertilizer substitute, of various options concerned with both selection of equipment for slurry spreading (strategic decisions) and procedures for managing the quantities and timing of applications (tactical decisions). In particular, the modeling procedures have been used to predict nitrogen losses and crop yields for several soils and locations in the cool temperate zones of Scotland and Ireland, where high winter rainfall and low soil moisture deficits are typical. The modeled predictions cover a range of options concerned with nutrient inputs from inorganic fertilizers and organic nitrogen in the form of animal manure slurry. The ultimate objective is to provide guidelines by which farm managers will base their decisions on equipment purchase, and nitrogen fertilizer and slurry applications for typical soil types, climates, and crops within this region. This work formed a component of a European Union-sponsored initiative on Slurry Waste and Agriculture Management (SWAMP, 1998), concerned with the development of a decision support system for whole-farms and the quantification of nutrients delivered by means of slurry applications.

An advanced manure management system must be capable of considering management options for a range of crops and climates, giving an almost infinite range of possible combinations for which computer simulation is the only feasible method of study. In this work, impacts on three cropping systems of a large number of nitrogen management practices representing tactical decision options about slurry spreading were considered for seven soils and three climates. Strategic decisions about equipment purchase were considered for a subset of these combinations. Physically based, weather-driven models were used for all simulations, as they could provide scenario predictions outside the model-calibrated boundaries. Such models are becoming more frequently used in agricultural management [e.g., APSIM (McCown et al., 1996) and RZWQM (Azevedo et al., 1997)].

In high rainfall climates such as in the United Kingdom and Ireland, effective installed field drainage plays an extremely important role because if it is absent or performing inefficiently, this can lead to high water tables and low crop yields. Therefore, in addition to models of nitrogen processes, it was also necessary to make use of a hydrological model that could accommodate inefficient drainage and hence reduced nitrate removal via field drains. The soil water and heat model SOIL (Jansson, 1996) had previously been demonstrated to simulate efficiently drained sites by McGechan et al., 1997, and as modified by Lewis and McGechan (1999) is capable of simulating inefficiently drained soils. The seven sites with various soil types were selected for this study because of their varying degrees of drainage efficiency. The decision to use the soil nitrogen model SOILN (Eckersten et al., 1996), in combination with an ammonia volatilization model from Hutchings et al. (1996), was made on the basis of the demonstration by Wu et al. (1998) that they could make good predictions of nitrate leaching losses and yields for a number of crop types. These validated models have been used in this chapter to evaluate the effects of slurry and nitrogen fertilizer management practices on nitrate leaching losses, gaseous nitrogen emissions, crop yields, and organic matter buildup in the soil.

DESCRIPTION OF SIMULATION MODELS USED

Soil Hydrological Model SOIL

The Swedish soil water and heat model SOIL (Johnsson and Jansson, 1991; Jansson and Halldin, 1979; Jansson, 1996) is a multilayer model and thus can indicate the soil water content and horizontal movement of water to field drain backfill at different depths, as well as deep percolation, with a range of drainage system options. The model therefore has a comprehensive treatment of processes that occur under wet soil conditions. Compared to any other soil water models, it combines the most sophisticated treatment of soil heat processes including freezing, and representation of falling and lying snow, which have a major influence on surface runoff and watercourse pollution from field spreading of wastes. Parallel simulation of soil heat processes alongside soil water movements is also important for a link to soil nitrogen dynamics models because soil nitrogen transformation processes are very dependent on both soil temperature and soil wetness. In addition, this model can readily be set up to carry out simulations over a long run of years (rather than the crop growth period in the middle of one year only). Most importantly, it includes representation of very fast movements of water by so-called "macropore flow" or "by-pass flow." Macropore flow is commonly associated with water movements through fissures or cracks in very dry soil, but also occurs in aggregated or structured soils at moisture contents near saturation, as studied extensively by Jarvis (1991) and Jarvis et al. (1991).

The version of the SOIL model that was used included classical drainage theory with representation of flows to drains from below as well as above the drain depth, according to the equations of Hooghoudt (1940) and Ernst (1956) as described further by Nwa and Twocock (1969) and Wesseling (1973). An adaptation of the Ernst equation to represent tile drains covered by soil layers rather than permeable backfill was implemented by Lewis and McGechan (1999) in a further revised version of the SOIL model to represent inefficient drainage systems that lead to high water tables and saturation surface runoff.

The SOIL model requires knowledge of soil physical properties for each layer (e.g., porosity and texture). Soil hydraulic properties are described in terms of parameters of the Brooks and Corey (1964) equation, usually obtained by measurement of the hydraulic conductivities and soil water contents at a range of matric tensions for each soil horizon. Subsurface drainage is estimated through knowledge of field parameters such as depth of the field drains, drain spacing, and permeable backfill resistance. These parameters had previously been selected using the adapted SOIL model to represent runoff-susceptible sites in Scotland and Ireland (Lewis and McGechan, 1999).

Soil Nitrogen Dynamics Model SOILN

Soil solute dynamics are very dependent on soil water movement, and hence models of soil solute dynamics are generally designed to be run in conjunction with a soil water model. The soil nitrogen model SOILN (Johnsson et al., 1987; Eckersten et al., 1996; and also described in Chapter 5 in this book) was designed to work in conjunction with SOIL. SOILN includes the major processes that describe the nitrogen cycle (inputs, transformations, transportation, and outputs) in agricultural soils. It is a multilayered model with layers determined by physical and biological characteristics.

Each layer has some representation of the inorganic and organic pools involved in the movement and transformations of nitrogen. Mineral nitrogen pools include ammonium and nitrate. Organic nitrogen is represented by fast cycling litter and manure-derived feces pools and a slow cycling humus pool. Mineralization and immobilization of nitrogen is controlled by carbon pools for litter and feces. The litter pool includes partially decomposed material such as dead roots, microbial biomass, and crop residues, while the humus pool is composed of stabilized decomposition products. Nitrogen is contained in plants, both above and below ground, with appropriate root distributions within the soil profile.

Nitrogen inputs to the uppermost soil layer can take the form of manure, inorganic fertilizer, and atmospheric deposition. As provided by its Swedish authors, representation of manure application in SOILN was slanted toward use of solid manure with plowing following application; thus, the model required slight adaptation to represent liquid slurry and application on grass with no plowing.

Soluble nitrate within the soil profile is transported with the water fluxes, and thus can be moved between soil layers, including upward movement during evaporation. Leaching of nitrate to drains with permeable backfill or semipermeable backfill takes place from each soil layer. An additional loss of nitrogen occurs through denitrification, which is also simulated from each layer.

SOILN has a mechanistic weather-driven sub-model representing crop growth. This interacts with the soil process routines in SOIL and SOILN to simulate growth in relation to radiation, soil water, and soil nitrogen availability, as well as indicating the daily uptake of nitrogen by the crop. As provided by its Swedish authors, the crop growth sub-model of SOILN (Eckersten and Jansson, 1991) represents a cereal crop and is therefore appropriate for considering spreading slurry on arable cropland. Work was also carried out on adapting a grass and grass/clover growth model (Topp and Doyle, 1996) to become a sub-model for SOILN for representation of spreading slurry on grassland. Developing and testing the linkage between SOILN and the monoculture grass growth sub-model is described by Wu and McGechan (1998a), and for the grass/clover mixed crop version of this by Wu and McGechan (1999).

Volatilization Model

Ammonia volatilization from applied slurry was estimated using the model of Hutchings et al. (1996). This model combines the partition of ammonia molecules in air and liquid through Henry's law, with aerodynamic resistance and surface boundary layer resistance terms. To calculate volatilization rates, knowledge of the atmospheric windspeed and pressure, and slurry temperature and pH is required. The volatilization rate also increases with the concentration of the slurry solid material as the slurry viscosity reduces the infiltration rate into the soil.

EXPERIMENTAL SITES AND CALIBRATION OF MODELS AT SITES

Experimental Sites and Hydrological Parameters

The experimental sites chosen reflected the range of typical soil types and drainage characteristics found in the main agricultural areas of Scotland and Ireland. These sites had previously been used in a complementary study on surface runoff (Lewis and McGechan, 1999) and are listed in Table 13.1. For all of these sites, hydraulic parameters for the SOIL model were available, with measurements made of the water release and hydraulic conductivity curves for the combination of soil matrix flow and macropore flow. Drainage characteristics of the fields are indicated in Table 13.1, with a profile conductivity defined as the maximum drainage rate from the field, which occurs when the water table is just at the soil surface.

Hydrological time series data were available for these fields, consisting of field drainage discharges from sites with good drainage, and watertable depths for sites with generally poor drainage. The site at Warren Field was monitored for both drainage and runoff flows. The time series data were available for calibration and validation of the SOIL model, as described by Lewis and McGechan (1999).

Table 13.1   Details of Soil Profiles, Arranged in Order of Increasing Leaching Ability

Location	Soil Type/ Cropping Regime	Drainage Characteristics	Profile Conductivity (mm/d)
Warren field, Wexford, Ireland	Silty clay loam/cut grass	Poor: 21.0 m separation, 0.5 m depth	2.0
Cowlands Wexford, Ireland	Sandy loam/grazed grass	Poor: 35.0 m separation, 0.5 m depth	0.5
Fence, Strathclyde, Scotland	Sandy silt loam/grazed grass	Poor: 10.0 m separation, 0.5 m depth	0.6
Beechgrove, Bush Estate, Penicuik, Scotland	Clay loam/grazed grass	Moderate: 9.0 m separation, 0.5 m depth	8.2
Crichton Royal Farm, Dumfries, Scotland	Silty clay loam/cut grass	Good: 7.0 m separation, 0.65 m depth	100.0
Glencorse Mains Farm, Penicuik, Scotland	Clay loam/arable cropping	Good: 7.5 m separation, 0.5 m depth	20.0
No. 3 Field, Bush Estate, Penicuik, Scotland	Sandy loam/arable cropping	Good: 7.0 m separation, 1.5 m depth	47.0

Hydrological Description of the Soil Profiles

A knowledge of actual field drainage conditions is important for the appraisal of any nutrient management plan for a given soil or site. The hydrological behavior of these fields could be subdivided into four main types, related to the soil properties and characteristics of the drainage system.

1. Fields with inefficient drainage systems due to drains covered with soil layers rather than permeable backfill. This is a common situation in Scotland and Ireland. Soil put into trenches above the pipes may have a high conductivity initially, but after about 5 years will revert to being similar to undisturbed soil. An inefficient drainage system can also be caused by drain spacings that are wider than in the initial installed drainage system. In effect, a doubling of spacing can occur in an old system when some drains are blocked. The fields at Beechgrove, Fence, and Cowlands (all grazed grasslands) exhibit poor drainage characteristics for the above reasons.
2. Fields that have efficient drainage systems and thus the profile drainage is primarily determined by the soil type. The drained plot sites at Glencorse (arable cropland), No. 3 Field (arable cropland), and Crichton (cut-grass) exhibit this type of behavior.
3. A thin permeable or highly conducting layer existing over an impermeable or low conductivity layer. This represents a common situation in which there is a thin topsoil layer above almost impervious subsoil or rock. The field at the Fence (grazed grassland) research farm site is a typical example; however, the drainage system is also inefficient at this site due to its large effective drain separation.

4. A high water table may exist in a field due to external influences from the catchment. This can be caused when groundwater sources outside the perimeter of the field contribute significantly to the water within the field. An example of this situation is presented by the Warren field site, into which a spring discharges, fed by groundwater from rising ground to the north of the field. In this case, recharge effects in both the low-lying plane of the site and from nearby hills produce an almost permanently high water table.

Nitrogen Cycling Data

A number of the fields discussed above have been used in long-term nitrogen cycling experiments. The last four fields listed in Table 13.1 have been monitored for a number of years for nitrate concentrations in drainflows, as well as harvested dry matter and crop nitrogen contents. These fields under various crops were used for calibration and validation of the SOILN model (Wu et al., 1998).

CLIMATIC DATA

Weather Data and Evaporation Model

To calculate daily transpiration, the SOIL model requires the daily climatic ("driving") variables, rainfall, mean temperature, vapor pressure, windspeed, and net radiation. These data sets were prepared from the daily weather records for the sites at Crichton and Bush in Scotland (Arnold, 1991), and termed here, respectively, West and East of Scotland climates, and for Warren and Cowlands in Ireland, termed the Southern Irish climate.

Global radiation was estimated from sunshine hours by the Ångström (1924) formula and daylength estimated from trigonometric relationships. Net radiation was estimated from global radiation, temperature, and vapor pressure by the Brunt (1932) formula. Evapotranspiration was then calculated using the Penman-Monteith equation (Monteith and Unsworth, 1990) for both bare soil and vegetation-covered soils.

Climatic Characteristics of Sites

In summary, the East of Scotland climate is generally dryer (by 130 mm average annual rainfall) and cooler (by 0.7°C annually) than that of the West of Scotland. There is a greater marked variation in monthly rainfall patterns for the West of Scotland climate, resulting in a higher average annual rainfall total (1100 mm), and with a slightly longer average daily sunshine. Comparing the Southern Irish climate data set with that of West of Scotland shows that annual rainfall is slightly less for the Irish data set (by 40 mm annually), but with the Southern Irish temperatures being approximately 1.2°C higher over the year.

There is a marked seasonal variation in average monthly rainfall for the West of Scotland and Southern Irish climates. The total rainfall in the months September to January usually reaches 40 to 50% of the annual total. This seasonal variation is less marked in the East of Scotland.

SIMULATION PROCEDURE

Simulations conducted in this work were carried out using model parameters calibrated for all sites for hydrology (McGechan et al., 1997; Lewis and McGechan, 1999), in which several years' experimental data on drainage flow, overland flow, and water table depths were generally used to validate the SOIL model. The processes involved in the nitrate transport model SOILN have been reviewed along with several other similar models by Wu and McGechan (1998b) and in Chapter 5 in this book. SOILN was modified to include a dynamic grass growth model and then calibrated for several fields under grass (Wu and McGechan, 1998a), using 3 years of data on nitrate losses with subsurface drainage flows and crop yields. Further calibration for nitrate uptake and losses using data on several fields under winter cereal has been described by Wu et al., 1998. The parameter set used for the dynamics of nitrogen in fields under spring cereals was taken from Johnsson et al. (1987).

Using these parameter sets, the impact of some equipment selection options, and a large number of nitrogen management scenarios involving fertilizer and slurry applications, were studied for combinations of soil/field and crop types. In particular, the effects of the nitrogen management scenarios on nitrate drainage flows, total gaseous nitrogen losses, and crop yields were investigated. The influence of different climates in this work was also studied by simulating conditions using two climatic data sets for the strategic study of equipment purchase, and all three climatic data sets for the tactical study of spreading quantities and timing. Long-term simulations (over 10 years) of continuous crop production were considered.

To make meaningful comparisons between all of the soil types, several model parameters were inferred for those fields that had no nitrate leaching validation data. Ideally, the model performance should be identified at all experimental sites, but a sensitivity analysis of the model (Wu et al., 1998), as well as other studies (Eckersten and Jansson, 1991), have shown that several parameters are not important to nitrate leaching while others may be inferred by soil type.

MANAGEMENT OPTIONS

Selection of Slurry Equipment and Spreading Strategy

The slurry management options addressed in the strategic study concern the choice of equipment in which capital must be invested, the size of store, and the method of spreading. Two store sizes and two spreading methods were considered.

A large slurry store enables spreading to be timed to coincide with a high rate of uptake by growing plants, whereas a small store requires extensive spreading in winter, which inevitably increases losses in various categories. Winter spreading can also aggravate trafficability and soil damage problems. Spreading during some winter months is prohibited in a number of European countries, including Sweden, Denmark, Germany, and The Netherlands, although it is allowed subject to some restrictions (MAFF, 1991; SOAFD, 1997) in the United Kingdom. The two store size options considered reflect farmers' differing attitudes to slurry. A farmer who regards slurry merely as an embarrassing waste to be disposed of will select the small store system, probably spreading slurry on fields nearest to the store whenever the store becomes full. In contrast, a farmer who regards slurry as a positive resource to be exploited will select the large store so that the slurry can be spread in spring when crops can use the nutrients, thereby maximizing the value of slurry as a fertilizer substitute. The large-sized store considered was adequate for a whole winter's production of slurry, compared with a small store that required emptying four times per year. A capacity of 1700 m³ was assumed for the large store option, sufficient for 1520 m³ annual production (see "Slurry Production and Composition") with 10% spare capacity, compared with a capacity of 500 m³ assumed for the small store.

Surface spreading equipment options considered were a vacuum tanker with a splash plate and a tanker-mounted shallow injector. Splash plate spreading leads to high ammonia volatilization losses, whereas injection reduces these losses to a very low level but at a cost of a more expensive machine with a higher power requirement and lower work rates. A 60-kW, 2WD tractor was assumed to be adequate for spreading slurry with a splash-plate vacuum tanker. For slurry injection with its higher power requirement, a 90-kW, 4WD tractor was selected. The same 7-m³ tanker was assumed for both spreading systems, with a four-tine winged injector unit mounted on the back when used for injection. These are the same equipment combinations as suggested by Warner et al. (1990), based on several years of operating slurry spreading equipment as described by Godwin et al. (1990).

Slurry Production and Composition

A common set of slurry parameters was selected for all the management options, based on typical approximate values reported by Dyson (1992). A herd of 100 dairy cows (mean weight 500 kg) plus 40 young cows (mean weight 250 kg), housed for 26 weeks per year, was assumed to produce 1520 m³ slurry, with mean composition 1.58 kg/m³ available N, 3.23 kg/m³ total N, 1.61 kg/m³ total P_2O_5, 3.01 kg/m³ total K_2O, and 7% dry matter.

Slurry Spreading

In the strategic options study, slurry was assumed to be spread at a rate of 50 m³/ha, the maximum application rate for surface spreading permitted by environmental protection regulations in the United Kingdom (MAFF, 1991; SOAFD, 1997). Although a higher rate (up to 140 m³/ha) is permitted for liquid wastes applied by injection, the same lower rate was selected for both application methods so that the results could be directly compared. In addition, for the slurry composition assumed here, the higher rate would have given nutrient applications in excess of crop requirements, a practice discouraged in the codes of practice (MAFF, 1991; SOAFD, 1997). A single application over 30.5 ha in spring was assumed with the large store, and four applications of 380 m³ over 7.6 ha for the small store distributed during the period for which the animals were housed. There were also two alternative options for the small store: one in which slurry was assumed to be spread repeatedly on the same 7.6-ha area of land (the worst case), the other in which slurry was spread on four occasions throughout the housed animal period but on different land each time. The worst case is permitted by farm waste disposal regulations in the United Kingdom, provided repeated applications are more than 6 weeks apart, and is common practice where a farmer regards slurry as a waste product to be disposed. Target dates for spreading were October 15, December 15, February 1, and March 15 for the small store, and March 15 for the large store. Spreading was simulated as actually taking place at a rate of 2.6 ha/d on the first available "spreading days" after the target date. A spreading day was defined as having to meet all the following conditions:

1. Soil water content in the upper topsoil layer < Field capacity (water content at 5 kPa tension) + 2%
2. Temperature in the upper topsoil layer > 0°C
3. No snow cover
4. Rainfall on current day < 2.5 mm

Spreading days were determined from weather-driven simulations with the SOIL model, with the upper topsoil layer defined as zero to 0.1 m depth. This is a similar procedure to that adopted for estimating the number of workdays for winter field operations by McGechan and Cooper (1994) and Cooper et al. (1997).

Nitrogen Inputs to SOILN Model

A simulation with the SOILN model requires an input file with information about dates and application rates of urinary N in slurry, feces N in slurry, and mineral fertilizer N, as well as the plowing date for arable cropland. The rate of urinary N was estimated from the available N in the slurry, depleted by a factor of 10% to represent losses in the store, as suggested by Dyson (1992) (Table 13.2); while feces

Table 13.2 Correction Factors for Calculation of Available Nitrogen from Slurry Between Store and Soil Incorporation

		Correction Factor
Step 1 In store	Aeration: short (2–3 days)	0.9
	Aeration: long (1 month)	0.75
	Agitation	0.9
Step 2 Time of application	Autumn	0.1
	November–December	0.4
	January–February	0.7
	Spring	1.0
Step 3 Method of application	Injection	0.9
	Low level, band	0.8
	Vacuum tanker, splash plate	0.7
Step 4 Arable land (plowed in within one day in cold weather)		1.0
Grassland		0.6

From Dyson, P.W. 1992. Fertiliser Allowances for Manures and Slurries. SAC Technical Note, *Fertiliser Series No. 14*. SAC, Edinburgh, Scotland.

N was calculated from the non-available N in the slurry without adjustment. Simulations with SOILN also require initial values for the soil nitrogen pools, which were estimated on the basis of soil organic matter levels of 7% for grassland and 3% for arable cropland.

Crop Management for Strategic Study

Assumed fertilizer applications to spring barley and grass crops were based on rates recommended by SAC (1996), listed as "Crop Requirements" in Table 13.3, reduced to allow for the contribution from available N and total P and K in the slurry using guidelines described (in the form of a farmer's advisory note) by Dyson (1992). The actual mineral fertilizer application rates are listed in Table 13.3, with those for N differing between each slurry management option when expected losses of available N from slurry by Dyson (1992; summarized in Table 13.2) were taken into account. For barley, fertilizer application was assumed to take place at sowing time; while for grass, the total quantities were split equally between three applications, taking place after the March slurry application and after the first and second silage cuts.

With four repeated slurry applications on the same area of arable cropland, the contribution from slurry exceeded the requirement, so no mineral N, P, or K was required. Similarly, no P or K was required with repeated applications on the same area of grassland, or with a single slurry application on arable cropland.

Plowing of arable cropland was assumed to take place on the first available plowing workday (day with soil water content in top layer < field capacity + 2%), as described by McGechan and Cooper (1994) and Cooper et al. (1997) after the following target dates: clay loam soil and large store, October 15; clay loam soil and small store, day following October slurry application; sandy loam soil, day following March slurry application. In simulations with SOILN, plowing distributes plant litter, from both the soil surface and the stubble from the previous crop, throughout the profile down to the plowing depth (specified in this case as 0.27 m).

Tactical Nitrogen Management Scenarios

Farmers are encouraged to comply with the Code of Good Practice "Prevention of Environmental Pollution from Agricultural Activity, PEPFAA" (SOAEFD, 1997). In this code, farmers are advised to apply rates of mineral nitrogen fertilizer to match crop requirements, taking into account soil nutrient status and any organic manure applied.

Table 13.4 gives the details of the management scenarios representing possible tactical decisions about quantity and timing of fertilizer spreading used in this work. These scenarios were derived from published fertilizer recommendations likely to be used by farmers (e.g., SAC, 1996; Younie et al., 1996; Dyson and Sinclair, 1993; Dyson, 1992). The scenarios cover three rates of slurry applications, with ten different possible spreading dates, depending on the crop type. The slurry can also be applied in one annual dose or split applications, and applied by either surface spreading or by injection. The three mineral fertilizer rates for each crop type reflect

Table 13.3   Available Nitrogen, Total Phosphorus, and Potassium, kg/ha from 50 m³/ha Slurry Application, and Adjustment to Mineral Fertilizer Application

| | Available N | | | | P₂O₅ | | K₂O | |
| | Arable | | Grassland | | | | | |
Crop Requirement	Injected 100	Surface 100	Injected 300	Surface 300	Arable 50	Grassland 115	Arable 50	Grassland 200
Slurry Nutrients								
Spring slurry application	63.8	49.6	38.3	29.8	80.5	80.5	151	151
Autumn slurry application	6.38	4.96	3.83	2.98	80.5	80.5	151	151
December slurry application	25.5	19.9	15.3	11.9	80.5	80.5	151	151
February slurry application	44.7	34.7	26.8	20.8	80.5	80.5	151	151
Four slurry applications	140	109	84.2	65.5	322	322	602	602
Mineral Fertilizer Required								
Spring slurry application	36.3	50.4	261	270	0	34.5	0	49.5
Autumn slurry application	93.6	95.0	296	297	0	34.5	0	49.5
December slurry application	74.5	80.2	284	288	0	34.5	0	49.5
February slurry application	55.3	65.3	273	279	0	34.5	0	49.5
Four slurry applications	0	0	215	234	0	0	0	0

From McGechan, M.B., and L. Wu. 1998. Environmental and economic implications of some animal slurry management options. *J. Agric. Eng. Res.*, 71:273-283. With permission.

Table 13.4   **Nitrogen Management Scenarios Used in the Nitrate Leaching Risk Assessment**

Variable	Option	Code	Value			Number of Options
Soil type	As Table 13.1	1 to 7				7
Crop	Grassland	1				3
	Winter cereal	2				
	Spring cereal	3				
Climatic conditions	East coast Scotland	1				3
	West coast Scotland	2				
	Southern Ireland	3				
Slurry N application (kg N/ha)			**Grassland**	**Winter Cereal**	**Spring Cereal**	
	Low	1	80	80	60	3
	Medium	2	140	120	90	
	High	3	190	160	120	
Fertilizer N application (kg N/ha)	Low	1	50	60	50	3
	Medium	2	100	100	80	
	High	3	150	140	110	
Spreading method	Surface spreading	1				2
	Injection or incorporation	2				
Slurry spreading date	Spring	1	3/4	14/3	3/4	10
	Summer	2	25/5	20/4	10/5	
	Autumn	3	10/10	10/10	10/10	
	Winter	4	14/2	14/2	14/2	
	Spring + autumn	5				
	Spring + summer	6				
	Spring + winter	7				
	Summer + autumn	8				
	Summer + winter	9				
	Autumn + winter	10				
Fertilizer spreading date	Spring	1	4/4	15/3	4/4	2
	Summer	2	26/5	21/4	11/5	
			Total number of options			21,630

typical average recommended rates, dependent on soil nutrient status and also whether any allowance has been made for applied slurry N. Fertilizer application dates selected for each crop are typical of agricultural practices in Scotland and Ireland and should be compared with Table 13.5, which identifies typical agronomic details.

Other details associated with the tactical decision scenarios were identical to those for the strategic study of equipment selection. These included details of equipment for surface spreading or injection, assumed composition of slurry, and selection of the first suitable workday after the target date for both slurry spreading and plowing.

Table 13.5   Typical Timings of Crop Husbandry Operations in Scotland and Ireland

Crop	Grassland	Winter Cereal	Spring Cereal
Sowing date	None	Early October	Early April
First fertilizer	Early April	Mid March	Early April
Second fertilizer	Late May	Mid April	Mid May
Spring slurry	Early April	Mid March	Early April
Summer slurry	End May	Mid April	Mid May
Harvest	(1) Late May (2) Early July (3) Late August	Mid August	Mid September
Autumn slurry[a]	Mid October	Mid October	Mid October
Winter slurry[a]	Mid February	Mid February	Mid February

[a] Autumn and winter applications are less efficient in providing available nitrogen to the next crop and could increase the risk of leaching losses.

RESULTS

Fate of Applied Nitrogen

Results from the strategic study are expressed in terms of the mean quantity of nitrogen recycled into biomass (crop and soil organic matter) over 10-year simulations, relative to the extent of losses that pollute the environment. These are presented in the form of input/output balances in Figure 13.1. The losses are categorized as nitrate leached to field drains, volatilized ammonia, and denitrified nitrogen. For the small store options involving spreading on different areas of land, results presented are the means of four simulations representing each area of land receiving slurry applications at different times. Results show substantial variations in the proportion of nitrogen passing by the different routes between the different slurry management options, between grass and arable cropland, and between the two soil types on arable cropland. Relatively small year-to-year variations were found for each option.

The levels of all the nitrogen pools in the soil profile, including litter, feces, ammonia, and nitrate, can differ from the beginning to the end of the year, and these are a significant part of the nitrogen balance when considered over an individual year. However, when averaged over a 10-year period, the imbalance in each of these pools, which tends to be positive in some years and negative in others, is small. The total imbalance from these pools (which are not shown in Figure 13.1) accounts for the small imbalance between each pair of total inputs and total outputs in Figure 13.1.

The proportions of nitrogen recycled for different times of slurry spreading (as with spreading on different areas of land for the small store) are plotted in Figure 13.2. This shows a rise in the proportion recycled with progressively later spreading from October to the spring. The proportion recycled is lower with repeated spreading on the same land than with a single application at any time. Separate figures for the proportion recycled to the harvested crop alone, and to the humus pool plus the crop, show that most of the variation with time of application is attributable to that recycled to the crop.

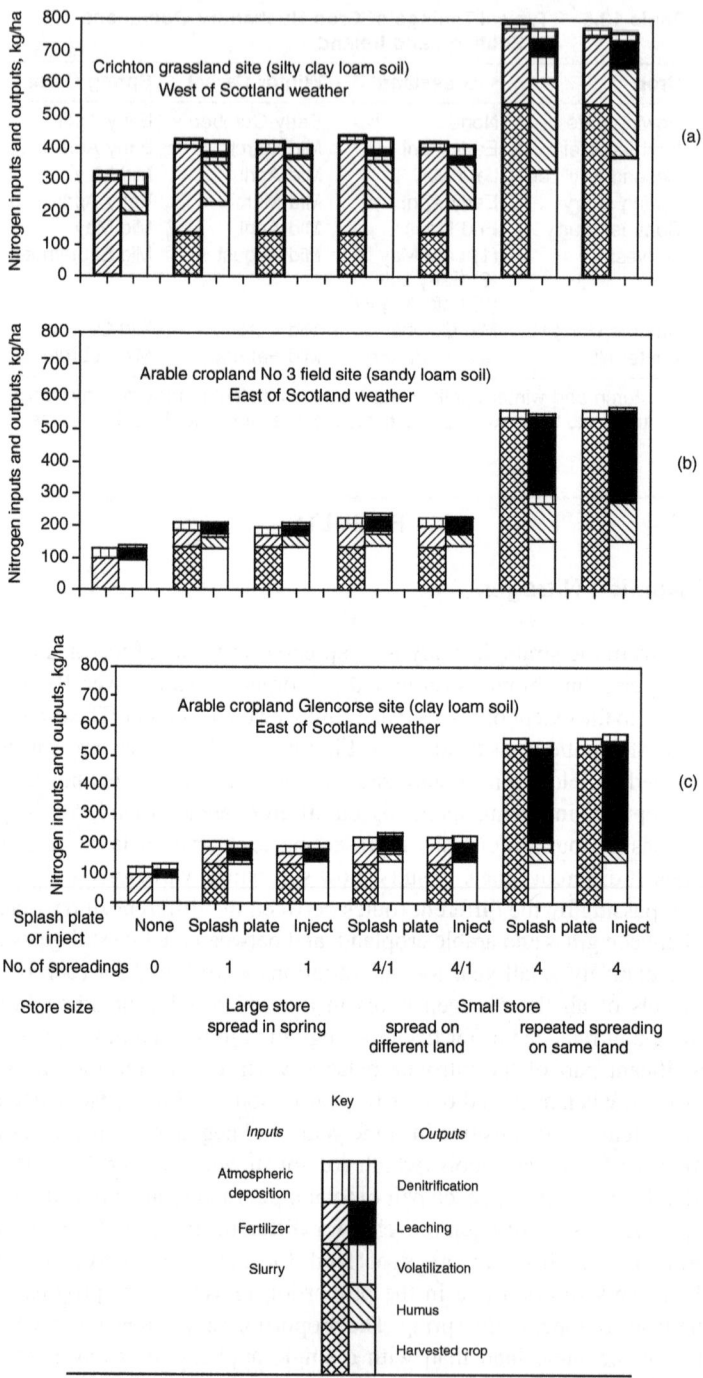

Figure 13.1 Mean annual nitrogen input/output balances for 10-year simulations. (From McGechan, M.B., and L. Wu. 1998. Environmental and economic implications of some animal slurry management options. *J. Agric. Eng. Res.*, 71:273-283. With permission.)

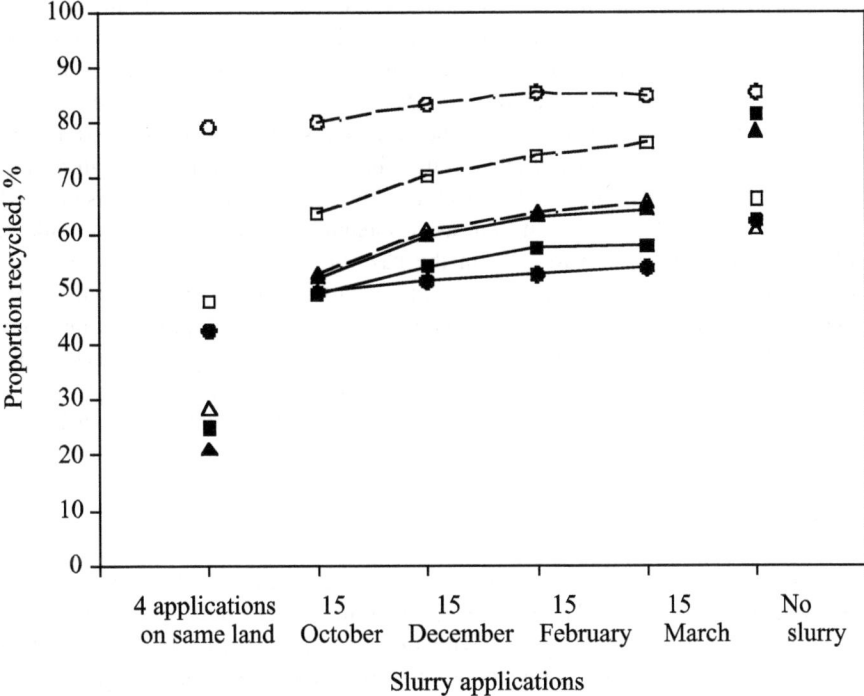

Figure 13.2 Proportion of applied nitrogen recycled with different numbers and times of slurry applications. ●, ■, ▲, ——, to harvested crop; ○, □, △, – – – –, to crop and soil humus; ●, ○, Crichton grassland site (silty clay loam soil); ■, □, No. 3 Field arable cropland site (sandy loam soil); ▲, △, Glencorse arable cropland site (clay loam soil). (From McGechan, M.B., and L. Wu. 1998. Environmental and economic implications of some animal slurry management options. *J. Agric. Eng. Res.*, 71:273-283. With permission.)

Effect of Store Size and Spreading Method

Results from the strategic study show about 10% of total N or 20% of available N lost by ammonia volatilization from slurry applied by a splash-plate system on arable cropland, while on grassland the figures rise to about 15% of total nitrogen or 30% of available nitrogen (Figure 13.1). These losses are higher with winter spreading, as required with the small store system, than when slurry is all spread in the spring, which can only be done with the large store. There is a substantial environmental benefit from an injection system, which cuts this gaseous emission to almost zero. The adjustment to mineral fertilizer suggested by Dyson (1992) to allow for higher available nitrogen from injected slurry (compared with surface spreading) is about right for arable cropland with the large store system. Repeated slurry spreading on the same land area (with the small store system) results in excess nutrients in the soil, particularly on bare soil or stubble with spring cereal land in winter. No mineral fertilizer whatsoever is required with such repeated slurry applications; thus, no adjustment can be made and injection contributes further to excess

nitrogen in the soil. For grassland, results indicate that the suggested mineral fertilizer reduction for injection could be increased slightly, as the current adjustment
only partially compensates for nitrogen lost by volatilization with surface spreading.

The effect of spreading method is also discernible in the results from the tactical
options study, and is in general agreement with the results of the strategic equipment
selection options study. However, this tends to be of secondary importance in relation
to leaching compared to the other routes for polluting losses. When slurry is injected,
almost no volatilization occurs, in contrast to spreading on the surface when available
nitrogen is reduced by up to 26% in the case of summer applications. Because
ammonia constitutes less than 20% of the nitrogen applied, an approximate 5%
change arises in slurry nitrogen input between the spreading options, which does
not manifest itself as large leaching effects. Frequent, repeated applications of slurry
would however increase the importance of the spreading method, as already shown.
Rees et al. (1993) found that nitrogen in injected slurry had a greater residual effect
on crop yield than surface-applied slurry nitrogen.

Nitrate Leaching

Nitrate leaching is the main route by which nitrogen is lost if it is in excess in
the system. Results from the strategic study show that leaching losses are always
highest for the small store system with repeated applications on the same land
(Figure 13.1) because four applications of slurry, of which some take place at an
inappropriate time of year, contribute to available nitrogen well in excess of requirements for crop growth. With this system, leaching losses are particularly high for
the arable cropland soils, and higher for the clay loam (Glencorse) than for the sandy
loam (No. 3 Field) soil. They are also higher with injection than with splash-plate
spreading, because retaining available nitrogen at the spreading stage further contributes to excess nitrogen in the system. With the large store system and spreading
only in the spring, leaching losses are higher for the sandy loam than for the clay
loam soil. With the large store system, leaching losses are very low from grassland;
this is the only case where denitrification losses exceed leaching losses.

Simulated nitrogen in harvested biomass and annual total leaching from the
tactical study with different slurry and fertilizer application rates are presented in
Figure 13.3. Results are shown for all 360 management scenarios at the clay loam
(Beechgrove) and sandy loam (No. 3 Field) sites under cut-grass, with East of
Scotland meteorological data. The general trend indicates that increasing the nitrogen
application (in slurry plus mineral fertilizer) leads to increased nitrate loss through
subsurface drains and increased total nitrogen in crop biomass. The nitrate leached
and the nitrogen in the crop increased nonlinearly with increased nitrogen application
rates for both spreading methods, and they were highly dependent on the spreading
date(s) and whether there were one or two slurry applications per year. For example,
nitrate leaching at the Beechgrove site increased by only 3% when a spring fertilizer
application was increased from 50 to 100 kg/ha, and a further 5% when this fertilizer
application rate was increased to 150 kg/ha, when accompanied by the lowest rate
of slurry application (80 kg/ha in the autumn). However, increasing the slurry nitrogen

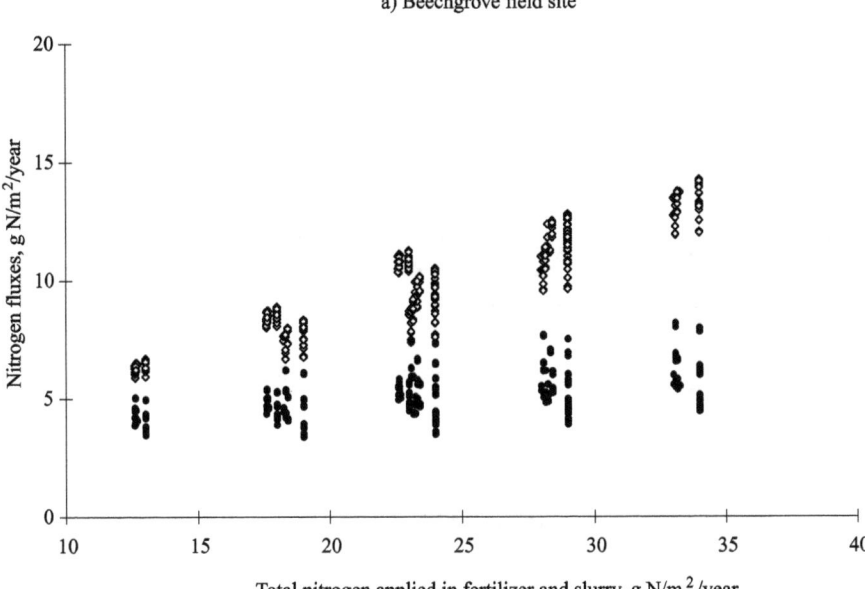

a) Beechgrove field site

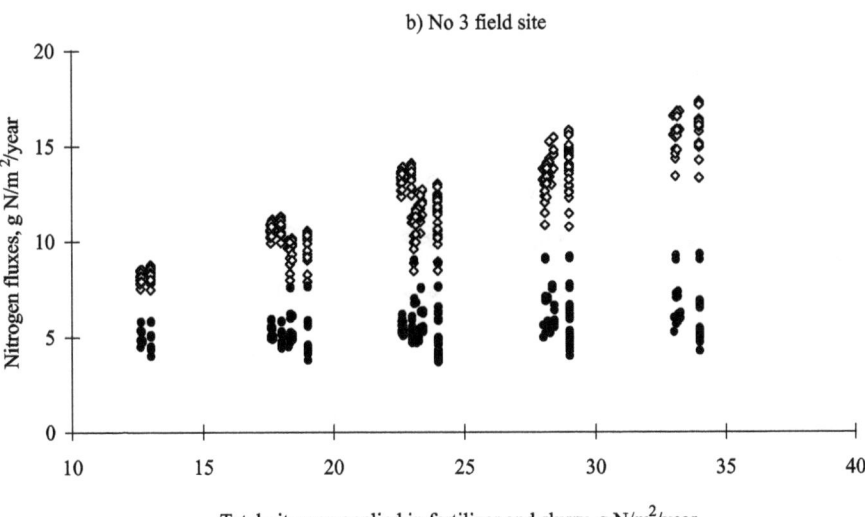

b) No 3 field site

◇ Harvest N   • Leached N

Figure 13.3   Nitrogen in harvested biomass and annual leached nitrogen against total nitrogen applied (in fertilizer plus slurry), for grass (two cuts) at the Beechgrove and No. 3 Field sites, with all management scenarios and the East of Scotland meteorological data set.

a) Fertilizer applied in spring, slurry applied in autumn

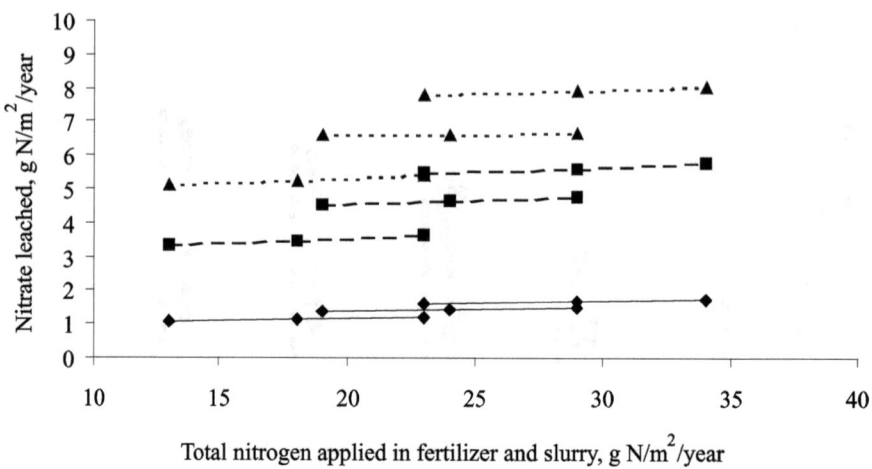

b) Fertilizer applied in spring, slurry applied in spring

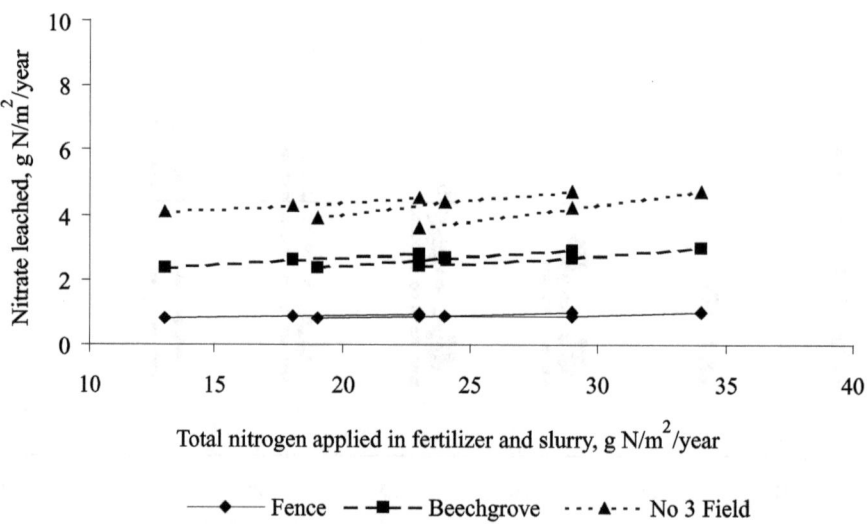

◆ Fence – ■ – Beechgrove ··▲·· No 3 Field

Figure 13.4 Comparison of nitrate leaching from several sites and two slurry spreading dates.

application rate to 190 kg/ha in conjunction with the high fertilizer application rate
led to an increase in leaching of 72%, and this was typical for all the soils investigated. This leaching effect for several sites under the nine possible fertilizer and
slurry application rates is shown in Figure 13.4, with data points joined by lines to
indicate a constant rate of slurry application. A spring slurry application produced

a marked decrease in overall leaching compared to an autumn application. Comparing the leaching pattern for these two spreading dates suggests that the feces component of slurry (the non-available nitrogen part which when incorporated into the soil becomes equivalent to litter material from dead plant components) has a significant influence on leaching. With an autumn slurry application, this represents additional leaching above that arising from a similar quantity of fertilizer nitrogen, whereas the converse is true for a spring slurry application. Similar findings have been reported in the literature from experiments on grassland (Unwin et al., 1986) and on arable cropland (Chambers et al., 1996; Beckwith et al., 1998).

Spreading dates and repeated slurry applications were found to have a critical effect on nitrate leaching. Figure 13.5 shows the leaching effects and other nitrogen losses for a given nitrogen application rate for the Beechgrove and No. 3 Field sites, with a wide spread in nitrate leaching and total nitrogen losses evident for the 20 surface spreading options. Repeated slurry applications, or a single slurry application in the autumn, gave rise to the most significant level of leaching, as already shown in relation to a small store (see "Effect of Store Size and Spreading Method"). This is also in agreement with findings from earlier studies that have led to recommendations that slurry spreading in the autumn should be avoided to minimize the risk of nitrate leaching (Chambers et al., 1996; SOAEFD, 1997; McGechan and Wu, 1998).

Denitrification

Results from the strategic study show denitrification losses more than twice as high from grassland compared with the arable cropland (Figure 13.1). This may reflect the larger nitrogen pools, particularly organic nitrogen, in the grassland soil. However, denitrification losses differed little between the different slurry management options and between the two arable cropland soils (Beechgrove and No. 3 Field). With the current state of development of the SOILN model, it is not possible to subdivide denitrified nitrogen into molecular nitrogen gas and nitrous oxide, but it is reasonable to assume that high levels of denitrification will be associated with high nitrous oxide emissions.

Results from the tactical study show denitrification increasing in wet soils, and the influence of this process on the total loss of nitrogen from the different sites is shown in Figure 13.6. By inspection of the full simulated output data set (not shown here), and also by subtracting leaching losses (Figure 13.4) from total losses (Figure 13.6) assuming volatilization to be small, it could be seen that denitrification accounted for a loss of nitrogen of approximately 1 g N/m for the well-drained and moderately drained soil profiles, with an increase to 2 g N/m for the poorly drained and often saturated Fence profile.

Soil Humus

In contrast to the other soil nitrogen pools, the one pool where there is a noticeable change over 10 years is soil humus. This is a very large pool, so the change is small as a percentage of the total pool size. However, it appears to be

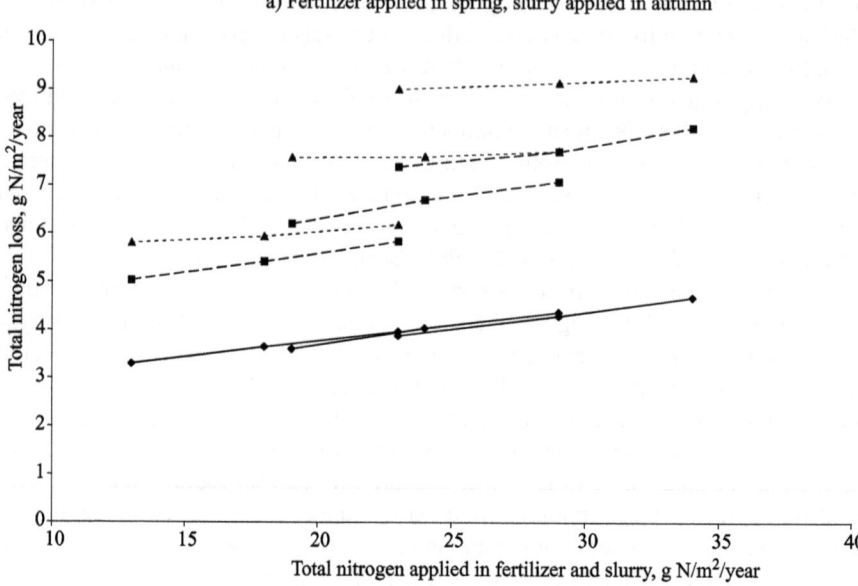

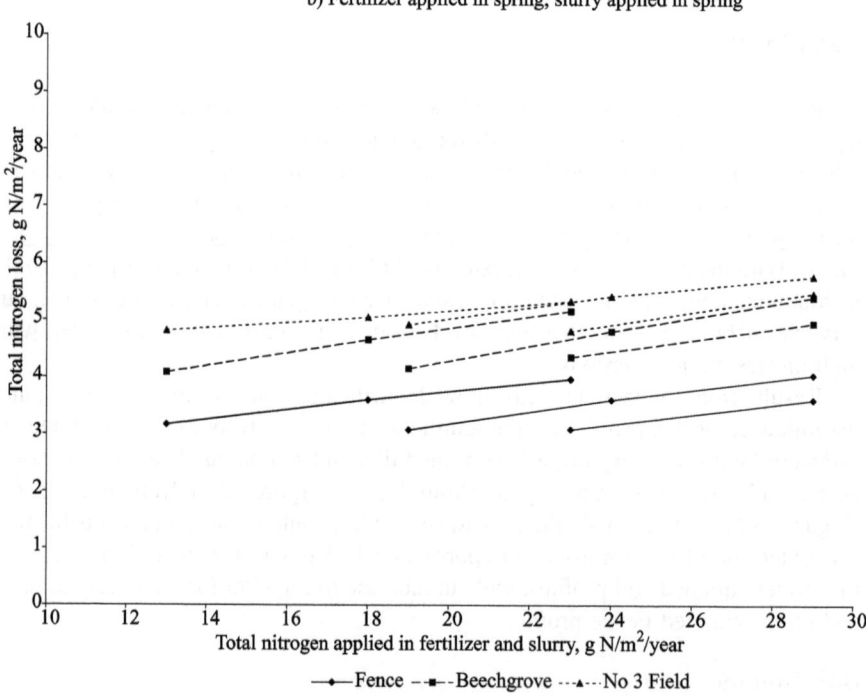

Figure 13.5 Comparison of total nitrogen loss from several sites and two slurry spreading dates: (a) fertilizer applied in spring, slurry applied in autumn; and (b) fertilizer applied in spring, slurry appied in spring.

a) Beechgrove field site

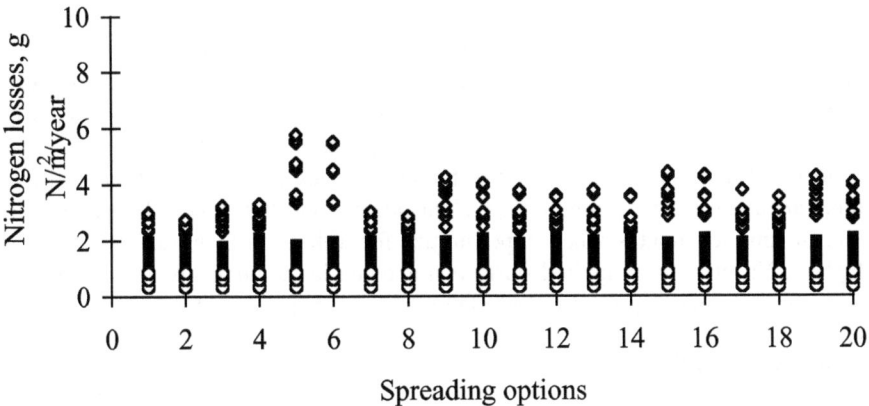

b) No 3 Field site

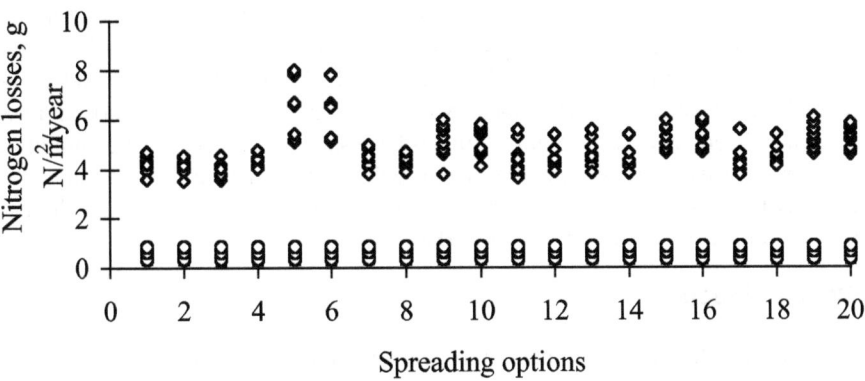

◇ N leached   ■ N Denitrified   ○ N Volatilized

Figure 13.6   Leached, denitrified, and volatilized nitrogen for grass (two cuts) with different surface spreading options and all nitrogen management scenarios, for the (a) Beechgrove and (b) No. 3 Field sites and the East of Scotland meteorological data set.

much larger when expressed as a percentage of the nitrogen added to and lost from the system, and is a significant part of the 10-year balance. For all options considered, there is an increase in the soil humus nitrogen pool over the years (Figure 13.1), due to the addition of non-available nitrogen from the feces component of slurry, which subsequently declines only slowly by mineralization. As would be expected, this increase is greater for four repeated slurry applications over the same area of land than for a single slurry application on each area. The smallest such increase occurs with the single surface spread application of slurry on the sandy loam arable cropland soil (No. 3 Field), where the rate of mineralization is almost as great (greater in some individual years) as the input of feces in slurry to the humus pool. An increase in soil humus can be regarded as one of the positive benefits of slurry application, because an increase in soil organic matter is generally associated with an improvement in soil structure for mineral agricultural soils. While there is little control of the timing of the release of humus nitrogen by mineralization to nitrate, mineralization is a temperature-dependent process and thus takes place at a faster rate in summer months when opportunities for utilization of nitrate by crops are highest. Nitrogen from applied slurry that ends up in the humus pool can therefore be regarded as a component of recycled nitrogen rather than a loss.

Soil Type Effects

The effects of soil type and field drainage characteristics were explored over several widely differing sites in the tactical study. Results presented in Figure 13.7 show nitrate leaching effects for three crops, with both slurry and fertilizer nitrogen applied at the highest rates in spring plus an autumn slurry application. Nitrate leaching is compared for all seven sites with results shown for the three climates, with groupings of three adjacent points referring to the East and West of Scotland and Ireland, and the soils ordered as in Table 13.1. Leaching was found to be generally greatest for the sandy loam No. 3 Field, and least for the poorly drained Irish sites. In contrast, surface runoff studies have shown that there is a high pollution risk for the Irish sites, which have a low field saturation hydraulic conductivity; whereas there was a low surface runoff pollution risk for the efficiently drained Scottish sites (McGechan and Lewis, 2000).

Climate Effects

Simulated harvested biomass nitrogen and nitrate leached are presented in Figure 13.8, for 20 surface spreading options, high application rates for both slurry and fertilizer nitrogen, all three climates, the Beechgrove soil, and a cut-grass crop. Crop growth simulations can be expected to be sensitive to temperature, and this manifests itself as crop yields approximately 4% larger, with the higher temperatures at the Irish compared to the Scottish sites. Nitrate leaching is very dependent on rainfall, and the simulations indicated this to be lowest in the East of Scotland, followed by Southern Ireland (5% higher relative to East of Scotland) and highest for the West of Scotland (a 17% increase). Under dry winter conditions, some studies have shown no effect of the timing of slurry application or nitrate leaching and nitrate nitrogen use efficiencies (Jackson and Smith, 1997). However, even in the East of Scotland, there is still a significant excess winter rainfall (over evaporation) and, thus, under these conditions, significant leaching is still likely to occur under particular management options such as autumn applications of slurry.

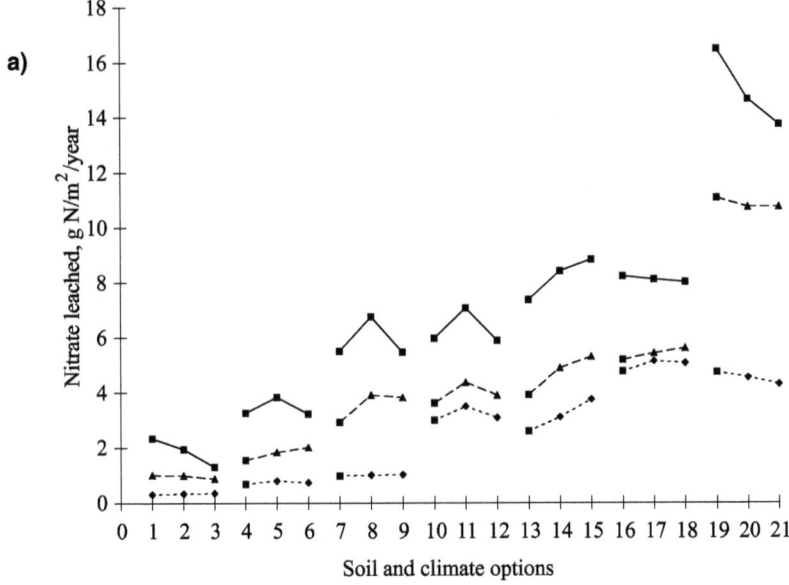

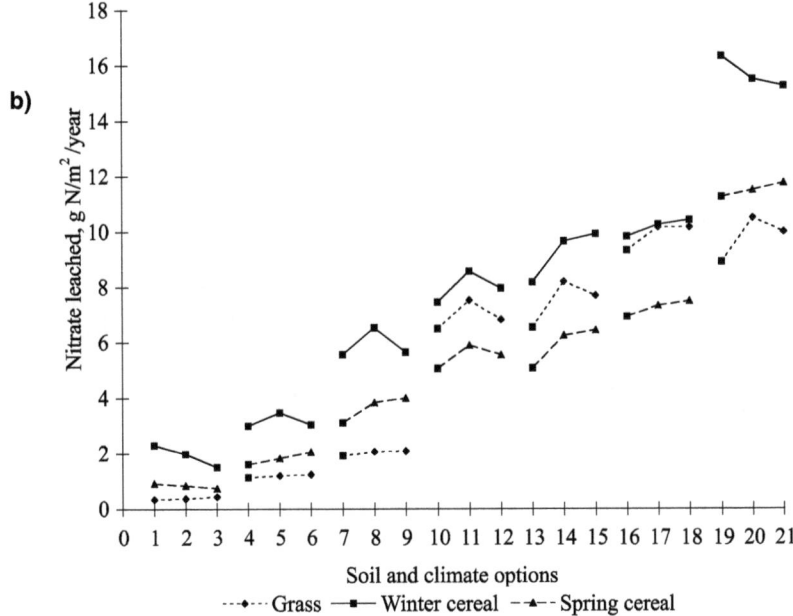

Figure 13.7   Leached nitrate with different sites (soil type) and climates, for the three crops with spring fertilizer and spring and autumn slurry applications all at the highest rates: (a) spring fertilizer, spring slurry; and (b) spring fertilizer, autumn slurry.

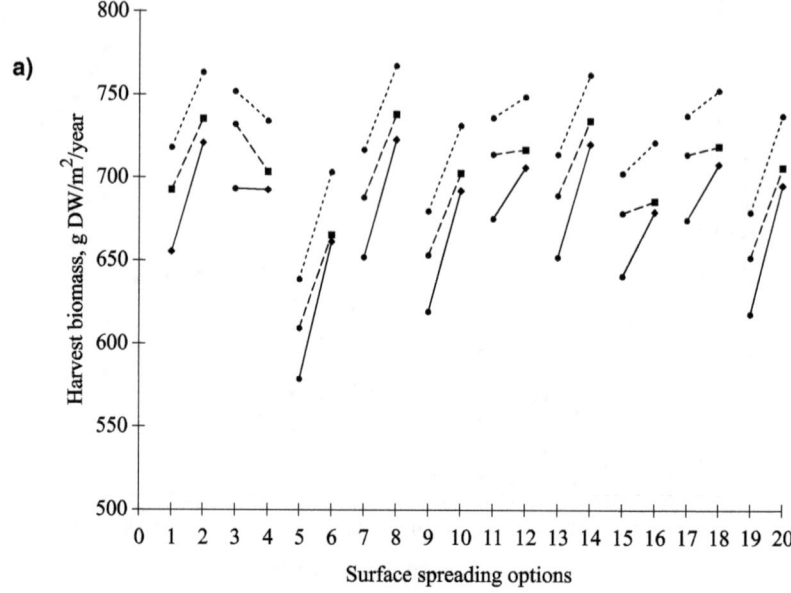

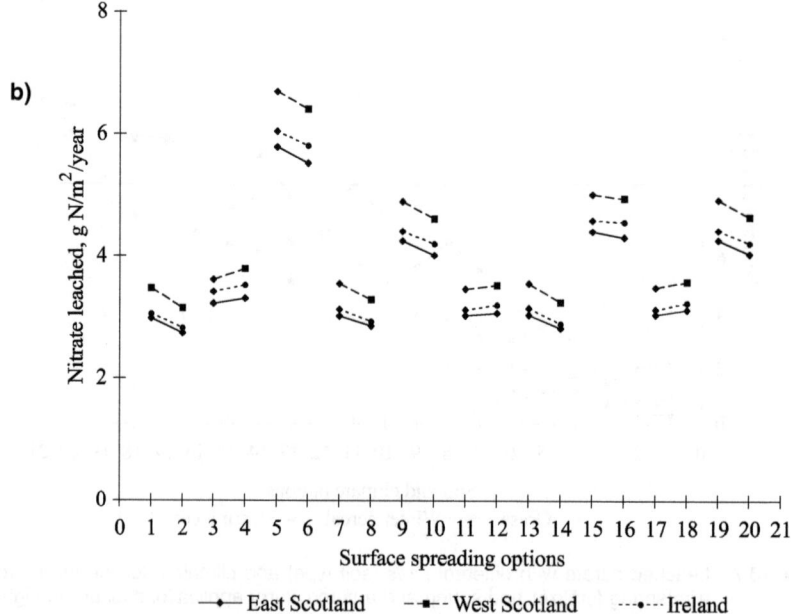

Figure 13.8   Harvested biomass (a) and leached nitrate (b) with different surface spreading options, for the three climates, the Beechgrove soil and fertilizer and slurry application at the highest rates.

Crop Comparisons

Results from the strategic study show a higher proportion of nitrogen recycled into crop biomass by grass than by spring barley (Figure 13.1), despite higher nitrogen fertilizer inputs on grassland. This suggests that grass has a greater potential to mop up excess nitrogen from the system, partly because it has a longer growing season and partly because annual dry matter biomass offtakes are higher for grass than for cereal crops. The higher total slurry application with the small store system and repeated applications on the same area of land, compared with the large store system, produces a response in terms of a higher yield in both crops, but this response is greater with grass than barley.

In the tactical study, a nonlinear trend in crop biomass yields and nitrate leaching in relation to nitrogen application rates was evident in each case (Figure 13.9). Overall losses from grass or a winter cereal were shown to be potentially greater than from a spring cereal, but these arose mainly as a result of the (typical) higher fertilizer nitrogen application rates.

CONCLUSION

Weather-driven simulations with ammonia volatilization, soil water, and soil nitrogen dynamics models have demonstrated the extent of losses of nitrogenous components that pollute the environment, from land spreading of slurry and mineral nitrogen fertilizer, with alternative management policies concerning storage, spreading method, application rates, and timing. The results of this work provide a sound scientific basis for decisions regarding environmentally friendly slurry management. They should be incorporated into decision support systems software, both for strategic decisions about capital purchase of equipment and for tactical day-to-day decisions about timing and quantities of slurry and fertilizer spreading. The work also demonstrates the value of simulation modeling in general, for testing the consequences of a wide range of decisions relating to weather-dependent agricultural processes that impinge on the environment, and incorporating the results into decision support systems. Related subject areas for future similar studies would be spreading FYM (farmyard manure) from straw-bedded animal systems, and testing the consequences of losses to the environment of phosphorus (the other potentially polluting plant nutrient) from fertilizer and manure.

The overall conclusions from the current study, which should be incorporated into decision support, are as follows:

1. Losses in the form of leached nitrate are high when slurry is spread during autumn or winter, compared with spreading in the spring. However, autumn or winter spreading may be unavoidable if the store is too small to hold all the slurry produced during the housing period.
2. Losses in the form of leached nitrate are extremely high, due to a gross overload of nutrients, where repeated slurry applications during the winter period take place

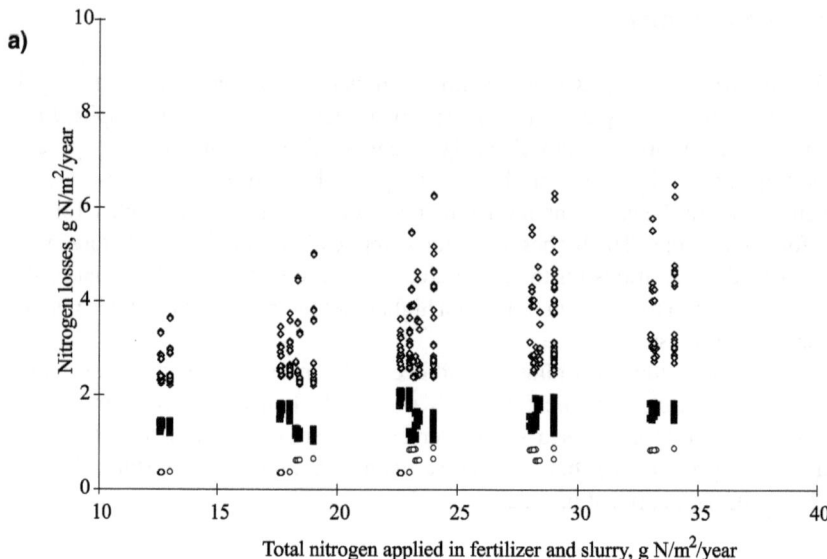

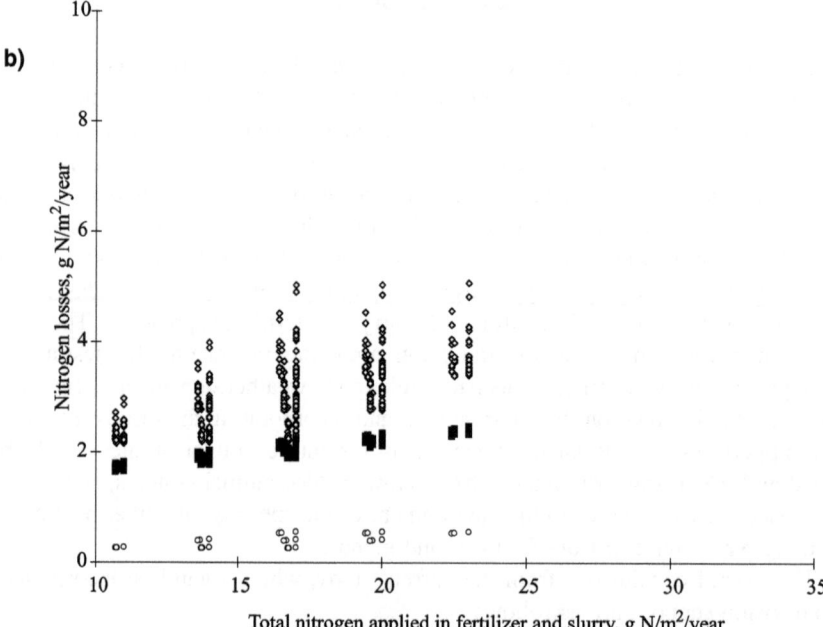

Figure 13.9 Nitrogen losses against total nitrogen applied (in fertilizer plus slurry), for the three crops at the Beechgrove site and the East of Scotland climate: (a) grass, (b) spring cereal, and (c) winter cereal.

on the same area of land. This is one option where the store is too small to hold all the slurry produced during the housing period, but management policies should be adopted to avoid this practice with only one application per year on each piece of land.

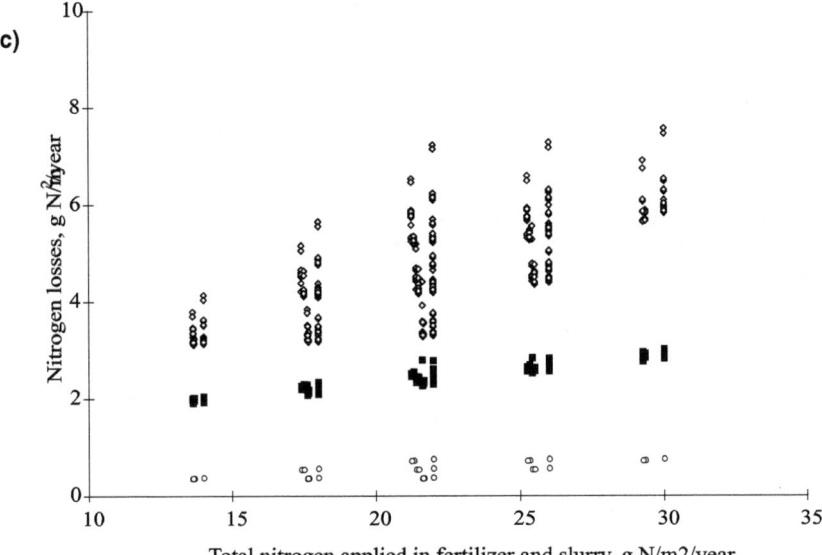

c)

Figure 13.9   (continued).

3. Slurry application by injection reduces ammonia volatilization losses to a low level, compared with surface spreading.

4. Slurry management measures have relatively little effect on nitrous oxide emissions from denitrification.

5. There is a greater potential for recycling nutrients in slurry by spreading on grassland than on arable cropland under cereals.

6. There are environmental benefits from reducing mineral fertilizer applications to allow for nutrients in slurry. There are also potential cost savings, contributing to the costs of storage and environmentally friendly spreading equipment (not discussed here, but presented in a study by McGechan and Wu, 1998). The results obtained can potentially be used as the basis for strategic management decisions regarding investment in storage and spreading equipment.

7. The most important decisions about slurry spreading concern the selection of spreading date and the selection of fields that are likely to minimize nitrate leaching.

8. There are significant differences in nitrate leaching (and crop nitrogen uptake) between different soil types (with Table 13.1 identifying the order of increasing leaching behavior among the sites and soils considered here), and also according to climate.

9. A wide range of spreading management scenarios have been simulated and compared to provide scientifically based environmental risk assessments regarding nitrogen losses, and to provide a basis for tactical decisions regarding quantity and timing of spreading.

REFERENCES

Ångström, A. 1924. Solar and terrestrial radiation. *Quarterly J. Roy. Meteorological Soc.*, 58, 389-420.

Arnold, C. 1991. *METDATA User Guide.* AFRC Computing Division, Harpenden, 32 pp.

Azevedo, A.S., P. Singh, R.S. Kanwar, and L.R. Ahuja. 1997. Simulating nitrogen managment effects on subsurface drainage water quality. *Agric. Syst.*, 55:481-501.

Beckwith, C.P., J. Cooper, K.A. Smith, M.A. Shepherd. 1998. Nitrate leaching loss following application of organic manures to sandy soils in arable cropping. I. Effects of application time, manure type, overwinter crop cover and nitrification inhibition. *Soil Use and Management*, 14:123-130.

Bergström L., H. Johnsson, and G. Torstensson. 1991. Simulation of soil nitrogen dynamics using the SOILN model. *Fert. Res.* 27:181-188.

Brooks, R.H., and A.T. Corey. 1964. Hydraulic properties of porous media. *Hydrology Paper No. 3,* Colorado State University, Fort Collins, CO. 27 pp.

Brunt, D. 1932. Notes on radiation in the atmosphere. 1. *Quarterly J. Roy. Meteorological Soc.*, 58:389-420.

Burt, T.P., A.L. Heathwaite, and S.T. Trudgill. 1993. *Nitrate — Processes, Patterns and Control.* John Wiley & Sons. Oxford. 456 pp.

Chambers, B.J., R.A. Hodgkinson, and J.R. Williams. 1996. Nitrate leaching losses from drained soils receiving farm manures. *Trans. 9th Nitrogen Workshop,* Braunschweig, Germany, 399-402.

Cooper G., M.B. McGechan, and A.J.A. Vinten. 1997. The influence of a changed climate on soil workability and available workdays in Scotland. *J. Agric. Eng. Res.*, 68:253-269.

Dyson, P.W. 1992. Fertiliser Allowances for Manures and Slurries. SAC Technical Note, *Fertiliser Series No. 14.* SAC, Edinburgh, Scotland.

Dyson, P.W., and A.H. Sinclair. 1993. Recommendations for Cereals: Nitrogen. *Fertiliser Series No. 5.* Technical Note T216, SAC, Edinburgh, Scotland.

Eckersten, H., and P.-E. Jansson. 1991. Modelling water flow, nitrogen uptake and production for wheat. *Fert. Res.,* 27:313-329.

Eckersten, H., P.-E. Jansson, and H. Johnsson. 1996. The SOILN Model User's Manual, Department of Soil Sciences, Swedish University of Agricultural Sciences, Uppsala.

Ellis, S., S. Yamulki, E. Dixon, R. Harrison, and S.C. Jarvis. 1998. Dentrification and N_2O emissions from a UK pasture soil following the early spring application of cattle slurry and mineral fertiliser. *Plant Soil,* 202:15-25.

Ernst, L.F. 1956. Calculation of the steady flow of groundwater in vertical cross sections. *Neth. J. Agricultural Sci.,* 4:126-131.

Godwin, R.J., N.L. Warner, and M.J. Hann. 1990. Comparison of umbilical hose and conventional tanker-mounted slurry injection systems. *Agri. Eng.,* 45:45-50.

Goss, M.J., E.G. Beauchamp, and M.H. Miller. 1995. Can a farming systems approach help minimize nitrogen losses to the environment? *J. Contam. Hydrol.,* 20:285-297.

Hooghoudt, S.B. 1940. Bepaling van den doorlaatfaktor van den grond met behulp van pompproeven (z.g. boorgatemethode). *Verslag Landbouw Onderzoek,* 42:449-541.

Hutchings, N.J., S.G. Sommer, and S.C. Jarvis. 1996. A model of ammonia volatilization from a grazing livestock farm. *Atmos. Environ.,* 30:589-599.

Jackson, D.R., and K.A. Smith. 1997. Animal manure slurries as a source of nitrogen for cereals: effect of application time on efficiency. *Soil Use and Management,* 13:75-81.

Jansson, P.-E. 1996. Simulation Model for Soil Water and Heat Conditions. Report 165 (revised edition). Department of Soil Sciences, Swedish University of Agricultural Sciences, Uppsala. 82 pp.

Jansson, P.-E., and S. Halldin, 1979. Model for annual water and energy flow in a layered soil. In S. Halldin (Ed.). *Comparison of Forest Water and Energy Exchange Models.* International Society for Ecological Modelling, Copenhagen, 145-163.

Jarvis, N. 1991. MACRO — A Model of Water Movements and Solute Transport in Macroporous Soils. Monograph 9, Swedish University of Agricultural Sciences, Department of Soil Sciences, Uppsala.

Jarvis, N.J., L. Bergström, and P.E. Dik. 1991. Modelling water and solute transport in macropore soil. II Chloride breakthrough under non-steady flow. *J. Soil Sci.,* 42:71-81.

Jarvis, S.C. 1996. Future trends in nitrogen research. *Plant Soil,* 181:47-56.

Jarvis, S.C., R.J. Wilkins, and B.F. Pain. 1996. Opportunities for reducing the environmental impact of dairy farming managements: a systems approach, *Grass Forage Sci.,* 51:21-31.

Johnsson, H., L. Bergström, P.-E. Jansson, and K. Paustian. 1987. Simulated nitrogen dynamics and losses in a layered agricultural soil. *Agric. Ecosyst. Environ.,* 18:333-356.

Johnsson, H., and P.-E. Jansson. 1991. Water balance and soil moisture dynamics of field plots with barley and grass ley. *J. Hydrol.,* 129:149-173.

Lewis, D.R., and M.B. McGechan. 1999. Watercourse pollution due to surface runoff following slurry spreading. 1. Calibration of the soil water simulation model SOIL for fields prone to surface runoff. *J. Agric. Eng. Res.,* 72:275-290.

MAFF. 1991. Agricultural Practice for the Protection of Water. Code of Good Practice, HMSO, 79 pp.

McCown, R.L., G.L. Hammer, J.N.G. Hargreaves, D.P. Holzworth, and D.M. Freebairn. 1996. APSIM: a novel software system for model development, model testing and simulation. *Agric. Syst.,* 50:255-271.

McGechan, M.B., and G. Cooper. 1994. Workdays for winter field operations. *Agri. Eng.,* 49:6-13.

McGechan, M.B., R. Graham, A.J.A. Vinten, J.T. Douglas, and P.S. Hooda. 1997. Parameter selection and testing the soil water model SOIL. *J. Hydrol.,* 195:312-334.

McGechan, M.B., and L. Wu. 1998. Environmental and economic implications of some animal slurry management options. *J. Agric. Eng. Res.,* 71:273-283.

McGechan, M.B., and D.R. Lewis. 2000. Watercourse pollution due to surface runoff following slurry spreading. 2. Decision support to minimize pollution model. *J. Agric. Eng. Res.,* 75:429-447.

Misselbrook, T.H., J.A. Laws, and B.F. Pain. 1996. Surface application and shallow injection of cattle slurry on grassland: nitrogen losses, herbage yields and nitrogen. *Grass Forage Sci.,* 51:270-277.

Monteith, J.L., and M.H. Unsworth. 1990. *Principles of Environmental Physics,* (2nd ed.). Edward Arnold, London, 291 pp.

Nwa, E.U., and J.G. Twocock. 1969. Drainage design theory and practice. *J. Hydrol.,* 9:259-276.

Powlson, D. 1993. Understanding the soil nitrogen cycle. *Soil Use and Management,* 9:86-93.

Rees, Y.J., B.F. Pain, V.R. Phillips, and T.H. Misselbrook. 1993. The influence of surface and subsurface application methods for pig slurry on herbage yields and nitrogen recovery. *Grass Forage Sci.,* 48:38-44.

SAC Farm Management Handbook 1996/97. The Scottish Agricultural College, Edinburgh.

Sieling, K., H. Schroder, M. Finck, and H. Hanus. 1998. Yield, N uptake, and apparent N-use efficiency of winter wheat and winter barley grown in different cropping systems. *J. Agric. Sci.,*131:375-387.

SOAFD. 1997. Prevention of Environmental Pollution from Agricultural Activity. Code of Good Practice. HMSO, Edinburgh, U.K., 110 pp.

Sommer, S.G., and N. Hutchings. 1995. Techniques and strategies for the reduction of ammonia emission from agriculture. *Water Air Soil Pollut.,* 85:237-248.

SWAMP. 1998. Optimal Use of Animal Slurries for Input Reduction and Protection of the Environment in Sustainable Agricultural Systems. Third Annual Progress Report (Consolidated and Individual). EU project: AIR-CT 94-1276.

Topp, C.F.E., and C.J. Doyle. 1996. Simulating the impact of global warming on milk and forage production in Scotland. 1. The effects on dry-matter yield of grass and grass — white clover swards. *Agric. Syst.,* 52:213-242.

Unwin, R.J., B.F. Pain, and W.N. Whinham. 1986. The effect of rate and time of application of nitrogen in cow slurry on grass cut for silage. *Agric. Wastes,* 15:253-268.

Warner, N.L., R J. Godwin, and M.J. Hann. 1990. An economic analysis of slurry treatment and spreading systems. *Agri. Eng.,* 45:100-105.

Wesseling, J. 1973. Subsurface flow into drains. In *Drainage Principles and Applications,* Volume II, Publication 16: International Institute for Land Reclamation and Improvement. Wageningen, The Netherlands.

Wu, L., and M.B. McGechan. 1998a. Simulation of biomass, carbon and nitrogen accumulation in grass to link with a soil nitrogen dynamics model. *Grass Forage Sci.,* 53:233-249.

Wu, L., and M.B. McGechan. 1998b. A review of carbon and nitrogen processes in four soil nitrogen dynamics models. *J. Agric. Eng. Res.,* 69:279-305.

Wu, L., M.B. McGechan, D.R. Lewis, P.S. Hooda, and A.J.A. Vinten. 1998. Parameter selection and testing the soil water model SOILN. *Soil Use and Management,* 14:170-181.

Wu, L., and M.B. McGechan. 1999. Simulation of nitrogen uptake, fixation and leaching in a grass/clover mixture. *Grass Forage Sci.,* 54:30-41.

Younie, D., G.E.J. Fisher, and G. Swift. 1996. Recommendations for Grazing and Conservation. *Fertiliser Series No. 4.* Technical Note T210, SAC, Edinburgh, Scotland.

CHAPTER 14

Parameterization of Soil Nitrogen Transport Models by Use of Laboratory and Field Data

E. Priesack, S. Achatz, and R. Stenger

CONTENTS

INTRODUCTION

During the past five decades, intensive agricultural land use has led to a strong increase in crop production. This was most often achieved by an increased application of nitrogen (N) fertilizers at the risk of soil and groundwater contamination, due to N leaching below the rooted soil. To avoid or decrease nitrate leaching,

N fertilizer application must be timed to match the period of highest N use by the crop and to account for the N mineralization-immobilization-turnover dynamics in the soil.

To improve N fertilizer management, soil-plant system models were applied to simulate adequate N supply for both, optimal crop growth and minimal N losses. Whereas most models correctly describe crop N demands and N uptake, description of the soil N turnover cycle remains difficult (De Willigen, 1991; Diekkrüger et al., 1995). In particular, data on the quality of organic fertilizers and of remaining plant residues, including decaying roots following crop harvest, is often incomplete. Hence it becomes difficult to test a soil-plant system model using complete, reliable data sets (e.g., Addiscott, 1993; Wu and McGechan, 1998). Most models of soil organic matter (SOM) turnover consider different conceptual SOM pools having different decomposition rates (Falloon and Smith, 2000), but model testing remains hindered by the lack of suitable methods for measuring these pools and fluxes (Magid et al., 1997). Model parameterizations therefore most often represent empirical relationships fitting the given model to available data, and which may not have any physical interpretation (Christensen, 1996; Falloon and Smith, 2000).

Adequate simulations of nitrate leaching require accurate modeling of soil water flow, because nitrate is mainly transported with the soil water in which it is dissolved. Determining soil hydraulic parameters, water retention curves (WRC), and unsaturated hydraulic conductivities needed for simulation of soil water flow in field conditions is complicated. To overcome this problem, pedotransfer functions (PTFs), which predict the soil water retention curve and saturated hydraulic conductivity K_{sat} from basic soil properties (i.e., texture, bulk density, and organic matter content), have been developed (Tietje and Tapkenhinrichs, 1993; Campbell, 1985).

Various PTFs to derive WRCs (water retention curves) from basic soil properties were applied in the simulation of water flow at different sites of a 1.5-km² research farm (Scheinost et al., 1997; Priesack et al., 1999). It was concluded from these simulations that a direct, straightforward, and automatic way to provide the necessary parameter values for water flow modeling at more than 400 grid points of the farm area was not possible. Consequently, the authors analyzed time series data on soil water and soil nitrate contents at different field sites and determined parameter values using indirect methods (i.e., inverse water flow modeling).

The objective was to test the usefulness of a newly developed inverse method for parameter estimation of N transport in agricultural soils, and to present examples for the estimation of soil hydraulic parameters and values for maximal rates of nitrification, N_2O production, and N mineralization.

DESCRIPTION OF EXPERT-N

Expert-N is a program package used for the simulation of water, nitrogen, and carbon dynamics in the soil-plant-atmosphere system. It comprises a number of modules describing various approaches to soil water flow, soil heat transfer, soil carbon and nitrogen turnover, crop processes, and soil management. Each module is composed of different sub-modules, which themselves contain different algorithms that can be selected to model each sub-process. The algorithms currently provided by Expert-N have been taken from published models, such as LEACHM (Hutson and Wagenet, 1992); CERES (Ritchie, 1991; Jones and Kiniry, 1986), HYDRUS (Simunek et al., 1998); SUCROS (van Laar et al., 1992); NCSOIL (Nicolardot and Molina, 1994); and DAISY (Hansen et al., 1991; Svendsen et al., 1995), or developed by the Expert-N team (Stenger et al., 1999). The possibility to choose between different algorithms allows the user to build a model that suits the specific purpose of the study and the availability of required input data. Comparison of differing algorithms for one sub-process under otherwise identical conditions should aid in finding the algorithms underlying the observed differences in simulation results, thus facilitating comparison and validation of different models. Expert-N also provides the means to enter user-programmed sub-modules by using a loading procedure for functions from external user-defined dynamic link libraries. In this manner, a step-wise development of a completely new N-model by the user is possible.

Soil Water Processes

Expert-N comprises different approaches for each of the following modules to simulate soil water flow: potential evapotranspiration, potential evaporation, actual evaporation, surface runoff, soil water leaching, snow accumulation and melting, and lower boundary condition for soil water flow. Calculation of potential evapotranspiration, for example, can be carried out by selecting one of several different methods, depending on the type of weather data input available. This process requires cautious and thorough work, as shown by the model comparison of Diekkrüger et al. (1995). In this study, the use of different approaches resulted in marked differences in simulated cumulative evapotranspiration and corresponding simulated cumulative leaching.

Expert-N can be used to simulate the vertical movement by applying a capacity model or the more mechanistic one-dimensional Richards equation. In the present study, non-steady-state water flow simulations were based on a numerical solution of the Richards equation:

$$\frac{\partial \theta}{\partial t} = \frac{\partial}{\partial z}\left[K(h)\cdot\left(\frac{\partial h}{\partial z}-1\right)\right] - S(t,z,h) \tag{14.1}$$

where t denotes time (d), z soil depth (positive downward) (mm), θ volumetric water content (mm^3/mm^3), $h = h(t,z)$ soil matric potential (kPa) converted to (mm) water head, $K(h)$ unsaturated hydraulic conductivity (mm/d), and $S = S(t,z,h)$ the water sink term for root water uptake (mm/mm/d).

Expert-N provides different numerical solvers for this flow equation. One such solver, according to the model HYDRUS 6.0 (Simunek et al., 1998), was selected for the inverse method. Required soil water retention curves for each soil horizon were given by van Genuchten-type curves (van Genuchten, 1980) and unsaturated conductivities were calculated according to Mualem (1976). Potential evapotranspiration was estimated by the Penman-Monteith equation (Smith, 1992), root water uptake was calculated according to Nimah and Hanks (1973), accumulation and melting of snow was determined by the method of Anderson (1973), and soil freezing and thawing was modeled by the approach of Flerchinger and Saxton (1989).

Solute Transport and N Turnover Processes

In contrast to capacity-type models, which consider most often only the convective transport of nitrates, the vertical transport of nitrogen in soils is usually simulated by the numerical solution of the one-dimensional advection-dispersion transport equation:

$$\frac{\partial}{\partial t}\left[\left(\theta+\rho\cdot k_d\right)\cdot c\right]=\frac{\partial}{\partial z}\left[\theta\cdot D(\theta,q)\frac{\partial c}{\partial z}-q\cdot c\right]+T(t,z,c) \qquad (14.2)$$

where t denotes time (d); z soil depth (mm), θ the volumetric water content (mm^3/mm^3), $c = c(t,z)$, the nitrogen species concentration of either urea, ammonium, or nitrate (kg/dm^3); $D = D(\theta,q)$, the dispersion coefficient (mm^2/d); $q = q(t,z)$, the Darcy water flow velocity (mm/d), $T = T(t,z,c)$ represents the sink resp. source term, including plant N uptake (kg/d), mineralized N, etc.; and finally $k_d = k_d(z)$ the adsorption constant for urea or ammonium (dm^3/kg). This equation was solved by applying the procedure based on the discretization scheme according to the LEACHN model (Hutson and Wagenet, 1992).

In addition to the model for vertical N movement, a complete N-model consists of sub-models for N mineralization, urea hydrolysis, nitrification, denitrification, and NH_3 volatilization. Expert-N provides two complete N models, which include the approaches used in both the LEACHN and CERES models. The more simplistic CERES model simulates N turnover independently of C turnover; whereas in LEACHN, as in most models, N dynamics is driven by C turnover. N transformations described by LEACHN follow the concept of Johnsson et al. (1987), who assume N mineralization to result from the decomposition of three different organic matter pools, namely, litter, manure, and humus.

For each pool N mineralization, respectively, N immobilization is determined from the carbon (C) decomposition and the C:N ratio of the pool using the following equations:

$$\frac{dN_{lit}}{dt}=\left[-\frac{N_{lit}}{C_{lit}}+\frac{f_e}{r_0}\left(1-f_h\right)\right]\cdot k_{lit}\cdot e_\theta\cdot e_T\cdot C_{lit} \qquad (14.3)$$

$$\frac{dN_{man}}{dt}=\left[-\frac{N_{man}}{C_{man}}+\frac{f_e}{r_0}\left(1-f_h\right)\right]\cdot k_{man}\cdot e_\theta\cdot e_T\cdot C_{man} \qquad (14.4)$$

$$\frac{dN_{hum}}{dt}=\frac{f_e\cdot f_h}{r_0}\left[\left(k_{lit}\cdot C_{lit}+k_{man}\cdot C_{man}\right)-k_{hum}\cdot N_{hum}\right]\cdot e_\theta\cdot e_T \qquad (14.5)$$

where the subscripts *lit* (litter), *man* (manure), and *hum* (humus) refer to the corresponding SOM pools; k is the reaction rate for C mineralization for the three soil organic matter pools; and the efficiency factor f_e denotes the fraction of the decomposed organic carbon converted to humus or to litter via incorporation into microbial biomass and not respirated as CO_2; the humification factor f_h defines the relative humus fraction that is produced, and r_0 is the C:N ratio of microbial biomass and humus.

Similar to the organic matter turnover rates, urea hydrolysis and nitrification rates are also described by first-order kinetics with the same correction functions e_θ and e_T accounting for the impact of water content and temperature (Johnsson et al., 1987). To calculate N_2O production and emission, different sub-models have been implemented in Expert-N. N_2O production during denitrification can be based on half-saturation or first-order kinetics, with subsequent N_2O reduction to N_2 on zero- or first-order kinetics. N_2O production during nitrification can be accounted for by assuming that produced N_2O-N levels represent a fixed fraction of the nitrified ammonium-N. N_2O emissions can be simulated by explicitly calculating gaseous N_2O transport within the soil profile and from the soil surface, or by simply assuming that the total N_2O amount produced is instantaneously emitted.

Plant Processes

Simulation of soil water and N uptake by plants occurs by selecting between complex crop growth models implemented in CERES (Ritchie, 1991; Jones and Kiniry, 1986), SUCROS (van Laar et al., 1992), SPASS (Wang, 1997), and the simple uptake functions used in LEACHM (Hutson and Wagenet, 1992; Nimah and Hanks, 1973; Watts and Hanks, 1978), all contained in Expert-N. The complex models can be applied to simulate the development and biomass growth of wheat, barley, maize, potato, and sunflower. The simple uptake functions do not predict crop growth but only describe water and nitrogen uptake. These functions can be modified by the following input data: start of plant growth, maximum crop cover, maximum rooting depth, maximal potential nitrogen uptake, and the dates these maxima are reached. Simulations of low water contents and lack of nitrogen in the rooting zone result in simulated lower actual water and nitrogen uptake than prescribed by the shapes of the uptake functions.

OPTIMIZATION METHODS

As opposed to direct methods, inverse modeling or inverse parameter estimation of soil water flow or solute transport optimize related model parameters by utilizing data from transient transport experiments. This is achieved by exploiting the model's ability to reproduce these transient data sets in terms of the dependence of the model on its parameters. Recently, inverse methods have become more popular (Kool and Parker, 1988; Yamaguchi et al., 1992; Romano, 1993; Hudson et al., 1996; Zijlstra and Dane, 1996; Lehmann and Ackerer, 1997; Abbaspour et al., 1997; Simunek et al., 1998; Pan and Wu, 1998; Schmied et al., 2000), due to their general applicability to various data types, improved simulation models, and the availability of more powerful computers (Pan and Wu, 1998). Inverse estimation of unsaturated hydraulic properties is a highly nonlinear optimization problem, and surfaces of the given objective functions plotted in parameter space are most often characterized by a complex geometry containing many local minima. In addition to the typically applied Levenberg–Marquardt algorithm (Press et al., 1992; van Genuchten et al., 1992), an alternative approach based on an annealing-simplex method has been proposed (Press et al., 1992; Pan and Wu, 1998). Both methods have been implemented into the Expert-N program package, which provides routines to simulate water flow, N-turnover, and N-transport in agroecosystems (Baldioli et al., 1995; Expert-N, 1999; Stenger et al., 1999).

The Levenberg–Marquardt method is a least-squares optimization approach, which is used for fitting if the model depends nonlinearly on the unknown model parameters. This method is widely used for parameter estimation and has become the standard nonlinear least-squares routine (Press et al., 1992). However, this method can be very sensitive to initial estimate (Hudson et al., 1996), and because it is a gradient-type method, it cannot guarantee the global optimal solution. In practice, if local optima are accepted, non-unique parameters can result from different initial estimates (Li et al., 1999). Simulated annealing is a technique particularly useful for finding the global optimum in between many local minima. Because it does not require calculation of derivatives, it can also be applied if the objective function is not smooth. The basic idea of simulated annealing comes from thermodynamics and statistical mechanics. The algorithm is in analogy to the physical process of slowly cooling a physical system (e.g., a piece of metal) until it reaches its minimum energy state (van Laarhoven and Aarts, 1987; Li et al., 2000). In combination with the downhill simplex method, simulated annealing can be applied to continuous variable problems by introducing a thermal fluctuation to the objective function that has to be minimized (Press et al., 1992; Pan and Wu, 1998).

To accelerate and stabilize the estimation procedure, both methods were combined into a multilevel approach (Chavent et al., 1994; Igler and Knabner, 1997). The resulting method consists of three different steps: (1) the optimization begins with the least possible number of degrees of freedom (if possible, n = 1); (2) the optimization result is interpolated for a parameterization with one or more degrees of freedom added, and is used as the next start value; (3) the second step is repeated as long as the optimization result is not significantly disturbed by oscillations. By this multilevel method, information on soil horizons can be used directly. Adequate parameter estimation can be achieved by starting with average parameter values constant throughout the entire soil profile, assuming a homogeneous soil, and continuing by dividing the profile successively to approximate the given division into soil horizons (Figure 14.1). This is especially true for cases with high numbers of degrees of freedom and highly nonlinear objective functions. The multilevel approach provides additional flexibility for the estimation procedures, in particular if layered soils are considered. To the authors' knowledge, in most publications on inverse methods, the estimation of unknown parameters has been limited to a single-layered, homogeneous soil (e.g., Kool and Parker, 1988; Pan and Wu, 1998; Schmied

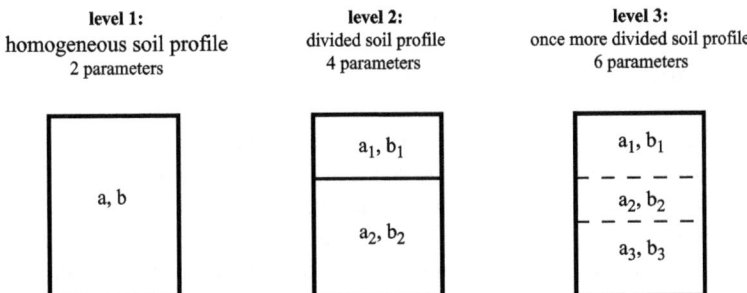

Figure 14.1 Multilevel optimization approach for soil profile parameters.

et al., 2000). Only in the study of Zijlstra and Dane (1996) have parameters been identified for a layered soil. To simultaneously estimate hydraulic parameters for three soil horizons, Zijlstra and Dane used an improved Levenberg–Marquardt algorithm, which was based on a quasi-Newton method. In contrast to this method, the multilevel approach iteratively increases the number of soil layers and successively provides more degrees of freedom for the path followed by the search algorithm to find the minimum of the objective function.

The optimization methods were applied to estimate parameters of the van Genuchten-type WRCs (water retention curves). The methods were also used to fit nitrification rates, N_2O production rates, and N mineralization rates from laboratory soil column data and an agricultural field experiment. A comparison of results obtained by the Levenberg–Marquardt and the annealing-simplex methods showed only minor differences. Most often, slightly smaller deviations from measured data (i.e., a lower minimum of the objective function [residuum]) was obtained by the simulated annealing-simplex method. The general form of the objective function is given by:

$$\sum_{k=1}^{m} \frac{1}{w_k} \cdot \left[\sum_{j=1}^{n} \sum_{i=1}^{p} \left(x_{kji,obs} - x_{kji,sim} \right)^2 \right]^{1/2} \tag{14.6}$$

where $x_{kji,obs}$ denotes the measured data, $x_{kji,sim}$ the corresponding simulated values; the index k stands for the m different types of data such as water contents, nitrate contents, or ammonium contents; the index $j = 1, ..., n$ enumerates the n different measured data (e.g., for different times of measurements), and the index i denotes the p different soil depths at which the measurements were taken. The weighting coefficient, w_k, for the data set of type k is defined by

$$w_k = \sum_{j=1}^{n} \sum_{i=1}^{p} x_{kji,obs}^2 \tag{14.7}$$

In the following applications, measured and simulated data h_{ji} for soil matric potentials were logarithmically transformed; in this case, $x_{ji} = \log(h_{ji})$ was used.

COLUMN AND FIELD EXPERIMENTS

Field Experiment

The field data sets used for the inverse parameter estimation are from a field experiment in which soil water flow, N transport, and N transformation processes were monitored. To assess, evaluate, and predict the effects of cultivation changes on N cycling in an agricultural landscape, the field experiment was carried out at the Klostergut Scheyern experimental farm to study N fluxes (net N mineralization, N leaching, root N uptake) and soil mineral N contents. The experimental farm is situated 45 km north of Munich (Germany), in a hilly landscape (450 to 490 m altitude) derived from tertiary sediments and partially covered with loess deposits. The average annual temperature is 7.4°C, and the yearly mean precipitation rate is 833 mm. Relevant agricultural management data are presented in Table 14.1 and basic soil properties in Table 14.2. Precipitation, air temperature, humidity, radiation, and wind speed were recorded every 10 minutes at a meteorological field station located at the experimental farm. Flessa et al. (1995), Stenger et al. (1996), and Dörsch (2000) give detailed information on the project aims and on the methods of sampling and chemical soil analyses. Here, only a brief overview of the experiment is given.

From August 1991 through December 1994, soil water contents and ammonium-N and nitrate-N amounts were determined every 3 to 4 weeks on a 1.48-ha field at 0 to 30, 30 to 60, and 60 to 90 cm depths (Stenger et al., 1996) and, additionally,

Table 14.1   Agricultural Management of the Field Site

Jan. 13, 1991	60 kg N/ha calcium ammonium nitrate
Mar. 13, 1991	50 kg N/ha calcium ammonium nitrate
May 15, 1991	50 kg N/ha calcium ammonium nitrate
Aug. 11, 1991	Harvest of winter wheat
Aug. 15, 1991	Tillage (10 cm)
Apr. 27, 1992	50 kg N/ha calcium ammonium nitrate
Aug. 17, 1992	1352 kg C/ha farmyard manure (68 kg N/ha, 6 kg NH_4/ha, C:N ratio = 19.9), tillage (10 cm), sowing of cover crop mixture including legumes
Feb. 4, 1993	Tillage (6 cm)
Apr. 23, 1993	Tillage (10 cm), sowing of sunflower
May 11, 1993	Mechanical weed control (5 cm)
May 24, 1993	Mechanical weed control (5 cm)
Sept. 21, 1993	Harvest of sunflower
Sept. 24, 1993	Tillage (10 cm)
Dec. 6, 1993	Tillage (10 cm)
Feb. 14, 1994	Tillage (15 cm)
Mar. 31, 1994	Sowing of lupine and harrowing (10 cm)
May 6, 1994	Mechanical weed control (5 cm)
May 13, 1994	Mechanical weed control (5 cm)
May 16, 1994	Mechanical weed control (10 cm)
June 1, 1994	Mechanical weed control (10 cm)
Aug. 23, 1994	Harvest of lupine
Aug. 30, 1994	Tillage (10 cm)
Sept. 20, 1994	Tillage (15 cm)

Table 14.2   Basic Soil Properties for the Field Site Profile

Soil depth (cm)	Clay (%)	Silt (%)	Sand (%)	Corg (%)	ρ (g/cm^3)	K_{sat} (mm/d)
0–20	19	25	56	1.15	1.49	6257
20–50	27	14	59	0.73	1.62	2710
50–90	22	11	67	0.21	1.63	3841
90–180	13	8	80	0.12	1.64	13,351

almost weekly at 0 to 30 cm depth during July 1992 to August 1994 (Flessa et al., 1995; Dörsch, 2000). Mineral N was extracted from field moist soils with 0.01 M CaCl$_2$ (soil:solution ratio = 1:2). NH$_4$-N and NO$_3$-N were determined colorometrically using a continuous flow analyzer (SA 20/40, Skalar Analytical, Erkelenz, Germany). Gravimetric water content was determined by drying at 105°C for 24 hours. From September 1994 through March 1997, volumetric water contents and pressure heads were monitored at six different depths (10, 20, 50, 90, 130, and 180 cm) using TDR probes (tensiometers) located in the same field approximately 20 m away from the site of mineral N measurements.

Column Experiment

For the soil column experiment, the reader is referred to the detailed description of Rackwitz (1997). For this experiment, undisturbed soil columns were sampled from topsoil (0 to 25 cm depth) and installed in a microcosm system (Hantschel et al., 1994a), which itself was installed in a constant-temperature room (14°C). The microcosm system was applied to keep the soil columns under a steady-state water flow regime by computer-controlled irrigation. It was also used to measure ventilation air flow rates, and air CO$_2$ and N$_2$O concentrations by online gas chromatography every 2 to 6 days. In the leachates, sampled at the same frequency, nitrate, ammonium, and chloride concentrations were determined to obtain breakthrough curves (BTCs).

After steady-state water flow conditions were achieved, an amount equivalent to 150 kg N/ha of ^{15}NH$_4$Cl (94 atom% ^{15}N) was added with the irrigation water. Following this application, increased N$_2$O emissions were observed; and after 8 weeks, most of the added ^{15}N was recovered as nitrate-^{15}N in the percolated solute.

In an additional batch experiment, a linear NH$_4$ sorption isotherm was determined and the distribution coefficient k_d = 0.92 cm^3/g was obtained.

MODEL APPLICATIONS (SIMULATION RESULTS AND DISCUSSION)

Parameter Estimation of Soil Hydraulic Property Functions

Mualem's theory (Mualem, 1976) was applied to estimate unsaturated hydraulic conductivities from water retention curves (WRCs). This was achieved using measured saturated hydraulic conductivities K$_{sat}$ (mm/d) and saturated water contents θ_s (mm^3/mm^3) for each soil horizon and assuming zero residual water contents θ_r (mm^3/mm^3) (Priesack et al., 1999). For the different soil horizons, therefore, only the WRC shape had to be estimated. In this particular case, the shape parameters α (mm^{-1}) and n (1) of the van Genuchten-type WRC (van Genuchten, 1980) derived from:

$$\theta(h) = \theta_r + \left(\theta_s - \theta_r\right)\left[1 + |\alpha \cdot h|^n\right]^{-1} \tag{14.8}$$

where θ (mm^3/mm^3) denotes the volumetric water content and h (mm) the matric potential. Necessary crop data to simulate water uptake included leaf area index, planting density, and rooting depth for different dates. Based on the TDR measurements of volumetric water contents and on soil matric potential measurements at six different depths of the field site for the time periods September 1, 1994 to December 31, 1995 (period 1995), and August 13, 1997 to March 31, 1997 (period 1997), the newly developed estimation procedure (Figure 14.1) was used to find the optimal α and n values for six different layers of the soil profile (Table 14.2). For reasons of comparison, both optimization algorithms — the Levenberg–Marquardt and the annealing-simplex methods — were applied. In the first step (level 1), an average α and an average n value were fitted for the whole soil profile. By subdividing the profile into halves and using the fitted average α and n value as start values for the next estimation, α and n values were obtained for the upper and lower halves of the soil profile (level 2). Using these values as start values to consider properties of the topsoil and to account for the bottom layer in the next step (level 4), the upper half was divided into an upper part (0 to 20 cm) and a lower part (20 to 90 cm), and the lower half into upper two thirds (90 to 150 cm) and a lower third (150 to 180 cm). Based on these newly obtained α and n values in the final step (level 6), the hydraulic parameters for six layers (0 to 10, 10 to 20, 20 to 50, 50 to 90, 90 to 150, and 150 to 180 cm) were estimated (Tables 14.3 and 14.4). Both optimization algorithms, in combination with the multilevel approach, led to stable water flow simulations, such that the numerical solver of the Richards equation converged in each case of the iterations during the search for optimal parameter values. Furthermore, the optimization routines were able to provide unique and stable parameter estimates, that is, values that were not sensitive to initial guess or noise originating from measured data. The α and n values obtained by either method are almost the same and, hence, represent retention curves of similar shapes. This leads to simulated water contents and matric potentials that are almost indistinguishable, as can be seen from the almost identical minimum of the objective function indicating the distance between observed and simulated water contents and matric potentials. In most cases, moreover, the 95% confidence intervals (Table 14.3) determined from the covariance of the parameter matrix for the Levenberg–Marquardt least-squares optimization contain parameter values obtained by the annealing-simplex method. This indicates that both methods were able to detect optimal parameter values, and that small deviations can be attributed to numerical errors of the flow simulator and optimization procedures.

To clarify if the global minimum is reached, further analysis of the surface geometry representing the response of the objective function in parameter space is needed, because the annealing-simplex method is able to find the global minimum only if it is significantly deeper than other local minima (Pan and Wu, 1998).

Table 14.3 Fitted van Genuchten Parameters for the Field Site Soil Profile for 1995 (Levenberg–Marquardt method)

Levels	1 Layer		2 Layers		4 Layers		6 Layers	
	α (cm⁻¹)	n	α (cm⁻¹)	N	α (cm⁻¹)	n	α (cm⁻¹)	n
Values	0.041	1.16	0.029	1.15	0.025	1.02	0.03	1.03
			0.041	1.18	0.024	1.16	0.016	1.23
					0.042	1.22	0.019	1.11
					0.040	1.17	0.045	1.21
							0.038	1.25
							0.040	1.17
Intervals	[0.040, 0.042]	[1.13, 1.17]	[0.028, 0.030]	[1.14, 1.17]	[0.024, 0.025]	[1.00, 1.03]	[0.029, 0.031]	[1.01, 1.04]
			[0.040, 0.042]	[1.17, 1.20]	[0.024, 0.025]	[1.13, 1.17]	[0.013, 0.017]	[1.21, 1.24]
					[0.024, 0.025]	[1.20, 1.23]	[0.019, 0.020]	[1.10, 1.12]
					[0.024, 0.025]	[1.16, 1.19]	[0.044, 0.047]	[1.19, 1.23]
							[0.037, 0.039]	[1.23, 1.27]
							[0.039, 0.041]	[1.16, 1.19]
Residuum	0.102		0.097		0.094		0.078	

Table 14.4 Fitted van Genuchten Parameters for the Field Site Soil Profile for 1995 (simulated annealing method)

Levels	1 Layer		2 Layers		4 Layers		6 Layers	
	α (cm⁻¹)	n	α (cm⁻¹)	N	α (cm⁻¹)	n	α (cm⁻¹)	n
Values	0.037	1.14	0.017	1.16	0.025	1.03	0.035	1.03
			0.041	1.20	0.020	1.16	0.014	1.25
					0.042	1.20	0.014	1.12
					0.050	1.17	0.047	1.17
							0.037	1.25
							0.046	1.16
Residuum	0.11		0.096		0.094		0.077	

Table 14.5   Fitted van Genuchten Parameters α and n for the Field Site Profile for 1997

	Levenberg–Marquardt Method		Annealing-Simplex Method	
	α (cm⁻¹)	n (–)	α (cm⁻¹)	n (–)
	0.072	1.01	0.086	1.01
	0.008	1.14	0.008	1.15
	0.013	1.11	0.013	1.10
	0.034	1.10	0.035	1.10
	0.102	1.12	0.123	1.13
	0.029	1.28	0.101	1.27
Residuum	0.137		0.133	

Table 14.6   Simulated Water Balances (mm)

Case	Runoff	Actual Evaporation	Actual Transpiration	Water Storage	Leaching
95/95	27	175	337	221	218
95/97	49	128	337	268	199
97/97	30	87	0	143	123
97/95	16	132	0	95	141

Both optimization procedures were also applied to the data period 1997, resulting in α and n values (Table 14.5) that deviated from those of period 1995. The estimated n values are often found for silty clay and clay, whereas the higher n values obtained for the 1995 period can be found for silty clay loam and sandy clay (Carsel and Parrish, 1988). This is in contrast to the observed textural data, in particular to the high sand fraction of the subsoil (90 to 180 cm), which is not represented by significantly higher n values. However, hydraulic property functions may not only depend upon soil texture. Soil structure can determine soil pore space geometry to a large extent and thus have a significant impact on WRCs and unsaturated conductivities (Roth et al., 1999). Water balances were calculated using the optimized hydraulic parameters derived from period 1995 data, for the simulation of periods 1995 (case 95/95) and 1997 (case 97/95) and are shown in Table 14.6. Similarly, simulations were performed with the optimal parameters from the period 1997 (cases 97/97 and 95/97). The main difference between the water balances determined with different hydraulic parameters occurs in the calculated actual evaporation. Using the parameters from period 1997, simulations are obtained with lower actual evaporation, higher water storage, and less cumulative leaching (e.g., the soil retains more water). The 10 to 15% difference observed when estimating cumulative leaching of soil water using the parameter values for the two different periods represents part of the inaccuracy and limitations of model prediction for nitrate loads leaving the rooted soil zone and contributing to potential groundwater pollution from the field site under consideration. The hydraulic parameters fitted to the period 1995 data appear useful for simulating volumetric water contents for the earlier period of August 15, 1991, to December 31, 1994, as simulated values compare well with measured ones (Figure 14.2). Because the data are an independent data set and not

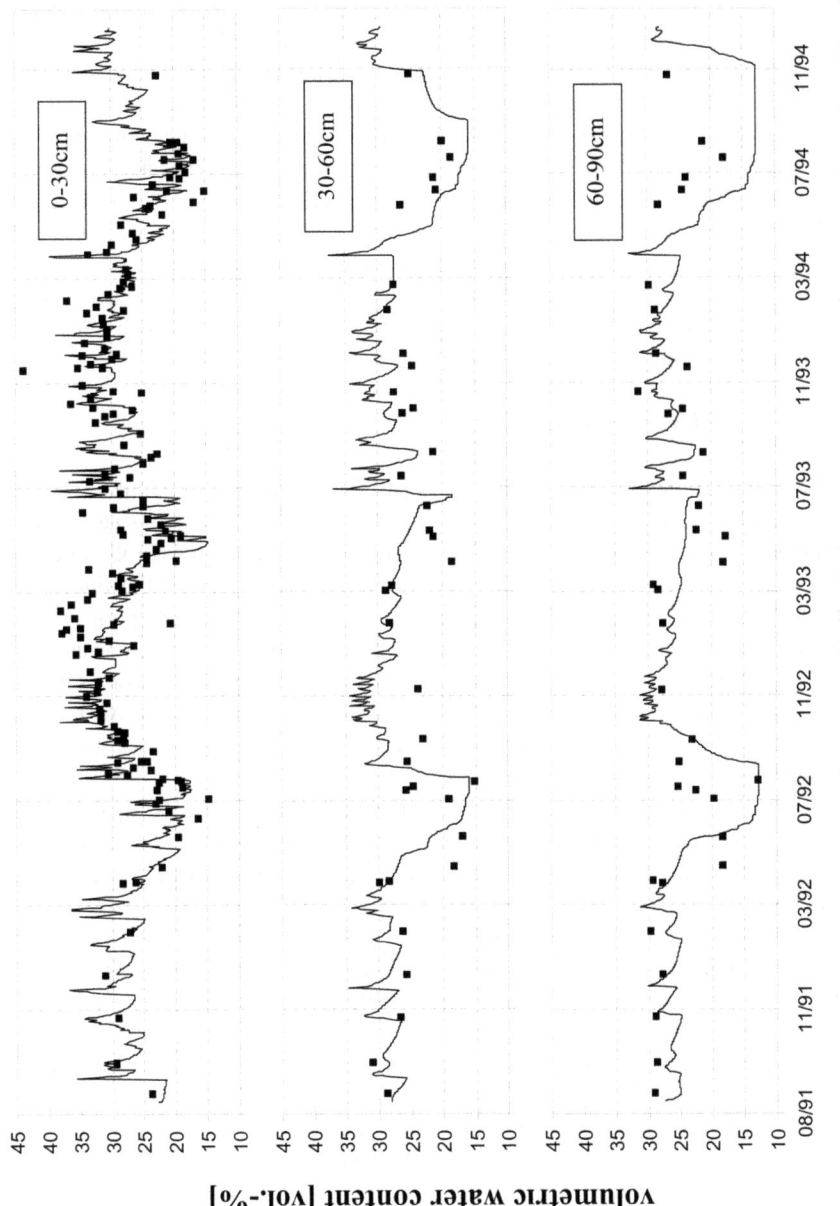

Figure 14.2 Measured and simulated volumetric water contents for three different depths (0–30 cm, 30–60 cm, and 60–90 cm) using fitted hydraulic properties.

used for parameter estimation, the simulated water contents, which often meet the measured values within a range of ±2.5 vol.%, demonstrate that the estimated parameters of the water flow model are adequate, in particular if the high spatial variability at the field site is taken into account (Stenger et al., 2000). Noticeable is the short period in February 1993 when the measured water contents of the upper soil (0 to 30 cm) were considerably higher than the simulated values. This period, characterized by repeated freezing and thawing, led to lower soil densities (i.e., to a higher soil porosity), a process which is not modeled by the Richards equation for water flow simulation. This short period gives an example where adjustment of unknown parameters to fit observed data becomes a futile process because the model does not account for observed soil porosity changes. The adjustment of inappropriate parameters does not necessarily lead to good fits if relevant processes are neglected. This case has been observed true by Schmied et al. (2000), with respect to the inverse modeling of field experiments that they have thus far considered. Therefore, inverse modeling may represent a potentially appropriate tool to identify processes either overlooked or inadequately represented by the model.

Estimation of Mineralization Rates

The N mineralization sub-model of Expert-N extended the definition of the manure pool to represent young soil organic matter (YSOM), because nitrate dynamics may be described better with a second, comparatively active pool (Frolking et al., 1998). Crop residues (above and below ground) are added both to the litter pool (with a mineralization rate of 0.045/d) and to the YSOM or manure pool (with a mineralization rate of 0.015/d). It is assumed here that plant residues consist of a large labile fraction and a more stable fraction of organic material. For example, cover crop residues were partitioned 90% to the litter pool and 10% to the manure pool, cereals straw 80 to 20%, sunflower straw 60 to 40%, and farmyard manure 10 to 90%, respectively.

To test if this partitioning of organic fertilizers to different pools is justified in the case of our field study, we applied the Levenberg–Marquardt procedure to estimate the mineralization rates (for litter, manure, and humus pools). The measured amounts of ammonium-N and nitrate-N (kg/ha) from August 1991 to December 1994 in the soil depths 0 to 30, 30 to 60, and 60 to 90 cm (Figure 14.3) were used to define the objective function. As additional model input to the weather, agricultural management, plant N uptake, and soil data (Table 14.3), the parameter values $f_e = 0.45$, $f_h = 0.3$ and $r_0 = 10$ for the mineralization model and the previously fitted soil hydraulic parameters for the water flow model were taken. The fitting of mineralization rates was stable for all different periods (Table 14.7) and unique optimal parameter values could be obtained for each period. Litter and manure mineralization rates differed only slightly (Table 14.7), and demonstrate that partitioning of crop residues into a second soil organic matter pool did not improve nitrate transport simulations for the field site considered. Therefore, in this case, if only soil ammonium-N and nitrate-N amounts are utilized in the estimation procedure, it is not possible to distinguish between different decomposition rates of freshly added soil organic material. Furthermore, mineralization rates appeared to be depth independent,

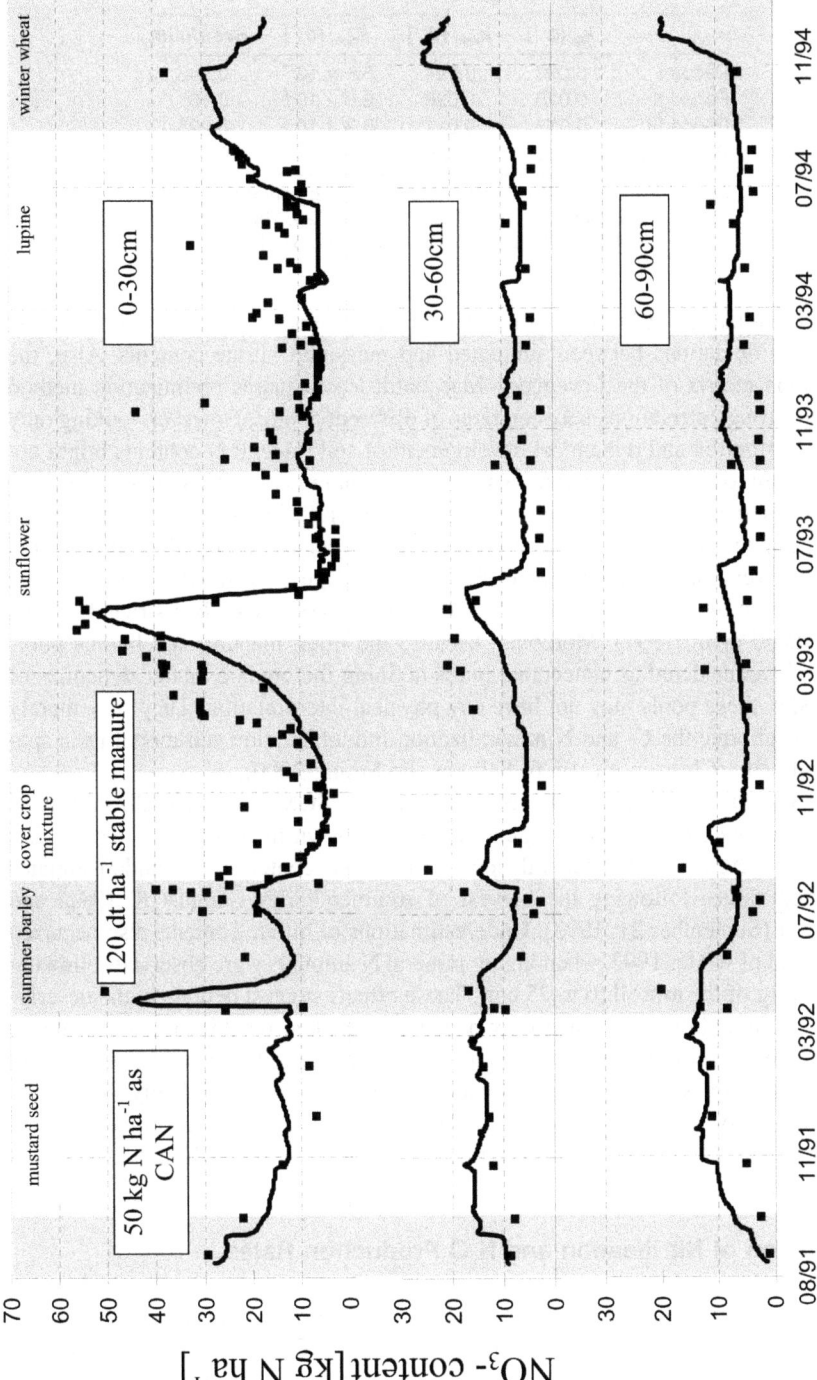

Figure 14.3 Measured and simulated nitrate N contents for three different depths (0–3 cm, 30–60 cm, and 60–90 cm) using fitted N mineralization rates.

Table 14.7   Fitted Mineralization Rates for Litter, Manure and Humus Pools

	k_{litt} [d^{-1}]	k_{man} [d^{-1}]	k_{hum} [d^{-1}]	Residuum
Period I	0.028	0.024	5.5×10^{-5}	0.143
Period II	0.028	0.026	6.0×10^{-5}	0.13
Period III	0.017	0.017	6.7×10^{-5}	0.22
Period I–III	0.025	0.025	5.5×10^{-5}	0.21

Note: Period I: Aug. 15, 1991–Aug. 5, 1992; period II: Aug. 6, 1992–Sept. 23, 1993; period III: Sept. 24, 1993–Dec. 31, 1994; period I–III: Aug. 15, 1991–Dec. 31, 1994.

as the optimization at higher levels for nonhomogeneous soil profiles did not lead to smaller deviations between simulated and measured nitrate contents. Also, the correlation matrix of the Levenberg–Marquardt least-squares optimization method showed strong correlations between rates of different depths. However, looking only at the distribution and dynamical development of soil mineral N contents might not be sufficient to discern differences between different mineralization rates of different soil organic matter pools at different depths. Therefore, without direct measurements of soil organic matter pools, the estimation of the related C and N decomposition rates by inverse methods remains difficult, as the degrees of freedom in the model interpretation of these field-scale processes may not be sufficiently restricted by data sets (Magid et al., 1997). Moreover, because the litter, manure, and humus pools have to be considered as conceptual pools defining the organic matter dynamics of the model, these pools may not have any physical interpretation. They may merely serve to calibrate the C- and N-mineralization-immobilization sub-model for a specific field site (Christensen, 1996; Falloon and Smith, 2000).

Simulation results for nitrate contents using the SOM pool mineralization rates fitted to the measured nitrate content values are shown in Figure 14.3. The model is able to reproduce the general course of nitrate contents, but underestimates measured values following the harvest of summer barley (August 8, 1992) and sunflower (September 21, 1993). Underestimations of nitrate contents also occurred during end of winter 1993, when higher mineral N amounts were observed following the thawing of the topsoil (0 to 25 cm). These effects suggest deficits in the description of N immobilization when only a part of the plant residues is plowed under and smaller mineral N amounts are immobilized than predicted by the simulation. During thawing periods, N mineralization seems to be high probably due to the mobilization of protected organic materials (Dörsch, 2000). The impact of soil freezing and thawing on N turnover rates therefore also needs a better model description.

Estimation of Nitrification and N$_2$O Production Rates

To accurately model N transport in soils, the decomposition rates of N species urea, ammonium, and nitrate (i.e., the rates of urea hydrolysis, of nitrification, and of denitrification) must be known, in addition to the rates describing C and N contents changes in soil organic matter pools. Moreover, transport parameters for the

N species, including coefficients of dispersion and adsorption, are needed, and may be obtained from laboratory soil column experiments (Hantschel et al., 1994b; Stenger et al., 1995; Rackwitz, 1997), *in situ* field experiments (Stenger et al., 1996), and through the application of inverse estimation methods (Jemison et al., 1994).

To further illustrate the inverse modeling technique, we analyzed the laboratory soil column experiment to estimate nitrification and N_2O production rates subsequent to mineral fertilizer application. For the transport and transformation of ammonium in soils, four processes must be considered: (1) sorption and desorption, (2) oxidation, (3) transport in the soil solution including convection, diffusion, and hydrodynamic dispersion, and (4) immobilization or mineralization by soil microbes. For each individual laboratory soil column, these processes can be analyzed by considering the breakthrough curves (BTCs) for fertilizer derived NH_4^+, NO_3^-, and Cl^-.

As the experimental recovery of ^{15}N added with the $^{15}NH_4^+$ fertilizer was high, it may be concluded that fertilizer N losses due to denitrification are almost negligible, and the immobilization of both $^{15}NH_4^+$ and $^{15}NO_3^-$ is low. For modeling purposes, we therefore neglect $^{15}NH_4^+$ and $^{15}NO_3^-$ immobilization and mineralization by soil microbes. If mineralization of soil organic matter or immobilization of mineral N by soil microbes cannot be neglected, these N turnover rates must be determined by additional measurements, or must be included as unknown parameters in the estimation procedure. We further assume that for steady-state water flow conditions, a constant volumetric water content θ and a constant Darcy flow velocity q adequately describe water flow in the soil column. The transport and transformation of fertilizer-derived NH_4^+ and NO_3^- in the soil solution of the soil columns is then described by the advection-dispersion equations for ammonium-N and nitrate-N, which then are only coupled by the sink term (for ammonium-N) and source term (for nitrate-N) representing the change in N concentration due to nitrification.

The constant dispersion coefficient D is estimated by fitting the chloride BTC. Applying the CXTFIT program for the linear equilibrium adsorption model (Parker and van Genuchten, 1984), a value of $D = 0.64$ (cm^3/g) was obtained. Using the fitted dispersion coefficient and the measured k_d value for NH_4^+ adsorption the nitrification rate $k = 0.48/d$ was obtained by simultaneously fitting the transport equations for ammonium-N and nitrate-N to the measured $^{15}NH_4^+$ and $^{15}NO_3^-$ BTCs. This was achieved by applying the Levenberg-Marquardt optimization algorithm.

Assuming that most of the emitted N_2O is $^{15}N_2O$, we applied two different models to estimate N_2O production, either during denitrification or during nitrification. The model that describes N_2O production rate during nitrification by:

$$\frac{dC_{N_2O}}{dt} = \gamma \cdot k_{nit,NO_3} \cdot f(t - \tau) \qquad (14.9)$$

and the model for the N_2O production rate during denitrification by:

$$\frac{dC_{N_2O}}{dt} = k_{den,N_2O,\max} \cdot \frac{\theta \cdot C_{NO_3}}{\theta \cdot C_{NO_3} + K_{NO_3}} \cdot f(t - \tau) \qquad (14.10)$$

Table 14.8   Fitted N$_2$O-Production Rates Using Different Models

Levels	Fitting of N$_2$O Production Rate During Denitrification $k_{den,N_2O,max}$				Fitting of N$_2$O Production Rate During Nitrification γ k_{nitNO_3}			
	1 Layer	2 Layers	4 Layers	8 Layers	1 Layer	2 Layers	4 Layers	8 Layers
	0.001	0.004	0.008	0.009	0.117	0.112	0.097	0.021
		0.000	0.000	0.008		0.078	0.566	1.995
			0.000	0.000			0.025	1.293
			0.000	0.000			0.017	0.271
				0.000				0.011
				0.000				0.012
				0.000				0.086
				0.000				0.001
Residuum	0.691	0.274	0.080	0.0790	0.889	0.887	0.848	0.325

Here, the N$_2$O-N concentration C_{N_2O} (g/cm^3) in the soil solution is assumed to be in instantaneous equilibrium with N$_2$O-N in soil air, γ is the proportionality coefficient between nitrification rate and N$_2$O production rate, f describes the delay time occurring before N$_2$O production begins subsequent to NH$_4^+$ application by:

$$
f(t-\tau) = \begin{cases} 1/2 \cdot \exp\left(\dfrac{t-\tau}{\tau}\right), & t \le \tau \\[2ex] 1 - 1/2 \cdot \exp\left(\dfrac{t-\tau}{\tau}\right), & t > \tau \end{cases}
\tag{14.11}
$$

$k_{den,N_2O,max}$ is the maximal N$_2$O production rate during denitrification, θ denotes the volumetric water content, and $K_{NO_3} = 0.01$ mg/cm^3 is the half-saturation constant for denitrification.

In the following, either the N$_2$O production rate model for nitrification was combined with the ammonium transport equation, or the N$_2$O production rate model for denitrification with the nitrate transport equation. It was assumed that N$_2$O is immediately emitted when it is produced. The N$_2$O emission data were used to fit the delay time and the coefficient γ for the N$_2$O production during nitrification, and the maximal N$_2$O production rate during denitrification $k_{den,N_2O,max}$. Using the multi-level fitting approach, depth-dependent values for $k_{den,N_2O,max}$ and γ can be obtained. The results shown in Table 14.8 illustrate that N$_2$O production occurs in the upper part of the soil columns. Both models for N$_2$O production can be applied to reproduce the observed time course of N$_2$O emissions (Figure 14.4).

SUMMARY AND CONCLUSION

The simulation of nitrate leaching from agricultural soils requires not only information on weather conditions and farmer management practices, but also data on soil parameters in order to quantify water flow and solute transport. The means by which these parameters are obtained is illustrated by inverse modeling of a laboratory soil column experiment and of a field experiment using time series of

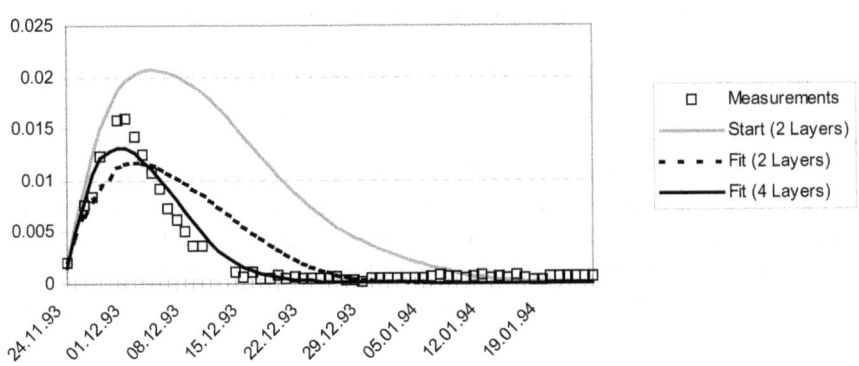

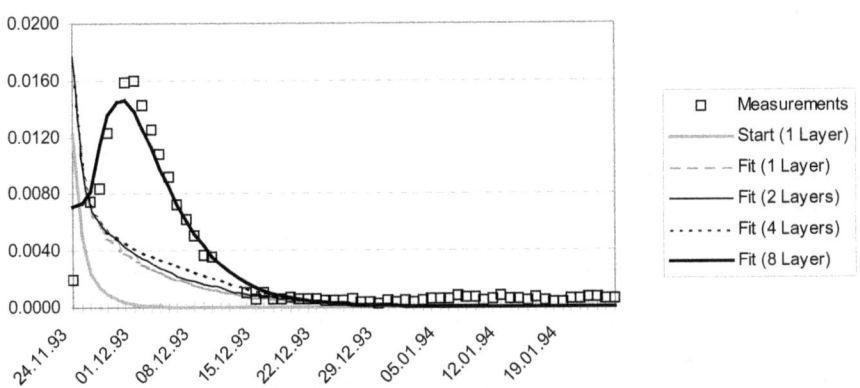

Figure 14.4 Fitting of observed N_2O emissions by two different modes: (a) N_2O production rate during denitrification with delay, and (b) N_2O production rate during nitrification with delay.

measured data for soil water and mineral nitrogen contents. Soil hydraulic parameters are needed to characterize water retention curves, soil hydraulic conductivities, and root water uptake. Required parameter values for nitrate transport include estimates for solute dispersion and for rates of nitrogen mineralization, nitrification, and denitrification.

For inverse simulation procedures, the Levenberg–Marquardt algorithm and a simulated annealing procedure were applied. These nonlinear optimization procedures minimize the different objective functions that quantify the deviation between measured and simulated soil water contents, or ammonium- and nitrate-N concentrations in the soil solution. These fitting algorithms were combined with a multilevel approach to accelerate and stabilize them, and were implemented into the open soil-plant-atmosphere model system Expert-N.

Results demonstrate the applicability of inverse modeling for estimating nitrate transport parameters. The newly developed fitting procedure of the Expert-N model was applied to estimate soil hydraulic properties and N turnover rates from laboratory soil column and agricultural field site experimental data. For these selected examples, it was possible to estimate parameters that enabled adequate water flow and N turnover simulations. The multilevel approach was shown effective in the analysis of dependency of parameters from soil depth. Thus, it was possible to identify the reaction space of N_2O production; albeit the process itself — either during nitrification or denitrification or both — could not be identified. Using time series data for the parameter estimation from different periods can lead to different parameter values. This illustrates the limitations of model prediction, for example, to estimate possible nitrate leaching to groundwater.

In summary, for the different models included in the program package Expert-N, the new inverse modeling method has been shown to be a useful and versatile tool to estimate model parameters, in particular if parameter values cannot be measured or measured values are not available. Moreover, inverse modeling can help to identify deficits in the model description of soil water flow and soil nitrogen turnover.

REFERENCES

Abbaspour, K.C., M.T. van Genuchten, R. Schulin, and E. Schäppi. 1997. A sequential uncertainty domain inverse procedure for estimating subsurface flow and transport parameters. *Water Resour. Res.,* 33:1879-1892.

Addiscott, T.M. 1993. Simulation modeling and soil behavior. *Geoderma,* 60:15-40.

Anderson, E.A. 1973. National Weather Service River Forecast System — Snow Accumulation and Ablation Model. National Oceanographic and Atmospheric Administration (NOAA), Tech. Memo., NWS-HYDRO-17, U.S. Department of Commerce, Silver Spring, MD, 217 pp.

Baldioli, M., T. Engel, E. Priesack, T. Schaaf, C. Sperr, and E. Wang. 1995. *Expert-N, ein Baukasten zur Simulation der Stickstoffdynamik in Boden und Pflanze. Version 1.0.* Benutzerhandbuch, Lehreinheit für Ackerbau und Informatik im Pflanzenbau TU München-Freising, Germany, 202 pp.

Campbell, G.S. 1985. *Soil Physics with BASIC.* Elsevier, New York.

Carsel, R.F., and R.S. Parrish. 1988. Developing joint probability distributions of soil water retention characteristics. *Water Resour. Res.,* 24:755-769.

Chavent, G., Jianfeng Zhang, and C. Chardaire-Riviere. 1994. Estimation of mobilities and capillary pressure from centrifuge experiments. In H.D. Bui, M. Tanaka, M. Bonnet, H. Maigre, E. Luzzato, and M. Reynier (Eds.). *Inverse Problems in Engineering Mechanics,* Balkema, The Netherlands, 265-272.

Christensen, B.T. 1996. Matching measurable soil organic matter fractions with conceptual pools in simulation models of carbon turnover: revision of model structure. In D.S. Powlsen, P. Smith, J.U. Smith (Eds.). *Evaluation of Soil Organic Matter Models Using Existing, Long Term Datasets.* NATO ASI Series No. 1, Vol. 38. Springer, Berlin, Heidelberg, New York, 143-160.

De Willigen, P. 1991. Nitrogen turnover in the soil crop system; comparison of fourteen simulation models. *Fert. Res.,* 27:141-149.

Diekkrüger, B., D. Söndgerath, K.C. Kersebaum, and C.W. McVoy. 1995. Validity of agro-ecosystem models. A comparison of results of different models applied to the same data set. *Ecol. Model.,* 81:3-29.

Dörsch, P., 2000. Nitrous Oxide and Methane Fluxes in Differentially Managed Agricultural Soils of a Hilly Landscape in Southern Germany. FAM-Berichtt 44, Munich, Germany, 226 pp.

Expert-N. 1999. Version 2.0. http://www.gsf.de/iboe/expertn/

Falloon, P.D., and P. Smith. 2000. Modelling refractory soil organic matter. *Biol. Fert. Soils,* 30:388-398.

Flerchinger, G.N., and K.E. Saxton. 1989. Simultaneous heat and water model of a freezing snow-residue-soil system. I. Theory and development. *Trans. ASEA,* 32(2):565-571.

Flessa, H., P. Dörsch, and F. Beese. 1995. Seasonal variation of N_2O and CH_4 fluxes in differently managed arable soils in southern Germany. *J. Geophys. Res.,* 100:23115-23124.

Frolking, S.E., A.R. Mosier, D.S. Ojima, C. Li, W.J. Parton, C.S. Potter, E. Priesack, R. Stenger, C. Haberbosch, P. Dörsch, H. Flessa, and K.A. Smith. 1998. Comparison of N_2O emissions from soils at three temperate agricultural sites: simulations of year-round measurements by four models. *Nutr. Cycl. Agroecosys.,* 52:77-105

Hansen, S., H.E. Jensen, N.E. Nielsen, and H. Svendsen. 1991. Simulation of nitrogen dynamics and biomass production in winter wheat using the Danish simulation model DAISY. *Fert. Res.,* 27:245-259

Hantschel, R.E., H. Flessa, and F. Beese. 1994a. An automated microcosm system for studying soil ecological processes. *Soil Sci. Soc. Am. J.,* 58:401-404.

Hantschel, R., E. Priesack, and R. Hoeve. 1994b. Effects of mustard residues on the carbon and nitrogen turnover in undisturbed soil microcosms. *Z. Pflanzenernähr. Bodenkd.,* 157:319-326.

Hudson, D., P.J. Wierenga, and R.G. Gills. 1996. Unsaturated hydraulic properties from upward flow into soil cores. *Soil Sci. Soc. Am. J.,* 60:388-396.

Hutson, J.L., and R.J. Wagenet. 1992. LEACHM: Leaching Estimation And Chemistry Model: A Process-based Model of Water and Solute Movement, Transformations, Plant Uptake and Chemical Reactions in the Unsaturated Zone. Version 3.0. Department of Soil, Crop and Atmospheric Sciences, Research Series No. 93-3, Cornell University, Ithaca, NY.

Igler, B., and P. Knabner. 1997. Structural identification of nonlinear coefficient functions in transport processes through porous media. Preprint, Universität Erlangen-Nürnberg, *Angewandte Mathematik I,* Germany, 20 pp.

Jemison, J.M., J.D. Jabro, and R.H. Fox. 1994. Evaluation of LEACHM: II. Simulation of nitrate leaching from nitrogen-fertilized and manured corn. *Agron. J.,* 86:852-859.

Johnsson, H., L. Bergstrom, P.-E. Jansson, and K. Paustian. 1987. Simulated nitrogen dynamics and losses in a layered agricultural soil. *Agric. Ecosys. Env.,* 18:333-356.

Jones, C. A., and J.R. Kiniry (Eds.). 1986. *CERES-Maize. A Simulation Model of Maize Growth and Development.* Texas A&M University Press, 194 pp.

Kool, J.B., and J.C. Parker. 1988. Analysis of the inverse problem for transient unsaturated flow. *Water Resour. Res.,* 24:817-830.

Lehmann, F., and P. Ackerer. 1997. Determining soil hydraulic properties by inverse method in one-dimensional unsaturated flow. *J. Environ. Qual.,* 26:76-81.

Li, L., D.A. Barry, J. Morris, and F. Stagnitti. 1999. CXTANNEAL: An improved program for estimating solute transport parameters. *Env. Modelling and Software,* 14:607-611.

Li, P., K. Löwe, H. Arellano-Garcia, and G. Wozny. 2000. Integration of simulated annealing to a simulation tool for dynamic optimization of chemical processes. *Chem. Eng. Process.,* 39:357-363.

Magid, J., T. Mueller, L.S. Jensen, and N.E. Nielsen. 1997. Modelling the measurable: interpretation of field scale CO_2 and N-mineralization, soil microbial biomass and light fractions as indicators of oilseed rape, maize and barley straw decomposition. In G. Cadisch and K.E. Giller (Eds.). *Driven by Nature: Plant Litter Quality and Decomposition.* CAB International.

Mualem, Y. 1976. A new model for predicting the hydraulic conductivity of unsaturated porous media. *Water Resour. Res.,* 12:513-522.

Nicolardot, B., and J.A.E. Molina. 1994. C and N fluxes between pools of soil organic matter: model calibration with long-term field experimental data. *Soil Biol. Biochem.,* 26:245-251.

Nimah, M.N., and R.J. Hanks. 1973. Model for estimation of soil water, plant and atmospheric interrelations. I. Description and sensitivity. *Soil Sci. Soc. Am. Proc.,* 37:522-527.

Pan, L., and L. Wu. 1998. A hybrid method for inverse estimation of hydraulic parameters: annealing-simplex method. *Water Resour. Res.,* 34:2261-2269.

Parker, J.C., and M.T. van Genuchten. 1984. Determining transport parameters from laboratory and field tracer experiments. *Vir. Agric. Ex. Stat. Bull.* 84-3. Vir. Poly. Inst. and State Univ., Blacksburg, VA.

Press, W.H., S.A. Teukolski, W.T. Vetterling, and B.P. Flannery. 1992. Numerical Recipes in FORTRAN., 2nd ed., Cambridge Univ. Press, USA.

Priesack, E., W. Sinowski, and R. Stenger. 1999. Estimation of soil property functions and their application in transport modelling. In M.Th. van Genuchten, F. Leij, and L. Wu (Eds.). *Proc. International Workshop on the Characterization and Measurement of the Hydraulic Properties of Unsaturated Porous Media.* Oct 22-24, 1997, Riverside, CA, 1121-1129.

Rackwitz, R. 1997. *Modelluntersuchungen zur Stickstoffdynamik in Schwarzerden mit unterschiedlicher Düngungsgeschichte.* Shaker Verlag, Germany, 164 pp.

Ritchie, J.T. 1991. Wheat phasic development. In J. Hanks and J.T. Ritchie (Eds.). *Modeling Plant and Soil Systems. Agronomy Monograph 31.* ASA-CSSA-SSSA, Madison, WI, 31-54.

Romano, N. 1993. Use of an inverse method and geostatistics to estimate soil hydraulic conductivity for spatial variability analysis. *Geoderma,* 60:169-186.

Roth, K., H.-J. Vogel, and R. Kasteel. 1999. A conceptual framework for upscaling soil properties. In J. Feyen, and K. Wiyo (Eds.). *Modelling of Transport Processes in Soils at Various Scales in Time and Space. International Workshop of EurAgEng's Field of Interest on Soil and Water.* Wageningen Pers., Wageningen, The Netherlands, 477-490.

Scheinost, A.C., W. Sinowski, and K. Auerswald. 1997. Regionalisation of soil water retention curves in a highly variable soilscape. I. Developing a new pedotransfer function. *Geoderma,* 78:129-143.

Schmied, B., K. Abbaspour, and R. Schulin, 2000. Inverse estimation of parameters in a nitrogen model using field data. *Soil Sci. Soc. Am. J.,* 64:533-542.

Simunek, J., M.T. van Genuchten, M.M. Gibb, and J.W. Hopmans. 1998. Parameter estimation of unsaturated soil hydraulic properties from transient flow processes. *Soil Tillage Res.,* 47:27-36.

Simunek, J., K. Huang, and M.Th. van Genuchten. 1998. The HYDRUS Code for Simulating the One-dimensional Movement of Water, Heat, and Multiple Solutes in Variably-Saturated Media. Version 6.0, Research Report No. 144, U.S. Salinity Laboratory, USDA, ARS, Riverside, CA, 164 pp.

Sinowski, W., A.C. Scheinost, and K. Auerswald. 1997. Regionalisation of soil water retention curves in a highly variable soilscape. II. Comparison of regionalization procedures using a pedotransfer function. *Geoderma,* 78:145-159.

Smith, M. 1992. Report on the Expert Consultation on Revision of FAO Methodologies for Crop Water Requirements. Land and Water Development Division. Food and Agricultural Organization of the United Nations, Rome, Italy.

Stenger, R., E. Priesack, and F. Beese. 1995. Rates of net nitrogen mineralization in disturbed and undisturbed soils. *Plant Soil,* 171:323-332.

Stenger, R., E. Priesack, and F. Beese. 1996. *In situ* studies of soil mineral N fluxes: some comments on the applicability of the sequential soil coring method in arable soils. *Plant Soil,* 183:199-211.

Stenger, R., E. Priesack, and F. Beese. 1998. Distribution of inorganic nitrogen in agricultural soils at different dates and scales. *Nutr. Cycl. Agroecosys.,* 50:291-297.

Stenger, R., E. Priesack, G. Barkle, and C. Sperr. 1999. Expert-N. A tool for simulating nitrogen and carbon dynamics in the soil-plant-atmosphere system. In M. Tomer, M. Robinson, and G. Gielen (Eds.). *NZ Land Treatment Collective Proceedings Technical Session 20: Modelling of Land Treatment Systems.* New Plymouth, New Zealand, 19-28.

Stenger, R., E. Priesack, and F. Beese. 2000. Spatial variation of nitrate-N and related properties within plots of 50m by 50m. *Geoderma,* in revision.

Svendsen, H., S. Hansen, and H.E. Jensen. 1995. Simulation of crop production, water and nitrogen balances in two German agro-ecosystems using the DAISY model. *Ecol. Modelling,* 81:197-212.

Tietje, O., and M. Tapkenhinrichs. 1993. Evaluation of pedo-transfer functions. *Soil Sci. Soc. Am. J.,* 57:1088-1095.

Vanclooster, M., P. Viaene, J. Diels, and K. Christiaens. 1994. WAVE: A Mathematical Model for Simulating Water and Agrochemicals in the Soil and Vadose Environment. Reference and User Manual (Release 2.0). Institute for Land and Water Management, Katholieke Universitteit Leuven, Leuven, Belgium.

Van Genuchten, M.T. 1980. A closed-form equation for predicting the hydraulic conductivity of unsaturated soils. *Soil Sci. Am. J.,* 44:892-898

Van Genuchten, M.T., F.J. Leij, and S.R. Yates. 1992. The RETC Code for Quantifying the Hydraulic Functions of Unsaturated Soils. U.S. EPA, Ada, OK.

Van Laar, H.H., J. Goudriaan, and H. van Keulen (Eds.). 1992. Simulation of crop growth for potential and water-limited production situations, as applied to spring wheat. *Simulation Reports 27.* 72 pp.

Van Laarhoven, P.J.M., and E.H.L. Aarts. 1987. *Simulated Annealing: Theory and Application.* Kluwer Academic, Dordrecht, The Netherlands.

Wang, E. 1997. *Development of a Generic Process-oriented Model for Simulation of Crop Growth.* Herbert Utz Verlag, Munich, 195 pp.

Watts, D.G., and J.R. Hanks. 1978. A soil-water-nitrogen model for irrigated corn on sandy soils. *Soil Sci. Soc. Am. J.,* 42:492-499.

Wu, L., and M.B. McGechan. 1998. A review of carbon and nitrogen processes in four soil nitrogen dynamics models. *J. Agric. Engng Res.,* 69:279-305.

Yamaguchi, T.P., P. Moldrup, D.E. Folston, and J.A. Hansen. 1992. A simple inverse model for estimating nitrogen reaction rates from soil column leaching experiments at steady state water flow. *Soil Sci.,* 154:490-496.

Zijlstra, J., and J.H. Dane. 1996. Identification of hydraulic parameters in layered soils based on a quasi-Newton method. *J. Hydrol.,* 181:233-250.

CHAPTER **15**

Application of the DAISY Model for Short- and Long-term Simulation of Soil Carbon and Nitrogen Dynamics

Lars S. Jensen, Torsten Mueller, Sander Bruun, and Søren Hansen

CONTENTS

INTRODUCTION AND SCOPE

A great number of soil C and N models have been developed by various research groups over the past two decades, as is evident from the contents of this book, but often with quite different objectives. Some models have been developed primarily for prediction of soil organic matter (SOM) changes over longer time scales, for example, the CENTURY (Parton et al., 1987), the RothC (Coleman and Jenkinson, 1996), and the ICBM (Andrén and Kätterer, 1997) models. Other models, for example, the DAISY (Hansen et al., 1991; Abrahamsen and Hansen, 2000; Chapter 6 in this book) and the CANDY (Franko, 1996) models, have been developed more with the aim of simulating short-term N dynamics in the soil, to facilitate evaluation of the environmental impact of different land use strategies, or the fertilization planning of agricultural crops. Obviously, models developed for long-term predictions will be somewhat inadequate for simulating short-term C and N dynamics; but as computing capacity and speed increases, a key question may be whether models developed for the short-term simulations of C and N dynamics also give adequate predictions of soil C and N in the long term. One problem often raised is the much larger number of parameters required by detailed models of short-term C and N dynamics. For long-term scenarios, these parameters may often be difficult to obtain or not available. However, if the parameter and input requirements are not excessive, or rigorous effective parameters are available (see Hansen et al., Chapter 16 in this book) and the capabilities for predicting long-term C and N trends have been successfully validated, the use of the same models for short- and long-term simulations should provide more rigid and valid predictions of output properties and enable scenario analyses of more complex system properties as, for example, soil quality, crop productivity and environmental impact in the long term.

As described by Hansen et al. in Chapter 16 in this book, the DAISY model has thus far mainly been validated on its ability to simulate nitrate leaching, crop dry matter production, and crop N uptake in the short term and has performed very well in several model comparison exercises (de Willigen, 1991; Vereecken et al., 1991; Diekkrüger et al., 1995). Although the major focus of the DAISY model never was to simulate soil organic matter (SOM) dynamics in a broad sense, soil C pools and turnover rates were taken into account in order to describe, mechanistically, the turnover of N in soil. The structure of the SOM sub-model thus resembles that of many of the more dedicated SOM models (e.g., Parton et al., 1987). Parameterization of the SOM sub-model was originally derived from both long-term experimental data (Broadbalk and Hoosfield plots at Rothamsted), as well as from short-term laboratory incubation data (Hansen et al., 1990). It is therefore of great importance to validate the DAISY model on detailed field data of both short-term temporal variability of C and N turnover and long-term soil C trends.

The objectives in this chapter are therefore to (1) to report the performance of DAISY with respect to the simulation of long-term trends in total soil C levels under different climatic and management conditions; (2) investigate how the internal partitioning of C between pools in the SOM sub-model structure is affected by differences in annual C inputs; (3) illustrate the general applicability of the model for long-term scenario analyses of land use strategies; and, finally, (4) describe how some of the SOM sub-model parameters can be calibrated on detailed field data and report on the validity of a parameter set derived in this way.

SOIL ORGANIC MATTER AND MINERAL NITROGEN SUB-MODELS

DAISY is a deterministic model that simulates the C and N dynamics in the soil-plant-atmosphere system (Hansen et al., 1990; 1991). It consists of sub-models for soil water (including solute movement), soil temperature, soil organic matter (SOM, including microbial biomass), soil mineral N, crop growth, and system management. Overall details on the sub-models are given in the model description in Chapters 15 and 16 of this book. In the following, some further details of the SOM sub-model, the mineral-N sub-model, and the corresponding input require-ments are described.

In the SOM sub-model (Figure 15.1), three organic pools are simulated: added organic matter (AOM), soil microbial biomass (SMB), and soil organic matter

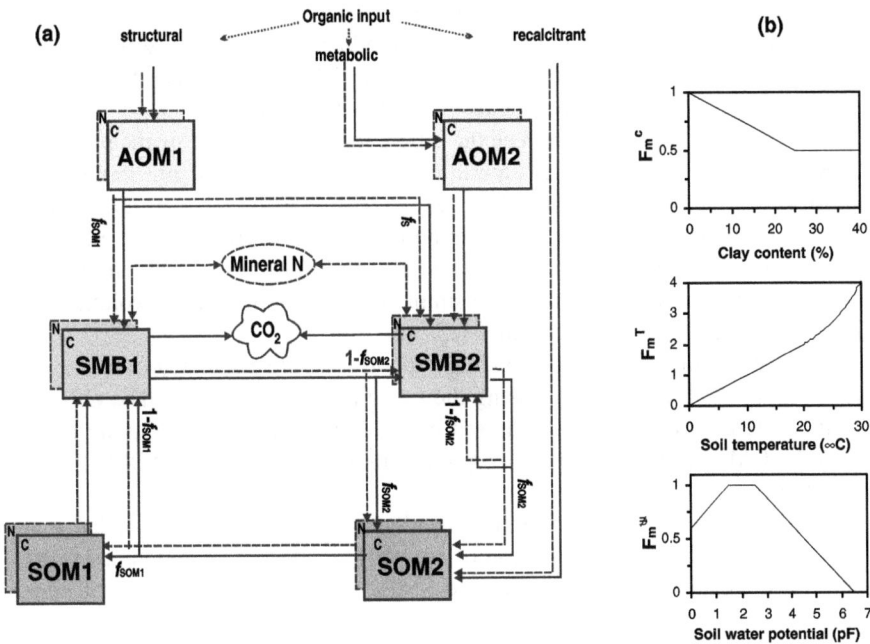

Figure 15.1   (a) Overview of the C and N fluxes between the different pools of organic matter, mineral N, and evolved CO_2 in the DAISY sub-model for soil organic matter (AOM = Added Organic Matter, SMB = Soil Microbial Biomass, SOM = native Soil Organic Matter, f_x = partitioning coefficients). (b) Turnover rate modifiers, F_m, for soil clay content, temperature, and water potential. Standard decomposition, death, and maintenance rate coefficients k_x^* (10°C, −10 kPa, 0% clay) are mul-tiplied with F_m to obtain actual rate coefficients k_x.

(SOM). The latter is native dead organic material not included in the AOM and SMB pools.

Each of the above-mentioned organic pools is divided into two sub-pools: one with a slow turnover (i.e., SOM1) and one with a fast turnover (i.e., SOM2). This discrete division is used as a very simple approximation of the corresponding continuum in nature. The division facilitates the description of the turnover of all the organic matter pools by first-order reaction kinetics, in agreement with the view that the rate-limiting step in the turnover is the rate at which a given pool dissolves into the soil solution (Nielsen et al., 1988). The division of soil microbes into SMB1 and SMB2 resembles the distinction between autochthonous and zymogenous soil microbial biomass, respectively (Jenkinson et al., 1987). For SMB1 and SMB2, the turnover rate is subdivided into a death rate and a maintenance respiration rate. Both rates are calculated separately with individual rate coefficients.

After crop harvest, turnover of dead root organic matter (ROM) is simulated analogously to AOM once soil tillage has been performed. If several incorporations of AOM occur, these are simulated as separate and identifiable AOM pools ($AOM1_a$ – $AOM2_a$, $AOM1_b$ – $AOM2_b$, etc.).

To determine the actual turnover rate coefficients, k_X, of each organic pool for a specific time step, turnover rate coefficients valid under standard conditions (10°C, –10 kPa, 0% clay), k_X^*, are multiplied with modifiers, F_m, for soil water potential, temperature, and clay content (the latter only for SOM1, SOM2, and SMB1 pools); see Figure 15.1b for functional values of the three modifiers.

Partitioning of C fluxes between the different pools is defined by the partitioning coefficients (f_X). Carbon fluxes entering the microbial biomass are multiplied by substrate utilization efficiencies (E_{SMB}, E_{SOM1}, E_{SOM2}, E_{AOM1}, E_{AOM2}). E_X defines the fraction of the substrate C coming from pool X (SMB1, SMB2, SOM1, SOM2, …) that is utilized for microbial growth. The remaining substrate C is respired as CO_2. Total CO_2 evolution can be calculated simply by summation of CO_2 evolution in the individual soil layers for every time step.

After every time step, corresponding N pools (N_{Xt}) are calculated from the actual amount of C in the pools (C_{Xt}) using the C:N ratios for each pool. The C:N-ratios of the different pools are constant and must be defined in the beginning (SMB1, SMB2, SOM1, SOM2) or when incorporated into the soil ($AOM1_i$, $AOM2_i$, $ROM1_i$, $ROM2_i$).

Net N-mineralization or N-immobilization is derived from the N-balance after each time step. If N-immobilization occurs, NH_4^+ is utilized in preference to NO_3^-, and the simulated growth of SMB1 and SMB2 may be limited by lack of mineral N in the soil.

With respect to the soil organic matter sub-model, the following parameters must be defined: (1) soil clay content for each soil horizon, (2) C:N-ratios of the different sub-pools of soil organic matter, (3) standard turnover rate coefficients of the pools of organic matter, (4) partitioning coefficients of the C fluxes, and (5) nitrification and denitrification parameters (default values exist for parameters 2–5). To calculate the N balance and hence the N losses from the soil, a balance depth to which all N pools are accumulated must be defined. The following initial conditions must be defined: (1) soil water potential, volumetric ice content, and soil temperature for the defined node points; (2) total organic C content and total org. C:N-ratio in the different soil horizons; (3) fraction of total organic C in SMB1 and SMB2; (4) subdivision of SOM-C in SOM1 and SOM2; and (5) NO_3-N and NH_4-N content for the defined node points. Every application of crop residue or organic fertilizer must be characterized by the (1) applied amount of wet weight, (2) dry matter (DM) content, (3) content of org. C, organic N, and NH_4-N in DM, (4) percent NH_4-N lost by volatilization, (5) fraction of AOM-C in AOM1 and AOM2, (6) C:N ratio of AOM1, (7) standard turnover rate coefficients of AOM1 and AOM2, and (8) partitioning coefficients. Default values for most of these parameters are available; and when crop residues are incorporated, these parameters are passed automatically from the respective crop module.

Further information is required on soil physical parameters, irrigation, climatic parameters, atmospheric N deposition, mineral fertilization, soil tillage, sowing, and harvest. Some details are given in Chapters 15 and 16 in this book; for further details, see Hansen et al. (1991).

LONG-TERM TRENDS IN SOIL ORGANIC C: VALIDATION AND SCENARIOS FOR FUTURE LAND USE MANAGEMENT

The problems of global climate change and the greenhouse effect have attracted public attention for a number of years and have thus been on the research agenda over the past decade. Soil organic matter represents a major C pool within the biosphere, estimated at about 1.4×10^{18} g globally, roughly twice that in the atmosphere. Changes in climate or in land use may affect this stock significantly through their effect on decomposition rates and residue inputs. However, several feedback mechanisms exist, and predictions about future trends in SOM and the possible effects on global change are therefore very difficult. Long-term experiments are very costly and the time scale involved to evaluate effects on SOM prohibits an exclusively experimental approach to this problem. As part of a GCTE (Global Change in Terrestrial Ecosystems) initiative, a global network (SOMNET) was initiated in 1995, with the aims of (1) establishing a meta-database of existing long-term experiments and current SOM models worldwide (currently containing 29 models and 76 long-term experiments) and (2) providing a forum for SOM experimentalists and researchers to share models and data sets (Smith et al., 1996; Powlson et al., 1998). Through such collaboration and synergy, it was envisaged that new knowledge on interactions between SOM and global change may evolve, enabling better predictions for the political stakeholders to act upon. The first joint effort of SOMNET was to carry out a model comparison exercise, in which 9 of the models in SOMNET were compared on their ability to simulate 12 data sets from 7 selected long-term experiments registered in SOMNET (Smith et al., 1997). The performance of the DAISY model in this comparison (Jensen et al., 1997b) will be reported here for some of these data sets.

Long-term Experimental Data: Availability, Quality, and Model Parameterization

Three of the SOMNET experiments are presented in more detail.

1. The *Prague-Ruzyne Plant Nutrition and Fertilization Management Experiment* (Klír, 1996) in the Czech Republic was started in 1956 on land that had been arable for centuries and consists of a two-course rotation (wheat-sugar beet), here represented with two treatments: either no amendments (nil treatment) or an average application of 100 kg fertilizer N (in a NPK fertilizer) and 11 metric tons farmyard manure (FYM) per year (termed the NPK+FYM treatment).
2. The *Tamworth Legume–Cereal Rotation* (Crocker and Holford, 1996) experiment in Australia was located on land that had been cultivated for about 100 years prior to the start of the experiment in 1966. It is represented with two treatments: a wheat-fallow rotation initiated with fallow from 1966 to 1969, and then a wheat crop every second year (or sorghum a few times) separated by fallow, or a lucerne-wheat-sorghum-clover-rotation, with lucerne from 1966 to 1969 and again from 1979 to 1983, subterranean clover from 1988 to 1990 and wheat or sorghum all remaining years; both treatments received no fertilizer N, only P, and the lucerne and clover were grazed by sheep.
3. The *Waite Permanent Rotation Trial* (Grace, 1996) in Australia was initiated in 1925 on a site previously under native grassland. Two of its treatments are represented: a wheat-fallow rotation and a wheat-oats-pasture-fallow rotation. Both treatments received no fertilizer N, only P, and the annual pasture was grazed by sheep.

Two additional experiments were simulated with the DAISY model, but will not be reported on in detail here: the *Bad Lauchstädt Static Fertilizer Experiment* (a long-term arable site) and the *Rothamsted Park Grass Experiment* (a permanent grassland site); see Jensen et al. (1997b) for details.

Experimental management information and measured properties, including total soil C values were all made available to the modelers. Modelers were not allowed to calibrate the models on data, for example, by adjusting decomposition rate parameters or internal model partitioning coefficients, but assumptions about initial total soil C values (often not available in the data) and varying the initial distribution of soil C between internal model pools were allowed. However, to do a true blind simulation, enabling evaluation of the predictive capabilities of the models, measured values were not known by the modelers prior to carrying out the simulation for one of the Waite rotations (wheat-oats-grass pasture-fallow).

Although the long-term experiments in SOMNET were selected for their data quality regarding soil organic matter, many of the experiments were either not designed for that particular purpose, or span such a long time period that the data quality does not meet modern standards. Therefore, the frequency of sampling was often not optimal for evaluating the models or the methods of sampling, or the analytical techniques differed both in time and between sites. Substantial variability in the SOM data thus often occurred. See, for example, Figure 15.2a for the Prague-Ruzyne data sets, where significant experimental scatter over time is evident. Considering the small changes in SOM that do occur under quite different management

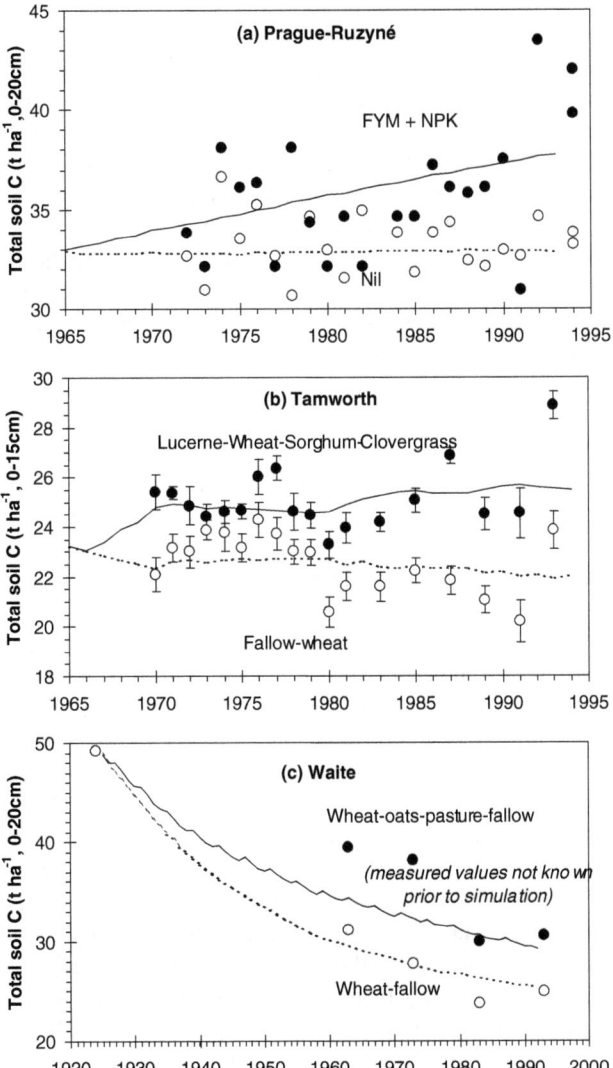

Figure 15.2 Measured (points) and simulated (lines) total soil C contents (not incl. AOM pools) for some selected SOMNET data sets: (a) the Prague-Ruzyné Plant Nutrition and Fertilisation Management Experiment (wheat-sugar beet rotation): either no amendments (nil treatment) or an average application of 100 kg fertilizer N (NPK) and 11 metric tons farmyard manure (FYM) per year; (b) the Tamworth Legume–Cereal Rotation experiment: a wheat-fallow rotation (fallow 1966–69, then every other year), or a lucerne(1966–69, 1979–83)-wheat-sorghum-clover (1988–90) rotation, no fertilizer N, only P; (c) the Waite Permanent Rotation Trial: a wheat-fallow rotation and a wheat-oats-pasture-fallow rotation; no fertilizer N, only P. For the latter rotation, measured values were not known by the modelers prior to carrying out the simulation. (From Jensen, L.S., Mueller, T., Nielsen, N.E., Hansen, S., Crocker, G.J., Grace, P.R., Klír, J., Körschens, M., and Poulton, P.R. 1997b. Simulating soil organic matter trends in long-term experiments with the soil-plant-atmosphere system model *Daisy. Geoderma*, 81:5-28. With permission from Elsevier Science.)

regimes and over reasonably long periods, this makes it very difficult to validate and compare the performance of models.

During the model comparison exercise, the difficulties arising from the different input requirements of the models became evident. As an example, DAISY requires daily values of global radiation, but these data were often missing or incomplete; hence, data of sun hours (which were recorded at nearly all sites for the full length of the experiments), together with the latitude of the site, were used to calculate global radiation (Rietveld, 1978). Similarly, DAISY requires rather detailed data on soil water characteristics (water release characteristics and hydraulic conductivity) to simulate the water content and water flow in the soil, but such detailed data were not available for any of the core sites. Therefore, water release and hydraulic conductivity curves had to be constructed from the information available on bulk density (from which water content at saturation can be calculated) and measured water content at field capacity (approx. −10 kPa) and at wilting point (−1500 kPa), and interpolation of the curves was then based on the textural class of the soil. Minor fine-tuning of hydraulic conductivity was done if necessary, judged from comparisons of simulated water content with the range of soil water contents given by the dataholders.

Assessment and Parameterization of Organic Matter Inputs

It also became evident through the comparison exercise that a long list of assumptions, differing somewhat between models, had to be made regarding some of the organic matter inputs to the experiments. Generally, soil tillage, sowing and harvest dates, atmospheric N deposition, inorganic N fertilization, as well as the amount and timing of organic fertilizer input were specified as model inputs according to the information provided by the dataholders (Smith et al., 1997). Plant materials deposited during the growth of the crops and crop residues incorporated after harvest were specified on a more individual basis (Table 15.1). Where available, data on amounts of crop residues after harvest (straw, stubble, sugar beet, or pasture top dry matter) were used as input; otherwise, the amounts simulated by the model were used. DAISY does not simulate dead root turnover or rhizodeposition, etc., but only incorporates simulated root dry matter when soil tillage takes place after harvest. For perennial and annual pastures, where no tillage takes place during the growing year, dead root turnover, rhizodeposition, and leaf litter inputs (Table 15.1) were therefore estimated by the dataholder and the modeler in collaboration. Estimates were based mainly on literature data (e.g., Jenkinson and Rayner, 1977; Jensen, 1993; Swinnen et al., 1994) on rhizodeposition from cereals, indicating that up to 1.9 metric ton C/ha/yr may enter the soil via rhizodeposition and root turnover. Hence, it was estimated that up to 2.5 metric ton C/ha/yr enters the soil in this way from perennial lucerne, somewhat less from clover and grass (see Table 15.1), and these inputs added to the soil on a monthly basis during the growing season. If pastures were grazed, animal excreta and urine returned to the soil were assumed to represent approximately 40% of N in the observed grass yields. When pastures were plowed under, the simulated dry matter of roots and above-ground plant material was incorporated. For cereals and root crops, no rhizodeposition or root

Table 15.1   Types and Amounts of Plant-derived Organic Inputs Used in the Simulation of the Individual Crops in Three of the Long-term Experiments

| | | Plant-derived Organic Inputs | | | |
| | | Above-ground | | Below-ground | |
Site	Crop	Type	Amount (t DM/ha/yr)	Type	Amount (t DM/ha/yr)
Prague-Ruzyné	S. Wheat	Stubble	1.5–2.0[a]	Roots	Sim.[c]
	S. Beet	Tops	0.5–1.0[a]	Roots	Sim.
Tamworth	S. Wheat	Stubble	0.2–0.5[a]	Roots	Sim.
	Lucerne pasture	Leaf litter	1.0–1.5[b]	Rhizodepos.	1.5[b]
	(perennial)	Excreta, graz. anim.	1.5–2.5[b]	Root turnover	1.0[b]
		Abovegr. at plow.	Sim.	Roots at plow.	Sim.
	Sorghum	Stubble	2.0[b]	Roots	Sim.
	Clover pasture	Leaf litter	1.0[b]	Rhizodepos.	1.0[b]
	(perennial)	Excreta, graz. anim.	1.5[b]	Root turnover	1.0[b]
		Abovegr. at plow.	Sim.	Roots at plow.	Sim.
Waite	S. Wheat	Stubble	0.1–0.7[a]	Roots	Sim.
	Oats	Stubble	0.1–0.9[a]	Roots	Sim.
	Grass pasture	Leaf litter	0.5[b]	Rhizodepos.	0.5[b]
	(annual, non-leg.)	Excreta, graz. anim.	0.4–1.7[b]	Roots at plow.	Sim.
		Abovegr. at plow.			

[a] Measured experimental values used.
[b] Measured values not available; estimated values used.
[c] Sim. = dry matter simulated by the DAISY crop submodel incorporated.

From Jensen, L.S., Mueller, T., Nielsen, N.E., Hansen, S., Crocker, G.J., Grace, P.R., Klír, J., Körschens, M., and Poulton, P.R. 1997b. Simulating soil organic matter trends in long-term experiments with the soil-plant-atmosphere system model *Daisy. Geoderma*, 81:5-28. With permission from Elsevier Science.

turnover was included; only the root dry matter simulated by the crop growth sub-model was incorporated after harvest.

The parameterization of these inputs, for example, the division of added organic matter C between the AOM1 and AOM2 pools, the C:N ratio of the added organic matter (if measured value was not available) and of the two sub-pools was based on standard values (see Table 15.4; see "Model Calibration" section), derived from incubation data used in the initial calibration of the model (Hansen et al., 1990). The parameterizations of the SMB and SOM pools, as well as nitrification and denitrification parameters, were set to default values (Hansen et al., 1991; Mueller et al., 1996). Initial total organic C content (Table 15.2) and total organic C:N ratio of the Ap horizon was taken or calculated from the observed data if available; otherwise, initial values were estimated. For each site, the initial total organic C was always assumed to be the same in all treatments. Distribution of total organic C down the soil profile was based on either observed data if available, or on estimates by the dataholders.

Simulation of Soil Organic Matter Trends and Partitioning Between Different Model Pools

Average annual total C inputs (the sum of C in organic manures, rhizodeposition, root turnover, leaf litter, and dead roots and crop residues incorporated after harvest)

Table 15.2   Initial (measured or estimated) and Final Soil Organic Matter Contents (simulated, see Figure 15.1), Average Yearly C Inputs and Their Effects on Partitioning of C Between SOM Pools in the DAISY Model (t C/ha: metric ton C per hectare)

Site	Treatment	Initial SOM (t C/ha) (year)	Average C Input[a] (t C/ha/yr)	Final Soil Organic Matter (year) (t C/ha)	Fraction of C_{tot} in Slow SOM1 Pool (%) Initial	Fraction of C_{tot} in Slow SOM1 Pool (%) End	Fraction of C_{tot} in Microbial Biomass[b]
Prague-Ruzyné	Nil inputs	32.9[c] (1965)	0.99	(1993) 32.9	80	80	0.57 ± 0.05
	FYM + NPK	(1965)	2.06	37.7	80	70	1.05 ± 0.10
Tamworth	Wheat-Fallow	23.2[c] (1965)	0.53	(1994) 22.0	80	85	0.51 ± 0.37
	Luc.-Wheat-Sorgh.-Clov.		1.54	25.5	80	72	0.86 ± 0.28
Waite	Wheat-Fallow	49.1 (1924)	0.24	(1992) 25.4	44	92	0.33 ± 0.18
	Wheat-Oat-Pasture-Fallow		0.93	29.3	44	83	0.59 ± 0.29

[a] Total C input = Organic fert. C + roots C + crop residue C (+ evt. rhizodeposition C + root turnover C + leaf litter C + excreta C).
[b] At the end of simulation. Initially 0.6% of total soil C is allocated to the microbial biomass pools (SMB1 + SMB2). ±: std. dev. of the mean indicates temporal variability.
[c] Estimated (not measured at the start of the experiment).

From Jensen, L.S., Mueller, T., Nielsen, N.E., Hansen, S., Crocker, G.J., Grace, P.R., Klir, J., Körschens, M., and Poulton, P.R. 1997b. Simulating soil organic matter trends in long-term experiments with the soil-plant-atmosphere system model Daisy. *Geoderma*, 81:5-28. With permission from Elsevier Science.

for the different experimental sites and treatments over the full length of the experiments can be seen in Table 15.2. It is evident that only relatively small changes in soil organic C contents occurred in both the Prague-Ruzyné and the Tamworth experiments, even though C inputs differed by a factor of 2-3. The cultivated native grassland of the Waite site lost substantial amounts of C, over the more than 60-year course of the experiment, soil C levels were more or less halved.

At the Prague-Ruzyné Plant Nutrition and Fertilization Management Experiment, the annually measured values of total soil C (Figure 15.2a) showed a relatively large year-to-year variability, but the general trend was for a somewhat increasing soil C level in the FYM+NPK treatment. This was confirmed by the simulations with DAISY, where an average C input of 0.99 metric tons C/ha/yr (Table 15.2) in the nil input treatment maintained the simulated soil C level, and 2.06 metric tons C/ha/yr in the FYM+NPK treatment produced an increase similar to the trend of the measured values.

The initial 4 years of lucerne pastures caused a marked increase in measured total soil C levels in the lucerne-clover rotation at the Tamworth Legume/Cereal Rotation on Black Earth (Figure 15.2b). The simulations predicted a general pattern similar to the measured values, with total soil C decreasing during fallow periods, maintained during annual crops, and increasing during perennial pastures.

With the Waite Permanent Rotation Trial data, modelers were only provided with measured total soil C for the wheat-fallow treatment (Figure 15.2c), but were asked to also simulate the wheat-oats-grass pasture-fallow treatment as a blind test, without knowing the measured levels. Prior to the start of the experiment in 1925, the site was under native grassland (Grace, 1996), and the cultivation led to a rapid decline in total soil C. Due to this pre-experimental history of the site, it was necessary to recalibrate the standard initial partitioning between the two native soil organic matter pools in the model (SOM1 and SOM2), which was reduced from the standard value of 80% in SOM1 to 44% (Table 15.2), with which the simulated total soil C decrease in the wheat-fallow treatment was similar to that measured (Figure 15.2c). This changed initial partitioning between SOM1 and SOM2 was then used also for the "blind" simulation of total soil C (measured values not known at time of simulation) in the wheat-oats-pasture-fallow treatment. Similar to the other pasture simulations, an assumed level of C input from root turnover, rhizodeposition, and leaf litter during years of pasture was input to the model, the average total C input being 0.93 metric tons C/ha/yr, compared to the low 0.24 metric tons C/ha/yr in the wheat-fallow treatment (Table 15.2). The simulated decrease in total soil C was within the range of measured soil C in the wheat-oats-grass-fallow treatment (Figure 15.2c), demonstrating the predictive capabilities of the model and confirming that the recalibrated initial partitioning between SOM1 and SOM2 was correct.

Comparing simulated and observed changes in total soil C can be difficult when the observed data has such large variability, as was the case for several of the data sets used in this study. However, average trends for simulated and measured data can be compared (Figure 15.3a), and this showed very good correspondence, except for the Bad Lauchstädt data where simulations were initialized with anomalously low values measured in 1956, which may not be valid (data not shown, see Jensen

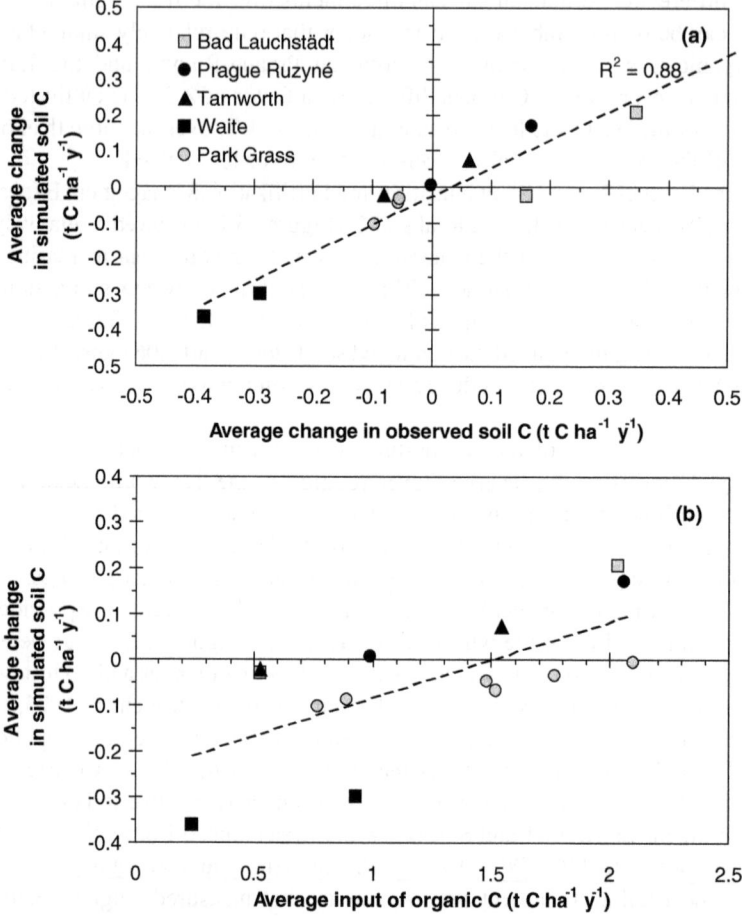

Figure 15.3 Comparison of average annual change in simulated total soil C for all investigated sites with (a) the average annual change in measured total soil C and (b) the average annual input of organic C to the soil used in the simulation, legend as in (a). (From Jensen, L.S., Mueller, T., Nielsen, N.E., Hansen, S., Crocker, G.J., Grace, P.R., Klír, J., Körschens, M., and Poulton, P.R. 1997b. Simulating soil organic matter trends in long-term experiments with the soil-plant-atmosphere system model *Daisy. Geoderma*, 81:5-28. With permission from Elsevier Science.)

et al., 1997b). However, the difference between the two treatments at Bad Lauchstädt was parallel to the others.

We have also compared the average trend in simulated total soil C against the average annual input of organic C (Figure 15.3b) to reveal whether total soil C changes were similarly related to C inputs at the different sites. As expected, clear differences between sites were evident, especially the Waite site, which showed a much larger decline in soil C. The sites differ widely with respect to climatic conditions and management histories; and this means that, for each site, a characteristic annual C input will be capable of maintaining the characteristic soil C level of the site. However, the average annual input capable of maintaining total soil C levels, approximately 1.5 metric tons C/ha/yr (Figure 15.3b, dotted line), compares well with estimates by Paustian et al. (1992), who also found that for a long-term arable site in Sweden, 1.5 metric ton C/ha/yr was capable of maintaining soil C levels. Similarly, for a long-term wheat-fallow site with different managements, Parton and Rasmussen (1994) estimated that 1.7 to 1.9 metric tons C/ha/yr was required to maintain total soil C.

The standard initial partitioning of native dead soil C between the two major pools in the model, SOM1 and SOM2 (80 and 20%, respectively; Table 15.2), was originally calibrated on the basis of some of the long-term arable experiments from Rothamsted, and may thus be valid only for soils that have been under long-term arable management. The validity of this partitioning was confirmed for all the sites included in this work except for the Waite site. The fraction of soil C remaining in the slow pool (SOM1, half-life of 700 y at 10°C, field capacity) at the end of the simulations was in the range from 70 to 92%, again depending on the average magnitude of C inputs (Table 15.2). As mentioned earlier, the Waite site had been native grassland prior to the start of the experiment, and the composition and decomposability of the soil organic matter developed under this vegetation was very different from that in soils under long-term arable management. Hence, a good simulation of the wheat-fallow rotation could be obtained by simply changing the initial fraction of native soil C in the slow SOM1 pool to 44%, and during the course of the simulation the amount of C in this pool actually increased (by 7%), whereas the native soil C pool with a faster turnover (SOM2, half-life of 13.6 yr at 10°C, field capacity) decreased dramatically (by 92%). After almost 70 years of cultivation the Waite soil seems to have reached a quasi-steady-state, with 83 to 92% of the native soil C in the slow SOM1 pool. This confirms that reasonable simulations of total soil C for land uses other than arable systems are possible with the default setup of the DAISY model, only by changing the initial partitioning of the native soil C, thus significantly facilitating a more general applicability of the model.

Scenario Example: Evaluating the Impact of Straw Removal on Soil Organic C Stores in Arable Land in Denmark

The Danish government has a very ambitious plan for sustainable development of energy consumption and utilization in Denmark, Energy 21, decided upon by the Danish parliament in 1996. The ultimate goal set for year 2030 is that 35% of the total energy consumption should be derived from renewable resources. This includes especially large increases in wind power and biomass for energy purposes, the latter to be more than doubled from the early 1990s to some 75×10^{15} Joules a year before the end of year 2000. Biomass would thus comprise almost 10% of the total consumption of fuel in the year 2000. The biggest expansion will be caused by the central heating and electrical power plants increasing use of straw and wood chips, and implies that, more or less, all excess straw from arable land countrywide (straw not used for animal feeding or bedding) would be used for energy. However, Danish topsoil C contents seem to show a declining trend already (Christensen and Johnston, 1997), and concern has been raised by both the research and agricultural communities regarding the soil quality impact of this strategy. Therefore, a research project was

Table 15.3 **Scenario Crop Rotations for Plant and Pig Production Farms, in Eastern (low precipitation climate 1) and Western (high precipitation, climate 3) Parts of the Country**

Plant and Pig Production Farms		Dairy Production Farms
Climate 1 (east):	Climate 3 (west):	Climate 3 (west)
Sp. barley	Sp. barley*	Maize (silage)
W. barley	Peas	Sp. barley (silage)
W. wheat*	W. wheat	Clovergrass (silage)
Sugar beet	W. rape	Clovergrass (pasture)
Sp. barley	W. wheat*	Sp. barley
W. wheat	Potatoes*	W. wheat*
W. wheat	Sp. barley	
Fallow (grass)	Fallow(grass)	

Note: For the scenario cattle farms, which predominate in the western parts, only one crop rotation was set up for climate 3 (west). *: denotes that a catch crop may be grown after this main crop.

initiated in 1999 to evaluate the impact of such land management changes on soil organic matter.

The approach taken has been to (1) use existing Danish long-term experimental data sets on soil organic matter to further validate and, if necessary, recalibrate the DAISY model; and (2) to set up different straw management scenarios for typical farm types (plant, pig, and dairy production) on different soil types (sandy loam and coarse sandy soil, predominant Danish soils) and subject to the natural climatic gradient in climate in Denmark (difference in average precipitation between eastern and western parts, 661 and 991 mm, respectively).

Based on agricultural statistics and management experience, crop rotations were set up for plant and pig production farms, similar for both farm types but differing slightly between the eastern (low precipitation climate 1) and western (high precipitation, climate 3) parts of the country. For the cattle farms, which predominate in the western parts, only one crop rotation was set up for climate 3 (west) (see Table 15.3).

When used as a pure plant production rotation, inorganic fertilizer N was applied to each crop according to the statutory N norms (Plantedirektoratet, 1999) issued annually by the Ministry of Food, Agriculture and Fisheries. When used as a pig production rotation or in the dairy production rotation, the maximum allowable amount of animal manure was applied, corresponding to approximately 170 kg total-N/ha, and any remaining demand for fertilization (assuming 55 to 60% efficiency of the applied total N in animal manure) was then supplemented as inorganic fertilizer.

Three scenarios for each climate-soil-farm type combination were run: (1) all straw from cereals, peas, and oilseed crops incorporated, regardless of farm type; (2) all straw from cereals, peas, and oilseed crops removed; and (3) as in (2), but with catch crops following all crops where possible (indicated by * in the rotations in Table 15.3) as a compensatory measure for the straw C removed.

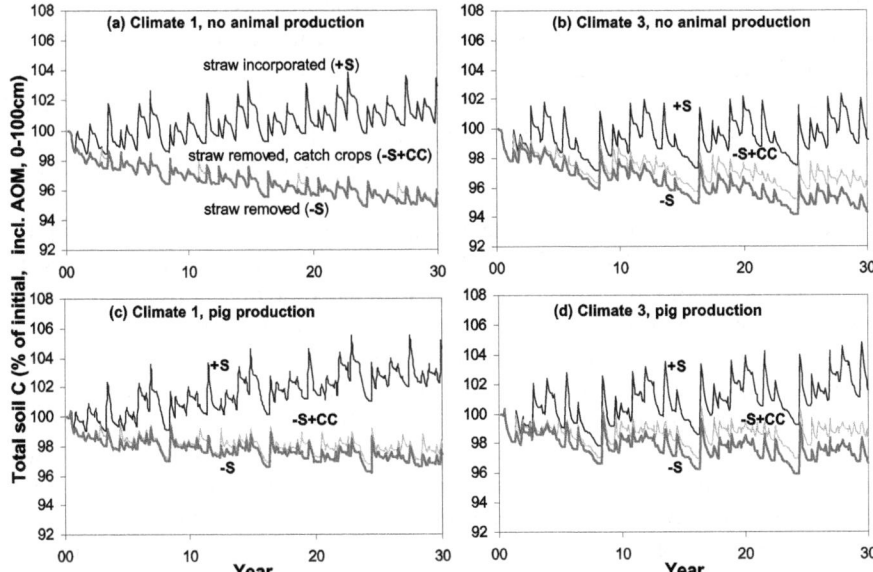

Figure 15.4 Simulated trends in total soil C (including plant and manure residue AOM pools) to 1-m depth, for a sandy loam soil, subject to either (a and b) plant production or (c and d) pig production, and to either (a and c) the low precipitation climate 1, or (b and d) the high precipitation climate 3. For each combination, three different straw management scenarios were applied: (i) all straw from cereals, peas, and oilseed crops incorporated, regardless of farm type; (ii) all straw from cereals, peas, and oilseed crops removed, and (iii) as in (ii), but with catch crops following all crops where possible. Total soil C referenced to initial level, which in this soil corresponds to ca. 100 metric tons C ha^{-1} to 1-m depth.

Simulations were run for 30 years with standardized weather data (same weather every year, containing natural day-to-day variations, but with monthly averages corresponding to long-term averages) and standard management operations (soil tillage, sowing, harvest).

Figure 15.4 shows some preliminary results of the simulated trends in total soil C (including plant and manure residue AOM pools) to 1-m depth for a sandy loam soil, subject to either pure plant or pig production and to either the high or the low precipitation climate and rotation. Total soil C to 1-m depth in this representative standard soil type corresponds, more or less, to 100 metric tons C/ha, and the overall impression is that the maximum treatment effects correspond to approximately ±5 metric tons C/ha. For the scenario with complete removal of the straw, soil C levels generally tend to decrease, especially with pure plant production; whereas with pig production, the manure input seems to compensate for most of the loss. On the other hand, with maximum incorporation of straw, total soil C tends to increase slightly in all combinations, although the annual variations are greater than the trend. Incorporation of all straw is somewhat unlikely in the real world, especially on the pig production farms, where some of the straw inevitably will be used for animal bedding, etc. Therefore, the expected difference in total soil C between incorporation and removal of straw may be somewhat less over time than that depicted in Figure 15.4. Only slight differences can be detected between the climate 1 and climate 3 rotations.

The scenario with catch crop use shows that this is obviously not a very efficient compensatory measure, primarily because catch crops can only be used in a few years of the crop rotation — 1 or 3 out of 8 years in the climate 1 and 3 rotations, respectively.

It is also evident that the intra-annual fluctuations in total soil C, including the plant and manure residues (AOM pools), are often of the same magnitude as the difference developed in total soil C between total incorporation and removal of straw over the simulated 30-year period. This underlines how difficult it would be to set up an experiment to verify these effects; if experimental variability and sampling uncertainty is added, significant differences would be very difficult to detect.

The scenario simulations of total soil C, as illustrated in this example, do not alone warrant judgments about management effects on soil quality. Evaluation will also have to include effects on the environment (e.g., N leaching, greenhouse gases) and on productivity (e.g., N mineralization capacity, crop productivity). Most of these parameters are also simulated by the DAISY model and can be used in the overall assessment, but the question still remains as to whether there are some critical thresholds of these parameters, including total soil C, below which soil productivity and quality decreases, eventually irreversibly. However, using dynamic simulation models for such scenario analyses as illustrated here provides a much better foundation for decisions than simple carbon balance models, like the ICBM model developed and applied by Kätterer and Andrén (1999).

SHORT-TERM DYNAMICS OF SOIL C AND N TURNOVER: CALIBRATION AND VALIDATION ON FIELD DATA

The validation and use of a complex model such as DAISY on long-term data sets and scenario analysis naturally raises the question: how adequate is the simulation of measurable model pools in the short term? We know that the model performs well with respect to crop production and nitrate leaching in the short term and with SOM in the long term as shown in the previous chapter section; but this may well be achieved with more or less inadequate simulations of short-term dynamics of the internal model variables, for example, the soil microbial biomass or the added residues decomposition. Furthermore, if data sets on such variables can be produced on a field scale and *in situ*, their use for overall calibration and validation of the model, may also prove to give a much more robust parameter set than the traditional calibration of selected parameters on experimental data derived from controlled conditions. We therefore carried out field experiments in the years 1993 to 1995, with the aim of producing detailed data on soil C and N pools and turnover after various crop residue inputs, and with great temporal resolution. One year's experimental data were used for calibration of selected parameters (Mueller et al., 1997), while the other data were used for independent validation and evaluation (Mueller et al., 1998b).

Experimental Data

Experimental data was obtained from two field experiments (Jensen et al., 1997a; Mueller et al., 1998a) conducted on a sandy loam soil, containing 13% clay, 1.3% C, and 0.14% N in the topsoil.

In the first experiment, 8 metric ton/ha of oilseed rape straw (*Brassica napus* L. cv. *Ceres*) was incorporated by rotavation to a depth of 15 cm after harvest in August. A control treatment was rotavated in the same way but without incorporation of rape straw (0 ton/ha).

In the second experiment (different year), 6 metric ton/ha of chopped barley straw (*Hordeum vulgare*), fresh chopped blue grass (*Poa pratensis*), or fresh chopped maize plants (*Zea mais*) were incorporated similar to the first experiment, and also here control plots without any plant material were established.

The soils were kept bare using herbicides if necessary. After incorporation, soil mineral N, soil surface CO_2 flux, soil microbial biomass (SMB) C and N, and light particulate soil organic matter (LPOM) were measured frequently. Soil surface CO_2 flux was measured using a chamber method with passive trapping of CO_2 in alkali over a period of 24 h (Jensen et al., 1996). SMB was measured by chloroform fumigation extraction (Brookes et al., 1985; Vance et al., 1987), using an f_{EC} factor of 2.22 (Wu et al., 1990) and an f_{EN} factor of 1.85 (Brookes et al., 1985; Joergensen and Mueller, 1996). LPOM contains the water-insoluble (particulate) part of the plant residues not yet decomposed at the sampling date and was separated by density ($\rho < 1.4$ g/cm^3) fractionation of particulate organic matter using a solution of Na-polytungstate as reported by Magid et al. (1997).

These variables were measured for a period of about 13 months in both experiments. However, in the first experiment, the last LPOM measurement was performed after about 18 months.

The driving variables (global radiation, precipitation, air temperature 2 m over ground, and potential evapotranspiration) were measured at a meteorological station placed next to the experimental fields. The model setup with regard to soil texture, water release characteristics, mineral N content, C_{org} content and C_{org}/N_t ratio of the various soil layers was based on measured values. In accordance with Petersen et al. (1995), rape roots and stubble from the previous crop were simulated as 1 metric ton/ha dry matter (0 to 15 cm). Simulations were carried out with and without incorporation of rape straw (8 metric tons DM/ha), barley straw, blue grass, or maize (all 6 metric tons DM/ha). For all treatments, a soil rotavation to 15-cm depth was simulated.

The measured contents of total C and N of the incorporated plant residues (AOM) were used as inputs in the simulation scenarios described below. Distribution of C and N in the residue inputs was performed as described below.

Model Calibration

Experiment 1 was used to calibrate the organic matter module of the DAISY model. First simulations with the default setup described by Hansen et al. (1990) (see Table 15.4) showed the following shortcomings (see thin lines in Figure 15.5): (1) SMB-C and SMB-N were simulated about 30 or 65% below the values measured

Table 15.4   Parameters and Initial Values Used in the Two DAISY-Model Simulation Setups

State Variable	Parameter		Default[a]	Modified[b]
			Setup Values	
SMB	Fraction of total C (initial)	SMB1 [%]	0.45	1.89
		SMB2 [%]	0.15	0.15
	C/N	SMB1	6	6.7
		SMB2	10	6.7
	Death rate coefficient[c]	SMB1 [d^{-1}]	0.001	0.000185
		SMB2 [d^{-1}]	0.01	0.01
	Maintenance respiration rate coeffic.	SMB1 [d^{-1}]	0.01	0.0018
		SMB2 [d^{-1}]	0.01	0.01
	Substrate utilization efficiency	SMB ≥ SMB [%]	60	60
AOM	Fraction of AOM C (initial)	AOM1 [%]	80	96[d]
		AOM2 [%]	20	4[e]
	C/N	AOM1	90	92[d]
		AOM2	72	19[e]
	Turnover rate coefficient	AOM1 [d^{-1}]	0.005	0.012
		AOM2 [d^{-1}]	0.05	0.05
	Substrate utilization efficiency	AOM1 ≥ SMB [%]	60	13
		AOM2 ≥ SMB [%]	60	69
SOM	Turnover rate coefficient	SOM1 [d^{-1}]	2.7×10^{-6}	2.7×10^{-6}
		SOM2 [d^{-1}]	1.4×10^{-4}	1.4×10^{-4}
	Substrate utilization efficiency	SOM1 ≥ SMB [%]	60	40
		SOM2 ≥ SMB [%]	60	50

Note: SMB = Soil Microbial Biomass, AOM = Added Organic Matter, SOM = native dead Soil Organic Matter.

[a] = default parameter set from Hansen et al.
[b] = modified parameter set from Mueller et al.
[c] = all rate coefficients are at standard conditions (10°C, optimal moisture, no clay).
[d] = water-insoluble, based on measured values for the rape straw
[e] = water-insoluble, based on measured values for the rape straw

From Mueller, T., Jensen, L.S., Magid, J., and Nielsen, N.E. 1997. Temporal variation of C and N turnover in soil after oilseed rape straw incorporation in the field: Simulations with the soil-plant-atmosphere model DAISY. *Ecol. Modelling,* 99:247-262. With permission from Elsevier Science.

after incorporation of 8 or 0 metric ton/ha rape straw, respectively; and (2) mineral N was considerably underestimated after incorporation of 8 metric ton/ha rape straw. However, mineral N was simulated reasonably without incorporation of rape straw. The latter indicates that the underestimated level of SMB is compensated by over-estimated death rates and maintenance rates of the SMB pools in the default setup. The described shortcomings led to a stepwise parameter modification. Table 15.4 compares the default parameters with the parameters resulting from this modification.

The **first step** of modification dealt with the control treatments without rape straw incorporation (0 ton/ha) only. C:N ratios of SMB1 and SMB2 were set to 6.7, which was the mean value measured in the 0 and 8 metric ton/ha rape straw treatments. Initial SMB1-C was increased to a level at which total SMB-C reached the measured level (Table 15.4). Then, death rate coefficient and maintenance res-piration rate coefficient of SMB1-C were diminished to fit the temporal pattern of SMB and mineral N.

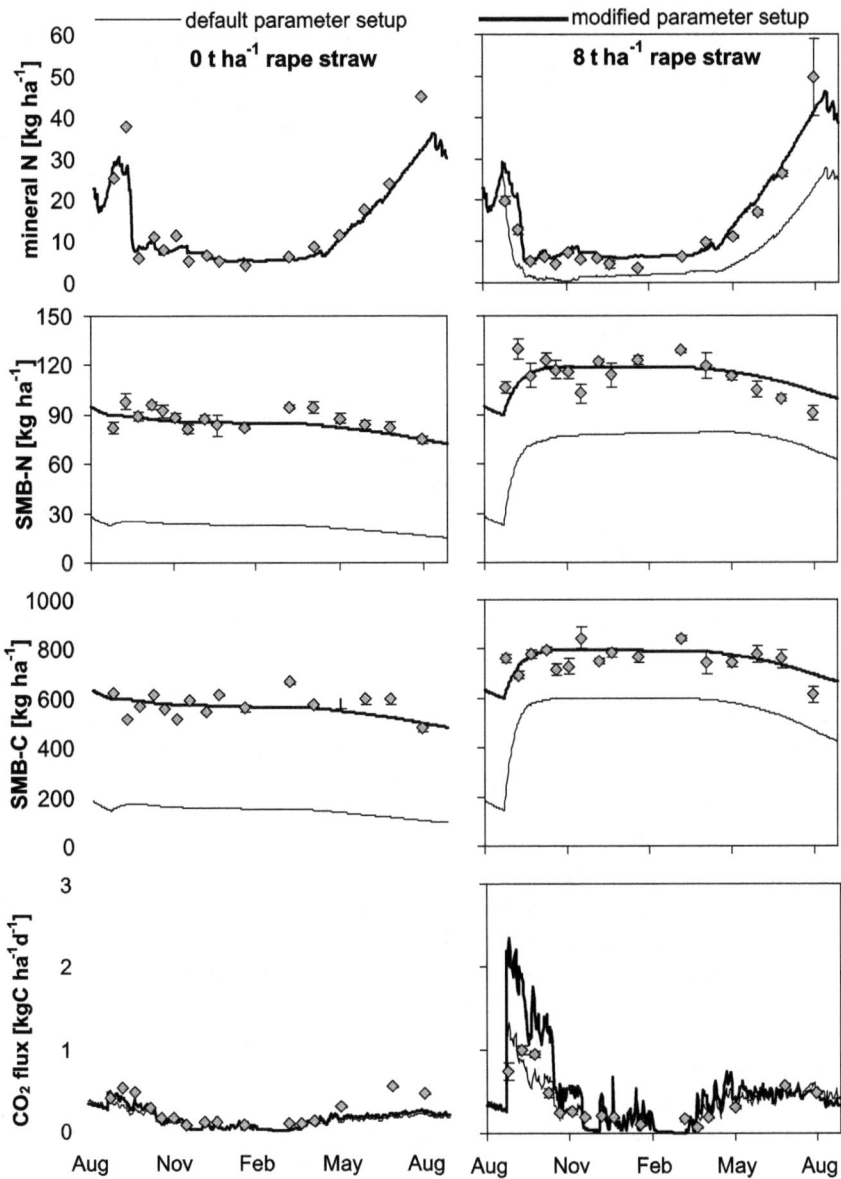

Figure 15.5   Simulated time courses of mineral N, SMB-N, and SMB-C (thick lines = modified
parameter setup; thin lines = default parameter setup), and the corresponding
measured values (bars show SE) in the treatments without rape straw incorporation (0 ton/ha) and with incorporation of 8 metric ton/ha rape straw. SMB = Soil
Microbial Biomass. (From Mueller, T., Jensen, L.S., Magid, J., and Nielsen, N.E.
1997. Temporal variation of C and N turnover in soil after oilseed rape straw
incorporation in the field: Simulations with the soil-plant-atmosphere model DAISY.
Ecological Modelling, 99:247-262. With permission from Elsevier Science.)

Table 15.5 Specification of the C Contents and the C/N Ratios of AOM (Added Organic Matter), and of the Partition into AOM1 (Water-insoluble) and AOM2 (Water-soluble) in the DAISY Simulations with the Modified Setup

		Added Residues			
		Rape Straw	Barley Straw	Blue Grass	Maize
Amount added (AOM)	C [kg/ha]	3969	2712	2682	2700
	N [kg/ha]	50	38	122	85
Fraction of AOM C in	AOM1 [%]	96	94	88	77
	AOM2 [%]	4	6	12	23
C/N of	AOM (overall)	80	72	22	32
	AOM1	92	110	25	37
	AOM2	19	12	12	23

Note: All values are based on experimentally determined data.

From Mueller, T., Jensen, L.S., Magid, J., and Nielsen, N.E. 1997. Temporal variation of C and N turnover in soil after oilseed rape straw incorporation in the field: Simulations with the soil-plant-atmosphere model DAISY. *Ecological Modelling*, 99:247-262; Mueller, T., Magid, J., Jensen, L.S., Svendsen, H.S., and Nielsen, N.E. 1998b. Soil C and N turnover after incorporation of chopped maize, barley straw and bluegrass in the field: evaluation of the DAISY soil-organic-matter sub-model. *Ecological Modelling*, 111:1-15. With permission from Elsevier Science.)

In a **second step**, the adjusted initial level of the SMB pools and turnover rate coefficients of SMB from the 0 ton/ha treatment were applied to the 8 metric ton/ha treatments. In addition, the initial subdivision of the AOM pool was based on the measured C and N contents in the water-insoluble fraction (= AOM1 with a relatively low turnover rate) and the water-soluble fraction (= AOM2 with a relatively high turnover rate) of the added rape straw (Jensen et al., 1997a) (see Table 15.5). After the modifications made in this step, predicted SMB increased markedly beyond the level measured in the field. This indicated that the added rape straw (AOM) was incorporated too effectively into SMB during the simulation.

In the default setup of DAISY, a universal substrate utilization efficiency (E_X) of 0.6 (approximately equal to the utilization efficiency of glucose in soil) is used for all C fluxes into SMB, independent of the pool from which they derive.

This led to the **third step** of modification, in which substrate utilization efficiencies (E_X) of each of the pools (SOM1, SOM2, SMB, AOM1, and AOM2) used as substrate by SMB were modified systematically in accordance with some predefined principles (Mueller et al., 1997). This resulted in a considerable decrease of E_{AOM1}, E_{SOM1}, and E_{SOM2} (see Table 15.4). At the same time as modifications of E_X-values were made, the turnover rate coefficient of AOM1 ($k_{AOM1}*$) was modified. Finally, the levels of SMB1-C and its turnover rate coefficients were fine-tuned.

The resulting simulations are shown in Figure 15.5 (thick lines). Compared to the simulations performed with the default setup, the root mean square error (RMSE) decreased and model efficiency (EF) increased markedly. This indicated a considerable improvement of the simulations (Loague and Green, 1991). Now, the modified turnover rates and the resulting metabolic quotient of SMB calculated from the modified setup were also found to be in much better accordance with literature data than for the default setup (Mueller et al., 1997). Measured soil surface CO_2 flux and

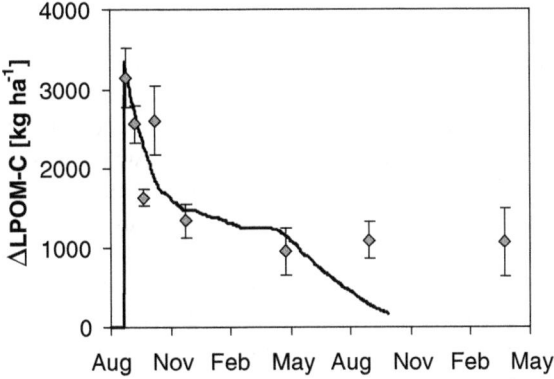

Figure 15.6   Time course of the difference between the straw amended (8 metric ton ha^{-1}) and non-amended treatments in water-insoluble plant residues (ΔAOM1-C) as simulated with the modified DAISY parameter setup (line) and the measured light particulate organic matter (ΔLPOM-C > 100 μm, ρ < 1.4 g/cm; points, bars show SE). (From Mueller, T., Jensen, L.S., Magid, J., and Nielsen, N.E. 1997. Temporal variation of C and N turnover in soil after oilseed rape straw incorporation in the field: Simulations with the soil-plant-atmosphere model DAISY. *Ecological Modelling,* 99:247-262. With permission from Elsevier Science.)

CO_2 evolution predicted by the model simulations agreed reasonably well (Figure 15.5). The apparent overestimation by the model can, to some extent, be explained by methodological problems of the CO_2 flux measurements as discussed by Jensen et al. (1996). Furthermore, total CO_2 evolution was simulated simply by summation of CO_2 evolution in the individual soil layers per day. The latter does not take into account slow gaseous diffusion at high water contents and dissolution of gaseous CO_2 in the aqueous phase as considered by Šimunek and Suarez (1993), for example. Due to these shortcomings, CO_2 evolution was not used as an indicator of the model performance during the modification of the parameter setup described above.

As shown in Figure 15.6, the turnover of the water-insoluble part of the incorporated rape straw, measured as the difference in LPOM-C between the straw amended and non-amended treatment, was reasonably simulated by the same difference in AOM1-C (more recalcitrant AOM-pool) during the first 8 months after incorporation. Later on, simulated AOM1-C decreased below the level of measured LPOM-C. This can be explained by a marked difference between the AOM1-pool in the DAISY model and the measured LPOM: during decomposition in the field, the C:N ratio of LPOM decreased considerably. Simultaneously, the relative lignin content of LPOM increased and the cellulose content decreased, altogether indicating a chemical modification (Magid et al., 1997). This is in contrast to the constant C:N ratio and standard turnover rate coefficient of the modeled AOM1 pool. Consequently, we conclude that modeled AOM1-C represented the measured LPOM-C well in the early stage of decomposition. However, in the later stage, the differences between AOM1-C and LPOM-C increased, with respect to both quality (C:N ratio) and turnover rates.

Model Evaluation

Data from the second experiment were used for an independent evaluation of the setup derived from the first experiment. All soil-related parameters were used without any further modification. The partitioning of added plant residue materials into more recalcitrant AOM1 and fast decomposing AOM2 was done according to the above-described concept, where AOM1 represents the particulate (water-insoluble) part and AOM2 represents the water-soluble part of the plant material (see Table 15.5).

The results of the simulations are shown in Figure 15.7 for the non-amended and plant residue amended treatments. Statistically significant differences between model predicted values and measured values of soil mineral N and soil microbial biomass N were observed in all treatments (F-test according to Addiscott and Whitmore (1987), followed by a multiple t-test). However, the model predictions of mineral N for the non-amended soil and for the soil receiving barley straw were more reliable than those for the two other treatments. This may be due to the fact that these two treatments were most similar to the rape straw experiment. As shown in Table 15.5, barley straw and rape straw were very similar in their quality parameters, whereas blue grass and maize clearly differ from these two materials. This was also true for the lignin and cellulose contents (Mueller et al., 1998a).

A distinct short-term pulse of SMB growth immediately after incorporation of the plant materials followed by a fast decay (Figure 15.7) was not predicted by the model simulations. However, the difference between the measured and predicted SMB pools did not induce a complementary difference for the mineral N pool. Soil microbial residues (SMR), temporarily protected against recycling via the microbial turnover and mineralization, may have been a sink for the N derived from the decaying SMB. Detailed calculations on the N budgets during this period support this hypothesis (Mueller et al., 1998b).

Measured soil surface CO_2 flux and CO_2 evolution predicted by the model simulations (data shown by Mueller et al., 1998b) generally agreed better than for the rape straw calibration data set, but significant differences were observed in all treatments at a few of the measuring dates.

The predicted dynamics of the more recalcitrant AOM1-pool (initialized as water-insoluble AOM) was correlated with the measured amounts of LPOM from the added plant materials (Figure 15.8). However, a slight overestimation of measured LPOM-C by AOM1-C in the initial period after incorporation of AOM was followed by a slight underestimation later on. Due to the high variability of the measured data (see differences between the two replicates for each sampling time in Figure 15.8), the differences between model predicted and measured values were not statistically significant. However, the differences occurred in all treatment according to the same pattern.

The discrepancy between the decay of AOM1-C and LPOM-C in the very early stages of decomposition indicates that a part of the water-insoluble (particulate) plant material was easily decomposable. The discrepancy is most pronounced in the maize treatment and most persistent in the blue grass treatment (Figure 15.8). Maize and blue grass represent fresh plant residues with relatively high contents of easily

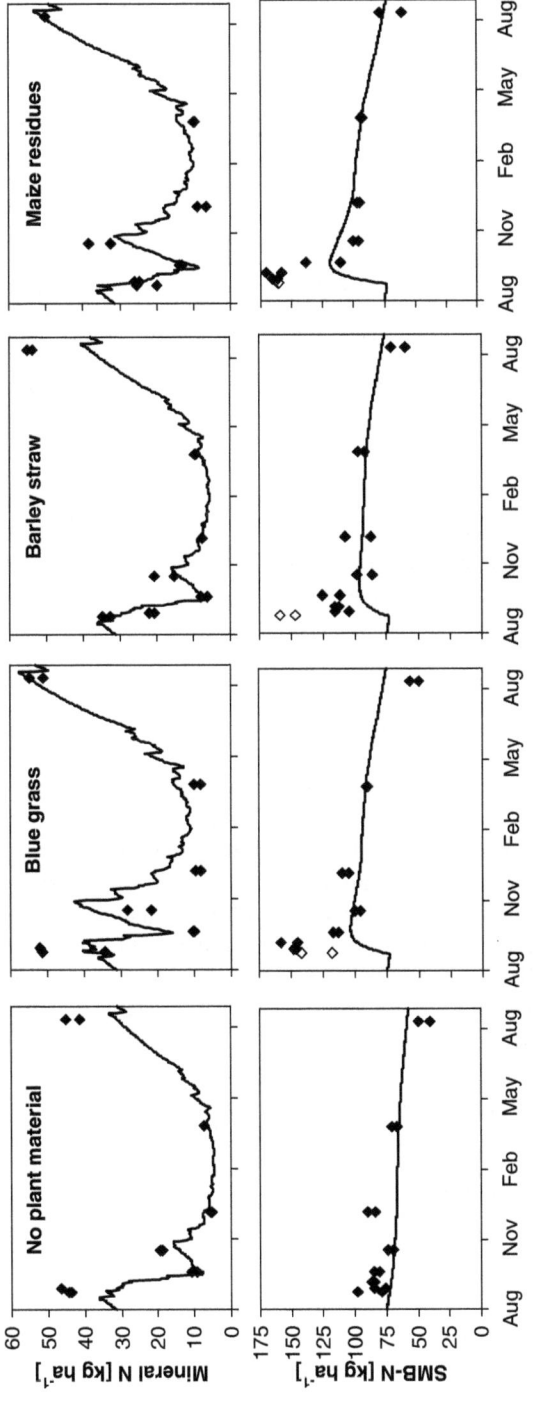

Figure 15.7 Time courses of model predicted (lines) and mean measured values (points) of mineral N, soil microbial biomass N (SMB-N) and soil microbial biomass C (SMB-C), and in the non-amended and amended treatments (6 metric ton/ha chopped blue grass, barley straw, or maize plants). Open symbols indicate methodological problems with this specific sample. (From Mueller, T., Magid, J., Jensen, L.S., Svendsen, H.S., and Nielsen, N.E. 1998b. Soil C and N turnover after incorporation of chopped maize, barley straw and bluegrass in the field: evaluation of the DAISY soil-organic-matter submodel. *Ecological Modelling*, 111:1-15. With permission from Elsevier Science.)

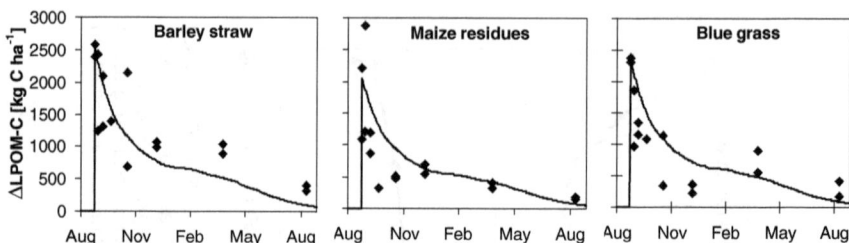

Figure 15.8   Time courses of the difference between the residue amended (6 metric ton/ha) and non-amended treatments in model predicted water-insoluble plant residues (ΔAOM1-C (lines)) and measured light particulate organic matter (ΔLPOM-C > 100 μm; ρ < 1.4 g/cm). (From Mueller, T., Magid, J., Jensen, L.S., Svendsen, H.S., and Nielsen, N.E. 1998b. Soil C and N turnover after incorporation of chopped maize, barley straw and bluegrass in the field: evaluation of the DAISY soil-organic-matter submodel. *Ecological Modelling*, 111:1-15. With permission from Elsevier Science.)

decomposable substances. The general assumption that water-insoluble plant material can be initialized as AOM1 (more recalcitrant added organic matter) may not be true for this type of organic material. This is supported by unpublished simulations of decomposition studies carried out with residues of green manure plant in our laboratories.

CONCLUSION

Overall, the DAISY model has performed reasonably well in simulating long-term trends in total soil C levels at most arable sites tested thus far, even under quite varying climatic conditions. However, a critical point, common to many other models, is the estimation of below-ground C inputs from rhizodeposition and root turnover under perennial crops (e.g., lucerne). As long as the models do not incorporate simulation of these C pathways, individual estimates of plant-derived C inputs will have to be used if the model is to be used for predictive scenario analyses. Although experimental data for calibration exists for cereals, it is still very sparse for perennial crops and, hence, model allocation of plant primary production to rhizodeposition and root turnover, as well as their turnover rates, remains rather hypothetical. Therefore, this must be a key subject in future research if soil organic matter models are to improve their general applicability for scenario analyses.

The overall calibration and validation in the short-term soil C and N dynamics have shown us that DAISY is able to reliably simulate the turnover of soil organic matter and microbial biomass if no organic matter is added and if organic matter with a relatively high content of recalcitrant material is added to the soil. However, the turnover of added organic matter with a high content of easily degradable substances is still not simulated satisfactorily. This may be due to the concept used for the subdivision of AOM into an easily degradable sub-pool and a more recalcitrant sub-pool based on water solubility; evidently, some of the non-soluble compounds may still be relatively easily degradable. Furthermore, microbial residues may play a role in the turnover processes, which is currently not reflected by the concepts behind the DAISY model.

REFERENCES

Abrahamsen, P. and Hansen, S. 2000. *Daisy*: An open soil-crop-atmosphere system model. *Environmental Modelling and Software*, 15:313-330.

Addiscott, T.M. and Whitmore, A.P., 1987. Computer simulation of changes in soil mineral nitrogen and crop nitrogen during autumn, winter and spring. *J. Agric. Sci.*, 109:141-157.

Andrén, O. and Kätterer, T., 1997. ICBM: the introductory carbon balance model for exploration of soil carbon balances. *Ecol. Appl.*, 7:1226-1236.

Brookes, P.C., Landman, A., Pruden, G., and Jenkinson, D.S. 1985. Chloroform fumigation and the release of soil nitrogen: a rapid direct extraction method to measure microbial biomass nitrogen in soil. *Soil Biol. Biochem.*, 17:837-842.

Christensen, B.T. and Johnston, A.E., 1997. Soil organic matter and soil quality: lessons learned from long-term field experiments at Askov and Rothamsted. In E.G. Gregorich and M.R. Carter (Eds.). *Soil Quality for Crop Production and Ecosystem Health*. Elsevier Science, Amsterdam, 399-430.

Coleman, K. and Jenkinson, D.S. 1996. RothC-26.3 — A model for the turnover of carbon in soil. In D.S. Powlson, P. Smith, and J.U. Smith (Eds.). *Evaluation of Soil Organic Matter Models Using Existing, Long-term Datasets*. Vol. NATO ASI Series I. Springer-Verlag, Heidelberg, 237-246.

Crocker, G.J. and Holford, I.C.R. 1996. The Tamworth legume/cereal rotation. In D.S. Powlson, P. Smith, and J.U. Smith (Eds.). *Evaluation of Soil Organic Matter Models Using Existing, Long-Term Datasets*. Vol. 38, Springer-Verlag, Heidelberg, 313-318.

Diekkrüger, B., Söndgerath, D., Kersebaum, K.C., and McVoy, C.W. 1995. Validity of agro-ecosystem models: a comparison of results of different models applied to the same data set. *Ecol. Modelling*, 81:3-29.

Franko, U. 1996. Modelling approaches of soil organic matter turnover within the CANDY system. In D.S. Powlson, P. Smith, and J.U. Smith (Eds.). *Evaluation of Soil Organic Matter Models Using Existing, Long-term Datasets*. Vol. 38, NATO ASI Series I, Springer-Verlag, Heidelberg, 247-254.

Grace, P, 1996. The Waite permanent rotation trial. In D.S. Powlson, P. Smith, and J.U. Smith (Eds.). *Evaluation of Soil Organic Matter Models Using Existing, Long-Term Datasets*. Vol. 38, Springer-Verlag, Heidelberg, 335-340.

Hansen S., Jensen N.E., Nielsen N.E., and Svendsen H., 1990. *Daisy* — Soil Plant Atmosphere System Model, *NPo-forskning fra Miljøstyrelsen*, A10.

Hansen S., Jensen N.E., Nielsen N.E., and Svendsen H., 1991. Simulation of nitrogen dynamics and biomass production in winter wheat using the Danish simulation model Daisy, *Fert. Res.*, 27:245-259.

Jenkinson, D.S. and Rayner, J.H. 1977. The turnover of soil organic matter in some of the Rothamsted classical experiments. *Soil Sci.*, 123:298-305.

Jenkinson, D.S., Hart, P.B.S., Rayner, J.H., and Parry, L.C. 1987. Modelling the turnover of organic matter in long-term experiments at Rothamsted. *Intecol Bulletin*, 15:1-8.

Jensen, B. 1993. Rhizodeposition by $CO_2{}^{+14}C$-pulse-labelled spring barley grown in small field plots on sandy loam. *Soil Biol. Biochem.*, 25:1553-1559.

Jensen, L.S., Mueller, T., Tate, K.R. Ross, D.J., Magid, J., and Nielsen, N.E. 1996. Measuring soil surface CO_2 flux as an index of soil respiration *in situ*: a comparison of two chamber methods. *Soil Biol. Biochem.*, 28:1297-1306.

Jensen, L.S., Mueller, T., Magid, J., and Nielsen, N.E. 1997a. Temporal variation of C and N mineralization, microbial biomass and extractable organic pools in soil after oilseed rape straw incorporation in the field. *Soil Biol. Biochem.*, 29:1043-1055.

Jensen, L.S., Mueller, T., Nielsen, N.E., Hansen, S., Crocker, G.J., Grace, P.R., Klír, J., Körschens, M., and Poulton, P.R. 1997b. Simulating soil organic matter trends in long-term experiments with the soil-plant-atmosphere system model *Daisy*. *Geoderma*, 81:5-28.

Joergensen, R.G. and Mueller, T. 1996. The fumigation extraction method to estimate soil microbial biomass: calibration of the k_{EN} value. *Soil Biol. Biochem.*, 28:33-37.

Kätterer, T. and Andrén, O. 1999. Long-term agricultural field experiments in Northern Europe: analysis of the influence of management on soil carbon stocks using the ICBM model. *Agric. Ecosyst. Environ.*, 72:165-179.

Klír, J. 1996. Long-term field experiment Praha-Ruzyne, Czech republic. In D.S. Powlson, P. Smith, and J.U. Smith (Eds.). *Evaluation of Soil Organic Matter Models Using Existing, Long-Term Datasets*. Vol. 38, Springer-Verlag, Heidelberg, 363-368.

Loague, K. and Green, R.E. 1991. Statistical and graphical methods for evaluating solute transport models: overview and application. *J. Contam. Hydrol.*, 7:51-73.

Magid, J., Jensen, L.S., Mueller, T., and Nielsen, N.E. 1997. Size-density fractionation for *in situ* measurements of rape straw decomposition — an alternative to the litterbag approach? *Soil Biol. Biochem.*, 29:1125-1133.

Mueller, T., Jensen, L.S., Hansen, S., and Nielsen, N.E. 1996. Simulating soil carbon and nitrogen dynamics with the soil-plant-atmosphere system model DAISY. In D.S. Powlson, P. Smith, and J.U. Smith (Eds.). *Evaluation of Soil Organic Matter Models Using Existing, Long-term Datasets*. Vol. 38, NATO ASI Series I, Springer-Verlag, Heidelberg, 275-281.

Mueller, T., Jensen, L.S., Magid, J., and Nielsen, N.E. 1997. Temporal variation of C and N turnover in soil after oilseed rape straw incorporation in the field: Simulations with the soil-plant-atmosphere model DAISY. *Ecol. Modelling*, 99:247-262.

Mueller, T., Jensen, L.S., Nielsen, N.E., and Magid, J. 1998a. Turnover of carbon and nitrogen in a sandy loam soil following incorporation of chopped maize plants, barley straw and blue grass in the field, *Soil Biol. Biochem.*, 30:561-571.

Mueller, T., Magid, J., Jensen, L.S., Svendsen, H.S., and Nielsen, N.E. 1998b. Soil C and N turnover after incorporation of chopped maize, barley straw and bluegrass in the field: evaluation of the DAISY soil-organic-matter submodel. *Ecol. Modelling*, 111:1-15.

Nielsen, N.E., Schjørring, J.K., and Jensen, H.E. 1988. Efficiency of fertilizer nitrogen uptake by spring barley. In D.S. Jenkinson and K.A. Smith (Eds.). *Nitrogen Efficiency in Agricultural Soils*. Elsevier Applied Science, London, 62-72.

Parton, W.J., Schimel, D.S., Cole, C.V., and Ojima, D.S. 1987. Analysis of factors controlling soil organic matter levels in Great Plains grasslands. *Soil Sci. Soc. Am. J.*, 51:1173-1179.

Parton, W.J. and Rasmussen, P.E. 1994. Long-term effects of crop management in wheat-fallow. 2. CENTURY model simulations. *Soil Sci. Soc. Am. J.*, 58:530-536.

Paustian, K., Parton, W.J., and Persson, J. 1992. Modeling soil organic matter in organic-amended and nitrogen-fertilized long-term plots. *Soil Sci. Soc. Am. J.*, 56:476-488.

Petersen, C.T., Jørgensen, U., Svendsen, H., Hansen, S., Jensen, H.E., and Nielsen, N.E. 1995. Parameter assessment for simulation of biomass production and nitrogen uptake in winter rape. *Eur. J. Agron.*, 4:77-89.

Plantedirektoratet, 1999. *Vejledning og skemaer. Mark og gødningsplan, gødningsregnskab, plantedække og harmoniregler 1999/2000.* (Field and fertilization plan, fertilizer accounting, crop-cover and harmony rules 1999/2000. In Danish). The Ministry of Food, Agriculture and Fishery, Copenhagen.

Powlson, D.S., Smith, P., Coleman, K., Smith, J.U., Glendining, M.J., Korschens, M., and Franko, U. 1998. A European network of long-term sites for studies on soil organic matter. *Soil Tillage Res.,* 47:263-274.

Rietveld, M.R. 1978. A new method for estimating the relative coefficients in the formula relating solar radiation to sun hours. *Agric. Meteorol.,* 19:243-252.

Šimunek, J. and Suarez, D.L. 1993. Modelling of carbon dioxide transport and production in soil. 1. Model Development. *Water Resources Res.,* 29:487-497.

Smith, P., Powlson, D., and Glendining, M. 1996. Establishing a Europea GCTE Soil organic matter network (SOMNET). In D.S. Powlson, P. Smith, and J.U. Smith (Eds.). *Evaluation of Soil Organic Matter Models Using Existing, Long-term Datasets.* Vol. 38, NATO ASI Series I, Springer-Verlag, Heidelberg, 81-98.

Smith, P., Smith, J.U., Powlson, D.S., Arah, J.R.M., Chertov, O.G., Coleman, K., Franko, U., Frolking, S., Gunnewiek, H.K., Jenkinson, D.S., Jensen, L.S., Kelly, R.H., Li, C., Molina, J.A.E., Mueller, T., Parton, W.J., Thornley, J.H.M., and Whitmore, A.P. 1997. A comparison of the performance of nine soil organic matter models using datasets from seven long-term experiments. *Geoderma,* 81:153-222.

Swinnen, J., van Veen, J.A., and Merckx, R. 1994. Rhizosphere carbon fluxes in field-grown spring wheat - Model calculations based on ^{14}C partitioning after pulse-labelling. *Soil Biol. Biochem.,* 26:171-182.

Vance, E.D., Brookes, P.C., and Jenkinson, D.S., 1987. An extraction method for measuring soil microbial biomass C. *Soil Biol. Biochem.,* 19:703-707.

Vereecken, H., Jansen, E.J., Hack-ten Broeke, M.J.D., Swerts, M., Engelke, R., Fabrewitz, S., and Hansen, S. 1991. Comparison of simulation results of five nitrogen models using different data sets. In *Soil and Groundwater Research Report II. Nitrate in Soils.* Final Report of Contracts EV4V-0098-NL and EV4V-00107C. Commission of the European Communities, Luxembourg, 321-338.

de Willigen, P. 1991. Nitrogen turnover in the soil-crop-systems: comparison of fourteen simulation models. *Fert. Res.,* 27:141-149.

Wu, J., Joergensen, R.G., Pommerening, B., Chaussod, R., and Brookes, P.C. 1990. Measurement of soil microbial biomass C by fumigation-extraction — An automated procedure. *Soil Biology and Biochemistry,* 22:1167-1169.

CHAPTER 16

Modeling Nitrate Leaching at Different Scales — Application of the DAISY Model

Søren Hansen, Christian Thirup, Jens C. Refsgaard, and Lars S. Jensen

CONTENTS

INTRODUCTION AND SCOPE

The first version of the DAISY model was developed in the late 1980s (Hansen et al., 1990; 1991a). Since then, it has been further developed to include more functionality as compared with the original version. Recently, the DAISY code was rewritten, and the new code is carefully designed to facilitate interaction with other models, either by replacing individual DAISY processes or by using DAISY as a part of a larger system, thus making DAISY an open software system (Abrahamsen and Hansen, 2000). However, the old, well-tested functionality of the model has been maintained.

The confidence, by which simulation results are viewed, strongly depends on the extent of model validation. During the 1990s, the DAISY model participated in a number of comparative model tests. The first of these comparative model tests was conducted in connection with an EU project entitled "Nitrate in Soils" (Thomasson et al., 1991). The second comparative test was performed at a workshop organized by The Institute for Soil Fertility Research in Haren, The Netherlands, as a part of the celebration of their centennial (Willigen, 1991). The third test was performed at a workshop organized by The Technical University of Braunschweig, Germany, as a part of a research program (Diekkrüger et al., 1995). Finally, the DAISY model was evaluated in connection with a NATO Advanced Research Workshop entitled "Evaluation of Soil Organic Matter Models Using Existing Long-Term Datasets" (Smith et al., 1997). In all the comparative validation tests, the evaluation of the performance of the DAISY model has been favorable. In this chapter, a few examples of the applications of the DAISY model in connection with these comparative tests are given, except for the DAISY contribution to "Evaluation of Soil Organic Matter Models Using Existing Long-Term Datasets," which is dealt with in Chapter 15.

The main concern regarding nitrate leaching is due to nitrogen pollution of surface and ground waters. Hence, the scale for which nitrate leaching is of special interest is the catchment scale. However, a catchment is composed of a number of agricultural fields subjected to various management practices, which to a large extent determine the leaching. The simulation of leaching at the catchment scale can be obtained by aggregation of the simulated leaching obtained at the field scale (a field being subject to only one management practice). The aggregation procedure may be deterministic in the sense that the nitrate leaching simulation is based on exact information on the management practices that take place within the watershed, or it may be statistical in the sense that the simulated management practices are based on statistical information. However, in both cases, the model must be able to predict leaching at field scale.

A mechanistic simulation model such as DAISY requires rather detailed information on the soil system. This information is often obtained at a scale much smaller than the field scale. The problem can be overcome by applying so-called effective parameters valid for the field scale, that is, by interpreting the field as an equivalent soil column (Jensen and Refsgaard, 1991). In the following, an example is given on how DAISY can be used in connection with effective soil parameters and how the model can be used in assessing nitrate leaching at the field scale and subsequently at the catchment scale.

MODEL DESCRIPTION

A brief review of the recent model is given in this section. Figure 16.1 shows an overall view of the model, including the "driving variables," viz. weather and management data, and its exchange with the environment of energy, water, CO_2, and

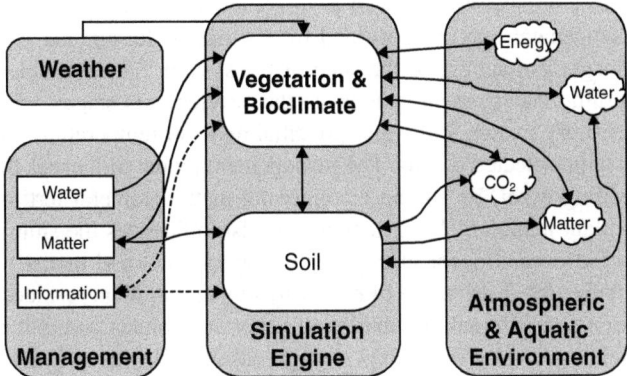

Figure 16.1   Schematic overview of the DAISY model. Flows of energy, matter, and carbon dioxide. The simulation engine is driven by weather and management data. The environment is represented by infinite sources and sinks.

matter. In the present context, the environment constitutes both the atmospheric and aquatic environments. The exchange of energy may include the exchange of radiation, sensible and latent heat at the surface. Furthermore, an exchange of heat at the bottom of the considered soil profile takes place. The exchange of water includes the evaporation of free water intercepted in the vegetation canopy, stored in a snowpack or ponding at the soil surface, as well as evaporation from the soil and transpiration from the vegetation. Furthermore, it includes surface runoff, deep percolation, and runoff through pipe drains and capillary rise at the bottom of the soil profile. The exchange of CO_2 includes the assimilation of CO_2 by vegetation and the respiration by vegetation and soil microbial biomass. The term "matter" in the diagram covers organic matter, solutes, and gaseous compounds. The loss of gaseous compounds includes losses due to denitrification and ammonia volatilization. Exchange of solutes includes movement of solutes with the surface runoff and movement of solutes with the water flow at the bottom of the profile.

Management Sub-model

The management model can be considered a programming language that allows for building rather complex management scenarios (Abrahamsen and Hansen, 2000; Abrahamsen, 1999). The management, in its simplest form, is just a list of management actions that are performed at a predefined date. This is adequate when simulating field experiments in which all the management actions are known in advance. However, in scenario studies, one often wants to investigate the effect of different management practices, where the performance of management actions are interrelated and where they are affected by the conditions simulated by the simulation engine. Such management scenarios can also be built using the management model. Management actions include soil tillage, sowing and harvesting, irrigation, fertilization, and spraying. Irrigation includes both overhead and surface irrigation. Fertilization comprises the use of mineral as well as organic fertilizers. Application methods comprise surface broadcasting as well as direct incorporation in the soil. The execution of a management may be governed by a simple date, number of days from a previous management action, or the state of the system. State variables that can be used in governing management actions are the development stage of the crop, shoot (leaf and stem) dry matter, soil water potential and soil temperature. The latter two are obtained at preselected depths. The management model influences the simulation engine by adding water, by adding or removing matter (mineral fertilizers; organic matter, e.g., plant material, slurry, manure). It also influences the simulation engine by transfer of information; for example, plowing may not add matter to the system but it may change the distribution of matter by plowing in plant residues and manure and by changing the internal distribution of water and solutes within the soil profile.

The simulation engine represents a single column. However, DAISY allows for running more simulation engines in parallel, each characterized by its own management. Each simulation may also have its own weather or, alternatively, a common or global weather for all columns may be defined. Each column is constituted by an aerial part (i.e., the vegetation and bioclimate model) and a below-ground part (i.e., the soil model). Extensive exchange of energy, water, CO_2, and matter takes place between the two parts. The soil surface constitutes an interface between the two sub-models.

The Vegetation and Bioclimate Sub-models

Figure 16.2 expands the vegetation and bioclimate model somewhat, viz. into separate vegetation and bioclimate models. The vegetation model comprises one or more crop growth models, which may all be an instance of the same generic crop model, but due to different parameterizations they simulate the growth of different crops. This feature makes it possible to simulate inter-cropping systems. The vegetation model accounts for the composite canopy of the present crops and it handles the communication with the bioclimate model. The individual instances of the crop growth model communicate with the soil model.

Crop Growth Model

The default crop growth model comprises a number of sub-modes describing main plant growth processes. Simple empirical relations describe most of these processes. Penning de Vries et al. (1989) have compiled information on such relations for several annual crops, and a number of these relations have been adopted in the present model. The crop growth model is split in two parts, viz. a shoot part and a root part. The shoot part of the growth is further divided into a phenological model, a net production model, and a canopy model.

The phenological model simulates the development stage of the crop on the basis of temperature and photoperiod (day length). The development stage plays a central role in the crop growth model as many of the sub-models are strongly influenced by the development stage, and many of the important parameters are functions of the development state (e.g., assimilate partitioning parameters, crop nitrogen demand parameters, parameters pertaining to senescence).

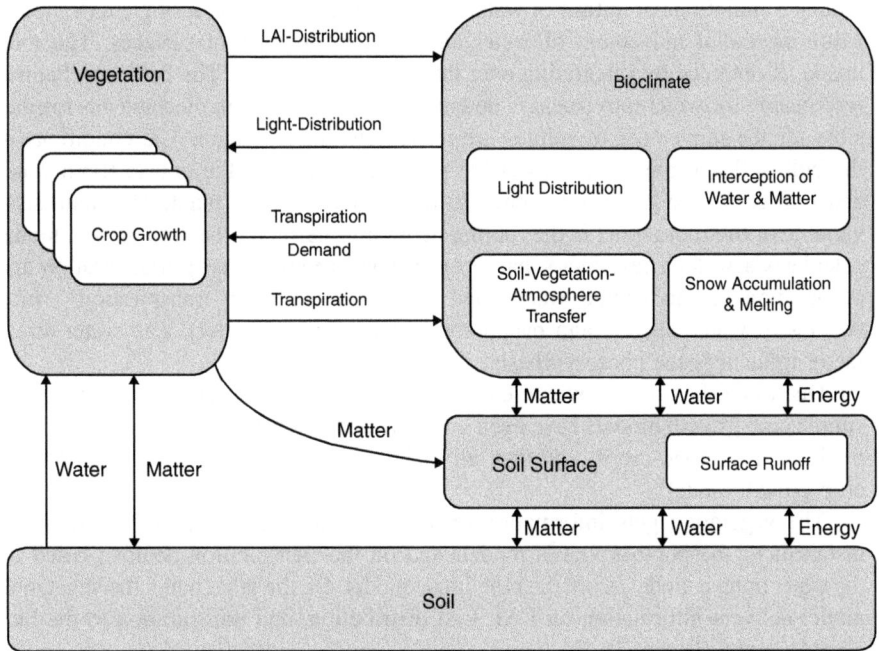

Figure 16.2 Schematic overview of flows of information, energy, water, and matter between the vegetation model, the bioclimate model, and the soil model. The vegetation model communicates directly with the soil model, while the bioclimate communicates through the soil surface component. The basic time step for the communication is 1 hour.

The net production model simulates photosynthesis or CO_2 assimilation; assimilate partitioning; growth and maintenance respiration; and loss of leaf and root dry matter due to senescence. Furthermore, low light interception in the lower part of the canopy may cause defoliation, and a high root:shoot ratio may cause loss of root dry matter. The net production model makes use of information on temperature and the light distribution within the canopy to calculate the photosynthesis. The main state variables of the net production model are the dry weight of root, stem, leaf, and storage organs (e.g., grain, seed, tubers, or beet).

The canopy model simulates the development of leaf area index (LAI), plant height, and leaf distribution. The simulation is based on information on development stage and the state variables of the net production model. The main state variables of the canopy model are LAI and plant height.

The crop nitrogen model simulates the crop nitrogen demand and nitrogen stress, and keeps track of the nitrogen stored in the crop (i.e., in the root, stem, leaf, and storage organ). The main state variables are the nitrogen content in the considered part of the crop.

The root part of the growth model simulates root penetration and root density distribution. Furthermore, it simulates uptake of water and nitrogen (ammonium and nitrate) in the soil solution. These calculations are based on information from the phenological model, the net production model, the crop nitrogen model, and the soil model. Simulations of uptake of water and solutes are based on a single root model, assuming radial movement of water and solutes toward root surfaces. The total uptake is obtained by integrating over the entire root system. The basic mechanism responsible for water movement is potential flow, and the main mechanisms responsible for the movement of solutes are mass flow and diffusion. The conditions in the bulk soil and the conditions at the root surface govern the uptake (both water and nitrogen). The latter is assumed to be regulated by the plant. The main state variable of the root model is the rooting depth. Furthermore, the crop growth model calculates a water stress factor based on demand (potential crop transpiration and evaporation of intercepted water) and supply (actual crop transpiration, which equates to water uptake, and evaporation of intercepted water). The water stress factor influences the photosynthesis.

Due to the flexible architecture of the DAISY software, a couple of other more simple crop growth models have been implemented. However, they can only be used in simulating mono-crops, and they all make use of the root model of the default crop growth model.

The vegetation gets information on the light distribution (used by the active instances of the photosynthesis model) and on the transpiration demand (used by the water uptake model) from the bioclimate model. On the other hand, the vegetation model delivers information on LAI, LAI distribution, and transpiration to the bioclimate model (Figure 16.2).

The Bioclimate Model

The bioclimate model comprises sub-models for light distribution in the canopy, interception of water and matter in the canopy, snow accumulation and melting, and soil-vegetation-atmosphere transfer (SVAT).

When water is allocated to the system either as precipitation or irrigation, then the snow accumulation and melting model is activated. The state of the allocated water (i.e., snow, rain, or a mixture of snow and rain) is simulated on the basis of air temperature. Freezing or melting of a possible snowpack depends on air temperature, soil heat flux (simulated by the soil model) and the received global radiation. The model allows liquid water to percolate or evaporate from the snowpack (if no snow is present, all the allocated water passes through the snowpack). The water percolating out of the snowpack is routed to the interception model where it is intercepted or routed to the surface as direct through-fall. Intercepted water is stored, allocated to the surface as canopy spill-off, or evaporated. The interception model is a simple capacity-type model, where the capacity depends on LAI.

The SVAT model simulates the evaporation fluxes or latent heat fluxes. More than one version of the SVAT model is implemented in the DAISY code. The simple version of the SVAT model is based on the concept of potential evapotranspiration, and it makes use of Beer's law in distributing the potential evapotranspiration between the soil surface and the canopy. The most comprehensive SVAT is based on resistance theory. It distributes the available energy between latent heat, sensible heat, and terrestrial long-wave radiation based on resistance against energy transfer between the various parts of the canopy-soil system (Keur et al., 2001). The simple version is the default model.

The soil surface model keeps track of water, solutes, and organic matter stored at the soil surface. Water received at the soil surface is routed to the surface runoff model. Here, it is infiltrated, stored, evaporated, or lost from the system as surface runoff. Infiltrability and maximum exfiltration rates, calculated by the soil water model, are key auxiliary variables in simulating infiltration and soil evaporation. Similarly, solutes received at the soil surface are routed to the surface runoff model. Here, it may move into the soil with infiltrating water, it may be stored, or lost from the system with the surface runoff (a part of the solute stored within the mixing sub-layer of the upper numeric layer may also be lost by surface runoff). Finally, organic matter stored at the surface may enter the soil due to tillage (the management model) or due to bio-incorporation (a new feature of the soil model).

The Soil Model

The soil model simulates the storage and movement of water, heat, and solutes (ammonium and nitrate). Furthermore, it simulates sorption and turnover processes of the considered solutes. It also simulates the storage, turnover, and bio-incorporation of organic matter. The considered turnover processes are mineralization of organic matter, nitrification, and denitrification. The vegetation model simulates plant uptake of solutes.

The soil model considers a soil profile composed of a number of soil horizons, each characterized by individual soil properties. The soil horizon is further divided into a number of numeric layers for the purpose of keeping track of the distribution of water, heat, solutes, and organic matter within the soil profile.

Default models for the simulation of the movement of water, heat, and solutes are based on the Richards equation, an extended Fourier equation (including freezing and melting of soil water), and the convection-dispersion equation, respectively. The equations are solved by finite-difference techniques. However, in the case of water movement, a simple model based on gravity flow is also implemented. A model that takes macropore flow into account is also implemented. Boundary conditions are very important when solving partial differential equations. The following lower boundary conditions are implemented: deep groundwater (assuming gravity flow at the bottom of the profile), fixed groundwater table, variable groundwater table read from file or supplied by another model through the API (application programmers interface), tile drains, and a lysimeter boundary condition. Sorption is simulated by Freundlich or Langmuir isotherms assuming instantaneous equilibrium.

The Soil Organic Matter Turnover Model

The soil organic matter turnover model simulates the decomposition of organic matter and the associated mineralization-immobilization turnover (MIT) of nitrogen. The model considers three distinguishable types of organic matter, viz. soil organic matter (SOM), soil microbial biomass (SMB), and added organic matter (AOM).

SOM constitutes a vast number of different organic compounds; however, the MIT model simulates the decomposition rate by splitting SOM into two separate pools and then applying first-order kinetics to each of the pools (SOM1 and SOM2). Similarly, the SMB and AOM are divided into two sub-pools. For further information on the MIT model, see Chapter 15.

Soil Mineral Nitrogen Model

The nitrification is simulated by Michaelis-Menten kinetics where the maximum nitrification rate is influenced by soil temperature and soil moisture expressed as soil water pressure potential.

A simple capacity model simulates the denitrification. The nitrification capacity or potential nitrification rate is assumed to be proportional to the CO_2 evolution simulated by the MIT model. The actual denitrification is simulated by reducing the potential denitrification according to the oxygen status mimicked by a function of the relative soil water content.

Data Requirements and Availability

The recent version of DAISY allows the user to define a large number of model parameters. However, based on the experience obtained with the model during the past decade, many parameters have been given default values or are estimated from transfer functions. Due to the special input language of DAISY, the user need not be concerned with these default parameters and transfer functions. However, information on these matters can be obtained from the DAISY user manual (Abrahamsen, 1999). To run the model, driving variables and model parameters are required.

The model can make use of weather data at different time scales. The most detailed simulations require hourly values of global radiation, air temperature, relative humidity, wind speed, and precipitation. However, for most purposes, daily values of global radiation, air temperature, potential evapotranspiration, and precipitation are sufficient. The minimum requirements are daily values of global radiation, air temperature, and precipitation.

The DAISY crop library contains the parameterizations of a number of different crops, and when one of these crops is sown, the model associates the parameters of the crop with its name. If a user wants to sow a crop that is not included in the crop library, a new set of parameters must be added to the library. However, the flexible nature of the input language of DAISY makes it possible to create new crops by calling up an existing crop and then overwriting the appropriate parameters. The minimum set of vegetation and bioclimate parameters required by the model are the names of the crops grown in the selected scenario, provided that the crop is included in the crop library.

The soil is constituted by a number of soil horizons and each of those must be defined. The minimum data required to describe a horizon include texture, organic matter content, SOM fractions, C:N-ratio in SOM1 and SOM2 (see Chapter 15), and soil hydraulic properties. The soil hydraulic properties can be described by a number of common models (e.g., Mualem-van Genuchten model, Burdine-van Genuchten model, Burdine-Brooks and Corey model, Mualem-Brooks and Corey model, etc.), and the appropriate parameters must be specified. The soil profile requires information on the size and location of the soil horizons as well as the depth of the numeric layers (the interface of two horizons must always be located at an interface between two numeric layers). Furthermore, the maximum rooting depth of the soil and the dispersivity of the considered soil profile must be specified. Finally, the type of lower boundary must be selected.

TEMPORAL DYNAMICS PERFORMANCE IN COMPARATIVE TESTS

A main objective of the EU project "Nitrate in Soils" (Thomasson et al., 1991) was to evaluate nitrate leaching models with respect to their ability to predict nitrate leaching from the root zone as a function of land use and fertilizer management. Basic information on the data sets used in the evaluation is shown in Table 16.1. Detailed information on the data sets is given by Breeuwsma et al. (1991) and on the simulations by Hansen et al. (1991b). It is noted that the data set covers quite a variety of fertilization practices, ranging from unfertilized to heavily fertilized plots and including both mineral fertilizers and animal manures. Simulated annual

Table 16.1   Datasets Used for Evaluation of the DAISY Model

Site, Soil Texture, and Period	Crop	Type of Fertilizer or Manure	Time of Application	Amount (kg N/ha), Year No. in Period 1	2	3	4	5
Jyndevad, sand, 1987–88	S. barley[a]	None		0	0			
	S. barley[a]	Mineral	Spring	120	120			
	S. barley[a]	Pig slurry	Spring	50	50			
	S. barley[a]	Pig slurry	Spring	100	100			
	S. barley[a]	Pig slurry	Autumn	100	100			
	S. barley w. ley[a]	Pig slurry	Spring	100	100			
Askov, sandy loam, 1987–88	S. barley[a]	None		0	0			
	S. barley[a]	Mineral	Spring	133	133			
	S. barley[a]	Pig slurry	Spring	100				
Ruurlo, loamy sand, 1980–84	Grass (plot 30)	None		0	0	0	0	0
	Grass (plot 37)	Mineral	Split	660	600			
	Grass (plot 19)	Cattle slurry	Split	81	83	100	66	49
	Grass (plot 39)	Mixed	Split	421	483			

[a] S. = spring.

From Breeuwsma, A., Djurhuus, J., Jansen, J., Kragt, J.F., Swerts, M.C.J.J., and Thomasson, A.J. 1991. Data sets for validation of nitrate leaching models. In *Soil and Groundwater Research Report II, Nitrate in Soils*. Final report of contracts EV4V-0098-NL and EV4V-00107-C, Commission of the European Communities, 219-235; Hansen, S., Jensen, H.E., Nielsen, N.E., and Svendsen, H. 1991b. Simulation of biomass production, nitrogen uptake and nitrogen leaching by the *DAISY* model. In *Soil and Groundwater Research Report II. Nitrate in Soils*. Final report of contracts EV4V-0098-NL and EV4V-00107-C, Commission of the European Communities, 300-309.

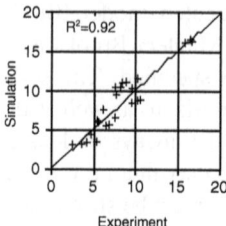 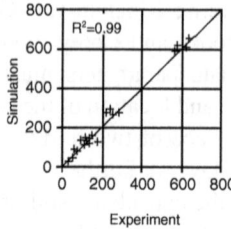 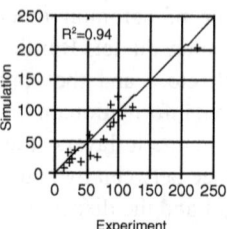

Figure 16.3 Scatter diagrams showing simulated vs. experimental values of above-ground dry matter yield (t DM/ha, left), above-ground nitrogen yield (kg N/ha, middle), and annual leaching (kg N/ha, right). For data, see Table 16.1. (From Hansen, S., Jensen, H.E., Nielsen, N.E., and Svendsen, H. 1991b. Simulation of biomass production, nitrogen uptake and nitrogen leaching by the *DAISY* model. In *Soil and Groundwater Research Report II. Nitrate in Soils.* Final report of contracts EV4V-0098-NL and EV4V-00107-C, Commission of the European Communities, 300-309.)

Table 16.2   Statistical Evaluation of the Performance of the DAISY Model

Statistical Criterion	Optimal Value	Dry Matter Production	Nitrogen in Shoot	Nitrate Leaching
Relative root mean square error	0	0.18	0.15	0.27
Coefficient of determination	1	0.86	0.92	1.09
Modeling efficiency	1	0.82	0.97	0.89
Coefficient of residual mass	0	−0.05	−0.08	0.05

Note: The coefficient of determination expresses the ratio of scatter of predicted values to that of the observed values. The modeling efficiency is always less or equal to 1. A perfect prediction would yield a modeling efficiency equal to 1. A negative modeling efficiency would indicate that the measured average would be a better estimate than the predicted values. Positive coefficients of residual mass indicate a tendency to underestimate the measurements, and negative ones indicate a tendency to overestimate.

Data from the Jyndevad, Askov, and Ruurlo sites. See Breeuwsma, A., Djurhuus, J., Jansen, J., Kragt, J.F., Swerts, M.C.J.J., and Thomasson, A.J. 1991. Data sets for validation of nitrate leaching models. In *Soil and Groundwater Research Report II, Nitrate in Soils.* Final report of contracts EV4V-0098-NL and EV4V-00107-C, Commission of the European Communities, 219-235.

dry matter production, nitrogen uptake in the shoot, and nitrate leaching were compared with corresponding experimental values, and the performance of the model was evaluated by statistical criteria as proposed by Loague and Green (1991). The considered statistical criteria were relative root mean square error, coefficient of determination, modeling efficiency, and coefficient of residual mass. The main results are shown as scatter diagrams in Figure 16.3 and as statistical criteria in Table 16.2.

The scatter diagrams (Figure 16.3) show an acceptable scatter around the 1:1 line; it should also be kept in mind that the experimental values are uncertain. The more rigorous evaluation performed by the statistical criteria indicates that the model has a small tendency to overestimate dry matter production and nitrogen uptake and a small tendency to underestimate nitrate leaching. Regarding leaching, the model explained about 90% (modeling efficiency 0.89) of the measured variability with a relative root mean square error of 27%. Furthermore, the coefficient of determination

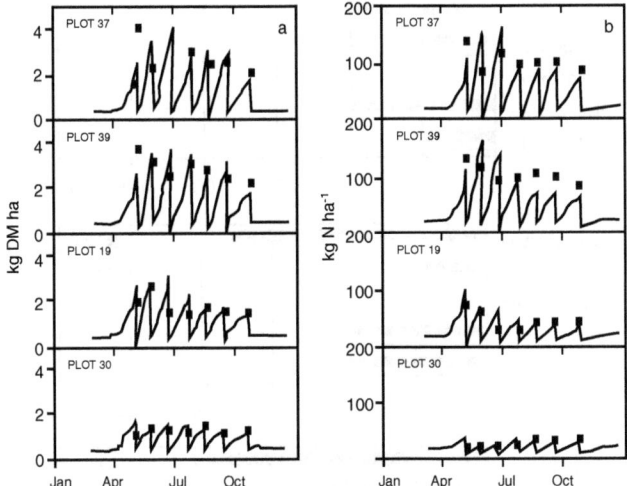

Figure 16.4   Comparison between measured and simulated values of dry matter production (t DM/ha, left) and above-ground nitrogen content (kg N/ha, right) at four different plots at Ruurlo, The Netherlands, 1980. (From Hansen, S., Jensen, H.E., Nielsen, N.E., and Svendsen, H. 1991b. Simulation of biomass production, nitrogen uptake and nitrogen leaching by the *DAISY* model. In *Soil and Groundwater Research Report II. Nitrate in Soils.* Final report of contracts EV4V-0098-NL and EV4V-00107-C, Commission of the European Communities, 300-309.)

(>1) indicates that the predicted scatter is less than the scatter found in the measured data. The simulation did overestimate the dry matter production. This is also what could be expected because the crop model included in DAISY does not take into account deficiencies other than water and nitrogen; that is, all other growth factors are assumed to be at an optimal level. Hence, very erroneous simulations can be expected when failure of the crop occurs due to reasons other than water or nitrogen stress. Hence, caution should be taken when large discrepancies between simulated and observed yields occur.

Simulation of grass for cutting poses special problems. Figure 16.4 shows a comparison between simulated and observed dry matter yield and the corresponding harvested nitrogen yields. It is noted that during the 1980 growth season, seven cuts were performed. A special difficulty pertaining to the simulation of grass for cutting is the simulation of regrowth after cutting because the model is very sensitive to the amount of active leaf area left in the field after cutting. In the present simulations, a common value of active leaf area after cutting was assumed throughout the entire simulation period. Another difficulty pertaining to the simulation of grass, and in fact all wintering crops, is the simulation of how growth is initiated in spring after the wintering. The crop model included in DAISY thus far does not account for the effects of damage during winter.

A major component in the mineral nitrogen balance is the nitrogen mineralization. In general, nitrogen mineralization was not measured in the present experiments. However, at all sites, unfertilized plots were included and such plots are well-suited to estimate net nitrogen mineralization. Figure 16.5 shows the main components

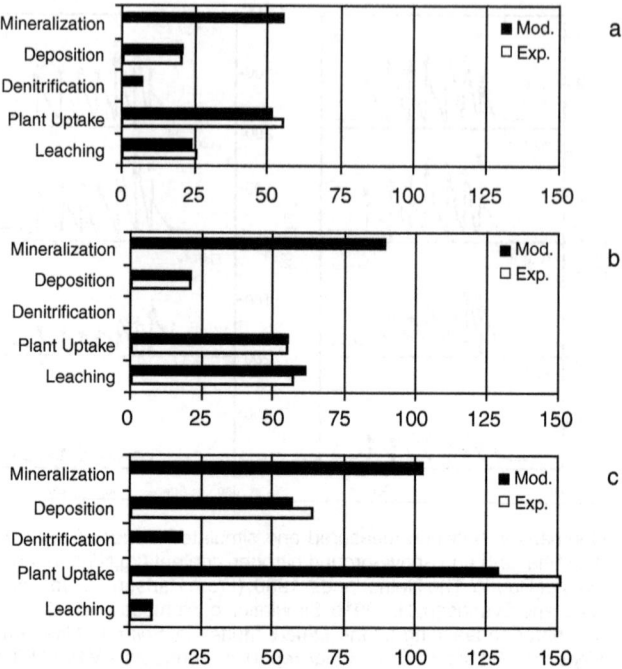

Figure 16.5   Main components in the annual nitrogen balance (kg N/ha) of unfertilized plots at Jyndevad, Denmark (a); Askov, Denmark (b); and Ruurlo, The Netherlands (c).

of the average annual nitrogen balance simulated at the three sites, together with the corresponding experimental values where available. At Jyndevad and Askov, Denmark, measurements of denitrification during the time of the experiment were performed, and these observations indicate that the simulated denitrification is of the right magnitude (Lind et al., 1990). Although the field measurements of denitrification are associated with considerable uncertainty, they do indicate that the simulation of net nitrogen mineralization is of the correct magnitude at the two sites. By use of the *in situ* method of Raison et al. (1987), Lind et al. (1990) measured the net nitrogen mineralization at the unfertilized plot at Jyndevad during the period 1988/1989 and they obtained a result, 95 kg N/ha, which exactly matches the result obtained with the simulation model. Similar measurements at the Askov unfertilized plot showed a mineralization of 70 kg N/ha, which is somewhat higher than the 60 kg N/ha simulated for the same period.

The parameterization of the mineralization model regarding rate and partitioning coefficient was based on a previous calibration (Hansen et al., 1990). These rate and partitioning coefficients are seldom changed and, in fact, they have been used unchanged in all simulations presented in this chapter. In general, only one parameter is used for calibration purposes, viz. the distribution of organic matter between the two SOM pools. A common distribution of organic matter between the two SOM pools was assumed for the Jyndevad, Askov, and Ruurlo simulations. Hence, the parameterizations only differ with respect to the total soil organic matter, which was measured at the sites (Breeuwsma et al., 1991), and for the 0 to 25 cm depth corresponding to 1.6%, 1.3%, and 3.6% for the three sites, respectively. The difference in soil organic matter content is the main cause responsible for the differences in net nitrogen mineralization.

From Figure 16.5 it appears that the simulation of the mineralization is satisfactory at the Jyndevad and Askov sites. At the Ruurlo site, the mineralization seems to be underestimated, provided that the denitrification is correctly simulated, which is quite uncertain. This indicates that the distribution of organic matter between the two SOM pools may not be universal and the simulation could have been improved through an individual calibration.

At the Haren workshop, celebrating the centennial of The Institute for Soil Fertility Research, 14 models of nitrogen turnover in the soil-crop system were compared. The model test was based on data sets originating from winter wheat experiments at three locations during 2 years (Groot and Verberne, 1990). Each experiment comprised three different nitrogen treatments. The observations included measurements of soil mineral nitrogen content, soil water content, groundwater table, dry matter production and dry matter distribution, nitrogen uptake, nitrogen distribution, and root length density. Results obtained by the DAISY model at the workshop were published by Hansen et al. (1991a) and further results obtained with the Haren data set can be found in Hansen et al. (1995).

General conclusions from the workshop were that (1) in general, simulation of the above-ground variables yielded better results than that of the below-ground variables (soil water and mineral nitrogen content); and (2) dry matter production and nitrogen uptake can be simulated satisfactorily in an independent way, that is, without parameter fitting as it is done in models such as DAISY and hence this approach seems to be an appropriate choice (Willigen, 1991). Furthermore, none of the participating models were able to account for the loss of mineral nitrogen occurring shortly after application of fertilizer in late spring and early summer.

Figure 16.6 shows the course of simulated and observed biomass production and nitrogen uptake for the Dutch experimental farm PAGV. The corresponding distribution of soil mineral nitrogen is shown in Figure 16.7. It is noted that very good simulations both with respect to the soil mineral nitrogen and the dry matter content and nitrogen content in the above-ground part of the plant were obtained in the case of fertilizer level N1 (140 kg N/ha). Acceptable results were obtained at fertilizer level N2 (240 kg N/ha, not shown), while not quite satisfactory results were obtained at fertilizer level N3 (300 kg N/ha). It is obvious that the model fails to account for the loss of soil mineral nitrogen occurring shortly after application of fertilizer in late spring and early summer, as stated in the general conclusion of the workshop. At the workshop, the reason for this loss was discussed, and although microbial immobilization was favored, no rational explanation could be given.

At the Braunschweig workshop, "Validation of Agroecosystem Models," the simulated data sets comprised 3-year sets of agricultural field measurement for two silt loam and one sandy location situated in Lower Saxony, Germany. A detailed description of the data sets is given by McVoy et al. (1995). Nineteen different modeling groups participated in the workshop, some using the same model. The simulation results obtained by the different modeling groups were delivered to the

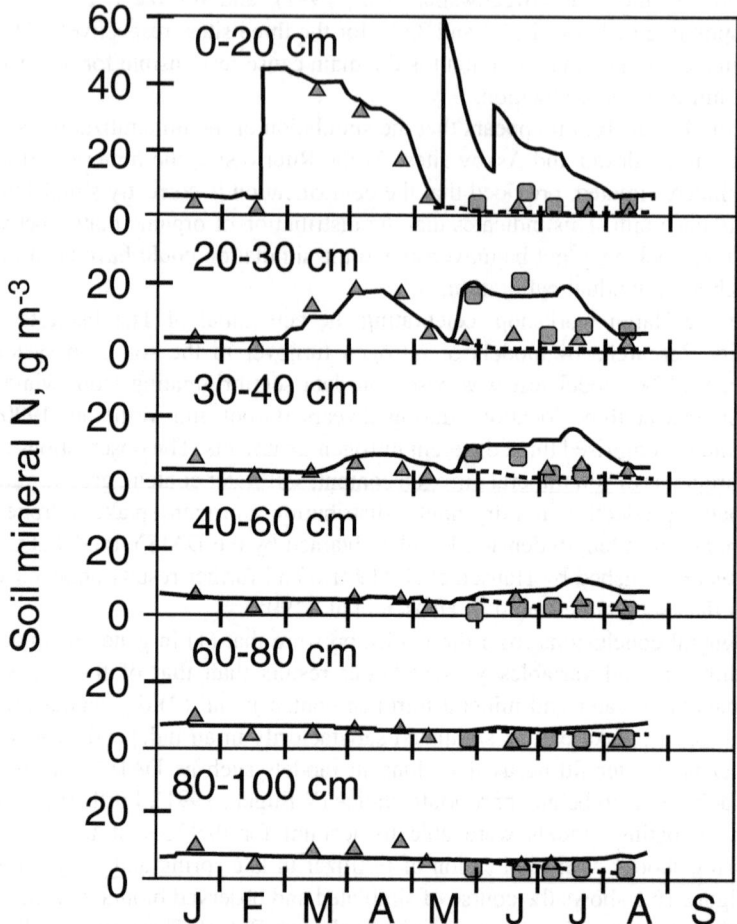

Figure 16.6   Comparison of measured (symbols) and simulated (lines) soil mineral nitrogen contents at PAGV 1984. Squares and solid lines represent high nitrogen fertilization (80+120+40 kg N/ha). Triangles and dashed lines represent low nitrogen fertilization (80 kg N/ha). (From Hansen, S., Jensen, H.E., Nielsen, N.E., and Svendsen, H. 1991a. Simulation of nitrogen dynamics and biomass production in winter wheat using the Danish simulation model DAISY. *Fert. Res.*, 27:245-259. With permission.)

workshop organizers, who compared and discussed the results. The main conclusions as stated by the organizers were (Diekkrüger et al., 1995):

1. The results differ significantly, although all models were applied to the same data set.
2. The experience of a scientist applying a model is as important as the difference between various model approaches.
3. Only one model simulated all main processes, such as water dynamics, plant growth, and nitrogen dynamics, with the same quality; the others were only able to partially reproduce the measured dynamics.

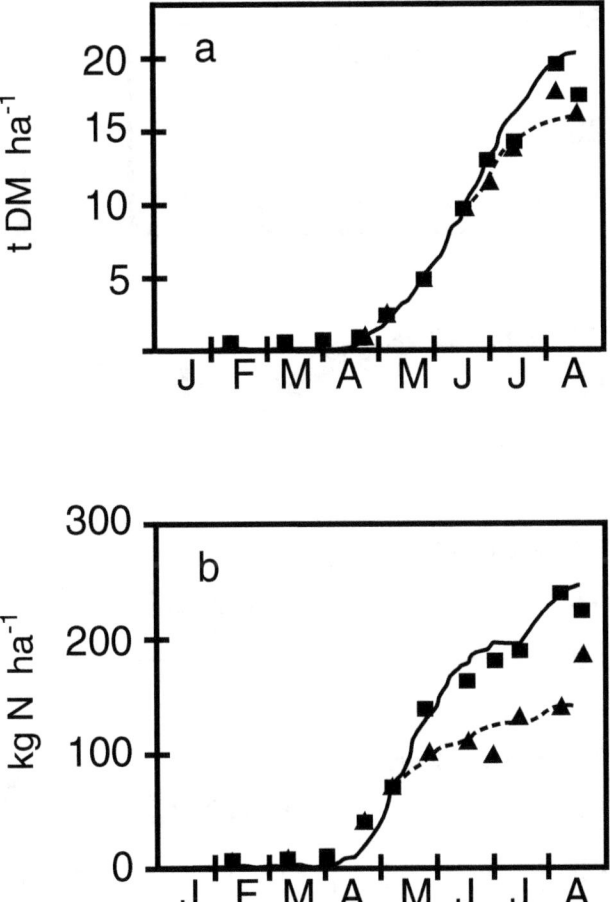

Figure 16.7   Comparison of measured and simulated above-ground biomass production (a) and nitrogen content in above-ground biomass (b). Legends as in Figure 16.6. (From Hansen, S., Jensen, H.E., Nielsen, N.E., and Svendsen, H. 1991a. Simulation of nitrogen dynamics and biomass production in winter wheat using the Danish simulation model DAISY. *Fert. Res.,* 27:245-259. With permission.)

The second conclusion strongly emphasizes the importance of proper training of model users because the user is responsible for the model parameterization. The model singled out in the third conclusion was the DAISY model as presented by Svendsen et al. (1995). Examples of the performance of DAISY at model validation exercise are shown in Figures 16.8 through 16.12, illustrating time courses in simulated and observed soil water content, soil water potential, soil temperature, shoot dry matter and leaf area index, and soil mineral nitrogen and shoot nitrogen contents. It should be noted that all the data are from the same site, the so-called Intensive Loam Site, and the same year, viz. 1989, where winter wheat was grown. Furthermore, it should be noted that parameters, which could be obtained from the data set, were used directly (e.g., soil organic matter content, soil water retention data,

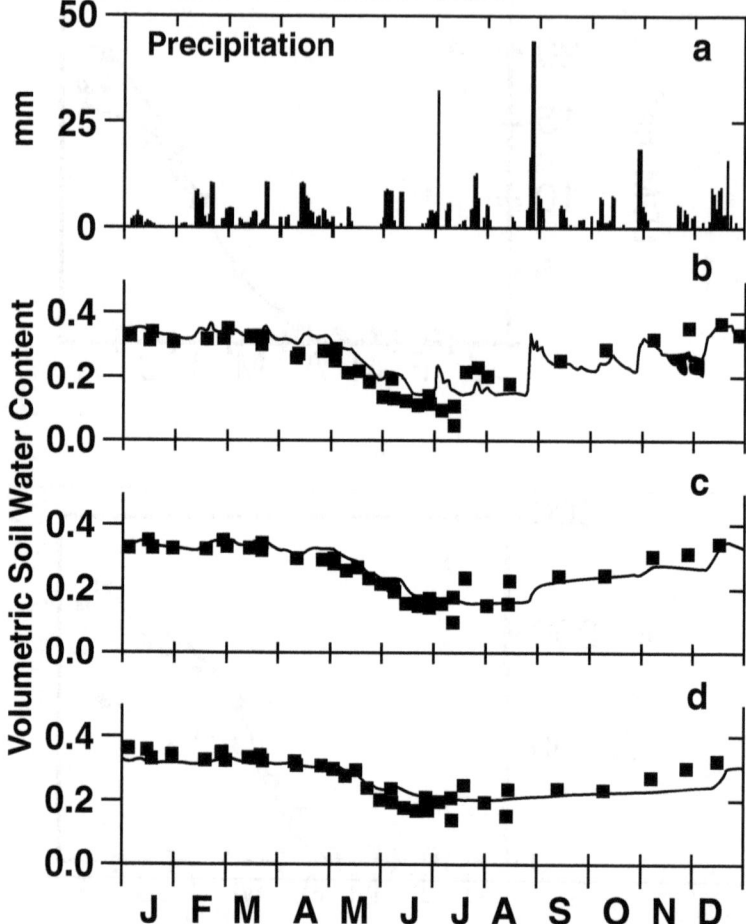

Figure 16.8 Precipitation and volumetric soil water content at the Intensive Loam Site 1989 with winter wheat. a: precipitation; b, c, d: volumetric soil water content, b: 0–30 cm, c: 30–60 cm, and d: 60–90 cm. Solid line is simulated liquid water plus ice content. (From Svendsen, H., Hansen, S., and Jensen, H.E. 1995. Simulation of crop production, water and nitrogen balances in two German agro-ecosystems using the DAISY model. *Ecological Modelling,* 81:197-212. With permission from Elsevier Science.)

etc.), while the parameters that were not measured at the sites were obtained from Hansen et al. (1990). Hence, regarding the parameterization of the nitrification and denitrification processes, default parameters were used; and regarding the mineralization model, the only calibration parameter used was the distribution of soil organic matter between SOM1 and SOM2 (see the model description and discussion earlier in this chapter).

Simulation of the soil water dynamics is illustrated in Figures 16.8 and 16.9 in terms of volumetric soil water content and soil water pressure potential (tension),

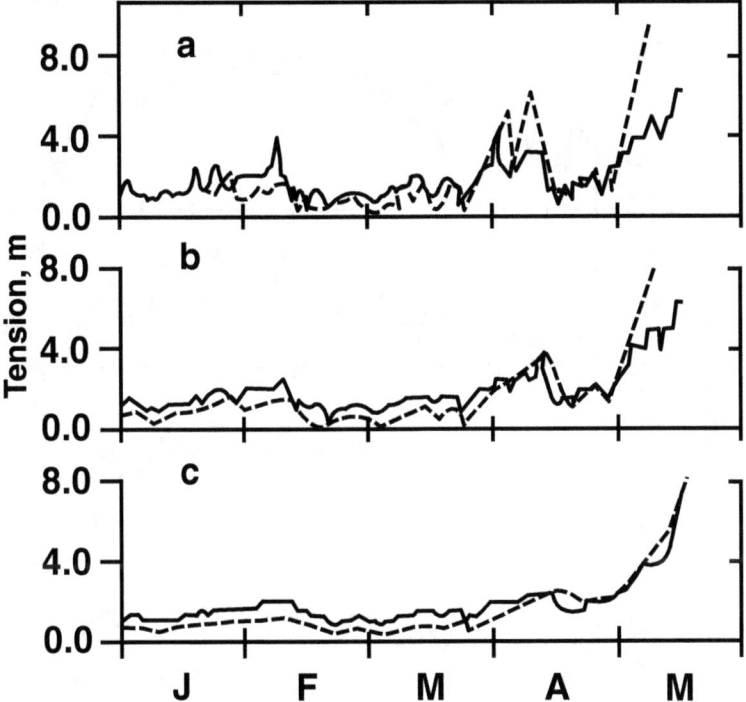

Figure 16.9 Soil water pressure potential for the Intensive Loam Site 1989. Depths are a: 10 cm, b: 20 cm, and c: 40 cm. Solid line is simulated pressure potentials; stippled line is experimental values. (From Svendsen, H., Hansen, S., and Jensen, H.E. 1995. Simulation of crop production, water and nitrogen balances in two German agro-ecosystems using the DAISY model. *Ecological Modelling*, 81:197-212. With permission from Elsevier Science.)

respectively. Discrepancies between simulated and observed values are most clearly identified in the more sensitive soil water pressure potential. Discrepancies especially develop in the upper soil layers in May. According to Figure 16.11, the winter wheat attains a closed canopy at that time, and hence the depletion of soil water in the upper soil is mainly due to water uptake by plants. At 40-cm depth, the simulation is very satisfactory, and in general the simulation of soil water dynamics is satisfactory.

Figure 16.10 compares simulated and measured values of soil temperature. Soil temperature is of importance because it influences biological processes in the soil. The agreement between simulated and measured values is very satisfactory, especially taking into account the extent to which the influence of temperature on biological processes is known.

Simulated and measured values of leaf area index (LAI) and accumulated shoot dry matter in the shoot part of the plant are shown in Figure 16.11. The version of the DAISY crop model used in these simulations does not simulate the proper LAI. Instead, the model simulated a crop area index (CAI, also including stem area, etc.).

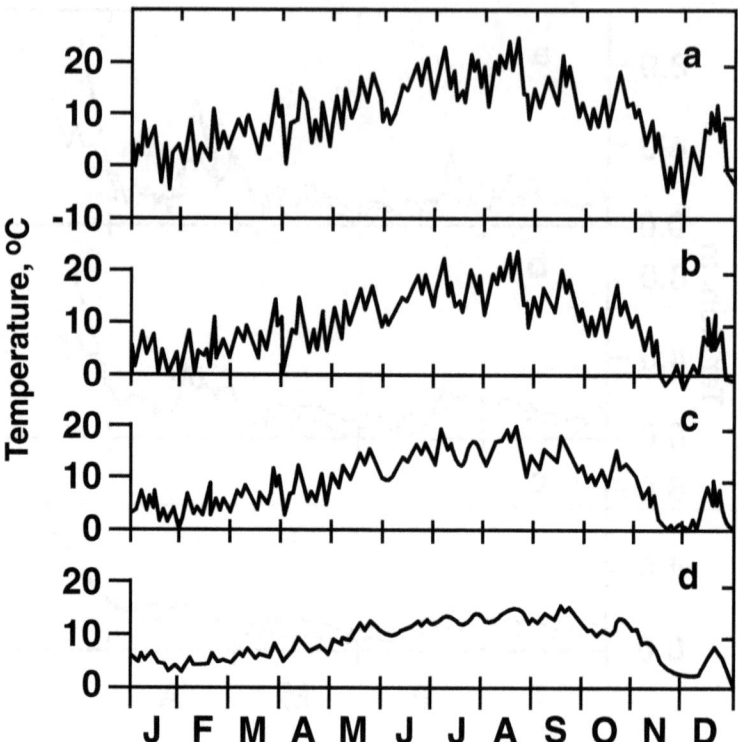

Figure 16.10   Air and soil temperatures for a winter wheat crop at the Intensive Loam Site 1989. a: air temperature, b, c, d: Solid line is simulated soil temperature. Depths are b: 5 cm, c: 20 cm, and d: 50 cm. (From Svendsen, H., Hansen, S., and Jensen, H.E. 1995. Simulation of crop production, water and nitrogen balances in two German agro-ecosystems using the DAISY model. *Ecological Modelling,* 81:197-212. With permission from Elsevier Science.)

However, in the first part of the growth period, leaves are the dominant contributor to the crop area index; thus, in this period, the measured LAI and simulated CAI can be compared. The figure shows a satisfactory agreement between measured and simulated values.

Figure 16.12 shows comparisons between simulated and measured values of accumulated nitrogen in the shoot and accumulated mineral nitrogen in the rooting zone, respectively. A satisfactory agreement is found for the accumulated nitrogen in the shoot, but a less satisfactory agreement is found for the accumulated mineral nitrogen in the rooting zone. Especially during late autumn, the model underestimates the soil mineral nitrogen content. The measured values were unexpectedly high during this period, as well as in 1990 (not shown). McVoy et al. (1995) comment that there is the possibility of unreported organic manure having been applied, but this is unsubstantiated.

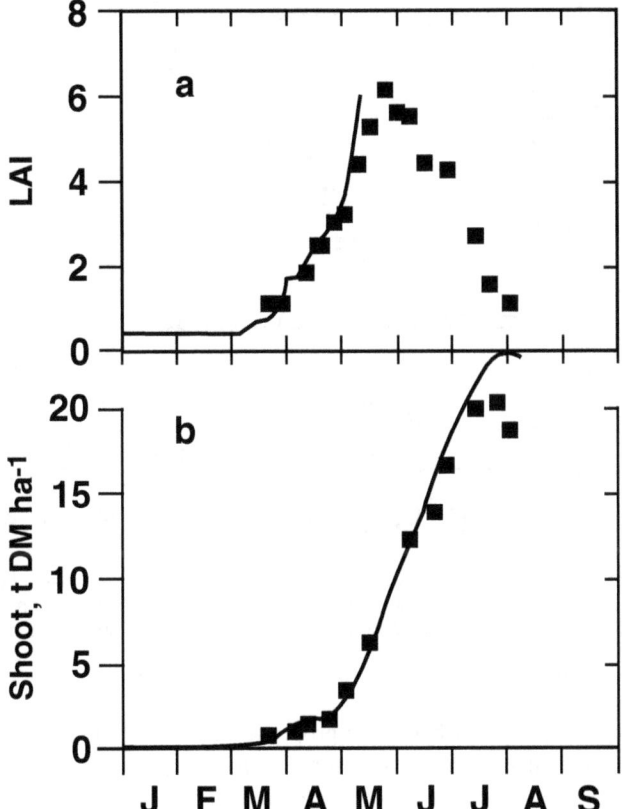

Figure 16.11 LAI development (a) and dry matter production (b) of winter wheat at the Intensive Loam Site 1989. (From Svendsen, H., Hansen, S., and Jensen, H.E. 1995. Simulation of crop production, water and nitrogen balances in two German agro-ecosystems using the DAISY model. *Ecological Modelling,* 81:197-212. With permission from Elsevier Science.)

APPLICATION OF EFFECTIVE PARAMETERS

A major problem in connection with model validation is spatial variability within the experimental plot or at field scale. This adds uncertainty to the experimental results. However, it also adds uncertainty to model parameters and, subsequently, model results. When simulating processes in an experimental plot, effective parameters are often used. However, many of the processes in the soil-plant-atmosphere system are strongly nonlinear in nature. One of the most nonlinear processes in the system is movement of water and the subsequent movement of solute in the soil. Djurhuus et al. (1999) have addressed this problem by applying different methods

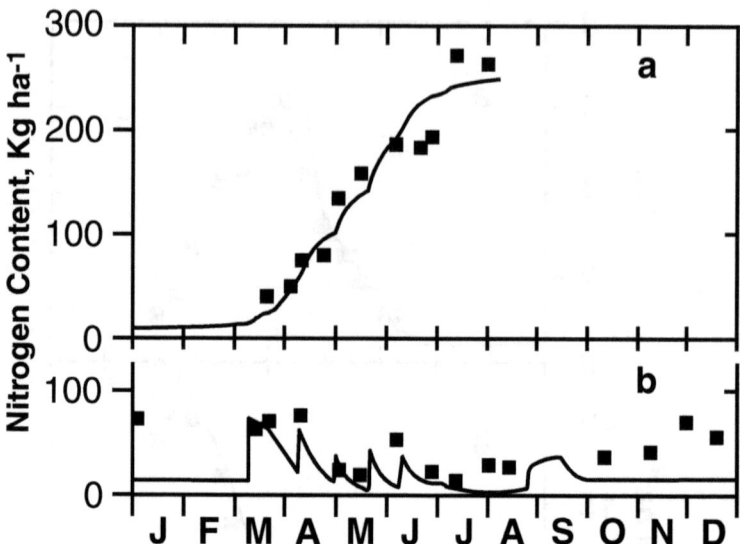

Figure 16.12   Nitrogen content of shoot of winter wheat (a) at the Intensive Loam Site 1989 and corresponding soil nitrogen content (b) (0–90 cm). (From Svendsen, H., Hansen, S., and Jensen, H.E. 1995. Simulation of crop production, water and nitrogen balances in two German agro-ecosystems using the DAISY model. *Ecological Modelling*, 81:197-212. With permission from Elsevier Science.)

of estimating effective soil hydraulic properties and using DAISY to simulate the temporal and spatial variation in soil nitrate concentrations. They measured the soil hydraulic properties (i.e., soil water retention and hydraulic conductivity) from 57 points located within an area of ca. 0.25 ha at two locations, Jyndevad and Rønhave. The former is a coarse sandy soil and the latter is a sandy loam soil. Furthermore, soil nitrate content was measured at the 57 points at 25- and 80-cm depths, respectively. The effective retention curve was estimated as an arithmetic mean, while the hydraulic conductivity was estimated as: (1) geometric mean, (2) arithmetic mean, (3) mean from a log-normal distribution, and (4) estimated from a stochastic large-scale model for water flow (Mantoglou, 1992).

First, DAISY was used to simulate the single 57 points at each location and statistics were calculated. The results are shown in Figures 16.13 and 16.14, along with average measured nitrate concentrations and error bars (±1 standard deviation). In general, the temporal variation agrees well with the observed variation. However, at Jyndevad at 80-cm depth, the predicted variation is overestimated in October, while it is underestimated especially in November and December (Figure 16.13). At Rønhave, the simulation at 80-cm depth agrees exceptionally well with the observed variation; but at 25 cm, a systematic underestimation is found. The discrepancy between measured and simulated concentrations developed especially between December 15 and 21, a period with heavy rain (71 mm) and may therefore be explained by water bypassing the suction cups. To describe this phenomenon the introduction of a preferential pathway is required. The Rønhave soil type is known to contain a considerable number of macropores, which can explain the bypass. The

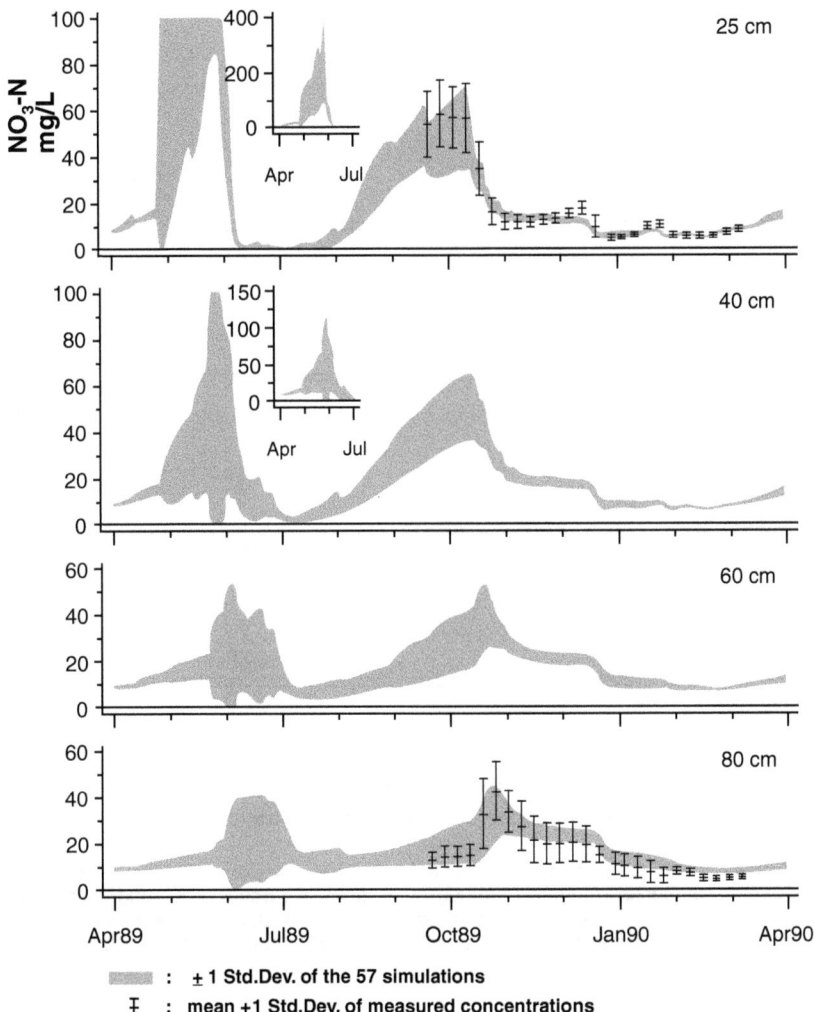

Figure 16.13 Comparison between simulated and measured concentrations of NO$_3$-N (mg/L) at Jyndevad. (From Djurhuus, J., Hansen, S., Schelde, K., and Jacobsen, O.H. 1999. Modelling the mean nitrate leaching from spatial variable fields using effective parameters. *Geoderma*, 87:261-279. With permission from Elsevier Science.)

version of DAISY used in the simulations did not include the option of simulating macropore flow.

The different ways of estimating effective parameters were tested by comparing simulations using effective parameters with the average of the 57 simulations. The comparison included the main elements of the water and nitrogen balances as well as the nitrate concentrations in the soil. It was concluded that effective parameters estimated as an arithmetic mean for the retention properties and a geometric mean for the conductivity properties gave satisfactory results. Hence, it was found that up-scaling by use of effective parameters was appropriate.

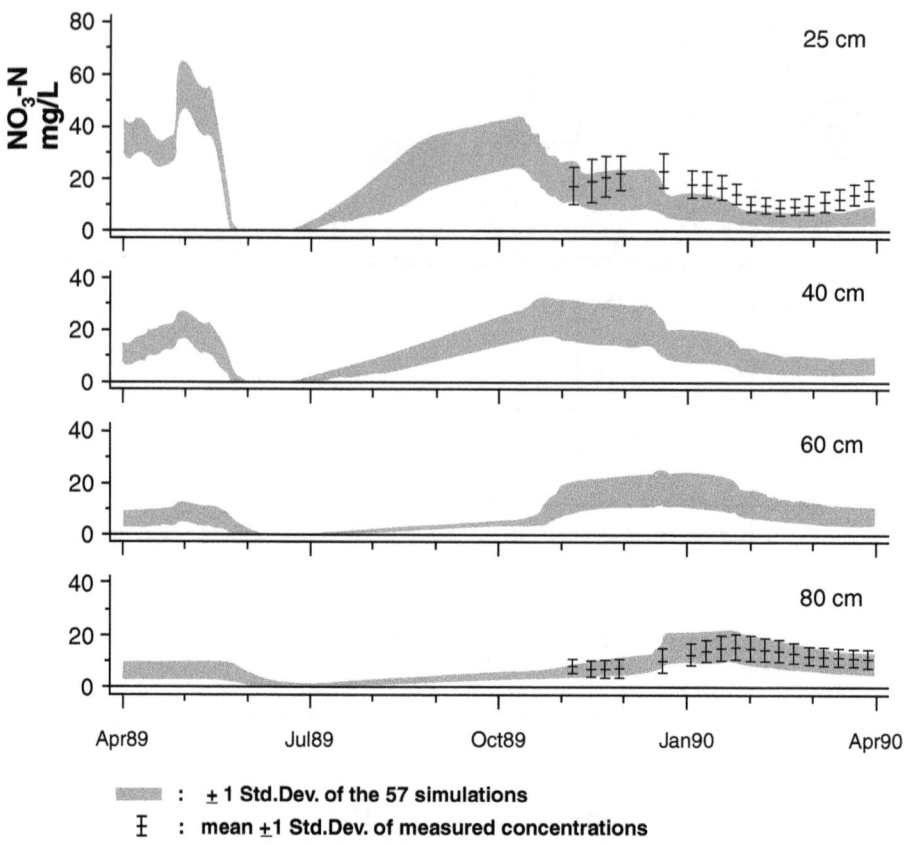

Figure 16.14 Comparison between simulated and measured concentrations of NO₃-N (mg/L) at Rønhave. (From Djurhuus, J., Hansen, S., Schelde, K., and Jacobsen, O.H. 1999. Modelling the mean nitrate leaching from spatial variable fields using effective parameters. *Geoderma,* 87:261-279. With permission from Elsevier Science.)

SIMULATION OF NITROGEN LEACHING AT LARGE SCALE

Large scale in this section is defined as a scale larger than the field scale, viz. the catchment scale ranging from a few km² to thousands of km². In large-scale modeling, the field can be considered the basic unit because it carries with it all the basic information on soil, vegetation, and management. As demonstrated in the previous sections, DAISY can be considered a model for this basic unit. In this section, two slightly different approaches are illustrated. The first approach is feasible at the 10 km² scale, and the second is feasible at the scale 100 km² or larger. Both examples take into account the legislative rules and regulations imposed on Danish agriculture. However, the general approach is flexible and can be adjusted to other rules and regulations.

Legislative rules and regulations have been imposed on Danish agriculture as a consequence of the first national Action Plan for the Aquatic Environment, APAE-I,

(Miljøstyrelsen, 1987), which was passed by the Danish parliament in 1987 to reduce the pollution of surface and groundwaters with organic matter, nitrogen, and phosphorus. The plan was very ambitious, and one of the goals was to reduce the nitrate losses from all of Danish agriculture to 50% of the level prevailing in the mid-1980s. One consequence of the APAE-I was that the Ministry of Food, Agriculture, and Fisheries each year issues statutory norms for nitrogen application to crops. These norms define the maximum allowable application of nitrogen (in the form of fertilizer or available animal manure nitrogen) to a specific crop adjusted to soil type, expected yield, and the previous crop in the crop rotation (e.g., Plantedirektoratet, 1997). If animal manures are used, minimum efficiencies of the total N content (expressed in terms of fertilizer equivalents) are required, with values for both first- and second-year efficiencies. However, as it became evident that the APAE-I did not secure the goals set forth, a second Action Plan for the Aquatic Environment (APAE-II) was passed by the Danish parliament in 1998. The consequence for agricultural management practices of this new action plan was mainly an intensification of existing regulations. However, the introduction of compulsory use of catch crops (at least 6% of the area) is new.

The following is an example of the application of DAISY to a groundwater pollution problem in a nitrate-sensitive area. The drinking water supply of a small rural community was threatened by pollution with nitrate of the aquifer where abstraction of groundwater for drinking water purposes takes place. The annual abstraction amounted to approximately 1.5×10^6 m^3. The abstraction takes place from an unconfined chalk aquifer covered by sandy quaternary deposits. The aquifer is generally assumed not to possess any substantial capacity for abiotic nitrate reduction. Hence, recharge should, on average, have drinking water quality, that is, the nitrate concentration should not exceed 50 mg/L according to EC Drinking Water Directive (80/778/EEC). The objective of this investigation was to analyze measures to secure the groundwater quality with respect to nitrate concentration. The analysis comprises the following scenarios:

Scenario 1. The simulation of the 1997-situation in order to define the basis of the overall analysis. The 1997-situation is a result of the APAE-I and succeeding legislation (e.g., Plantedirektoratet, 1997).

Scenario 2. The simulation of the effect of the implementation of APAE-II, except for the catch crop requirement.

Scenario 3. As scenario 2, but including catch crops on the maximum feasible area.

Scenario 4. As scenario 2, but with reduced use of animal manure.

Scenario 5. As scenario 2, but with a further reduction of the statutory N application norms (to 80% of APAE-II).

The first problem to be solved in this context is to identify the recharge area contributing to the groundwater abstracted for drinking water purposes. A potential recharge area was identified by the regional water authority based on groundwater level maps of the regional aquifer. Part of the area is riparian, dominated by peat and gyttja, the latter having a low hydraulic conductivity that hampers the exchange of water between surface and aquifer. Hence, the riparian areas were excluded from the considered recharge area. The size of the revised recharge area was approximately

950 ha, and the predominant land use was agriculture (approximately 73% of the total area). The remaining area was mainly forest (14%), town (7%), and road (2%).

The next step was to gather information from the local extension service about soil properties and agricultural management within the area. Information about the soil comprised the soil classification of the individual fields. The classification was in accordance with the Danish soil classification system. Information about agricultural management included cropping in 1996, 1997, and 1998; planned use of mineral fertilizers and animal manures, including amount and timing; whether straw incorporation took place; the expected yield; and whether or not the field could be irrigated. Regarding animal manure, the information also included expected utilization percent, the type of animal manure, and its total nitrogen content. The utilization percent converts animal manure nitrogen into the equivalent amount of mineral nitrogen. All information was given on an individual field basis. It was possible to gather this information from nine different farms in the area covering approximately two thirds of the agricultural area. Figure 16.15 shows the individual fields within the area.

The parameterization of the soil part of DAISY was based on the information on soil type. Five different soil types were identified within the area of consideration, viz. JB1 (coarse sandy soil) through JB5 (sandy loam) (Ministry of Agriculture, 1976). However, JB4 covered only a very minor part of the area and was aggregated

Figure 16.15   Nitrate concentrations (mg/L) in percolating water: scenario 1.

with JB5 and simulated as a JB5 soil. Typical values for the required model parameters for each soil type were constructed on the basis of 1300 soil profile descriptions
stored in the Danish Soil Profile Database. Maximum rooting depth was based on
Madsen (1985). The adopted maximum rooting depths were 60, 62, 68, and 75 cm
for JB1, JB2, JB3, and JB5, respectively. The corresponding root zone capacities
were 74, 92, 102, and 147 mm plant available water. The dominant soil types were
JB1 (coarse sand), and JB3 (coarse loamy sand) covering 42 and 48% of the area,
respectively. The amount of soil organic matter in the 0 to 20 cm depth was set to
2.6 and 3.0% for the sandy and the loamy soils, respectively, whereas the distribution
of soil organic matter on SOM1 and SOM2 was based on Børgesen et al. (1997);
the ratio 82:18 was used. Other soil parameters were obtained from Hansen et al.
(1990).

Based on agricultural management information, crop rotations were constructed
for each farm. The crop rotations were dominated by winter wheat (21% of farmed
area), winter barley (22% of farmed area), and spring barley (11% of farmed area)
for the farms with pigs or no animal production and by fodder crops (total 11% of
farmed area) (e.g., grass, maize, and fodder beet) on the cattle farms.

The 1996/1997–1997/1998 average application of fertilizer and manure nitrogen
amounted to 216 kg N/ha. Mineral fertilizer accounted for 120 kg N/ha and animal
manure for 96 kg total-N/ha. With an average efficiency of 48% and 10 to 15% for
the animal manure in the first and second year after application, this gave an average
application rate of 165 kg available-N/ha. The amount of animal manure nitrogen
available in the area was well below the maximum application of 170 kg N/ha set
by the EC Nitrate Directive (91/676/EEC). Based on the agricultural management
information, fertilizer application to the individual crops in the selected crop rotations was estimated. It was agreed that the selected fertilization practice did comply
with rules and regulations stated by APAE-I. Composition of the animal manures
was obtained from the Danish Agricultural Advisory Centre (1998) and Iversen et al.
(1998). The adopted values are shown in Table 16.3. It is noted that a lower nitrogen
content in urine and slurry is assumed in the scenarios representing the situation
after the implementation of APAE-II, as compared to the scenario representing the
present situation. This reduction in nitrogen content is based on the assumption that
improved utilization of fodder is going to occur during the time of implementation
of APAE-II (Iversen et al., 1998). In animal production systems, the urine contributes
33% and 53% of the total manure nitrogen for pig farms and cattle farms, respectively
(Danish Agricultural Advisory Centre, 1998).

Loss of ammonium nitrogen at application of animal manure depends on the
method of application. If the fertilizer is incorporated directly (banding), it is
assumed that 6% of the ammonium nitrogen is lost as volatilization. If the fertilizer
is placed at the surface by hoses, it is assumed that 10% is lost; and if it is surface
broadcasted, a loss of 14% is assumed (Sommer and Thomsen, 1993).

Conventional tillage practices were assumed. Plowing takes place in the middle
of March for spring-sown crops and soon after harvest of the previous crop for
autumn-sown crops. Irrigation is assumed to take place when less than 10 mm of
the plant available water is left in the root zone in the irrigated rotations. The scenarios
showed an annual average irrigation of 87 mm, corresponding to consumption of

Table 16.3   Composition of Animal Manures

Parameter	Cattle			Pig		
	Farmyard Manure	Urine	Slurry	Farmyard Manure	Urine	Slurry
Parameters Used in Scenario 1 (APAE-I)						
Nitrogen content, kg N/t	5.1	5.2	4.6	7.0	5.5	5.9
Ammonium content, kg N/t	1.3	4.2	2.8	2.4	4.2	4.2
Dry matter content, kg DM/t	200	40	79	230	40	66
C:N ratio	19.6	3.9	8.6	16.4	3.9	5.6
Parameters Used in Scenarios 2 through 5 (after APAE-II)						
Nitrogen content, kg N/t	5.1	4.4	4.2	7.0	4.4	5.3
Ammonium content, kg N/t	1.3	3.4	2.4	2.4	3.4	3.6
Dry matter content, kg DM/t	200	40	79	230	40	66
C:N ratio	19.6	4.5	9.4	16.4	4.5	6.2

From Danish Agricultural Advisory Centre. 1998. Håndbog i Plantedyrkning. Landskontoret for Planteavl, Landbrugets Rådgivningscenter, Skejby, 200 pp; Iversen, T.M., Grant, R., Blicher, Mathiesen, G., Andersen, H.E., Skop, E., Jensen, J.J., Hasler, B., Andersen, J., Hoffmann, C.C., Kronvang, B., Mikkelsen, H.E., Waagepetersen, J., Kyllingsbæk, A., Poulsen, H.D., and Kristensen, V.F. 1998. Vandmiljøplan II — faglig vurdering. Danmarks Miljøundersøgelser, 44 pp.

approximately 167×10^3 m³, which is well below the permit of 220×10^3 m³. At harvest, it is assumed that a part of the yield is lost. For cereals, it is assumed that 5% of the grain and 25% of the straw are lost (Andersen et al., 1986). For other crops, a common loss of 5% is assumed.

The simulations were performed for a 10-year period — April 1987 to May 1998. The simulations were performed in such a way that all combinations of crop and year were simulated and the results were taken as average annual values (April 1 to March 31). The variation in precipitation encountered by the simulation period includes years with drought as well as relatively ample water supply.

The crop model included in DAISY is calibrated on experimental data sets where deficiencies of different kinds are controlled. In this case, the model is applied to conditions of practical agriculture; hence, the model can be expected to overestimate the production. Therefore, the crop model was recalibrated to give yields in terms of nitrogen uptake that correspond to norm yields as given with the fertilizer norms (Plantedirektoratet, 1997).

The percolation and leaching were estimated as weighted averages of the percolation and leaching of the individual land use. It was a basic assumption that the approximate two thirds of the agricultural area where information on management practices was available were representative for the entire agricultural area. Nitrate concentrations were calculated as flux concentrations. For the non-agricultural area, percolation and nitrogen leaching were estimated on the basis of literature values (see Table 16.4).

Table 16.4 Percolation and Leaching for the Different Land Uses Before the Implementation of APAE-II.

Land Use	Area (ha)	Percolation (mm/year)	Leaching (kg N/ha/year)	Concentration (mg NO₃/L)
Agriculture	668	312	90	128
Set aside	28	350	10	13
Forest	131	260	5	9
Town	67	320	10	14
Road	24	350	10	13
Farmstead	19	300	15	22
Hedge	16	260	10	17
Moor	11	350	10	13
Area	953	310	66	94

Values for non-agricultural areas: Callesen et al., 1996; Gundersen et al., 1998; Magid et al., 1995; Renger et al., 1986; Windolf et al., 1998.

Main results of the nitrate leaching situation before the implementation of APAE-II are shown for the different land uses in Table 16.4, and for the different fields of the investigated farms in Figure 16.15. It is noted that agriculture is responsible for most of the nitrate leaching in the considered area. If it is assumed that the percolation does not change, then the leaching from the agricultural area should be reduced to approximately 46 kg N/ha to comply with the required 50 mg NO₃/L. The average annual leaching of 90 kg N/ha for the agricultural area is a result of a considerable variation between years (range: 39 to 134 kg N/ha) and between crop rotations (range 38 to 186 kg N/ha).

The individual components of the annual mineral nitrogen balance for the agricultural area is shown in Figure 16.16. It is noted that the model predicts an agricultural system that is in approximately steady state regarding soil organic matter, because the annual change in soil nitrogen is only 2 kg/ha. Furthermore, it is noted that the primary nitrogen source is fertilization, and that only approximately two thirds of the applied nitrogen is removed from the fields as yield.

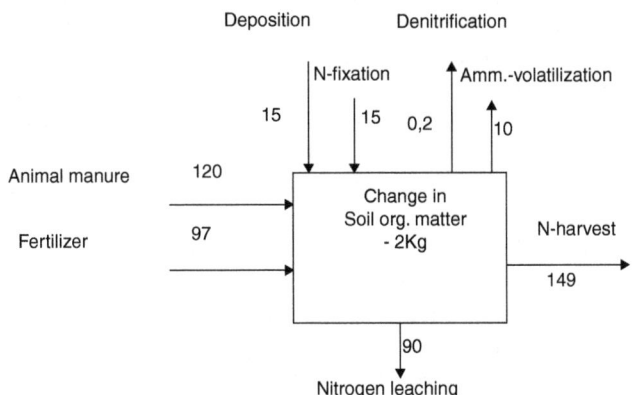

Figure 16.16 Main components in the nitrogen balance: scenario 1.

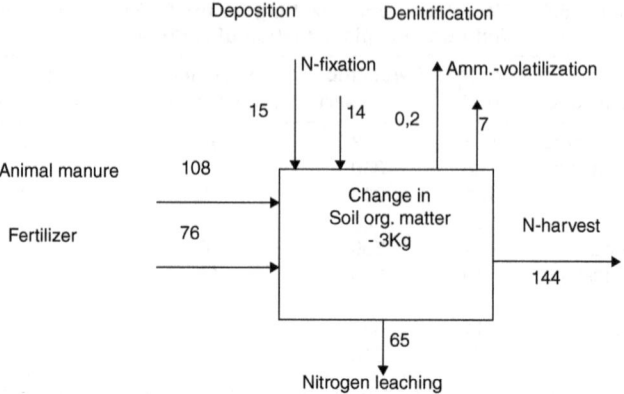

Figure 16.17 Main components in the nitrogen balance: scenario 2.

Scenario 2, representing the situation after the implementation of APAE-II, was based on the new reduced (10% compared to APAE-I) statutory nitrogen application norms, improved utilization of the animal manure nitrogen, and a lower content of nitrogen in the animal manure, due to improved utilization of the fodder. Introduction of catch crops is not included in this scenario. The results are shown in Figure 16.17. The improved feeding practice reduced the available amount of animal manure by 12 kg N/ha/year; and furthermore, the annual application of mineral fertilizer was reduced by 21 kg/ha. Improved handling of the animal manure, in combination with reduced application, reduced the ammonium volatilization by 3 kg N/ha/year. The annual nitrogen yield decreased 5 kg/ha, while the main effect of the reduced input to the system was a reduction in annual leaching from 90 kg N/ha to 65 kg N/ha (i.e., a reduction of 25 kg N/ha). Furthermore, the range of spatial variation in annual leaching was also reduced, from 38 to 186 kg N/ha to 51 to 86 kg N/ha. The range in nitrate concentration in the percolating water was reduced to the interval from 69 to 125 mg/L. However, the measures taken in scenario 2 still do not sufficiently reduce the leaching. Hence, further measures must be taken.

Scenario 3 includes growth of catch crops wherever in the crop rotation this was practically feasible, regardless of implications for pests and diseases. This resulted in the establishment of catch crops on 24% of the area, which was much higher than the 6% required by APAE-II, and in a reduction in annual nitrogen leaching of 7 kg N/ha as compared to scenario 2. These 7 kg were due to an increase in nitrogen yield of 1 kg/ha, a change in the storage term corresponding to 1 kg/ha, and finally a reduction in the annual fixation corresponding to 5 kg N/ha. It may come as a surprise that the reduction in leaching in this scenario is primarily due to a decrease in nitrogen fixation. However, the effect of catch crop is mainly due to an immobilization of mineral nitrogen when the catch crop is growing in autumn and winter and a subsequent release of mineral nitrogen after the catch crop is plowed in, often prior to sowing of a nitrogen-fixating crop (e.g., pea). In general, an appreciable decrease in leaching is observed subsequent to the crop after which the catch crop is established. Nevertheless, scenario 3 still did not reduce the leaching sufficiently.

Three of the rotations were characterized by high load of animal manure nitrogen (132, 202, and 176 kg N/ha) and concomitant high simulated leaching (79, 68, and 86 kg N/ha, respectively). In scenario 4, the animal manure N loads of all crop rotations were reduced to 100 kg N/ha, which corresponds to one livestock unit per hectare according to the EC Nitrate Directive (91/676/EEC). Because statutory nitrogen norms remained the same, a concomitant increase in fertilizer nitrogen application occurred. This scenario resulted in a 2 to 6 kg N/ha reduction in the leaching, and a corresponding 2 to 4 kg N/ha reduction in the crop uptake.

Scenario 5 represents a reduction of the statutory nitrogen application norms to 80% of APAE-II level, corresponding to 72% of the APAE-I level. This scenario resulted in a reduction of 14 to 22 kg N/ha as compared with the APAE-II scenario (scenario 2), corresponding to a 28% reduction. If catch crops were also included, the reduction would be approximately 37%. The corresponding annual average concentration in the percolation would be 52 mg N/L and 46 mg N/L for the scenario without and with catch crops, respectively, which is the required level necessary to secure the drinking water quality in the area.

Moving from the 10-km² scale to a larger scale (i.e., 100 km² scale), it becomes very laborious to collect information on soil and agricultural management on an individual field basis. In fact, in many cases, such a collection of data is not feasible; thus, one would normally have to rely on other sources of information such as soil maps and agricultural statistics. It is obvious that resorting to this kind of information adds uncertainty to predicted nitrogen leaching. Hansen et al. (1999) investigated this problem. The main objectives of this study were:

1. To calculate regional nitrate leaching from farmland based on publicly available information, taken as much as possible from standard European databases.
2. To quantify the uncertainty of the nitrate leaching associated with the uncertainty attributed to the information on which the calculations are based.

The study was carried out as a case study where the Karup region located in Jutland, Denmark, was chosen as the model area.

To quantify uncertainty, Monte Carlo analysis was adopted. However, in this case (a complex leaching model), the traditional Monte Carlo technique, wherein realizations are drawn from a simple random field, is prohibitively costly in terms of computer time. Hence, to keep within the computational requirements, the number of parameters included in the analysis was limited and an effective sampling technique (Latin hypercube) enabling small sampling sizes was adopted (McKay et al., 1979). The following model parameters were included in the Monte Carlo analysis: clay content (representing soil texture and derived parameters), SOM2 size (representing active soil organic matter and mineralization), slurry dry matter content and total nitrogen content (representing agricultural management and manuring practices), and precipitation as the most spatially variable of the weather parameters.

In the study, most model parameters were adopted from previous studies (e.g., Hansen et al., 1990; 1991a; Svendsen et al., 1995; Jensen et al., 1997). Model parameters, which were specific for this study, include soil texture, soil hydraulic properties, soil organic matter content, and partly the characterization of animal manures (slurry). Soil hydraulic properties were estimated by pedotransfer functions as proposed by Cosby et al. (1984). These pedotransfer functions require information on soil texture (i.e., clay, silt, and sand content). This reduces the required soil information to information on soil texture and soil organic matter content.

Information on soil texture was obtained from an FAO soil map of Denmark. All soil types in the Karup region fall within one FAO texture class (coarse), which covers soils with less than 18% clay and more than 65% sand. Hence, estimates on soil texture as well as hydraulic properties derived from information on texture are associated with considerable uncertainty. By assuming that the clay and the silt content are correlated, Hansen et al. (1999) transferred all uncertainty on soil texture and soil hydraulic properties to the uncertainty on the clay content, which was assumed to follow a uniform probability density function (pdf) within the range 0 to 18% clay.

The European databases did not provide information on soil organic matter in agricultural soils. As the estimate of the most important quantity — the active SOM2 pool — was based solely on an informed guess, a considerable uncertainty is associated with this estimate. Hansen et al. (1999) reduced all uncertainty associated with the organic matter content and the parameterization of the MIT model to uncertainty on the estimates of SOM2. They assumed that the pdf SOM2 could be described by a truncated normal distribution characterized by the mean 0.5% carbon, the standard deviation 0.22, and the range 0.06 to 0.94.

Agricultural management scenarios were constructed on the basis of information in the agricultural statistics and expert knowledge on typical management practices. The information obtained from the agricultural statistics comprised agricultural land use (crop type distribution) and livestock densities (cattle and pigs). Based on the statistical information and general knowledge, three farm types associated with typical crop rotations were constructed (Table 16.5). It was a simplifying assumption that the agriculture in the region can be represented by only three different farm

Table 16.5   Assumed Farm Types (Rotation and Livestock Densities)

Year in Rotation	Farm Type			
	Arable Farm	Cattle Farm 1	Cattle Farm 2	Pig Farm
1	Pea	Fodder beet	Spring barley	Winter barley
2	Winter wheat	Spring barley with ley	Spring barley with ley	Winter rape
3	Spring barley	Grass (4 cuts)	Grass (4 cuts)	Winter wheat
4		Grass (3 cuts)	Grass (3 cuts)	Spring barley
5		Winter wheat	Winter wheat	
Rel. Distribution (%)	18	25	25	32
LU/ha[a]	0	1.4	1.4	1.9

[a] LU = livestock unit.

Modified from Hansen, S., Thorsen, M., Pebesma, E.J., Kleeschulte, S., and Svendsen, H. 1999. Uncertainty in simulated leaching due to uncertainty in input data. A case study. *Soil Use and Management*, 15:167-175.

systems, based on cattle, pigs, or on arable crops without livestock. In the simulations, it was assumed that the farmers applied amounts of fertilizer and manure corresponding to the statutory nitrogen norms (Plantedirektoratet, 1996). Due to the high livestock density in the region, slurry was a substantial source of nitrogen in the region. Hence, the management of slurry is of prime importance for the leaching losses in the region. A main problem in management of slurry is the large variability found in the composition of the slurry. This variability may cause the actual nitrogen application in slurry to differ from the planned one, and therefore introduces considerable uncertainty. Hansen et al. (1999) coped with this by introducing uncertainty in the dry matter content and the nitrogen content of the slurry. It was assumed that slurry composition could be described by truncated normal probability density functions (pdf). The adopted cattle slurry parameters — mean, standard deviation, and range — were 7.5%, 2.5%, and 1.89 to 14.35% for dry matter content, and 0.5%, 0.12%, and 0.24 to 1.02% for total N content, respectively. The corresponding parameters for pig slurry were 4.9%, 2.5%, and 0.82 to 13.79% for dry matter content, and 0.61%, 0.18%, and 0.24 to 1.02%% for total N content, respectively.

To "validate" their simulations, Hansen et al. (1999) compared simulated yields with data obtainable from the agricultural statistics (Table 16.6). It is noted that the yields obtained by the simulations correspond very well with the yields obtained from the agricultural statistics. Yields of grass cannot be obtained from the agricultural statistics because grass is normally used on the farm for feeding.

Using the Latin hypercube sampling scheme, it was found that 25 Monte Carlo runs were sufficient to describe the uncertainty with respect to the means and standard deviations of the simulated output. Figure 16.18 shows the average annual leaching, with uncertainty represented as error bars for the four selected crop rotations. It is noted that no substantial differences exist between the two cattle farm rotations, and that both the magnitude of the leaching and the uncertainty increases in the order arable farm, cattle farm, and pig farm. This may come as no big surprise because the livestock density increases in the same order (Table 16.5). The area

Table 16.6   Annual Economic Yield (in kg DM/ha), Aggregated Values for all Farm Types

	Agricultural Statistics 1989–1993	Deterministic Simulation[a]	Monte Carlo Simulation	
			Mean	CV'%
Spring barley	3.9[b]	3.9	3.8	11
Winter barley	4.9[b]	4.8	4.6	11
Winter wheat	5.9[b]	5.9	5.8	10
Winter rape	2.6[c]	2.6	2.4	11
Pea	3.0[d]	3.3	3.4	13
Fodder beet	12.9[e]	13.2	13.0	6
Grass		16.9	16.5	7

[a] Based on mean parameters.
[b] Converted assuming 16% water.
[c] Converted assuming 9% water.
[d] Converted assuming 14% water.
[e] Converted assuming 20% dry matter content.

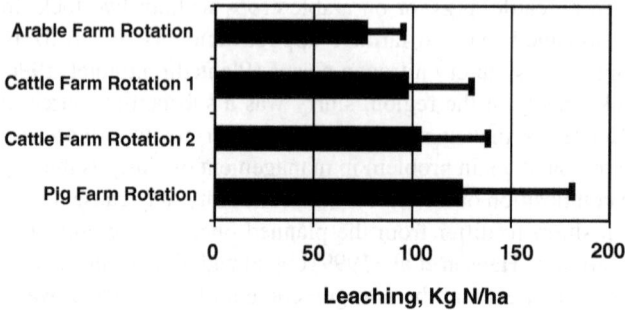

Figure 16.18   Average annual nitrate leaching for four different crop rotations. Uncertainty is represented by error bars (1 standard deviation).

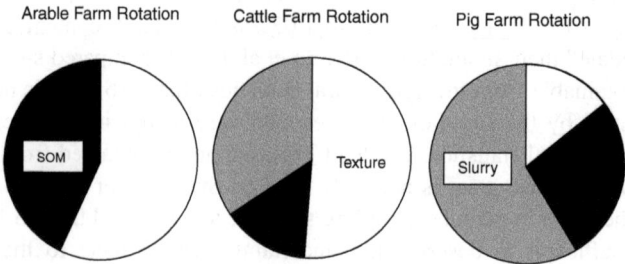

Figure 16.19   Breakdown of total uncertainty into contributions from the error sources: texture (white), soil organic matter (black), and slurry (grey). Total uncertainty is estimated as total variance.

weighted leaching was 106 kg N/ha. This value was in good agreement with the findings of Jensen et al. (1996). Considering the uncertainty expressed in terms of a coefficient of variation (CV), this was approximately 20%, 30%, and 40% for the arable farm rotation, the cattle farm rotations, and the pig farm rotation, respectively. Hansen et al. (1999) did a breakdown of the total uncertainty (expressed as variance) in contributions arising from the considered individual error sources. The results are illustrated in Figure 16.19. First it is noted that the contribution from uncertainty on precipitation was negligible. The contributions from the soil texture (clay content) and soil organic matter (SOM2) were significant in all cases. In the rotations receiving slurry, the contribution from slurry composition is important; and, especially in the pig farm rotation, this contribution is very important. However, Hansen et al. (1999) concluded that the contribution to the overall uncertainty stemming from the uncertainty on the slurry component might be overestimated. Furthermore, they concluded that no single, all-important error source could be identified.

Refsgaard et al. (1999) combined DAISY with a distributed, physical-based hydrological catchment model, MIKE SHE (Abbott et al., 1986; Refsgaard and Storm, 1995). The MIKE SHE model describes overland flow, channel flow, flow in the unsaturated zone, and flow in the saturated zone (a 3-D groundwater model). Based on Hansen et al. (1999), this model system was used to simulate the nitrate

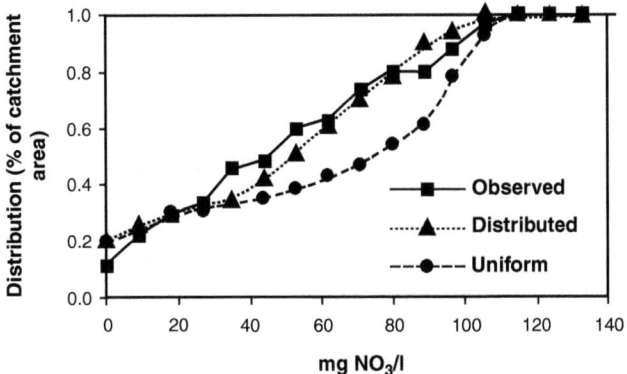

Figure 16.20 Comparison of the statistical distribution of nitrate concentrations in groundwater for the Karup catchment predicted modeling based on distributed agricultural representation, modeling based on uniform agricultural representation, and measured nitrate concentrations (35 wells).

concentration in the upper part of the Karup aquifer. In this connection, the cropping pattern in the catchment was represented by 18 crop rotation schemes that were distributed randomly over the area in such a way that the statistical distribution was in accordance with the agricultural statistics. The simulation results were compared to the concentrations actually measured in the aquifer (Figure 16.20). Because the location of the individual fields within the 518-km^2 large catchment is unknown, the comparison can only be done on a statistical basis. The nitrate concentrations simulated by the model system (the curve denoted "distributed") are seen to match the observed data remarkably, both with respect to the average concentrations and the statistical distribution of concentrations within the catchment. It may be noted that the critical nitrate concentration level of 50 mg/L (EC Drinking Water Directive [80/778/EEC]) is exceeded in about 60% of the area.

Furthermore, for evaluating the importance of the adopted up-scaling method ("distributed"), another model run has been carried out for the Karup catchment with another up-scaling method. This alternative method is based on up-scaling of soil/crop types all the way from point scale to catchment scale. This implies that all of the agricultural area is described by one representative ("uniform") crop instead of the 18 cropping schemes used in the "distributed" method. This representative crop has been assumed to have the same characteristics as the dominant crop (namely, winter wheat) and further to be fertilized by the same total amount of organic manure as in the other simulations, supplemented by some mineral fertilizer up to the nitrate amount prescribed in the norms defined by Plantedirektoratet (1996).

The differences between the two up-scaling methods are illustrated in Figure 16.20. The "uniform" method results in a lower average concentration and a less-smooth areal distribution as compared to the distributed agricultural representation. Thus, in the case of the "uniform" representation, the nitrate concentrations fall into two main groups. Approximately 30% of the area, corresponding to the natural areas with no nitrate leaching, has concentrations between 0 and 20 mg/L, while the remaining 70%, corresponding to the agricultural area with the "uniform" crop, has concentrations between 70 and 90 mg/L. In the "distributed" agricultural representation, the area distribution curve is much more smooth, in accordance with the measured data.

SUMMARY AND CONCLUSIONS

It is concluded that the DAISY model is capable of simulating crop yields and nitrate leaching over a wide range of management conditions as demonstrated by the high performance in the many comparative validation studies reported in this chapter. It has been demonstrated that DAISY is able to simulate main processes in the soil-plant-atmosphere system on a time scale of a growing season, as well as the main temporal variation within a growing season and on the spatial scale of an experimental plot or field. Furthermore, we have documented that the problems of spatial variability may be overcome by the use of effective parameters, enabling the up-scaling from column to plot or field scale.

The applicability of the DAISY model at larger scales has also been illustrated. Different legislative initiatives with the purpose of reducing nitrate leaching and securing groundwater quality were analyzed for a small catchment (1000 ha). In this case, detailed information on agricultural practices was available. The analysis showed that the present legislation was not able to secure the groundwater quality in this particular case. In another study, the leaching of a 500-km² catchment was simulated on the basis of statistical information from agricultural practices. In this case, DAISY was combined with a fully distributed, physical-based, hydrological catchment model, and the result was evaluated by comparing simulated values of nitrate concentrations in the upper groundwater with corresponding measured values. Two different up-scaling procedures were adopted. A commonly used up-scaling procedure, in which up-scaling is used all the way from point scale to catchment scale by selecting the dominant crop type in each grid, resulted in one uniform crop representing all the agricultural area. The results showed that this method is not satisfactory for simulation of nitrate leaching and groundwater concentrations. However, an alternative procedure, in which the up-scaling procedure preserved the statistical distribution of agricultural management practices, gave satisfactory results. This indicates that the predictive capability of the DAISY model that has been demonstrated at the field scale can be transferred to catchment scale if the real agricultural management is adequately preserved in the catchment modeling.

If only limited or general information on agricultural management and soil is available, Monte Carlo analysis revealed that considerable uncertainty on simulation results could be expected. However, the Monte Carlo analysis did not isolate single, all-important information, which always should be collected in order to reduce the uncertainty. The type of information of most importance did vary with the agricultural system.

It should be noted that the successful applications of DAISY reported in this chapter confine themselves to agricultural systems characterized by high productivity. That is, agricultural systems where, for example, deficiencies of phosphorus, potassium, or micronutrients are of little or no importance; the influence of weeds, pests, and diseases is kept at a low level; and inadequate emergence of the crop due to adverse physical, chemical, or biological conditions in the soil does not occur. In fact, the model in its present state only considers water and nitrogen stress. Hence, caution must always be taken when applying the model to an agricultural system in which stress factors other than these may occur. Furthermore, simulation of new cultivars of crops not already included in the DAISY crop library may make it necessary to re-parameterize the crop model if satisfactory simulation results are to be obtained.

REFERENCES

Abbott, M.B., J.C. Bathurst, J.A. Cunge, P.E. O'Connell, and J. Rasmussen. 1986. An introduction to the European Hydrological System — Système Hydrologique Européen "SHE". 1. History and philosophy of a physically based distributed modeling system. 2. Structure of a physically based distributed modeling system, *J. Hydrol.*, 87:45-77.

Abrahamsen, P. 1999. DAISY Program Reference Manual. Dina notat no. 81. 187 pp. <URL:http:www.dina.kvl.dk/~abraham/daisy/guide/guide.html>.

Abrahamsen, P. and Hansen, S. 2000. DAISY: an open soil-crop-atmosphere system model. *Environmental Modelling and Software,* 15:113-330.

Andersen, A., Hhaar, V., and Sandfærd, J. 1986. Det tidsmæssige forløb af stofproduktion og næringsstofoptagelse i vinter- og vårformer af kornarter. Beretning S-1854. Statens Planteavlsforsøg.

Børgesen, C.D., Kyllingsbæk, A., and Djurhuus, J. 1997. Modelberegnet kvælstofudvaskning fra landbruget — Betydningen af reguleringer i gødningsanvendelsen og arealanvendelsen indført fra midten af 80érne og frem til august 1997. SP-rapport nr. 19. Danmarks Jordbrugsforskning, 66 pp.

Breeuwsma, A., Djurhuus, J., Jansen, J., Kragt, J.F., Swerts, M.C.J.J., and Thomasson, A.J. 1991. Data sets for validation of nitrate leaching models. In *Soil and Groundwater Research Report II, Nitrate in Soils.* Final report of contracts EV4V-0098-NL and EV4V-00107-C, Commission of the European Communities, 219-235.

Callesen, I., Thormen, A., Raulund, Rasmussen, K., and Østergaard, H.S. 1996. Nitratkoncentrationen i jordvand under danske skove. *Dansk Skovforenings Tidsskrift,* 81:73-94.

Cosby, B.J., Hornberger, M., Clapp, and Ginn, T.R. 1984. A statistical exploration of relationships of soil moisture characteristics to the physical properties of soils. *Water Resources Research,* 20:682-690.

Danish Agricultural Advisory Centre. 1998. Håndbog i Plantedyrkning. Landskontoret for Planteavl, Landbrugets Rådgivningscenter, Skejby, 200 pp.

Diekkrüger, B., Söndgerath, D., Kersebaum, K.C. and McVoy, C.W. 1995. Validity of agroecosystem models. A comparison of results of different models applied to the same data set. *Ecol. Modelling,* 81:3-29.

Djurhuus, J., Hansen, S., Schelde, K., and Jacobsen, O.H. 1999. Modelling the mean nitrate leaching from spatial variable fields using effective parameters. *Geoderma,* 87:261-279.

Groot, J.J.R. and Verberne, E.L.J. 1990. Response of wheat to nitrogen fertilization, a data set to validate simulation models for nitrogen dynamics in crop and soil. *Fert. Res.,* 27:349-383.

Gundersen, P., Wright, R.F., and Rasmussen, L. 1998. Effects of enhanced nitrogen deposition in a spruce forest at Klosterhede, Denmark, examined by moderate NH_4NO_3 addition. *For. Ecol. Manage.,* 101:251-268.

Hansen, S., Jensen, H.E., Nielsen, N.E., and Svendsen, H. 1990. DAISY: Soil Plant Atmosphere System Model. NPO Report No. A 10. The National Agency for Environmental Protection, Copenhagen, 272 pp.

Hansen, S., Jensen, H.E., Nielsen, N.E., and Svendsen, H. 1991a. Simulation of nitrogen dynamics and biomass production in winter wheat using the Danish simulation model DAISY. *Fert. Res.,* 27:245-259.

Hansen, S., Jensen, H.E., Nielsen, N.E., and Svendsen, H. 1991b. Simulation of biomass production, nitrogen uptake and nitrogen leaching by the *DAISY* model. In *Soil and Groundwater Research Report II. Nitrate in Soils.* Final report of contracts EV4V-0098-NL and EV4V-00107-C, Commission of the European Communities, 300-309.

Hansen, S., M.J. Shaffer, and H.E. Jensen. 1995. Developments in modeling nitrogen transformations in soil. In Bacon, P. (Ed.). *Nitrogen Fertilization and the Environment.* Marcel-Dekker, New York, 83-107.

Hansen, S., Thorsen, M., Pebesma, E.J., Kleeschulte, S., and Svendsen, H. 1999. Uncertainty in simulated leaching due to uncertainty in input data. A case study. *Soil Use Manage.,* 15:167-175.

Iversen, T.M., Grant, R., Blicher, Mathiesen, G., Andersen, H.E., Skop, E., Jensen, J.J., Hasler, B., Andersen, J., Hoffmann, C.C., Kronvang, B., Mikkelsen, H.E., Waagepetersen, J., Kyllingsbæk, A., Poulsen, H.D., and Kristensen, V.F. 1998. Vandmiljøplan II — faglig vurdering. Danmarks Miljøundersøgelser, 44 pp.

Jensen, K.H.J. and Refsgaard J.C. 1991. Spatial variability in physical parameters and processes in two field soils. II. Water flow at field scale. *Nord. Hydrol.,* 22:303-326.

Jensen, C., Stougaard, B. and Østergaard, H.S. 1996. The performance of the Danish simulation model *DAISY* in prediction of N_{min} at spring. *Fert. Res.,* 44:79-85.

Jensen, L.S., Mueller, T., Nielsen, N.E., Hansen, S., Crocker, G.J., Grace, P.R., Klir, J., Körschens, M., and Poulton, P.R. 1997. Simulating trends in soil organic carbon in long-term experiments using the soil-plant-atmosphere model *DAISY. Geoderma,* 81(1/2):5-28.

Keur, P. van der, Hansen, S., Schelde, K., and Thomsen, A. 2000. Modification of DAISY SVAT model for use of remotely sensed data. *Agric. For. Meteorol.,* 106:215-231.

Lind, A.-M., Debisz, K., Djurhuus, J., and Maag, M. 1990. Kvælstofomsætning og — udvaskning I dyrket ler — og sandjord. NPO Report No. A 9. The National Agency for Environmental Protection, Copenhagen, 94 pp.

Loague, K. and Green, R.E. 1991. Statistical and graphical methods for evaluating solute transport models: overview and application. *J. Contam. Hydrol.,* 7:51-73.

Madsen, H.B. 1985. Distribution of spring barley roots in Danish soils of different texture and under different climatic conditions. *Plant Soil,* 88:31-43.

Magid, J., Skop, E., and Christensen, N. 1995. Set-aside areas as strategic tools for surface and groundwater production. In J.F.T. Schoute, P.A. Finke, F.R. Veeneklas, and H.P. Wolfert (Eds.). *Scenario Studies for the Rural Environment.* Kluwer Academic, The Netherlands, 15-21.

Mantoglou, A. 1992. A theoretical approach for modeling unsaturated flow in spatially variable soils: effective flow models in finite domains and non-stationarity. *Water Resour. Res.,* 23:37-46.

McKay, M.D., Conover, W.J., and Beckman, R.J. 1979. A comparison of three methods for selection values of input variables in the analysis of output from a computer code. *Technometrics,* 2:239-245.

McVoy, C.W., Kersebaum, K.C., Arning, M., Kleeberg, P., Othmer, H., and Schröder, U. 1995. A data set from north Germany for the validation of agroecosystem models: documentation and evaluation. *Ecol. Modelling,* 81:265-297.

Miljøstyrelsen. 1987. Handlingsplan mod forurening af det danske vandmiljø med nærings-salte. Miljøstyrelsen, 20 pp.

Ministry of Agriculture. 1976. Den danske Jordklassificering. Teknisk redegørelse. 88 pp.

Penning de Vries, F.T.W., Jansen, D.M., Berge, H.F.M. ten, and Bakema, A. 1989. Simulation of ecophysiological processes of growth in several annual crops. *Simulation Monographs.* PUDOC, Wageningen, The Netherlands, 271 pp.

Plantedirektoratet. 1996. Vejledninger og skemaer 1995/1996. Ministry for Food, Agriculture and Fishery, 38 pp.

Plantedirektoratet. 1997. Vejledning og skemaer. Ministeriet for Fødevarer. Landbrug og Fiskeri. 38 pp.

Raison, R.J., Connel, M.J., and Khanna, P.K. 1987. Methodology for studying fluxes of soil mineral-N in-situ. *Soil Biol. Biochem.,* 19:521-530.

Refsgaard, J.C. and Storm, B. 1995. MIKE SHE. In V.P. Singh (Ed.). *Computer Models of Watershed Hydrology.* Water Resources Publication, 809-846.

Refsgaard, J.C., Thorsen, M., Birk Jensen, J., Kleeschulte, S., and Hansen, S. 1999. Large scale modeling of groundwater contamination from nitrogen leaching. *J. Hydrol.,* 221:117-140.

Renger, M., Strebel, O., Wessolek, G., and Duynisveld, W.M. 1986. Evapotranspiration and groundwater recharge — a case study for different climate, crop patterns, soil properties and groundwater depth conditions. *Zeitschrift fur Pflanzenernahrung und Bodenkunde,* 149:371-381.

Smith, P., Smith, J.U., Powlson, D.S., Arah, J.R.M., Chertov, O.G., Coleman, K., Franko, U., Frolking, S., Gunnewiek, H.K., Jenkinson, D.S., Jensen, L.S., Kelly, R.H., Li, C., Molina, J.A.E., Mueller, T., Parton, W.J., Thornley, J.H.M., and Whitmore, A. P. 1997. A comparison of the performance of nine soil organic matter models using data sets from seven long-term experiments. *Geoderma,* 81(1/2):153-222.

Sommer, S.G. and Thomsen, I.K. 1993. Loss of nitrogen from pig slurry due to ammonia volatilization and nitrate leaching. In *Nitrogen Flow in Pig Production and Environmental Consequences: Proceedings of the First International Symposium,* Wageningen, The Netherlands, 353-367.

Svendsen, H., Hansen, S., and Jensen, H.E. 1995. Simulation of crop production, water and nitrogen balances in two German agro-ecosystems using the DAISY model. *Ecol. Modelling,* 81:197-212.

Thomasson, A.J, Bouma, J., and Lieth, H. (Eds.). 1991. *Soil and Groundwater Research Report II. Nitrate in Soils.* Contract NOS EV4V-0098-NL and EV4V-00107-C-AM. Commission of the European Communities, Luxembourg.

Willigen, P. de. 1991. Nitrogen turnover in the soil-crop system; comparison of fourteen simulation models. *Fert. Res.,* 27:141-149.

Windolf, J., Svendsen, L.M., Ovesen, N.B., Iversen, H.L., Larsen, S.E., Skriver, J., and Erfurt, J. 1998. Ferske vandområder — Vandløb og kilder. Vandmiljøplanens Overvågningsprogram 1997. Faglig rapport fra DMU nr. 253. Danmarks Miljøundersøgelser.

CHAPTER 17

Performance of a Nitrogen Dynamics Model Applied to Evaluate Agricultural Management Practices

K.C. Kersebaum and A.J. Beblik

CONTENTS

INTRODUCTION

Agricultural production in the 20th century is often linked to various problems, such as nutrient pollution to water resources, trace gas emissions, or decreased soil fertility. The efforts to evaluate agricultural management and land use for sustainable production systems is increasingly being supported by the use of agroecosystem models in which the processes of C/N dynamics in the soil-crop-atmosphere system play a central role. Although various models of different complexity have been developed, their application for practical purposes is limited because site-specific input requirements of research models are often not fulfilled by commonly available data. Additionally, there is a scaling conflict between the regional assessments required for environmental management decisions and the punctual nature of state-of-the-art models. Spatial variability, scarcity of regional input data, and limitations to regional validation further exacerbate this conflict.

The following examples of model application should give an idea about the performance of a relatively simple model approach for nitrogen dynamics applied for different practical purposes.

MODEL DESCRIPTION

Genealogy

Based on a simple model approach from Richter et al. (1978) describing nitrate transport assuming quasi-stationary flux conditions during winter and experimental work on nitrogen mineralization of Richter et al. (1982), a first preliminary model has been developed by Nuske (1983) to simulate nitrogen dynamics in loess soils over the winter period. The aim was to create a tool that allows one to calculate mineral nitrogen content in early spring, which is usually determined by soil sampling as a basis for fertilizer recommendation. Extending this concept with a complete water balance sub-model for transient water flux and a module for the dynamic simulation of crop growth for winter wheat, a new model (HERMES) has been built by Kersebaum (1989) that is able to simulate nitrogen dynamics in the soil-crop system for the entire growing season. The simulated deficiency between nitrogen supply from soil and the demand of the growing crop was used to calculate recommendations for fertilizer application for the entire growing season. In the following years, the model was subsequently modified for other crops and the integration of a module for denitrification (Kersebaum, 1995). With the use of the model for regional nitrogen leaching studies, the requirements for data handling increased and led to the development of MINERVA (Beblik, 1996), which is portable to different hardware platforms (ANSI-C) and allows the direct use of modern relational databases and GIS.

Fundamentals

The basic process formulations of both the HERMES and MINERVA models are identical. The models were developed to be used for practical purposes, which implies that relatively simple model approaches were chosen which operate under restricted data availability. The main processes considered are nitrogen mineralization, denitrification, transport of water and nitrogen, and crop growth and nitrogen uptake by crops.

Water Balance

The simulation of soil water dynamics is necessary to describe transport of nitrate as well as biological nitrogen transformations such as mineralization and denitrification. Because soil information is often very limited under practical conditions, the chosen capacity approach appeared appropriate to be used on a field, as well as on a regional scale, because parameters can be easily derived from rough basic soil information. The capacity parameters required by the model are attached to the model by external data files that are consistent with the German soil texture classification and their capacity parameters (AG Bodenkunde, 1995). The basic values are modified by organic matter content, bulk density, and hydromorphic indices. For groundwater affected sites, capillary rise is calculated dependent on soil texture and the distance to the groundwater using tabulated values of AG Bodenkunde (1995). These flux rates are defined for a water content of 70% of crop-available water and are used by the model as a steady-state flux up to the lowest soil layer with less than 70% of crop-available water.

Potential evapotranspiration is calculated using the empirical method of Haude (1955) and modified by Heger (1978) for different plant covers. The method requires the vapor pressure deficit at 2 a.m., which corresponds well with its daily maximum. The calculation of actual evapotranspiration considers the soil water status and time-variable root distributions of plants over depth.

Nitrogen Transformations

The sub-model for nitrogen mineralization follows the concept of net-mineralization and simulates release of mineral N from two pools of potentially decomposable nitrogen according to first-order reactions.

$$N_{min}(t) = N_{slow} \times \left(1 - e^{-k_{slow}(T,\Theta)*t}\right) + N_{fast} \times \left(1 - e^{-k_{fast}(T,\Theta)*t}\right) \qquad (17.1)$$

with N_{min} (kg N/ha) as nitrogen, mineralized at time t; N_{slow} and N_{fast} as decomposable fractions; and k_{slow} and k_{fast} as reaction coefficients (day^{-1}), depending on temperature (T) and soil moisture (Θ). Pool sizes are derived as a fixed percentage of soil organic matter nitrogen (13%) (Nuske, 1983) and the composition of different crop residues and manure. Nitrogen in crop residues recycled to the soil is calculated automatically, using the simulated N-uptake and a crop-specific relation to the N-export with the yield. Daily mineralization coefficients are calculated, dependent on mean air temperature using two Arrhenius functions from Nuske (1983) and Nordmeyer and Richter (1985) (Figure 17.1):

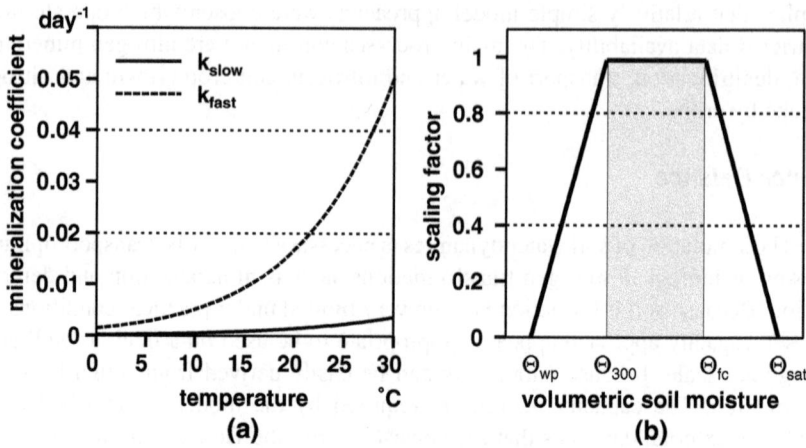

Figure 17.1 Functions used in the HERMES and MINERVA models to describe the dependency of net mineralization on (a) temperature and (b) soil moisture.

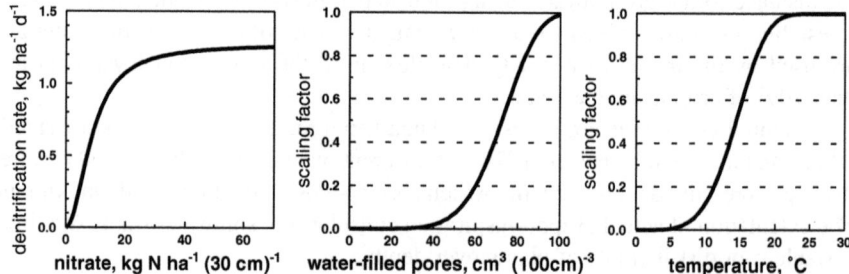

Figure 17.2 Functions used in the HERMES and MINERVA models to describe denitrification rates dependent on nitrate content, water-filled pore space, and temperature.

$$k_{slow} = 4.0 \times 10^9 \times e^{-8400/(T+273)} \tag{17.2}$$

$$k_{fast} = 5.6 \times 10^{12} \times e^{-9800/(T+273)} \tag{17.3}$$

Soil moisture effects are considered according to Myers et al. (1982) (Figure 17.1). Mineralization is limited to the top 30 cm of the soil (at maximum). Denitrification is calculated by an approach taken from Richter and Söndgerath (unpublished, cited in Schneider, 1991). Daily denitrification N_{den} (kg N/ha) is simulated for the topsoil using Michaelis-Menten kinetics modified by reduction functions dependent on water-filled pore space (Θ_r) and temperature (T) (Figure 17.2):

$$N_{den} = \frac{V_{max} \times (NO_3)^2}{(NO_3)^2 + K_{NO_3}} \times f(\Theta_r) \times f(T) \tag{17.4}$$

with a maximum denitrification rate V_{max} (kg N/ha/day), the soil nitrate content NO_3 (kg N/ha) and the Michaelis-Menten coefficient for nitrate (kg N/ha) and functions for water content and temperature:

$$f(\Theta_r) = 1 - e^{-\left(\frac{\Theta_r}{\Theta_{crit}}\right)^6} \qquad (17.5)$$

$$f(T) = 1 - e^{-\left(\frac{T}{T_{crit}}\right)^{4.6}} \qquad (17.6)$$

Critical values T_{crit} for temperature and Θ_{crit} for water-filled pore space are set to 15.5°C and 76.6 cm^3 (100 cm^{-3}), respectively.

The maximum denitrification loss per time unit is related to the total carbon content of the plowing layer.

Crop Growth and Nitrogen Uptake

The sub-model for crop growth was developed on the basis of the SUCROS model of van Keulen et al. (1982). Driven by the global radiation and temperature, the daily net dry matter production by photosynthesis and respiration is simulated. The partitioning of the assimilates to different crop organs is determined by the phenological development of the crops, which is calculated by a cumulative biological time based on thermal time (°C * day) modified by day length and vernalization. The root dry matter is distributed over depth by an empirical function according to Gerwitz and Page (1974), with an increase of rooting depth with the biological time. Dry matter production will be reduced if water or nitrogen is limiting.

Water stress is estimated by the ratio between actual and potential transpiration. To consider nitrogen stress, a crop-specific function of a critical nitrogen content in the crop depending on the stage of development is defined. Figure 17.3 shows typical functions used for cereal crops.

For nitrogen uptake, the demand of the crop is calculated from the difference between the actual nitrogen content of the crop and a maximum N content (see Figure 17.3) which is also related to the biological time. Actual nitrogen uptake is limited by the supply of soil mineral nitrogen by convective and diffusive transport to the roots and by a maximum uptake rate per centimeter root length.

Different parameter sets stored in external data files allow simulation of different crop covers. However, the estimation of crop parameters requires intensive research, which has been done mostly for several main crops. Therefore, the background of the parameter sets for crops is quite different. A procedure to estimate crop growth parameters from field experiments is described in the following section.

Data Requirements

Data required by the model can be separated into three parts: weather data, soil information, and management data. An overview of the data requirements and usual data sources is given in Table 17.1.

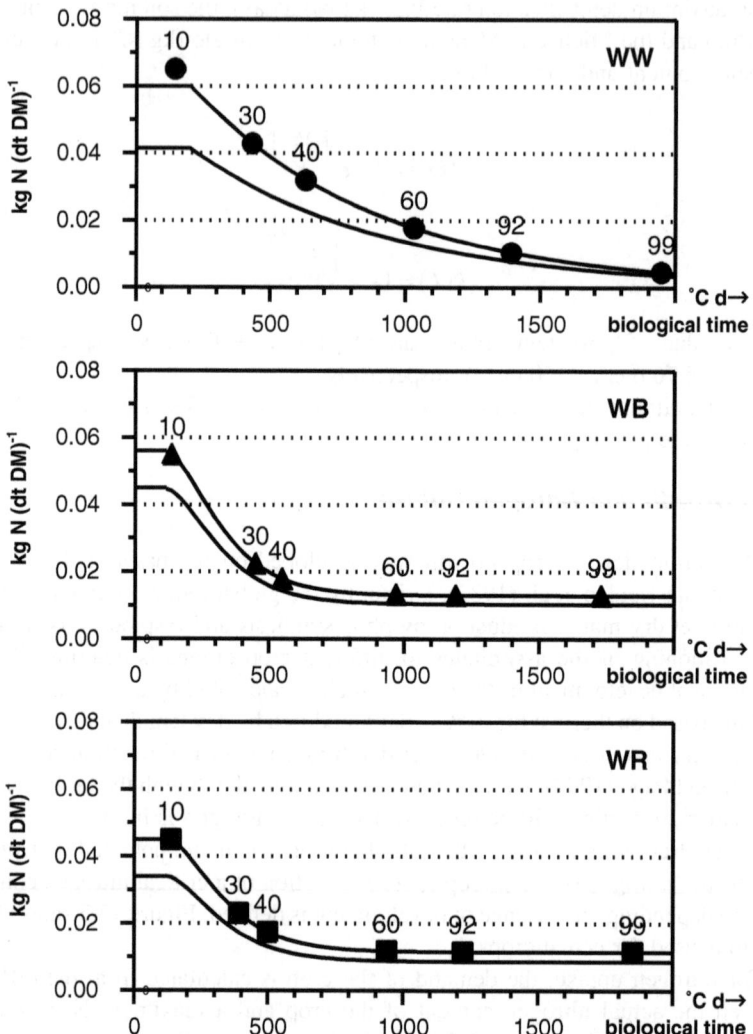

Figure 17.3 Model relations between calculated biological time (expressed as degree days, °C d) and crop phenology and corresponding functions for maximum and critical nitrogen contents in the crop biomass for winter wheat (WW), winter barley (WB), and winter rye (WR) (lines are marking functions, while markers/numbers represent phenological stages).

The model operates on a daily weather data basis. Input data required include precipitation, average air temperature, global radiation, or alternatively sunshine duration and the vapor pressure deficit at 2 a.m. Soil information is required at a resolution of 10 cm for the profile. For the plowing layer, the organic matter content and its C:N ratio should be given. The soil texture classes of AG Bodenkunde are the most important soil information required. Parameterization of water capacity parameters according to AG Bodenkunde (1995) is attached in an external file that allows one to modify default values or to create new user-defined classes. Information

Table 17.1 Input Data Requirements of the HERMES and MINERVA Models and Possible Data Sources for Different Scales

Scale Purpose	Field Scale: Management/Fertilizer Recommendations	Catchment Scale: Management and Land Use Evaluation	Regional Scale: Risk Assessment and Scenario Analysis
Required Input		Data Source	
Meteorological data: (Daily) Precipitation Air temperature Vapor pressure deficit Global radiation	Neighboring met. station Online weather data, Typical weather scenario	Neighboring met. station Weather data time series	Representative met. station Long-term time series Weather generator
Soil data: Soil texture class	Mapped soil profiles Texture, hydromorphy	Large-scale soil map units Representative soil profiles	Medium-scale soil map Representative soil profiles for aggregated soil map units
Humus (C:N) Groundwater table	Measurements Average and amplitude	Soil and land use-specific local estimates Hydrographical map	Soil, climate, and land use-specific estimates Hydrographical map
Management data:	Field data from farmer's database	Land use survey Field database Farm and crop-specific fertilization and management schemes	Statistical data of field crop distribution, livestock density, average fertilization, and yield level
Previous crop (1. crop): Yield/harvest date (start of simulation) *Crop (resp. rotation):* Sowing/harvest date Tillage (date/depth) Crop residue management Nitrogen fertilization (kind/date/amount) Irrigation (date/amount)			Expert knowledge of regional crop-specific management schemes

on groundwater depth is used to calculate capillary rise. Bulk density and percentage of stones are used to modify capacity parameters.

Mandatory data for management are dates of sowing, harvest and soil tillage, nitrogen fertilizer and water application (kind of, quantity, and date of application/incorporation). Default initial vertical distributions of mineral nitrogen content and soil moisture are used, dependent on previous crop and season. At any time during a simulation run, model state variables (mineral nitrogen, soil moisture, phenology) can be optionally actualized by observed values (e.g., to exactly reflect measured states after periods with less confident data).

Data Structure

All user interfaces are using the same core modules as described in "Genealogy." Model parameters are stored in external databases and can be extended or customized by users to fit their needs. They contain general expert knowledge and define IDs for all kinds of soil matrices, crops, and fertilizers used to describe specific simulation cases. Site-specific input data files (climate, soil profile, management) are stored with daily resolution in relational databases and are read synchronous to simulation. On the regional scale, we use the polygon-attribute-table of GIS (Arc/Info, GRASS, etc.) to direct the process simulation based on thematic maps. The output of state variables can be chosen by the user and simultaneously presented on-screen and written into files. Descriptions of database format and scripting syntax for MINERVA are documented online at the model homepage (http:://www.hydrologie.tu-cottbus.de/meson/).

Optimization of Crop Parameters

Nitrogen uptake and water consumption heavily depend on specific properties of crops. Storage of nitrogen in crop biomass also works as an important sink for soil mineral nitrogen and leads to different amounts of residual nitrogen after harvest and different potentials of nitrogen losses. Especially for leaching studies, a good agreement with phenological development, dry matter production, and N concentration in the canopy are of substantial interest. Physiological measurements from experiments (field, greenhouse) are time-consuming and related to specific years and sites. A combined strategy based on good estimates, a multi-site, multi-year but relatively sparse data set, and an automatic parameter optimization was developed to increase the crop spectrum of the model describing N uptake from soil for different crop covers.

By combining MINERVA with an external tool (NPAR), an arbitrary collection of simulation cases can be used for parameter optimization. NPAR does a multivariate, nonlinear fit without derivates (Jacob, 1982) by means of a hierarchic algorithm with six levels of complexity. Parameters are thematically grouped into sets. Some basic parameters are related to general properties of the crop (e.g., maximum assimilation rate, soil moisture demand for germination), while others determine phenological development (e.g., temperature sums, day length), partitioning of daily dry matter production to plant organs (roots, leaves, stems, storage organs), or describe the function of N concentration in plant dry matter vs. phenology.

phenological temperature		dry matter + nitrogen		lev.	+ quality
temp.sums	*n o t*	*n o t*	*n o t*	1	stage
l o w	day length vernalization	*n o t*	*n o t*	2	LAI
f i xed	*l o w*	assim., LAI, N content	*n o t*	3	dry mass N content
f i xed	*f i xed*	*l o w*	distribution of daily prod. dry matter	4	mineral N
all parameters in the set are varied simultaneously				5 / 6	soil moist.

Figure 17.4 Strategy for optimization with six levels and overlapping parameter sets (not used, low range, fixed results from last level; +quality are cumulative error components).

In the initial phase, the data set was built with crop-specific data from the literature and completed on the basis of a suitable existing data set of similar crops. The cases collected for an overall fit should cover a wide spectrum of different climatic conditions, intensities of agricultural practice, and soil types. Optimization is divided into six levels with overlapping subsets of parameters (see Figure 17.4).

After fitting thermal time to phenological stages (level 1, controlled by the deviation between observed and simulated phenological stages), the procedure is repeated, adding light efficiency and vernalization factors. Keeping results from levels 1 and 2 in a narrow range, in level 3 the specific leaf area, growth efficiency, and maximum assimilation rate are optimized with a mixed error criteria (stage × mass × N content). At level 4, partitioning rates of daily dry matter production are varied and improvements are calculated on the product of all errors (stage × mass × N content × mineral N × soil moisture, scaled). Finally, overall optimization, based on results of levels 1 to 4, is done to reflect all interactions in the parameter set.

Parameter sets resulting in fatal errors of the simulation procedure get a penalty error to avoid unsuccessful directions of the optimization process. Improvement of each step is internally qualified and the optimization procedure continues up to a given count of iterations or until further improvements fall below a certain threshold. Details of the procedure are given by Beblik et al. (1996).

MODEL VALIDATION

For practical purposes, it is important that models can be used with a sparse data set and without intensive investigations for parameter adjustments. Therefore, standardized parameter deviation procedures operating on basic soil and management information are used to obtain reproducible parameter sets for specific input situations. To ensure that models are accepted by users such as farmers and political decision-makers, it is necessary to validate them for a wide spectrum of different sites.

Time Series Measurements

A common method to validate models is to compare the results of different state variables to a time sequence of measured values. Both models have been tested on various data sets without changes of the basic parameter sets or the internal transfer algorithms (e.g., Kersebaum and Richter, 1991; Kersebaum, 1995). Only crop parameters were adopted if the model was applied for crop rotations with insufficient background for single crops.

Figure 17.5 depicts a typical example of a model validation on a farmer's field in northwestern Germany. Basic soil and management information from the farmer and weather data from a neighboring station of the German weather service were used to model water and nitrogen dynamics in a cereal rotation of winter rye following silage maize. Soil samples for mineral nitrogen and water content were taken after harvest and in early spring before fertilization. The primary aim of the investigators was to monitor mineral nitrogen changes over winter to get a rough estimate of nitrogen losses by leaching in this typical sandy soil. Data were suitable to test the model performance under practical conditions. In the given example, the simulation fits well the measured data for water and nitrogen content in the soil, except for a higher deviation in mineral nitrogen content in summer 1994 shortly after manure

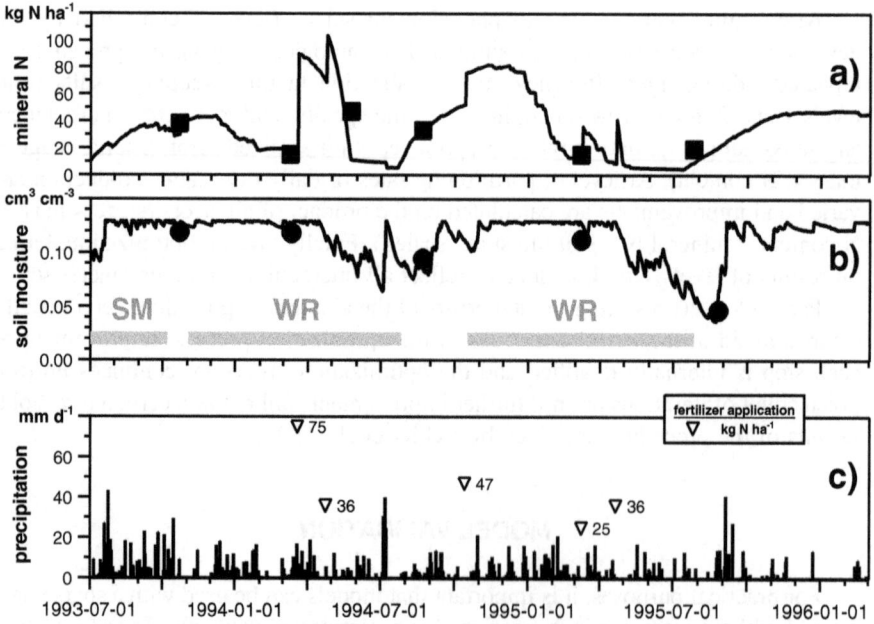

Figure 17.5 Example of a time series of water and mineral nitrogen content; observations on a farmer's field in northwestern Germany used to validate the MINERVA model (SM: silage maize; WR: winter rye).

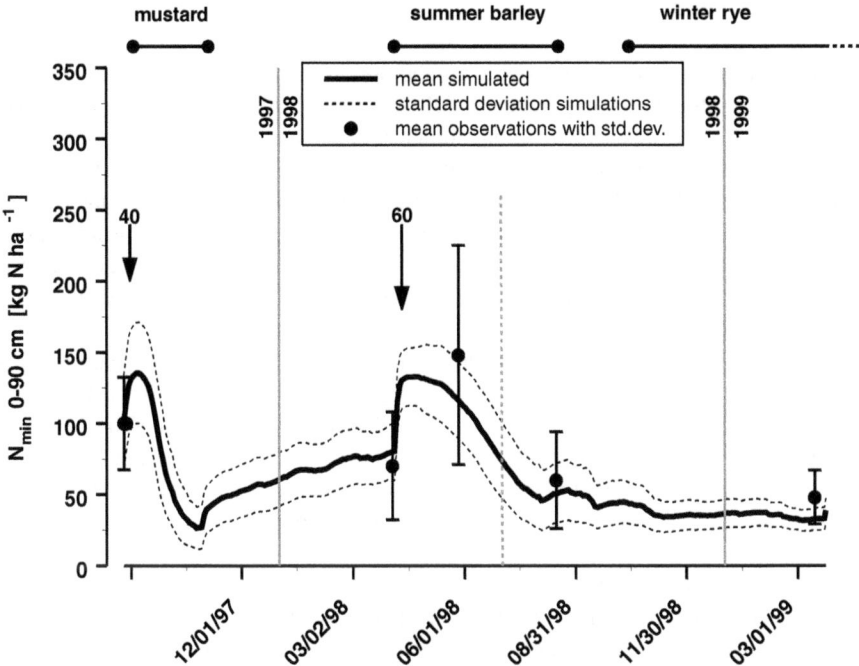

Figure 17.6 Measured and simulated average and standard deviations of soil mineral nitrogen in the root zone (0–90cm) for 14 points of a field in Luettewitz, Saxony (arrows mark fertilizer applications in kg N/ha).

application. However, in many cases, the agreement between measured and simulated values is not as good as shown in this figure. Often, it is not clear whether the model really fails, because accuracy of input data, especially for management data, is sometimes doubtful and the confidence limits of the measured values due to spatial variability are unknown. Figure 17.6 shows an example of the standard deviations of mineral nitrogen content in the root zone measured on a 30-ha loess field in Saxony (Germany) at 14 different locations (five soil samples mixed at each point) throughout the growing season of spring barley. Although texture did not vary significantly within the field, the mineral nitrogen content after harvest of triticale showed standard deviations within these points of 33 kg N/ha (coefficient of variation = 0.33). A nested sampling at five locations of the field at 25 points in a 6 × 6 m grid showed that mineral nitrogen contents varied even within small distances in the same order of magnitude (coefficients of variation = 0.22–0.4) as in the whole field. Details of the spatial analysis were given by Wendroth et al. (2001).

The model was applied to the 14 locations separately with their individual mineral nitrogen content and their corresponding soil profile data. The average time course of these individual simulations and the corresponding standard deviations are also plotted in Figure 17.6. Although deviations between the average values of 28 kg N/ha (= 20%) occurred in May 1998, the difference is not significant regarding the error of the observation.

Lysimeter Data

Lysimeter measurements provide a good opportunity to test model performance, especially for the flux components which are difficult to measure in the field. The HERMES model was tested among other models on a 13-year data set of the lysimeter station in Brandis/Saxony. The lysimeters were taken as undisturbed soil monoliths of 300-cm depth from different typical sites in eastern Germany. Water and nitrogen flux were measured in 300-cm depth, and changes in soil water storage were estimated by continuous weighting of the monolith. A detailed description of the physical and chemical properties of the lysimeters is given in Keese et al. (1997) and Knappe and Keese (1996).

Simulations were done for two different soil types, which were taken as three replicate lysimeter monoliths. The lysimeter group 5 represents a silty sand with an average stone content of 21% in the upper 200 cm of the profile. The lysimeter group 7 represents a loamy sand over sandy loam with low stone content (2 to 3%). Soil parameters for simulation were derived by the internal standard procedure using texture and organic matter measurements from neighboring locations of their origin fields.

Table 17.2 shows the comprehensive results of 1982 to 1992 (11 years) for the main flux components and the simulated nitrogen export with harvested crop material compared to the measured values. For lysimeter group 7, only two monoliths are considered because the observed results of the third lysimeter differed significantly from the others. For the cumulative values, the relative deviation for actual evapotranspiration and seepage are fairly low (maximum deviation 5 and 12%, respectively). The relative deviations for nitrogen leaching are a little higher for the sandy soil (20%), but simulation is within the range of observed values. Also, the cumulative exported nitrogen was simulated satisfactorily. Nevertheless, there are significant differences, especially for the water fluxes and nitrogen uptake by crops in

Table 17.2   Comparison Between Observed and Simulated Results of Water and Nitrogen Fluxes Summarized for the Period 1982–1992 for Undisturbed Lysimeters of Two Different Soil Types

	Lysimeter 5			Lysimeter 7		
	Measured Average Min Max	Simulated	Average Absolute Annual Deviation	Measured Average Min Max	Simulated	Average Absolute Annual Deviation
Actual evapo-transpiration (mm)	5236 5170 5327	5066	51	6034 5999 6070	5733	38
Seepage in 300 cm depth (mm)	1705 1526 1853	1724	46	1037 1032 1041	1163	38
N leaching in 300 cm depth (kg N/ha)	442 349 492	350	12	173 166 180	166	8
N-uptake (kg N/ha)	1318 1209 1411	1362	32	1848 1772 1912	1664	33

Note: Lysimeter group 5 (silty sand) = 3 replicates; lysimeter group 7 (loamy sand – sandy loam) = 2 replicates.

single years, mostly due to insufficient crop parameterization. The absolute deviations given in Table 17.2 refer to hydrologic seasons (November to October), instead of calendar years, because high deviations can occur at the end of the year due to the time lag of observed seepage compared to simulated water fluxes by a capacity approach. The average absolute deviations of single years for nitrogen leaching are relatively low for both soil types. The differences between both soil types were well reflected by the simulation results.

APPLICATIONS

Model Application for Fertilizer Recommendations

For about 20 years, best management practices for nitrogen fertilization have been developed in Germany for different crops on the basis of soil mineral nitrogen in early spring. Because soil sampling is time- and labor-consuming and soil mineral status may change very quickly depending on soil properties and weather conditions, we focused on the development of an applicable nitrogen model that should be able to calculate fertilizer recommendations based on site-specific simulations of the most important processes of nitrogen dynamics. The main goal was to minimize the amount of fertilization without significant losses in crop yield.

The basic concept for calculating fertilizer recommendations is illustrated in Figure 17.7. Starting from an initial mineral nitrogen content in soil (e.g., after harvest of the previous crop), the model simulates the nitrogen dynamics in the soil crop system using actual weather data until the day of consultation. At this time (point 1), the model predicts nitrogen uptake and soil mineral nitrogen changes operating with typical site-specific weather scenarios until the model indicates that the next relevant development stage for fertilization (e.g., stem elongation) has been reached (point 3). Weather scenarios are based on records of former seasons at the same or similar sites. Regarding the depletion of mineral soil nitrogen in the root zone, the model accumulates the potential nitrogen uptake during phases of nitrogen deficiency (beginning at point 2), considering also the amount of mineralized nitrogen. The total deficiency is recommended to be applied in subsequent applications ahead of critical phases. In its present status, the model can be used to recommend nitrogen fertilization, primarily for cereals. Extensions for oilseed rape, corn, sugar beets, and potatoes are in an experimental stage.

To test our concept, we contributed to different field experiments in Germany that should evaluate several methods for nitrogen best management practices. An experimental evaluation of different methods with long-term scope was started in 1997 by the official agricultural federal state agency of Hanover in Lower Saxony with various soil and climate conditions. Soil and management information from the sites were provided by the investigators, and actual weather data for the specific sites were available online. All experiments were designed as Latin squares with 100 m^2 plots in four replicates. Although the experiments included up to 13 different management schemes, we selected the non-fertilized control (Z) and the two most relevant methods for comparison with the model results (M).

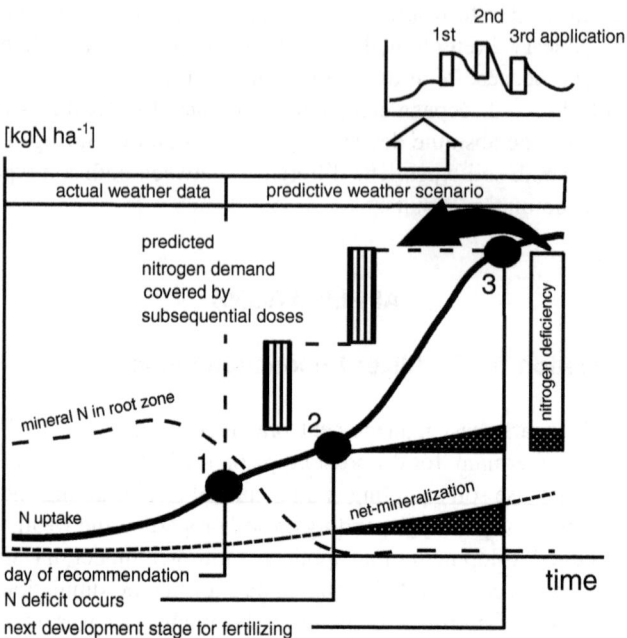

Figure 17.7 Scheme basic of the concept for calculating fertilizer recommendations with HERMES and MINERVA models.

The most commonly used standard method (N_{min}-method) was developed by Wehrmann and Scharpf (1986). It considers the mineral nitrogen of the root zone (0 to 90 cm) in early spring and calculates the required fertilizer application from the difference to a crop-specific optimum that was experimentally determined from a large number of field experiments. The estimated requirement is modified by correction factors, which should consider the influence of different soils, climate, and special management treatments (e.g., manure application). The results of this method are used in our comparison as reference (REF = 100%).

The second method (C) for comparison is a combination of the N_{min}-method and the estimation of the nitrogen demand of the crops by measuring the intensity of green color of crop leaves by a chlorophyll meter at defined development stages. Measured values are used to calculate the required nitrogen application from tabulated results of field experiments performed for the most common crop varieties of cereals.

Figure 17.8 compares the results of 3 years for winter wheat (ww), winter rye (wr), and winter barley (wb) for a total of 41 experimental sites. Each bar represents the average of a variable number of experimental sites (1997: 5/4/3; 1998: 8/2/5; 1999: 7/3/4 for ww/wr/wb, respectively). Column A relates the observed yield of the non-fertilized plot (Z), the chlorophyll method (C), and model-based management (M) to the yield of the standard N_{min}-method (REF = 100). The absolute level of harvested yields for the reference method is tabulated for each year in the plots. Although simulations are primarily based on initial measurements of the previous

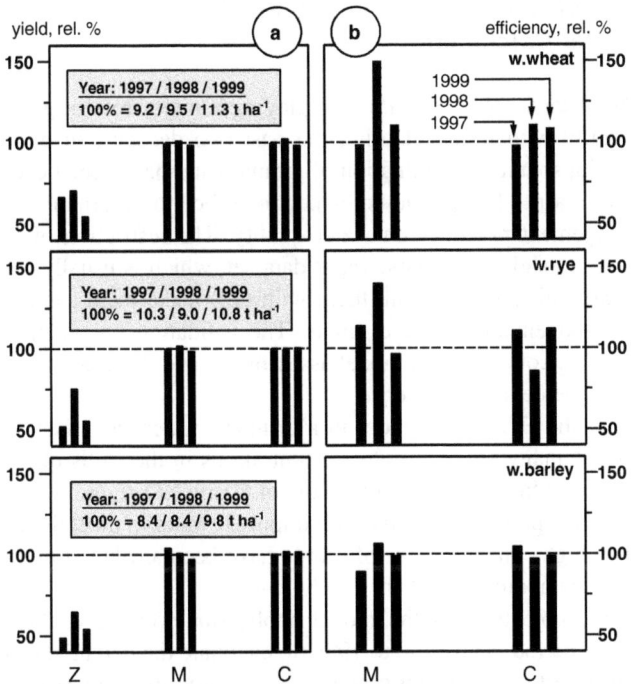

Figure 17.8 Summarized results (41 plots) of different methods for nitrogen fertilizer recommendations referred to the standard method (N_{min} = 100%) for three crops in 3 different years for (a) crop yield and (b) nitrogen efficiency (increase of yield compared to non-fertilized plot per kilogram fertilizer applied). (Z = no fertilizer application (control), M = model-based recommendation (MINERVA), and C = chlorophyll measurement.)

years and the data supplied by the investigators, the achieved yields are not significantly different from those based on intensive measurements. To consider the different amounts and distribution schemes, we calculated a nitrogen efficiency factor (yield response per kilogram N fertilized), which relates the amount of given nitrogen fertilizer to the increase in yield compared to the non-fertilized control (Z). Column B of Figure 17.8 shows the efficiency of the chlorophyll method and the model relative to the standard N_{min}-method (= 100%). Results show, on average, a beneficial effect of the model recommendation compared to the standard N_{min}-method and also to the chlorophyll method for winter wheat (M = 123%, C = 106%) and winter rye (M = 114%, C = 105%). For winter barley, no significant difference between the efficiencies of the three methods can be estimated. The chosen indicator does not consider differences in grain protein, which is a relevant quality criterion at least for winter wheat. Nevertheless, the results do indicate that the model can be applied for practical purposes if site and management are well-documented. The results of multi-season simulations (see "Time Series Measurements") suggest that it should be possible to operate the model for advisory purposes over a couple of years without frequent soil sampling.

Assessment of Nitrate Pollution in a Small Drinking Water Catchment

Agricultural land use is identified as playing a major role in groundwater pollution, especially with nitrate. During the past two decades, several strategies to reduce non-point source contamination of groundwater have been developed. In that context, there is a need to predict site-specific effects of alternative forms of land use and management practices on water quality. The restriction of the HERMES and MINERVA models on a sparse input data set, which is usually available at the farm level, and their ability to link them with GIS make them suitable to evaluate agricultural management for larger areas. The validation on practically managed fields supports the confidence of model assessments and the acceptance of proposed alternatives by the decision-makers.

We applied the HERMES model in a drinking water catchment of 2.8 km² in northwestern Germany where nitrate concentrations in the wells of the water plant were about twice as high as the EU threshold of 50 mg NO_3/L (= 11.3 mg NO_3-N/L). In fact, the water supplier has to mix with imported water to meet the drinking water standards. The high concentrations of nitrate have been attributed to the arable land use, which covers about 75% of the catchment.

The specific objectives of the model application were to analyze the present situation for the period from 1990 to 1996, to calculate the potential of water quality improvement if best management practices are applied to the existing land use, and to quantify the requirements for additional changes in land allocation to meet, on average, the drinking water standard. The study is described in detail by Kersebaum (2000). Here, we focus only the aspect of present land use evaluation.

Simulations were carried out for the intersected polygons of a field map and a digitized soil map with daily weather data from a neighboring meteorological station. Crop distribution was collected from 1990 to 1996 for the fields within the official protection zone. The crops of an extended area that was estimated to contribute to the catchment of the wells were mapped only in 1996, and crop distribution from 1990 to 1995 was extrapolated back using typical crop rotations.

Detailed management information (sowing, harvest, date and amount of fertilizer) was available for the period 1990 to 1991. For the following period, the total amount of mineral and organic nitrogen fertilizer was given by the official advisory bureau for each crop and year using the average recommendation based on the observed soil mineral nitrogen content in early spring.

Field-specific observations of the soil mineral nitrogen content in 0 to 90 cm depth and the organic matter content of the plow layer (0 to 30 cm) were used to initialize the simulation in fall 1990.

Mineral nitrogen was also measured three times a year (spring, after harvest, fall) during 1990 to 1995. Figure 17.9 shows a comparison of the simulated mineral nitrogen content in fall 1991, which was the only year with field-specific management data. Looking at the relationship between measured and simulated values for single fields (Figure 17.9a), the coefficient of determination is only 0.29. Nevertheless, there were in this year uncertainties within the information coming from the farmers regarding the exact amounts of fertilizer and their application dates.

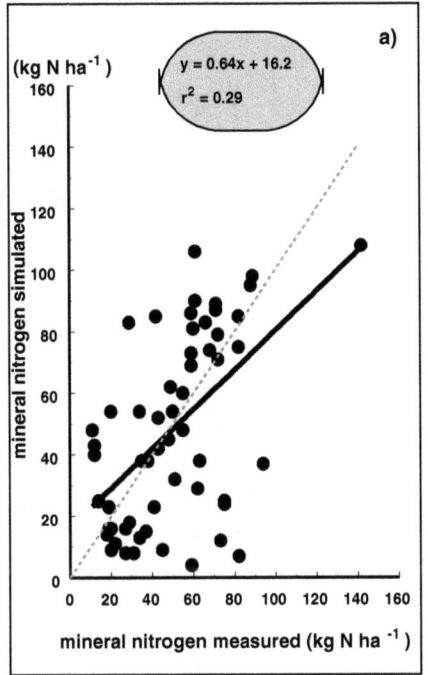
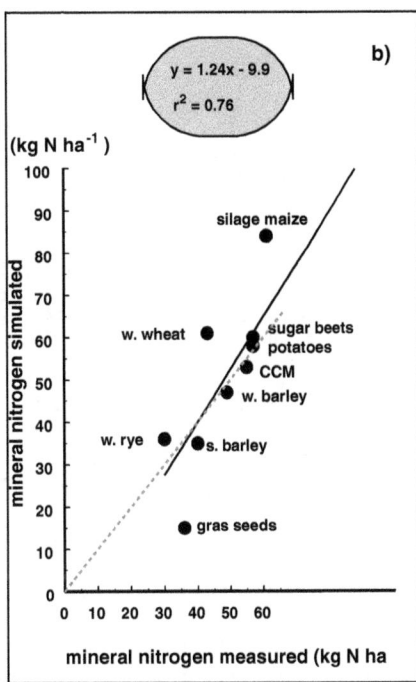

Figure 17.9 Comparison of measured and simulated mineral nitrogen content in the root zone on (a) single fields of a water catchment in fall 1991, and (b) their average values separated for the previous crops.

Additionally, the generalization of representative soil profiles for mapping units leads to deviations due to spatial variability within the unit. Another problem is that observations represented mixed samples from unknown locations within the fields, whereas the soil map shows up to five mapping units within one field. Aggregating the results for different crops, the crop-specific averages after harvest show a close relationship (Figure 17.9b) between simulated and observed values ($r^2 = 0.76$). Even in the years with uncertain, nonspecific information, the average mineral nitrogen contents observed in different years at different times were reflected relatively well by the model results ($r^2 = 0.63$). Due to a relatively shallow water table and a coarse-textured soil, the travel time between root zone and aquifer is short; thus, the simulated nitrate concentration of the seepage water can be related to observed nitrate concentrations in the upper aquifer as an additional indicator that simulation results are in the right order of magnitude. Figure 17.10 is a map of the average nitrate concentrations during 1990 to 1996 in the catchment. The added bar plots show measured nitrate concentrations and their temporal variation during the investigation period, which are compared to simulations of corresponding neighboring areas assuming a lateral shift of about 50 m with the westward groundwater flow. Most of the simulated concentrations are within the observed variation of the corresponding monitored values. Only for the observation point in the north (Figure 17.10a), the simulation underestimates the observed concentrations; this

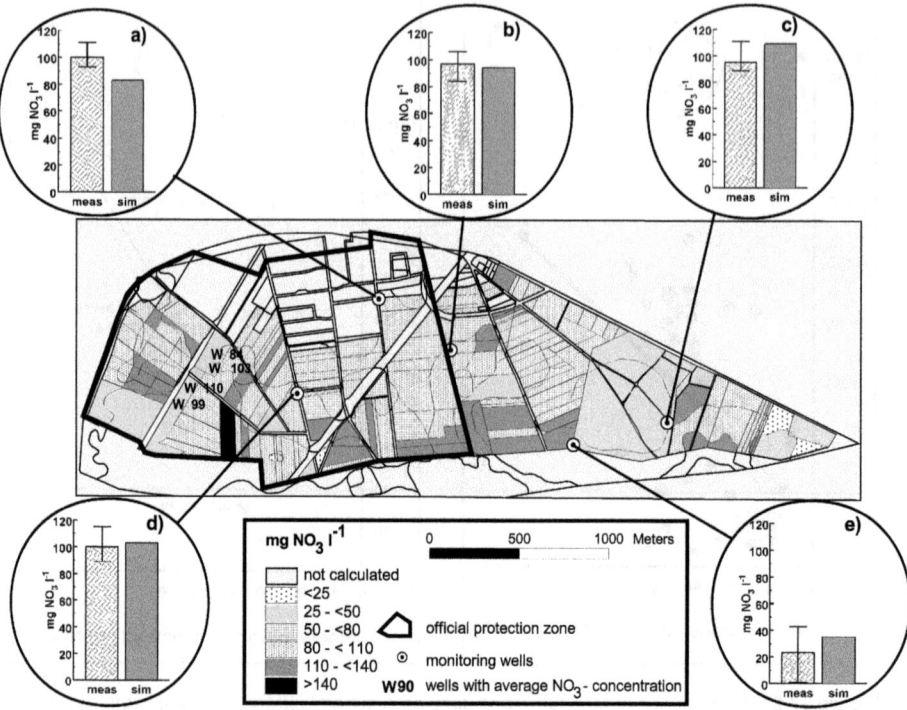

Figure 17.10   Spatial distribution of simulated annual nitrate concentrations of seepage water in a drinking water catchment in northwestern Germany (average 1990 to 1996) and comparison with observed nitrate concentrations of the upper aquifer at five monitoring wells (a through e). (See text for more details.)

might be an effect of the neighboring settlement area with gardens, something not considered in the calculation. It can also be seen that the groundwater entering the official protection zone at its eastern border (Figure 17.10b) shows concentrations similar to the observations within the protection zone (Figure 17.10a and d). Only at the downstream side of the forested area can nitrate concentrations below the threshold be observed (Figure 17.10e), whereas the concentration at the eastern border of the forest is obviously related to the incoming groundwater flux from the neighboring arable land (Figure 17.10c). Because the model application is restricted to agricultural land, the concentration under forest must be roughly assessed from the literature. For the region, an annual nitrogen deposition of 40 kg N/ha in forests (Meesenburg et al., 1994) can be considered. For such deposition rates, a review (Dise and Wright, 1995) of nitrogen leaching and deposition in different European forest ecosystems indicated annual leaching losses under different European forests of about 20 kg N/ha for similar deposition rates. Combining this amount with the estimated groundwater recharge of 250 mm/year, an average nitrate concentration of 35.8 mg NO_3/L (= 8 mg N/L) of the percolation water can be assessed for the forested area.

Integrating simulated nitrogen leaching and seepage for the entire area produces a mixed nitrate concentration of groundwater recharge of 86 mg NO_3/L, which is within the range of observed average concentrations in the wells of the water supply plant (84 to 110 mg NO_3/L; see also Figure 17.10).

CONCLUSION

Application of a model to a real-world situation often means that it must operate under conditions of restricted or uncertain input data, and there is usually no possibility of calibrating model parameters. Thus, it is necessary to validate the model on data sets apart from the sites where they have been developed. To ensure that the model can be used for real applications, the model should be operated for validation with the normal input data even if the description of an investigation site provides much more detailed data. Comparison of model results with measured state variables must consider the accuracy of the observations. High temporal and spatial variability of the studied variables often limit the coefficient of determination that can be achieved on a field or regional scale.

When a model is applied to solve a certain problem, the validity of the model can rarely be evaluated directly. Under these circumstances, the model performance must be controlled indirectly by observations that are strongly related to the problem to be solved.

As an example, results of model-based fertilizer recommendations were compared to alternative methods based on soil and crop measurements in terms of crop yields and nitrogen efficiency. Results are encouraging and indicate that the model approach is suitable for this purpose, at least under middle European conditions. Nevertheless, the current technical trend to consider spatial variability within fields in the frame of precision farming strategies provides a higher quality of input data for the future. Further investigations will have to prove if the present model approach is sufficiently sensitive to reflect the variation within fields.

The second example has shown model application for a small catchment that addresses the problem of nitrate pollution of groundwater resources. Here, the problem in validating the model with local observations becomes more evident because soil data must be generalized to representative soil profiles for soil map units. Aggregating the results for specific crops, soils, or years shows that the model can reflect these general trends and allows for separation of the effects of different crop rotations and fertilization schemes. Although travel time of nitrate from the root zone to the aquifer would require long-term observations and simulations to determine the effect of land use and management on groundwater quality, currently observed nitrate concentrations of the groundwater can be used as an indirect indicator of model performance if land use and management remains fairly constant over the observed period, as in our example.

Although the model does not explicitly simulate carbon dynamics, the chosen approach seems to be a robust and suitable solution for the demonstrated purposes. Nevertheless, the model has clear limitations to reflect short-term net immobilization of mineral nitrogen and long-term effects of land use on soil organic matter dynamics and related soil properties. Furthermore, the possibilities for alternative model scenarios remain limited by the spectrum of crops that can be simulated dynamically. Particularly for organic farming systems, there is a need for model approaches linking crop growth to N_2 fixation for a better site-specific estimation of nitrogen balances.

REFERENCES

AG Bodenkunde. 1995. *Bodenkundliche Kartieranleitung*, 4th ed. Schweizerbart, Stuttgart. 392 pp.

Beblik, A.J. 1996. Description of the N-dynamics simulation model "MINERVA." In *Nitrate Leaching from Arable Soils into the Groundwater of Differently Contaminated Catchments Typical for Lower Saxony Germany*. Final Report, BEO No. 0339121 C Ministry of Science and Technology. 235 pp.

Beblik, A.J., T. Lickfett, and T. Harden. 1996. A multicase optimization procedure for fitting plant growth parameters in the SUCROS based N dynamics model MINERVA. *Transact. 9. Nitrogen Workshop*, Braunschweig, Technical University, Braunschweig, 79-82.

Dise, N.B., and R.F. Wright. 1995. Nitrogen leaching from European forests in relation to nitrogen deposition. *Forest Ecol. Manage.*, 71:153-161.

Gerwitz, A., and E.R. Page. 1974. An empirical mathematical model to describe plant root systems. *J. Appl. Ecol.*, 11:773-781.

Haude, W. 1955. Zur Bestimmung der Verdunstung auf möglichst einfache Weise. *Mitteilgn. Dtsch. Wetterdienst*, 11.

Heger, K. 1978. Bestimmung der potentiellen Evapotranspiration über unterschiedlichen landwirtschaftlichen Kulturen. *Mitteilgn. Dtsch. Bodenkundl. Ges.*, 26:21-40.

Jacob, H.G. 1982. Rechnergestützte Optimierung statischer und dynamischer Systeme. *Fachber. Messen, Steuern, Regeln*, 6:229.

Keese, U. et al. 1997. Vergleichende bodenphysikalische Untersuchungen zwischen Lysimetern und ihren Herkunftflächen am Beispiel von drei typischen Böden Mitteldeutschlands unter landwirtschaftlicher Nutzung. *Arch. Acker-Pfl. Boden*, 41:209-231,

Kersebaum, K.C. 1989. Die Simulation der Stickstoffdynamik von Ackerböden, Ph.D. thesis, University Hannover. 141 pp.

Kersebaum, K.C. 1995. Application of a simple management model to simulate water and nitrogen dynamics. *Ecol. Modelling*, 81:145-156.

Kersebaum, K.C. 2000. Model-based evaluation of land use strategies in a nitrate polluted drinking water catchment in North Germany. In R. Lal (Ed.). *Integrated Watershed Management in the Global Ecosystem*, CRC Press, Boca Raton, FL, 223-238.

Kersebaum, K.C., and J. Richter. 1991. Modelling nitrogen dynamics in a plant-soil system with a simple model for advisory purposes. *Fert. Res.*, 27:273-281.

Keulen, H. van, F.W.T. Penning de Vries, and E.M. Drees. 1982. A summary model for crop growth. In F.W.T. Penning de Vries and H.H. van Laar (Eds.), *Simulation of Plant Growth and Crop Production*. PUDOC, Wageningen, 87-97.

Knappe, S., and U. Keese. 1996. Untersuchungen zu ausgewählten chemischen Eigenschaften langjährig landwirtschaftlich genuzter Boeden von Lysimetern im Vergleich zu Profilen auf deren Herkunftflächen. *Arch. Acker-Pfl. Boden*, 40:431-451.

Meesenburg, H., K.J. Meiwes, and R. Schultz-Sternberg. 1994. Entwicklung der atmogenen Stoffeinträge in niedersächsische Waldbestände. *Forst und Holz*, 49:236-238.

Myers, R.J.K., C.A. Campbell, and K.L. Weier. 1982. Quantitative relationship between net nitrogen mineralization and moisture content of soils. *Can. J. Soil Sci.,* 62:111-124.

Nordmeyer, H., and J. Richter. 1985. Incubation experiments on nitrogen mineralization in loess and sandy soils. *Plant Soil,* 83:433-445.

Nuske, A. 1983. *Ein Modell für die Stickstoff-Dynamik von Acker-Lössböden im Winterhalbjahr — Messungen und Simulationen.* Ph.D. thesis, University Hannover. 164 pp.

Richter, J., H.C. Scharpf, and J. Wehrmann. 1978. Simulation der winterlichen Nitratverlagerung in Böden. *Plant and Soil,* 49:381-393.

Richter, J., A. Nuske, W. Habenicht, and J. Bauer. 1982. Optimized N-mineralization parameters of loess soils from incubation experiments. *Plant Soil,* 68:379-388.

Schneider, U. 1991. Messungen von Denitrifikations- und Nitratauswaschungsverlusten in einem landwirtschaftlich genutzten Wassereinzugsgebiet, Ph.D. thesis, University Bonn, 86 pp.

Wehrmann, J., and H.C. Scharpf. 1986. The N_{min}-method — an aid to integrating various objectives of nitrogen fertilization. *Z. Pflanzenernaehr. Bodenk.,* 149:428-440.

Wendroth, O., P. Jürschik, K.C. Kersebaum, H. Reuter, C. van Kessel, and D.R. Nielsen. 2001. Identifying, understanding, and describing spatial processes in agricultural landscapes — four case studies. *Soil Till. Res.,* (accepted).

CHAPTER 18

Modeling Nitrogen Behavior in the Soil and Vadose Environment Supporting Fertilizer Management at the Farm Scale

Juan David Piñeros Garcet, Amaury Tilmant, Mathieu Javaux, and Marnik Vanclooster

CONTENTS

INTRODUCTION

Globally, 30 to 50% of the Earth's surface is believed to be adversely affected by non-point source pollutants (Duda, 1993). The agricultural sector is considered to play a major role in non-point source pollution because agricultural activities result in the movement of fertilizer residues, agrochemicals, and soil particles from the soil surface into rivers and streams via runoff and erosion, and into subsurface soil and groundwater via leaching (Corwin et al., 1999). Nitrate contamination of surface and groundwater bodies from agricultural origin is a major issue in Europe, as well as in the United States (Kolpin et al., 1999).

The nitrate problem continues to attract the attention of many scientists at the international level. The International Association of Hydrological Sciences, for example, has addressed the issue of nutrient loads to surface and groundwater (Heathwaite, 1999). In the United States, the American Geophysical Union and the Soil Science Society of America organized in 1997 a joint outreach conference entitled "Application of GIS, Remote Sensing, Geostatistics and Solute Transport Modeling, to the Assessment of Non-Point Source Pollution in the Vadose Zone" contributing significantly to a better understanding of the issues related to non-point source pollution (Corwin et al., 1999). In Europe, EURAQUA addressed the issue of farming without harming (Kraats, 2000). Notwithstanding all these initiatives, research related to a better understanding and control of the nitrate problem continues to rank very high on the research priority list, an example being the Fifth Framework Program of Directorate General Research of the European Commission entitled "Combating Diffuse Pollution."

The nitrate problem is also of concern for groundwater managers in Belgium. At the national level, nearly 80% of the drinking water in Belgium is extracted from groundwater, which is very often adversely affected by nitrate contamination. Three vulnerable zones have been identified in the country:

1. The Flemish Region, characterized by intensive agriculture, horticulture, and animal breeding, shallow groundwater overlain by sandy and loamy-sandy soils.
2. The vulnerable zone of the Brusselean sandy aquifer in the Walloon Region, characterized by extensive agriculture in the quaternary loess, central loamy belt.
3. The vulnerable zone of the Hesbaye limestone, an area also characterized by an extensive agriculture in the central loamy belt; yet the loess in this case is overlying a karstic aquifer.

Control of the nitrate problem in the vulnerable areas is a complicated task, given the typical Belgian heterogeneous and patchy land use. Both the agricultural and the non-agricultural sectors contribute to the nitrate pollution problem. For the vulnerable area of the Brusselean region, for example, it has been estimated that

16% of the nitrate contamination is of domestic origin (De Becker et al., 1985). Measures to reduce the nitrate load should therefore focus on both sources. Actions to control the nitrate pollution of domestic origin include the implementation of a more integrated sewage plan, now receiving high priority from regional environmental authorities. Nitrate pollution of agricultural origin, on the other hand, should be regulated by the EU/91/676 directive stipulating that for vulnerable areas, good agricultural practices must be implemented. Regional authorities are now responsible for the definition and the implementation of these good agricultural practices. As an example, the Flemish authorities recently approved the second "Manure Action Plan" MAP-2 (2000). With this decree, the adoption of actual fertilizer practice by the agricultural sector enters its third phase.

Given the complexity of the system, and the space-time variability of the involved processes, multiple-scale modeling is needed for assessing the role of agriculture in environmental quality, and in particular groundwater quality in terms of nitrogen pressure (Vanclooster et al., 1996). As compared to field studies, modeling studies provide an opportunity to analyze the possible impact of a given practice for a wide range of environmental settings. In the past, many modeling studies were realized in a purely academic context, supporting the unraveling of different N processes in a complex system. Currently, however, models are also needed as operational engineering and decision support tools, thereby creating a range of possibilities to refine the nitrogen management strategies at different levels.

Initial modeling studies of nitrogen fate and transport in agroecosystems addressed the processes on the local scale (e.g., Van Veen, 1977; Groot et al., 1991). Some successful studies have been reported in the literature (e.g., Groot et al., 1991; Diekkrüger et al., 1995). In a further step, as more detailed databases and appropriate information technology became available, larger-scale modeling approaches have been presented. The availability of digital soil maps and land use maps now makes it possible to use process-oriented local scale models in a spatially distributed way (e.g., Styczen and Storm, 1993; Christiaens et al., 1996; Birkinshaw and Ewen, 2000). This distributed methodology is now used for evaluating and spatially refining land use management strategies (e.g., Groenendijk and Boers, 1999; Pudenz and Nützmann, 1999; Brenner et al., 1999). However, the potential advantages of information technologies and integrated larger-scale nitrogen models are currently offset by the difficulty in moving geographic data, and model inputs and outputs between different structures and resolutions. In addition, the lack of knowledge of statistical methods for summarizing spatial patterns, as well as the difficulty and cost of large-scale model validation, are other major issues. Finally, the difficulty in placing an economic value on current and alternative land use practices does not fill the gap between researchers and decision-makers (Wilson, 1999). These off-settings should be addressed in current and future research programs.

This chapter presents a case study in which a detailed local-scale nitrogen model was spatially integrated, allowing simulation of a larger-scale nitrogen balance. Such a large-scale modeling approach was conceived for the entire area of the Brusselean aquifer, characterized by a deep aquifer and a deep vadose zone. The model supports the development of medium- and long-term fertilizer management strategies that must be implemented for this vulnerable zone in the framework of the EU/91/676 nitrate directive. This chapter also illustrates how the integrated model allows one to discriminate between farm types and fertilizer strategies in terms of medium- to long-term nitrate pollution of the deep aquifer.

MATERIAL AND METHODS

The Study Area

The Region of the Brusselean Aquifer

The region of the Brusselean aquifer covers an area of 134,000 ha in the central part of Belgium, between Brussels, Gembloux, and Charleroi. A moderate maritime climate with a mean annual temperature of 9.8°C and a mean annual rainfall of 780 mm characterizes the region. The aquifer, which is exploited by different water supply companies, is situated within tertiary sandy deposits. The groundwater level is at a variable depth, outcropping at the surface within the valleys, but situating at depths sometimes exceeding 50 m below the plateaus, its mean depth being 25 m. In most places, the aquifer is overlain by a permeable quaternary loamy deposit, within which fertile loamy soils develop. Major field crops in the area are winter wheat, spring wheat, sugar beets, and grassland, intersected with minor field crops such as barley, chicory, potatoes, and other diverse crops.

Given the unconfined nature of the vulnerable aquifer (Laurent, 1977), it is subjected to a range of non-point source pollution hazards. Previous studies elucidated the nitrate problem of the aquifer with nitrate contamination levels sometimes exceeding the 50 ppm drinking water limit (Bogaert, 1996; Goovaerts et al., 1993). Also, agricultural and domestic sources are suspected to contribute to the contamination (De Becker et al., 1985; Bocken, 1986).

Realizing the vulnerable character of the aquifer and the requirements imposed by the EU/91/676/CEE directive, the regional authority (the Région Wallonne) is currently in charge of defining good agricultural practices. However, given the variable hydrogeological settings and the expected large reaction time of the system for alternative agricultural practices and fertilizer strategies, scope exists for developing a large-scale modeling approach.

The Two Analyzed Farms

Model calculations were performed at the farm-scale level for two distinct farms. The land use characteristics of the two farms are given in Table 18.1. Both farms are mixed, non-intensive, and representative of most of the farm types in the area. However, these farms differ significantly in pedological and hydrogeological settings. Based on a detailed GIS (see Figures 18.1 to 18.4), it was shown that the first farm was very heterogeneous in terms of soil type. Parcels were situated on plateaus as well as in valleys. Parcels on the valley slopes are on a sandy soil type. The depth

Table 18.1   Actual and Simulated Land Use of the Two Selected Farms (as a percentage of total farm area)

| | Farm 1 | | Farm 2 | |
	Actual	Simulated	Actual	Simulated
Grassland	35	21	60	61
Maize	21	24	21	21
Beet	13	15	15	6
Winter wheat	25	25	4	6
Winter barley	0	15	0	6
Other (fallow crops and chicory)	6	0	0	0

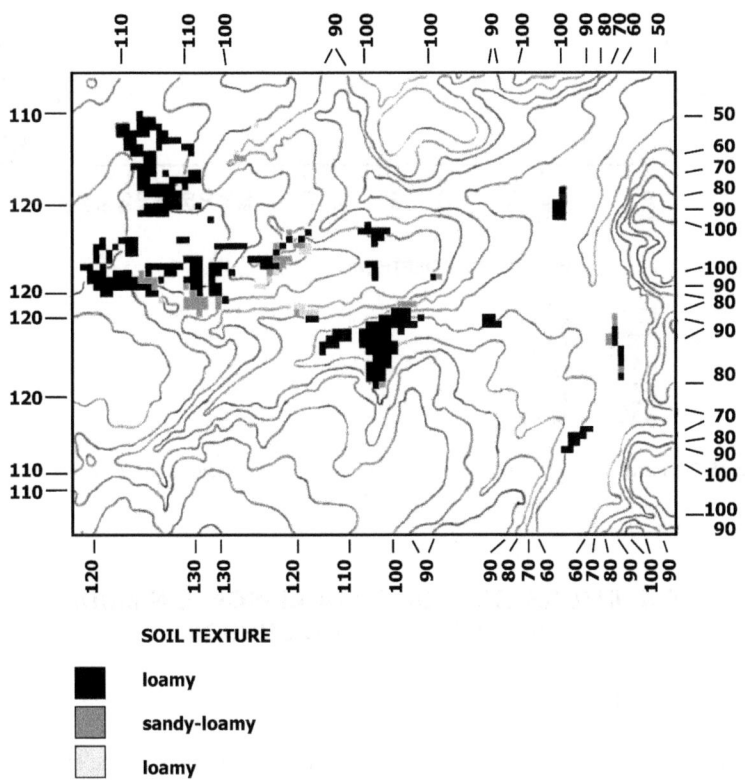

SOIL TEXTURE

■ loamy

▨ sandy-loamy

□ loamy

Figure 18.1 Farm 1: height above sea level and soil texture.

to the aquifer is variable. Within the valley, a shallow aquifer pierces within the topsoil profile. The second farm is more homogeneous in terms of soil type, because most of the field parcels are situated on the plateaus, dominated by a homogeneous loamy soil type. The mean aquifer depth is shallower, and its maximum aquifer depth is also smaller as compared to the first farm.

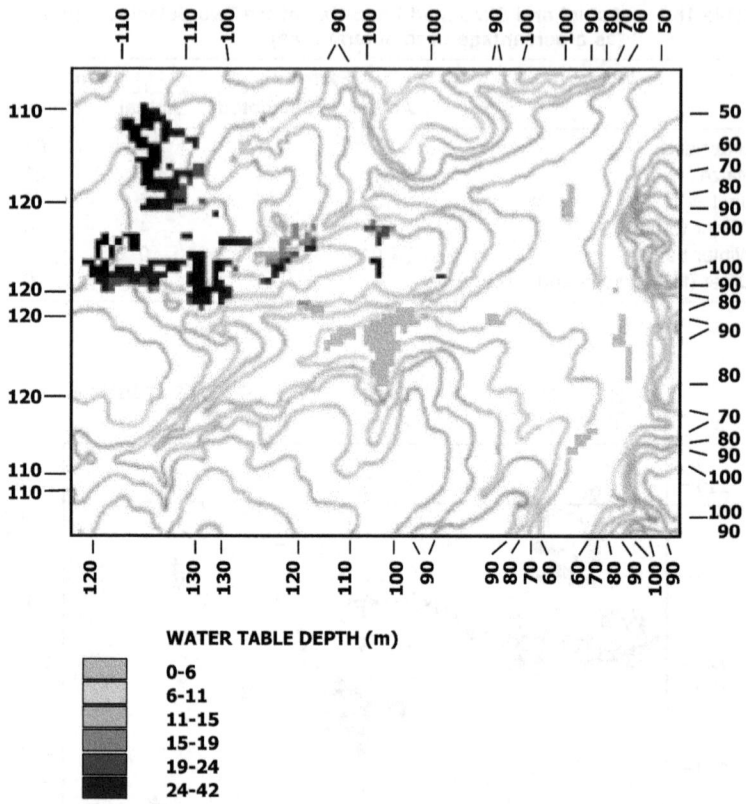

Figure 18.2 Farm 1: height above sea level (m) and water table depth (m).

THE INTEGRATION OF A LOCAL-SCALE N MODEL
FOR THE TOP- AND SUBSOIL

Scale Considerations

To comply with the regional and hydrogeological settings of the study area and the long-term reaction time of the system, the pursued modeling method must allow for evaluation of long-term agriculture practices at the regional scale.

To meet the spatial requirements, a distributed modeling approach was adopted. In particular, the three-dimensional heterogeneous unsaturated hydrogeological formation was decomposed into a set of vertical heterogeneous columns of variable thickness. The horizontal variability of the physicochemical properties within the column was ignored; that is, effective parameters were considered, but those properties were assumed to vary from column to column. The surface resolution was arbitrarily set at 50×50 m, which is a trade-off between the heterogeneity and the covered area.

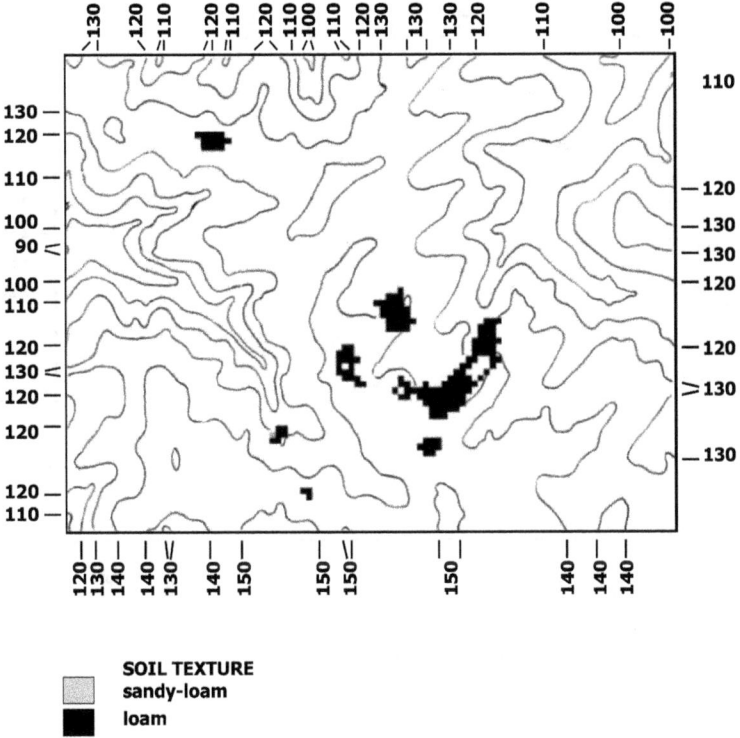

Figure 18.3 Farm 2: height above sea level (m) and soil texture.

Each vertical column was decomposed into a topsoil part and a subsoil part. The topsoil part extended up to a depth of 1.5 m, while the subsoil was delineated by the bottom of the topsoil and the top of the aquifer. Nitrogen behavior in the topsoil was modeled with the detailed deterministic N model WAVE (Vanclooster et al., 1994). The subsoil N transport and behavior in the vadose zone was modeled using a simple transfer function for either a one- or two-layer system, depending on the presence or absence of loamy parent material under the topsoil. A different modeling approach for the topsoil and the subsoil is justified by the differences in sensitivity of some key processes. For example, the N fluxes in the topsoil are largely influenced by the unsteady-state boundary conditions at the soil surface and the turnover of N within the soil organic biomass and rhizosphere, inducing unsteady-state leaching fluxes at the bottom of the topsoil profile. However, in the subsoil, both processes are becoming unimportant, resulting in a merely steady-state N flux dispersing in the subsoil. The linking of the topsoil and subsoil models was facilitated by the availability of a GIS of the physical variables of the system, inferred from pedological and geological maps and databases.

The Topsoil Local-Scale N Model

Within the study, use was made of the local-scale deterministic WAVE (Water and Agrochemicals in the soil and Vadose Environment) model. A detailed description

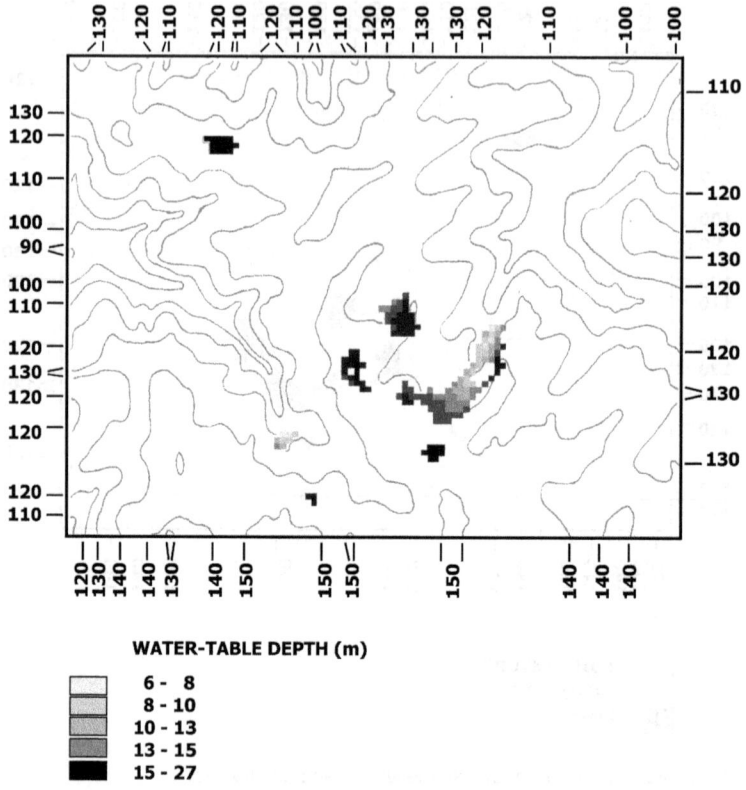

WATER-TABLE DEPTH (m)

	6 - 8
	8 - 10
	10 - 13
	13 - 15
	15 - 27

Figure 18.4   Farm 2: height above sea level (m) and water table depth (m).

of the WAVE model can be found in Vanclooster et al. (1994). The WAVE model combines different *ad hoc* state-of-the-art models such as SWATRER (Belmans et al., 1983), SUCROS (Spitters et al., 1988), and modules of LEACHM (Wagenet and Hutson, 1989). The model is a revised version of the SWATNIT model (Vereecken et al., 1990). It is programmed in a modular way and can easily be extended to model the fate of other agrochemicals in the soil-crop environment. The WAVE model is mechanistic and deterministic. It can handle different soil horizons, which are divided into equidistant soil compartments. Water, heat, and solute mass balance equations are developed for each compartment, taking into consideration different sink/source terms. Physical transport equations (1-D Richards equation for water flow, 1-D non-equilibrium convection dispersion equation) are implemented, which are solved numerically using finite-difference techniques.

The nitrogen module used in the WAVE model was originally developed by Vereecken et al. (1990; 1991) and named SWATNIT (Simulating WATer and NITrogen). The nitrogen module describes the transformation processes for the organic and inorganic nitrogen in the soil. Also, the uptake of nitrogen by plants is described by means of a sink term added to the transport equation. The potential transformation rates, which are model inputs, are reduced for temperature and moisture content in the soil profile. Different levels of complexity exist for describing the nitrogen transformation processes. The complexity level of a selected model will strongly depend on the desired level of explanation (short term/long term). A simple, first-order decay model for the mineralization of the organic nitrogen may give a good description of the mineralization processes shortly after organic matter addition. However, such a model is likely to fail in describing the long-term mineralization process involved in the decomposition of the organic matter fraction. This is of particular concern for our case study, for which long-term sustainable measures need to be implemented. A too-detailed model description, on the other hand, is offset by intensive parameter requirements and limited model applicability. Generic parameter databases, describing the N transformation processes for a wide range of soil and land uses, remain lacking.

The WAVE model has been evaluated using data for a range of field studies. A summary of the model validation and application studies is given in Table 18.2. The model parametrization in our study was based on field experiments realized within our study area (Marriage et al., 1999; Ducheyne et al., 2000).

The Subsoil Local-Scale Nitrate Transport Model

Preliminary experiments on subsoil materials revealed that N turnover within the unsaturated subsoil was unimportant. As a result, the nitrate that leached from the topsoil formation was considered as a tracer injected within the subsoil unsaturated soil column. Only physical transport parameters will affect the tracer travel time and dispersion. Transport model identification experiments, similar to those presented in Vanclooster et al. (1995b) and Vanderborght et al. (1997), were carried out on undisturbed soil monoliths, showing that nitrate transport within the subsoil unsaturated sandy formation behaves as a classical convection-dispersion process. The identified mobile to total moisture content and hydrodynamic dispersivities were further used to predict the nitrate load at the top of the aquifer by means of a CDE transfer function model formulation (Jury and Roth, 1990):

$$Js(t) = v \cdot \theta \cdot \int_{0}^{t} Cin(t') \cdot \frac{z}{2 \cdot \sqrt{\pi \cdot D \cdot (t-t')^3}} \cdot \exp\left(-\frac{(z-v(t-t'))^2}{4D(t-t')}\right) \cdot dt' \quad (18.1)$$

with Js, the nitrate loading flux at the top of the aquifer ($M/L^2/T$); v, the mean pore water velocity (L/T); set equal to Jw/θ, with Jw the mean Darcian soil water drainage flux (L/T); θ, the mean mobile moisture content (L^3/L^3); D, the hydrodynamic dispersion coefficient (L^2/T); and Cin(t) the daily nitrate concentration at the bottom of the topsoil profile (M/L^3), as simulated with the WAVE model. The mean mobile moisture content was obtained by multiplying the mobile:total moisture ratio, as identified in the transport experiments, with observations of total moisture content within the subsoil (Tilman et al., 1978). The apparent dispersion coefficient was obtained by multiplying the identified hydrodynamic dispersivity with the mean pore water velocity.

Table 18.2 Evaluation Studies of the WAVE Model

Ref.	Water	Solute	Heat	Crop	Nitrogen	Pesticide	Field Testing (T)/ Application (A)	Remarks
Burauel et al. (1995)	x	x	x			Benazolinethyl	T	Lysimeter
Chang et al. (1996)	x	x	x	Maize	x		A	Field
Chang et al. (1995)	x	x	x	Maize	x		A	Field
Christiaens et al. (1996)	x	x	x	Maize	x		A	Region
Diels (1994)	x			Grass			T	Lysimeter, Field
Droogers and Bouma (1997)	x			Potato			T/A	Field
Droogers and Bouma (1996)	x			Potato, cereals, sugar beets, meadow			A	Field
Droogers (1998)	x			Grass			A	Field
Droogers et al. (1997)	x	x	x	Potato, grass	x		A	Field
Droogers (1997)	x	x	x	Potato, cereals, grass	x		A	Field
Ducheyne et al. (1998a)	x	x	x	Sugar beet, winter wheat, winter barley maize, potato, grass	x		T	Field
Ducheyne et al. (1998b)	x	x	x	Sugar beet, winter wheat, winter barley, maize, potato, grass	x		T	Field
Ducheyne (2000)	x	x	x	Sugar beet, cereals	x		T/A	Field
Dust (1995)	x	x	x	Oilseed rape		Clopyralid	T	Lysimeter
El-Sadek et al. (1999)	x			Cereals			T/A	Field
Espino et al. (1995)	x						T	Field
Green et al. (1996)	x			Orchard trees			T	Field
Hubrechts (1998)	x	x	x	Maize	x		T	Field

Reference				Crop	Notes		T/A	Study type
Kim et al. (1993)	x	x	x				T/A	Laboratory
Kosmas et al. (1996)	x	x	x	Tobacco				Field
Mallants et al. (1996)	x	x					T	Lysimeter / Laboratory columns
Mallants et al. (1998)	x	x					T	Laboratory columns
Marriage et al. (1999)	x	x	x	Winter wheat		x	T	Lysimeters
Meiresonne et al. (1999)	x	x		Poplar trees			T/A	Field
Normand (1996)	x	x	x	Maize			T	Field
Riga and Charpentier (1999)	x	x	x	Apple trees			T	Field
Vachaud et al. (1997)	x	x	x	Maize, meadow			A	Catchment
Vanclooster et al. (1992)	x	x	x	Maize			T	Field crop
Vanclooster (1995)	x	x	x	Sugar beet, winter wheat			T	Field
Vanclooster et al. (1995)	x	x	x	Sugar beet, winter wheat			T	Field
Vanclooster et al. (1996)	x	x	x	Maize			T/A	Field
Tiktak (2000), Vanclooster and Boesten (2000)	x	x	x	Grass	Ethoprophos, bentazon		T	Field
Van Uffelen et al. (1997)	x	x	x	Potato			A	Field
Vereecken et al. (1991)	x	x	x	Winter wheat			T	Field crop
Vereecken and Kaiser (1999)	x	x		Bare soil			T	Soil column
Verhagen et al. (1995)	x	x	x	Potato, cereals, sugar beets, meadow			A	Farm
Verhagen (1997a)	x	x	x	Potato			A	Field
Verhagen (1997b)	x	x	x	Potato			A	Field
Xu Di (1998)	x	x	x	Cereals			A	Field

The Scenario Definition

To represent combinations of crop rotation, soil, subsoil, climate, agronomic, and land use factors to be used in modeling at the farm scale, one can define five types of scenarios:

1. Fertilization scenario: temporal distribution, quantity, and type of nitrogen fertilization for a given column in the farm's distributed spatial model
2. Climatic scenario: time series of climatic factors (identical for all columns)
3. Agricultural scenario: for each column in the farm's distributed spatial model, it is possible to choose a combination of crop rotations, fertilization scenario, and soil type; each one of such a combination being called an "agricultural scenario"
4. Land use scenario: describes the spatial distribution of the agricultural scenarios in a particular farm
5. Topsoil nitrate leaching scenario: time series of leaching at 1.5-m depth, for each column of the distributed spatial model, obtained using the topsoil model and used as input for the subsoil transport model

Realistic long-term (i.e., 30 years) fertilization, climatic, and agricultural scenarios were established and processed. For the climatic scenarios, available climatic time series from the Belgian Royal Meteorological Institute were used. Further, 77 different agricultural scenarios were constructed, thereby combining different crop types, crop rotations, fertilizer, and soil types. The parameterization of the scenarios was based on a questionnaire that was sent to a range of farmers in the area so that current agricultural practices could be identified (Meiers et al., 1998).

Description of the Agricultural Scenarios

A range of 77 agricultural scenarios were defined, combining crop rotations with realistic fertilization scenarios for a given soil type. When defining the agricultural scenarios, the following constraints were adopted: the modeled rotation and soils must be representative for the study area; the modeled fertilization strategies must correspond to (1) the current fertilization doses and timings; (2) the maximum allowed fertilization dose as imposed by the current legislation; and (3) the envisaged fertilization practice, as proposed in an indicative Good Agricultural Practice Plan.

Among the 77 agricultural scenarios, 56 were developed for representing the agriculture on the plateaus. Two crop rotations were considered: (1) beets, winter wheat, barley and winter wheat; and (2) beets, winter wheat, potatoes, and winter wheat, each having either a winter catch crop or not. Three periods for amending the organic fertilizer before the sugar beet crop were considered: (1) application between August 4 and 31; (2) application between November 1 and 30; and (3) application between February 1 and 29. Six fertilization doses were used (Table 18.3). The remaining 21 agricultural scenarios were developed for representing other agricultural practices on the loamy and sandy-loam to sandy soil types. For these scenarios, five crop rotation types were identified: (1) monoculture of maize; (2) sugar beets-winter wheat-barley rotation; (3) maize-winter wheat-winter

Table 18.3   Fertilization Doses for the 56 Scenarios Representing the Agriculture on the Plateaus

Dose	Beets			Winter Wheat in Crop Rotation 1			Winter Wheat in Crop Rotation 2			Barley			Potatoes		
	Org[a]	Mi[b]	Tot[c]	Org[a]	Mi[b]	Tot[c]	Org[a]	Mi[b]	Tot[c]	Org[a]	Mi[b]	Tot[c]	Org[a]	Mi[b]	Tot[c]
1	210	120	330	0	223	223	0	170	170	0	223	223	210	120	330
2	210	120	330	0	170	170	0	170	170	0	170	170	185	145	330
3	188	133	321	0	173	173	0	173	173	0	144	144	186	131	317
4	0	139	139	0	173	173	0	173	173	0	144	144	0	151	151
5	0	80	80	0	119	119	0	119	119	0	85.7	85.7	0	80	80
6	357	124	481	0	211	211	0	211	211	0	193	193	275	180	455

a Organic fertilization
b Mineral fertilization
c Organic + mineral fertilization

barley rotation; (4) potato-winter wheat and winter barley rotation; and (5) permanent grassland.

Description of the Land Use Scenarios

The agricultural scenarios were combined to form land use scenarios, in order to represent the land use in the two analyzed farms. The agricultural scenarios were conceived to represent agriculture in the region of the Brusselean aquifer, and not in a particular farm; thus, a specific assignment procedure was developed that optimizes the similarity between the actual and modeled land use on the two farms. The adopted procedure is a multi-objective optimization procedure using the Levenberg–Marquardt algorithm to approximate at best the fertilization doses and the percentage of the area occupied by each crop. The results of the assignment procedure are shown in Table 18.1. At the scale of the two selected farms, two land use scenarios were constructed, taking into consideration the evolution of the system for a period of 30 years. The first scenario considers a *status quo* of the actual N fertilizer practice. The second scenario corresponds to an optimized one, considering the region-specific N fertilizer recommendations (Van Bol, 2000) as suggested to be implemented in the Good Agriculture Practice plan of the region.

Description of the Leaching Scenarios

For each agricultural scenario, the topsoil N model was processed so as to generate topsoil nitrate leaching scenarios. In total, 77 different leaching scenarios were generated with the local-scale topsoil N model, and related to the 77 agricultural scenarios. Those 77 leaching scenarios are spatially distributed over the two farms following the land use scenarios.

RESULTS

Topsoil Simulation Results for Variable Agricultural Scenarios

For a statistical analysis of simulation results, the 56 scenarios representing the agriculture on the plateaus were chosen, and daily simulation results were aggregated on a decade basis.

The Impact of the Crop Rotation

Figure 18.5 displays the mean simulated nitrate leaching for two contrasting crop rotations on a loamy soil. The mean for a given rotation type and a given decade is calculated over the six fertilization doses presented in Table 18.3 and the three periods for amending. Obviously, nitrate leaching for both rotations is more pronounced during the winter period than during the summer period. The small increase in summer leaching in decades 20 and 21 is a result of some extreme summer storm events, which were present in the adopted 30-year climatic time series. The crop

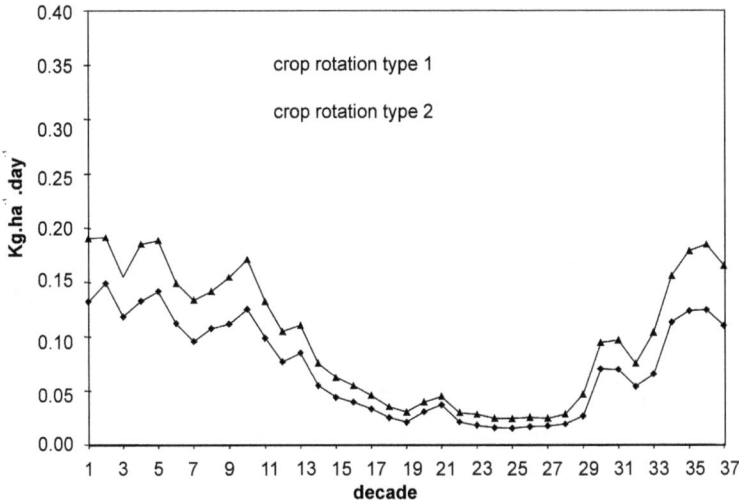

Figure 18.5   Simulated nitrate leaching on a loamy soil for two crop rotation types without winter catch crop (1: sugar beet-winter wheat-winter barley-winter wheat rotation; 2: sugar beet-winter wheat-potato-winter wheat rotation). The mean for a given rotation type and a given decade is calculated over six fertilization doses and three periods for amending.

rotation encompassing a potato crop is significantly more vulnerable than the crop rotation encompassing a barley winter crop, which indicates a less appropriate N scheduling for potato crops as compared to a fertilizer schedule adopted for winter barley.

The Impact of Timing of the Organic Fertilizer Application Before the Sugar Beet Crop

Significantly different N loadings from the topsoil layer were also simulated for scenarios with different organic fertilizer schedules. Organic amendments are applied before the sugar beet crop and can either be applied shortly after harvesting the winter wheat, during autumn or during spring, just before the planting of the sugar beets. When there is no winter catch crop, the summer application after harvest is likely to create high N loadings during the subsequent winter periods. Disposing organic fertilizer on a bare soil for a prolonged time interval seems to be environmentally unsound. On the other hand, when a winter crop is present, the spring application seems to be untimely. Our results illustrate the complexity of the analyzed systems, as well as the importance of establishing correct scenarios of crop sequences.

The Role of a Winter Catch Crop

Our simulation results showed an increase in the long-term N load when winter catch crops are inserted into the crop rotation without adapting the fertilizer rates.

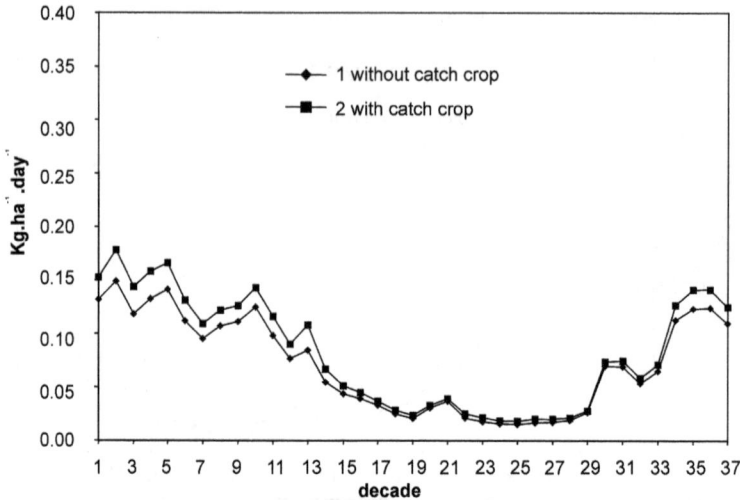

Figure 18.6   Simulated nitrate leaching on a loamy soil for the sugar beet-winter wheat-winter barley-winter wheat rotation crop rotation without (1) and with (2) a winter catch crop. The mean for a given rotation type and a given decade is calculated over six fertilization doses and three periods for amending.

This is illustrated, for example, in Figure 18.6, where the mean simulated nitrate leaching for the same crop rotation with and without a winter catch crop is shown. This unexpected result can be explained as follows. If a winter catch crop is used, the nitrate content within the topsoil profile will be reduced during the winter period, especially in the soil layer close to the soil surface, as illustrated in Figure 18.7. However, if the catch crop is reincorporated into the soil, for example, before the sugar beet crop, then it will increase the N content of the labile organic litter pool. Given the low C:N content of the incorporated fresh catch crop, a mineralization flush of N will take place, which will continue during the following seasons at a rate dependent on the C and N turnover constants of the organic pool. Mineralization flushes of the freshly incorporated soil organic matter have also been identified in field experiments subjected to similar agricultural practices under similar edapho-climatic conditions (Clotuche and Peeters, 2000; Clotuche et al., 1998). Following our simulations, parts of the mineralized N will not meet the N requirements of the sugar beet crop on a timely basis, and will therefore be partially lost to the environment. In addition, following our simulations, N denitrification during winter was reduced when a catch crop was present. This can be explained by the different positions of the center of mass of the nitrate profile for rotations with and without a catch crop. Because more nitrate will be available in the deeper humid soil layers when no catch crop is present, more denitrification will take place. Given the fact that the process of denitrification can be considered a net loss within the N balance, it will be inversely proportional to the risk of N leaching. However, a cautionary note should be added. In our agricultural scenarios, the fertilization rate was maintained constant with and without catch crops. In practice, farmers will reduce their

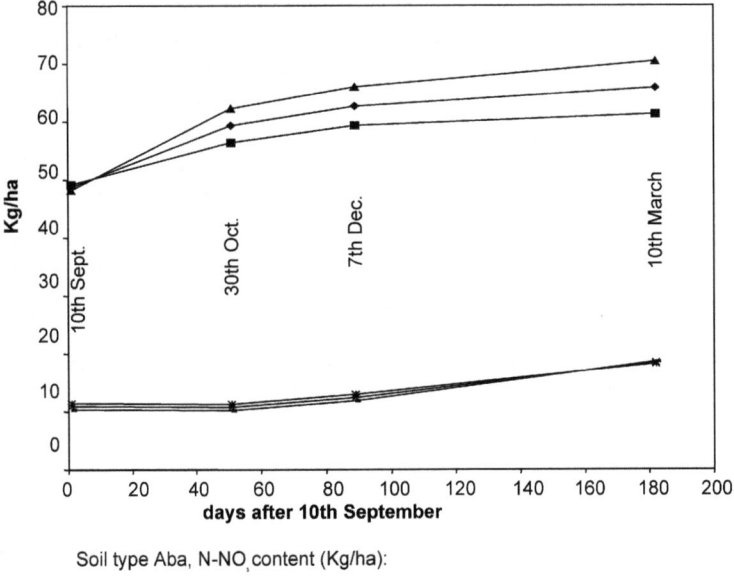

Figure 18.7   Simulated N profiles during the winter period for the topsoil (0 to 150 cm) and the layer 100 to 150 cm, for all scenarios with and without winter catch crop. The presented means are calculated over six fertilization doses and three periods for amending.

fertilization rates proportional to the expected amount of N coming from the catch crops. This could probably offset, in practice, the negative impact of catch crop use.

The efficiency of a winter catch crop for reducing nitrate leaching is a matter of debate in the agronomical literature. Also, positive (Hansen and Djurhuus, 1997; Schroder et al., 1996; BrandiDohm et al., 1997) as well as negative effects (Beckwith et al., 1998; Goss et al., 1998; Ritter et al., 1998) of catch cropping have been reported. We conclude that the efficiency in reducing nitrate leaching will be strongly influenced by the C:N ratio of the catch crop and the turnover rate constants of the different organic matter pools. However, given the uncertainty in these soil parameters in actual C:N models, we do not believe that we yet have a solid base for rejecting catch cropping as an appropriate measure for reducing N leaching. Albeit, even if the efficiency of catch cropping in terms of N leaching reduction is doubtful, it will definitely have beneficial impacts on soil erosion reduction because the soil surface is covered during winter time.

Table 18.4   Basic Soil Physical Properties of the Selected Soil Profiles

Soil Type	Horizon	Thickness (cm)	Clay (%)	Loam (%)	Sand (%)	C (%)
Aba	Ap	25	14	77	8	0.9
	Bt	74	21	74	5	0.2
	C	65	17	78	5	0.1
Sbx	A1	15	8	35	58	1.9
	G	185	4	12	85	1.6
Zbx	A1	15	4	17	81	0.8
	G	185	3	6	90	0.1

Note: The soil parameters have been identified from the Belgian soil database AARDEWERK.

From Van Orshoven, J., J. Maes, H. Vereecken, J. Feyen, and R. Dudal. 1988. A structured data base of Belgian soil profile data. *Pédologie*, 38:191-206. With permission.

The Role of the Soil Type

The soil type of a given field plot will significantly influence N leaching from the topsoil. This can be clearly illustrated with a grassland agricultural scenario receiving 225 kg N/ha, and this for three different soil types: a loamy soil, a sandy soil, and sandy loamy soil. The physical properties of the considered soils are given in Table 18.4. The concentration of the leaching water at a depth of 1.50 m is given in Figure 18.8. As expected, the purely sandy soil exhibits a significantly larger N

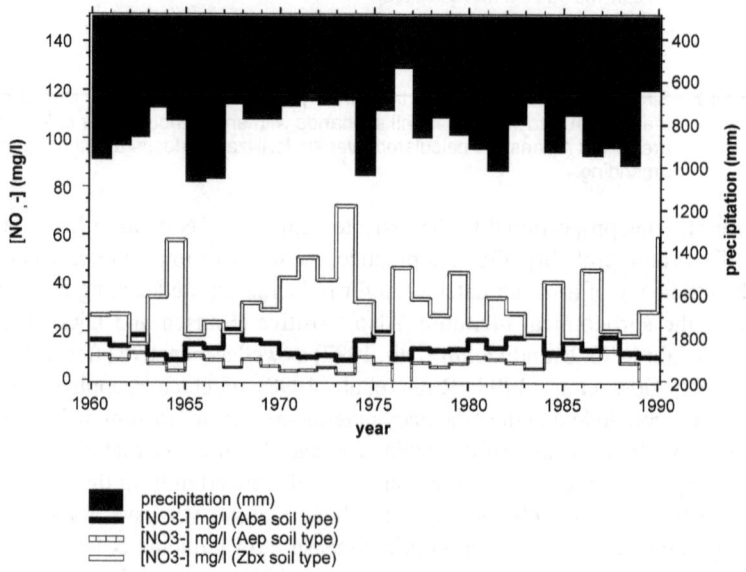

Figure 18.8   Simulated annual mean nitrate concentration of the leaching water leaving the topsoil profile for a grassland agricultural scenario on three different soil types. The fertilization rate is 255 kg N/ha/year.

leaching potential as compared to the loamy and sandy loamy soils. The larger vulnerability of sandy soils is due to the higher transport velocities of N within the root zone of crops, and to the lower moisture content, thereby reducing the N transformation processes.

Farm-Scale Simulation Results

Figure 18.9 yields the expected mean evolution of the nitrate concentration at the top of the aquifer for a period of 30 years, given a *status quo* of the actual N fertilizer practice. The results were obtained by linking the leaching scenarios for the topsoil, generated with the WAVE model, with the subsoil transport model. The given values are the non-weighted mean of the calculated concentrations for all grid cells of both farms, for all land use scenarios. It is observed that the nitrate concentrations at the top of the aquifer below the two selected farms is expected, on average, to increase and to pass the limit of 50 mg/L. However, the variability between the calculated grid cells at farm level is extremely high, as illustrated by the standard deviation bars. This shows that scope exists to optimize the field plots within the farms, based on their geographical position. The steady increase in nitrate concentrations (on average, 0.7 mg/L/year) corresponds to many observations of long-term evolution of nitrate concentrations as observed in many drinking water wells of the region.

Figures 18.10 and 18.11 illustrate the evolution of mean nitrate concentration at the top of the aquifer for the two selected farms, and thus for the two land use evolution scenarios. It is first shown that both farms behave significantly differently in terms of absolute contamination levels and exhibit a different sensitivity toward the optimized land use scenarios. The larger contamination level of the second farm

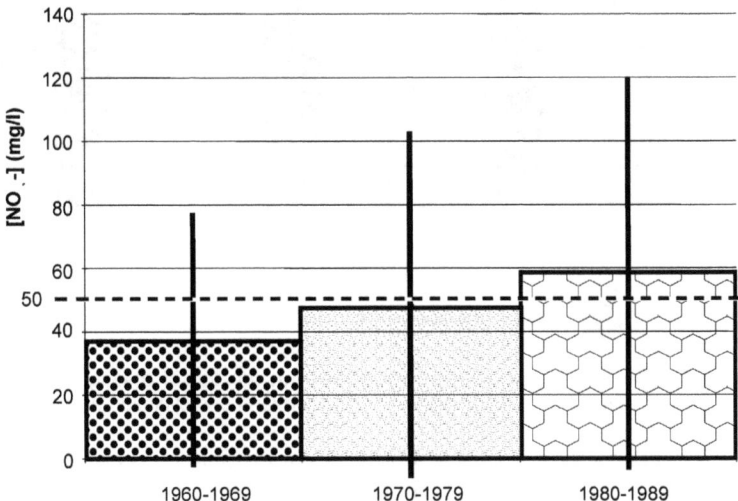

Figure 18.9   Mean and standard deviation of the simulated nitrate concentration at the top of the aquifer (non-weighted mean of the calculated concentrations for all grid cells of both farms, for all land use scenarios).

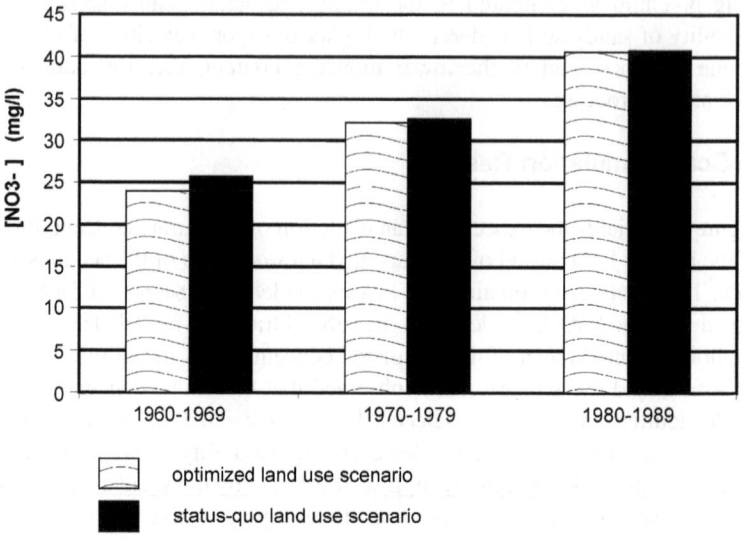

Figure 18.10   Evolution of the nitrate concentration at the top of the aquifer for farm 1, subjected to a *status quo* and optimized land use evolution scenario.

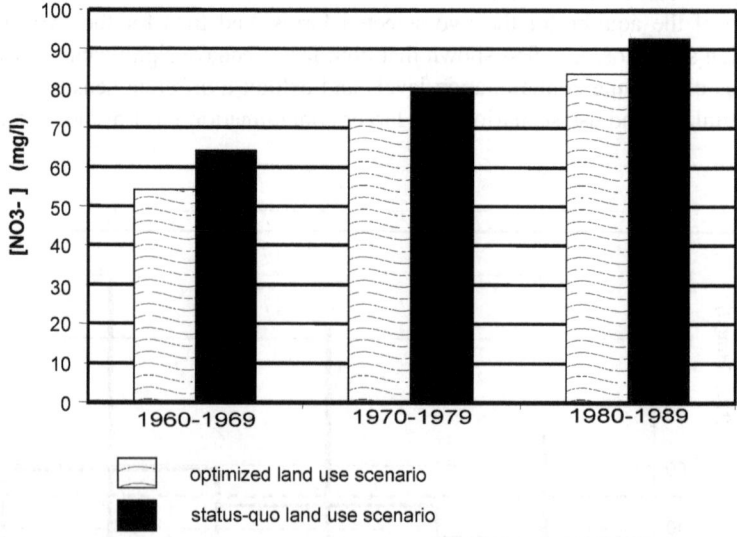

Figure 18.11   Evolution of the nitrate concentration at the top of the aquifer for farm 2, subjected to a *status quo* and optimized land use evolution scenario.

can be explained by the geographical positions of the field plots and the associated different crop scenario. It is further shown that, following our simulations, the proposed modification of the fertilizer practice will not be offset by a decrease in nitrate concentration at the top of the aquifer. A statistical test confirmed that the observed decrease is insignificant. Moreover, a more elaborate statistical analysis showed that the simulated nitrate concentration at the top of the aquifer was extremely sensitive to the thickness of the unsaturated sandy layer.

DISCUSSION AND CONCLUSION

In this study, an attempt was made to link a detailed soil crop N model with a subsurface nitrate transport model within a GIS framework to predict medium- and long-term evolutions of the nitrate content recharging a deep but vulnerable sandy aquifer. The method was implemented for two pilot farms in the area, and allowed us to analyze the impact of land use and fertilizer management strategies on the potential evolution of subsurface water quality. Notwithstanding that we are convinced that the approach results in a powerful analysis tool, we recommend that caution should be given to the following points.

First, the model was only implemented for two selected farms, representing a spatial coverage less than 0.1% of the total area. Any regional extrapolation of the present results is therefore unjustified. However, given the current evolution in information technology, we expect that the method could easily be applied to a regional assessment. This, however, is subject to the availability of digital soil maps, geological maps, land use maps, and Digital Terrain models at the appropriate spatial resolution.

Further, notwithstanding that the topsoil model was subjected to a range of validation studies in the past, and that the subsoil transport model was validated using detailed flow experiments on large undisturbed soil cores, validation of the integrated farm-scale model cannot be stated. The lack of validation at the farm-scale level is primarily due to a lack of experimental data on nitrate behavior in the subsoil and at the top of the aquifer. No long-term time series of experimental data with the appropriate spatial resolution is presently available. It is believed that major shortcomings of the actual farm-scale modeling concept are related to the following issues, which we recommend be addressed in future research programs:

1. *Impact of the point pollution.* The actual model only considers the nitrate recharge from non-point source pollution. However, as has already been addressed in previous studies (De Becker et al., 1985), point pollution from, for example, sewage pits can contribute to nitrate load at the regional level.
2. *Impact of the surface boundary condition.* The actual model does not consider surface hydrological processes such as run off and erosion. Obviously, some nutrients will be lost by surface runoff and particulate bounded processes.
3. *Uncertainty regarding horizontal movement of nutrients within the subsoil.* The current model only considers vertical transport, adopting a parallel column concept. Horizontal fluxes within the unsaturated zone can be important, especially at larger scales.
4. *Uncertainties regarding the degradation potential within the subsoil.* The current model version does not consider any degradation of the nitrate in the subsoil. However, denitrification in locally saturated aggregates in the subsoil cannot be à priori excluded.
5. *Uncertainties with respect to long-term organic matter degradation constants.* This can be very important, especially if the long-term dynamics of the organic matter plays a role. This was illustrated by the ambiguous role of the catch crop.

REFERENCES

Beckwith, C.P., J. Cooper, K.A. Smith, and M.A. Sheperd. 1998. Nitrate leaching loss following application of organic manures to sandy soils in arable cropping. 1. Effects of application time, manure type, overwinter crop cover and nitrification inhibition. *Soil Use Manage.*, 14:123-130.

Belmans, C., J.C. Wasseling, and R.A. Feddes. 1983. Simulation of the water balance of a cropped soil: SWATRE. *J. Hydrol.*, 63:271-286.

Birkinshaw, S.J. and J. Ewen. 2000. Nitrogen transformation component for SHETRAN catchment nitrate transport modeling. *J. Hydrol.*, 230:1-17.

Bocken, P. 1986. La composition chimique des sources du bassin de la Dyle. 2. Analyse des variations des teneurs en NO_3^-, Cl^-, Ca^{2+} et conductivité. *Revue de L'Agriculture*, 6:1299-1312.

Bogaert, P. 1996. Geostatistics Applied to the Analysis of Space-Time Data. Ph.D. thesis. Université Catholique de Louvain. 176 pp.

BrandiDohm, F.M., R.P. Dick, M. Hess, S.M. Kaufman, D.D. Hempill, and J.S. Selker. 1997. Nitrate leaching under cereal rye crop. *J. Environ. Qual.*, 26:181-188.

Brenner, A.J., P.L. Richards, M. Barlage, and P. Sousounis. 1999. The impact of land use changes 1820-2020 on water quality and quantity in southeast Michigan. In L. Heathwaite (Ed.). *Impacts of Land-Use Change on Nutrient Loads from Diffuse Sources*. IAHS Publication 257, 229-233.

Burauel, P., K. Hücker, M. Dust, G. Reinken, and F. Führ. 1995. The fate of (Benzene- U14-C) benazolin-ethyl in a lysimeter study supported by detailed laboratory experiments and model calculation. *BCPC Monograph*, 62:211-216.

Chang, Y., M. Vanclooster, L. Hubrechts, and J. Feyen. 1996. Multicriteria decision analysis in irrigation scheduling. In *Evaporation and irrigation scheduling. Proceedings of the International Conference,* Nov. 3-6, 1996, San Antonio, Texas, 1128-1133.

Chang, Y., M. Vanclooster, P. Viaene, and J. Feyen, 1995. A multi-objective irrigation scheduling procedure maximising economic return and minimising environmental impact. In *Advances in Hydro-Science and Engineering (ICHE'95)*. Tsinghua University Press, Beijing, China,136-143.

Christiaens, K., K. Vander Poorten, M. Vanclooster, and J. Feyen. 1996. Combining GIS and a dynamic simulation model to assess nitrogen leaching susceptibility on a regional scale: a case study. In O. Van Cleemput et al. (Ed.). *Progress in Nitrogen Cycling Studies.* Kluwer Academic, 407-411.

Clotuche, P. and A. Peeters. 2000. Nitrogen uptake by Italian ryegrass after destruction of non-fertilised set-aside covers at different time in autumn and winter. *J. Agron. Crop Sci.*, 184:121-131, 2000.

Clotuche, P., B. Godden, V. Van Bol, A. Peeters, and M. Penninckx. 1998. Influence of set-aside on the nitrate contents of soil profiles., *Environ. Pollut.*, 102:501-506.

Corwin, D.L., K. Loague, and T.R. Ellsworth (Ed.). 1999. *Assessment of Non-point Source Pollution in the Vadose Zone*, Geophysical Monograph 108, AGU, Washington, 369 pp.

De Becker, E., G. Billen, V. Rousseau, and E. Stainier, 1985. *Etude de la contamination des eaux souterraines et de surface dans le bassin de la Dyle à Wavre*. Rapport final. Ministère de la Région Wallonne pour L'eau, L'environnement et la Vie Rurale.

Diekkrüger, B., D. Söndgerath, K. Kersebaum, and C. McVoy. 1995. Validity of agro-ecosystem models: a comparison of results of different results applied to the same data set. *Ecol. Modelling,* 81:3-29.

Diels, J., 1994. A Validation Procedure Accounting for Model Input Uncertainty: Methodology and Application to the SWATRER Model. Ph.D. Thesis, Leuven University.

Droogers, P., and J. Bouma. 1996. Biodynamic versus conventional framing effects on soil structure as expressed by simulated potential productivity. *Soil Sci. Soc. Am. J.*, 60:1552-1558.

Droogers, P., and J. Bouma. 1997. Soil survey input in exploratory modeling of sustainable soil management practices. *Soil Sci. Soc. Am. J.*, 61:1704-1710.

Droogers, P. 1997. *Quantifying Differences in Soil Structure Induced by Farm Management.* LU Wageningen, 134 pp.

Droogers, P. 1998. Time aggregation of nitrogen leaching in relation to critical thresholds values. *J. Contam. Hydrol.* 30:(3-4)363-373.

Droogers, P., F.B.W. van der Meer, and J. Bouma. 1997. Water accessibility to plant roots in different soil structures occurring in the same soil type. *Plant Soil*, 188:83-91.

Ducheyne, S. 2000. Derivation of the Parameters of the WAVE Model Using a Deterministic and Stochastic Approach. Doctoraatsproefschrift nr. 434, Faculteit Landbouwkundige en Toegepaste Biologische Wetenschappen. Leuven, Belgium. 127 pp.

Ducheyne, S., L. van Ongeval, H. Vandendriessche, and J. Feyen. 2000. Application of the deterministic calibration and validation procedure to a limited number of field sites. *Agric. Water Manage.*, submitted.

Ducheyne, S., N. Schadeck, and J. Feyen. 1998a. Optimisation of the nitrogen fertilizer package for field crops using experimental data in conjunction with numerical modeling. In *Proceedings of the AgEng98 Conference* (also on CD-Rom), 24-27 August, Oslo, Norway.

Ducheyne, S., N. Schadeck, and J. Feyen. 1998b. Modellering van de migratie van nutriënten in de bodem. Interne publicatie nr. 52. Instituut voor Land- en Waterbeheer, K.U. Leuven, Belgium.

Duda, A.M. 1993. Addressing non point sources of water pollution must become an international priority, *Water Sci. Technol.*, 28:1-11.

Dust, M. 1995. Comparison of the Results from a Two Year Lysimeter Experiment on Degradation and Transport of Clopylarlid in Soil with Results of Calculations Using WAVE and PELMO Models. Inaugural dissertation zur Erlangung des Grades Koktor in Agrarwissenschaften der Hohen Landwirtschaftlichen Fakultat der Rheinischen Friedrich Wilhelms, Universitat zu Bonn, 154 pp.

El-Sadek, J. Feyen, and J. Berlamont. 1999. Analysis of the drainage component of the simulation models WAVE, SWAP and DRAINMOD. In J. Feyen and K. Wiyo (Eds.). *Modelling of Transport Processes in Soils*. Wageningen Pers, Wageningen, The Netherlands, 390.

Espino A., D. Mallants, M. Vanclooster, J. Diels, and J. Feyen. 1995. A cautionary note on the use of pedotransfer functions for the estimation of soil hydraulic properties. *Agric. Water Manage.*, 29:235-253.

Goovaerts P., Ph. Sonnet, and A. Navarre. 1993. Factorial kriging analysis of springwater contents in the Dyle River Basin, Belgium. *Water Resources Res.*, 29:2115-2125.

Goss M.J., K.R. Howse, D.G. Christian, J.A. Catt, and T.J. Pepper. 1998. Nitrate leaching: modifying the loss from mineralized organic matter., *Eur. J. Soil Sci.*, 49:649-659.

Green S.R., B.E. Clothier, M. Vanclooster, and J. Feyen. 1996. Water uptake by orchard trees: measurements and a model. In L. Currie and P. Loganathan (Eds.). *Recent Developments in Contaminant Transport in Soils*. Fertilizer and Lime Research Institute, Massey University, New Zealand, 143-152.

Groenendijk, P., and P. Boers. 1999. Surface water pollution from diffuse agricultural sources at regional scale. In L. Heathwaite (Ed.). *Impacts of Land-use Change on Nutrient Loads from Diffuse Sources*. IAHS Publication 257. 235-244.

Groot, J.J.R., P. de Willigen, and E.L.J. Verberne. 1991. Nitrogen turn-over in the soil-crop systems. *Fert. Res.*, 27:141-386.

Hallberg, G.R., and D.R. Keeney. 1993. Nitrate. In W.M. Alley (Ed.). *Regional Ground Water Quality.* Van Nostrand Reinhold, New York, 297-321.

Hansen, E.M., and J. Djurhuus. 1997. Nitrate leaching as influenced by soil tillage and catch crop. *Soil Tillage Res.*, 41:203-219.

Heathwaite L. (Ed.). 1999. *Impacts of Land-use Change on Nutrient Loads from Diffuse Sources.* IAHS Publication 257, 271 pp.

Hubrechts, L. 1998. Transfer Function for the Generation of Thermal Properties of Belgian Soils. Ph.D. thesis. Faculteit Landbouwkundige en Toegepaste Biologische Wetenschappen. 236 pp.

Jury, W.A., and K. Roth. 1990. *Transfer Functions and Solute Movement Through Soil. Theory and Applications.* Birkhauser Verlag, Basel.

Kim, D.J., H. Vereecken, J. Feyen, M. Vanclooster, and L. Stroosnijder. 1993. A numerical model of water movement and soil deformation in a ripening clay soil. *Modelling Geo-Biosphere Processes,* 1:185-203.

Kolpin, D., M. Burkart, and D. Goolsby. 1999. Nitrate in groundwater of the midwestern United States: a regional investigation on relations to land use and soil properties. In L. Heathwaite (Ed.). *Impacts of Land-use Change on Nutrient Loads from Diffuse Sources.* IAHS Publication 257, 111-116.

Kosmas, C., N. Danalatos, E. Ntzanis, and N. Yassaglou. 1996. The application of pedotransfer functions in predicting ground water recharge at regional scale. In A. Bruand, C. Duval, H. Wösten, and A. Lilly (Eds.). *The Use of Pedotransfer in Soil Hydrology Research in Europe.* INRA, Orléans, 111-119.

Kraats, J.A. (Ed.). 2000. Farming without harming: the impact of agricultural pollution on water systems. *Fifth Scientific and technical review of EurAqua.* 232 pp.

Laurent, E. 1977. *Monographie du Bassin de la Dyle.* Ministère de la Santé et Publique et de L'environnement, Belgique.

Mallants, D., D. Jacques, M. Vanclooster, J. Diels, and J. Feyen. 1996. A stochastic approach to simulate water flow in macroporous soil. *Geoderma,* 70:299-234.

Mallants, D., P.H. Tseng, M. Vanclooster, and J. Feyen. 1998. Predicted drainage for a sandy loam soil: sensitivity to hydraulic property description. *J. Hydrol.,* 206:136-147.

MAP-2. 2000. Decreet houdende wijziging van het decreet van 23 Januarie 1991 inzake de bescherming van het leefmilieu tegen de verontreiniging door meststoffen. *Belgisch Staatsblad,* 30/03/2000:10004-10032.

Marriage, Q., L. Pussemier, and M. Vanclooster. 1999. Evaluation of soil hydraulic pedotransfer functions for modeling the soil water drainage in cropped lysimeters. In J. Feyen and K. Wiyo (Eds.). *Modelling of Transport Processes in Soils.* Wageningen Pers, Wageningen, The Netherlands. 297.

Meiers, P., V. van Bol, and A. Peeters. 1998. *Convention Prop'Eau Sable.* Annual Report. Laboratoire d'Ecologie des Prairies. Université Catholique de Louvain.

Meiresonne, L., N. Nadezhdina, J. Cermak, J. Van Slycken, and R. Ceulemans. 1999. Measured sap flow and simulated transpiration from poplar stand in Flanders (Belgium). *Agricultural and Forest Meteorology,* 96:165-179.

Normand, B. 1996. *Etude expérimental et modélisation du devenir de l'azote dans le système sol-plante-atmosphère.* UJF, Grenoble, INPG-UJF-CNRS. 190 pp.

Pudenz, S., and G. Nützmann. 1999. Scenario calculations of regional subsurface transport if phosphorus in a sub-basin of the Spree River near Berlin. In L. Heathwaite (Ed.). *Impacts of Land-use Change on Nutrient Loads from Diffuse Sources.* IAHS Publication 257. 213-219.

Riga, P., and S. Charpentier. 1999. Simulation of nitrogen dynamics in an alluvial sandy soil with drip fertigation of apple trees. *Soil Use Manage.,* 15:35-40.

Ritter, W.F., R.W. Scarborough, and A.E.M. Chimside. 1998. Winter crop as a best management practice for reducing nitrate leaching. *J. Contam. Hydrol.,* 34:1-15.

Schroder, J.J., W. van Dijk, and W.J.M. De Groot. 1996. Effects of cover crops on the nitrogen fluxes in silage maize production systems. *Neth. J. Agric. Sci.*, 44:239-315.

Spitters, C.J.T., H. Van Keulen, and D.W.G. Van Kraailingen. 1988. A simple but universal crop growth simulation model, SUCRO87. In R. Rabbinge, H. Van Laar, and S. Ward (Eds.). *Simulation and Systems Management in Crop Protection.* Simulation Monographs, PUDOC, Wageningen, The Netherlands.

Styczen, M., and B. Storm. 1993. Modelling of nitrogen movements on a catchment scale. A tool for analysis and decision making. *Fert. Res.,* 36:1-6.

Tiktak A., 2000. Application of pesticide leaching models to the Vredepeel dataset. II. Pesticide fate. *Agric. Water Manage.*, 44:119-134.

Tilman E., L.W. De Backer, R. Klein, and J.P. Kuypers. 1978. Sonde automatique des mesure d'humidité de grande profondeur, synthèse des essais réalisés jusqu'ne juillet 1977. Évaluation des ressources en eau du bassin hydrographique de la Dyle dans la région de Louvain-la-Neuve. Université catholique de Louvain, 1978.

Vachaud G., B. Normand, F. Bouraoui, M. Vanclooster, F. Moreno, J.E. Fernandez, N. Yassaglou, and K. Kosmas. 1997. Evolution des ressources en eau souterraines en zone de culture irrigué du bassin Méditerraneen en cas de changements climatiques. In *Agricultural and Sustainable Development in Mediterranean Countries. Proceedings.* Montpellier 10-12 mars 1997.

Van Bol, V. 2000. Rapport d'activités 1 avril 1999–30 mars 2000. *Convention Prop'Eau Sable.* Laboratoire D'écologie des Prairies, Université catholique de Louvain, 147 pp.

Van Orshoven, J., J. Maes, H. Vereecken, J. Feyen, and R. Dudal. 1988. A structured data base of Belgian soil profile data. *Pédologie*, 38:191-206.

Van Uffelen, C.G.R., J. Verhagen, and J. Bouma. 1997. Comparison of simulated and crop yield patterns for site specific management. *Agric. Syst.,* 54:207-222.

Van Veen, J.A. 1977. Ph.D. thesis. Vrije Universiteit Amsterdam, 164 pp.

Vanclooster, M., 1995. Nitrogen Transport in Soil: Theoretical, Experimental and Numerical Analysis. Ph.D. Thesis, Faculteit Landbouwkundige en Toegepaste Biologische Wetenschappen, 220 pp.

Vanclooster, M., and J.J.T.I. Boesten. 2000. Application of nine pesticide leaching models to the Vredepeel dataset. I. Water, solute and heat transport. *Agric. Water Manage.,* 44.

Vanclooster, M., H. Vereecken, J. Diels, F. Huysmans, W. Verstraete, and J. Feyen. 1992. Effect of mobile and immobile water in predicting nitrogen leaching from cropped soils. *Modelling Geo-Biosphere Processes*, 1:23-40.

Vanclooster, M., J. Feyen, and E. Persoons. 1996. Nitrogen transport in soils: what do we know. In L. Currie and P. Loganathan (Eds.). *Recent Developments in Contaminant Transport in Soils*. Fertilizer and Lime Research Centre, Massey University, New Zealand, 127-141.

Vanclooster, M., P. Viaene, J. Diels, and J. Feyen. 1995a. A deterministic validation procedure applied to the integrated soil crop model WAVE. *Ecol. Modelling*, 81:183-195.

Vanclooster, M., D. Mallants, J. Diels, J. Vanderborght, J. Van Orshoven, and J. Feyen. 1995b. Monitoring solute transport in a multilayered sandy lysimeter using time domain reflectometry. *Soil Sci. Soc. Am. J.*, 59(2):341-355.

Vanclooster, M., P. Viaene, J. Diels, and K. Christiaens. 1994. *WAVE — A mathematical model for simulating water and agrochemicals in the soil and vadose environment. Reference and user's manual (release 2.0).* Institute of Land and Water Management, Katholieke Universiteit Leuven, Belgium.

Vanderborght, J., C. Gonzalez, M. Vanclooster, D. Mallants, and J. Feyen. 1997. Effects of soil type and water flux on solute transport. *Soil Sci. Soc. Am. J.*, 61:372-389.

Vereecken, H., M. Vanclooster, and M. Swerts. 1990. A simulation model for the estimation of nitrogen leaching with regional applicability. In R. Merckx and H. Vereecken (Eds.). *Fertilization and the Environment*. Leuven Academic Press, Belgium, 250-263.

Vereecken H., M. Vanclooster, M. Swerts, and J. Diels, 1991. Simulating water and nitrogen behavior in soil cropped with winter wheat. *Fert. Res.*, 27:233-243.

Vereecken, H., and R. Kaiser, R. 1999. Analysis of multi-step outflow data and pedotransfer functions to characterize soil water and solute transport at different scales. In M.Th. van Genuchten, F. Leij, and L. Wu (Eds.). *Characterization and Measurement of the Hydraulic Properties of Unsaturated Porous Media*. University of California, 1089-1102.

Verhagen J., 1997a. Spatial Soil Variability as a Guiding Principle in Nitrogen Management. Ph.D. Thesis. LU Wageningen, 107 pp.

Verhagen J., 1997b. Site specific fertilizer application for potato production and effects on N-leaching using dynamic simulation modeling. *Agriculture Ecosystems and the Environment*, 66:(2)165-175.

Verhagen J., H. Booltink, and J. Bouma. 1995. Site specific management: balancing production and environmental requirements at farm level. *Agric. Syst.*, 49:369-384.

Wagenet, R., and J. Hutson. 1989. *LEACHM, a process-based model of water and solute movement, transformations, plant uptake and chemical reactions in the unsaturated zone*. Centre for Environ. Res., Cornell University, Ithaca, NY, 147 pp.

Wilson, P. 1999. Current and future trends in the development of integrated methodologies for assessing non-point source pollutants. In D.L. Corwin, K. Loague, and T.R. Ellsworth (Eds.), *Assessment of Non-point Source Pollution in the Vadose Zone*. Geophysical Monograph 108, AGU, Washington, 343-361.

Xu, Di. 1998. Evaluation of soil management practices through experiments and soil water simulation. Application to the Xiongxian area (China). Ph.D. Thesis 1837, EPFL, Lausanne, Switzerland, 152 pp.

CHAPTER **19**

Modeling Transformations of Soil Organic Carbon and Nitrogen at Differing Scales of Complexity

Robert F. Grant

CONTENTS

THE CONTRIBUTION OF MODELING SOIL C AND N TRANSFORMATIONS TO UNDERSTANDING ENVIRONMENTAL ISSUES

Soil C and N models are important components of ecosystem models that are finding growing use in the testing of hypotheses for ecosystem behavior under documented site conditions and in the prediction of ecosystem behavior under hypothesized site conditions. These predictions are of growing importance in the management of environmental issues associated with land use practices. Some of these issues, and the key modeling issues associated with them, are discussed below.

Soil C Sequestration and Soil Quality

The long-term stabilization of soil organic carbon (SOC) has important implications for the sequestration of atmospheric CO_2 and for the maintenance of soil quality. Land management effects on stabilization can be estimated from changes in SOC with different land use practices under site-specific conditions of soil and climate. Such estimates require long-term experiments and are therefore confined to a limited number of sites. Insights gained from these experiments can be extended to a wider range of conditions through the use of ecosystem models that include C and N transformations in soils and plants. Such models have been used to simulate changes in SOC with grazing intensity (Parton et al., 1994), fertilizer and organic amendments (Jenkinson et al., 1987; Li et al., 1994; Parton and Rasmussen, 1994), and tillage and rotations (Grant, 1997; Grant et al., 1998; Paustian et al., 1998). These models are now being used to project changes in SOC under hypothesized management practices as part of regional and national studies of CO_2 emissions.

Because changes in SOC represent the difference between C inputs from net primary productivity (NPP) and C losses from heterotrophic respiration (R_h), ecosystem models used to simulate C sequestration must be capable of simulating land use effects on both NPP and R_h. The modeling of NPP must be clearly distinguished from that of phytomass growth by which NPP is represented in many models. NPP is the difference between gross primary productivity (GPP, or gross CO_2 fixation) and autotrophic respiration (R_a) expended in maintenance and growth of phytomass. NPP includes both standing phytomass and litterfall, of which a large part occurs in the root system. An accurate simulation of both above- and below-ground litterfall is therefore a critically important requirement of an ecosystem model used to estimate changes in SOC. The testing of modeled litterfall is difficult because litterfall remains in a measurable state for only a brief period of time, especially below ground. However, NPP can also be estimated from the difference between gross primary productivity (GPP) and autotrophic respiration (R_a) deduced from continuous CO_2 flux measurements using eddy covariance techniques. Such measurements are becoming a more important part of ecosystem model testing (Amthor et al., 2001).

A key process in the modeling of R_h is the rate at which labile C from litterfall becomes protected from rapid decomposition, thereby reducing R_h. This rate is commonly considered to be a function of soil clay content (e.g., van Veen and Kuikman, 1990) and residue lignin content (Paustian et al., 1997), which determine the capacity of a soil to protect labile C. Thus, clay soils typically have lower R_h than sandy soils with the same C inputs. Hassink and Whitmore (1997) presented an interesting development of this function in which the protection of labile C and the decomposition of protected C are determined by the degree to which the capacity for labile C protection is already exploited. The accurate modeling of the rate at which labile C is protected is important because this rate determines long-term C sequestration in soils. The accurate modeling of the maximum capacity for C protection, if such a capacity exists, is also important because it will determine the maximum C accumulation of which a soil is capable.

Trace Gas Emissions and Atmospheric Quality

Because N_2O and CH_4 are radiatively active in the atmosphere, there is much interest in estimating net exchange of these gases between terrestrial ecosystems and the atmosphere as part of global climate studies. These estimations are needed to evaluate management options for reducing N_2O and CH_4 emissions as part of national commitments to reduce the release of greenhouse gases to the atmosphere. Mathematical models of these emissions are currently being used to estimate agricultural contributions to regional and national greenhouse gas emissions (e.g., Desjardins and Keng, 1999).

Nitrous oxide is an intermediate respiration product of facultative anaerobes that is generated rapidly with the onset of O_2 deficits during nitrification and denitrification. High NO_3^- concentrations favor the intermediate product N_2O, while low O_2 and high labile SOC concentrations favor the terminal product N_2 (Smid and Beauchamp, 1976). The onset of O_2 deficits is determined by the degree to which O_2 reduction is constrained by O_2 transport and solubility, each of which is differently affected by soil temperature and water content. Methane is a terminal product of anaerobic respiration under prolonged O_2 deficits in saturated soils and a substrate of aerobic respiration in unsaturated soils. It is produced from acetate, H_2, and CO_2, which are the terminal products of a syntrophic community of anaerobic fermenters and H_2-producing acetogens that oxidize and reduce organic C. Acetate is transformed into CH_4 and CO_2 by acetotrophic methanogens, and H_2 and CO_2 are transformed into CH_4 by hydrogenotrophic methanogens. Methane emitted in saturated soils can move upward into unsaturated soils where much of it may be oxidized to CO_2 by autotrophic methanotrophs (Cicerone and Oremland, 1988; Fechner and Hemond, 1992; Sass et al., 1990; Schütz et al., 1989), or absorbed into root systems through which it may be transferred directly to the atmosphere.

Rates of N_2O and CH_4 emissions are therefore not simple functions of substrate concentrations and soil environment (Blackmer et al., 1982; Christensen, 1983; Flessa et al., 1995; Robertson, 1994), but rather complex functions of interacting transformation and transport processes and the environmental conditions under which they occur. These environmental conditions include temperature (Bremner and Shaw, 1958; Dunfield et al., 1993; Frolking and Crill, 1994; Nommik, 1956), O_2 and water table depth (Allison et al., 1960; Fechner and Hemond, 1992; Martikainen et al., 1993), and labile C from NPP (Aselmann and Crutzen, 1990; Smid and Beauchamp, 1976; Whiting and Chanton, 1992), such that temporal and spatial patterns of emissions under site-specific conditions are highly variable and complex. Consequently, the predictive value of short-term flux measurements for long-term estimates of these emissions is confined to sites with similar soil and climate (Blackmer et al., 1982). Mathematical models used to estimate these emissions should explicitly simulate the processes from which these emissions arise, including nitrification, denitrification, methanogenesis, and methanotrophy. They should also explicitly simulate spatial variability in soil environmental conditions, including O_2, water, heat, SOC, and roots. These models should be tested for their ability to represent the large spatial and temporal variability that characterizes N_2O and CH_4 flux measurements.

Soil Nutrient Uptake and Plant Growth

The fixation and stabilization of C in terrestrial ecosystems are often limited by the availability of soil nutrients (most commonly, N and P) for uptake by microbes and roots. Soil-plant models have long been used to simulate increases in plant N uptake and soil C accumulation when this limitation is removed by N fertilizer application (e.g., de Willigen 1991). A key objective of these simulations is to evaluate alternative strategies for improving the efficiency with which fertilizer and other soil amendments such as manure are used to increase plant growth and soil C storage.

Uptake of nutrients by roots depends on the convective and diffusive movement of nutrients through the soil solution to the root surface and on the active uptake of nutrients at the root surface. This movement is driven by nutrient concentrations in solution that are controlled by reactions with organic (mineralization-immobilization) and inorganic (adsorption-desorption, precipitation-dissolution) forms of the nutrient in soil. These reactions are influenced by soil OM quality, temperature, water content, pH, ionic strength, and mineral composition, so that effects of soil quality on nutrient uptake may be site and time dependent. The active uptake of nutrients depends on the density, surface area, and uptake kinetics of the root system (Anghinoni and Barber, 1980). Root density and surface area are influenced by plant growth, and by soil temperature, water content, pH, bulk density, and nutrient concentrations so that association between root growth and nutrient uptake is highly site and time dependent. The extension of root density and surface area by mycorrhizae is necessary for adequate rates of P uptake in most soils.

Mathematical models used to calculate plant nutrient uptake should couple a model of the soil transformations by which soluble nutrient concentrations are determined (e.g., Grant and Heaney, 1997) with a model of root and mycorrhizal growth by which active uptake of soluble nutrients is driven (e.g., Grant and Robertson, 1997). The robustness of the nutrient uptake model will be directly related to the diversity of conditions under which the soil transformation and root growth models function. Such diversity requires that a comprehensive range of soil transformations and root growth responses be represented in the model (e.g., Jones et al., 1990).

Soil Nutrient Loss and Water Quality

The effect of land use practices such as fertilization and manure management on water quality (e.g., nitrate and phosphate concentrations) has become an important environmental issue to which ecosystem models have been applied (e.g., Hansen et al., 1990; Hutson and Wagenet, 1992; Shaffer and Larson, 1987). However, it has been generally observed that these models require further refinement before they can confidently be used for predicting nutrient losses to ground and surface water under different land use practices.

Some key limitations to the modeling of land use effects on water quality are:

- The inability of most functional and process models to simulate the apparent immobilization of fertilizer N (de Willigen, 1991; Jaynes and Miller, 1999; Johnson et al., 1999), indicating that the accurate simulation of biological processes that control solution N concentration remains a problem.
- Uncertainty in modeling gravity-driven water flow and associated solute transport through soil macropores as affected by soil water content and tillage. Failure to represent macropores in SOM models leads to underestimation of losses from seepage and overestimation of those from runoff (Ghidey et al., 1999). There is also uncertainty about the modeling of solute transfer between macropore and micropore water fractions during macropore flow.
- The large spatial variation in relationships among soil water content, water potential, and hydraulic conductivity. This variation can cause large differences in convective solute transport to emerge across small distances within an apparently homogeneous soil profile (Addiscott and Wagenet, 1985).

Soil Disturbance and Land Management

Land use practices that involve soil disturbance and vegetation removal have been widely observed to cause losses of soil C to the atmosphere. These losses account for about one third of the atmospheric CO_2 that has accumulated since pre-industrial times, and indicate widespread soil degradation. Some land management practices that reduce these losses include:

1. Greater cropping frequency (Bremer et al., 1994; Campbell et al., 1995), which causes C inputs through plant fixation to be raised comparatively more than C losses through soil respiration.
2. Reduced tillage, which causes plant residues to accumulate on the soil surface (Mielke et al., 1986). These residues decompose more slowly than those incorporated by tillage because of reduced contact with soil microorganisms (Reicosky et al., 1995).

3. Use of fertilizers (Nyborg et al., 1995), which raises primary productivity and hence inputs of plant C and provides N required for C stabilization.
4. Application of organic amendments such as manure, which may contribute C directly to the soil (although Janzen et al. (1998) regard this as a C transfer rather than a gain), and which may also improve nutrient availability and thereby primary productivity and C inputs.
5. Use of perennial legumes and grasses (Bremer et al., 1994; Campbell et al., 1991), which maintain continuous inputs of C to soil through perennial root systems that are more prolific than those of annual crops. Nitrogen fixation by legumes also provides N required for C stabilization.

Ecosystem models have been used to simulate changes in SOC with different land management practices as described above. Such models must be capable of representing the processes by which these land management practices affect SOC. For example, the effect of tillage practices on SOC should be modeled from an explicit simulation of a surface residue layer with its own microbial populations and microclimate (temperature and water content) by which microbial activity and hence decomposition rate in the residue are determined. This residue layer should also affect surface energy balances and thereby subsurface heat fluxes and temperatures. Tillage practices in the model could affect the degree and depth to which surface residues are incorporated into the underlying soil profile. Incorporation would increase residue contact with soil microbial populations and microclimate, thereby increasing its decomposition rate. Incorporation would also reduce residue heat fluxes, leading to more rapid soil heating and cooling. This modeling approach can simulate the more rapid soil warming (Grant et al., 1995) and efflux of CO_2 (Grant, 1997) following tillage to which much of the soil C loss caused by tillage is attributed (Reicosky et al., 1995).

Tillage has additional effects on CO_2 efflux by improving contact between soil C and soil microbial populations that also need to be considered in models. In some models (e.g., Molina et al., 1983; Li et al., 1992), this effect is simulated by redistributing some C from resistant to labile C pools. Tillage also alters soil surface properties such as surface roughness, residue coverage, and depressional water storage, and subsurface properties such as bulk density that have indirect effects on CO_2 efflux. These alterations are included in some ecosystem models (e.g., Shaffer and Larson, 1987; Williams et al., 1989).

Modeling the effects of perennial legumes and grasses vs. annual crops on SOC requires increases in the below- vs. above-ground allocation of plant C that arise from the perennial growth habit. This allocation is greatly affected by climate, including irradiance (Aguirrezabal and Tardieu, 1996), temperature (Vincent and Gregory, 1989), and atmospheric CO_2 concentration (Chaudhuri et al., 1990; Kimball et al., 1995), likely through its effects on CO_2 fixation (Aguirrezabal et al., 1993), and by soil properties, including water (Hoogenboom et al., 1987) and nitrogen (Ennik and Hofman, 1983). Therefore, ecosystem models must simulate root NPP, including root turnover and rhizodeposition, separately from shoot NPP in a way that represents site effects on root C allocation and NPP without the need for site-specific calibration (e.g., Grant et al., 2001b). Modeling the effects of legumes on SOC will require the simulation of biological N_2 fixation by root nodules and the exchange of fixed C and N between nodules and roots. Changes in below- vs. above-ground allocation of plant C are not currently represented well enough in ecosystem models to allow site-specific effects on SOC through root NPP to be modeled (Parton and Rasmussen, 1994).

THE PROCESS OF MODEL DEVELOPMENT

Defining Model Scope and Objectives: General vs. Specific Applications

Most ecosystem models are constructed with a specific objective related to one of the environmental issues described above. Development is considered complete when all processes believed to affect model results in relation to its objective are represented with the minimum parameterization required to simulate the environmental sensitivity of each process. Examples of how minimum parameterization can be established for specific processes are given by Sinclair et al. (1976) and Stockle (1992). They demonstrated that the spatial aggregation of leaf surfaces within a canopy into sunlit and shaded fractions does not adversely affect the accuracy with which canopy photosynthesis and transpiration are calculated. Consequently, most ecosystem models currently use this spatial aggregation when simulating these processes.

When model objectives expand to include a wider range of ecosystem conditions, the level of parameterization appropriate to those objectives becomes more detailed. The spatial aggregation of leaf-level processes such as photosynthesis and transpiration described above is valid for the mono-specific crop stands that the models of Sinclair et al. (1976) and Stockle (1992) were intended to simulate. However, should model objectives be expanded to include competition for irradiance among different species within a complex biome, then greater spatial resolution would be needed in the modeling of leaf area used in simulating photosynthesis and transpiration (e.g., Grant, 1994a; Spitters and Aerts, 1983). A corresponding need for greater spatial resolution in the modeling of root growth arises when model objectives include competition for water and nutrient uptake among different species within a complex biome.

Parameterizing Model Algorithms at Appropriate Temporal and Spatial Scales According to Model Scope and Objectives

There are two basic requirements for the parameterization of robust ecosystem models:

1. Parameterization must be at temporal and spatial scales smaller than those at which prediction is to occur; otherwise, the key model function of spatial and temporal integration is not accomplished. For example, daily phytomass growth in many models is still calculated from daily solar radiation using a parameter called radiation use efficiency (RUE in g MJ^{-1}). However, RUE is evaluated by regressing seasonal accumulation of phytomass (g m^{-2}) on that of absorbed solar radiation

(MJ m^{-2}). The temporal and spatial scales at which RUE is parameterized (season and canopy) are thus larger than those at which it is used (daily and leaf). Its value varies with leaf age, temperature, irradiance, nutrient concentration, and water status, all of which change during parameterization. The value of RUE thus incorporates the integrated effects of these changes at the site at which parameterization is conducted, and therefore is often found not to be robust (Goudriaan, 1996). Furthermore, above-ground litterfall may or may not be accounted for during the parameterization of RUE, and below-ground litterfall is not accounted for, so that the relationship between RUE and NPP is not clearly defined. Consequently, a more biochemically based model of CO_2 fixation (Farquhar et al., 1980) at smaller temporal and spatial scales (usually hour and leaf) has been adopted in many ecosystem models to drive phytomass growth. This model can be parameterized from basic research conducted independently from site-specific measurements of CO_2 fixation, and can be used under a wide range of environmental conditions without reparameterization. This model simulates GPP so that if R_a is calculated from the energetics of plant C transformations (Penning de Vries, 1983), credible estimates of NPP needed for SOC modeling can be made.

2. Parameters must be capable of evaluation independently of the model in which they are used. There is frequent recourse to the evaluation of model parameters by comparing results from the model with those from an experiment under a common set of site conditions. The parameterized value becomes an artifact of model formulation at the time of comparison, and unless this value can be corroborated by independent experimentation, the algorithm in which the parameter is used should be replaced. An example of such parameterization appears in Hanson et al. (1999), where key physiological parameters controlling plant growth assumed values that varied by a factor of two or more following evaluation at different sites. However, no attempt was made to establish whether these values were consistent with independent experimental evidence, or even whether such variation was biologically realistic. A clear objective of model development should be the avoidance of algorithms that require the use of the model in their parameterization.

MODELING SPECIFIC SOIL PROCESSES

Primary Productivity

The accurate modeling of above- and below-ground NPP is required of any model used to predict the effects of land use or climate change on SOC. NPP is simulated at varying levels of complexity in different models:

1. From prescribed C inputs, estimated at monthly time steps from annual above-ground phytomass (e.g., Coleman and Jenkinson, 1996). Because land use practices affect NPP and its distribution between above- and below-ground components, results from these models cannot be independent of site-specific assumptions about land use effects on NPP. The use of prescribed inputs compromises the independence of model testing where such inputs have not been accurately measured.

2. From simple associations with soil water and precipitation adjusted for mineral nutrients, typically at monthly time steps (e.g., Parton and Rasmussen, 1994). These associations require site-specific calibration against seasonal data for plant growth,

although once calibrated they can be quite portable within zones of similar climate and soil types (Smith et al., 1997). However, it is important that such calibration not compromise the independence of model testing. There is a need for improved methods to estimate below-ground NPP from calibrated plant growth in these models.

3. From radiation use efficiency modified by temperature, humidity, and atmospheric CO_2 concentration (C_a), usually at daily time steps (e.g., Kiniry and Bockholt, 1998). Values of radiation use efficiency can be derived from basic energetics of CO_2 fixation, but are usually derived from seasonal data for plant growth so that different values are used in different models. Again, the estimation of NPP from calibrated plant growth remains a problem for RUE models.

4. From the biochemistry of CO_2 fixation (Farquhar et al., 1980) and respiration (Penning de Vries, 1983). The basic nature of this approach confers a generality and robustness of model performance that is necessary for the simulation of diverse ecosystems. Consequently, this approach to the modeling of NPP is expanding in use.

Below-ground NPP is an important determinant of SOC because, under some conditions, as much as 40% of plant C can be transferred to roots. Below-ground NPP in many ecosystem models is a prescribed fraction of above-ground NPP that may be constant (e.g., Parton and Rasmussen, 1994) or phenology dependent (Hansen et al., 1990). The spatial distribution of below-ground NPP is sometimes modeled from a prescribed logarithmic function of soil depth (e.g., Ritchie et al., 1988). However, below-ground NPP is a larger fraction of above-ground NPP and its spatial distribution changes in perennial vs. annual plant species and in water or nutrient limited soil conditions. Deeper, more prolific root systems contribute to the higher rates of C accumulation frequently measured under forages vs. cereals (Bremer et al., 1994; Campbell et al., 1991). Therefore, changes in root:shoot C allocation caused by growth habit or soil conditions need to be simulated in eco-system models used to estimate changes in SOC (Parton and Rasmussen, 1994). One approach to modeling root:shoot C allocation implements the functional equi-librium hypothesis of Thornley (1995) in which transport of C and N occurs along concentration gradients generated within the plant by C fixation and N uptake vs. C and N consumption in roots and shoots. This hypothesis allows root vs. shoot growth to adapt to changing environmental conditions and can be parameterized independently of site-specific data for root growth (Grant, 1998; Grant et al., 2001b). However, the extent to which below-ground NPP in models can be tested is strongly limited by a lack of well-constrained data (Smith et al., 1997).

Decomposition and Respiration

Many ecosystem models share some common approaches to the simulation of R_h (Paustian, 1994):

1. First-order kinetics, based on the assumption that metabolic demand of the soil biomass exceeds substrate supply. However, this assumption becomes invalid when substrate supply exceeds biomass metabolic demand, which is most readily apparent

when soils are amended with a C substrate under suboptimal temperatures or water contents (e.g., Sørensen, 1981; Stott et al., 1986).

2. Discrete soil and litter C fractions with differing rates of decomposition, usually based on age and chemical composition. Although these fractions can be based on measurable components of C in some models, in others they are artifacts of model formulation. Values for rate constants used with these fractions depend on model-specific definitions of the fractions (e.g., "labile," "stable," "inert"), and are there-fore model dependent (Paustian et al., 1997). The allocation of total soil C to these fractions may be arbitrary, which poses a problem for models that use first-order kinetics because this allocation directly affects C oxidation rates according to the first-order rate constants. One way to address this problem is to allocate larger portions of total SOC to inactive fractions in soils with higher SOC contents. However, this must be accomplished without reference to test data from the site at which model testing is to be conducted.

3. Coupled transformations of C and N based on observed C:N stoichiometry, so that N can affect C cycling. However, P and S also exert important constraints on C and N transformations (Gijsman et al., 1996; McGill and Cole, 1981) that are often overlooked or poorly represented in ecosystem models.

In a review of current ecosystem models, McGill (1996a) observed that their future development would benefit from a more mechanistic treatment of soil organisms. Such a treatment is necessary because C and N transformation rates may depart from the first-order kinetics commonly assumed in many of these models. An alternative model to first-order kinetics is one in which microbial activity is explicitly represented as the agent of C and N transformations using Monod kinetics (e.g., McGill et al., 1981; Smith, 1982). The relationship between amounts and transformation rates of soil C in this model changes from first order at low soil C to zero order at high soil C. Such a change is consistent with observations that soil respiration rates do not increase in proportion to SOC concentrations (Rutherford and Juma, 1989). C oxidation rates in microorganism-based models are therefore less sensitive to the initial allocation of soil C among different kinetic fractions than are those in first-order models. Instead, C oxidation rates are more determined by microbial biomass and specific activity as affected by access to organic C and nutrients.

A further development in microorganism-based models is their parameterization from the known energetics and kinetics of the C and N oxidation-reduction reactions conducted by different microbial populations (McGill, 1996b). For example, a model of methanogenesis has been parameterized from the energetics and kinetics of fermentation, methanogenesis (acetotrophic and hydrogenotrophic), and microbial growth, based entirely on chemostat studies, and tested against methane emissions from soil cores (Grant, 1998). With no changes, this model gave methane emissions from natural wetlands that were comparable to those measured with flux towers and surface chambers (Grant et al., in preparation). The extension of this approach to the parameterization of other SOC transformations would reduce dependence on temporally and spatially aggregated, site-specific data for changes in soil C and N that are currently used in the parameterization of first-order models. The adoption of this approach will require closer collaboration between microbiologists and eco-system modelers than has occurred in the past.

Humification

The greater stabilization of C in soils with higher clay content was clearly demonstrated by Sørenson (1981). Consequently, soil clay content is increasingly used in ecosystem models to slow organic C decomposition and to allocate C decomposition products among compartments of differing decomposition rates (McGill, 1996a). Verberne et al. (1990) partitioned the products of microbial decomposition into protected vs. non-protected organic fractions using coefficients ranging from 0.3 for sandy soils to 0.7 for clayey soils. Van Veen and Kuikman (1990) and Whitmore et al. (1991) proposed that efficiency of substrate utilization increased with clay content. However, studies of clay effects on substrate utilization have not substantiated this hypothesis (Filip, 1979; van Veen et al., 1985). Hansen et al. (1990) and Li et al. (1992) proposed that first-order decomposition rates of organic substrates were reduced by clay content. However, this hypothesis does not explain the increase in labeled C recovered as amino acids (Sørenson, 1981), or as microbial biomass (van Veen et al., 1985), from soils with higher clay content. It appears more likely that microbial products and metabolites (Stevenson and Elliott, 1989) and products of lignin degradation (Shulten and Schnitzer, 1997) are stabilized by clay surfaces. Parton et al. (1987) thus use residue lignin content to allocate decomposition products to slow vs. active pools. Grant et al. (2001b) coupled some of the hydrolysis products of lignin with those of protein and carbohydrate according to the stoichiometry proposed by Shulten and Schnitzer (1997) and allocated the resulting compound to a particulate organic matter complex according to soil clay content. Rates of particulate organic matter formation were thus a function of residue lignin and soil clay contents and of heterotrophic microbial activity. These rates contributed to long-term changes in soil C modeled under different land management practices that were corroborated by field measurements. The parameterization of functions for humification at appropriate spatial and temporal scales remains a problem.

Nitrification

Nitrification is an important process in the control of mineral N transformation between the less mobile NH_4^+ and more mobile NO_3^- forms. The accurate representation of nitrification in ecosystem models is therefore necessary for the accurate simulation of N loss by leaching and denitrification. Nitrification is currently receiving more attention in ecosystem models because it is an important cause of N_2O emissions following fertilization. However, nitrification is influenced by several environmental factors, including substrate (NH_4^+ and CO_2) concentration, aeration, temperature, and pH. In simpler models, nitrification rates have been simulated as zero-order (Addiscott, 1983; Flowers and O'Callaghan, 1983; Sabey et al., 1969) and as first-order (Baldioli et al., 1994; Gilmour, 1984; McLaren, 1969; Parton et al., 1996) functions of NH_4^+ based on field studies of NO_3^- formation rates. In more detailed models, first-order functions based on Michaelis-Menten kinetics are coupled to nitrifier biomass and specific activity (Ardakani et al., 1974; McGill et al., 1981; McInnes and Fillery, 1989; van Veen and Frissel, 1981). In all models, these functions are modified by the product of dimensionless functions for temperature, pH, and water-filled pore space, the last of which may be texture dependent (Parton et al., 1996). In all current models, CO_2 is assumed to be non-limiting, although sensitivity of nitrification to CO_2 concentration has been demonstrated (Kinsbursky and Saltzman, 1990). The effect of aeration on nitrification is usually modeled from linear functions of water content (Clay et al., 1985; Gilmour, 1984) or potential (Sabey et al., 1969), although it is a nonlinear function of O_2 concentration (Amer and Bartholomew, 1951). The sensitivity of nitrification to temperature has been modeled from linear (Clay et al., 1985) or Arrhenius (Gilmour, 1984) functions, the parameters of which may be influenced by climate (Malhi and McGill, 1982). The sensitivity of nitrification to pH has been modeled from linear functions of pH (Gilmour, 1984; McInnes and Fillery, 1989), although it may arise from pH effects on substrate NH_3 concentration (Suzuki et al., 1974).

There is a wide range of published values for parameters used in the above relationships, suggesting complex interactions among the environmental factors influencing nitrification. Such interactions are currently modeled by multiplying constraints imposed by individual factors (Clay et al., 1985; Gilmour, 1984) upon rates of product formation, although the validity of this approach has not been clearly established. Furthermore, it is difficult to parameterize functions for nitrification independently of those for related functions that also affect NH_4^+ concentrations in soils (e.g., mineralization, volatilization, adsorption). An alternative approach to the modeling of nitrification was proposed by Grant (1994b) based on the energetics of NH_3 oxidation and microbial growth and driven by microbial kinetics constrained by O_2 uptake and pH effects on substrate NH_3 concentration. This approach was entirely parameterized from well-constrained chemostat studies conducted independently of field measurements and avoids assumptions about how environmental factors interact with nitrification. This model also avoids the use of site-specific constraints imposed on the nitrification rates of some other models (e.g., Clay et al., 1985; Parton et al., 1996) that may limit their application.

Denitrification

Denitrification is an important cause of N loss by gaseous emission, notably of N_2O, which is an important greenhouse gas. Denitrification has been represented in models by first-order functions of NO_3^- and available carbon that are multiplied by the product of dimensionless functions of soil temperature and water content (Baldioli et al., 1994; Li et al., 1992; Parton et al., 1996; Ritchie et al., 1988; Sharpley and Williams, 1990; Stockle and Nelson, 1995; Xu et al., 1998). The functions of soil water content may have arbitrary constraints that can cause underestimates of the frequency of denitrification events (Marchetti et al., 1997). It is apparent that short-term temporal variation in the emission of N_2O is too large to be explained by simple functions of soil water content, temperature, or N and C substrates (Blackmer et al., 1982; Christensen, 1983; Flessa et al., 1995; Robertson, 1994). This variation suggests that denitrification is determined by complex interactions among N transformation and transport processes and the environmental conditions under which they function. Such interactions need to be more fully represented in these models if they are to simulate N_2O emissions reliably. However, it is difficult to parameterize functions for denitrification independently of those for related functions that also affect NO_3^- concentrations in soils (e.g., nitrification, leaching, immobilization). Therefore the kinetics of the NH_4^+ oxidation and the NO_3^- reduction pathways, which have been simulated under controlled laboratory conditions (Grant, 1994b; Leffelaar and Wessel, 1988; McConnaughey and Bouldin, 1985), should be linked to simulations of water, heat (including freezing and thawing), and O_2 transfer through heterogeneous soil layers if they are to be used to estimate denitrification and N_2O emissions under field conditions. This linkage is especially important in the estimation of emissions during spring thaw when transfer processes are affected by phase changes of water. The accurate simulation of water transfer from soil hydraulic functions is a key requirement for the modeling of N_2O emissions from denitrification (Frolking et al., 1998).

The dissimilatory reduction of NO_3^- is understood to proceed through a sequence of reaction products that include NO_2^-, NO^-, N_2O and N_2, the last three of which may be emitted as gases. The accurate partitioning of emissions among these gases remains a challenge to ecosystem models (Frolking et al., 1998). The reduction of NO_3^- is suppressed by O_2, and reduction of N_2O is suppressed by NO_3^- (Blackmer and Bremner, 1978; Firestone et al., 1979; Weier et al., 1993). These reaction kinetics suggest a declining preference for electron acceptors of $O_2 > NO_3^- > NO_2^- > N_2O$ by the facultative anaerobes responsible for denitrification (Cady and Bartholomew, 1961). A direct inhibition by NO_3^- on N_2O reduction has been used in some models to simulate denitrification reaction sequences (Arah and Vinten, 1996; McConnaughey and Bouldin, 1985), although such inhibition has not been observed experimentally (Betlach, 1979). This preference scheme for electron acceptors has been coupled to heat, water, and O_2 transport algorithms and to the energetics of NO_x oxidation-reduction reactions and microbial growth (Grant, 1991; 1995). The combined scheme has been used to simulate hourly and daily changes in denitrification products and their ratios under laboratory conditions (Grant et al., 1993c), and in N_2O emissions under field conditions (Grant et al., 1993d; Grant and Pattey, 1999) without imposing site-specific maximum limits to process rates required in some first-order models (e.g., Baldioli et al., 1994; Shaffer et al., 1991).

Volatilization and Deposition of Ammonia

Volatilization of NH_3 can cause important losses of N from heavily fertilized or manured ecosystems, although it is rarely represented in ecosystem models. Fleisher et al. (1987) simulated volatilization of soil NH_3 from ammonia partial pressure determined by ammonia solubility and ammonia-ammonium solute equilibrium coupled with ammonium-calcium equilibrium by Gapon exchange. The ammonia-ammonium solute equilibrium was also used by Li et al. (1992) to simulate soil NH_3 volatilization. Solution and exchange equilibria are the most appropriate processes for simulating soil pH, water content, and cation exchange effects on soil NH_3 volatilization. Significant volatilization of NH_3 can also occur through plant stomates in heavily fertilized ecosystems, although such volatilization is not currently included in ecosystem models.

Deposition of atmospheric NH_3 is not currently included in most ecosystem models although it can account for as much as 10% of total crop N requirements under normal atmospheric concentrations (Hutchinson et al., 1972), and more under higher concentrations caused by local NH_3 emissions (Whitehead and Lockyer, 1987). Ammonia deposition may be of particular importance in modeling long-term NPP and C storage in N-limited ecosystems. The modeling of NH_3 exchange between plants and the atmosphere should be based on NH_3 fluxes through explicitly calculated stomatal and aerodynamic conductances regulated by an internal NH_3 compensation point derived from plant N concentrations. The absence of plant-atmosphere NH_3 exchange is an important oversight in ecosystem models.

Transport and Leaching of Nitrate

As for other ecosystem processes, there are functional and mechanistic approaches to the modeling of solute transport. In functional models, the calculation of water fluxes is based on a vertical displacement hypothesis in which soil water is displaced downward within prescribed maximum and minimum volumetric contents, with provision in some cases for mobile and immobile fractions (Addiscott and Whitmore, 1987; Whitmore and Parry, 1988). Convective solute transport is thus calculated as the vertical water flux divided by the maximum soil water content. Rose et al. (1982) developed this approach further by accounting for diffusion and dispersion about the convective solute front.

Mechanistic models use the Richards equation for micropore coupled with convective-dispersive solute transport of solutes (Hansen et al., 1990; Hutson and Wagenet, 1992), the concentrations of which are also determined by adsorption-desorption, precipitation-dissolution, and mineralization-immobilization reactions (Grant and Heaney, 1997; Hutson and Wagenet, 1992). This approach has been further developed by adding gravity-driven macropore flow that functions in parallel with micropore flow (Hanson et al., 1999).

Both functional and mechanistic approaches have been adapted from their basic formulations to represent nonsteady-state water flows and vertical inhomogeneity in bulk soil hydraulic properties. However, neither approach has been fully adapted to represent horizontal inhomogeneity, which can greatly affect vertical convective transfer at small spatial scales (Addiscott and Wagenet, 1985). Further development of solute transport modeling will likely require extension of one-dimensional models to two and three dimensions in which a range of soil hydraulic properties are stochastically allocated to different cells at an appropriate spatial scale. Such models could explicitly represent spatial variation in solute transport that could be tested against experimental measurements.

Methanogenesis and Methanotrophy

Methane is an important greenhouse gas that is emitted from wetlands and rice paddies, so the accurate simulation of CH_4 emissions is a necessary feature of ecosystem models used in climate change studies. Frolking and Crill (1994) regressed surface CH_4 fluxes measured over a peat soil on soil temperature, depth to water table, and precipitation. This regression was used with a peat soil climate model to simulate seasonal changes in daily CH_4 fluxes. CH_4 oxidation was not explicitly represented in their model. However, parameterization of such regression models is conducted on a seasonal time scale using the model itself, and so site- and model-specific values result. In a more mechanistic approach, Cao et al. (1995) simulated daily CH_4 fluxes over rice paddies from simulated rates of organic matter decomposition and rhizodeposition, and from soil redox potential, pH, temperature, surface water depth, and mineral fertilizer. Methane oxidation was simulated from CH_4 production using a ratio that increased with rice dry matter on the assumption that soil oxygenation increased with root growth. Carbon transformations in this model were first-order functions of substrate mass, with no explicit simulation of oxidation kinetics.

A more biologically based approach to the modeling of CH_4 emission has been used to study methanogenesis in bioreactors (e.g., Mosey, 1983; Shea et al., 1968). This approach is based on the stoichiometries, kinetics, and yields of the microbial populations involved in methanogenesis. This approach, if adapted to soils, could lead to a model of CH_4 emissions that would be applicable under a wide range of soil and climate conditions because values of model parameters could be derived from well-constrained bioreactor tests conducted independently of the larger eco-system model in which the parameters are to be used. The adaptation of a biologically based approach to the simulation of CH_4 transformations could be based on the hypothesis that methanogenesis can be represented from the interrelated activities of four microbial communities defined by functional type: (1) anaerobic fermenters and H_2-producing acetogens, (2) acetotrophic methanogens, (3) hydrogenotrophic methanogens, and (4) autotrophic methanotrophs (Grant, 1998; 1999). Each of these communities, in fact, consists of several interacting populations that are aggregated in the model to maintain simplicity. The validity of this hypothesis has been tested against CH_4 emissions reported by Tsutsuki and Ponnamperuma (1987) from a soil amended with different organic materials and incubated under anaerobic conditions at 20 and 35°C. These emissions clearly departed from first-order kinetics, indicating the need for a higher-order model.

Methane oxidation is carried out by obligate and facultative methanotrophic bacteria that use C_1 compounds as carbon and energy sources. These bacteria are aerobic, so that oxidation is strongly controlled by soil water content through its control of O_2 transfer (Adamsen and King, 1993; Lessard et al., 1994). Methane oxidation is not strongly controlled by soil temperature (Dunfield et al., 1993; King and Adamsen, 1992; Lessard et al., 1994), suggesting that it is constrained by diffusion of CH_4 to localized oxidation sites. A biologically based model of CH_4 oxidation should therefore represent the transformation of CH_4 into organic C and CO_2 as controlled by the kinetics of methanotrophy and by the transport of CH_4 and oxygen. Such a model was coupled to other models of C transformation, including methanogenesis (Grant, 1998), C oxidation (Grant and Rochette, 1994), microbial growth (Grant et al., 1993a,b,c,d), and soil gas transfer (Grant, 1993; 1999). The model was used to test the hypothesis that the sensitivity of CH_4 oxidation to soil temperature and water content is determined by diffusion constraints to the transfer of CH_4 and O_2 through soil. This sensitivity was tested with data for CH_4 oxidation rates measured over an oxic podzol at different temperatures and water contents by Adamsen and King (1993) and King and Adamsen (1992).

Plant Nutrient Uptake

The complex behavior of nutrients in soil-root systems has led to the use of simulation modeling as a means of synthesizing existing knowledge of, and developing a predictive capability for, nutrient uptake by root systems. In simpler models of soil-plant systems (e.g., Hanson et al., 1999; Jones et al., 1984; Parton et al., 1988), nutrients are allocated to different organic and inorganic fractions among which transfer is calculated from first-order rate constants. However, the inorganic fractions are defined by their kinetic, rather than chemical, behavior, and their sizes and rate constants must therefore be estimated empirically for each soil type. Root uptake in these models is calculated from plant demand (the product of plant nutrient concentrations and growth rates), constrained by soil labile nutrient concentrations and water contents, without explicitly representing the transport and uptake processes by which uptake occurs. Nutrient uptake algorithms in these models cannot therefore be parameterized independently of the ecosystem model in which they are to be used. The reliability of these models may therefore be confined to the sites at which they are parameterized because the effects of soil environment on nutrient transport and root activity are incorporated into their parameters and state variables, rather than treated explicitly.

In the mechanistic model of Barber and Cushman (1981), nutrient transport to root surfaces is controlled by soil buffer capacity, defined as the ratio of changes in solid vs. solution forms of the nutrient (Barber and Silberbush, 1984). However, this ratio is soil and site specific (Morel et al., 1994), and is influenced by biological activity, pH, ionic strength, mineral composition, and the kinetics of nutrient transformations. This model therefore needs to be combined with one in which soluble nutrient concentrations are explicitly calculated from reactions with solid phases. In the model of Barber and Cushman (1981), nutrient uptake at the root surface is determined by root length and surface area, for which measured or estimated values must be provided, and uptake by mycorrhizae is not considered. Some parameters in this model, such as maximum root uptake rates, may require site-specific adjustment when used under field conditions (Lu and Miller, 1994), although such adjustment may only be necessary when mycorrhizae are ignored. This model therefore needs to be combined with one for root and mycorrhizal growth for calculation (e.g., Grant and Robertson, 1997). However, the model of Barber and Cushman (1981) is capable of independent parameterization from well-constrained tests conducted at small temporal and spatial scales, and thus is well-suited for use in larger ecosystem models.

THE PROCESS OF MODEL TESTING

Well vs. Poorly Constrained Tests of Model Algorithms

Concern is frequently expressed about the over-parameterization of so-called "complex" models when simple models with fewer parameters can simulate a given data set with equal, and sometimes better, accuracy (e.g., Schulz et al., 1999). However, the testing of model hypotheses enabled by these data sets is often poorly constrained and therefore inconclusive because the data sets consist of highly aggre- gated data. These data are usually the net result of many different processes and hence are often unable to distinguish among alternative model hypotheses. An inference is frequently drawn from such testing that simpler models are preferable because they are less likely to contain offsetting errors in their parameters that allow accurate simulation of results from inaccurate simulation of processes. A more useful inference from such testing is that the data set has failed to discriminate among alternative model hypotheses and therefore better-constrained data sets are needed that allow testing of component processes. These better-constrained data sets are typically collected at smaller temporal and spatial scales.

For example, modeled changes in SOC concentration are often tested against measurements taken at large temporal scales (e.g., every 5 to 10 years in Smith et al., 1997). However, many ecosystem models used to simulate changes in SOC are based on algorithms in which interactions among temperature, irradiance, nutrients, CO_2, and water on the processes controlling ecosystem C exchange are represented using basic biophysical principles. These interactions are known from experimentation to be highly dynamic and nonlinear, with pronounced variation at an hourly time scale and at a sub-meter (e.g., plant leaf or soil aggregate) spatial scale. Testing of modeled C exchange should therefore first be conducted for component processes at appro- priate temporal and spatial scales under defined, orthogonal changes in environmen- tal conditions. Such tests are well-constrained because test results arise from a limited number of component processes (e.g., leaf C fixation, soil C respiration) with unique responses to independently controlled changes in boundary conditions (e.g., tem- perature, water, irradiance, CO_2). In the case of leaf C fixation, such tests often use chamber experiments in which leaf CO_2 fixation is measured under controlled, independent changes in irradiance, temperature, and CO_2 concentration. Experimen- tal results for CO_2 fixation rate are compared with model results generated under the same combinations of irradiance, temperature, and CO_2 concentration to test model sensitivity to changes in atmospheric boundary conditions (e.g., Grant, 1989; 1992). These tests can be conducted on intact leaves *in situ* (e.g., Grant et al., 1999) for closer comparison with other model tests at larger temporal and spatial scales (Figure 19.1). Corresponding tests for soil respiration often use results from labo- ratory experiments to ensure controlled inputs of C and N to homogeneous soils under controlled changes in temperature or water content. Experimental results for CO_2 emission and C stabilization are compared with model results over time periods that range from hours to months. Examples of such tests are given in Grant et al. (1993a,b) and in Grant and Rochette (1994).

Once confidence in model hypotheses for C exchange at the process level has been established from well-constrained tests conducted under controlled laboratory conditions, the spatial scale of testing can then be extended from that of the com- ponent processes to that of the integrated ecosystem. Such extension can be best achieved through comparison of modeled ecosystem C and energy exchange with measurements of ecosystem C and energy fluxes using eddy covariance techniques at flux towers. These measurements are typically aggregated to half-hourly or hourly

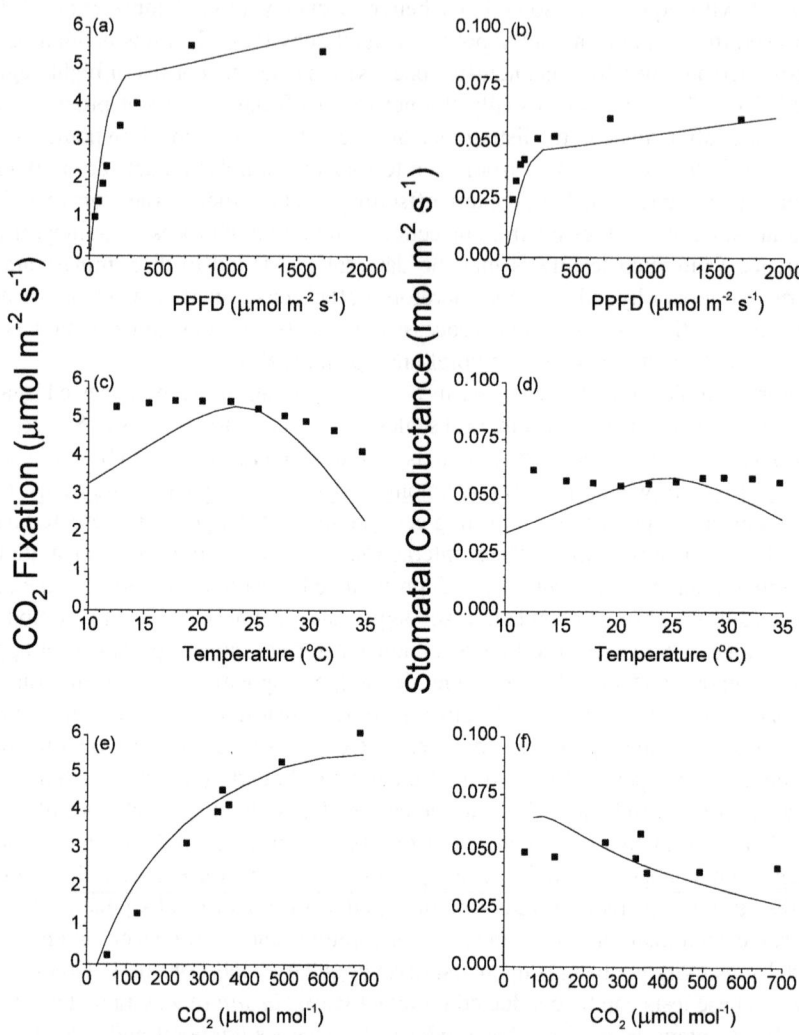

Figure 19.1   Simulated (lines) and measured (symbols) responses of CO_2 fixation and stomatal conductance by needles in the upper part of a 115-year-old spruce canopy in the BOREAS southern study area to changes in (a, b) irradiance (C_a = 355 µmol mol^{-1}, T_a = 15°C, H_a = 1.3 kPa); (c, d) air temperature (C_a = 345 µmol mol^{-1}, I = 1800 µmol m^{-2} s^{-1}, H_a increased with temperature from 1.1 to 2.0 kPa), and (e, f) CO_2 concentration (I = 1000 µmol m^{-2} s^{-1}, T_a = 11°C, H_a = 1.0 kPa.). Measured data from Berry et al. (1998) using portable controlled chambers on intact needles in the black spruce canopy. (From Grant, R.F., P.G. Jarvis, J.M. Massheder, S.E. Hale, J.B. Moncrieff, M. Rayment, S.L. Scott, and J.A. Berry. 2001a. Controls on carbon and energy exchange by a black spruce — moss ecosystem: testing the mathematical model *ecosys* with data from the BOREAS experiment. *Global Biogeochem. Cycles* (in press). With permission.)

averages to reduce the effects of short-term fluctuations. Comparisons of modeled vs. measured fluxes are less well-constrained than those conducted under controlled conditions at smaller spatial scales because test data are the net product of several interacting processes (e.g., GPP vs. R_a and R_h), each of which responds differently to changes in boundary conditions. These tests are also less well-constrained because changes in individual boundary conditions are correlated rather than independent (e.g., diurnal changes in temperature follow those of irradiance). These tests are therefore less able to discriminate among alternative model hypotheses (e.g., the accuracy of alternative hypotheses for the sensitivity of CO_2 fixation to irradiance vs. temperature may not be distinguished). Depending on the height of the tower at which they are measured, eddy covariance fluxes also integrate spatially over sometimes diverse landscapes, so that the association between fluxes and specific landscape patches may not be clear (e.g., Amthor et al., 2001). The accuracy of eddy covariance techniques is reduced by low wind speeds, and in some cases these techniques may underestimate evapotranspiration, so that some screening of eddy covariance data used in model tests is necessary. Nonetheless, model tests against eddy covariance data are an important link between model tests at smaller and larger scales, both temporal and spatial, used in the evaluation of model performance. An example of a model comparison against eddy covariance fluxes is given in Figure 19.2. This comparison is a direct extension of that in Figure 19.1.

The degree of constraint in model tests against eddy flux data can be improved by supplementing tower flux measurements using eddy covariance with soil flux measurements using automated surface enclosures (e.g., Goulden and Crill, 1997), thereby resolving ecosystem fluxes into above- and below-ground components. Amthor et al. (2001) presented results from comparative tests of several models against concurrently measured tower and surface fluxes. Further resolution in modeled vs. measured flux testing may be achieved in the future by using ^{13}C isotopic signatures in eddy covariance fluxes to separate GPP from R_a and R_h (Flanagan and Ehleringer, 1998). By testing against diurnal changes in the magnitude and amplitude of ecosystem C fluxes, the sensitivity of modeled GPP, R_a, R_h, and hence NEP, to short-term changes in surface boundary conditions (irradiance, temperature, humidity, wind speed, precipitation) and longer term changes in soil conditions (water and nutrient status) can be evaluated. Hourly tests against these fluxes are usually confined to periods of 1 to 2 weeks over the course of the year (e.g., Figure 19.2), although they may extend throughout the year if flux data are available. These tests are an important current and future direction in ecosystem modeling.

Tests against tower and surface flux measurements of C and energy exchange may then be extended to longer time periods (e.g., daily to yearly) by using temporally and spatially aggregated flux data (e.g., daily sums of C and water exchange at the ecosystem scale). Amthor et al. (2001) present a comparative test of several models against net ecosystem C exchange aggregated to 4-day means over 3 years. An example of a model test vs. yearly aggregates of ecosystem C exchange estimated from eddy covariance and other techniques is given in Table 19.1. This test is a direct extension of the model test vs. hourly net ecosystem C exchange in Figure 19.2. Data at these temporal scales aggregate ecosystem responses to boundary conditions that may change substantially within aggregation periods. Model tests at these scales can be used to corroborate those at an hourly scale, but are not substitutes for them because they are less well-constrained. Models that function only at daily or monthly temporal scales are driven by temporally aggregated weather data (e.g., daily or monthly sums

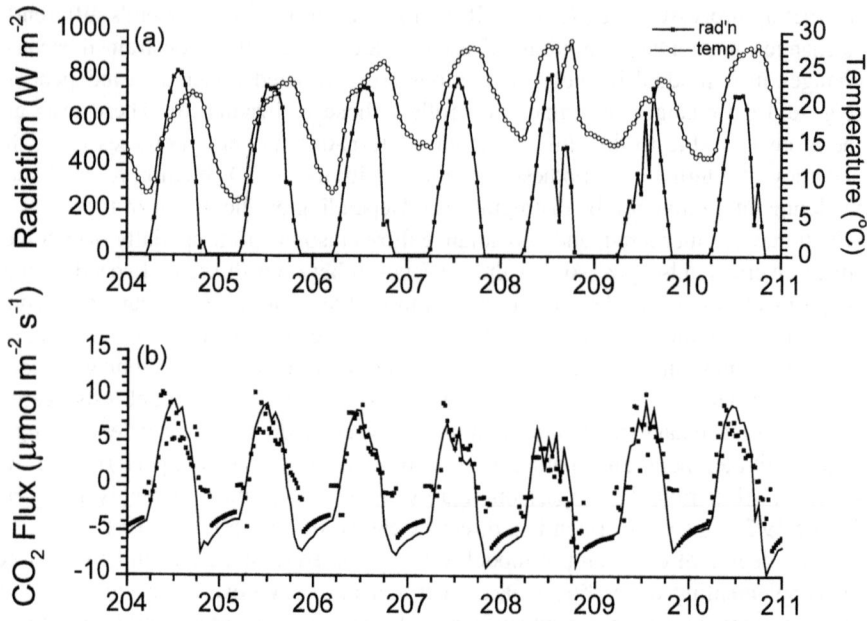

Figure 19.2   (a) Radiation and air temperature, and (b) CO_2 fluxes simulated (lines) and
measured (symbols) over a 115-year-old spruce canopy in the BOREAS southern
study area from July 24 (DOY 205) to July 30 (DOY 211) 1994. Model results
for canopy CO_2 fixation are aggregated spatially from those for needle CO_2 fixation
demonstrated in Figure 19.1. Measured data from eddy covariance measurements of Massheder et al. (1998). Note warmer temperatures during DOY 207
to 209 raise $R_a + R_h$ (larger nighttime effluxes represented as negative values)
without raising GPP (differences between daytime influxes and nighttime effluxes
similar to those under cooler temperatures). Warmer temperatures, especially
with low radiation, can thus cause this ecosystem to change from an aggrading
to a degrading state. (From Grant, R.F., P.G. Jarvis, J.M. Massheder, S.E. Hale,
J.B. Moncrieff, M. Rayment, S.L. Scott, and J.A. Berry. 2001a. Controls on carbon
and energy exchange by a black spruce — moss ecosystem: testing the mathematical model *ecosys* with data from the BOREAS experiment. *Global Biogeochem. Cycles* (in press). With permission.)

or averages of irradiance, temperature, humidity, wind speed, and precipitation).
Results from these models cannot be tested directly against C and energy fluxes
measured at half-hourly to hourly time scales, but rather against indices of these fluxes
that have been aggregated to the same spatial and temporal scales as those at which
the models function. Such models are therefore not amenable to better constrained tests.

Model testing vs. C fluxes described above may then be further extended temporally by comparing modeled vs. measured changes in SOC at a decadal time scale
(e.g., Grant et al., 2001b; Jenkinson et al., 1987; Parton and Rasmusson, 1994). An
example of such a test vs. wood C accumulation is given in Figure 19.3, which is a
direct extension of that vs. annual C exchange in Table 19.1. Such tests are of great
ecological interest and are important to model development because they impose
constraints to the long-term performance of models tested at smaller time scales.
However, these tests are very poorly constrained because the test data are of low

Table 19.1   **Annual Carbon Balance of a Black Spruce — Moss Forest Simulated by the Ecosystem Simulation Model *ecosys* (described in Chapter 6), and Estimated from Flux Measurements and Allometric Techniques at the BOREAS Southern Old Black Spruce (Modeled data is aggregated temporally from net C exchange in Figure 19.2.)**

		Simulated	Estimated
		(g C/m²)	
Spruce	Gross Fixation	660	
	Respiration[a]	429	
	Net Primary Productivity (NPP)	231	
	Senescence	128	53[d] + 120[g,] 91,54[e]
	Net growth: wood	64	80[d]
	foliage	16	8[d]
	roots	3	
Moss	Gross fixation	288	
	Respiration[a]	184	
	Net Primary Productivity (NPP)	104	
	Senescence	103	37[e], 50-150[h]
	Net growth	0	12[d]
Soil	Respiration[b,c]	294	368,283[e]
Ecosystem	Gross Fixation	948	1090[f]
	Total respiration: autotrophic	613	785[f]
	heterotrophic	294	
	Net Primary Productivity (NPP)	335	266[d,g],307[f]
	Net Ecosystem Productivity	41	

[a] Includes root respiration·
[b] Excludes root respiration.
[c] Includes CO_2-C and CH_4-C.
[d] Gower et al. (1997) Litterfall above-ground only.
[e] Nakane et al. (1997) Well and poorly drained sites. Litterfall, above-ground only. Soil respiration includes root respiration.
[f] Ryan et al. (1997).
[g] Steele et al. (1997) from below-ground NPP.
[h] Harden et al. (1997).

precision, and may be explained by a wide range of alternative model hypotheses, not all of which may be valid. It is therefore imperative that such tests be supported by better constrained tests conducted at higher levels of temporal and spatial resolution before ecosystem models are used for predicting long-term changes in SOC. An example of combined model testing that extends from temporal and spatial scales of the leaf and the hour to the ecosystem and the century is given in Grant et al. (1999).

Quantitative Tests of Model Performance

Objective criteria are needed by which to assess whether the extent of disagreement between modeled and measured results warrants rejection of model hypotheses. A good description of such criteria is given in Smith et al. (1997). For replicated test data, an F test can be used to compare the mean square for differences between mean measured and modeled values with the mean square for error in the measured values.

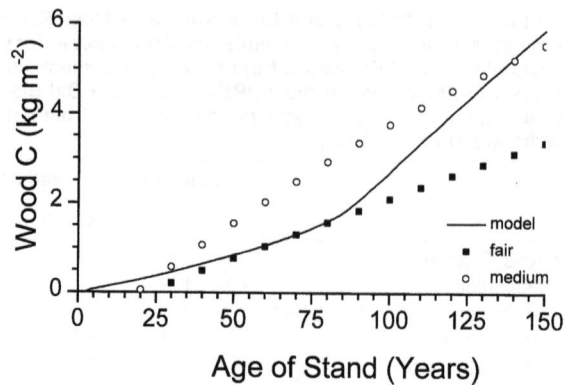

Figure 19.3   Spruce wood C simulated in a black spruce — moss stand in the BOREAS southern study area during 160 years (line), and spruce wood C at fair and medium sites calculated from wood volume measurements by the Alberta Forest Service (1985) (symbols). Model results for wood C are aggregated temporally from net C exchange in Figure 19.2 and Table 19.1. Small gains in soil C were also simulated. (From Grant, R.F., P.G. Jarvis, J.M. Massheder, S.E. Hale, J.B. Moncrieff, M. Rayment, S.L. Scott, and J.A. Berry. 2001a. Controls on carbon and energy exchange by a black spruce — moss ecosystem: testing the mathematical model *ecosys* with data from the BOREAS experiment. *Global Biogeochem. Cycles* (in press). With permission.)

F ratios less than the critical value at 5% indicate that the model hypotheses cannot be confidently rejected. For unreplicated test data, the root mean square for differences between mean measured and modeled values can be used to evaluate the comparative accuracy of different models (e.g., Amthor et al., 2001). If an estimate of the standard error of the measured values can be made, then it may be used to establish the 95% confidence interval of the measured values. If this interval is larger than the root mean square for differences between mean measured and modeled values, then the model hypothesis cannot be confidently rejected. If no estimate of the standard error can be made, then the coefficient of determination can be used to calculate the fraction of the variance in the measured values that is explained by covariance between measured and modeled values. The significance of the coefficient of determination may be established by comparison with its standard error. Evidence of bias in modeled results can be assessed by calculating the relative difference between modeled and measured values from the average difference between modeled and measured values relative to the measured values. Values larger or smaller than zero indicate persistent overpredictions or underpredictions, respectively, by the model. Evidence of model bias can be further assessed by regressing modeled values on measured values and determining whether the regression coefficient and intercept deviate significantly from 1 and 0, respectively.

CURRENT TRENDS AND FUTURE DIRECTIONS IN MODELING ECOSYSTEM C AND N TRANSFORMATIONS

The degree of constraint with which ecosystem models could be tested in the past was limited by the low temporal and spatial resolution at which data could be recorded. Recent developments in field data acquisition are rapidly removing these constraints and allowing model testing to be conducted at more appropriate temporal and spatial scales. The measurement of ecosystem C exchange using the eddy covariance technique (Baldocchi et al., 1988) is a striking example of these developments. This technique has now been used for several years to measure net C exchange by terrestrial ecosystems (e.g., Wofsy et al., 1993) and has now achieved considerable refinement. Although measurement uncertainties persist (Baldocchi et al., 1996), eddy covariance data sets are becoming key tools for testing the response of modeled C and energy exchange to short term changes in weather under diverse site conditions (e.g., Sellers et al., 1997b). Improved sensors such as tunable diode lasers are being used to adapt eddy covariance techniques to the measurement of other gas fluxes such as N_2O and CH_4. This adaptation will greatly improve the constraints with which modeled rates of denitrification and methanogenesis can be tested (e.g., Grant and Pattey, 1999). Fluxes measured at eddy covariance towers may in the future be resolved into those of atmospheric vs. terrestrial origin through isotopic analysis of eddy fluxes (Bowling et al., 1999; Flanagan and Ehleringer, 1998). The resulting ability to resolve model testing into components of C fixation vs. respiration will be an important advance in ecosystem modeling.

The need for more site diversity in model testing is driving the ongoing expansion of tower flux networks such as EUROFLUX and AmeriFLUX, in which this and related micrometeorological techniques are being used to monitor mass and energy exchange between terrestrial ecosystems and the atmosphere. Associated biogeochemical data (mostly within-ecosystem distributions of C, N, and water) collected at flux tower sites provide further constraint to model testing. The combined collection of flux and biogeochemical data requires interdisciplinary teams of scientists to work within well-coordinated projects such as the Boreal Ecosystem Atmosphere Study (BOREAS) in northern Canada or the Large-scale Biosphere Atmosphere Study (LBA) in Brazil. These projects are an important development in ecosystem research and vital to ecosystem model development.

Most current ecosystem models require considerable development before they are capable of being tested at the temporal scale of eddy flux measurements. These developments may be partially accomplished by reducing the time scale of the rate constants by which they are currently driven, without changing basic model hypotheses. In some cases, a more basic approach may be taken in which model hypotheses are based on unifying principles of ecosystem energetics (e.g., Odum, 1971) that link those of energy exchange, heat transfer, CO_2 fixation, autotrophic respiration, and heterotrophic respiration in the simulation of ecosystem behavior. Examples of how energetics can be used to simulate specific C transformation processes were given above. Parameterization of these hypotheses will lead to closer integration of ecosystem modeling with the basic sciences than has occurred to the present.

Constraints imposed on long-term model testing are being improved by the development of data set networks in which measurements of changes in SOC at decadal to century time scales are recorded under diverse site and management conditions (e.g., Smith et al., 1996). These data sets can be used to identify systematic biases in ecosystem models (Smith et al., 1997) that may not have been apparent from earlier tests at smaller time scales. The causes of these biases must then be identified from further model tests at smaller time scales designed to reveal these biases. Models tests of net C exchange against eddy covariance measurements at hourly time scales and against SOC measurements at century time scales will thus become complementary, each strengthening the other, in future model development.

During the past decade, there has been a growing awareness of the importance of topography in determining C and nutrient dynamics in terrestrial ecosystems. Watershed models have been used for many years to simulate water movement through complex landscapes. However, they have not been coupled with full eco-system models to simulate the impact of water movement on nutrient transfers and NPP. The extension of ecosystem models from the one-dimensional formulations common today to two- and three-dimensional formulations in the future will become an important trend in ecosystem modeling. These formulations will require the concurrent solution of all water, energy, gas, and solute fluxes across four or six boundaries around each landscape unit such that model results are independent of solution structure. Multidimensional ecosystem models will have important appli-cations in the elucidation and eventual prediction of landscape effects on ecosystem NPP and nutrient accumulation vs. loss. Such prediction will eventually make valuable contributions to nutrient management strategies in agricultural landscapes.

The testing of multidimensional models will require changes in research practices from their current emphasis on homogeneous field plots to a future emphasis on heterogeneous landscapes. These landscapes are often represented by slope transects in which changes of soil water, temperature, and nutrient concentrations, and of plant growth and nutrient uptake, are monitored at field sites designated by slope position (e.g., shoulder, upper slope, lower slope, footslope, depressional area). These measured changes will be used to test corresponding changes modeled con-currently for each slope position as affected by downslope movement of water and solutes determined by the inclination and aspect of the modeled slope. Three-dimensional ecosystem models will eventually be coupled with soil and topographic data in geographic information systems (GIS) to model soil and plant dynamics at a landscape level. Supporting these developments is the explosive growth in GIS technology used in soil, topography, and yield mapping.

Ecosystem models that include coupled soil-plant C and N dynamics are now replacing the land surface schemes (e.g., Dickenson et al., 1986; Verseghy et al., 1993) formerly used to represent mass and energy exchange between land surfaces and the atmosphere within general circulation models (GCMs). These models, included as so-called third-generation land surface schemes in GCMs (Sellers et al., 1997a), improve the simulation of terrestrial effects on atmospheric composition and energy status, and thus are considered necessary to improved modeling of global climates (Dickenson, 1995; Sellers et al., 1997a). The key function of the ecosystem models used as third-generation land surface schemes in GCMs is the accurate simulation of mass and energy exchange under diverse conditions of soil quality, land management, and climate. These models must therefore function at sub-hourly to hourly time steps and be rigorously tested against eddy covariance measurements from flux towers under diverse site conditions. The need for such testing is one of the main reasons for the expansion of flux tower networks.

The coupling of ecosystem models with GCMs will eventually enable regional and even global assessments of how hypothesized changes in atmospheric composition and associated changes in climate will impact ecosystem function, and how, in turn, these impacts will affect atmospheric composition and hence climate. When the effects of soil disturbance and land management (rotation, tillage, harvesting, fire) are included in these third-generation land surface schemes, GCMs can be used with soil and land management databases for integrated impact assessments of climate change on agriculture, forestry, and natural ecosystems. Such assessments will be advances on those currently based on macro-scale ecosystem models (e.g., VEMAP, 1995) with highly aggregated expressions of C exchange that function at low temporal resolution independently of climate (e.g., Melillo et al., 1993). The credible scaling of ecosystem models within GCMs is critically dependent on the continued development of spatially referenced databases containing soil attributes used to model water and nutrient movement (e.g., density, texture, organic C content, CEC, water retention at field capacity, and wilting point).

REFERENCES

Adamsen, A.P.S., and G.M. King. 1993. Methane consumption in temperate and subarctic forest soils: Rates, vertical zonation, and responses to water and nitrogen. *Applied Env. Microbiol.,* 59:485-490.

Addiscott, T.M. 1983. Kinetics and temperature relationships of mineralization and nitrification in Rothamsted soils with differing histories. *J. Soil Sci.,* 34:343-353.

Addiscott, T.M., and R.J. Wagenet. 1985. Concepts of solute leaching in soils: a review of modeling approaches. *J. Soil Sci.,* 36:411-424.

Addiscott, T.M., and A.P. Whitmore. 1987. Computer simulation of changes in soil mineral nitrogen and crop nitrogen during autumn, winter and spring. *J. Agric. Sci. Camb.,* 109:141-157.

Aguirrezabal, L.A.N., S. Pellerin, and F. Tardieu. 1993. Carbon nutrition, root branching and elongation: can the present state of knowledge allow a predictive approach at a whole-plant level? *Env. Exp. Bot.,* 33:121-130.

Aguirrezabal, L.A.N., and F. Tardieu. 1996. An architectural analysis of the elongation of field-grown sunflower root systems. Elements for modeling the effects of temperature and intercepted radiation. *J. Exp. Bot.,* 47:411-420.

Alberta Forest Service. 1985. *Alberta Phase 3 Forest Inventory: Yield Tables for Unmanaged Stands.* Alberta Energy and Natural Resources. Edmonton, Alberta.

Allison, F.E., S.N. Carter, and L.D. Sterling. 1960. The effect of partial pressure of oxygen on denitrification in soil. *Soil Sci. Soc. Am. Proc.,* 24:283-285.

Amer, F.M., and W.V. Bartholomew. 1951. Influence of oxygen concentration in soil air on nitrification. *Soil Sci.,* 71:215-219.

Amthor, J.S., J.M. Chen, J.S. Clein, S.E. Frolking, M.L. Goulden, R.F. Grant, J.S. Kimball, A.W. King, A.D. McGuire, N.T. Nikolov, C.S. Potter, S. Wang, and S.C. Wofsy. 2001. Boreal forest CO_2 exchange and evapotranspiration predicted by nine ecosystem process models: intermodel comparisons and relationships to field measurements. *J. Geophys. Res.* (in press).

Anghinoni, I., and S.A. Barber. 1980. Phosphorus influx and growth characteristics of corn roots as influenced by phosphorus supply. *Agron. J.,* 72:685-688.

Arah, J.R.M., and A.J.A. Vinten. 1996. Simplified models of anoxia and denitrification in aggregated and simple-structured soils. *Eur. J. Soil Sci.,* 46:507-517.

Ardakani, M.S., R.K. Schulz, and A.D. McLaren. 1974. A kinetic study of ammonium and nitrate oxidation in a soil field plot. *Proc. Soil Sci. Soc. Amer.,* 38:273-277.

Aselmann, I., and D.J. Crutzen. 1990. A global inventory of wetland distribution and seasonality, net primary productivity, and estimated methane emissions. In A.F. Bouwman (Ed.), *Soils and the Greenhouse Effect.* John Wiley, New York, 441-449.

Baldioli, M., T. Engel, B. Klöcking, E. Priesack, T. Schaaf, C. Sperr, and E. Wang. 1994. Expert-N: ein baukasten zur simulation der stickstoffdynamik in boden und pflanze. *Prototyp. Benutzerhandbuch, Lehrenheit für Ackerbau und Informatik im Planzenbau.* TU München, Friesing. 1-106.

Baldocchi, D., R. Valentini, S. Running, W. Oechel, and R. Dahlman. 1996. Strategies for measuring and modeling carbon dioxide and water vapor fluxes over terrestrial ecosystems. *Global Change Biol.,* 2:159-168.

Baldocchi, D.D., B.B. Hicks, and T.P. Meyers. 1988. Measuring biosphere-atmosphere changes of biologically related gases with micrometeorological methods. *Ecology,* 69:1331-1340.

Barber, S.A., and J.H. Cushman. 1981. Nitrogen uptake model for agronomic crops. In I.K. Iskander (Ed.). *Modeling Waste Water Renovation — Land Treatment.* Wiley Interscience, New York, 382-409.

Barber, S.A., and M. Silberbush. 1984. Plant root morphology and nutrient uptake. In D.M. Kral (Ed.). *Roots, Nutrient and Water Influx, and Plant Growth.* ASA Spec. Publ. No. 49. Amer. Soc. Agron. Madison, WI, 65-87.

Berry, J.A., J. Collatz, J. Gamon, W. Fu, and A. Fredeen. 1998. BOREAS TE-04 Gas Exchange Data from Boreal Tree Species. Available online at [http://www-eosdis.ornl.gov/] from the ORNL Distributed Active Archive Center, Oak Ridge National Laboratory, Oak Ridge, Tennessee.

Betlach, M.R. 1979. Accumulation of Intermediates During Denitrification — Kinetic Mechanisms and Regulation of Assimilatory Nitrate Uptake. Ph.D. thesis. Michigan State Univ. *Diss. Abstr.* 80:06083.

Blackmer, A.M., and J.M. Bremner. 1978. Inhibitory effect of nitrate on reduction of N_2O to N_2 by soil microorganisms. *Soil Biol. Biochem.,* 10:187-191.

Blackmer, A.M., S.G. Robbins, and J.M. Bremner. 1982. Diurnal variability in rate of emission of nitrous oxide from soils. *Soil Sci. Soc. Am. J.,* 46:937-942.

Bowling, D.R, D.D. Baldocchi, and R.K. Monson. 1999. Dynamics of isotopic exchange of carbon in a Tennessee deciduous forest. *Global Biogeochem. Cycles,* 13:903-922.

Bremer, E., H.H. Janzen, and A.M. Johnson. 1994. Sensitivity of total, light fraction and mineralizable organic matter to management practices in a Lethbridge soil. *Can. J. Soil Sci.,* 74:131-138.

Bremner, J.M., and K. Shaw. 1958. Denitrification in soils. II. Factors affecting denitrification. *J. Agric. Sci.,* 51:40-52.

Cady, F.B., and W.V. Bartholomew. 1961. Influence of low pO_2 on denitrification processes and products. *Soil Sci. Soc. Am. Proc.,* 25:362-365.

Campbell, C.A., V.O. Biederbeck, R.P. Zentner, and G.P. Lafond. 1991. Effect of crop rotations and cultural practices on soil organic matter, microbial biomass and respiration in a thin Black Chernozem. *Can. J. Soil Sci.,* 71:363-376.

Campbell, C.A., B.G. McConkey, R.P. Zentner, F.B. Dyck, F. Selles, and D. Curtin. 1995. Carbon sequestration in a Brown Chernozem as affected by tillage and rotation. *Can. J. Soil Sci.,* 75:449-458.

Cao, M., J.B. Dent, and O.W. Heal. 1995. Modeling methane emissions from rice paddies. *Global Biogeochem. Cycles,* 9:183-195.

Chaudhuri, U.N., M.B. Kirkham, and E.T. Kanemasu. 1990. Root growth of winter wheat under elevated carbon dioxide and drought. *Crop Sci.,* 30:853-857.

Christensen, S. 1983. Nitrous oxide emission from a soil under permanent grass: seasonal and diurnal fluctuations as influenced by manuring and fertilization. *Soil Biol. Biochem.,* 15:531-536.

Cicerone, R.J., and R. Oremland. 1988. Biogeochemical aspects of atmospheric methane. *Global Biogeochem. Cycles,* 8:385-397.

Clay, D.E., J.A.E. Molina, C.E. Clapp, and D.R. Linden. 1985. Nitrogen — tillage — residue — management. II. Calibration of potential rate of nitrification by model simulation. *Soil Sci. Soc. Am. J.,* 49:322-325.

Coleman, K., and D.S. Jenkinson. 1996. RothC-26.3 — A model for the turnover of carbon in soil. In D.S. Powlson, P. Smith, and J.U. Smith (Eds.). *Evaluation of Soil Organic Matter Models.* NATO ASI Series, Vol. I 38. Springer-Verlag, Berlin, 237-246.

Desjardins, R.L., and J. Keng. 1999. Nitrous oxide emissions from agricultural sources in Canada. In R.L. Desjardins, J. Keng, and K. Haugen-Kozyra (Eds.). *Proceedings of the International Workshop on Reducing Nitrous Oxide Emissions from Agroecosystems.* Agriculture and Agri-Food Canada; Alberta Agriculture, Food and Rural Development. Banff, Alberta. 3-5 March, 1999, 51-56.

de Willigen, P. 1991. Nitrogen turnover ion the soil-crop system; comparison of fourteen simulation models. *Fert. Res.,* 27:141-149.

Dickenson, R.E. 1995. Land surface processes and climate modeling. *Bull. Am. Met. Soc.,* 76:1445-1448.

Dickenson, R.E., A. Henderson-Sellers, P.J. Kennedy, and M.F. Wilson. 1986. Biosphere-Atmosphere Transfer Scheme (BATS) for the NCAR Community Climate Model. NCAR Tech. Note TN+275+STR.

Dunfield, P., R. Knowless, R. Dumont, and T.R. Moore. 1993. Methane production and consumption in temperate and subarctic peat soils: response to temperature and pH. *Soil Biol. Biochem.,* 25:321-326.

Ennik, G.C., and T.B. Hofman. 1983. Variation in the root mass of ryegrass types and its ecological consequences. *Neth. J. Agric. Sci.,* 31:325-334.

Farquhar, G.D., S. von Caemmerer, and J.A. Berry. 1980. A biochemical model of photosynthetic CO_2 assimilation in leaves of C_3 species. *Planta,* 149:78-90.

Fechner E., and H.F. Hemond. 1992. Methane transport and oxidation in the unsaturated zone of a *Sphagnum* peatland. *Global Biogeochem. Cycles,* 6:33-44.

Filip, Z. 1979. Wechselwirkungen von Mikro-organismen und tonmineralen — eine Ubersicht. *Zeitschrift fur Pflanzenernahrung und Bodenkunde,* 142:375-386.

Firestone, M.K., M.S. Smith, R.B. Firestone, and J.M. Tiedje 1979. The influence of nitrate, nitrite and oxygen on the composition of the gaseous products of denitrification in soil. *Soil Sci. Soc. Am. J.,* 43:1140-1144.

Flanagan, L.B., and J.R. Ehleringer. 1998. Ecosystem — atmosphere CO_2 exchange: interpreting signals of change using stable isotope ratios. *Trends Ecol. Evol.,* 13:10-14.

Fleisher, Z., A. Kenig, I. Ravina, and J. Hagin. 1987. Model of ammonia volatilization from calcareous soils. *Plant Soil,* 103:205-212.

Flessa, H., P. Dörsch, and F. Beese. 1995. Seasonal variation of N_2O and CH_4 fluxes in differently managed arable soils in southern Germany. *J. Geophys. Res.,* 100:23115-23124.

Flowers, T.H., and J.R. O'Callaghan. 1983. Nitrification in soils incubated with pig slurry or ammonium sulfate. *Soil Biol. Biochem.,* 15:337-342.

Frolking, S., and P. Crill. 1994. Climate controls on temporal variability of methane flux from a poor fen in southeastern New Hampshire: measurement and modeling. *Global Biogeochem. Cycles,* 8:385-397.

Frolking, S., A.R. Mosier, D.S. Ojima, C. Li, W.J. Parton, C.S. Potter, E. Priesack, R. Stenger, C. Haberbosch, P. Dörsch, H. Flessa, and K.A. Smith. 1998. Comparison of N_2O emissions from soils at three temperate agricultural sites: simulations of year-round measurements by four models, *Nutr. Cycling Agroecosys.,* 55:77-105.

Ghidey, F., E.E. Alberta, and N.R. Kitchen. 1999. Evaluation of the Root Zone Water Quality Model using field-measured data from the Missouri MSEA. *Agron. J.,* 91:183-192.

Gijsman, A.J., A. Oberson, H. Tiessen, and D.K. Friesen. 1996. Limited applicability of the CENTURY model to highly weathered tropical soils. *Agron J.,* 88:894-903.

Gilmour, J.T. 1984. The effects of soil properties on nitrification and nitrification inhibition. *Soil Science Soc. Am. J.,* 48:1262-1266.

Goudriaan, J. 1996. Predicting crop yield under global change. In B.H. Walker and W. Steffen (Eds.). *First GCTE Science Conference.* Woods Hole, MA. 23-27 May 1994.

Goulden, M.L., and P.M. Crill. 1997. Automated measurements of CO_2 exchange at the moss surface of a black spruce forest. *Tree Physiol.,* 17:537-542.

Gower, S.T., J.G. Vogel, J.M. Norman, C.J. Kucharik, S.J. Steele, and T.K. Stow. 1997. Carbon distribution and aboveground net primary production in aspen, jack pine and black spruce stands in Saskatchewan and Manitoba, Canada. *J. Geophys. Res.,* 102:29,029-29,041.

Grant, R.F. 1989. Test of a simple biochemical model for photosynthesis of maize and soybean leaves. *Agric. For. Meteorol.,* 48:59-74.

Grant, R.F. 1991. A technique for estimating denitrification rates at different soil temperatures, water contents and nitrate concentrations. *Soil Sci.,* 152:41-52.

Grant, R.F. 1992. Simulation of carbon dioxide and water deficit effects upon photosynthesis of soybean leaves with testing from growth chamber studies. *Crop Sci.,* 32:1313-1321.

Grant, R.F. 1993. Simulation model of soil compaction and root growth. I. Model development. *Plant Soil,* 150:1-14.

Grant, R.F. 1994a. Simulation of competition between barley (*Hordeum vulgare* L.) and wild oat (*Avena fatua* L.) under different managements and climates. *Ecol. Model.,* 71:269-287.

Grant, R.F. 1994b. Simulation of ecological controls on nitrification. *Soil Biol. Biochem.,* 26:305-315.

Grant, R.F. 1995. Mathematical modeling of nitrous oxide evolution during nitrification. *Soil Biol. Biochem.,* 27:1117-1125.

Grant, R.F. 1997. Changes in soil organic matter under different tillage and rotation: mathematical modeling in *ecosys. Soil Sci. Soc. Am. J.* 61:1159-1174.

Grant, R.F. 1998. Simulation in *ecosys* of root growth response to contrasting soil water and nitrogen *Ecol. Model.,* 107:237-264.

Grant, R.F. 1999. Simulation of methanotrophy in the mathematical model *ecosys. Soil Biol. Biochem.,* 31:287-297.

Grant, R.F., N.G. Juma and W.B. McGill. 1993a. Simulation of carbon and nitrogen transformations in soils. I. Mineralization. *Soil Biol. Biochem.,* 27:1317-1329.

Grant, R.F., N.G. Juma, and W.B. McGill. 1993b. Simulation of carbon and nitrogen transformations in soils. II. Microbial biomass and metabolic products. *Soil Biol. Biochem.,* 27:1331-1338.

Grant, R.F., M. Nyborg, and J. Laidlaw. 1993c. Evolution of nitrous oxide from soil. I. Model development. *Soil Sci.,* 156:259-265.

Grant, R.F., M. Nyborg, and J. Laidlaw. 1993d. Evolution of nitrous oxide from soil. II. Experimental results and model testing. *Soil Sci.,* 156:266-277.

Grant, R.F., and P. Rochette. 1994. Soil microbial respiration at different temperatures and water potentials: theory and mathematical modeling. *Soil Sci. Soc. Am. J.,* 58:1681-1690.

Grant, R.F., R.C. Izaurralde, and D.S. Chanasyk. 1995. Soil temperature under different surface managements: testing a simulation model. *Agric. For. Meteorol.,* 73:89-113.

Grant, R.F., and D.J. Heaney. 1997. Inorganic phosphorus transformation and transport in soils: mathematical modeling in *ecosys. Soil Sci. Soc. Am. J.,* 61:752-764.

Grant, R.F., and J.A. Robertson. 1997. Phosphorus uptake by root systems: mathematical modeling in *ecosys. Plant Soil,* 188:279-297.

Grant, R.F., R.C. Izaurralde, M. Nyborg, S.S. Malhi, E.D. Solberg, and D. Jans-Hammer-meister. 1998. Modelling tillage and surface residue effects on soil C storage under current vs. elevated CO_2 and temperature in *ecosys.* In R. Lal, J.M. Kimble, R.F. Follet, and B.A. Stewart (Eds.). *Soil Processes and the Carbon Cycle.* CRC Press, Boca Raton, FL, 527-547.

Grant, R.F., and E. Pattey. 1999. Mathematical modeling of nitrous oxide emissions from an agricultural field during spring thaw. *Global Biogeochem. Cycles,* 13:679-694.

Grant, R.F., T.A. Black, G. den Hartog, J.A. Berry, S.T. Gower, H.H. Neumann, P.D. Blanken, P.C. Yang, and C. Russell. 1999. Diurnal and annual exchanges of mass and energy between an aspen-hazelnut forest and the atmosphere: testing the mathematical model *ecosys* with data from the BOREAS experiment. *J. Geophys. Res.,* 104:27,699-27,717.

Grant, R.F., P.G. Jarvis, J.M. Massheder, S.E. Hale, J.B. Moncrieff, M. Rayment, S.L. Scott, and J.A. Berry. 2001a. Controls on carbon and energy exchange by a black spruce — moss ecosystem: testing the mathematical model *ecosys* with data from the BOREAS experiment. *Global Biogeochem. Cycles* (in press).

Grant R.F., N.G. Juma, J.A. Robertson, R.C. Izaurralde, and W.B. McGill. 2001b. Long term changes in soil C under different fertilizer, manure and rotation: testing the mathematical model *ecosys* with data from the Breton Plots. *Soil Sci. Soc. Am. J.,* 65:205-214.

Grant, R.F. N.T. Roulet, and P.M. Crill. In preparation. Methane efflux form boreal wetlands: theory and testing of the ecosystem model *ecosys* with chamber and tower flux measurements.

Hansen, S., H.E. Jensen, N.E. Nielsen, and H. Svensen. 1990. *DAISY: Soil Plant Atmosphere System Model.* Ministry of the Environment. Copenhagen, Denmark, 270 pp.

Hanson, J.D., K.W. Rojas, and M.J. Shaffer. 1999. Calibrating the Root Zone Water Quality Model. *Agron. J.,* 91:171-177.

Harden, J.W., K.P. O'Neill, S.E. Trumbore, H. Veldhuis, and B.J. Stocks. 1997. Moss and soil contributions to the annual net carbon flux of a maturing boreal forest. *J.Geophys. Res.,* 102:28,805-28,816.

Hassink, J., and A.P. Whitmore. 1997. A model of the physical protection of organic matter in soils. *Soil Sci. Soc. Am. J.,* 61:131-139.

Hoogenboom, G., M.G. Huck, and C.M. Peterson. 1987. Root growth rate of soybean as affected by drought stress. *Agron. J.,* 79:607-614.

Hutchinson, G.L., R.J. Millington, and D.B. Peters. 1972. Atmospheric ammonia: absorption by plant leaves. *Science,* 175:771-772.

Hutson, J.L., and R.J. Wagenet. 1992. LEACHM: Leaching Estimation and Chemistry Model: Research Series 92-3, Water Resources Institute, Cornell Univ., Ithaca, NY.

Janzen, H.H., C.A. Campbell, E.G. Gregorich, and B.H. Ellert. 1998. Soil carbon dynamics in Canadian agroecosystems. In R. Lal, J.M. Kimble, R.F. Follet, and B.A. Stewart (Eds.). *Soil Processes and the Carbon Cycle.* CRC Press, Boca Raton, FL, 57-80.

Jaynes, D.B., and J.G. Miller. 1999. Evaluating the Root Zone Water Quality Model using data from the Iowa MSEA. *Agron. J.,* 91:192-200.

Jenkinson, D.S., P.B.S. Haet, J.H. Rayner, and L.C. Perry. 1987. Modelling the turnover of organic matter in long-term experiments. *INTECOL Bull.*, 15:1-8.

Johnson, A.D., M.L. Cabrera, D.V. McCracken, and D.E. Radcliffe. 1999. LEACHN simulations of nitrogen dynamics and water drainage in an Ultisol. *Agron. J.*, 91:597-606.

Jones, C.A., C.V. Cole, A.N. Sharpley, and J.R. Williams. 1984. A simplified soil and plant phosphorus model: I. Documentation. *Soil Sci. Soc. Am. J.*, 48:800-805.

Jones, C. A., W.L. Bland, J.T. Ritchie, and J.R. Williams. 1990. Simulation of root growth. In J. Hanks, and J.T. Ritchie (Eds.). *Modeling Plant and Soil Systems*. Amer. Soc. Agron. No. 31. Madison, WI, 91-123.

Kimball, B.A., P.J. Pinter Jr., R.L. Garcia, R.L. LaMorte, G.W. Wall, D.J. Hunsaker, G. Wechsung, F. Wechsung, and T. Kartschall. 1995. Productivity and water use of wheat under free-air CO_2 enrichment. *Global Change Biol.*, 1:429-442.

King, G.M., and P.S. Adamsen. 1992. Effects of temperature on methane consumption in a forest soil and in pure cultures of the methanotroph *Methylomonas rubra*. *Appl. Env. Microbiol.*, 58:2758-2763.

Kiniry, J.R., and A.J. Bockholt. 1998. Maize and sorghum simulation in diverse Texas environments. *Agron. J.*, 90:682-687.

Kinsbursky, R.S., and S. Saltzman. 1990. CO_2-nitrification relationships in closed soil incubation vessels. *Soil Biol. Biochem.*, 22:571-572.

Leffelaar, P.A., and W.W. Wessel. 1988. Denitrification in a homogeneous, closed system: experiment and simulation. *Soil Sci.*, 146:335-349.

Lessard, R., P. Rochette, E. Topp, E. Pattey, R.L. Desjardins, and G. Beaumont. 1994. Methane and carbon dioxide fluxes from poorly drained adjacent cultivated and forest sites. *Can. J. Soil Sci.*, 74:139-146.

Li, C., S. Frolking, and R. Hariss. 1994. Modeling carbon biogeochemistry in agricultural soils. *Global Biogeochem. Cycles*, 8:237-254.

Li, C., S. Frolking, and T.A. Frolking. 1992. A model of nitrous oxide evolution from soil driven by rainfall events: I. Model structure and sensitivity. *J. Geophys. Res.*, 97:9759-9776.

Lu, S., and M.H. Miller. 1994. Prediction of phosphorus uptake by field-grown maize with the Barber-Cushman model. *Soil Sci. Soc. Am. J.*, 58: 852-857.

Malhi, S.S., and W.B. McGill. 1982. Nitrification in three Alberta soils: effect of temperature, moisture and substrate concentration. *Soil Biol. Biochem.*, 14:393-399.

Marchetti, R., M. Donatelli, and P. Spallacci. 1997. Testing denitrification functions of dynamic crop models. *J. Environ. Qual.*, 26:394-401.

Martikainen, P.J., N. Hannu, P. Crill and J. Silvila. 1993. The effect of changing water table on methane fluxes at two Finnish mire sites. *Suo*, 43:237-240.

Massheder, J.M., J. B. Moncrieff, and M.B. Rayment. 1998. BOREAS TF-09 SSA-OBS Tower Flux, Meteorological, and Soil Temperature Data. Available online at [http://www-eosdis.ornl.gov/] from the ORNL Distributed Active Archive Center, Oak Ridge National Laboratory, Oak Ridge, Tennessee.

McConnaughey, P.K., and D.R. Bouldin. 1985. Transient microsite models of denitrification. I. Model development. *Soil Sci. Soc. Am. J.*, 49:886-891.

McGill, W.B. 1996a. Review and classification of ten soil organic matter (SOM) models. In D.S. Powlson, P. Smith and J.U. Smith (Eds.). *Evaluation of Soil Organic Matter Models Using Existing Long-Term Datasets*. NATO ASI Series I, Vol. 38, Springer-Verlag. Heidelberg, 111-132.

McGill, W.B. 1996b. Soil sustainability: microorganisms and electrons. *Solo/Suelo'96. XIII Congresso Latino Americano de Ciéncia do Solo*. 4-8 Aug. 1996. Sao Paulo, Brazil.

McGill, W.B., and C.V. Cole. 1981. Comparative aspects of cycling of organic C, N, S and P through soil organic matter. *Geoderma*, 26:267-286.

McGill, W.B., H.W. Hunt, R.G. Woodmansee, and J.O. Reuss. 1981. Phoenix, a model of the dynamics of carbon and nitrogen in grassland soils. In F.E. Clark, and T. Rosswall (Eds.). *Terrestrial Nitrogen Cycles. Ecological Bulletins,* 33, 49-115.

McInnes, K.J., and I.R.P. Fillery. 1989. Modeling and field measurements of the effect of nitrogen source on nitrification. *Soil Sci. Soc. Am. J.,* 53:1264-1269.

McLaren, A.D. 1969. Steady state studies of nitrification in soil: Theoretical considerations. *Proc. Soil Sci. Soc. Am.,* 33:273-275.

Melillo, J.M., A.D. McGuire, D.W. Kicklighter, B. Moore III, C.J. Vörösmarty, and A.L. Schloss. 1993. Global climate change and net primary production. *Nature,* 363:234-240.

Mielke, L.N., J.W. Doran, and K.A. Richards. 1986. Physical environment near the surface of plowed and no-tilled soils. *Soil Till. Res.,* 7:355-366.

Molina, J.A.E., C.E. Clapp, M.J. Shaffer, F.W. Chichester, and W.E. Larson. 1983. NCSOIL, a model of nitrogen and carbon transformations in soil: description, calibration and behavior. *Soil Sci. Soc. Am. J.,* 47:85-91.

Morel, C., H. Tiessen, J.O. Moir, and J.W.B. Stewart. 1994. Phosphorus transformations and availability under cropping and fertilization assessed by isotopic exchange. *Soil Sci. Soc. Am. J.,* 58:1439-1445.

Mosey, F.E. 1983. Kinetic descriptions of anaerobic digestion. In *Third International Symposium on Anaerobic Digestion.* Boston Univ., Cambridge, MA, 37-52.

Nakane, K., T. Kohno, T. Horikoshi, and T. Nakatsubo. 1997. Soil carbon cycling at a black spruce (*Picea mariana*) forest stand in Saskatchewan, Canada. *J.Geophys. Res.,* 102:28,785-28,793.

Nommik, H. 1956. Investigations on denitrification, *Acta Agric. Scand.,* 6:195-228.

Nyborg, M., E.D. Solberg, S.S. Malhi, and R.C. Izaurralde. 1995. Fertilizer N, crop residue and tillage alter soil C and N content in a decade. In R. Lal, J.M. Kimble, R.F. Follet, and B.A. Stewart (Eds.). *Soil Management and Greenhouse Effect.* Lewis Publishers, Boca Raton, FL, 93-99.

Odum, E.P. 1971. *Fundamentals of Ecology,* 3rd ed. W.B. Saunders Co., Toronto. 574 pp.

Parton, W.J., D.S. Schimel, C.V. Cole, and D.S. Ojima. 1987. Analysis of factors controlling organic matter levels in Great Plains grasslands. *Soil Sci. Soc. Am. J.,* 51:1173-1179.

Parton, W.J., and P.E. Rasmussen. 1994. Long term effects of crop management in wheat-fallow. II. CENTURY model simulations. *Soil Sci. Soc. Am. J.,* 58:530-536.

Parton, W.J., J.W.B. Stewart, and C.V. Cole. 1988. Dynamics of C, N, P and S in grassland soils: a model. *Biogeochemistry,* 5:109-131.

Parton, W.J., D.S. Schimel, C.V. Cole, and D.S. Ojima. 1994. Analysis of factors controlling soil organic matter in great plains grasslands. *Soil Sci. Soc. Am. J.,* 51:1173-1179.

Parton, W.J., A.R. Mosier, D.S. Ojima, D.W. Valentine, D.S. Schimel, K. Weier, and A.E. Kulmala. 1996. Generalized model for N_2 and N_2O production from nitrification and denitrification. *Global Biogeochem. Cycles,* 10:401-412.

Paustian, K. 1994. Modeling soil biology and biochemical processes for sustainable agriculture research. In Z.E. Pankhurst, B.M. Doube, V.V.S.R. Gupta, and P.R. Grace (Eds.). *Soil Biota Management in Sustainable Farming Systems.* CSIRO Information Services, Melbourne, 182-193.

Paustian K., G.I. Ågren G, and E. Bosatta. 1997. Modelling litter quality effects on decomposition and soil organic matter dynamics. In G. Cadisch, and K.E. Giller (Eds.). *Driven by Nature: Plant Litter Quality and Decomposition.* CAB International, 313-335.

Paustian, K., E.T. Elliot, and K. Killian. 1998. Modeling soil carbon in relation to management and climate change in some agroecosystems in central North America. In R. Lal, J.M. Kimble, R.F. Follet, and B.A. Stewart (Eds.). *Soil Processes and the Carbon Cycle.* CRC Press, Boca Raton, FL, 459-.

Penning de Vries, F.W.T. 1983. Modeling of growth and production. In O.L. Lange, P.S. Nodel, C.B. Osmond, and H. Ziegler (Eds.). *Physiological Plant Ecology IV.* Springer Verlag. Berlin, 117-150.

Reicosky, D.C., W.D. Kemper, G.W. Langdale, C.L. Douglas Jr., and P.E. Rasmussen. 1995. Soil organic matter changes resulting from tillage and biomass production. *J. Soil Water Cons.,* 50:253-261.

Ritchie J.T., D.C. Godwin, and S. Otter-Nacke. 1988. *CERES-Wheat: A Simulation Model of Wheat Growth and Development.* Texas A&M Univ. Press. College Station, TX.

Robertson, K. 1994. Nitrous oxide emission in relation to soil factors at low to intermediate moisture levels. *J. Environ. Qual.,* 23: 805-809.

Rose, C.W., F.W. Chichester, J.R. Williams, and J.T. Ritchie. 1982. Application of an approximate analytic method of computing solute profiles with dispersion in soils. *J. Environ. Qual.,* 11:151-155.

Rutherford, P.M., and N.G. Juma. 1989. Dynamics of microbial biomass and fauna in two contrasting soils cropped to barley (*Hordeum vulgare* L.). *Biol. Fertil. Soils,* 8:144-153.

Ryan, M.G., M.B. Lavigne, and S.T. Gower. 1997. Annual cost of autotrophic respiration in boreal forest ecosystems in relation to species and climate. *J.Geophys. Res.,* 102:28,871-28,883.

Sabey, B.R., L.R. Frederick, and W.V. Bartholomew. 1969. The formation of nitrate from ammonium nitrogen in soils: IV. Use of the delay and maximum rate phases for making quantitative predictions. *Proc. Soil Sci. Soc. Am.,* 33:276-278.

Sass, R.L., F.M. Fisher, and P.A. Harcombe. 1990. Methane production and emission from a Texas rice field. *Global Biogeochem. Cycles,* 4:47-68.

Schulz, K., K. Beven, and B. Huwe. 1999. Equifinality and the problem of robust calibration in nitrogen budget simulations. *Soil Sci. Soc. Am. J.,* 63:1934-1941.

Schütz, H., W. Seiler, and R. Conrad. 1989. Processes involved in formation and emission of methane in rice paddies. *Biogeochemistry,* 7:33-53.

Sellers, P.J., R.E. Dickinson, D.A. Randall, A.K. Betts, F.G. Hall, J.A. Berry, G.J. Collatz, A.S. Denning, H.A. Mooney, C.A. Nobre, N. Sato, C.B. Field, and A. Henderson-Sellers. 1997a. Modelling the exchanges of energy, water and carbon between continents and the atmosphere. *Science,* 275:502-509.

Sellers, P.J., F.G. Hall, and BOREAS members (1997b) BOREAS in 1997: experiment review, scientific results, and future directions. *J.Geophys. Res.,* 102:28,731-28,769.

Shaffer, M.J. and W.E. Larson. 1987. NTRM, a soil-crop simulation model for nitrogen, tillage and crop-residue management. USDA Conserv. Res. Rep. 34-1. National Technical Information Service, Springfield, VA.

Shaffer, M.J., A.D. Halvorson and F.J. Pierce. 1991. Nitrate leaching and economic analysis package (NLEAP): model description and application. In Follet R.F. et al. (Eds.). *Managing Nitrogen for Groundwater Quality and Farm Profitability.* Proc. Symp. ASA, SSSA and CSSA. Anaheim, CA. SSSA Madison, WI, 285-322.

Sharpley, A.N., and J.R. Williams. 1990. EPIC — Erosion/Productivity Impact Calculator: 1) Model Documentation. USDA-ARS Tech. Bull. 1768.

Shea, T.G., W.E. Pretorius, R.D. Cole, and E.A. Pearson. 1968. Kinetics of hydrogen assimilation in the methane fermentation. *Water Res.,* 2:833-848.

Shulten, H.-R. and M. Schnitzer. 1997. Chemical model structures for soil organic matter and soils. *Soil Sci.,* 162:115-130.

Sinclair, T.R., C.E. Murphy Jr., and K.R. Knoerr. 1976. Development and evaluation of simplified models for simulating canopy photosynthesis and transpiration. *J. App. Ecol.,* 13:813-829.

Smid, A.E., and E.G. Beauchamp. 1976. Effects of temperature and organic matter on denitrification in soil. *Can. J. Soil Sci.,* 56:385-391.

Smith, O.L. 1982. *Soil Microbiology: A Model of Decomposition and Nutrient Cycling*. CRC Press, Boca Raton, FL. 273 pp.

Smith, P., J.U. Smith, D.S. Powlson, W.B. McGill, J.R.M. Arah, O.G. Chertov, K. Coleman, U. Franko, S. Frolking, D.S. Jenkinson, L.S. Jensen, R.H. Kelly, H. Klein-Gunnewiek, A.S. Komarov, C. Li, J.A.E. Molina, T. Mueller, W.J. Parton, J.H.M. Thornley, and A.P. Whitmore. 1997. A comparison of the performance of nine soil organic matter models using datasets from seven long-term experiments. *Geoderma*, 81:153-225.

Smith, P., J.U. Smith, and D.S. Powlson (Eds.). 1996. Soil Organic Matter Network (SOMNET): 1996 Model and Experimental Metadata. GCTE Report 7. GCTE Focus 3 Office, Wallingford. 255 pp.

Sørenson, L.H. 1981. Carbon-nitrogen relationships during the humification of cellulose in soils containing different amounts of clay. *Soil Biol. Biochem.*, 13:313-321.

Spitters, C.R.T., and R. Aerts. 1983. Simulation of competition for light and water in crop-weed associations. *Asp. Appl. Biol.*, 4:467-483.

Steele, S.J., S.T. Gower, J.G. Vogel, and J.M. Norman. 1997. Root mass, net primary production and turnover in aspen, jack pine and black spruce forests in Saskatchewan and Manitoba, Canada. *Tree Physiol.*, 17:577-587.

Stevenson, F.J., and E.T. Elliott. 1989. Methodologies for assessing the quantity and quality of soil organic matter. In D.C. Coleman, J.M. Oades, and G. Uehara (Eds.). *Dynamics of Soil Organic Matter in Tropical Ecosystems*. Univ. of Hawaii Press, 173-241.

Stockle, C.O. 1992. Canopy photosynthesis and transpiration estimated using radiation interception models with different levels of detail. *Ecol. Model.*, 60: 31-44.

Stockle, C.O., and R. Nelson. 1995. CropSyst. Cropping Systems Simulation Model. User's Manual (V. 1.04). Wash. State Univ. Pullman, WA.

Stott, D.E., L.F. Elliott, R.I. Papendick, and G.S. Campbell. 1986. Low temperature or low water potential effects on the microbial decomposition of wheat residue. *Soil Biol. Biochem.*, 18:577-582.

Suzuki I., U. Dular, and S.C. Kwok. 1974. Ammonia or ammonium ion as substrate for oxidation by *Nitrosomonas europaea* cells and extracts. *J. Bacteriol.*, 120:556-558.

Thornley, J.H. 1995. Shoot:root allocation with respect to C, N and P: an investigation and comparison of resistance and teleonomic models. *Ann. Bot.*, 75:391-405.

Tsutsuki, K., and F.N. Ponnamperuma. 1987. Behavior of anaerobic decomposition products in submerged soils: effects of organic material amendment, soil properties and temperature. *Soil Sci. & Plant Nutr.*, 33:13-33.

van Veen, J.A., and M.J. Frissel. 1981. Simulation model of the behavior of N in soil. In M.J. Frissel, and J.A. van Veen (Eds.). *Simulation of Nitrogen Behaviour of Soil-Plant Systems*. Centre for Agricultural Publishing and Documentation. Wageningen, 126-144.

van Veen, J.A., and P.K. Kuikman. 1990. Soil structural aspects of decomposition of organic matter by micro-organisms. *Biogeochemistry*, 11:213-233.

van Veen, J.A., J.N. Ladd, and M. Amato. 1985. Turnover of carbon and nitrogen through the microbial biomass in a sandy loam and a clay soil incubated with $[^{14}C(U)]$ glucose and $[^{15}N](NH_4)_2SO_4$ under different moisture regimes. *Soil Biol. Biochem.*, 17:747-756.

VEMAP. 1995. Vegetation/ecosystem modeling and analysis project (VEMAP): comparing biogeography and biogeochemistry models in a continental-scale study of terrestrial ecosystem responses to climate change and CO_2 doubling. *Global Biogeochem. Cycles*, 9:407-437.

Verberne, E.L.J., J. Hassink, P. de Willigen, J.J.R. Groot, and J.A. van Veen. 1990. Modelling soil organic matter dynamics in different soils. *Neth. J. Agric. Sci.*, 38:221-238.

Verseghy D.L., N.A. McFarlane, and N. Lazare. 1993. CLASS — A Canadian land surface scheme for GCMs. II. Vegetation model and coupled runs. *Int. J. Climatol.*, 13:347-370.

Vincent, C.D., and P.J. Gregory. 1989. Effects of temperature on the development and growth of winter wheat roots. II. Field studies of temperature, nitrogen and irradiance. *Plant Soil,* 119:99-110.

Weier, K.L., J.W. Doran, J.F. Power, and D.T. Walters. 1993. Denitrification and the dinitrogen/nitrous oxide ratio as affected by soil water, available C and nitrate. *Soil Sci. Soc. Am. J.,* 57:66-72.

Whitehead, D.C., and D.R. Lockyer. 1987. The influence of the concentration of gaseous ammonia on its uptake by the leaves of Italian Ryegrass, with and without an adequate supply of nitrogen to the roots. *J. Exp. Bot.,* 38:818-827.

Whiting, G.J., and J.P. Chanton. 1992. Plant-dependent CH_4 emission in a subarctic Canadian fen. *Global Biogeochem. Cycles,* 6:225-231.

Whitmore, A.P., and L.C. Parry. 1988. Computer simulation of the behavior of nitrogen in soil and crop in the Broadbalk continuous wheat experiment. In D.S. Jenkinson, and K.A. Smith (Eds.). *Nitrogen Efficiency in Agricultural Soils,* Elsevier, London, 418-432.

Whitmore, A.P., K.W. Coleman, N.J. Bradbury, and T.M. Addiscott. 1991. Simulation of nitrogen in soil and winter wheat crops: modeling nitrogen turnover through organic matter. *Fert. Res.,* 27:283-291.

Williams, J.R., P.T. Dyke, W.W. Fuchs, V.W. Benson, O.W. Rice, and E.D. Taylor. 1989. *EPIC — Erosion Productivity Impact Calculator.* USDA Tech. Bull. 17.

Wofsy, S.C., M.L. Goulden, J.W. Munger, S.-M. Fan, P.S. Bakwin, B.C. Daube, S.L. Bassow, and F.A. Bazzaz. 1993. Net exchange of CO_2 in a mid-latitude forest. *Science,* 260:1314-1317.

Xu, C., M.J. Shaffer, and M. Al-kaisi. 1998. Simulating the impact of management practices on nitrous oxide emissions. *Soil Sci. Soc. Am. J.,* 62:736-742.

Index

Milton Keynes UK
Ingram Content Group UK Ltd.
UKHW021934071024
449327UK00022B/1800